Insect Pests of Cotton

INSECT PESTS OF COTTON

Edited by

G.A. Matthews

International Pesticide Application Research Centre,
Imperial College at Silwood Park, Ascot, UK

and

J.P. Tunstall

Formerly Research Entomologist in Central Africa
and
Director of the Agricultural Research Council of Malawi

CAB INTERNATIONAL

CAB INTERNATIONAL
Wallingford
Oxon OX10 8DE
UK

Tel: Wallingford (0491) 832111
Telex: 847964 (COMAGG G)
Telecom Gold/Dialcom: 84: CAU001
Fax: (0491) 833508

A catalogue entry for this book is available from the British Library.

ISBN 0 85198 724 9

Typeset by Create Text Ltd, Bath
Printed and bound in the UK at the University Press, Cambridge

Contents

Contributors

J.P. Bournier: *Institut Recherches du Coton et Textiles Exotiques, Départment du Centre de Coopération Internationale en Recherche Agronomique pour le Développement, BP 5035, 34032 Montpellier Cedex, France.*

S.W. Broodryk: *South African Development Trust Corporation Ltd, PO Box 213, Pretoria 0001, South Africa.*

G.D. Butler, Jr.: *United States Department of Agriculture, Agricultural Research Service, Western Cotton Research Laboratory, 4135 E Broadway, Phoenix, Arizona 85040, USA.*

D.G. Campion: *Natural Resources Institute, Central Avenue, Chatham Maritime, Chatham, Kent ME4 4TB, UK.*

R. Couilloud: *Département du Centre de Cooperation Internationale en Recherche Agronomique pour le Développement, BP 5035, 34032 Montpellier Cedex, France.*

J.P. Deguine: *Centre de Coopération Internationale en Recherche Agronomique pour le Développement, Institut la Recherche Agronomique, BP 22, Mároua, Cameroon.*

J.R. Gannaway: *Texas Agricultural Experiment Station, PO Box 219, Lubbock, Texas 79401–9757, USA.*

D.J. Greathead: *International Institute of Biological Control, Silwood Park, Buckhurst Road, Ascot, Berkshire SL5 7TA, UK.*

J. Gutierrez: *Institut Français de Recherche Scientifique pour le Développement en Coopération (ORSTOM), BP 5045, 34032 Montpellier Cedex 1, France.*

F.A. Harris: *Mississippi State University, Delta Research and Extension Center, Stoneville, Mississippi 38776, USA.*

T.J. Henneberry: *United States Department of Agriculture, Agricultural Research Service, Western Cotton Research Laboratory, 4135 E Broadway, Phoenix, Arizona 85040, USA.*

W.R. Ingram: *Orchards, Alhampton, Shepton Mallet, Somerset BA4 6PZ, UK.*

K.A. Jones: *Natural Resources Institute, Central Avenue, Chatham Maritime, Chatham, Kent ME4 4TB, UK.*

A.B.S. King: *Natural Resources Institute, Central Avenue, Chatham Maritime, Chatham, Kent ME4 4TB, UK.*

F. Leclant: *Ecole Nationale Supérieure Agronomique, 2 Place Pierre Viala, 34060 Montpellier Cedex 1, France.*

T.F. Leigh (deceased): *University of California, Shafter Research Station, 17053 Shafter Avenue, Shafter, California 93263, USA.*

J.W.M. Logan: *Natural Resources Institute, Central Avenue, Chatham Maritime, Chatham, Kent ME4 4TB, UK.*

G.A. Matthews: *International Pesticide Application Research Centre, Imperial College at Silwood Park, Buckhurst Road, Sunninghill, Ascot, Berkshire SL5 7PY, UK.*

J.D. Mumford: *Department of Biology, Imperial College at Silwood Park, Buckhurst Road, Sunninghill, Ascot, Berkshire SL5 7PY, UK.*

J.M. Munro: *Flat 33, 20 Craiglea Place, Morningside Grove, Edinburgh EH10 5QA, UK.*

G.A. Norton: *Cooperative Research Centre for Tropical Pest Management, University of Queensland, Brisbane, Queensland 4072, Australia. Formerly at: Imperial College, Silwood Park, Ascot, Berkshire SL5 7PY, UK.*

G.K.C. Nyirenda: *University of Malawi, Bunda College of Agriculture, Lilongwe, Malawi.*

W. Reed: *Waterside, Sherborne Street, Bourton-on-the-Water, Gloucestershire GL54 2BY, UK.*

J.W. Smith: *United States Department of Agriculture, Boll Weevil Research Unit, PO Box 5367, Mississippi 39762, USA.*

J.P. Tunstall: *'Austers', Woodmans Green, Whatlington, Nr Battle, Sussex TN33 0NJ, UK.*

Preface

In his book *Insect Pests of Cotton in Tropical Africa*, published in 1958 by the Commonwealth Institute of Entomology, E.O. Pearson provided a unique account of the biology and control of the vast complex of insects found in African grown cotton. Pearson's experience was gained while working in several countries before synthetic insecticides were made widely available. He considered very carefully how to optimize yields using all the biological and cultural options of insect control. However, insecticides were beginning to be used commercially on cotton, notably in the USA and the Sudan, and with the prospect of their increased widespread application he sought the assistance of R. Maxwell-Darling in adding a brief note that pointed out many of the possible dangers of insecticide use, especially in small-scale peasant farming. Over the last 35 years much has been achieved in increasing cotton yields with chemical control, but the problems foreseen by Pearson and Maxwell-Darling have also occurred. Use of insecticides on cotton has undoubtedly increased yields, but today there is growing and justified concern about the effect of insecticides and other pesticides on the environment, the selection of pests resistant to insecticides, new pests due to the destruction of natural enemies and other undesirable effects.

There is now an increasing trend to develop an integrated system of Insect Pest Management (IPM), in which all possible control tactics are harmonized into a programme which minimizes the reliance on insecticides. There have been successes, but in many cases the adoption of a more sensible and sustainable use of chemical control has only been accepted after a crisis point has been reached. Thus the loss of the cotton industry in the Ord Valley of Australia, due to insecticide resistance, stimulated the industry to cooperate in the development of an insecticide resistance management strategy in Queensland and New South Wales.

The implementation of IPM requires close collaboration of farmers, extension staff, research scientists, the agrochemical industry and seed producers – in fact everybody concerned with cotton production. Many of the insects attacking cotton also affect other major crops so decisions need to be based on the agroecosystem, rather than an individual crop. To achieve more effective and widespread use of IPM technology, a broader look at each area is needed to decide on an appropriate package of tactics.

This book is not strictly a revision of Pearson's work as it covers the world's cotton areas with additional insects, while condensing and omitting sections of the original work. A revision implies a replacement whereas the present book, although different from Pearson, is a logical 'follow on' and brings together more recent works on the biology and behaviour of cotton insects with those of the past, and examines current developments in control programmes. Even with the specialist knowledge of the authors, there is so much literature available now that no book can cover the subject completely, especially as progress continues to be made. The appearance of transgenic cottons is at much the same stage as were many insecticides in the 1950s, when Pearson wrote his book. The impact of transgenic cotton on production has yet to be calculated, but like insecticides it will not be a simple and effective step forward; there will be problems and their use will need to be part of an IPM programme.

Acknowledgements

We would like to acknowledge the Overseas Development Administration (UK), the International Cotton Advisory Committee (Washington DC), and Shell South Africa, for financial assistance towards the costs of including colour plates in this book.

I THE COTTON PLANT

1 Cotton and its Production

J.M. Munro

Flat 33, 20 Craiglea Place, Morningside Grove, Edinburgh EH10 5QA, UK

The Commercial Product

The Market for Cotton

Cotton is the most important natural textile fibre in the world. World consumption of textile fibres including wool, linen, silk and man-made fibres has risen steadily since the end of the Second World War, partly because of a growing population and partly because consumption per head is increasing. Fibre consumption correlates well with income, and while the more developed countries consume 15–20 kg per head, many of the developing countries use less than 5 kg. There is ample scope for further growth as world population increases and incomes in the developing countries improve.

The production of wool, silk and linen together amounts to only about one tenth of the production of cotton, but man-made fibres have taken an increased share of the market especially in the developed countries. Cotton now supplies less than 50% of the fibre market but in spite of this, consumption of cotton has risen steadily since 1946 at about 3% per annum, thus world demand has increased from six million tonnes in 1947–1948 to 18 million tonnes in 1986–1989 (Fig. 1.1, Table 1.1).

Man-made fibres are of two kinds, cellulosic and synthetic. Cellulosic fibres are derived from natural cellulose, obtained originally from cotton waste and linters (the short fibres left on most cotton seeds after ginning), but now the main source of cellulose is wood pulp. The cellulose polymers, long chains of identical molecules, are extracted, purified and formed into a viscous mass by various treatments, and new fibres produced by extrusion through a perforated plate or spinneret. Synthetic fibres are polymer chains made from entirely new synthetic chemicals, each of which confers its own special properties on the resulting fibres. Man-made fibres may be spun as long filaments into filament yarn, or chopped up into staple fibre of any desired length. Staple fibre is commonly used

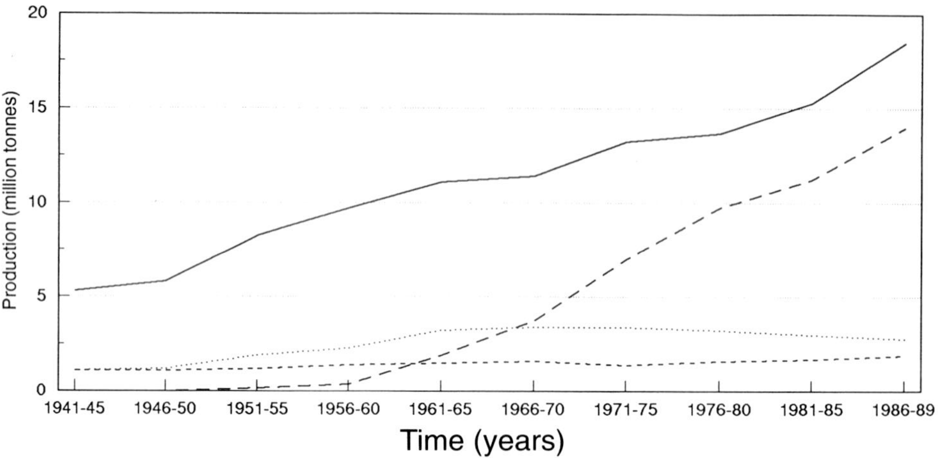

Fig. 1.1. World production of textile fibres. — Cotton; --- Wool; ...... Cellulosic; — — — Synthetic.

in blends with wool or cotton, and is spun on the same machines as the natural fibres.

The competition for use in clothing between cotton and man-made gets most attention from the press and public, but only about half of the cotton produced is used in this way. The National Cotton Council of America divides the textile fibre market into three main groups: apparel, household and industrial, and publishes figures for the end use of cotton and man-made fibres in the USA

Table 1.1. World production of textile fibres.

	4–year average 1986–1989	
Type of fibre	Thousand metric tonnes	% of world total
Man-made fibres		
Cellulosic	2,868	7.7
Synthetic	13,940	37.5
Total	16,808	45.2
Natural fibres		
Raw cotton	18,432	49.6
Raw wool	1,860	5.0
Raw silk	60	0.2
Total	20,352	54.8
World total	37,160	100.00

Source: *Textile Outlook International*, The Economist Intelligence Unit, 40 Duke Street, London, UK.

Table 1.2. Fibre consumption in textile end-use manufacturing in the USA in 1987.

	Cotton	Other fibres	Total	% Cotton
	'000 metric tonne equivalents			
Apparel				
Men's wear	687	493	1180	58
Women's wear	311	455	766	41
Children's wear	95	98	193	49
Total	1093	1046	2139	51
House furnishings				
Towels	220	9	229	96
Sheets and pillow cases	104	79	183	57
Drapery and upholstery	111	226	337	33
Other	125	1637	1762	7
Total	560	1951	2511	22
Industrial uses	143	750	893	16
Grand total	1796	3747	5543	32

Source: *Cotton Counts its Customers*, National Cotton Council of America, Memphis, Tennessee, USA.

(Table 1.2). Similar figures are difficult to find in some countries, but a comparison between the statistics for yarn and fabric production gives some indication of how much goes into textiles (Fig. 1.2).

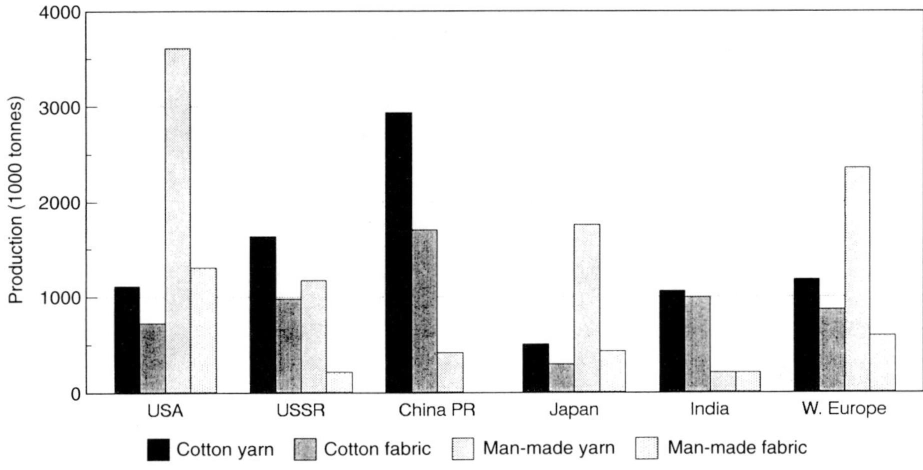

Fig. 1.2. Production of yarn and fabrics (in '000 metric tonnes) (ICAC figures, Washington, DC, 1983).

The Cotton Fibre

Ripe cotton picked from the plant is termed seed cotton, which is processed at a ginnery to separate the fibre from the seed. The fibre is called cotton lint or raw cotton, and is compressed into bales for transport to the spinning mills or other users.

Each cotton fibre is a tubular outgrowth from the seed coat, formed by the elongation of a single cell. After the cell has grown to its full length, additional cellulose is deposited on the inside of the cell wall in a spiral pattern. Just before the boll ripens, the cell contents are withdrawn leaving a hollow or lumen in the centre of the fibre. When the boll opens, the fibres dry out and the tube collapses to form a ribbon; the spiral pattern of the secondary thickening causes this ribbon to twist. When the fibres are spun into thread, this twist prevents the individual fibres from slipping over each other, and adds to the strength of the spun yarn.

The fibres are distributed fairly evenly over the seed surface, but are not all of the same length; those at the distal or rounded end of the seed tend to be longer than those at the basal or hilum end. This means that the staple length, the commercial measure of the length of the fibre, has no exact meaning, as each sample of cotton contains a range of fibre lengths seen in the Baer diagram (Fig. 1.3). A trained cotton classer can estimate the staple length by eye; more objective measures include the effective length (Fig. 1.3) calculated from the Baer diagram, and the span length measured on a fibrograph.

The two most important measures of a commercial sample are staple length and grade. Staple length is genetically controlled, and each variety has its own potential staple length modified by growing conditions. Staple length is measured in thirty-seconds of an inch or in millimetres; similar lengths are grouped into classes ranging from extra long staple (35 mm and above) through long and medium to short staple (below 21 mm) (Table 1.3). The longest staple is found in

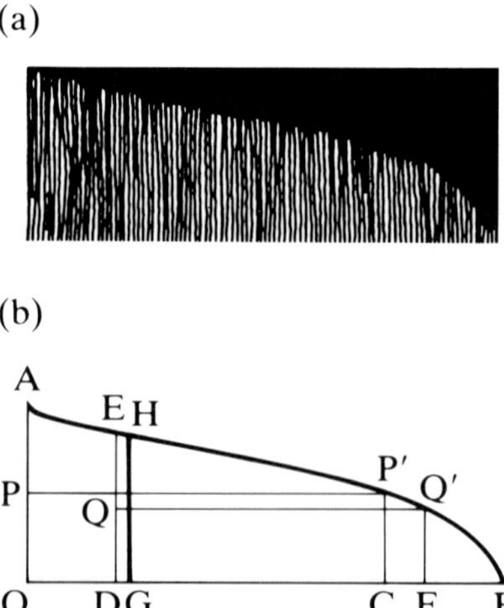

Fig. 1.3. Baer diagram and effective length (after Clegg, 1931): (a) drawing of typical Baer diagram; (b) treatment of trace from Baer diagram. OP = ½OA. OD = ¼OC. DQ = ½DE. OG = ¼OF. Effective length = GH; Percentage of short hairs = 100 × FB/OB.

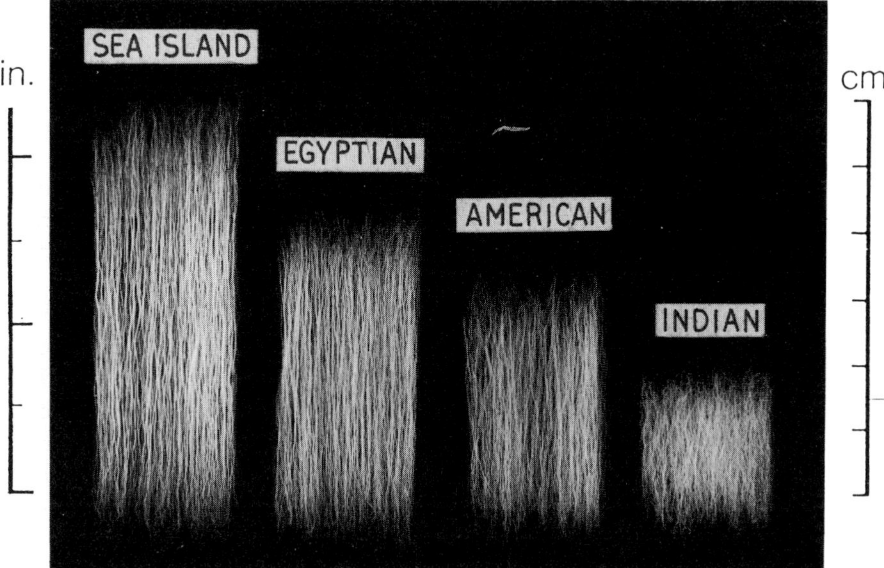

Fig. 1.4. Range of cotton fibre lengths.

the *barbadense* cottons such as Sea Island and Egyptian, and the shortest in the *desi* cotton of the Old World species (Fig. 1.4). Long staple cotton needs a regular water supply as the fibres are usually fine and prone to neps if subjected to water stress; in general the harsher the conditions the shorter the staple length.

Grade depends on the cleanliness of the sample, taking trash, dirt and discoloration into account, and depends largely on conditions during and after harvest. Hand-picked cotton is generally one or two grades better than machine-picked. Pest damage will often lower the grade: the lint may be discoloured or damaged by direct infection with a fungus transmitted by insects such as cotton stainers; by secondary infection of tissue damaged by bollworm or boll weevil; or by diseases like bacterial blight and anthracnose which can attack otherwise undamaged bolls.

Other fibre characters used in assessing lint quality are maturity, fineness and fibre strength. Immature cotton lacks the secondary thickening of the cell walls

Table 1.3. Classification of staple length.

	Staple length	
Description	Millimetres	Inches
Short	below 21	below 13/16
Medium	21–25	13/16–1
Medium long	26–28	1 1/32–1 3/32
Long	28–34	1 1/8–1 11/32
Extra long	35 and above	1 3/8 and above

either because the boll has opened prematurely or conditions have interfered with normal development and ripening. As a result the fibre is weak and may cause breakages in spinning, or it may form neps, clumps of immature fibres, which take up dyes differently from mature fibres; the woven fabric will be uneven in colour and texture, and lack durability. Fibre fineness can be measured directly as weight per unit length, but is usually measured in micronaire units. The micronaire measures the resistance to air flow of a fibre plug of standard size and weight; the pressure needed to force air through the plug is proportional to the surface area and hence to the fineness of the fibres. It is important to remember that the micronaire measures both fineness and maturity, as immature fibres are finer than mature ones, and it is sometimes necessary to carry out further tests to distinguish between these two characters. Fibre strength is measured on special instruments, the Pressley and the Stelometer, which measure the force needed to break a bundle of fibres of standard weight. The clamps holding the fibres may be close together at 'zero gauge' or separated by 1/8 of an inch at '1/8 in gauge'; the zero gauge Pressley test is commonly used in the USA, while the 1/8 gauge Stelometer is more widely used in Europe.

Recently high volume instruments (HVI) have been devised for the rapid measurement of fibre characters: length and length uniformity, micronaire, fineness and maturity, strength and elongation. In the HVI systems manual operation is replaced by automated instruments controlled by a microprocessor, which then prints out the data in the required format (Sasser, 1981).

The Cotton Plant

Taxonomy

Cotton belongs to the genus *Gossypium* in the family Malvaceae. The family was divided into two groups or tribes by Reichenbach (1828) according to the type of fruit – one group in which the carpels, each containing a single seed, separate from each other when ripe (schizocarpous); while in the other the capsular fruits split into locules leaving the seeds free and exposed. The latter group contains two series centred on *Hibiscus* and *Gossypium* respectively, but until recently botanists considered them to be so weakly differentiated that they included them in a single tribe. However, Fryxell (1979) has collected sufficient evidence to justify their separation into two distinct tribes, the Hibisceae and the Gossypieae, the main diagnostic character of the latter being the gossypol glands or 'punctae', as they were called originally by Alefeld in 1861.

Genera of both groups may be of interest to entomologists as alternative hosts to some of the important cotton pests when their distribution overlaps areas where cotton is grown commercially. More distantly related genera, such as *Bombax*, *Sterculia*, *Adansonia* (Baobab) and *Theobroma* (Cocoa), can also harbour some of the pests of cotton.

There are some 33 diploid species ($2n = 26$) of *Gossypium*, which can be divided into a number of groups according to the fertility of crosses between them. Crosses between species in the same group show a high degree of chromosome pairing and give fertile hybrids: the chromosomes are therefore closely related, and are said to belong to the same genome. The genomes have been designated A to G, and to a large extent the differences are the result of geographical isolation: the C genome is confined to Australia (10 species) and the D genome to America (12 species), while genomes A, B and E are found in Africa and Asia. Genomes F and G comprise one species each which do not fit into the original five groups.

The A genome is of particular interest in that it is the only one which produces cotton lint. It is found in the wild species *Gossypium herbaceum* var. *africanum* in southern Africa, which seems to be the nearest existing species to the wild ancestor of the cultivated cottons (Fryxell, 1979). The tetraploid cottons (2n = 52) which originated in America also owe their lint to the A genome. Cytological and genetic evidence has shown that half of their chromosomes are similar to those of the A genome from the Old World, and the other half to those of the D genome of the American wild cottons, none of which produces true lint. How and where the two species met and crossed is still a matter for debate, but it must have happened long before 3000 BC, as cotton fabrics of that date have been found in Peru.

The varieties grown commercially for their lint all belong to four species of *Gossypium*, the diploid Old World species *G. herbaceum* and *G. arboreum*, and the tetraploid New World species *G. hirsutum* and *G. barbadense.* The first two are cultivated almost exclusively in Asia, and produce the *desi* cottons of India and Pakistan, while smaller quantities are grown in Thailand and China. *Gossypium barbadense* provides the long staple cottons of the Nile valley, Sea Island cotton in the West Indies, Tanguis in Peru and Pima cotton in New Mexico and Arizona.

By far the most widely cultivated species is *G. hirsutum*, often called Upland cotton, which produces the bulk of the crop in the major producing countries as well as providing nearly all the commercial cotton from the minor producers worldwide.

The ancestors of the present day cultivated varieties were perennial shrubs or small trees, in which the period of vegetative growth was more or less prolonged, and fruiting was controlled by the influence of day length and temperature on the production of fruiting branches, or of rainfall on the retention of flowerbuds. The annual habit has been developed by selection for early flowering types which are not affected by day length. Almost without exception, the tendency in cotton breeding has been towards early flowering varieties, resulting in a shorter growing season and a longer interval between harvest and the sowing of the next crop. This has enabled cotton production to spread into increasingly higher latitudes, notably in Russia and China, characterized by a short hot summer season. Before the use of insecticides the ability of a variety to recover from an insect attack and produce a late crop was considered an advantage, but a shorter season means that less insecticide is needed. Large bolls make hand picking easier, but where mechanical harvesting is general, boll size is unimportant and bolls have tended to be smaller in modern varieties.

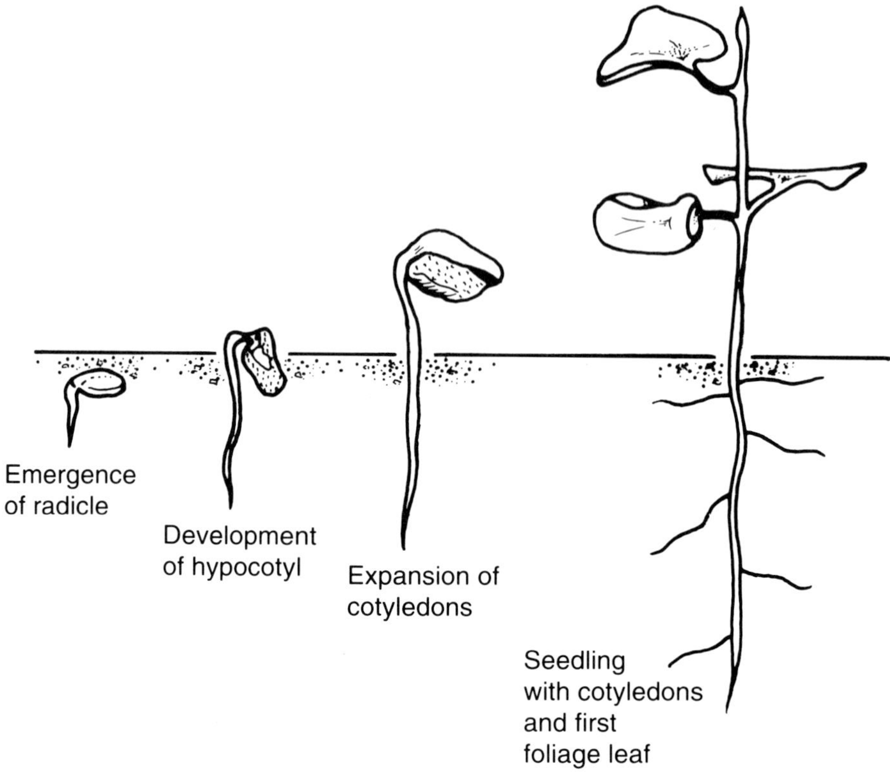

Fig. 1.5. Stages of seedling development.

Morphology

The cotton plant is typically a small bush about 1.0–1.5 m high, with a vertical mainstem and more or less horizontal branches giving a pyramidal shape when free from competition. The flowers are showy with cream to yellow petals; the fruit is green, and splits open when ripe to expose the white fluffy seed cotton, which usually remains hanging from the open bolls until it is picked.

Cotton is nearly always grown as an annual crop from seed, but if left to itself or ratooned it can survive into a second season. This provides a vehicle for the carry-over of pests and diseases from one season to the next, and one of the first measures in pest control is the establishment and enforcement of a close season between harvest and sowing when no live cotton plants are permitted, either as stand-over cotton or as volunteer seedlings.

Given sufficient moisture the seed will germinate in four to ten days depending on the temperature. The radicle emerges first, and as it grows downwards the hypocotyl elongates dragging the whole seed up above the soil surface; the cotyledons unfold, pushing off the ruptured seed coat, and the neck of the hypocotyl straightens so that the cotyledons are horizontal and the apical bud between them is facing upwards. The radicle grows downwards forming a taproot, from which branch roots grow sideways and downwards absorbing water and nutrients (Fig. 1.5).

The main stem of the cotton plant grows from the apical bud between the cotyledons. The cotyledons are opposite each other, but subsequent mainstem leaves are alternate and spirally arranged. Each leaf subtends a node with two or three buds which may remain dormant or produce lateral branches; even if they remain dormant, their positions are clearly indicated by the petioles of the mainstem leaves.

The cotyledons are heart-shaped with short petioles; mature leaves are about as broad as they are long, three to five lobed for about half their length, usually with a nectary on the underside of each main vein; the petiole is about the same length as the leaf. The leaves vary in size, shape, texture and hairiness according to the species or variety.

Branches from the mainstem are of two different types, monopodial (vegetative) and sympodial (fruiting). The monopodial branches are similar to the mainstem as growth continues through the apical bud, and the leaves and nodes are arranged in a spiral. The sympodial branches arise from the monopodia; the apical bud produces a flower while growth of the branch continues from a lateral bud at the same node, giving a zigzag appearance to the branch, while the monopodial branches are relatively straight.

The first branches on the mainstem above the cotyledons are monopodial, but at a certain point, which depends mainly on variety and temperature, the mainstem starts to produce sympodial branches. This point is called the node of the first fruiting branch (NFB), and every node from then on will produce a sympodium unless the plant is injured in some way.

These branches form the main fruiting structure of the plant, and about 60% of the crop is normally set on the first two nodes of the first ten fruiting branches from the mainstem. A pest management programme aims to protect as many of these bolls as possible. An early, well matured crop is easier to pick, and checks the vegetative growth of the plant.

The NFB is an important character and is recorded by counting the mainstem nodes upwards from the cotyledons, which are usually counted as zero. The NFB of the average Upland cotton is between four and six, but is much higher in the diploid Old World cottons and in *G. barbadense*: Tanguis cotton has an NFB of 12 or more. The lower the NFB the earlier the crop and the fewer the possible number of monopodial branches. The number of nodes below the NFB which actually produce branches depends on the vigour of the plant – in dry or poor conditions, or with very close spacing they may all remain dormant, while a vigorous, well watered and well spaced crop can produce monopodia at every node down to and including the cotyledonary nodes. Vigorous monopodia can grow as tall as, or taller than, the mainstem, and in stormy weather such branches are liable to break away at their junction with the mainstem.

While each node from the NFB upwards produces one fruiting branch, a second and rarely a third branch can develop from the same node; these branches may be sympodial or monopodial, but the most common is a short sympodium bearing a single flower. In a really vigorous plant, all sorts of buds which are normally dormant will develop, producing an exceedingly complex pattern of branching, as well as a second flower at many of the nodes on the fruiting branches.

The flowerbuds are enclosed in three triangular leaf-like bracts, usually deeply toothed, forming a three-sided pyramid paradoxically called a 'square'. This remains closed until pushed open by the elongating bud a day or so before flowering. The flower is about 50 mm in diameter with five creamy-white to yellow petals, which turn pink within a day and later red to purple before being shed along with the staminal column and style. There is a purple spot near the base of each petal in all species except *G. hirsutum*.

The basic pattern of flowering is fairly simple and surprisingly regular. Assuming that no bud shedding takes place before the flowers open, the first flower is always at the first node of the first fruiting branch, that is the branch which develops from the NFB. Flowering then proceeds upwards from branch to branch with a vertical flowering interval (VFI) of about three days, and outwards along each sympodium with a horizontal flowering interval (HFI) of about six days. It follows that the second flower on the first branch will open on the same day as the first flower on the third branch (six days after the first flower). Although this ratio of 2:1 between HFI and VFI may increase slightly as flowering progresses, it is a very useful approximation: any flower is contemporary with that two branches below and one node further from the mainstem.

The flowering zone on each plant can be visualized as a hollow cone which gradually moves upwards and outwards from the first flower: the spiral pattern illustrated by Balls (1915) and copied by others is incorrect. The actual flowering intervals depend on temperature, with the HFI ranging from 5.26 days in the hotter parts of the USA to 11 days in Uganda at an altitude of 1150 m just north of the equator; the VFI is approximately half of these values.

Only a small proportion of the flowerbuds produced by the cotton plant reach maturity as open bolls, and this proportion decreases as the plant gets older. This can be partially explained by postulating that the developing bolls have priority in the demand for nutrients, the supply of which is limited by radiation and photosynthesis; as the size and number of bolls increase, the demand for nutrients exceeds the supply, when surplus buds and small bolls are shed. Physiological shedding occurs for this and other reasons, and may reach quite high proportions. Counts of pest damage among fruiting bodies collected from the ground below the crop must be assessed in relation to similar losses which would have occurred in any case. Insect damage to the buds causes the surrounding bracts to flare open, producing what is called a 'flared square'. Shedding is particularly common at two stages: when the buds are very small, called 'matchhead shedding' from the size of the buds; and a few days after the flower opens, the stage at which an unfertilized flower will shed. Once the boll is set it can sustain considerable insect damage without shedding.

When the corolla is shed after flowering, the bracts close again round the young boll. The boll takes about two months to ripen, but this period may be longer or shorter depending on the temperature. In about half this time the boll grows to full size while the lint hairs inside it are elongating; while in the second half of the boll period there is no further increase in size, but secondary thickening is being laid down in the lint cells.

While the loss of buds and small bolls by shedding delays the time when the demand for carbohydrates exceeds the supply, the loss of leaves has the opposite effect, reducing the supply. The original mainstem leaves are lost by senescence before the crop matures, but a new leaf is produced at every node while growth continues. A severe drought at any stage in the growth cycle will stop vegetative growth and cause partial or complete shedding of the leaves. Leaves damaged by pests or diseases are not always shed, but they lose their effectiveness for respiration and photosynthesis, reducing the supply of carbohydrates for growth and fruit production. Artificial removal of leaves has shown that defoliation right up to the time when the first bolls open can have an adverse effect on yield.

Although the fibre or lint is worth eight to ten times as much as the seed in a batch of seed cotton, the seed does produce a number of useful by-products. Linters are the short fibres left on the seed after ginning and are a source of pure cellulose; when cleaned and sterilized they produce cotton wool for medical and other uses. The husks of the seed have no great value, but may be used as fuel for the ginnery or oil mill. The kernels provide an edible oil and a cattle cake rich in protein. These are heated or cooked to rupture the oil-bearing cells enabling the oil to flow freely, which is then extracted in a screw press or with a solvent. The screw press leaves at least 5% of oil in the cake or meal, while solvent extraction leaves less than 1%. In China stalks are processed to make hardboard.

Ecology

Soils

Cotton grows well on a wide range of soil types from almost pure sand to heavy clay, but a sticky clay is difficult to manage in wet weather and tractor operations may be impossible. A good crop needs a well-developed root system, and any soil condition which restricts root growth will prevent the crop from reaching its full potential. Restrictions to look for are a hard pan developing below the cultivated top soil, compaction caused by the use of tractors and farm machinery, especially when the soil is wet, and a high water table or waterlogging which prevents air reaching the roots.

Cotton tolerates a fairly wide range of soil acidity and alkalinity, but while it will grow on soils as acid as pH 4.0 the best crops are grown at pH 6.0 or more. The reaction to alkalinity depends on the kind of salts and the soil type: black alkali containing sodium carbonate is more destructive than white alkali containing sulphates and chlorides, which may reach pH 8.0 or more with no ill effects on the cotton crop. The alkali is more destructive in sandier soils.

Temperature

Broadly speaking cotton growth occurs between 14°C and 38°C, and in this range the higher the temperature the faster the growth (Fig. 1.6). According to Purseglove (1968) 'the optimum temperature for germination is 34°C, for growth

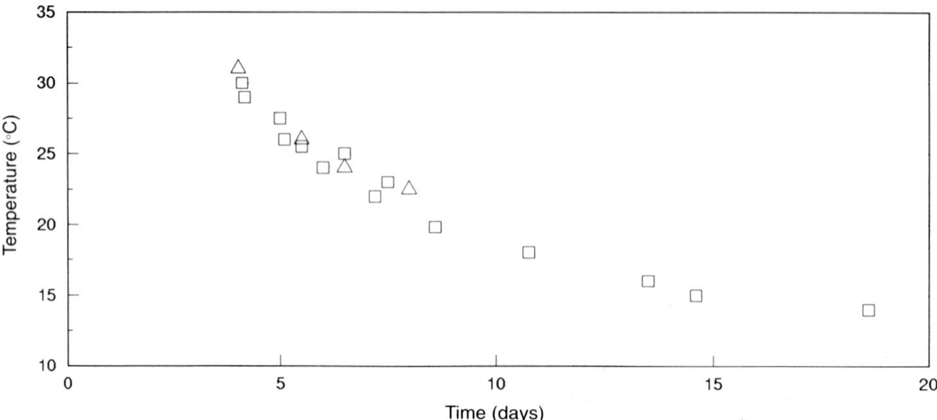

Fig. 1.6. Horizontal flowering interval in days. Two isophases: □ (Zaitzev, 1928); △ (Mauney, 1966).

of seedlings 24–29°C and for later continuous growth 32°C'. But temperature in the field is subject not only to seasonal changes, but also to the daily cycle of day and night. The effect of this diurnal variation is different from that of a constant temperature with the same mean, and this is especially important at high temperatures, where growth stops at about 38°C. Cotton can withstand considerably higher air temperatures during the day if they alternate with cool nights, but no growth takes place around midday under direct sunlight in really hot climates like that of Egypt.

Low temperatures and slow growth mean that the crop takes longer to mature, giving more time for pest populations to build up and damage the crop. On the other hand a long growing season may allow the crop time to recover from damage by early season pests, and still produce a late crop. In temperate latitudes, sowing has to wait for the temperature to rise in the spring; in the USA the sowing date is later in the north, following roughly the movement of the 16°C isotherm, and the rate of growth accelerates as the temperature rises in spring and summer. In the tropics and subtropics growth is possible all the year round, and the sowing date depends on water supply and pest incidence rather than on temperature.

Sunshine

Light is essential for photosynthesis, but the process do not require direct sunlight; it can proceed quite satisfactorily in cloudy weather. In a sunny climate, periods of cloudy weather can cause the shedding of buds and small bolls (Dunlap, 1945; Goodman, 1955), but where cloudy weather is normal in the growing season, as in Ecuador and Peru, flowering and fruit setting do not seem to be affected. Sunlight has a direct effect on temperature, and the effect of these two factors may be difficult to separate in the field.

Wind

High winds can affect the cotton crop directly in two ways. The first of these is lodging, where the growing crop is flattened with either partial uprooting or permanent bending of the mainstem. This results in a tangled mass of vegetation, the crop is difficult to penetrate for spraying and picking, and is liable to bollrot in the moist conditions under the leaves. Some varieties are more susceptible to lodging than others, and rank growth makes lodging more probable. Topping the plants before they get too tall may reduce the susceptibility to lodging. The second effect occurs after the bolls have opened, when high winds may blow the ripe cotton out of the open bolls on to the ground where it collects dirt and trash. This has led to the development of storm-proof varieties in which the ripe cotton is held more tightly in the husk; this does not affect mechanical picking, but makes hand picking more difficult.

Wind speed is an important factor in the evaporation of moisture from the crop, and hence affects the water requirement. Wind can also assist the movement of flying insects, allowing them to spread beyond their normal range of flight.

Water

Most crops, including cotton, produce their best yields when the water requirements of the crop are balanced by the water available to it, either from rainfall or irrigation. If we are to know the extent to which this balance is being achieved, it is necessary to measure or estimate both supply and demand. Supply is not too difficult: a rain-gauge to measure rainfall and metering devices in the irrigation canals are accurate enough for most purposes, but losses by seepage from irrigation canals, by deep drainage and by surface runoff are less easy to measure. On the demand side, crop water requirements are almost always estimated. Direct measurement is both difficult and expensive, and is used mainly as a research tool at a limited number of sites to check the accuracy of various methods of estimation.

Four methods of estimating crop water requirements are described in detail by Doorenbos and Pruitt (1977), who also discuss the accuracy and limitations of the different methods, and of the instruments used. Estimates are usually made in three stages: first the potential evapotranspiration from a 'reference crop', a well watered uniform grass cover, is calculated from climatic data; secondly this is multiplied by a 'crop factor' which depends on the crop being grown, its stage of growth and the length of the growing season; thirdly this figure is adjusted for the effect of certain local conditions and agricultural practices in the area.

Between supply and demand there is the buffering effect of water storage. The quantity that can usefully be stored in the soil itself will depend on the physical properties of the soil and the depth of the root zone of the crop. Most of the water held in the root zone between wilting point and field capacity is available to the growing crop, and may range from sufficient water for the whole season to a few days supply in a light sandy soil. Storage capacity can be supplemented by dams and reservoirs.

Supply, demand and storage can be combined in a water balance sheet which can be used in a variety of ways. For rainfed crops the best sowing date, other things being equal, is that which matches the crop water requirement with the average expectation of rainfall. In areas where cotton has not been grown before, the same matching can be used to determine what are its chances of success. For irrigated crops, the application of irrigation water can be based on a water balance sheet, which will show when a critical water deficit is reached. In planning an irrigation scheme, the canal system must be capable of delivering sufficient water to meet peak water requirements, water must be available when required throughout the growing season, and the capacity of any storage works must match the extent and timing of the demand for water. The water balance sheet provides essential data for the water engineers.

PRODUCTION

The Producers

Cotton is grown by a large number of individuals whose cotton holdings range in size from less than one to over one thousand hectares; by contrast, man-made fibres are produced by large chemical companies in a relatively small number of factories. The individual grower can decide each year whether he wants to grow cotton or not, so that the world cotton production tends to be more volatile than that of man-made fibres, which require long-term capital investment.

The major cotton producing countries are the United States of America, the former Union of Soviet Socialist Republics and the People's Republic of China, each of which produces more than two million tons of cotton annually. The rest of the world crop comes from more than 70 different countries, which produce appreciable quantities of cotton either for export or increasingly for home consumption. The size of the figure for cotton production gives no indication of how important the crop is to the country's economy: in an industrialized country like the USA, cotton production contributes only 0.1 per cent of the gross national product (GNP); but in an agricultural country like Zimbabwe, cotton is the third most important crop after tobacco and maize, contributing 2.6% to the GNP. Table 1.4 shows cotton production data for the main producing countries of the world.

The cotton growing areas of the world lie in a belt between 42°N and 33°S; outside these limits the summers are either not long enough or not hot enough for the cotton plant to complete its growth cycle, although quick maturing varieties are being bred to extend these limits in the former USSR, Europe and China. The cotton belt can be divided into three climatic zones: temperate, subtropical and tropical. Desert conditions are found in parts of the first two

Table 1.4. Production of cotton by countries producing over 50,000 tonnes 1987–1990 (expressed in thousand tonnes).

	1987	1988	1989	1990	Average
Africa					
Chad	48	53	58	55	54
Egypt	352	322	293	310	319
Ivory Coast	93	114	128	106	110
Mali	79	75	97	97	87
South Africa	59	69	54	56	60
Sudan	163	136	142	125	142
Tanzania	63	85	84	59	73
Zimbabwe	90	116	94	79	95
America					
Argentina	100	282	195	270	212
Brazil	552	837	625	660	668
Colombia	134	97	108	122	115
Mexico	220	312	162	168	216
Paraguay	84	194	220	230	182
Peru	95	93	103	93	96
USA	3214	3355	2655	3353	3144
Asia and Australia					
Australia	214	284	286	305	272
China	4245	4149	3788	4166	4087
India	1085	1477	1612	1497	1418
Iran	104	121	117	144	122
Pakistan	1430	1426	1456	1530	1460
Syria	123	180	155	180	160
Turkey	537	650	585	628	600
Former USSR	2460	2762	2686	2613	2630
Europe					
Greece	174	235	255	265	232
Spain	83	120	62	77	86

Source: *FAO Quarterly Bulletin of Statistics*, Rome.
1 metric tonne = 4.5929 bales of 480 lb.

zones, where cropping is only possible with irrigation; a hot irrigated desert probably provides the ideal climate for cotton growth and yield.

The three major producers of cotton (USA, the former USSR and China) lie in the temperate zone, where the winters are too cold for cotton to grow, and sowing takes place in spring when the temperature reaches about 16°C, usually between March and May. The crop is harvested in the autumn, when ginneries in America have occasionally had to cope with cotton harvested after a fall of snow. The length of the season is limited by the date when spring temperatures are sufficiently high for germination, and at harvest by the onset of winter. Low temperatures will stop growth, and any bolls remaining on the plant will either remain green and unopened until the spring, or if there is a frost they will open prematurely, even though the lint is immature and thin walled.

Much of the cotton in the temperate zone is grown in desert conditions with irrigation – California and Arizona in the USA, all of the former USSR and an increasing proportion in China. In the Middle East and Europe the crop is also mainly irrigated – Spain, Greece, Turkey, Syria and the northern region of Iran adjoining the former USSR. Desert conditions and cold winters affect the incidence of cotton pests: pink bollworm only occurs in the western states of the USA, while the boll weevil migrated from Mexico into the southern and eastern states, but has not spread into the west. The land area with a temperate climate in the southern hemisphere is limited, but Argentina and Paraguay lie outside the tropics, and irrigated crops are grown in South Africa and Australia where the climate is more subtropical than temperate.

In the subtropical zones growth of cotton is possible all the year round and water supply rather than temperature is the limiting factor. There is a single wet season in the year: north of the equator this lasts approximately from May to September, while in the south it lasts from November to March. Annual rainfall is often less than 1000 mm per annum, and a well-distributed 500 mm is usually sufficient to produce a good crop. Raingrown crops are sown at the start of the wet season and harvested in the dry season when the rains have finished. Picking therefore takes place from November to January in the northern hemisphere and from May to July in the southern, but the time when the cotton comes on to the world market depends on how quickly it can be transported and ginned.

Most of the world's deserts lie along the tropics of Cancer and Capricorn, the largest being the Sahara in Africa, which continues eastward through Arabia as far as Pakistan. Irrigated cotton is grown in some of these deserts: in the north of Mexico, in the Nile valley and Israel, in Pakistan and in parts of India; in the southern hemisphere on the coastal plain of Peru, along the Orange River in South Africa, and in New South Wales and Queensland in Australia. The Ord River scheme on the north coast of Western Australia was abandoned because of lack of water, high production costs, insect infestations and problems with fruit setting at high temperatures.

The growing season for irrigated crops in the desert regions depends on when water is available. This may depend on the rainfall hundreds of miles from the irrigated crop, and is affected by competing claims for water by the various users. In the case of the Nile, the Sudan guarantees the traditional flow of water to Egypt where the cotton crop is sown from February to April. The water from the Blue Nile flows freely through the sluices of the Sennar Dam from January to July, but as the river flow increases the flood water is surplus to the needs of Egypt and is retained in the dam to supply the Gezira with its water. Cotton in the Gezira is sown in August and September, and matures through the winter months. July and August are also the period of maximum rainfall in the Sudan, when 70% of the year's rain usually falls; some 400 mm is the average for the year at the Gezira Research Station, being drier to the north and wetter to the south. This 400 mm was sufficient for the *dura* (sorghum) crop which was grown before the Gezira scheme started in 1925, but is not reliable and usually insufficient for a cotton crop; in fact anything more than average rainfall, even if it falls before the cotton is sown, has been shown to reduce the yield of irrigated cotton.

The soils of the Abyan delta in Yemen, and Tokar and Gash deltas in the Sudan, must be almost unique in that they can store sufficient water from one flood irrigation to carry the cotton crop right through from sowing to harvest. The rivers flood twice a year: the *seif* floods in April, May and June are used to grow crops other than cotton, and the much heavier *kharif* in July, August and September is used mainly for cotton. Most of the flood water is controlled and distributed through canals to the cotton fields, replacing the field to field watering used originally in the deltas. The average but very erratic rainfall is of the order of 50 mm per annum.

The problems of irrigation are the same for cotton as for other crops: increasing soil salinity, lowering of the water table and increasing demand for limited water resources.

The tropical or moist equatorial zone is characterized by two rainy seasons in the year, and has been defined as having a mean annual rainfall exceeding 1000 mm and not more than four months during the year with a mean monthly rainfall of less than 50 mm. Much of this zone is occupied by tropical rain forest, but round the perimeter of the forest is often an area of tall grass savannah and it is in this peripheral region that cotton is grown.

The West Indies, although well north of the equator, have a classical tropical climate and this enables them, without irrigation, to grow Sea Island cotton with the longest staple length in the world.

Cotton crops near the equator are grown under a range of conditions. In Colombia there are two distinct cotton production zones with different growing seasons: the crop in the central zone is sown in March at the start of the first rains or in late January if irrigation is available; in the north of the country sowing takes place from late July to September. In Ecuador, within a few degrees of the equator, there is only one rainy season starting in January. A single rainy season also occurs in Indonesia due to the trade winds striking the high central spine of the western leg of Sulawesi. Westerlies deposit the peak of their rainfall in January on the west side; while within a few miles, in the eastern half of the peninsula, the peak changes to May–June when the easterly trade winds are blowing.

With two rainy seasons, the food crops are usually sown at the start of the longest rainy season, while cotton is sown at the end of this season and matures through the second rains. In Kenya the two seasons are called the long and the short rains, but the so-called short rains are often longer and heavier than the long.

In the northern and central zones of India, raingrown cotton is sown in June–July and the irrigated crop rather earlier in May. In the southern zone, however, some of the irrigated cotton is sown in the other half of the year: December–January in the coastal region of Andhra Pradesh, and February–March in parts of Tamil Nadu. Two overlapping cotton seasons in the same area facilitate the carry-over of pests from year to year.

Production Methods

The conditions under which cotton is grown vary enormously, not only in the size of holding but in the sophistication of the methods employed. A peasant farmer using hand cultivation may have between two and four hectares of arable crops, of which perhaps half a hectare is cotton (Prentice, 1972), while a fully mechanized farm of 120 to 160 hectares may have 40 hectares of cotton. Some machinery may be owned and operated by the farmer, but some operations are carried out by specialized contractors. Estate production may require management of over 1000 ha of cotton. The wide range of production practices and their various combinations make any broad classification difficult, especially as the motivation may be different: the peasant farmer may not use any fertilizer or insecticide because he is not convinced that they are economic for him, or because he lacks the money or credit to purchase them; another farmer having access to inputs may apply increasing quantities of fertilizer and insecticide until the costs become prohibitive and the government is forced to control their application; while others are limiting their use of chemicals partly to slow down the development of resistance in a pest species, and partly to reduce the contamination of rivers and water supplies, and preserve the environment for future generations. It is often worth while examining whether a low input, low yielding crop may be more profitable to peasant farmers than a high yielding one requiring a high input. Such farmers do adapt recommendations to fit their own economic circumstances by reducing the number of pesticide sprays (Farrington, 1977).

Many of the operations on the cotton crop are the same as those employed for growing other arable crops and, in the same community, the methods used for cotton will in general show the same level of sophistication as those used for other crops.

Agronomy

Land Preparation

Methods of land preparation are much the same for all arable crops and will depend in any one area on the general level of agricultural development. The aim is to encourage rapid growth of the crop but to check the growth of weeds; cotton in its young stages is particularly intolerant of weed competition. With hand hoeing, large weeds are first removed and burned, and the soil is then loosened to produce a smooth seed bed, leaving any small weeds on the surface to dry out.

Cultivation is not as a rule very deep, but it may be necessary to break up a hard pan which is impeding root development. Some farmers in fact favour zero tillage, where the soil is left undisturbed and weeds are controlled by herbicides. Ploughing with oxen or tractors enables the farmer to cultivate a much larger area than hand hoeing, but this is only worth while if he has sufficient resources to cope with the subsequent operations: weed control, fertilizers, pest control and harvesting. Ploughing is followed by surface cultivation with a tined or disc harrow to kill off any seedling weeds and to leave a smooth seed bed. Cotton does best when the soil under the seed is fairly firm, and a rotavator tends to leave the soil too loose and pulverized.

There is considerable controversy over whether cotton should be grown on ridges or on the flat. Ridging is essential for furrow irrigation and contour ridging may be used to control soil erosion and assist rainfall penetration. On fairly flat land ridging may, or may not, increase cotton yields. Experimental results suggest that tie-ridging (basin listing) increases yields in a dry season, while cotton growing on ridges may survive temporary flooding better than on the flat.

Sowing

Cotton seed as it comes from the gin is usually fuzzy, and the individual seeds stick together making it difficult to sow uniformly either by hand or by machine. To facilitate sowing, the fuzz can be reduced by mechanical delinting, a process similar to saw ginning, and income from the linters more or less pays for the delinting. Alternatively seed is completely delinted with acid, either by the relatively simple wet acid process using sulphuric acid, or the more complex dry acid process using vapour from hydrochloric and nitric acids. The wet acid process allows insect damaged seed to be floated off when the seed is washed, leaving only high quality seed. The dry process involves much greater capital expenditure and needs to be closely supervised, but avoids the problems of providing for and disposing of large quantities of washing water.

Cotton is generally sown in rows between 60 cm and 100 cm apart either as a continuous ribbon or as clumps of four to six seeds at 20–50 cm spacing (hill-drop planting). Either method may be used with machine seeders, but hill-drop is preferred for hand sowing. There is considerable latitude in plant population after thinning, which may range from 30,000 to 60,000 plants per hectare, with little or no effect on yield.

Sowing should be at a uniform depth of 20 to 50 mm and the seed covered with loose soil; the wetter the soil the shallower the depth of sowing. Fuzzy seed is usually sown at the rate of 25–35 kg per ha but when seed is provided free, as in many African countries, much higher rates have been reported; acid delinted seed is sown at 12 to 16 kg per ha. A reserve of seed should be kept especially for raingrown cotton, in case resowing is necessary. Filling in gaps is generally considered unnecessary.

Time of planting will depend on a number of factors, including temperature, water supply, pest incidence and availability of labour. In temperate climates cotton must be sown in the spring as soon as temperature permits, in order to benefit from the warm summers (in parts of China the seed is sown under polythene to ensure the soil warms up more rapidly). In the subtropics where there is unimodel rainfall, sowing is timed so that the crop water requirements match the pattern of rainfall; in tropical areas with two wet seasons, the time of sowing will depend more on pest incidence and labour requirements. With peasant cultivation, food crops take priority over cotton, so with limited labour, the cotton crop is often sown much later than the optimum date for high yields. Here the possibility of sowing in dry soil before the start of the rains should be considered to avoid a clash with the sowing of food crops. However, this may result in competition for labour to weed the cotton and food crops. Problems of water supply are much simpler when irrigation is available, but other factors remain the same.

Thinning

Ideally the amount of seed sown should be just enough to produce the desired stand, but this is rarely possible because of the likelihood of unforeseen losses. Even in the most highly developed systems of agriculture it is common practice to sow an excess of seed and to thin out the seedlings to the spacing and population required. Seed should have a minimum of 80% germination in laboratory tests, but in areas subject to attack by *Dysdercus* spp. germination of fuzzy seed can be very much lower. The combined force of several seedlings will usually give better emergence than widely spaced single seeds, especially on soil prone to capping. Thinning is usually done when the seedlings show two or three true leaves, about two or three weeks after sowing, and preferably when the soil is damp so that it causes the least possible disturbance to the remaining seedlings. Late thinning must be avoided. Hand thinning allows some selection to be made for the most vigorous seedlings and extra plants can be left to compensate for gaps in the stand. No completely satisfactory machine has been devised for thinning, but machinery can be used to remove most of the unwanted seedlings and so reduce the amount of hand labour. Hand thinning is universal among peasant farmers and is still common practice even when other operations are highly mechanized.

Rotations

Rotation of cotton with other crops should be followed in the interests of sound crop husbandry, although in practice the nature of the rotation is likely to be governed by economic as well as agronomic considerations. Factors requiring attention when planning a cropping sequence are:

1. Soil erosion is likely to be increased on lands with slopes greater than about 3% if successive cotton crops are sown.
2. Cotton should follow a short season crop or an early ploughed fallow to conserve soil moisture; this assists in the establishment of cotton especially in seasons of light early rains.
3. Care should be taken to avoid following a highly fertilized crop with cotton as this can cause excessive early vegetative growth.

4. Successive cotton crops are liable to increase the occurrence of nematodes, *Rhizoctonia*, *Fusarium* and *Verticillium* wilts.
5. Economic considerations.

Fertilizers

Fertilizer requirements depend to a large extent on the amount of nutrients already present in the soil and so will vary widely from one locality to another; soil analysis and local fertilizer trials will be needed before precise recommendations can be made. Of the major nutrients, some added nitrogen (N) is nearly always beneficial, potash (K) is rarely needed, while phosphate (P) will depend less on total P than on how much is available for plant use. Here again the economics of a low input/low yield regime must be compared with those of high input/high yield. Excess N, especially when combined with heavy watering, can cause rank growth leading to bud and boll shedding and possibly lodging of the crop. Some soils may benefit from additional P even though a soil analysis shows high levels of total P, because the P is fixed in an unavailable form in the soil complex. It is quite common for a small application of P to depress yields, while applications above a certain level give responses once the capacity for fixation is satisfied.

Phosphate, potash and some of the nitrogen are usually applied as a basal treatment about the time of sowing, and mechanical seeders can often deliver both fertilizer and seed. The remainder of the N is delayed to guard against rank growth and is only applied if responses can be expected or where symptoms of N deficiency appear.

Of the minor elements, the absence of boron or sulphur is the most likely cause of deficiency symptoms in cotton.

Weed Control

Weed control in cotton differs little from that in other crops and the methods used are the same: hoeing, inter-row cultivation and the use of herbicides. Weed competition is most critical in the first six or eight weeks, before the crop is sufficiently established to provide ground cover and shade. Pre-emergent selective herbicides are ideal for controlling weeds during this period and are sprayed after sowing but before the emergence of the cotton seedlings. Sprays may be applied to the whole area or confined to a band along the row where it will become increasingly difficult for hoes and cultivators to reach without damaging the young cotton plants. Late season weeds can be a problem, in particular those which shed seeds on the open cotton, and bindweeds which climb over the cotton plants and tangle with the branches, preventing access between the rows for weeding, spraying and harvesting.

Some perennial weeds are especially difficult to control because no satisfactory chemicals have been found which do not also damage the cotton crop: the most important are nut-grass (*Cyperus* spp.), Johnson grass (*Sorghum halepense*) and Bermuda grass (*Cynodon dactylon*). Herbicides can sometimes be applied to the growing crop using shielded nozzles to direct the spray on to the soil between the rows of cotton without touching the stems and leaves of the cotton plants.

Insect Control

Specific methods of control are considered for each of the major insect pests included in this book. Overviews of cultural, biological and chemical methods are also given in later chapters.

Apart from a few areas in which inorganic insecticides such as calcium arsenate were used, cotton was grown without insecticide until 1945. Since then an increasingly large area of cotton has been sprayed, although there are still areas where profitable crops are grown under a low input, low yield system.

Harvesting

Much of the world's cotton is picked by hand, and will continue to be so as long as labour is relatively cheap; hand picking gives a cleaner product which commands a premium on the world market.

The peasant farmer is seldom concerned about the efficiency of harvesting as the whole family helps using whatever is available – baskets, bags, cloth sheets – to collect the picked cotton, and spending the afternoons and evenings separating out any stained or damaged locules (locs) from the clean first-grade seed cotton. On some farms storage of large amounts of seed cotton, due to delays in transport or marketing facilities, can be a major problem.

With a larger farm requiring hired labour, the same care and cleanliness cannot be expected, but much can be done to improve efficiency. Specially designed picking bags leave both hands free for picking, and have a separate pocket for any cotton seen to be damaged or dirty. Moving between two rows, the picker takes cotton from the nearest half of the bushes on both sides; damaged bolls are left on the plant until the last pick; the bottom of the bag may be tied with cord which is released for easy emptying. The most efficient picking was seen in the work of picking gangs in the USA before the introduction of mechanical harvesting.

Nowadays all the cotton in the USA and Australia is picked by machine, as is about three quarters of the crop in the former USSR; the proportion is increasing in Greece, Mexico and Argentina. Mechanical harvesters need large level fields in which to operate satisfactorily, as well as a network of service stations for spares and maintenance. There are two main types of machine, spindle pickers and strippers, with spindle pickers being used for good quality cotton. Each unit consists of a pair of belts which enclose a single row of cotton plants; the spindles projecting from the belts rotate independently and tangle with the seed cotton in the open bolls, pulling it away from the husks. Strippers are nonselective and harvest husks, green bolls and some leaves and branches as well as ripe cotton; they do not harvest individual rows, so are suitable for narrow row cotton; they need much more elaborate cleaning machinery at the ginnery to separate the cotton from the trash. Mechanical harvesting therefore requires a specialized production system with appropriate cultural practices and plant characteristics.

Marketing

The organization required to market and gin the seed cotton from a large number of individual farms differs from country to country, and ranges from the sale by the growers at local markets set up by a cotton marketing organization, to the ginning of individual farmers' crops on contract by privately owned ginneries.

The peasant farmer's crop is too small to warrant individual handling, so most governments set up marketing boards responsible for buying, ginning and marketing cotton on a country-wide scale. This enabled the government to fix prices and to maintain stabilization funds to even out fluctuations in world prices. Increasingly, primary marketing has been taken over by cooperative societies which do not always buy directly from the farmers. They employ buyers to operate small markets on commission, while sometimes independent middlemen buy cotton from the farmers at less than the official price, collecting and transporting it for resale in larger lots to the board or cooperative. In the USA privately owned ginneries contract to gin the cotton from large individual farmers, and deliver the baled lint to merchants or wholesalers as directed by the farmer. The ginner may take over the cotton seed as part payment for ginning, and sell it independently. Each country has evolved its own organization which is continually being modified to meet changing needs, resulting in a great variety of systems.

Crop Residues

Prompt disposal of crop residues after harvest is important to prevent the carry-over of pests and diseases from one season to the next, and is often enforced by the declaration of an official close season. Where cultivation is largely by hand, the cotton stalks are uprooted or chopped off just below ground level, and piled in heaps to dry before burning them. Where machinery is available, the parts of the plant above ground are first chopped up by a rotary mower or flail and then ploughed in; the ploughing also dislodges the stumps of the cotton plants, preventing their regeneration. The aim is to bury all the plant residues, and a mouldboard plough does this more efficiently than a disc plough. In China stalks are also processed into hardboard.

Loose seed cotton spilled around the fields and along the roadsides should be collected and burned. The same applies to cotton stores and ginneries, which may require fumigation to destroy any pests which can overwinter there. Volunteer cotton seedlings which emerge in and around the cotton fields should continue to be destroyed before the start of the close season.

References

Alefeld, F.G.C. (1861) Ueber die Stellung der Gattung Gossypium und mehrer Anderer. *Botanische Zeitung* 19, 229–301.

Balls, W.L. (1915) *The Development and Properties of Raw Cotton.* Black, London.

Clegg, G.G. (1931) The stapling of cottons, *Shirley Institute Memoirs – Series A 3,* 33–62.

Doorenbos, J. and Pruitt, W.O. (1977) Crop water requirements. *FAO Irrigation and Drainage paper No. 24* (revised). FAO, Rome.

Dunlap, A.A. (1945) *Fruiting and shedding of cotton in relation to light and other limiting factors.* Bulletin No. 677. Texas Agricultural Experimental Station, College Station, Texas.

Farrington, J. (1977) Research-based recomendations versus farmers' practices: some lessons from cotton spraying in Malawi. *Experimental Agriculture* 13, 9–15.

Fryxell, P.A. (1979) *The Natural History of the Cotton Tribe.* Texas A & M University Press, College Station, Texas.

Goodman, A. (1955) The effect of cloudiness upon the shedding of fruiting points from cotton at Tokar Delta. *Empire Cotton Growing Review* 33, 24–34.

Hector, J.M. (1936) *Introduction to the Botany of Field Crops,* Vol. 2. Central News Agency, Johannesburg.

Mauney, J.R. (1966) Floral initiation of Upland cotton in response to temperature. *Journal of Experimental Botany* 17, 452–459.

Prentice, A.N. (1972) *Cotton.* Longmans, London.

Purseglove, J.W. (1968) *Tropical Crops: Dicotyledons.* Longman, London.

Sasser, P.E. (1981) The basics of high volume instruments for fibre testing. *Beltwide Cotton Production Research Conference 1981.* National Cotton Council, Memphis, Tennessee.

Zaitzev, G.S. (1928) Effect of temperature on the development of the cotton plant (Quoted by Hector, 1936).

II Insects and Mites

2 Insect and Mite Pests: General Introduction

G.A. Matthews

International Pesticide Application Research Centre, Imperial College at Silwood Park, Buckhurst Road, Sunninghill, Ascot, Berkshire SL5 7PY, UK

Introduction

Hargreaves (1948) listed 1326 species of insects recorded on cotton. The great majority of these are of little or no economic importance so far as cotton is concerned, and some, although collected on cotton, have not been observed to feed on it. Many are sporadic, casual or accidental visitors to the crop. Fortunately only a relatively small number of insects are of major economic importance, but individually or by their combined effects they can cause significant losses in yields. Economic entomologists have paid considerable attention to these insects because agricultural departments have given emphasis to fostering cash crops to provide revenue for development, and the need to provide a reliable source of fibre for the textile industries. The following chapters provide an updated review of the current knowledge of the main pest species and the control tactics that are being adopted.

Origins of the Pests of Cotton

The pests of cotton fall into three classes according to the host plants on which they occur in nature: those that are virtually restricted to the genus *Gossypium* or its close relative *Cienfugosia*; those that occur also on other members of the order Malvales; and those that feed freely or preferentially on other orders of plants.

Pests Restricted to *Gossypium* and its Immediate Relatives

As Pearson (1958) pointed out, the most important pests of a crop plant are often insects that are associated with the plant species in its uncultivated, wild state. However, cultivated cottons of the genus *Gossypium* do not occur in the wild state, with the possible exception of *G. herbaceum* var. *africanum*. The boll weevil *Anthonomus grandis* was thought to have originated on the wild species of *Gossypium* in southern Mexico, possibly *G. gossypiodes* and *G. trilobum*, but recent research has suggested that *A. grandis* evolved on plants of the genus *Hampea* and subsequently shifted to cotton (Burke *et al*., 1986; Cate and Summy, 1991). This may have coincided with the origin of the tetraploid species of *Gossypium*, and the subsequent spread of the cultivated cottons by man in close contact with the wild hosts. The recent realization of the importance of *Hampea* has significant implications for pest management as there is a complex and diverse guild of parasitoids associated with *A. grandis* on *Hampea*. Apart from generalist parasitoids these have not shifted to a cotton habitat. Cate and Summy (1991) suggest that specific plant substances in *Hampea* are essential for one of the parasitoids *Bracon compressitarsis* to find its host, so biological control depends on these key chemical compounds.

It is interesting that *A. grandis* which has spread into South America can also breed on *Cienfugosia affinis* which is indigenous there. Furthermore the smaller boll weevil *A. vestitus* that occurs in the coastal regions of Peru and Ecuador apparently evolved on a form of *G. barbadense*, the 'native Peruvian variety of cotton' which may have arisen from the hybridization of the Old World diploid cotton being brought into contact by man with the truly wild species *G. raimondii*, that occurs there.

In Africa, *Diparopsis* spp. and in South America, *Sacadodes pyralis* both appear to have originated similarly from wild cotton *Gossypium* or *Cienfugosia* and subsequently spread to the cultivated cottons. Certain less important pests may have similar origins. Thus *Bucculatrix thurberiella* occurs on *G. thurberi* and in Africa *Apion soleatum* is only known on cultivated cotton.

Control of each of these pest species has been most successful where there has been rigid enforcement of crop destruction over an entire agroecosystem. Wherever there has been ratoon or standover cotton populations have survived or increased between seasons, and subsequent attempts at control with insecticides early the following season have resulted in disruption of the effects of natural enemies and often exacerbated the pest problems. Selection of cultivars which mature more rapidly so that harvest can be completed before diapause populations can increase is clearly an important component of such a strategy.

Pests Associated Chiefly with the Malvales

Many important pests of cotton stainers are species that otherwise feed chiefly, or exclusively, upon other indigenous plants belonging to the Malvales, and have fed on cultivated cotton when this has become available. *Pectinophora gossypiella* is perhaps the most important in this group as it has become of major importance worldwide, but, like those species confined to cotton it has been effectively controlled by a rigid close season. The primary hosts of *Earias* spp., the cotton stainers *Dysdercus* spp. and several minor pests are indigenous Malvales. The Malvales include four families which have a mainly tropical distribution. These are the Bombacaceae, Malvaceae, Sterculiceae and Tiliaceae. Among the most important alternative hosts of cotton stainers are the baobab tree *Adansonia digitata*, the Kapok tree *Ceiba pentandra*, species of *Sterculia* and numerous shrubs, including okra *Abelmoschus esculentus*, widely grown as a vegetable crop and other closely related *Hibiscus* spp. The relative importance of these host plants for cotton pests varies considerably. In some cases the presence of these hosts acts as trap crops and reduces the infestation on cotton, whereas in other areas their availability when cotton is not grown provides a means of survival between seasons.

Polyphagous Pests

The third and largest group of cotton pests consists of those with a wide range of alternative host plants. Some are the polyphagous cutworms, armyworms and leaf-eating larvae that do not have any special attachment to cotton, but attack it, along with other cultivated crops, as they happen to encounter it. Where irrigation extends the growing season and allows a range of crops to be grown throughout the year there has been a significant increase in the severity of some pest species. In Egypt the leafworm *Spodoptera littoralis* survives through the winter on berseem clover *Trifolium alexandrinum* grown as a fodder crop. *Helicoverpa* spp. and *Heliothis* spp. have become the most important insect pests of cotton in almost every cotton growing country because of the wide range of cultivated hosts as well as wild hosts. Devastating infestations on cotton often occur in association with maize, especially in monsoonal areas as cotton remains attractive even after the maize has matured. In irrigated areas a sequence of maize plantings can act as trap crops alongside cotton, but out-of-season irrigated wheat and other hosts enable populations to survive in areas where there would otherwise be a natural decline.

A similar situation exists with jassids, aphids and whiteflies that had to survive on a diversity of wild hosts during a dry season, but with continuous production of vegetables, these pest populations can continue even in the absence of cotton. Even if only a small number migrate to cotton, populations can build up to serious proportions as a result of rapid and prolific breeding in the cotton crop.

Insects Attacking the Various Parts of the Cotton Plant

Short notes on the species that attack each part of the cotton plant are given below as an introduction to the detailed information given in subsequent chapters that describe the most important insect pests. Notes on diseases are included, with mention of some of the very minor insect pests that do not merit detailed consideration.

Sown Seed and Seedlings

The germinating seeds and the seedlings of cotton, like those of other field crops, are liable to be attacked by a range of polyphagous insects, including the soil-dwelling larvae of beetles, such as wireworms (*Elateridae*), *Tenebrionidae* (which tend to replace the former in the tropics), and white grubs (*Melolonthidae*); and, above ground-level, by cutworms (the larvae of Noctuid moths) and many kinds of grasshoppers. In Africa, in addition, millepedes attack the germinating seeds and harvester termites are particularly troublesome pests of seedlings. Insect attack at this stage usually kills the plant, and a sweeping attack results in loss of stand and consequently in some reduction of yield, although some compensatory growth often occurs if the plants were closely spaced. Frequently, however, insect attack at this stage results in little more than normal thinning and, as a general rule, seedling pests are not important in Africa.

Diseases of seedlings include sore shin, caused by the fungus, *Rhizoctonia solani*, and the blackarm phase of bacterial blight, caused by *Xanthomonas malvacearum*. The former tends to occur in patches and is associated with poor soil drainage and unseasonably cool weather; it is not prevalent in tropical Africa. In contrast, the latter disease can be of first importance causing outright death of the seedlings, or partially crippling them so that the development of the plant is impaired; infected seedlings also serve as sources of secondary spread of the disease in the field. The disease can be much diminished by clean-up of infected debris from the previous season, by seed disinfection, effected by delinting with sulphuric acid or by dressing with chemical fungicides, or by the use of resistant strains of cotton.

Root

In comparison with many other field crops, cotton suffers little from root-feeding insects. The most important are the larvae of certain Eumolpid beetles of the genus *Syagrus*. The few species of root-feeding Aphids and Coccids recorded are mainly from the more temperate cotton-growing regions, with the exception of a Coccid, belonging to a recently described genus, which is associated with an obscure disorder of cotton in Zaire. Mole-crickets (Gryllotalpidae) are occasionally troublesome, and termites are always liable to attack plants that have suffered any root injury, and under some circumstances act as primary pests.

The roots of the cotton plant are apt to suffer much more severely from diseases than from pests. A disorder of full-grown plants that results in slow wilting, following rotting of the cortex of the hypocotyl and taproot, has been observed particularly in irrigated or low-lying areas subject to waterlogging. It has been associated with *Rhizoctonia* fungi (including *R. bataticola*) and usually results in the death of the plant. Two widespread diseases of the utmost importance that manifest themselves as a wilt, involving the disorganization of the vascular system, and thus affecting the whole plant, are caused by the fungi, *Fusarium vasinfectum* and *Verticillium dahliae*. Although both can be carried by the seed, they are primarily soil-borne and enter the plant through the root system.

Stem

After the main stem or main branches of the cotton plant have become woody, they are liable to attack by the larvae of woodboring beetles belonging to the Bostrychidae, Buprestidae, weevils (Curculionidae) and longhorns (Lamidae). Such attacks are rare on annual crops as there is insufficient time for a serious infestation to build up after suitable tissues have become available. Where, however, cotton is allowed to stand over or is grown in the neighbourhood of abandoned cotton plantations, these pests are able to cause some damage.

The stem is more susceptible to damage while it is still green and tender. In South America *Eutinobothrus* spp. is an important stem borer. In Africa, the young larvae of species of the Noctuid moths, *Diparopsis* and *Earias*, which are normally bollworms, often attack the growing point of the stem and bore a short distance down inside it, particularly if flowerbuds are not available; elsewhere, larvae of certain Pyralid moths act as tip-borders. Many of the species of Hemiptera that suck the foliage of cotton plants also feed upon the green stem, notably the cotton aphid *Aphis gossypii*, and the whitefly, *Bemisia tabaci* (Aleyrodidae), but they do not exert any localized effect on the stem. A number of species of scale insects and mealybugs (Coccoidea) are recorded from the stems of cotton, but they seem invariably to be associated with plants that are in poor condition and to become numerous only towards the end of the plant's life; they presumably accelerate senescence, but none of them is known to transmit disease. Direct and localized injury of a serious nature is caused to the stem by only two kinds of Hemiptera and in both cases seems due to the injection of toxic saliva: one of these insects, *Anoplocnemis curvipes* (Coreidae), attacks and kills the tip of the main stem, but is never abundant; the other, *Helopeltis schoutedeni* (Miridae), can cause crippling damage and is at times a major pest in certain regions.

The major disease affecting the stem is the blackarm phase of bacterial blight (*Xanthomonas malvacearum*). This takes the form of a lesion spreading back from, usually, the tip of the main stem or a side branch, and generally initiated on a leaf vein or petiole, although the bacteria appear able to gain entry also via regions of mechanical injury, such as the scars caused by hail. The disease may cripple or kill the stem concerned, and thus reduce the structure on which the crop can be formed.

Leaf

The leaf is the part of the plant from which, as might be expected, by far the greatest number of species of insects have been recorded. The great majority are chewing insects, mostly beetles (nearly all of which attack the leaves as adults), Lepidopterous larvae and Orthoptera. Those of importance are, principally, insects especially associated with the Malvaceae, although it so happens that the Egyptian cotton worm *Spodoptera littoralis*, the only African species whose depredations are sufficiently extensive or regular to warrant special measures against it, is not so restricted. Apart from this Noctuid, none of the defoliating insects that occurs in Africa attains the status reached by the neotropical cotton leafworm,

Alabama argillacea (Noctuidae). Several members of the Miridae, a family of sucking bugs whose saliva is toxic to plant tissues, are responsible for malformation of the leaf; these include *Helopeltis*, which also attacks the stems, *Lygus* and *Campylomma*. Apart from other and probably more serious effects that follow the attacks of these insects, they can cause a substantial loss of effective leaf area. The leaves are also attacked by Homoptera, of which jassids, e.g. *Jacobiasca, Amrasca* (Cicadellidae), the cotton aphid, *Aphis gossypii*, and whitefly, *Bemisia tabaci*, are the most important. The effect of these three kinds of insects is debilitatory, but the first, in addition, appears to disrupt transport in the vessels, and the last two excrete copious honeydew, on which moulds develop that still further impair the leaf function. A peculiar disorder of the whole plant follows the feeding of a Psyllid, *Paurocephala gossypii.* Thrips (Thysanoptera) cause severe damage in limited regions of Africa by scarifying the epidermal layers of the leaf.

The most important disease affecting the foliage of cotton is angular leaf-spot, another phase of bacterial blight (*Xanthomonas malvacearum*); the lesions, if numerous and confluent, may result in extensive leaf-tattering, but this is perhaps less important than the blackarm phase to which severe leaf attack can give rise. Premature senescence of the leaf, followed by defoliation, results from infection by *Alternaria* fungus, but attacks by this usually indicate a physiological imbalance of the whole plant.

Leaf-curl, a virus transmitted by whiteflies (*Bemisia* spp.) causes the leaves of *G. barbadense* to become crinkled, leathery and curled-up at the edges, the stems to become whip-like and the flower buds to shed. Upland cottons (*G. hirsutum*) are scarcely affected and some varieties of *G. barbadense* are strongly resistant. Leaf-curl disease is restricted to those parts of Africa south of the Sahara where *barbadense* cottons are grown, and may have spread to the Sudan, where it was for a time a very serious threat to cotton-growing, from West Africa; it is not found in Egypt.

Flowerbuds

There are remarkably few species of insects recorded as attacking specifically the flowerbuds of cotton, although they include some of the most important pests of the plant. The cotton weevils (*Anthonomus* spp., Cuculionidae), which, together with the cotton flower midge (*Contarinia gossypii* (Felt.), Cecidomyiidae), and the cotton flea hopper, *Psallus seriatus* (Miridae), are essentially New World species. The flower bud is subject to attack, usually when it is at least part-grown, by each of the major kinds of bollworm (*Helicoverpa, Heliothis, Earias, Pectinophora and Diparopsis*). Three kinds of Mirid bugs feed on the bud, causing shedding: *Creontiades*, which attacks the buds approaching full size; and *Campylomma* and *Lygus*, which, in contrast, attack the bud when it is extremely small. Attack by these last two Mirids also results in the truncation of the sympodial branches: the mechanism of this is not fully understood, but it is associated with the loss of the small flower buds, and thus perhaps with the interruption of supplies to the vegetative axillary bud.

Diseases affecting the flowerbuds are dealt with in connection with green bolls.

Flowers

The flower of the cotton plant only lasts, effectively, for one day, that on which it opens. Being large and showy, the blooms attract large numbers of species of flower-eating beetles, mainly Meloidae, but also some Rutelidae and Cetoniidae. These attack mainly the corolla and may not always prevent boll-setting; they are probably not of economic importance.

Green Boll

The most important boll-feeding insects are Lepidopterous larvae, the boll weevil and sucking bugs. The latter are by far the more numerous, as regards number of species involved, all of which belong to the Heteroptera. Members of this group are well adapted because of their relatively long mouthparts, to pierce the wall of the boll and suck the contents.

The bollworm genera previously mentioned, together with *Cryptophlebia leucotreta*, comprise virtually all the Lepidoptera whose larvae feed on the green boll. Those of the last-named species tend to tunnel into the boll wall before entering the cavity; the rest go straight through and feed on the developing seeds. Their attacks cause the very young bolls to be shed; the older bolls may produce some seed cotton from the undamaged locules, but more often the boll cavity is invaded by secondary fungi and the whole boll is rotted.

The sucking bugs that feed on the green boll belong to five principal families. Amongst the Mirids, species of *Lygus*, *Adelphocoris* and some other genera feed on the very young bolls, and *Helopeltis* on the older ones. The extent to which they puncture the boll cavity is uncertain, but attacks by the former cause shedding, and in the case of *Helopeltis*, at any rate, the lesions are caused by the feeding of the bugs on the boll contents. Those of the Pentatomidae include the common green shield bug (*Nezara viridula*), *Calidae dregii* and species of several other genera, such as *Acrosternum*, *Agonoscelis* and *Piezodorus*; they do not breed to any marked extent on cotton, but reach the crop as immigrant adults from their numerous wild and cultivated plant hosts, on which they feed on the fruits and seeds. A number of Coreidae and Lygaeidae behave in the same way. With the exception of *Calidea dregii*, none of the species concerned is of much importance. The fifth family, the Pyrrhocoridae, comprises the cotton stainers (*Dysdercus* spp.). These breed in profusion on the later stages of the cotton crop, the younger nymphs feeding on the seed in the recently opened bolls, and the older nymphs and adults on both open and green bolls. They damage the latter in two ways; by killing the young embryos of the seeds on which they feed, and by introducing the fungi of stigmatomycosis (principally *Ashbya gossypii*), which kills the seed and the developing lint hair, thus causing the condition known as staining. Some of the other Hemiptera that feed on the interior of the boll can also carry the infection.

Apart from the internal boll disease caused by fungi, the boll-disease phase of bacterial blight (*Xanthomonas malvacearum*) is extremely important, both as a source of loss of flowerbuds and young bolls, and of damage to the older bolls. Bacterial lesions develop on the bracteoles of the flowerbud or boll and travel thence to the peduncle, which is killed, thus causing the bud or young boll to abort or to be shed. The bacterial infection also frequently enters at the base of the boll, via moisture held between the bracteoles and the boll wall, the infection spreading thence up the central columella and causing the rotting of the loculi.

Open Bolls and Cotton Seed

When the cotton boll has opened, the lint is fully mature and is no longer subject to damage by insects. The only insects that attack previously undamaged cotton seed in the field, on any scale, are the common stainer bugs, *Dysdercus*, and the cotton seed bugs, *Oxycarenus* (Lygaeidae). All active stages of both genera feed on the seed in the open bolls and are capable of lowering the germinating capacity and reducing the oil-content. *Oxycarenus* is particularly well able to breed rapidly and build up very large populations in the field. The principal bollworms cease to attack the seeds when these reach maturity, although the pink bollworm (*Pectinophora gossypiella*) may complete its feeding inside an attacked seed.

The majority of the insects found in open bolls that have been damaged, apart from those mentioned, are scavengers that are of no economic importance; the more notable of these, because they are sometimes mistaken for the pink bollworm, are the larvae of another Gelechiid moth, *Mometa zemiodes*, and of *Sathrobota* spp. (Cosmopterygidae). Many of the common pests of stored products have been recorded from stored cotton seed; these lie outside the scope of this book, but it may be remarked that, as a general rule, stored cotton seed is remarkably free of pests of this kind.

One of the results of insect attack on the developing green boll is that, at or before the normal maturity date, the carpels split apart but fail to reflex. The boll contents are thus exposed to infection by fungus spores, but tend to dry only slowly, and considerable secondary rotting thus takes place. Should unseasonably wet weather accompany boll opening, the carpels themselves are attacked by rots and moulds, and the same happens with bolls that have aborted, without opening, because of some primary infestation or infection. All these diseases of mature boll are strictly secondary, although they often give the impression that they are primary causes of a loss of crop.

References

Burke H.R., Clarke, W.E., Cate, J.R. and Fryxell, P.A. (1986) Origin and dispersal of the boll weevil. *Bulletin of the Entomological Society of America* 32, 228–238.

Cate, J.R. and Summy, K.R. (1991) Ecology of the cotton bollweevil: the key to current successful controls and clues for future strategies. Abstract of paper presented at XII International Plant Protection Congress, Rio de Janeiro, August 1991.

Hargreaves, H. (1948) *List of Recorded Cotton Insects of the World.* Commonwealth Institute of Entomology, London, 50pp.

Pearson, E.O. (1958) *The Insect Pests of Cotton in Tropical Africa.* Commonwealth Institute of Entomology, London.

3 *Heliothis/Helicoverpa* (Lepidoptera: Noctuidae)

A.B.S. King

Natural Resources Institute, Central Avenue, Chatham Maritime, Chatham, Kent ME4 4TB, UK

Introduction

The Noctuid genera *Heliothis* and *Helicoverpa* contain pests of worldwide agricultural importance. Of the 80 or so species recorded by Todd (1978), only *Helicoverpa armigera*, *Helicoverpa zea*, *Helicoverpa punctigera* and *Heliothis virescens* are of consistently major pest status on cotton (Hardwick, 1965; Fitt, 1989).

Their larvae are the most important bollworms worldwide, with *H. armigera* distributed throughout the Old World, *H. zea* and *H. virescens* confined to the New World and *H. punctigera* limited to Australia. Other species in the genera *Helicoverpa* and *Heliothis* of local or occasional importance on cotton include *H. peltigera* and *H. viriplaca* (= *dipsaceae*) from southwest Asia, the Middle East and southern Europe (Table 3.1).

Until recently all of the above species were assigned to the genus *Heliothis*; but as a result of a taxonomic revision by Matthews (1987) the original genus has been split into *Heliothis* and *Helicoverpa*, as originally advocated by Hardwick (1965).

The four major species of *Heliothis* are found on a very wide range of wild and cultivated host plants, with the larvae, particularly the later instars, preferentially feeding upon the reproductive organs, although there are exceptions. All fruiting stages of cotton are attacked, and in many cotton growing areas losses can be very severe in the absence of chemical control. Control of *Heliothis* and *Helicoverpa* on cotton has depended almost exclusively upon insecticides; initially DDT, then endosulfan and latterly the pyrethroids, which until recently, have been outstanding in their efficacy. However, resistance to insecticides has resulted in diminishing returns and crop failures in some areas, and has led to a renewed interest in management strategies placing less reliance on pesticide inputs.

Table 3.1. *Heliothis* and *Helicoverpa* species of major and minor agricultural importance, their distribution, common names and principal crop hosts. (From Matthews, 1987; King and Coleman, 1989; Fitt, 1990)

Species	Common name	Distribution	Crop hosts
HELICOVERPA			
H. armigera (Hübner)	American bollworm, gram pod borer	Africa, southern Europe, Asia, Australia	Maize, sorghum, cotton, sunflower, pigeonpea, chickpea, groundnut, tomato, soyabean, okra
H. zea (Boddie)	Corn earworm, cotton boll worm, tomato fruitworm	Americas	Maize, sorghum, cotton, tomato, sunflower, soyabean
H. punctigera (Wallengren)		Australia	Cotton, sunflower, lucerne, soyabean, chickpea, safflower
H. assulta (Guenée)	Oriental tobacco budworm	Australia, SE Asia, China, India, Africa	Tobacco, *Physalis* and other Solanaceae
H. fletcheri (Hardwick)		Sahel Africa	Sesame, sorghum
H. geletopoeon (Dyar)		Southern South America	Cotton, maize, tobacco, soyabean, flax
HELIOTHIS			
H. virescens (Fabricius)	Tobacco budworm	Americas	Tobacco, cotton, tomato, sunflower, soyabean
H. peltigera (Denis & Schiffermüller)		Europe, SW Asia, Middle East, Africa	Cotton, safflower
H. viriplaca (Hufnagel) (=dipsacae (L.))		SW Asia, Middle East, southern former USSR	Cotton, legumes

Distribution

Helicoverpa armigera is the most widely distributed of the group, occurring throughout Africa, India, central and southeast Asia to Japan and the Philippines, the Middle East, southern Europe, eastern and northern Australia, New Zealand and many eastern Pacific Islands (Pearson, 1958; Fitt, 1989). Irregular migratory flights can extend this range into more northerly temperate latitudes (Pedgley, 1985). *Helicoverpa zea* is a closely related species which together with *H. virescens* is restricted in distribution to the New World, with permanant populations between latitudes 40°N and 40°S. *Helicoverpa zea* occurs from Canada to Argentina, the West Indies and Hawaii; *Heliothis virescens* is similarly, but less widely distributed and is less abundant in the cooler regions (Neunzig, 1969). Again, migratory movement is likely to be responsible for temporary northward extensions of the ranges of both species (e.g. Hardwick, 1965; Raulston *et al.*, 1986). *Helicoverpa punctigera* is native to Australia and has not been recorded outside that continent.

The distribution of the minor *Heliothis* and *Helicoverpa* species is much more restricted (Table 3.1). Additional details and bibliography on the distribution of economic Heliothinae are given by Widmer and Schofield (1983).

Taxonomy

For a long time there has been considerable controversy concerning the correct nomenclature of the most notorious bollworms, *H. armigera* and *H. zea*. Up to 1953, the Old and New World bollworms were considered to be the same species, when the names *Heliothis armigera* or *Chloridea obsoleta* Fabricus (and others) were used for both, although Hampson (1903) considered that *C. obsoleta* was correct. The name *Heliothis armigera* was later generally accepted after the Old and New World species had been distinguished by Common (1953). However, for some time after this, combinations of the two generic and two specific names have appeared in the literature, and it was not until 1955 that the name *Heliothis zea* became widely accepted for the New World species (Todd, 1955).

In 1965, Hardwick (1965) introduced the name *Helicoverpa* to distinguish members of the corn earworm complex which he considered to be morphologically distinct and phylogenetically separate from *Heliothis* species. However, Hardwick's conclusions were not widely accepted by entomologists, largely for reasons of potential confusion and convenience. In the opinion of Nye (1982), 'since generic boundaries are generally a matter of opinion ... it is better to have a broad-based genus *Heliothis* for field workers and to divide it into subgenera, such as *Helicoverpa* for the convenience of taxonomists ...' and this opinion has held until recently. However, as a result of recent taxonomic revision of the Heliothinae, Hardwick's nomenclature is now generally accepted (Matthews, 1987).

The Australian members of the genus *Heliothis* were revised by Common (1953) who distinguished four species; and more recently a fifth (Common, 1985). Of these, only *H. armigera* and *H. punctigera* have economic importance.

In addition to the controversy over nomenclature, the gender of the name *Heliothis*, with which the species name has to agree, has also been at issue. Traditionally treated as feminine, hence justifying *armigera* as the agreeing adjectival specific name, Steyskal (1971), and later Todd (1978) asserted that *Heliothis* was in fact masculine, with the result that *armigera* should then become *armiger*. In order to resolve this issue it was reviewed by Nye (1980) who requested that the Commission of Zoological Nomenclature rule that the gender of the generic name *Heliothis* should be feminine, and this has now been accepted (Knutson, 1989).

A brief history of the nomenclature of *Helicoverpa armigera* and *H. zea* is given by Nye (1982) and Poole (1989); a more detailed one, in his revision of the Heliothinae, is provided by Matthews (1987).

Description of the Stages

The biology of *Heliothis/Helicoverpa* spp. is typical of the Noctuidae. The various stages of *Helicoverpa armigera* have been fully described by Pearson (1958), Jayaraj (1982); those of *Helicoverpa zea* and *Heliothis virescens* by Neunzig (1964 and 1969); of *Helicoverpa punctigera* by Common (1953) and Zalucki *et al.* (1986); and all by Hardwick (1965). Here, the various stages are described, taking *H. armigera* as typical.

Egg

Subspherical with a flattened base; apical area surrounding the micropyle smooth, the rest of the surface sculptured with approximately 24 longitudinal ribs, the alternate ones being slightly shorter, with numerous much finer transverse ridges (carinae) between them. Colour whitish or creamy-white shortly after being laid, developing a central, reddish-brown band as the embryo develops. This gradually darkens, together with the rest of the egg which becomes a uniform greyish brown before hatching. Infertile eggs are white or yellowish, becoming increasingly yellow and cone-shaped as they desiccate.

The eggs of *H. zea* tend to differ from those of *H. virescens* by their size (average of 0.51 mm and 0.55 mm in height and 0.57 and 0.56 mm in width respectively), the number of ridges (21–31, average 24.7 for *H. zea* and 18–35, average 21.3 for *H. virescens*), and by the greater prominence of the transverse carinae in *H. zea*. However, the extent of intraspecific variation in these characters precludes their use as a reliable key to the eggs of the two species (Neunzig, 1969). The eggs of *H. punctigera* are indistinguishable from those of *H. armigera* (Kirkpatrick, 1961), but may be separated by electrophoretic techniques (Daly and Gregg, 1985).

Larva

Newly hatched larvae are a translucent yellowish white, with faint darker longitudinal lines and brown to black head capsules. The thoracic and anal shields, thoracic legs, setae and their tubercle bases and spiracles are also brown to black, giving the larva a spotted appearance. Prolegs are present on the third to sixth and tenth abdominal segments. The second instar is essentially similar but with some slight darkening of the ground colour and some lightening of the scleritized head capsule, thoracic and anal shields and thoracic legs. In *H. virescens* these are yellowish rather than the reddish-brown of *H. zea*. Neunzig (1969) also describes a key to separate first and second instar *H. zea* and *H. virescens* based on arrangement of the setae.

The third instar is characterized by two colour types having either a green or a red-brown ground colour, with respectively a greenish-fawn or cream to fawn coloured head capsule. The brown form predominates in *H. armigera*, *zea* and *punctigera* (Hardwick, 1965).

The characteristic patterning becomes more prominent and colouring generally darker in later instars, although this depends to some extent upon the nature of the host diet. The colours and patterns always lose intensity prior to each moult.

The full-grown larva is about 35 to 42 mm long, the integument having a characteristic granular appearance, consisting of close-set minute tubercles. The head capsule is mottled light brown to reddish-brown, the prothoracic and anal plates pale brown, setae dark and the spiracles and claws black. The colour pattern consists essentially of a single or pair of dark, median mid-dorsal bands, flanked on each side by a wider pale band then by a broad dark band and finally, on the lateral line, a broad, unbroken white or yellowish band on which the row of spiracles shows up clearly. The underside is uniformly pale. A number of narrow, somewhat wavy longitudinal lines is superimposed on the basic dorsal pattern.

The ground coloration of the last instar is highly variable, ranging from blue-green to yellow, pink and reddish-brown, with the density of the longitudinal lines and pigmentation of the setal tubercles more or less expressed. The reason for these considerable colour differences has been the subject of conjecture. Carshatt (1964) demonstrated that an interaction of light, temperature, food and heredity determined the colour of *H. virescens* larvae; with light green or pinkish larvae, having little cuticular pigmentation, developing under more exposed and warmer conditions. Ramos and Morallo-Rejesus (1976) demonstrated that larval colouring in *H. armigera* was largely due to nutritional factors, of which the carotinoids were of major importance. From the author's observations, the background colour of the host, and possibly its nutritional content appear to influence late instar colour: larvae on sunflower heads were often a pale yellow, whereas those on chickpea were more frequently a uniform green. Preferential predation of contrasting colour types may also be involved. However, such correlations are by no means fixed as larvae of all colour forms may be found on all types of host. All four species exhibit this colour variability (Plate I).

The later larval instars of *H. zea* and *H. virescens* are described in numerous early publications, some of which show how the two species may be distinguished (Garman, 1920; Brazzel *et al.*, 1953). The colour of the dorsal anterior cervical shield hairs may be used to distinguish between sixth instar *H. armigera* (white) and *H. punctigera* (black) (Zalucki *et al.*, 1986). However the only reliable method for separating younger stages is by electrophoretic techniques (Daly and Gregg, 1985).

The number of larval instars in all four species is variable from five to seven (Neunzig, 1969); although six instars is most common (Hardwick, 1965). Prolonged development under low temperatures was associated with an extra moult in southern Africa (Pearson, 1958).

Pupa

The pupa is normally 14–22 mm long, 4.5–6.5 mm in width across the thorax. Typical of the Noctuidae, it is mahogany-brown, smooth-surfaced, rounded both anteriorly and posteriorly and with two tapering parallel spines at the posterior tip. The distance between the bases of these spines can be used as a criterion to separate *H. armigera* from *H. punctigera* (>0.22 mm, *armigera*; <0.20 mm, *punctigera*) (Zalucki *et al.*, 1986). The pupae of *H. virescens* are usually smaller than those of *H. zea*, averaging 18.2 mm in length and 4.7 mm in width. The two species may be separated by the smaller abdominal spiracles and absence of maxillary palpi in *H. virescens* (Neunzig, 1969). The considerable variation in size is generally the result of larval food quality. Females are on average heavier than males.

Adult

The adults are stout-bodied moths, typical of the Noctuidae, with a wingspan of 35–40 mm and body length of 18–19 mm. Smaller specimens are not infrequent. The general colour of *H. armigera*, *zea* and *punctigera* varies from dull greenish-yellow, buff to olive-grey and light brown, sometimes with a pinkish suffusion, especially on the underside, and with darker brown to blackish markings on the wings. Females are darker than males and grey-brown in *H. armigera*. The forewings usually have a dark kidney-shaped spot half way between the base and apex of the forewing; this and the darker submarginal bands on both fore and hindwings are more distinct on the underside. There is a broad, irregular band, which may be broken into a series of indistinct dots or crescents, across the wing parallel to its distal margin. The hindwing is pale with a more or less distinct darker brown or black patch along the distal margin. The female is generally darker in colour, most notably in *H. armigera*. Male *H. armigera*, almost uniformly pale cream in colour have been found in India; pale coloration was associated with high temperatures during pupal development. The coloration of *H. virescens* is light olive to brownish-olive, with three oblique darker bands, usually with adjacent whitish bands, on the forewings. The hindwings are pearly white with a fuscous border along the outer margin; this is usually indistinct in males, but distinct and often suffused with red in the female.

Helicoverpa punctigera adults may be separated from *H. armigera* by the presence of indistinct transverse lines on the forewings of both sexes and an uninterrupted black terminal fascia in the hindwing; in *H. armigera* the black fascia has a pale patch between veins M3 and Cu2 (Zalucki *et al.*, 1986). For rubbed specimens, which have lost most of their wing scales, genital characters have to be used. Further distinguishing features are described and keys to the adult characters given in Common (1953), Kirkpatrick (1961), Hardwick (1965) and Matthews (1987).

Adults of the economically important species of *Heliothis* and *Helicoverpa* are illustrated in Plate II.

Adult Behaviour

The life history and reproductive activities of the four species are broadly similar.

Emergence

The emergence of *H. armigera* moths has a circadian rhythm; starting at dusk it continues until midnight, after which it virtually ceases (Tripathi and Sharma, 1984; Riley *et al.*, 1992). Peak emergence takes place between 20.00 and 22.00 hours (Roome, 1975; Singh and Singh 1975); between 19.00 and 23.00 hours in *H. zea* (Callahan, 1958; Lingren *et al.*, 1988). After emerging from the ground, moths climb up a nearby vertical surface and go through three phases of wing drying: the crumpled wing stage, butterfly stage and the quiescent moth stage. The first two stages generally last about an hour, the quiescent moth stage rather longer; the overall time of inactivity increased in length with lateness of emergence in the night (Riley *et al.*, 1992). A similar pattern of behaviour is described for *H. zea* by Callahan (1958) and Lingren *et al.* (1988). There is no emergence during the daytime.

Calling and Mating

Female *H. armigera* and *H. virescens* moths release pheromone (call) in the early hours of the morning, from about 02.00 to 04.30 hours (Ramaswamy, 1990; A.B.S. King, unpublished), during which time males search for females in the crop and mating takes place. Loganathan (1981) observed peak mating in *H. armigera* at 04.00 hours, but Hardwick (1965) recorded it to peak between 10.30 and 11.30 hours. Calling in *H. armigera* began two to five days after emergence, 322 minutes after the scotophase (or 00.22 hours, assuming this starts at 19.00 hours), advancing to 134 minutes on day 5 (21.14 hours) (Kou and Chow, 1987). Mating effectively terminated sex pheromone production in *H. zea*, although the physical presence of a spermatophore or sperm was not responsible for the response (Raina, 1989). Pheromone production resumed on subsequent nights. The duration of successful copulation (transfer of a spermatophore) varied from 85–115 minutes in *H. zea* (Raina, 1989), with a mean of 1 hour 29 minutes (± 22 minutes) (Hardwick, 1965). Mating was never observed in *H. zea* on the night of emergence nor on the following night (Callahan, 1958). Mating began about four days after emergence, but only if moths had fed (Pearson, 1958). Hardwick (1965) noted that mating in *H. armigera* began on the third night after emergence, peaking on the fourth night, but that males could start mating on the second night. The preoviposition period ranged from one to four days in *H. armigera* (Jayaraj, 1982).

Differences in the time and age at which calling starts in *H. armigera* suggests that there is some genetic variation in the onset of calling, or of the preoviposition period. This may relate to differing propensities for migration, and hence delayed reproduction, between moths from different geographical origins (Fitt, 1989; Colvin, 1990).

Temperature and humidity were critical to optimal mating in *H. zea* (Callahan, 1962), *H. armigera* (Hackett and Gatehouse, 1982a) and *H. virescens* (Henneberry and Clayton, 1991). Roome (1975) observed mating in *H. armigera* only during the coolest and most humid period of the night but never when conditions were warm and dry. Oviposition is greatly stimulated by mating, although female *H. armigera* and *H. zea* may lay infertile eggs before mating (Pearson, 1958; Hardwick, 1965; Armes 1989). However, Callahan (1958) maintains that unfertilized *H. zea* moths reabsorb infertile eggs before they mature and that infertile eggs are never laid. While this is certainly untrue for *H. armigera* in captivity, infertile eggs are seldom found under natural conditions.

Up to seven matings by a single pair of *H. armigera* moths have been recorded in captivity over 11 days, although of the 35 pairs examined by Hardwick (1965) 14 did not mate. Callahan (1958) and Hardwick (1965) describe in detail the reproductive behaviour of *H. zea* and *H. armigera* respectively.

Feeding

According to most workers, feeding is an essential prerequisite to mating and egglaying in *H. armigera* (Parsons *et al.*, 1938; Parsons and Marshall, 1939; Hardwick, 1965) and in *H. zea* (Callahan, 1962); however other workers (e.g. Vanderzant *et al.*, 1962; Lukefahr and Martin, 1964) observed both mating and oviposition in unfed females. Their results may have been influenced by the nature of the larval diet (Lozina-Lozinskii, 1939; Callahan, 1962). Unfed adult *H. zea* and *H. virescens* reared from larvae fed on corn or artificial diet mated and produced eggs, whereas adults reared from larvae fed on cotton squares failed to do so unless given a diet of sucrose, when fecundity increased by about 50%. Fertility was not significantly affected (Lukefahr and Martin, 1964). *Helicoverpa punctigera* females mated within one to two days of emergence although lack of sugar greatly decreased fecundity and mating frequency (Cullen, 1969).

Moths ingest sugars and amino acids, usually in the form of nectar. *Helicoverpa zea* moths have also been observed to feed on aphid honeydew in maize and sorghum, on the exudate produced by the sphacelial stage of ergot fungus (*Claviceps* spp.), pigeonpea sap-flows from buds damaged by larval feeding and on the extrafloral nectaries of cotton (Phillips and Whitcombe, 1962; Adler, 1987). Similar feeding habits have been observed in *H. armigera* in India by A.B.S. King (unpublished) where adults were observed engorged on the honeydew of *Dalbulus maidis* (Delong and Wolcott) and ergot exudates on sorghum.

The feeding habits of adult moths in the field are, however, poorly understood and there is little information on the species of plants preferred (but see list for *H. zea* by Callahan, 1958), alternative food sources, frequency of flower visits and on the seasonal availability of food sources, or if indeed food is at any time a limiting factor under natural conditions. Clearly, if much energy has to be expended in finding food sources, this will be at the expense of reproductive activities. Foraging patterns of *H. zea* were conditioned by reward, time for mating activities and range of diet (Adler, 1987), and in *H. punctigera* by night-time temperatures (Coombs, 1992). In cotton the extrafloral nectaries provide a source of food and may be important (Maxwell *et al.*, 1976), and was the rationale behind the development of nectariless cotton varieties to reduce oviposition

(Lukefahr, 1982). Adult feeding has been observed both during the day and night, although it was most frequent at dusk (Zalucki *et al.*, 1986; Adler, 1987; Coombs, 1992).

Fecundity

In South Africa, the average number of eggs laid by *H. armigera* was 730, with a maximum of 1600 over an oviposition period which varied from 10–23 days (from Pearson, 1958). In *H. punctigera*, the mean number of eggs per female lifetime was 1395 (± 160) and 1437 (± 229) at 19 and 24°C respectively, with a maximum (at 24°C) of 2899 and a maximum egg production in one day of 691. The daily average number of eggs per female was 112 at 24°C over a lifspan of 12.8 days, and 83 at 19°C over a lifespan of 16.8 days (from Zalucki *et al.*, 1986). The average number of eggs laid by *H. zea* was 1075 (± 565, max. 2240); by *H. armigera* , 1702 (max. 4394) and by *H. punctigera* 1823 (max. 2752) (Hardwick, 1965). Many more estimates of fecundity and longevity are given in the literature, but measured under laboratory conditions they may bear little resemblance to actual fecundity in the field, of which no estimates exist (Fitt, 1989).

The preoviposition period of *H. virescens* and *H. zea* varied from one to eight days, depending largely on temperature (Isley, 1935; Nadgauda and Pitre, 1983; Ellington and El-Sokkari, 1986). In *H. armigera*, the first eggs were usually deposited on the fourth or fifth night after emergence; the mean nightly numbers of eggs deposited then increased to a peak on the ninth day for *H. armigera* and the fourteenth day for *H. zea*, with oviposition most strongly peaked in *H. armigera* (Hardwick, 1965). While it is assumed that these tests were carried out at the cited rearing temperature of 25° ±0.5, they seem excessively long compared with more recent data from Armes (1989). He found that at 24°C, oviposition by sucrose diet-fed *H. armigera* started on day three and peaked on day six, but lack of diet delayed oviposition. A delay in the provision of the sucrose diet by five days after emergence delayed peak oviposition by three days. However, delays in diet availability shorter than five days and temperatures between 20 and 35°C did not significantly affect the pattern of oviposition (Armes, 1989).

Fecundity is influenced by the quantity and nutritive value of the adult and larval diet (Lukefahr and Martin, 1964; Pretorius, 1976; Nadgauda and Pitre, 1983; Armes, 1989), adult longevity (Hardwick, 1965; Armes, 1989), temperature (as a function of longevity), humidity (Armes, 1989) and time of year; factors which are often interrelated. Both fecundity and longevity in *H. armigera* and *H. virescens* moths are a function of the adult diet (Willers *et al.*, 1987; Armes, 1989), which has to contain amino acids (Hackett, 1980; Topper, 1981). Fecundity in *H. armigera* was significantly correlated with pupal weight (Pretorius, 1976), although Pearson (1958) and Reed (1965) found this link with body size to be weak. In *H. zea* and *H. virescens* there was no correlation between pupal size and longevity or fecundity, although *H. virescens* was more dependent on adult feeding than *H. zea* to realize its potential fecundity (Lukefahr and Martin, 1964).

Flight Activity

Moths are generally active only after dark, although daytime flight is not uncommon in *H. punctigera* (Cullen, 1969; Coombs, 1992) and, in general, may be greater when population densities are high and late in the season at higher latitudes (Hardwick, 1965). Moths normally start to fly at, or just before, dusk when they begin searching for, and feeding on nectar and other food sources. Most moth activity has been recorded during dusk and the early part of the night (Cullen, 1969; Roome, 1975; Persson, 1976; Ramaswamy, 1990; Riley *et al.*, 1990, 1992), with most female activity earlier, and male activity later (Persson, 1976). A small dawn flight peak is also sometimes observed (Topper, 1987a; Fitt, 1989; Ramaswamy, 1990). Diurnal flight and feeding by *H. punctigera* was a function of low air temperatures at dusk of the previous night (Coombs, 1992).

Records of the periodicity of moth activity will differ with the method of monitoring. Light trap catches are influenced by moonlight and meteorological conditions (Dent and Pawar, 1988) and by the phenology of surrounding crops (A.B.S. King, unpublished). Pheromone traps measure male activity only during the natural calling time of females. Direct night-time observation (Drake and Fitt, 1990; Riley *et al.*, 1990, 1992) should give the most accurate measure of local activity but may be too crop or location specific.

The flight and dispersal activity of *H. armigera* has been extensively studied by Farrow and Daly (1987), Riley *et al.* (1990, 1992) and others, and is reviewed by Fitt (1989). Hartstack *et al.* (1982, 1986), Lingren *et al.* (1982), Raulston *et al.* (1982 and 1986) and others have studied flight behaviour in *H. zea* and *H. virescens.*

Oviposition

Oviposition in *H. armigera* usually started some hours after dusk, initially alternating with feeding, later becoming the predominant activity until soon after midnight (from Pearson, 1958), however, Topper (1987a) found it to be virtually complete by midnight. Persson (1974) found that oviposition by Australian *H. armigera* exhibited a bimodal distribution, with the initial peak larger than the later one and with peak female activity coinciding with peak oviposition, and that oviposition was strongly depressed by moonlight. Phillips and Whitcombe (1962) recorded most oviposition by *H. zea* in the field between 19.00 and 21.30 hours, whereas Callahan (1958) recorded it throughout the night, with only slightly more during the early hours. *H. virescens* also oviposited throughout the night in a number of 'bouts' (Lingren *et al.*, 1977; Ramaswamy, 1990). Moths did not oviposit at 10°C and more eggs were laid at 21 and 27°C than at 15 and 32°C (Henneberry and Clayton, 1991).

Moths are highly selective in their choice of hostplant in a suitable condition of development (Hardwick, 1965). All four species lay their eggs singly, often on or near the flowering or fruiting parts of the host, with peak oviposition often coinciding with, or just before, peak flowering or nectar production (Parsons,

1940; Cullen, 1969; Roome, 1975; Wardaugh *et al.*, 1980; Alvarado-Rodriguez *et al.*, 1982; Adjei-Maafo and Wilson, 1983 a,b; Firempong and Zalucki, 1990b). This may be seen as an adaptation to ensure that nitrogen-rich reproductive tissues are available to developing larvae. However, there are frequent exceptions to this and, usually in the absence of flowers, eggs may be laid on any part of the plant (Cullen, 1969; Passlow, 1973; Wilson, 1976; Broadley, 1980; Wardaugh *et al.*, 1980), and not always on the most appropriate site for the initiation of larval feeding (Fitt, 1990).

Geographic differences in preference for oviposition sites have sometimes been noted and host preference has been seen to change over time. *Helicoverpa zea* prefers to lay eggs on the silks of maize (as does *H. armigera* in India, although it is not a major problem there), whereas in South Africa it lays most of its eggs on the stalks (Pearson, 1958). *H. armigera* eggs are found mostly on the upper part of the cotton crop canopy (Beeden, 1974; Patel *et al.*, 1974; Wilson *et al.*, 1983; Fitt, 1990), with moths often preferring the upper leaf surfaces, young vegetative growing points and squares. In Australia, few *H. armigera* or *H. punctigera* eggs are laid on the terminals, although egg distribution differs significantly between the species and between cotton varieties having different leaf characters (Fitt, 1990).

Beeden (1974) noted that the preferred oviposition sites on cotton changed with development, from leaf surfaces and stems early in the season to bolls and peduncles later on. *H. zea* is reported to lay a high proportion of eggs directly on the growing tips (Farrar and Bradley 1985a), although, according to Pearson (1958, and references quoted), one third of the eggs laid by *H. armigera* in southern Africa were on the flower buds and the rest scattered all over the plant. Farrar and Bradley (1985b) noted that egg location on the plant had little effect on subsequent establishment of larvae.

On pigeonpea in India most of the eggs of *H. armigera* are laid on the flowers and flowerbuds, only sparingly on the leaves and then only during the vegetative phase of the host. On chickpea the eggs are laid on the leaves, mostly on the underside, and from the time the plants are still very small. In contrast to other hosts, oviposition on chickpea declines from the onset of flowering (author's observations). On tomato, the eggs of *H. zea* were laid mostly on the periferal leaflets and near the flowers (Wilcox *et al.*, 1956; Alvarado-Rodriguez *et al.*,1982). In sunflower, eggs are laid on both the upper surfaces of the leaves and on the calyx of the flower bud. In sorghum and millet, *H. armigera* lays most of its eggs between the spikelets of the head during the few days from preanthesis to just after full anthesis (A.B.S. King, unpublished). In Queensland, Australia, eggs were laid by *H. armigera* on the leaves of sorghum with larvae developing on the vegetative tissues (I.J. Titmarsh, personal communication), and by *H. zea* on the whorls of young maize (Phillips and Whitcombe, 1962).

Moths tend to prefer hairy over glabrous leaf surfaces on which to oviposit (Cullen, 1969; Lukefahr *et al.*, 1971), a factor which has been utilized for breeding less preferred cotton varieties (Thomson and Lee, 1980; Lukefahr, 1982) (see later). The likely selective advantages of this habit are discussed by Fitt (1990).

Host Selection

The highly polyphagous nature of the economic Heliothinae has been amply recorded (e.g. Pearson, 1958; Hardwick, 1965; Neunzig, 1969; Bilapate, 1984; Stadelbacher *et al.*, 1986; Zalucki *et al.*, 1986; Matthews, 1987). Hostplants differ considerably in their relative acceptability to ovipositing females (Johnson *et al.*, 1975; Firempong and Zalucki 1990b) and in their suitability for larval development (Pearson, 1958; Gross and Young, 1977; Stadelbacher *et al.*, 1986; Zalucki *et al.*, 1986; Topper, 1987a). The extent of this polyphagy in relation to natural hostplants is clearly one key factor to the success of these species. However, relative differences in acceptability for oviposition are not necessarily reflected in suitability for larval development (Fitt, 1990). Some chemical extracts of tomato which were attractive to adult moths were toxic to young larvae (Fitt, 1990).

Host selection in the Heliothinae has been comprehensively reviewed by Fitt (1990), and for moths in general by Ramaswamy (1988). There is considerable information on the distribution or pattern of host plant utilization, but little is known about the actual processes of host location and recognition. Puzzling geographical differences in host preference have often been noted (Reed and Pawar, 1982), and the existence of races or subspecies, or even species differentiation, within a population, specific to or greatly preferring much restricted host ranges, has been suggested (Bhattacherjee and Gupta, 1972; Fox and Morrow, 1981). However, the absence of evidence for moths selecting the host species on which they had fed as larvae, the generally short periods during which most hosts are attractive to ovipositing adults or palatable to larvae and the considerable, although not precisely known dispersal capability of adult moths, all tend to contra-indicate such intrinsic specificity (Fitt, 1990). There is, hovever, experimental evidence that early adult experience positively influences host plant selection (Firempong and Zalucki, 1991). This implies that moths would tend to oviposit on the same or neighbouring plants to those on which they developed as larvae, provided they were still attractive. This behaviour could account for the rapid increase of insecticide resistance on long season crops such as cotton and pigeonpea.

Specific attractants and oviposition stimulants have been identified for *Heliothis* and *Helicoverpa* spp. (Jackson *et al.*, 1984; Tingle and Mitchell, 1984; Juvic *et al.*, 1988). Volatiles attractive to moths have been identified from the silks and tassels of maize (Cantelo and Jacobson, 1979), one of which, phenyl acetaldehyde, was attractive to both *H. zea* and *H. armigera* (C.S.Pawar, unpublished). However whole plant extracts generally elicited the strongest responses, indicating that host recognition and acceptance was mediated by a complex of both volatile, as well as probably non-volatile, chemical stimuli (Fitt, 1990).

There is, however, little concrete evidence that volatiles from crops attract moths from more than a few metres; for example, no upwind flight towards a cotton crop by *H. virescens* was observed (Ramaswamy *et al.*, 1987). However, Drake and Fitt (1990) and this author have noted moths doubling back into an attractive crop, after having overshot the boundary. Clearly some mechanism must exist to account for initial attraction into the crop and by which moths can perceive boundaries, whether by visual or olfactory cues. Apparent indifference to the presence of attractive hosts has also been noted (Fitt, 1990), and both *H. armigera* and *H. punctigera* have been observed to fly downwind out of an attractive crop and over fallow ground. The physiological status of these moths was thought likely to have differed from those which remained within the crop, although this was not determined (Fitt, 1990). Differences in egg densities in large, adjacent blocks of alternative hosts (Haggis, 1982; Topper, 1987a,b) suggests that moths can, however, effectively select the most suitable among several hosts in an area.

How long moths remain in the selected area depends upon the state of the crop (Roome, 1975), although this was based on light trap records. Topper (1987a), using night-time observations, demonstrated a diel cycle of movement: from resting on groundnuts, to feeding in lentils or sorghum and ovipositing in cotton (and sorghum). Data from pheromone traps indicated that *H. armigera* was more sedentary in an attractive crop than *H. punctigera* (Fitt *et al.*, 1989).

Duration of Stages and Larval Behaviour

Egg

The duration of the egg incubation period depends upon temperature. It has been recorded as 2–5 (usually 3) days in India (Singh and Singh, 1975; Jayaraj, 1982; Rajagopal and Channa Basavanna, 1982) and western Tanganyika (Reed, 1965); and for *H. zea* in the USA (Ewing *et al.*, 1947). Egg incubation periods of as long as 17 days have been recorded for *H. zea* in the US, and of 14 days for *H. armigera* (from Pearson, 1958).

Larva

On eclosion the young larva usually eats some or all of the empty egg shell before feeding on the plant. Newly hatched larvae may often wander for some distance, with occasional surface feeding before settling down at a preferred site: in cotton this is usually on a flower bud or flower, which is eventually hollowed out if it does not first absciss. Older larvae prefer buds and young bolls, although leaves are also eaten, as when plants are small or few bolls are present. Although a large larva normally feeds with only the front portion of its body inside the boll, with the remainder protruding, it may be entirely hidden inside the boll, with only a hole, often lightly spun over with silk threads, to indicate its presence. However, most buds and bolls which have been attacked show an accumulation of faeces between the surface and the enclosing bracts. This may render the larva more prone to discovery and attack by natural enemies reponding to the semio-chemicals given off.

Moulting often takes place on the upper surface of a leaf in full sunlight, possibly to hasten drying of the new cuticle (Reed, 1965).

Larvae tend to move between feeding sites and contiguous plants, especially when preferred food is limiting and population densities are high; it is this habit of partially damaging many fruiting structures which make the Heliothinae so destructive. Encounters between older (>3rd instar) larvae usually result in antagonism, which may be followed by cannibalism in *H. armigera* and *H. zea*: thus a maize cob on which several eggs had been laid may have only a single large larva. A mathematical model in which the extent of cannibalism may be estimated and related to larval density has been derived for *H. zea* (Stinner *et al.*, 1974). *Helicoverpa punctigera* larvae tend not to be cannibalistic (G.P. Fitt, personal communication).

The number of larval instars varies from five to seven with six being most common (Hardwick, 1965). It varies with the species, the individual and is influenced by the nutritive value of the larval diet and temperature (Hardwick, 1965; Nadgauda and Pitre, 1983).

The duration of larval development is dependent not only on the temperature but also on the nature and quality of the host. Larval development in *H. virescens* took from 17.5 to 25.0 days in June–July and August–September respectively, on tobacco in North Carolina (Neunzig, 1969). In *H. armigera*, it varied from 12.8 to 21.3 days on maize at an average temperature of between 27.2 and 24.0°C, and varied from 15.2 days on maize to 23.8 days on tomato (at an average temperature of 24.3°C) (Rajagopal and Channa Basavanna, 1982). Reed (1965) recorded an average larval period on cotton flower buds of 21.1 ± 0.32 days (including a prepupal period of 2.72 days), at a temperature which varied between 21 and 27°C. In Uganda on a diet of sorghum, the mean larval period over the whole season was 24.8 days (Coaker, 1959).

According to laboratory studies by Dhandapani and Balasubramanian (1980) the larval period ranged from 17 to 20 days, with a minimum and maximum on pigeonpea and tomato respectively. However, Singh and Singh (1975) recorded 8–12 days on tomato, whereas on cotton and maize the larval period was 18.3 and 18.0 days, comparable with the 18.2 and 17.4 days recorded by Pretorius (1976). Larval durations on soybean, maize, cotton and tomato were 15, 19, 21 and 24 days respectively (Doss, 1979). The weights of fully-grown larvae differed considerably with diet; the heaviest larvae had fed on cotton, the lightest ones on tomato and sorghum (Jayaraj, 1982).

Comparisons between authors' data are, however, not possible where temperatures and other rearing conditions are not standardized or given, and it is not known whether estimates of larval (or pupal) development times include or exclude the prepupal stage.

Pupa

On completion of growth the fully fed larva drops or crawls to the ground and enters the soil to pupate. The depth at which the pupal cell is formed varies considerably, being dependent upon the hardness and wetness of the soil, presence of cracks and surface litter. It is most often formed at a depth of 2.5 to 17.5 cm (Jayaraj, 1982). Distinct differences between *H. zea* and *H. virescens* were recorded by Neunzig (1969); *H. virescens* pupated within the first 5 cm of soil

Plate I

Colour forms of the larvae of economically important species of *Heliothis* and *Helicoverpa*: **(1)**, **(2)**, **(3)** *Helicoverpa armigera* on cotton, Gambia (W. King); **(4)** *Helicoverpa zea* on tomato, Costa Rica; **(5)** *Helicoverpa armigera* on millet, India; **(6)** *Helicoverpa zea* on maize; **(7)** *Heliothis virescens* on pigeonpea, Costa Rica (A.B.S. King) (see Chapter 3).

Plate II

Adult moths of economically important species of *Heliothis* and *Helicoverpa*: **(1)** *Helicoverpa armigera*; **(2)** *Helicoverpa zea*; **(3)** *Heliothis peltigera*; **(4)** *Helicoverpa punctigera*; **(5)** *Helicoverpa fletcheri*; **(6)** *Helicoverpa geletopoeon*; **(7)** *Helicoverpa assulta*; **(8)** *Heliothis virescens*; **(9)** *Heliothis viriplaca*, females below (see Chapter 3).

Plate III

(1) *Pectinophora gossypiella* Saund larva (ZENECA Agrochemicals) (see Chapter 4). **(2)** *Earias* larva (G.A. Matthews) (see Chapter 5). **(3)** *Diparopsis castanea* Hmps larva (G.A. Matthews) (see Chapter 6). **(4)** *Cryptophlebia leucotreta* (Meyrick) (IRCT): (a) egg, (b) larva, (c) pupa, (d) adult (see Chapter 7).

Plate IV

(1) *Spodoptera littoralis* Boisduval adult (see Chapter 8). **(2)** *Alabama argillacea* damage on cotton in Brazil (see Chapter 8). **(3)** *Anthonomus grandis* adult (see Chapter 9). (4) *Apion soleatum* Wagner adult (see Chapter 10). **(5)** *Dysdercus intermedius* Distant adult (G.A. Matthews) (see Chapter 11). **(6)** *Dysdercus* nymphs (Cillier). **(7)** *Phonoctonus nigrofasciatus* Stål, a predator of stainers (*Dysdercus*) (G.A. Matthews) (see Chapter 11). **(8)** *Calidea dregii* Germ. adult (see Chapter 11).

1 2 3a 3b 3c 3d

Plate V

Aphids and their predators (IRCT) (see Chapter 12). **(1)** Apterous viviparous female cotton aphid, *Aphis gossypii* Glover. **(2)** Alate. **(3)** Coccinellid predators of aphids: (a) *Cheilomenes* sp. nymph; (b) *Cheilomenes* sp. adult; (c) *Exochomus* sp. nymph; (d) *Scymnus* sp. nymph.

Plate VI

(1) Aphids parasitized by: (a) Aphelinides; (b) Aphidiides (IRCT) (see Chapter 12). **(2)** *Bemisia tabaci*: (a) nymphs; (b) adults (ZENECA Agrochemicals) (see Chapter 13).

1 2 3 4 5 6

Plate VII

(1) Jassid, *Jacobiella facialis* (Jacobi), nymph (ZENECA Agrochemicals) (see Chapter 14). **(2)** Jassid damage on susceptible glabrous cotton in variety trial in Malawi (G.A. Matthews). **(3)** *Taylorilygus vosseleri* damage in Africa (G.A. Matthews) (see Chapter 16). **(4)** *Helopeltis* on cotton (G.A. Matthews). **(5)** Scale insects on cotton in India (G.A. Matthews) (see Chapter 16). **(6)** Damage caused by thrips – cotyledons thickened and blistered (IRCT) (see Chapter 17).

1 2 3 4 5 6

Plate VIII

(1) *Zonocerus elegans*, the elegant or stinking grasshopper (G.A. Matthews) (see Chapter 18). **(2)** Red spider mites (*Tetranychus*), eggs and adults (ZENECA Agrochemicals) (see Chapter 20). **(3)** An example of biological control – parasitoids on a noctuid egg (IIBC) (see Chapter 24). **(4)** Pheromone trap for *Spodoptera littoralis* (see Chapter 26). **(5)** Knapsack spraying (see Chapter 27). **(6)** Using a spinning disc ULV sprayer in Africa (see Chapter 27).

whereas most *H. zea* pupated between 2.5 and 7.5 cm, a few at greater depths. In *H. zea*, the larva may make a U-shaped tunnel, burrowing down then up, stopping just short of the soil surface, and pupating at the lower end of the tunnel (Pearson, 1958). Very occasionally pupation takes place inside a tunnel in a maize cob (Reed, 1965), or on the soil or leaf surface. A loose web of silk is usually spun by the larva before pupating.

The prepupal stage is shorter and stouter, a more uniform colour and less active than the fully grown larva. It generally lasts for 1–4 days.

The pupae of all four species undergo a facultative diapause: the duration of the pupal stage therefore depends upon whether or not diapause has been induced during the earlier life stages, and its intensity. In non-diapause *H. armigera*, the pupal period ranges from about six days at 35°C to over 30 days at 15°C, lasting for about 10–14 days under field conditions in central India. It may also vary slightly with the larval host (Jayaraj, 1982). In diapausing individuals the pupal period may last several months. Pupal temperatures of over 30°C produced paler coloured adults.

Adult

The length of adult life is largely determined by the availability of food, in the absence of which depletion of the fat body is rapid and death occurs in only a few days (Pearson, 1958; Armes, 1989). While access to water prolonged lifespan, honey or sugar solution gave significantly greater longevities than water alone (Callahan, 1961; Topper, 1987b; Armes, 1989). The benefits of a sucrose diet over water were related to the nature of the larval foodplant, probably in terms of resulting adult fat content, such that adult feeding effectively offset any detrimental effects of poor larval diet (Topper, 1987b). Longevity was also related to initial (= pupal) weight and temperature (Armes, 1989).

Female moths are generally longer lived than males (Pearson, 1958), however Callahan (1958) and Armes (1989) found male and female longevities to be comparable in captive *H. zea* and *H. armigera*. Longevity is also a function of activity; immobilized *H. zea* lived longer (Callahan, 1958), and flight activity greatly reduced longevity in unfed *H. virescens*, although much less so when moths were fed with sucrose (Willers *et al.*, 1987). The mean survival times of unfed-flown female moths was 6.5 days compared with 10.0 days for unfed-unflown, 14.2 days for fed-flown and 16.8 for fed-unflown (Willers *et al.*, 1987). Callahan's (1958) assertion that mated females lived longer was not supported by Hardwick (1965), who found there to be no significant difference between the longevities of mated and unmated females of both *H. zea* and *H. armigera*.

Moth longevity in captivity varied from 7.5–14.0 days for male and 9.2–18.0 days for female *H. zea* in the USA; and from 1–23 days for male and 5–28 days for female *H. armigera*, in southern Africa (Pearson, 1958). Mean and maximum longevity at 'room temperature' (21–27°C in winter and 27–32°C in summer) was similar for *H. armigera* (1 and 33 days), *H. zea* (17 and 37 days), *H. punctigera* (14.5 and 36 days) and *H. assulta* (16 and 24 days) (Hardwick, 1965).

The relevance of laboratory data on longevities to those under natural conditions is questionable. While predation, disease and abiotic factors will naturally reduce longevity, the ability of moths to find conditions optimal to their survival in the field may not be met in caged conditions where stress-induced mortality may be higher. No data on field mortality have been recorded.

Pupal Diapause

Winter Diapause

In common with many temperate noctuids, *Helicoverpa zea*, *H. armigera*, *H. punctigera* and *Heliothis virescens* exhibit a facultative diapause which enables populations to survive adverse conditions. While for the greater part these are of low winter temperatures, in three out of the four species there is evidence for a similar, but less well defined, response to hot, dry summer conditions (summer diapause). The induction of (winter) diapause is typically by exposure of the larvae to short photoperiods and low temperatures, with the optimal response occurring within the range of 11.5–12.5 hours daylength and 19–23°C (Komarova, 1959; Hardwick, 1965; Phillips and Newsom, 1966; Roach and Adkisson, 1970; Cullen and Browning, 1978; Roome, 1979; Hackett and Gatehouse, 1982a,b). Exposure of adults and eggs to increasing photoperiods greater than ten hours, followed by larval exposure to photoperiods less than 13 hours (at a constant 21°C) was necessary to induce diapause in *H. zea* (Wellso and Adkisson, 1966; Adkisson and Roach, 1971).

High temperatures (26–29°C) during the late larval and prepupal stages, prior to pupal ecdysis, counteract earlier diapause-inducing conditions (Browning, 1979; Roome, 1979; Hackett and Gatehouse, 1982b), however, low temperatures during both larval and pupal stages and shortening photoperiods during egg and larval development increased the incidence and intensity of diapause in *H. zea* (Wellso and Adkisson, 1966) and *H. punctigera* (Cullen and Browning, 1978; Browning, 1979).

In response to a need to adapt to the highly unpredictable environment in which it is endemic, *H. punctigera* has evolved a complex diapause strategy (Fitt, 1989). As well as responding to an extremely narrow range of photoperiods (12–12.5 h daylength at 19°C), the induction of diapause is also strongly affected by the temperature and/or photoperiod experienced by the eggs and the pharate adult (Cullen and Browning, 1978). In the field, diapausing and non-diapausing populations may coexist virtually throughout the year in subtropical and temperate Australia (Fitt, 1989).

Under natural conditions, there is no (Coaker, 1959; Nel, 1961) or very little (Reed, 1965; Hackett and Gatehouse, 1982a,b; ICRISAT, unpublished) diapause in *H. armigera* populations in tropical regions, although laboratory tests have shown that diapause can be induced in these populations by exposure to low temperatures and short (*c.* 12 h) photoperiods (Roome, 1979; Hackett and Gatehouse, 1982a,b; A.B.S. King, unpublished).

The incidence and particularly the intensity of diapause in *Heliothis* and *Helicoverpa* spp. show considerable variation (Fye and Carranza, 1973; Wilson *et al.*, 1979; Fitt, 1989). This may reflect the effects of uncontrolled conditions such as larval diet (Goryshin, 1958; Phillips and Newsom, 1966) and genetic variation in the response to diapause-inducing stimuli (Hardwick, 1965; Holtzer *et al.*, 1976; Roome, 1979). It was also observed that diapause in *H. armigera*, *H. zea* and *H. virescens* was more intense in the relatively small proportion induced early in the season than later on (Pearson, 1958; Reed, 1965; Lopez and Hartstack, 1985), and may be a mechanism to assure synchrony of emergence.

In general, the expression of diapause, both as the proportion of the population entering this state and its intensity, varies with the latitude, such that the moderating influence of temperature on photoperiod becomes less, and the length of the pupal period at a given temperature becomes greater with increasing latitude (Roome, 1979). This was quantified by Danilevskii (1965) who found that the critical photoperiod falls with successive increments of temperature at a rate of one hour per 5°C, and led Hardwick (1965) to postulate the existence of a cline of genetic susceptibility to diapause among populations with distance from the equator. The existence of a genetic basis of geographic variation in response to diapause-inducing stimuli, as discussed by Reed (1965), Roome (1979) and Wilson *et al.* (1979), still needs to be verified (Fitt, 1989).

Diapause development typically follows two phases (Pearson, 1958; Wilson *et al.*, 1979): the completion of development at a low temperature (typically 12–13.5°C; phase 1) (Fitt, 1989) is a necessary prerequisite to an ability to respond to higher temperatures (phase 2) with a threshold of about 17°C for *H. armigera* (Wilson *et al.*, 1979) and 18°C for *H. zea* (Lopez and Hartstack, 1985). Diapause termination, or phase 2 development, proceeded more rapidly in *H. virescens* than in *H. zea* (Lopez and Hartstack, 1985), and in most of the USA, *H. virescens* emerges earlier in the spring than *H. zea* (Neunzig, 1969; Gross and Young 1977; Lopez *et al.*, 1984; Lopez and Hartstack 1985).

Summer Diapause

An ability to aestivate, or undergo diapause to survive prolonged hot, dry conditions has been reported for a number of noctuids (Masaki, 1980) including some species of *Helicoverpa* and *Heliothis*. Hackett and Gatehouse (1982b) recorded the prolongation at high temperatures (35°C) of a 'winter' diapause induced by short photoperiods and low temperatures in *H. armigera* in the Sudan Gezira. This diapause, which lasted over 80 days, was broken by a fall in temperature, as would be expected with the rains following an extended dry period. While this would offer a mechanism for surviving the dry season when few hosts are availiable, its field existence has not been confirmed. However *H. fletcheri*, which is endemic to this region, exhibits a summer diapause lasting up to 9.5 months (Hackett and Gatehouse 1982a,b).

In Arizona and California *H. virescens* enters a true summer diapause when larvae are exposed to very high temperatures (43°C for 8 h daily), although the proportion of females entering diapause was about half that of males. Non-diapause males were sterile at these temperatures (Butler *et al.*, 1985). A similar response is suspected in some populations of *H. armigera* and *H. punctigera* in

Australia (Fitt, 1989; D.A.H. Murray, unpublished; G. Daglish, unpublished); however in Hyderabad, India, summer diapause could not be induced by exposure to high temperatures in *H. armigera* (W. Reed, unpublished).

The diapause condition may be recognized by the presence of the larval eye-spots in the pupae. These disappear some four days after pupal ecdysis in non-diapausing pupae and on completion of diapause in diapausing pupae (Phillips and Newsom, 1966). The fat body of the newly formed pupa consists of firm, rounded lobes and remains in this condition throughout diapause. In the developing pupa, histolysis breaks this down into a mass of minute free granules, which eventually reform into the adult fat body (Pearson, 1958).

Flight and Dispersal

The importance and success of the economic Heliothinae is, in large measure, due to their highly mobile nature, enabling them to exploit widely separated food sources and to escape natural enemies. The importance of understanding moth movement to the development of effective management strategies has stimulated detailed studies in Australia, the USA and India, in which a wide range of approaches and techniques have been used (Drake, 1990).

In *Heliothis* and *Helicoverpa* spp., trivial and so-called migratory movements essentially form a poorly-defined continuum, with their scale depending upon atmospheric conditions, the spatial distribution of favourable habitats and the behaviour of the migrants (Farrow and Daly, 1987). The types of dispersal activity may thus be conveniently divided according to the distance travelled as they affect the infestation of local or distant crops. However, the environmental stimuli and innate conditions which control the extent of local or migratory flight are, as yet, not clearly understood (Fitt, 1989).

Long Distance Movement

Like other Noctuidae, the Heliothinae make use of two strategies to survive or take advantage of the seasonality of their habitats or host plants: diapause through periods of drought or cold; and spatial redistribution by migratory or dispersive flight (Raulston *et al.*, 1982, 1986; Pedgley, 1985; Farrow and Daly, 1987; Pedgley *et al.*, 1987).

Although long-distance movement is undertaken by *Heliothis virescens*, *Helicoverpa zea* and *H. armigera* moths, they have seldom been observed to display typical migratory behaviour (Schaefer, 1976; Drake and Fitt, 1990; Riley *et al.*, 1992; R. A. Farrow, personal communication), and differ from obligate migrant noctuids like *Spodoptera exempta* (Rose *et al.*, 1985). Rather, they behave like facultative migrants, undertaking long-distance flight in response to deteriorating local conditions such as shortage of nectar sources and larval hosts. Flight is in directions largely governed by the passage of prevailing weather systems (Fitt, 1989). The four major species may be ranked in decreasing order of migratory activity: *H. punctigera* >> *H. zea* > *H. virescens* > *H. armigera* (Farrow and Daly,

1987). There is strong evidence that *H. punctigera* is an obligate migrant (Drake and Fitt, 1990; Fitt and Pinkerton 1990).

Most long-range flight falls into the non-migratory class of movement defined by Farrow and Daly (1987). Short-range or trivial movements are largely appetitive, involving feeding, mating and oviposition, whereas long-range movement generally involves downwind flight between crops (Topper, 1987a,b; Fitt, 1989). Consistent with Johnson's (1969) 'oogenesis-flight syndrome', in which migratory flight is confined to the period between emergence and attainment of sexual maturity, any factor affecting the onset of maturity would also influence the tendency to migrate. In *H. armigera*, Hackett (1980) demonstrated that maturation was delayed in moths derived from larvae fed on poor quality food and starved as adults, and that they flew for longer than well-fed controls. Conversely, feeding reduced the pre-maturation period and mating reduced flight some 20-fold in female moths (Armes and Cooter, 1991). Colvin (1990) showed that the pre-maturation (= reproduction) period (PRP) of *H. armigera* from Malawi was significantly longer than that of an Indian strain. He suggested that this may be related to the greater importance of migration in the more transient habitat in Malawi, where there has been selection for a longer PRP.

Much of the evidence for long-distance and migratory movement is circumstantial and based upon the appearance of moths in geographical locations where they should not otherwise have occurred at either that time (for example, in trap catches before the emergence of local diapausing populations), or in the numbers recorded (Fitt, 1989). Thus there is evidence for migration of *H. armigera* in Europe (Pedgley, 1985), the Middle East (Pedgley, 1986), India (Pedgley *et al.*, 1987), for *H. armigera* and *H. punctigera* in Australia (Daly and Gregg, 1985; Farrow and Daly, 1987); and for *H. zea* and *H. virescens* in North America (Hardwick 1965; Hartstack *et al.*, 1982, 1986; Lopez *et al.*, 1984; Mueller *et al.*, 1984; Raulston and Houghtaling, 1986).

There are many records in the literature of sudden increases in the numbers of moths in light or pheromone traps, indicative of migratory movements (Widmer and Schofield, 1983). These were often associated with particular synoptic weather patterns, usually warm winds preceding cold frontal systems (Muller and Tucker, 1986; Drake and Farrow, 1988). Large increases in egg densities on cotton in the Sudan Gezira were associated with outflows from approaching rainstorms (Haggis, 1982).

More direct evidence of migration exists from mark and recapture experiments with *H. virescens* (Hendricks *et al.*, 1973; Raulston *et al.*, 1982) and *H. zea* (Lopez *et al.*, 1979). The presence of pollen attached to trapped *H. zea* moths, identified their origin to be some 750–1000 km distant to the south (Hendrix *et al.*, 1987). Similar evidence for long distance movement of *H. armigera* and *H. punctigera* from inland Australia has been demonstrated by Gregg *et al.* (1990). The appearance of insecticide resistant *H. armigera* in areas where the insecticide in question had not been used (Goodyer and Greenup, 1980; McCaffery *et al.*, 1989) is a good indicator of long-distance dispersal. Identification of the source area can be done by establishing trajectories by 'backtracking' along the previous nights' prevailing winds at over 500 m above ground level (Pedgley *et al.*, 1987; Westbrook *et al.*, 1990). Evidence from backtracking and unaccountably high levels of pyrethroid resistance strongly indicated that *H. armigera* moths had flown from cotton in coastal Andhra Pradesh, India into food crops some 250 km inland (McCaffery *et al.*, 1989), and some 330 km by *H. virescens* in Texas (Westbrook *et al.*, 1990). Wind-assisted movements of a few hundred kilometres have been recorded for *H. punctigera* in Australia (Drake *et al.*, 1981; Drake and Farrow, 1988). Electrophoretic analyses showed that the patterns of genetic variation between geographically separated populations of *H. armigera* and *H. punctigera* in Australia were consistent with gene flow resulting from long-distance movement (Daly and Gregg, 1985).

Long-range movement in the economic Heliothinae is discussed more widely by Farrow and Daly (1987).

Local Movement

The flight behaviour and local movement of *Heliothis* and *Helicoverpa* spp. have been studied extensively with night-time observational techniques, including infrared, X-band radar and 3D video tracking (Fitt, 1989; Riley *et al.*, 1990, 1992) and mark-recapture experiments (Graham *et al.*, 1978; Lopez *et al.*, 1979; Hayes and Hopper, 1987; King *et al.*, 1990).

Most activity by *H. armigera* was recorded during the early part of the night and immediately above attractive crops. Longer range appetitive dispersal was also undertaken at low elevation; flying populations above 10 m were very low by comparison (Drake and Fitt, 1990; Riley *et al.*, 1992). Mark-capture experiments indicated that there was a rapid exodus of *H. armigera* moths from the emergence site (King *et al.*, 1990), and although a proportion of moths remained within the trapping area, moving into nearby crops if these were in a favourable condition, most left the area altogether (Lopez *et al.*, 1979; Fitt and Pinkerton, 1990; King *et al.*, 1990), a behaviour with important implications for control strategies. The proportion of *H. punctigera* leaving the study area was higher than *H. armigera*, confirming that *H. punctigera* is the more mobile species of the two (Drake and Fitt, 1990; Fitt and Pinkerton, 1990).

In the laboratory, flight swings (Hackett, 1980) and mills (Armes and Cooter, 1991) have been used to assess the flight potential of *H. armigera* under various conditions. These confirmed the diel pattern of flight observed in the field, the substantial depressing effect of mating on flight activity, especially on prolonged flight and that flight activity in unmated females increased to a peak four days after emergence (Armes and Cooter, 1991). With a mean flight speed of about 4.8 km h^{-1} and a median distance flown of 40 km per night, clearly flights of over 200 km per night are possible even with low to moderate wind speeds. Mated females exhibited a large number of short periods of activity (Armes and Cooter, 1991); this is consistent with typical host seeking and oviposition behaviour, as has been observed in the field.

Sex Pheromone

Composition

In all four species a sex pheromone is produced by the female moth when she is calling to attract males. The attractant chemical consists of two or more components in an optimal blend. The chemical identities of these components have been worked out for *H. armigera* by Piccardi *et al.* (1977); Nesbitt *et al.* (1979, 1980); for *H. punctigera* by Rothschild (1978), Rothschild *et al.* (1982); and for *H. zea* and *H. virescens* by Klun *et al.* (1979, 1980). The major component for *H. zea* and *H. virescens* is (Z)-11-hexadecenal, with minor components of hexadecanal, (Z)-9-hexadecenal and (Z)-7-hexadecenal for *H. zea*, and with the addition of (Z)-9-tetradecenal, tetradecenal and (Z)-11-hexadecen-1-ol for *H. virescens*. The main component for *H. armigera* is (Z)-11-hexadecenal, with minor components of (Z)-9-hexadecenal and (Z)-11-hexadecenal-1-ol and for *H. punctigera* a 50:50:1 mixture of (Z)-11-hexadecenal, (Z)-11-hexadecenal acetate and (Z)-9-tetradecenal (Nesbitt *et al.*, 1980; Rothschild *et al.*, 1982).

Use of Sex Pheromones for Trapping

The main use to which *Heliothis* and *Helicoverpa* pheromones have been put is detection and population monitoring with traps (Lopez *et al.*, 1990), where their species specificity and simplicity of servicing are major advantages over black light traps. Optimal ratios and formulations of pheromones in plastic or rubber dispensers have been developed by different workers (Rothschild *et al.*, 1982; Lopez *et al.*, 1990), and have been used extensively in specially designed traps to monitor (male) moth populations. The types of traps used have included open pan water traps, sticky traps and various designs of cage and funnel traps (Pawar *et al.*, 1988; Lopez *et al.*, 1990), which varied greatly in efficiency. For practical reasons capture efficiency has had to be compromised by design simplicity, availability of materials and durability in the field (Lopez *et al.*, 1990); the dry plastic funnel trap developed at ICRISAT was adopted because of these latter features. Its inability to capture any *H. zea* and *H. virescens* compared with other traps (wire cone, plastic cone and sticky traps) (Lopez *et al.*, 1990) contrasts markedly with its relative efficiency for *H. armigera* in India (Pawar *et al.*, 1988), and suggests fundamental differences in behaviour between the Old and New World species. Dry funnel traps of comparable design are in use in Australia (Wilson, 1985) and Israel (Kehat *et al.*, 1982).

Population monitoring with pheromones implies estimating changes in adult male density over time; for these data to be useful, there must be some consistent relationship with field infestations. On a field to field basis, the relationship, appropriately offset in time, between trap catches and densities of immature stages or damage, has been variable to poor (Harstack *et al.*, 1978; Kehat *et al.*, 1982; Rothschild *et al.*, 1982; Johnson, 1983; Leonard *et al.*, 1989; ICRISAT, unpublished), although it was closest when moth densities are low and at the beginning of the seasonal cycle. High levels of inter-trap variation, as observed for *H. armigera* at ICRISAT, India, negate the usefulness of catch data from individual traps as indices of action threshold (ICRISAT, 1986). Trap position, both absolute and in relation to the crop, wind speed (and direction), relative humidity and temperature, can all significantly affect catch (Rothschild *et al.*, 1982; Dent and Pawar, 1988; ICRISAT, unpublished). Moonlight and barometric pressure have also been shown to affect captures of *H. virescens* and *H. zea* in the USA (Harstack *et al.*, 1978; Hendricks and Hartstack, 1978), although in India the effect of moonlight on pheromone trap catch was minimal (Dent and Pawar, 1988).

As an indicator of the start of an infestation, as for example after spring emergence (Hartstack *et al.*, 1978) or a migratory 'wave front', a qualitative measure is all that is necessary. Predictive models, incorporating pheromone trap catches have been developed in the USA to provide early warnings of probable oviposition patterns on cotton crops, or as aids to scheduling scouting and other pest management activities on an area-wide basis (Hartstack *et al.*, 1978; Hendricks and Hartstack 1978; Hartstack and Witz, 1981; Harstack *et al.*, 1983; Lopez *et al.*, 1990).

There is considerable potential for the use of pheromone trapping to monitor *Heliothis* and *Helicoverpa* populations, but for this to be exploited a better understanding of the key biotic and environmental factors influencing captures is necessary (Lopez *et al.*, 1990). The use of pheromones for mating disruption as part of IPM packages has met with some success for other cotton pests such *Pectinophora gossypiella* and *Earias* spp. (Baker *et al.*, 1990; McVeigh *et al.*, 1990). This technique has not, so far, shown much promise for heliothine bollworms, probably because of their highly mobile and polyphagous nature (D.G. Campion, personal communication).

Problems associated with a stable formulation of the aldehyde components (McLaughlin and Mitchell, 1982) and the release of the various components in their optimal proportions over time are added complications. Even so, while investigators in the USA and Australia have shown reductions in captures of males, egglaying and mating of tethered female moths in small scale field trials (Sparks *et al.*, 1982; R.A. Vickers, personal communication), mating inhibition was minimal and egglaying scarcely affected following applications of *H. zea* and *H. virescens* pheromone constituents to the crop on a field scale (Sparks *et al.*, 1982).

The recent discovery of the pheromone biosynthesis activating neuropeptide (PBAN) in *H. zea* has opened the possibility of interference with pheromone production and may provide the basis for developing future management strategies (Menn *et al.*, 1989).

Host Range

All four species of agricultural importance are highly polyphagous, their larvae feeding on a wide and diverse range of cultivated and wild host plants in a large number of different plant families – 49 in Australia (Zalucki *et al.*, 1986). Matthews (1987) lists records for crops alone of: *H. armigera*, 63 species in 27 families; *H. zea*, 49 species in 16 families; *H. punctigera*, 18 species in 11 families; and *H. virescens*, 55 species in 14 families. Lists of both cultivated and wild hosts found supporting populations of the different species exist for: *H. armigera* in India (Bilapate, 1984; Manjunath *et al.*, 1989), Pakistan (Moyhuddin, 1989), Southeast Asia (Napompeth, 1989), Africa (Pearson, 1958; Greathead and Girling, 1989), Egypt (Ibraham *et al.*, 1989); for *H. armigera* and *H. punctigera* in Australia (Zalucki *et al.*, 1986; Twine, 1989); for *H. armigera*, *H. viriplaca* and *H. peltigera* in Western Europe (Meirrose *et al.*, 1989), and for *H. zea* and *H. virescens* in North, Central and South America (Isley, 1935; Martin *et al.*, 1976; Hallman, 1984; Stadelbacher *et al.*, 1986; Kogan *et al.*, 1989). The predominant host families include the Leguminosae, Solanaceae, Malvaceae, Compositae and Gramineae.

Despite a capability to breed on a wide range of host plants, species-linked preferences exist (Johnson *et al.*, 1975; Roome, 1975; Zalucki *et al.*, 1986), although this is modified by seasonal availability of hosts in the preferred developmental stage. Whereas all four species attack legumes, maize and sorghum are often preferred over other crops by *H. armigera* and *H. zea* (Neunzig, 1969; Johnson *et al.*, 1975; Roome, 1975), but are not, or very rarely, attacked by *H. punctigera* and *H. virescens* (Fitt, 1990).

Cotton often supports the highest populations, especially in many areas after alternative hosts have senesced (Fitt, 1989, 1990), but is not considered to be the most preferred host plant of any species (Johnson *et al.*, 1975; Firempong and Zalucki, 1990a).

Individual females will often oviposit on a range of host plants, occasionally including plants, or substrates, unsuitable for larval development (Juvic *et al.*, 1988), whereas, by contrast, larval development can sometimes proceed satisfactorily on plants which are not selected for oviposition (Isley, 1935; Parsons 1940). The mechanisms of host selection are however poorly understood; the patterns of host use and the behavioural processes leading to host selection and oviposition are reviewed by Fitt (1990).

Seasonal Activity

As a result of the large geographical ranges of the economic Heliothinae, their cycles of seasonal activity vary considerably. In those regions where low winter temperatures and short photoperiods induce diapause, emergence is brought about by increasing spring temperatures, as in *H. zea* and *H. virescens* in the USA, *H. armigera* in southern Europe and *H. armigera* and *H. punctigera* in southern Australia. However, immigrants from lower latitudes may appear before local populations emerge, as in the case of *H. zea* in the USA (Hardwick, 1965; Stadelbacher and Pfimmer, 1972; Slosser *et al.*, 1975; Hartstack *et al.*, 1982), and crops will become infested by two separate populations.

The population dynamics of the four species may be complex. Generations may overlap due to the mixture of immigrant and local populations and differential development rates, and only become synchronized when forced through the time restriction of a short period of host suitability or adverse environment.

In a pest whose behaviour is adapted to the exploitation of ephemeral hosts, seasonal phenology is largely host or crop mediated. This is clearly illustrated in the descriptions of population dynamics of the four species: for *H. armigera* in western Tanzania by Reed (1965) and Nyambo (1988) and in Botswana by Roome (1975); in Sudan by Topper (1987b); in India by Bhatnagar *et al.* (1982), Pawar *et al.* (1986b) and Srivastava *et al.* (1992); for *H. zea* and *H. virescens* in North Carolina by Neunzig (1969), and California by Pearson *et al.* (1989); and for *H. armigera* and *H. punctigera* in Australia by Wardaugh *et al.* (1980).

The number of generations per season is directly influenced by temperature and host availiability. While up to 11 generations per year are possible where hosts are continuously availiable in the tropics, 3–5 are more usual in temperate and subtropical environments (Fitt, 1989). Generation size or abundance is largely governed by the interaction between initial population size and timing, host suitability, abundance and sequencing. Rainfall indirectly influences population size through its effect on host abundance and timing, or directly through the redistribution of ovipositing populations (Haggis, 1982). Prediction of population change will therefore depend very much upon the simplicity or consistency of the environment. While Hartstack *et al.* (1983) could predict the size of later, economically important generations on the basis of early season trap catches of *H. zea* and *H. virescens* in the relatively simple crop environment of central Texas, this was not possible for *H. punctigera* in Australia (Wardaugh *et al.*, 1980; Fitt *et al.*, 1989) or for *H. armigera* in Zimbabwe (Gledhill, 1982).

Breeding on wild and volunteer crop host plants early in the season, before crops have reached the susceptible stage or are widely available, is important to the population development of *Heliothis* and *Helicoverpa* spp. in many regions (Pearson, 1958; Reed, 1965; Mueller and Phillips, 1983; Wardaugh *et al.*, 1980; Stadelbacher *et al.*, 1986). These populations are usually at low density and widely dispersed, but may be important in maintaining populations until suitable host crops become availiable (Fitt, 1989). The extent to which populations in local refugia, such as weeds and gardens, relative to migrant populations from further afield, are important to crop infestation is not clear, as it is difficult to quantify either. The strong correlations sometimes found between populations on spring hosts and those on crops later in the season has resulted in strategies for area-wide management of natural habitats harbouring the pest (Knipling and Stadelbacher, 1983; Mueller *et al.*, 1984; Harris and Phillips, 1986). However, the level and times of immigration and the extent to which population responses are density dependent are both critically important to the validity of this strategy (Mueller *et al.*, 1984). The strategy, moreover, requires a discrete, relatively circumscribed agroecosystem to which it can be applied, as may exist in the Mississippi Delta region. However, in Australia early spring populations of *H. punctigera* occur on widespread native host plants both within and outside the cropping area, where the possibilities of suppression are limited (Fitt, 1989), a situation which also exists for *H. armigera* in southern India.

Population Regulation

Naturally occurring predators, parasites and pathogens are important in regulating numbers of economic *Heliothis* and *Helicoverpa* (Quaintance and Brues, 1905; van den Bosch and Hagen, 1966; Lingren *et al.*, 1968; van den Bosch *et al.*, 1969; Bell, 1982; Carner and Yearian, 1989). In Texas, cotton is unprotected by chemical pesticides for about 95% of the growing season, a fact attributed to the efficiency of natural mortality agents, in the widest sense of the meaning (Sterling, 1989). Recent emphasis and increased interest in developing this component as part of integrated pest management (IPM) strategies has largely been stimulated by deteriorating confidence in insecticides. This has been the result of increasing resistance (Sparks, 1981; Wolfenbarger *et al.*, 1981; Gunning *et al.*, 1984; Luttrell *et al.*, 1987; Sawicki and Denholm, 1987; Daly and Murray, 1988; McCaffery *et al.*, 1989), resurgence following the destruction of natural enemies and the upsurge of secondary pests (Newsom and Brazzel, 1968; Smith, 1970; Knipling, 1979; King, 1986; King and Coleman, 1989), high cost (Jayaraj *et al.*, 1989) and greater awareness of environmental considerations, including pesticide residues in food (Jayaraj *et al.*, 1989; King and Coleman, 1989; Menn *et al.*, 1989; Morton and Collins 1989).

Ecology and Natural Control

The importance of both biotic and abiotic control agents on the seasonal abundance of *Heliothis* and *Helicoverpa* spp. is still poorly understood. Few detailed life tables for field populations, in which the various mortality factors are quantified and partitioned, have been published (Zalucki *et al.*, 1986; Fitt, 1989; Room *et al.*, 1990) although some more or less complete analyses do exist for specific locations and crops (e.g. Room, 1979a; Titmarsh, 1985; Ma Shijun and Ding Yanquin, 1989; Nanthagopal and Uthamasamy, 1989; Kyi *et al.*, 1991; A.B.S. King, unpublished). From these analyses, early stage mortality is invariably the most severe, although its causes and extent may vary greatly and comparable data sets are too few to identify the factors responsible for population regulation (Room *et al.*, 1990).

Many isolated estimates of parasitism and predation of *Heliothis* and *Helicoverpa* spp. have been recorded from all over the world (e.g. Coaker, 1959; Reed, 1965; Neunzig, 1969; Shepard and Sterling, 1972; Stadelbacher *et al.*, 1984; King *et al.*, 1985; Pawar *et al.*, 1986a; Steward *et al.*, 1990). Observations have been made of predation and parasitism of immature stages to identify the agents concerned (Bishop and Blood, 1977; Room, 1979b), to assess their potential as control agents (Samson and Blood, 1980), or to infer potential impact in the field through correlations between natural enemy and prey populations (Bishop and Blood, 1977, 1980, 1981; Room, 1979b; Cock *et al.*, 1989, 1991). Despite the volume of information on parasitism and, to lesser extent predation, it is often fragmentary. Comparisons between the extent of control exercised are often complicated by differences in methodology and environmental circumstances, and their impact is seldom expressed, or can be interpreted, in terms of effect on seasonal patterns of abundance (Zalucki *et al.*, 1986; Fitt, 1989; Seymour and Jones, 1990). Quantification of the impact of parasitism has, however, been studied in some detail for *H. armigera* and *H. punctigera* by Titmarsh (1985), and for *H. virescens* and *H. zea* by Hopper (1989) using models. The shortfalls of existing data and methodologies to be used to assess the impact of parasites and predators are reviewed by Seymour and Jones (1990).

The characteristic traits of mobility and rapid colonization of new habitats, which make *Heliothis* and *Helicoverpa* such successful r–strategists, usually overwhelm the capacity of natural enemies to respond functionally and numerically to regulate populations. This asynchrony or delayed response between a host and its natural enemies, common in r–type insects, is a major factor limiting the effectiveness of natural control (Fitt, 1989). Severe dry seasons, moreover, can break the synchrony between parasite and host (Greathead and Girling, 1982), although local populations may be kept at a stable low level in ecosystems where breeding is continuous throughout the year, as indicated by Coaker (1959) in Uganda. The overall abundance of *Heliothis* and *Helicoverpa* spp. is more probably regulated by region-wide climatic factors, acting either directly or through the abundance and quality of host plants, than by biotic factors (van Emden and Williams, 1974; Fitt, 1989). As evidence for this, yearly changes in total pheromone trap catches of *H. armigera* at widely separated locations and ecosystems over the Indian subcontinent were remarkably consistent (Srivastava *et al.*, 1992; W. Reed, personal communication).

Early Stage Mortality

Mortality has frequently been recorded to be highest during the egg and first instar larval stages (Hardwick, 1965; Room, 1979b; Goodenough *et al.*, 1986; Ma Shijun and Ding Yanquin, 1989; Seymour and Jones, 1990; Kyi *et al.*, 1991; A.B.S. King, unpublished), although it may become less with progression of the season (Wilson and Greenup, 1977). Kyi *et al.* (1991) recorded losses of 33–88% eggs and 93–100% first instar *H. armigera* larvae on cotton. The principal causes of mortality have been variously attributed to adverse climatic conditions (Wilson, 1982), dislodgment of eggs (Titmarsh, 1985; Kyi *et al.*, 1991) and larval incompatibility with the host (Titmarsh, 1985), cannibalism of eggs by larvae (Twine, 1971, 1974; Twine *et al.*, 1983), predation and parasitism (Room, 1979b; Twine *et al.*, 1983; Evans, 1985) and competition (Room, 1979b). Ridgway and Lingren (1972) estimated that at least 50–90% of *H. virescens* and *H. zea* eggs may be destroyed by natural enemies. Arguably, the most important among these are the egg parasites, *Trichogramma* spp. and *Telenomus* spp., which can sometimes destroy large numbers of eggs. However, Pearson (1958) disregards the importance of egg parasitism compared with the more severe effects of predation and abiotic factors, a conclusion also reached by Kyi *et al.* (1991). In Kenya, Cock *et al.* (1989) recorded 20% egg parasitism by the Scelionid *Telenomus ullyetti* Nixon. Parasitism of *H. armigera* eggs, chiefly by *Trichogramma* sp., was density dependent (ICRISAT, 1988), and, in *H. zea*, varied with the spatial distribution of eggs on the host plant (Morrison *et al.*, 1980). Estimates of the destruction of eggs and young larvae by predators, cannibalism and abiotic factors is much more difficult to determine, and few reliable estimates exist.

Mid Stage Larval Mortality

During the mid larval instars, biotic mortality is mainly caused by parasitism by Braconid and Ichneumonid wasps. In the USA, the most important and specific parasite of *H. zea* and *H. virescens* is *Microplitis croceipes* (Cresson), although another common parasite, *Cardiochiles nigriceps* Viereck, shows a marked specificity to *H. virescens* (Neunzig, 1969; Goodenough *et al.*, 1986; van den Berg *et al.*, 1988). King *et al.* (1985) reported that in 1981 and 1982, respectively, 31% and 50% of *H. virescens* and *H. zea* larvae, collected from sprayed and unsprayed cotton in SE Arkansas, had been parasitized by *M. croceipes*, comprising over 90% of all parasitic insects reared. *Microplitus croceipes* is active on a wide range of *H. virescens* hosts (Burleigh and Farmer, 1978), although it is rarely found on maize (E.G. King, personal communication).

The ichneumonid *Campoletis chlorideae* Uchida is probably the most important larval parasite in India, but, unlike *M. croceipes*, parasitism is affected by larval host plant, and varies from 46% on sorghum to 3% on pigeonpea (Pawar *et al.*, 1986a). In Australia, *Microplitis* sp. is generally regarded as the predominant parasite of larval *Helicoverpa* spp, particularly on sunflower and tobacco (Twine, 1989), although levels of parasitism did not exceed 30% (Wilson and Greenup, 1977). In the Ord Valley, larvae were heavily parasitized by the Braconid *Microgaster* sp. (Michael, 1973). In East and South Africa, the most important hymenopteran larval parasites of *H. armigera* are *Meteorus laphygmae* Brues and *Apanteles* spp. (van den Berg *et al.*, 1988; Cock *et al.*, 1989).

The later larval instars are generally more subject to parasitism by *Tachinidae*, so that little direct reduction in larval damage results. In the USA, *Archytas marmoratus* (Townsend) and *Eucelatoria bryani* Sabrosky are the principal species; in India *Carcelia illota* Curran and, to a lesser extent, *Goniophthalmus halli* Mes. and *Paleorixa laxa* (Curran) parasitize up to 22% of *H. armigera* larvae on pigeonpea (Bhatnagar *et al.*, 1983), and up to 54% on chickpea (A.B.S. King, unpublished), although Bhatnagar *et al.* (1983) recorded only 3% on chickpea. In Tanzania, Reed (1965) recorded up to 13% parasitism by *Drino inberbis* (Wied.) (=*Paleorixa laxa*), but in East and South Africa, respectively *Linneamyia longirostris* (Marquart) and *P.laxa* gave the highest rates of parasitism (up to 42% by *P. laxa*) (van den Berg *et al.*, 1988; Cock *et al.*, 1989).

Pupal Mortality

There are few reliable estimates of prepupal and pupal mortality of *Heliothis* and *Helicoverpa*. Although there is evidence that it may be as high as 80% in *H. armigera* (D.A.H. Murray, personal communication; A.B.S. King, unpublished), Fitt and Daly (1990), through extensive soil sampling in various crops, showed that as many as 50% survived under cotton. The fate of pupae is however difficult and laborious to determine under natural conditions, and the relevance of survival rates measured under cage conditions is questionable (Fitt, 1989). The estimation of prepupal and pupal mortality in diapausing/overwintering generations, often with a view to predicting emergent spring populations, has occupied much attention in the USA (Stadelbacher and Pfimmer, 1972; Stadelbacher and Martin, 1980; Rummel *et al.*, 1986; Stadelbacher *et al.*, 1986; Rummel and Neece, 1989). Pupal mortalities of *H. zea* and *H. virescens*, caused mostly by adverse environmental factors, ranged from 60% to >99% (see summary in Fitt, 1989). However, in Australia and Botswana, higher survival rates (*c.* 40–50%) were recorded for *H. armigera* (Roome, 1979; Fitt and Daly, 1990), possibly because the cold, wet winter conditions, so inimical to survival (Eger *et al.*, 1983) were not experienced there.

Fitt and Daly (1990), working in northern New South Wales in Australia, recorded over six parasitic species emerging from field collected pupae of *H. armigera*, the most important of which was *Heteropelma scaposum* (Morely). Overall rates of pupal parasitism under cotton varied from 10% to 42%. Predation of prepupae and pupae by ants may be significant. Pearson (1958) and Cock *et al.* (1991) note that ants, particularly *Pheidole* sp., were important predators of *H. armigera* in eastern and southern Africa, and *Camponotus* sp. has been observed to take *H. armigera* prepupae on the soil at ICRISAT Center, India (A.B.S. King, unpublished). Physical factors impeding adult emergence may also account for an additional 10–20% mortality (Fitt, 1989).

Effect of Host Plant on Parasitism

Host-related differences in the activity of heliothid parasites have often been recorded (Quaintance and Brues, 1905; Bhatnagar *et al.*, 1982; Pawar *et al.*, 1986a; Zalucki *et al.*, 1986; van den Berg *et al.*, 1988, 1990; Manjunath *et al.*, 1989). The influence of host plant is often more pronounced on egg parasitism, which is generally higher on cereals. In India, the eight year average rates of parasitism of the eggs of *H. armigera* (mainly by *Trichogramma chilonis* Ishii) were 33% on sorghum, 15% on groundnut and 0.3% on pigeonpea (Pawar *et al.*, 1986a). In Gujurat state, *Trichogramma chilonis* parasitized up to 98% of *H. armigera* eggs on tomato, potato and lucerne, but no egg parasitism was recorded from chickpea, probably because of the acid exudate secreted by the leaves (Manjunath *et al.*, 1989), and only 0.1% egg parasitism was recorded from pigeonpea (Bhatnagar *et al.*, 1982). Similar differences in egg parasitism, mainly by the Scelionid, *Telenomus* sp. nr. *triptus* Nixon have been recorded from Australia (Zalucki *et al.*, 1986), where parasitism rates of 60% on unsprayed cotton (Waite, 1981) and up to 99% on sorghum were recorded (Michael, 1973; Robertson, 1977). Goodenough *et al.* (1986) recorded 60% and 30% parasitism of *H. zea* eggs on corn and sorghum respectively. In southern Africa, egg parasitism by *Trichogrammatoidea lutea* Girault varied with the crop from <1% on citrus to 75% on maize (from van den Berg *et al.*, 1988), and in Senegal, *Trichogrammatoidea* sp. parasitized from 17% *H. armigera* eggs on *Acanthospermum* to 80% on tomato (Bhatnagar, 1987).

In India, the parasites attacking *H. armigera* on sorghum in a sorghum/

pigeonpea intercrop (mainly *Trichogramma* sp. and *Campoletis chorideae*) did not parasitize *H. armigera* on the later maturing pigeonpea, where parasitism was chiefly by the tachinid *Carcelia illota* (Bhatnagar *et al.*, 1982). On *H. armigera* infesting pigeonpea, up to 22% larvae were parasitized by Diptera and less than 5% by Hymenoptera; on chickpea, only 3% larvae were parasitized by Diptera and 17% by Hymenoptera (Bhatnagar *et al.*, 1983). In Africa, as in India, parasitism of *H. armigera* on sorghum was higher and by more parasitic species, than on other crops (van den Berg *et al.*, 1988, 1990). When parasitism rates were analysed for *H. armigera* feeding on four crops in Tanzania, the parasitoid species, *Palexorista laxa*, *Apanteles diparopsidis* Lyle and *Chelonus curvimaculatus* Cameron were all most strongly associated with sorghum, whereas *Cardiochiles* sp. and *Charops* sp. were, respectively, more closely associated with cotton and the weed-crop *Cleome* sp., and virtually absent on sorghum (van den Berg *et al.*, 1990).

Parasitism of *H. armigera* larvae by the nematode *Hexamermis* sp. in southern India was much higher on groundnut, tomato and some low growing weeds than on chickpea, sorghum and pigeonpea, where it was virtually absent (Bhatnagar *et al.*, 1985). In Andhra Pradesh, the nematode was much more prevalent on Alfisols than on Vertisols. The relatively high (up to 52%) rates of *H. armigera* parasitism by nematodes (*Hexamermis* sp.) on low-growing crops and weeds on the more sandy textured Alfisols sustaining low, early season populations of *H. armigera* may be important in reducing numbers.

Predation

In general, predators have received less attention than parasites as natural control agents of the economic Heliothinae. The impact of predators has usually been based on indirect evidence: as in the induced build-up of *H. zea* and *H. virescens* resulting from their elimination by insecticides (King, 1986), their effect on known prey populations in cages (Lingren *et al.*, 1968; Ridgway and Jones, 1968, 1969; van den Bosch *et al.*, 1969; Lopez *et al.*, 1976; Thead *et al.*, 1987); by exclusion (van den Bosch *et al.*, 1969; Pedigo *et al.*, 1983) and, in *H. armigera*, by exclusion cages (Cock *et al.*, 1991), and direct observation of activity and association studies (Bishop and Blood, 1977, 1980, 1981; Room, 1979b; Cock *et al.*, 1989, 1991).

McDaniel *et al.* (1981), using the method of sprayed/unsprayed crop comparisons (King and Coleman, 1989), estimated high levels of natural predation of early stage *H. virescens* and *H. zea* in unsprayed cotton.

The most common predators of *Heliothis* and *Helicoverpa* include *Chrysopa* spp., *Chrysoperla* spp., *Nabis* spp., *Geocoris* spp., *Orius* spp., *Polistes* spp. and various Pentatomidae, Reduviidae, Coccinellidae, Carabidae, Formicidae and Araneida (Bishop and Blood, 1980; Greathead and Girling, 1982; King *et al.*, 1982; Zalucki *et al.*, 1986; van den Berg *et al.*, 1988; Cock *et al.*, 1989, 1991). Some predators have been used in augmentative release studies, notably *Chrysopa carnea* Steph. (Ridgway *et al.*, 1977). Although effective in large numbers, the high cost of large scale production of this predator precluded its economic use as an applied control measure (King *et al.*, 1986).

Very little attention has been given to the role of vertebrate predators, other than by casual observation, and even this is infrequent because of their shyness and predominantly nocturnal or early morning activity. Their impact on mortality is almost certainly underestimated. Larger larvae are frequently taken by birds and pupae by rodents; adults may be taken by bats. *Helicoverpa armigera* populations on chickpea were drastically reduced by predation by cattle egrets (*Bubulcus ibis*) and river tern (*Sterna aurantia*) in Orissa, N. India (Ghode *et al.*, 1988). At ICRISAT Centre, India, a large population of fourth to sixth instar larvae on a two hectare field of chickpea were virtually annihilated by a flock of small birds (author's observations). The impact of these predators may therefore be high under certain conditions.

Pathogens

The contribution of diseases to the mortality of immature *Heliothis* and *Helicoverpa* varies seasonally, geographically and with the host plant (Zalucki *et al.*, 1986). Stress, induced by any of a range of conditions, such as crowding, poor food quality, starvation and adverse climatic conditions all encourage the expression of diseases. In general, diseases reach substantial levels in a population only when conditions are humid and densities high. Transmission and survival of the disease organism is then enhanced and epidemics can cause populations to crash. Few accurate estimates of the mortality caused by naturally occurring pathogens exist, although disease induced mortalities of 60–80% on lucerne (Bishop, 1984) and up to 60% on cotton (Wilson and Greenup, 1977) have been recorded in Australia. Pathogens are generally not specific to a particular instar, although Daoust (1974) found that the younger instars were more susceptible to nuclear polyhedrosis virus (NPV). The impact of NPV is therefore difficult to assess, because death is rapid and the remnants of small larvae not easily discernible (Teakle, 1989).

There are a number of naturally occurring generalist pathogens which commonly cause mortality of heliothine larvae, most of which have been evaluated as potential biocontrol agents (Bell, 1982). The most common of these are nuclear polyhedrosis (NPV) and granulosis viruses, *Bacillus thuringiensis* (Berliner) and fungal pathogens *Nomuraea* (*Spicaria*) *rileyi* (Farlow) Samson, *Metarhizium anisopliae* (Metch.) Sorokin, *Entomophthora aulicae* and *Beauvaria bassiana* (Bals.) Vuillemin. The microsporidian protozoa *Nosema heliothidis* and *Vairimorpha necatrix* are frequently present in populations in the USA, causing a chronic disability but rarely direct mortality (Yearian *et al.*, 1986).

Details of the extent of natural control are scarce: *B. bassiana* caused 20% mortality of overwintering pupae in Namoi, Australia (Wilson and Greenup, 1977); fungal infection by *N. rileyi* varied with the crop from 20% in tomato and 28% in field beans to 37% in vegetable pigeonpea in southern India (Gopallakrishnan and Narayanan, 1989). Epizootics of *E. aulicae* causing 48–100% mortality of *H. zea* larvae on sorghum have been recorded (Hamm, 1980), but in general the mortality caused by these organisms is much lower. Some of the above pathogens have been used as control agents in commercial formulations.

Abiotic Factors

The effects of abiotic mortality factors, most likely to operate at the vulnerable egg and early larval stages, have tended to be underestimated or disregarded (Fitt, 1989), although recent studies by Kyi *et al.* (1991) have demonstrated their importance on cotton. Many biotic and abiotic factors are closely interrelated and their impact strongly dependent upon the nature of the host. The concentration of eggs at one site, as on the silks of maize, may lead to increased levels of predation, parasitism or cannibalism. The distance or time a neonate larva has to travel before finding a suitable feeding site will increase exposure and the probability of its dying from starvation, predation or desiccation; exposure of neonate larvae to mortality factors was increased because locomotion was impeded by hairiness in cotton (Ramalho *et al.*, 1984). Early survival of *H. armigera* was always higher on determinate than on indeterminate pigeonpea cultivars and on compact-headed than on open-headed sorghum, probably as a result of the proximity of suitable feeding sites and/or cover (A.B.S. King, unpublished), and cotton variety had a highly significant effect on egg mortality (Kyi *et al.*, 1991). How well eggs adhere to a leaf surface, or are protected by virtue of their position, will affect losses caused by dislodgement by wind and rain and discovery by predators; desiccation and dislodgement were major causes of mortality, particularly on young plants (Wilson, 1982; Kyi *et al.*, 1991). Physical mortality factors have been estimated to account for up to 95% mortality of eggs and small larvae in tobacco and cotton (Titmarsh, 1985; Wilson, 1982; Kyi *et al.*, 1991).

High mortalities may be sustained by pupae, particularly those overwintering for prolonged periods at low temperatures and damp conditions (Slosser *et al.*, 1975; Eger *et al.*, 1983; Rummel and Neece, 1989), and by adults whose exit from the pupal cell is impeded, as by the effect of rainfall or cultivation (Roach and Hopkins, 1979; Roach, 1981; Murray and Zalucki, 1990a). Waterlogging for 2–3 days killed over half non-diapause *H. punctigera* and *H. armigera* pupae (Murray and Zalucki, 1990b). Land management practices are sometimes employed to increase pupal mortality (see later).

Natural Enemies Worldwide

Pearson (1958) lists the parasites of *H. armigera* recorded in Africa; this has been updated in a comprehensive review of all the natural enemies of *H. armigera* in Africa by van den Berg *et al.* (1988). The worldwide distribution, abundance and potential for biocontrol of the natural enemies of *Heliothis* and *Helicoverpa* have been reviewed (King *et al.*, 1982; King and Coleman, 1989; King and Jackson, 1989).

The principal parasites and predators of *Heliothis* and *Helicoverpa* have been listed: worldwide by King *et al.* (1982) and in King and Jackson (1989); in the USA by Goodenough *et al.* (1986), Nordlund *et al.* (1986); for eastern and southern Africa by Greathead (1966), van den Berg *et al.* (1988); for Australia by Zalucki *et al.* (1986). The use of cultural and biological control tactics in the control of *H. zea* and *H. virescens* in the USA is overviewed by Johnson *et al.* (1986); the worldwide use of biological control as a means of suppressing *Heliothis* and *Helicoverpa* populations is addressed in the workshop proceedings edited by King and Jackson (1989).

Pest Status

The four species, *Helicoverpa zea*, *H. armigera*, *H. peltigera* and *Heliothis virescens* are pests of major importance in most areas where they occur, damaging a wide variety of food, fibre, oilseed, fodder, commodity and horticultural crops. The major pest status of the four species is rooted in the essentials of their biology. High mobility, polyphagy, rapid and high reproductive rate and diapause make them particularly well adapted to exploit transient habitats like man-made agroecosystems. Their predilection for the harvestable parts of high value crops, like cotton, tomato, tobacco and sweetcorn, confers a high economic cost to their depredations. Their damage to important subsistence crops, like maize, sorghum and the pulses, and small producers' cash crops can be severe in socioeconomic terms.

Monetary losses result from the direct reduction of yields and from the cost of monitoring and control, particularly the cost of insecticides. In the USA, Hardwick (1965) puts the annual cost of damage by *H. zea* at hundreds of millions of dollars, with the cost of insecticides just on cotton at over $50 million (Ignoffo, 1973); other estimates (Agricultural Research Service, 1976) put the nationwide cost of *H. zea* and *H. virescens* damage at over one billion dollars annually, despite the use of insecticides costing a further $250 million (King and Coleman, 1989). In Queensland, Australia, the cost of crop loss and control was estimated at over A$16 million in 1979 (Alcock and Twine, 1981). Wilson (1982) estimated total Australian loss at A$23.5 million. With increases in insecticide prices and replacement of the relatively cheaper pyrethroids with more expensive alternatives to counter pyrethroid resistance, Twine (1989) has estimated that present costs in Queensland alone would now have increased to some A$25 million annually. In the irrigated Ord valley, cotton production had to be abandoned because of the destruction of natural enemies and resistance to insecticides (Reed and Pawar, 1982).

In India, where *H. armigera* commonly destroys more than half the yield, crop losses in the pulses, chickpea and pigeonpea were estimated at over $300 million per annum (Reed and Pawar, 1982). Mehrotra (unpublished) estimates that the total losses in both pulses and cotton exceeded $530 million per annum and that insecticides, to the cost of nearly $120 and $7.2 million are used annually on cotton and pulses respectively. With the recent crop failures and increased insecticide use resulting from pyrethroid resistance in *H. armigera* (Dhingra *et al.*, 1988; McCaffery *et al.*, 1989), these figures may need to be revised upwards. In Africa, few estimates of crop loss exist. However, Reed and Pawar (1982) estimate that the loss of cotton in Tanzania must exceed $20 million in most years.

Regional, and even relatively local differences in host plant preference can give rise to differences in pest status on particular crops. For example, in southern India, while *H. armigera* is abundant in light trap catches and damages legumes and other crops throughout the region, it is a major pest of cotton in only certain locations (Reed and Pawar, 1982), although this is now changing. In northern India, *H. armigera* is unimportant on cotton (this has now changed), although infestation of chickpea in the same area may be severe, and cause up to 90% pod damage (Sehgal and Ujagir, 1990). In southern Uganda, *H. armigera* was not a major pest of cotton, compared with northern Uganda and Tanzania. Coaker (1959) concluded that this was because climatic conditions in southern Uganda allowed the continuous development of both *H. armigera* and its natural enemies, whereas in northern Uganda and Tanzania populations of both were disrupted by a prolonged dry season. Reinvading *H. armigera* populations were then able increase to damaging proportions before natural enemies could catch up.

The level of infestation and damage of cotton may be influenced by its phenology, the presence and phenology of other crop and wild hosts, and climatic conditions. The presence of other cotton bollworms, such as *Diparopsis* spp., *Pectinophora gossypiella* (Saund.) and *Earias* spp. occupying or competing for the same niches, may also affect the relative importance of heliothine bollworms. The pest status of *H. armigera* on cotton in western Tanzania depended on good early season rainfall which allowed infesting populations to build up on early-sown sorghum and maize (Reed, 1965; Nyambo, 1988). However, later sown maize and sorghum, with their susceptible stage coincident with that of cotton, tended to divert or reduce infestations (Nyambo, 1988). This tactic, known as trap cropping, although widely applied and recommended in the past, has seldom been successful (Pearson, 1958; Fitt, 1989). The main drawback for protecting cotton in this way is the protracted flowering period of cotton and the generally short attractive periods, and sometimes augmentative effect on the pest, of potential diversionary crops like maize (Pearson, 1958; Fitt, 1989).

While there is much evidence that the pest status of *Heliothis* and *Helicoverpa* on cotton has been largely pesticide-induced, particularly of *H. virescens* in the USA, there are many records of the species having been a major problem on cotton before insecticides were widely used, particularly in Africa (Reed and Pawar, 1982). Moreover, agronomic factors, such as high yielding varieties, increased use of irrigation, fertilizers, scale of production and plantings of alternate crop hosts will all have contributed towards the greater prevalence and pest status of the economic Heliothinae in recent years (Reed and Pawar, 1982; Fitt, 1989).

Management and Control

In the absence of pesticides natural control of *Heliothis* and *Helicoverpa* spp, in all but a relatively few cases, is insufficient to prevent economic damage. Their control on cotton and other high value crops therefore depends heavily upon chemical insecticides, although integration with other control tactics and increased rationalization of insecticide use has been the subject of much recent research. A major constraint on moderating chemical control of the heliothine bollworms on cotton has been the need to deal with a complex of pests, often with irreconcilable control needs. It was therefore not feasible to forgo early season insecticide applications to preserve natural enemies in some states in the USA, because of the need to suppress boll weevil (Plapp and Bull, 1989).

In the development of integrated pest management (IPM) strategies for *Heliothis* and *Helicoverpa* spp., different control tactics may be combined to suppress numbers below a threshold, which depends upon the cost of implementation and the expected value of the yield resulting from it, and is dependent upon a knowledge of the relationship between population density and economic loss. It is often difficult to obtain precise data on this relationship because it is rarely simple, and many extraneous factors, both environmental and socioeconomic, may influence it. In practical terms, thresholds, if used at all, are often by rule of thumb, and it is only in the more sophisticated agricultural societies that models have been used to derive optimum control, usually insecticide application, tactics. Moreoever, the level of grower sophistication and the presence of other pests in the complex will affect the control strategy employed.

Control tactics which have been evaluated against *Heliothis* and *Helicoverpa* include cultural manipulation of the crop and its environment, biological control including the use of microbial preparations, sex pheromones for population monitoring, mass trapping or confusion, sterile back cross techniques, host plant resistance, tolerance or lack of preference and chemical control. All of these tactics require some ecological understanding of the relevant heliothine species, preferably on a regional basis, to be effective (Fitt, 1989).

Chemical Control

Management of *Heliothis* and *Helicoverpa* in cotton and other high value crops, still relies heavily on insecticides, often to the exclusion of other methods. With the increasing restraints upon insecticide use due to pest resistance and environmental considerations, the integration of several management tactics has become necessary to reduce reliance and prolong the utility of important insecticides. In cotton, pest management strategies have to cope with a complex of pests, so that the choice of insecticides and other tactics will depend upon the pests concerned and their relative importance as components of this complex; a situation which varies from region to region and may often change over time.

There is a substantial literature on the comparative efficacies of insecticides against *Heliothis* and *Helicoverpa* on cotton and other crops; some of the more recent work is reviewed and publications listed by Zalucki *et al.* (1986) and Matthews (1989); it is beyond the scope of this work to review these in detail. The development of integrated pest management (IPM) strategies and their component tactics have been reviewed for cotton in Green and Lyon (1989) and by Matthews (1989) and for other crops, chiefly pulses, in Reed and Kumble (1982) and by Reed *et al.* (1989a).

Most insecticide applications are targeted at the larval stages, although Joyce (1982) argues that larvicides have only a limited role to play in heliothine management, and favoured adults and eggs (and neonate larvae) as the principal target stages. The advantages of ovicides like chlordimeform and methomyl are their minimal effects on non-target organisms and, because most heliothine eggs are laid on the upper surfaces of leaves (Beeden, 1974; Wilson *et al.*, 1980; Hassan *et al.*, 1990) and on the upper part of the plant (Joyce, 1982; Farrar and Bradley, 1985a), they are accessible to spraying. Spray decisions based on egg counts would destroy both invading adults and eggs, and leave a residue to kill future eggs and neonate larvae (Joyce, 1982). Early application would also allow for delays in response time (Morton and Collins, 1989). Additionally, resistant and susceptible larvae exposed to a commercial application of pyrethroid did not show differences in mortality before they were four days old; targeting of neonates is therefore essential in areas where resistant populations are present (Daly *et al.*, 1988).

Action thresholds based on egg numbers have been used successfully as the basis for control decision making in cotton since 1961 in Malawi and Zimbabwe. Spraying was recommended at an average of one half an egg per plant on the basis of twice-weekly counts (Matthews and Tunstall, 1968). In the Sudan Gezira over 2 eggs or larvae per 18 plants was the threshold (Haggis, 1982), while in Australia 2 eggs per metre row (Wilson, 1981) or 5–20 brown (about to hatch) eggs per metre, was used depending upon the stage of the crop (Shaw and Colton, 1987). In the USA, emphasis on scouting has been for the boll weevil on fruiting structures (Morton and Collins, 1989). Alternative methods of scouting include thresholds based on damage, as recommended by Croft *et al.* (1984) in Texas and by Kabissa (1989) in Tanzania. Damaged buds are easier to detect and sample than eggs or larvae (Kabissa, 1989), but may be poorly correlated with crop damage (Keerthisinghe, 1982). Kabissa (1989) also argues that some damage may actually increase yields; however this will only be true if larval populations are low (less than one per plant).

Small producers in many less developed countries have limited resources, so are often unwilling to spend money on control until damage is inevitable and/or large larvae are seen. At low populations this may be good policy. However, when infestation is heavy, by the time spraying commences the low dosages of insecticides often used fail to adequately control the larger larvae; additionally much damage will already have been done.

Insecticide Resistance

The four major species of *Heliothis* and *Helicoverpa* have probably been responsible for the use and misuse of more insecticides than any other insect species. A wide variety of insecticides have been used to control the larvae and in many areas applications have been, and still are, heavy and frequent. The considerable selection pressure which the species have experienced has resulted in the development of resistance to the major classes of insecticides in many of the areas where these have been used (Wolfenbarger *et al.*, 1981).

A notable exception to the appearance of resistance problems in the control of *H. armigera* applies to cotton in Zimbabwe, despite the entire crop having been treated with insecticides since the early 1960s. DDT and endosulfan were used extensively, but were replaced later by a restricted use of synthetic pyrethroids. The absence of severe resistance problems may be due to strict control in the use of insecticides on cotton involving timing of applications on the basis of egg counts, targeting of the young larval stages with minimal dosage rates, effective spray deposition and the absence of 'cocktail' treatments, as well as virtually no insecticides being applied to maize and other host plants.

Although *H. armigera* has always been an important cotton pest in parts of southern Africa, the need to control heliothine bollworms in other areas such as south India is relatively recent; infestations have often been induced by applications against other pests. In the 1950s, 1960s, 1970s DDT and methyl parathion were widely used on cotton and food crops; these wide spectrum insecticides were initially effective, but gave rise to increasingly severe attacks by a wide range of secondary pests as their natural control elements were destroyed. Resistance to them also developed, as in the Ord Valley, Australia (Wilson, 1974). In many countries DDT was phased out during the 1970s for environmental reasons, and the high mammalian toxicity of parathion led to curtailment of its use in favour of less acutely toxic insecticides. The most important of these was endosulfan, which was very widely recommended and used against heliothine bollworms from the mid 1970s and early 1980s. Severe resistance to endosulfan has only been reported from Australia (Forrester *et al.*, 1993), where it was one of the factors responsible for the abandonment of cotton production in the Ord Valley. Some tolerance to endosulfan is still present but this has tended to remain stable under current management practices. Although it continues to be used, endosulfan was widely overtaken during the late 1970s and 1980s by the synthetic pyrethroids.

From the late 1970s, the relatively inexpensive synthetic pyrethroids were increasingly relied upon to control heliothine bollworms, as they are very effective against lepidopterous larvae and forgiving of poor managment. Applications were often made without reference to economic thresholds, and based on a calendar or crop stage basis, as exemplified by the 'toxic carpet' approach (Morton *et al.*, 1981). This reversion to the 'bad old ways', and excessive and widespread reliance on the pyrethroid 'panacea' very rapidly led to the development of resistance in *H. armigera*, later in *H. virescens*, and only recently in *H. zea*, although not in *H. punctigera* (Wilson, 1974; Fitt, 1989). Consequently, control failures in many cotton growing regions were reported throughout the 1980s. Field failures resulting from *H. armigera* resistance to pyrethroids have been reported from Australia (Gunning *et al.*, 1984; Daly and Murray, 1988; Forrester *et al.*, 1993), Thailand and Turkey (Ahmad and McCaffery, 1988), India (Dhingra *et al.*, 1988; McCaffery *et al.*, 1989) and Indonesia (McCaffery *et al.*, 1988). In the USA, pyrethroid resistance, associated with control failures, was found in *H. virescens* in Mississippi and Texas in 1985 and 1986 (Luttrell *et al.*, 1987; Roush and Luttrell, 1989) and later in both *H. virescens* and *H. zea* in several other states (Leonard *et al.*, 1988). Pyrethroid resistance in *H. virescens* has also been reported from Columbia (Jackson, 1989).

Because of their economic advantages, low toxicity to mammals and to some parasites and predators (King and Coleman, 1989), much effort has been directed towards developing management strategies aimed at prolonging the use of synthetic pyrethroids. This has included the formation of manufacturers' associations, such as the Pyrethroid Efficacy Group (PEG), the Insecticide Resistance Action Committee (IRAC) and international organizations such as the International Organization of Resistant Pest management (IOPRM) with this objective.

Insecticide resistance management (IRM) strategies have been aimed either to prevent the development of resistance, or to contain it, as in the case of *H. armigera* in eastern Australia. All rely on a strict temporal restriction in the use of the pyrethroids and their alternation with other insecticide groups to minimize selection for resistance (Daly and McKenzie, 1986; Sawicki *et al.*, 1989). In Australia, the IRM strategy is dictated by the cotton growing season which is divided into 3 stages: 1, early growth; 2, the peak squaring, flowering and boll formation; and 3, end of season. The use of pyrethroids is restricted to stage 2, which was initially a 42 day period corresponding to a single *H. armigera* generation, when no more than 3 applications should be made. Endosulfan should not be used during stage 3, to decrease the chances of reselecting for cyclodiene resistance (Forrester and Cahill, 1987). Management of pyrethroid and endosulfan resistance in Australia has recently been reviewed in detail (Forrester *et al.*, 1993).

The strategy showed initial success, but in the fourth year high levels of resistance were present in early season populations and, although it declined again in the following two years, resistance in late season populations in the sixth year were the highest recorded (Forrester, 1989). The strategy depends for its success on the dilution of resistance through interbreeding with immigrant populations of susceptible insects from unsprayed crops and wild hosts (Forrester and Cahill, 1987). However, if resistance is present in these, as indicated by Gunning and Easton (1989), and a large proportion of the local overwintered population is resistant, as shown by Fitt and Forrester (1988) and Daly and Fitt (1990), and contibutes significantly to early season infestation, then this strategy is unlikely to be effective in the long term. There is some uncertainty as to whether insecticide resistance confers biological constraints in *H. armigera* and *H. virescens*. There is evidence both for the absence (Payne *et al.*, 1988; Gunning and Easton, 1989; Sawicki and Denholm, 1989), and presence (Campanhola *et al.*, 1991) of reduced fitness in pyrethroid resistant strains. If, as is indicated, this is principally associated with metabolic resistance, the increased incidence of the nerve insensitivity resistance mechanism, as noted by West and McCaffery (1992) and Gunning *et al.* (1991) in Indian and Australian *H. armigera*, which does not confer the disadvantage of reduced fitness, implies a major threat to current resistance management strategies.

Counter to the benefits derived from diluting influxes of susceptible insects, the capacity of *Heliothis* and *Helicoverpa* spp. for long range movement has serious implications on the spread of resistance, particularly into previously unsprayed areas. Based on their work on gene flow between populations of *H. armigera* in Australia, Daly and Gregg (1985) strongly suspected this to occur. The sudden appearance of pyrethroid resistant insects (and poor results on crops where previously normal rates of pyrethroid applications had given satisfactory control) in Texas 330 km downwind (Westbrook *et al.*, 1990), and in Andhra Pradesh,

India 250 km downwind from highly resistant populations in cotton (McCaffery *et al.*, 1989), largely confirms this.

The underlying factors influencing the development of resistance are genetic, ecological, behavioural and agronomic (Fitt, 1989). The importance of ecological and behavioural factors has already been shown, but is particularly well illustrated in the case of *H. punctigera*, which has not developed resistance to any chemical although subjected to the same intense selection on cotton as *H. armigera*. However, *H. punctigera* is more mobile, more widely distributed and has a much wider natural host range than *H. armigera* in its native Australia, so that it maintains large enough populations in unsprayed refugia to adequately dilute selection for resistance in cropping areas (Fitt, 1989).

Field failures are the result of a complex interrelation of genetic, environmental and management factors, rather than resistance alone (Daly and Murray, 1988). The relevance of differential mortality/ selection with larval age, insecticide decay, poor application and of adult moths is discussed by Daly *et al.* (1988) who stress the importance of targeting applications against very small larvae, as already recommended much earlier by Matthews and Tunstall (1968) for cotton in Zimbabwe.

The biochemical basis of resistance has also received recent attention. Pyrethroid resistance in *H. armigera* may be conferred through 3 mechanisms: detoxication by mixed-function oxidases, nerve insensitivity and delayed penetration (Daly, 1988; Ahmad *et al.*, 1989; Sawicki and Denholm, 1989; West and McCaffery, 1992). The detoxication mechanism is inhibited by piperonyl butoxide, so that synthetic pyrethroids synergized with it may be one way of prolonging their effectiveness against populations in which detoxication is the principal resistance mechanism (Sawicki and Denholm, 1989).

Cultural Control

Cultural manipulations of the crop, cropping system and land management have been used as tactics in the management of *Heliothis* and *Helicoverpa*. The use of short season cotton cultivars has proved highly effective in Texas. Plant growth regulators have been used on cotton to increase accessibility of larvae to spray droplets and to shorten crop time to maturity (Bradley *et al.*, 1986). Trap cropping, and the use of diversionary hosts was widely applied and recommended in the past, but has seldom been successful (Pearson, 1958; Fitt, 1989). Although infestation of cotton by *H. armigera* was reduced by late planted maize and sorghum flowering coincidentally with it (Nyambo, 1988), their comparatively short attractive periods, and the potential of earlier planted crops to augment or create infesting populations, were major disadvantages.

Post harvest cultivation to destroy pupae, mainly of the overwintering generation, has received considerable attention in the USA as a means of strategic control (Fife and Graham, 1966; Young and Price, 1977; Roach, 1981; Bradley *et al.*, 1986; Luttrell *et al.*, 1986; Rummel and Neece, 1989). Although the natural survival of overwintering pupae may be as low as under 5% (Slosser *et al.*, 1975; Stadelbacher and Martin, 1980; Fitt, 1989), there is still controversy over whether these, or immigrants, are principally responsible for infestations of the following season's crops in southern USA (Luttrell *et al.*, 1986). The importance of ploughing cotton stubble to reduce overwintering populations of pyrethroid resistant *H. armigera* in the Namoi Valley, Australia is stressed by Fitt and Forrester (1988).

The close season, aimed to deny suitable hosts for one or more generations, is legally enforced in some countries principally to reduce *Pectinophora* infestations, and is considered useful in relation to the cotton pest complex as a whole. Its benefit for heliothine bollworms is, however, questionable in view of the migratory potential of these species.

Based on observations in the Mississippi Delta that infestations on cotton tended to be higher near extensive early-season host habitats, a strategy for area-wide suppression of the first generation of *H. virescens* and *H. zea* feeding on wild hosts was proposed by Knipling and Stadelbacher (1983). Suppression of these populations, using a variety of techniques including mass releases of predators and parasites, applications of pathogens, autocidal (release of sterile males) techniques and destruction of spring hosts, would reduce the level of damage to cotton and soybeans in the next generation to an acceptable level and, according to his model, significantly reduce populations in subsequent years (Stadelbacher, 1982). The rationale was criticized by Mueller *et al.* (1984), who pointed out that the strategy would be invalid if moth immigration was the principal origin of populations invading crops and that the measures could be more detrimental to natural enemies than to heliothine populations. Using a simulation model of insect movement Schneider (1989) estimated that suppression of *H. virescens* populations over an area 25 km in diameter would be necessary to prevent significant recolonization of the area within two generations.

The key issue of moth immigration, and movement in general, is of general relevance to the long-term effectiveness of any control strategy aimed to suppress more than one generation; it may have little point if crop infestation is mainly by immigrant moths of distant origin.

Biological Control

There has been much interest in classical biological control as a tactic in the management of *Heliothis* and *Helicoverpa*, particularly in the USA. This has been comprehensively reviewed in Johnson *et al.* (1986), King and Coleman (1989) and by King and Jackson (1989). There has also long been interest in exploiting entomophagous pathogens such as *Bacillus thuringiensis* and *Heliothis* NPV (Ignoffo, 1973; Yearian *et al.*, 1986; Carner and Yearian, 1989; Jayaraj *et al.*, 1989), although to date these have not provided a viable alternative to insecticides. Currently, research is underway to develop transgenic crop varieties expressing the active *Bt* toxin, a development which has already shown promise in tomato (Menn *et al.*, 1989).

Probably the most frequently tried method of achieving control with natural enemies in annual crops has been by augmentative releases of artificially reared parasites or predators (King *et al.*, 1986; King and Coleman, 1989). While the technical feasibility of suppressing heliothine bollworm populations in cotton by this method have been demonstrated, the results have not always been consistent and the economic feasibility dubious at best, when compared with insecticides (King *et al.*, 1982; King and Coleman, 1989).

Because of the dependence of this technique on the capability of producing large numbers of parasites or predators simultaneously and economically, emphasis has been placed on *Trichogramma* spp., which are most amenable to mass rearing and are now the most commonly augmented entomophagous arthropods in the world (King *et al.*, 1985). However, releases in cotton have not been consistently effective in reducing heliothine populations (King and Coleman, 1989). Cotton production, relying heavily on chemicals to control a range of pests, is not a favourable environment for entomophagous arthropod activity, although some parasites of *H. virescens* and *H. zea* in the USA are less vulnerable to synthetic pyrethroids (Plapp and Vinson, 1977; King *et al.*, 1985; Elzen *et al.*, 1987; Plapp and Bull, 1989), now the most commonly used insecticide group in cotton. The most effective of these, *Microplitis croceipes* (Cresson) (King *et al.*, 1985), is also relatively tolerant to organophosphate (OP) insecticides (Powell *et al.*, 1986; King and Coleman, 1989) and strategies involving its augmentative release are currently being developed (Menn *et al.*, 1989). However, the increasingly widespread occurrence of resistance in the economic Heliothinae to the synthetic pyrethroids and the destruction of natural enemies through the need for early season control of other pests with insecticides, severely limit the scope for economic control strategies using beneficial insects in intensive production systems (King *et al.*, 1982; Menn *et al.*, 1989).

Microbial Control

Their relative specificity, potential activity and safeness to non-target organisms and the environment have made microbial pesticides a favoured component for IPM strategies. Since 1959, considerable efforts have been made to develop the most promising agents, *Bacillus thuringiensis* (*Bt*) and *Heliothis* NPV, into commercially viable biological insecticides (Ignoffo, 1973; Bell, 1982; Yearian *et al.*, 1986). However, with the arrival of the pyrethroids in the 1970s research on microbials and other alternatives, particularly in the USA, received less emphasis.

Improved strains of microbial agents (e.g. *Bt* var. *kurstaki* and commercial formulation of *Heliothis* NPV known as Elcar®) were, however, inconsistent and marginal at best in controlling moderate to high *Heliothis* or *Helicoverpa* populations, compared with standard insecticides (Rogers *et al.*, 1983; Yearian *et al.*, 1986). Poor performance was attributed to rapid degradation by UV light and insufficient quantities being ingested by larvae to cause mortality. Formulations containing gustatory stimulants, which increased ingestion of the pathogen, were more successful (Andrews *et al.*, 1975; Carner and Yearian, 1989). More recently, formulations with increased persistence and including feeding stimulants or adjuvants like Coax® and Gustol®, have significantly increased the effectiveness of *Bt* and *Heliothis* NPV (Bell and Romine, 1980; Bell, 1982; Smith and Hostetter, 1982; Yearian *et al.*, 1986; Carner and Yearian, 1989). However, under heavy pest pressure, adequate crop protection may still not be provided (Luttrell *et al.*, 1983; Yearian *et al.*, 1986).

Mixtures with sublethal doses of some insecticides have also been tested. While results have generally been no better than either the pathogen or the insecticide applied individually, some combinations, for example *Bt* or NPV with chlordimeform, particularly with the addition of a gustatory adjuvant, have been comparable with the insecticide standard (Yearian *et al.*, 1980). The *Bt* formulation Dipel® combined with endosulfan or monocrotophos had a synergistic effect on larval mortality, compared with carbaryl, malathion or phenthoate (Dabi *et al.*, 1989). However, pathogen adjuvant mixtures, with or without addition of insecicides, are more expensive than insecticides alone, so that their use on a regular basis does not appear to be economically practical (Yearian *et al.*, 1986), at least while current insecticide use remains economical and environmentally tolerable.

Sterile Backcross

Laster (1972) found that sterile male offspring are always produced when *Heliothis subflexa* females mate with *H. virescens* males; although the female hybrids are fertile, they also produced sterile male offspring. This fact has been exploited in the development of population management by means of releasing sterile males, a tactic originally proposed by Knipling (1960), and modelled on the successful screw-worm eradication programme (Menn *et al.*, 1989). The technique was evaluated on the island of St Croix, in the Virgin Islands, where, after a three year release programme, suppression was achieved (Proshold *et al.*, 1982).

The main drawbacks to such a programme are the cost of producing the large numbers of backcross insects to be introduced, and that suppression is only likely to be successful in isolated or contained ecosystems, where population dynamics are not greatly influenced by emigration and immigration.

Host Plant Resistance (HPR)

The development of crop cultivars resistant or tolerant to damage by *Heliothis* and *Helicoverpa* spp. has major potential in their management (Wiseman, 1982: Kennedy *et al.*, 1987; Fitt, 1989), particularly for communities with few resources (Lukefahr, 1982). Many crops possess some genetic potential which can be exploited by breeders to produce varieties less subject to pest damage.

Resistance may take three basic forms or 'modalities', antixenosis, antibiosis and tolerance, each of which can be expressed to different degrees (Kennedy *et al.*, 1987). Antixenosis, can be expressed as non-preference or reduced attractiveness to ovipositing females; antibiosis reduces the rate of population increase through survival and reproduction, and is often expressed in slower development rates and increased larval mortality (Wiseman, 1982). Tolerance effectively raises the action threshold for control (Kennedy *et al.*, 1987). The impact of HPR therefore depends upon the modality or type of resistance, the behaviour and ecology of the pest and the complexity of the cropping system (Fitt, 1989). The management consequences may not always be desirable; in a diverse agroecosystem, antiaxenic-based resistance may effectively shift the focus of infestation to other crops, and increased tolerance allow pest populations to reach higher than normal levels (Kennedy *et al.*, 1987).

A notable example of antiaxenic resistance in cotton, and some other crops, is the glabrous leaf character. Significantly fewer eggs were laid by *H. zea* and *H. virescens* on glabrous compared with pubescent leaved cotton (Lukefahr *et al.*, 1971; Robinson *et al.*, 1980); but the pubescent surface adversely affected mobility and survival of young *H. virescens* larvae (Ramalho *et al.*, 1984). Oviposition by heliothines was also less on nectariless than on cultivars with extra-floral nectaries (Lukefahr *et al.*, 1971; Adjei-Maafo and Wilson, 1983b), and when both glabrous and nectariless characters were combined, egg numbers were reduced by up to 80% (Lukefahr *et al.*, 1971). However, the benefits of smooth leaf and nectariless characters have been difficult to demonstrate on a field scale (Lukefahr, 1982; Watson, 1989). Nectariless cotton also supported fewer predators (Adjei-Maafo and Wilson, 1983a), although the compensatory effect on heliothine populations has not been quantified (Schneider *et al.*, 1986).

Antibiosis is expressed in cotton varieties having high levels of terpenoids, particularly gossypol and other heliocides. These varieties were more resistant to attack by *Heliothis* and *Helicoverpa* (Lukefahr, 1982), and in combination with the smooth leaf character, could significantly suppress larval numbers (Lukefahr *et al.*, 1975). Although the presence of resistance has always been at the expense of yield, resistant, high gossypol lines have recently been developed which are competitive in yield with non-resistant varieties when insects are controlled with insecticides (Jenkins, 1989). The interactions of *Heliothis* and *Helicoverpa* with their hostplants, sources of resistance, its implication and durability are reviewed and discussed by Schneider *et al.* (1986) and Kennedy *et al.* (1987).

Varieties of other crop hosts showing resistance to *Heliothis* or *Helicoverpa* have been identified or developed in chickpea (Lateef, 1985; Ujagir and Khare, 1987), pigeonpea (ICRISAT, 1986, 1987), soybean, tomato, maize, sorghum, millet and tobacco (Wiseman *et al.*, 1977; Lukefahr, 1982; Wiseman and Isenhour, 1990; Javaid *et al.*, 1991).

Apparent resistance may result from physiological characters: for example, determinate cotton varieties reduce the period at risk to infestation (Matthews, 1989), while indeterminate pigeonpea increases early larval losses from predation and desiccation (A.B.S. King, unpublished). The frego bract and okra leaf characters and an open plant structure, while not necessarily conferring any resistance directly, enhance penetration and coverage by insecticides (Matthews 1989; Thomson and Lee, 1980).

In cotton, the development of resistance has often to include a complex of pests, often peculiar to a region, and involve plant characters which may be mutually exclusive. For example, pubescence is inimical to jassids but favoured by heliothine bollworms (Matthews, 1989). It cannot, moreover, be assumed that host plant resistance will be compatible with natural control; there is relatively little published work on the interactions between the predators and parasites of *Heliothis* and *Helicoverpa* and host plant resistance (Wiseman, 1982; Reed *et al.*,

1989b). Adjei-Maafo and Wilson (1983a) found that both *Helicoverpa* spp. and the numbers of many predatory and parasitic arthropods were significantly less on nectariless than on nectaried cotton. Predation of *H. zea* by the anthocorid *Orius insidiosus* was greater on resistant than on susceptible genotypes of maize (Isenhour *et al.*, 1989), but rates of *H. armigera* parasitism were consistently lower on resistant than on susceptible chickpea and pigeonpea cultivars, although this was more than compensated by the effects of resistance (Reed *et al.*, 1989b). In general, one would expect density dependence to reduce the impact of natural enemies on resistant crops.

Modelling

Models are conceptual or mathematical devices which aim to describe or simulate natural processes. They can be used to predict the outcome of hypothetical eventualities and, as management tools, to predict or establish the optimal tactics required to achieve a particular result, within the constraints of the model.

The population models developed for *Heliothis* and *Helicoverpa* spp., mostly in cotton, have as their principal aim the improvement of pest management strategies, such as optimal timing of insecticide applications. Models have been developed in the USA to aid in the management of *H. virescens* and *H. zea* in different cropping systems; most notable amongst these are HELSIM-2 for *H. zea* in North Carolina (Stinner *et al.*, 1974) and MOTHZV for *H. zea* and *H. virescens* in Texas (Hartstack *et al.*, 1976). In Texas, BUGNET, which incorporated MOTHZV, was developed as a state-wide extension tool to provide management advice for a range of pests of cotton and sorghum (Hartstack, 1982). Later versions of MOTHZV incorporated detailed population and physiology algorithms as well as crop models for corn, cotton and sorghum. The models are driven by initial pheromone trap catches and a day-degree concept to calculate the development of both insects and crop (Hartstack, 1982). Elaborations, such as a submodel relating larval infestation to yield for different crop varieties, have been incorporated later. In general, models are highly specific to the cropping system and location.

In Australia, SIRATAC, a computer-based pest management system, was developed to rationalize insecticide use on cotton (Room, 1979a, 1983; Hearn *et al.*, 1981). Based on a temperature driven development model for *H. armigera* and *H. punctigera*, SIRATAC has been progressively updated as research results became available. The on-line system incorporates a temperature driven cotton development model, including the natural fruiting habit of the plant, and submodels to incorporate damage relationships, the impact of natural enemies and pre-determined or dynamic thresholds for pests. The system gives management options, and the outcomes of using 'soft' or 'hard' insecticides (Zalucki *et al.*, 1986; D. Larsen, personal communication). Although the operating company, Siratac Ltd, now no longer exists, the programme still operates through a user group, using individual PC packages which have been developed from it. Although these lack the sophistication of the original system, Cox *et al.* (1991) have shown that significant improvements could be obtained by adjusting the threshold at specific times during the crop growth period, which roughly corresponds to the three phases of the Resistance Management Strategy (Forrester and Cahill, 1987; Forrester, 1990).

A regional population model, aimed to aid management of *H. armigera* and *H. punctigera* in cotton by predicting infestation levels, has recently been developed by CSIRO in Australia. The model, HEAPS, incorporates modules stimulating adult movement, oviposition, development, survival and host phenology, and estimates populations in each of a grid of simulation units into which a cotton producing region is divided, taking account of both bollworm species in cotton as well as other crop and non-crop hosts in the region (Dillon and Fitt, 1990; Hamilton and Fitt, 1990).

A model has recently been developed by Hopper and Stark (1987) for *H. zea* and *H. virescens* management in cotton in Massachussetts, USA. The model, DEMHELIC, emphasizes the impact of natural enemies on mortality and larval feeding, and the differential effects of insecticide types on natural enemies and heliothine larvae. Although having crop and pest development parameters in common with HEAPS, DEMHELIC is primarily a management model based on scouting and crop protection economics.

A relatively simple simulation model of *H. armigera* on pigeonpea in India has recently been developed by Holt *et al.* (1990) to optimize insecticide use for the control of susceptible and resistant larvae on pigeonpea. The driving variable is the flowering phenology of the crop, on which oviposition time and larval survival strongly depend. The optimal time and application frequency to control the larval progeny of a wholly immigrant population were most sensitive to the time and duration of immigration, flowering time, moth age at immigration and the development time of young larvae. Like the other models described, this model is also highly specific to the local ecology of the pest and the particular cropping situation.

Conclusion

The complexity of *Heliothis* and *Helicoverpa* population dynamics and host relations and the large differences between the agroecosystems and socioeconomic environments in which they are pests, make generalized statements on management largely meaningless. The pests and their impact need also to be understood in a regional context and management strategies developed on this basis. While this wider outlook is being addressed in some of the more sophisticated economies and large-scale agriculture, particularly in cotton, a similar approach may not be feasible for small producers having scarce resources, farming in often highly heterogeneous, remote and sometimes unpredictable environments. In these situations emphasis has to be placed on the integration of those elements from an array of tactical options which are most suited to the local situation. A future prospect must therefore be to increase the number, and availability, of these options; models can play an important role in suggesting the most appropriate combination of these.

Acknowledgements

The author would like to thank T.G. Wood, N.J. Armes and D.J.deB. Lyon of NRI; G.P. Fitt of CSIRO; E.G. King of USDA/ARS and W. Reed, formerly of ICRISAT, for commenting upon the original manuscript. The work for this chapter was supported by funding from ODA.

References

Adjei-Maafo, I.K. and Wilson, L.T. (1983a) Factors affecting the relative abundance of arthropods on nectaried and nectariless cotton. *Environmental Entomology* 12, 349–352.

Adjei-Maafo, I.K. and Wilson, L.T. (1983b) Association of cotton nectar production with *Heliothis punctigera* (Lepidoptera: Noctuidae) oviposition. *Environmental Entomology* 12, 1160–1170.

Adkisson, P.L. and Roach, S.H. (1971) A mechanism for seasonal discrimination in the photoperiodic induction of pupal diapause in the bollworm *Heliothis zea* (Boddie). In: Menaker, M. (ed.), *Symposium on Biochemistry*. National Academy of Science, Biological and Agricultural Division, Washington, DC, pp. 272–280.

Adler, P.H. (1987) Temporal feeding patterns of adult *Heliothis zea* (Lepidoptera: Noctuidae) on pigeonpea nectar. *Environmental Entomology* 16, 424–427.

Agricultural Research Service (1976) ARS National *Heliothis* Planning Conference, New Orleans, Louisiana. USDA, Washington, DC, 36 pp.

Ahmad, M. and McCaffery, A.R. (1988) Resistance to insecticides in a Thailand strain of *Heliothis armigera* (Hübner) (Lepidoptera: Noctuidae). *Journal of Economic Entomology* 81, 45–48.

Ahmad, M., Gladwell, R.T. and McCaffery, A.R. (1989) Decreased nerve sensitivity is a mechanism of resistance in a pyrethroid resistant strain of *Heliothis armigera* from Thailand. *Pesticide Biochemistry and Physiolology* 35, 165–171.

Alcock, B. and Twine, P.H. (1981) The cost of *Heliothis* in Queensland crops. Presented at a Workshop on the Biological Control of *Heliothis* spp., 23–25 September 1980. Queensland Department of Primary Industries, Toowoomba, Queensland, Australia, pp. 1–10.

Alvarado-Rodriguez, B., Leigh, T.F. and Lange, W.H. (1982) Oviposition site preference by the tomato fruitworm (Lepidoptera: Noctuidae) on tomato, with notes on plant phenology. *Journal of Economic Entomology* 75, 895–898.

Andrews, G.L., Harris, F.A., Sikorowski, P.P. and McLaughlin, R.E. (1975) Evaluation of *Heliothis* nuclear polyhedrosis virus in a cotton seed oil bait for the control of *Heliothis virescens* and *Heliothis zea* on cotton. *Journal of Economic Entomology* 68, 87–90.

Armes, N.J. (1989) Summary of life-table experiments on *Heliothis armigera* (Hübner) (Lepidoptera: Noctuidae). Natural Resources Institute, Chatham, Kent (Physiology and Behaviour Section). Internal Publication. 57pp.

Armes, N.J. and Cooter, R.J. (1991) Effects of age and mated status on flight potential of *Helicoverpa armigera* (Hübner) (Lepidoptera: Noctuidae). *Physiological Entomology* 16, 131–144.

Baker, T.C., Staten, R.T. and Flint, H.M. (1990) Use of pink bollworm pheromone in pest management. In: Ridgeway, R.L., Silverstein, R.M. and Inscoe, M.N. (eds) *Behavior-Modifying Chemicals for Insect Management.* Marcel Dekker, New York, pp. 417–436.

Beeden, P. (1974) Bollworm oviposition on cotton in Malawi. *Cotton Growing Review* 5, 52–61.

Bell, M.R. (1982) The potential use of microbials in *Heliothis* management. In: Reed, W. and Kumble, V. (eds) *Proceedings of the International Workshop on* Heliothis *Management*, 15–20 November 1981, ICRISAT, Patancheru, A.P., India. ICRISAT, Patancheru, A.P., India, pp. 137–145.

Bell, M.R. (1990). Use of baculovirus for the control of *Heliothis* spp. in area-wide pest management programs. In: *Proceedings and Abstracts of the 5th International Colloquium on Invertebrate Pathology and Microbial Control*, 20–24 August 1990, Adelaide, Australia, pp. 486–490.

Bell, M.R. and Romine, C.L. (1980) Tobacco budworm: Field evaluation of microbial control in cotton using *Bacillus thuringiensis* and a nuclear polyhedrosis virus with a feeding adjuvant. *Journal of Economic Entomology* 73, 427–430.

Bhatnagar, V.S. (1987) *Rapport de Synthese (1981–86) et Recommandations, Sous-programme de Lutte Biologique, Projet CILISS de Lutte Integree, Senegal.* FAO, Rome, 168pp.

Bhatnagar, V.S., Lateef, S.S., Sithanantham, S., Pawar, C.S. and Reed, W. (1982) Research on *Heliothis* at ICRISAT. In: Reed, W. and Kumble, V. (eds) *Proceedings of the International Workshop on* Heliothis *Management*, 15–20 November 1981, ICRISAT Center, Patancheru, A.P., India. ICRISAT, Patancheru, A.P., India, pp. 385–396.

Bhatnagar, V.S., Sithanantham, S., Pawar, C.S., Jadhav, D.S., Rao, V.K. and Reed, W. (1983) Conservation and augmentation of natural enemies with reference to IPM in chickpea and pigeonpea. In: *Proceedings of the International Workshop on Integrated Pest Control in Grain Legumes*, 4–9 April 1983. EMBRAPA, Goiania, Brazil, pp. 157–180.

Bhatnagar, V.S., Pawar, C.S., Jadhav, D.R. and Davies, J.C. (1985) Mermithid nematodes as parasites of *Heliothis* spp. and other crop pests in Andhra Pradesh, India. *Proceedings of the Indian Academy of Science* (Animal Science) 94, 509– 515.

Bhattacherjee, N.S. and Gupta, S.L. (1972) A new species of *Heliothis* Ochsenheimer (Noctuidae, Lepidoptera) infesting cotton and tur (*Cajanus indicus*) in India with observations on three other common species of the genus. *Journal of Natural History* 6, 147–151.

Bilapate, G.G. (1984) *Heliothis* complex in India – a review. *Agricultural Review London* 5, 13–26.

Bishop, A.L. (1984) *Heliothis* spp. and *Merophyas divulsana* (Walker) in the seasonal damage of lucerne in the Hunter Valley, New South Wales. *General and Applied Entomology* 16, 36–49.

Bishop, A.L. and Blood, P.R. (1977) A record of beneficial arthropods and insect diseases in southeast Queensland cotton. *PANS* (*Pest Articles and News Summaries*) 23, 384–386.

Bishop, A.L. and Blood, P.R. (1980) Arthropod ground strata composition of the cotton ecosystem in southeastern Queensland and the effect of some control strategies. *Australian Journal of Zoology* 28, 693–698.

Bishop, A.L. and Blood, P.R. (1981) Interactions between natural populations of spiders and pests in cotton and their importance to cotton production in southeastern Queensland. *General and Applied Entomology* 13, 98–104.

Bradley, J.R., Herzog, G.A., Roach, S.H., Stinner, R.E. and Terry, L.I. (1986) Cultural control in southeastern U.S. cropping systems. In: Johnson, S.J., King, E.G. and Bradley, J.R. (eds) *Theory and Tactics of* Heliothis *Population Management: 1 – Cultural and Biological Control.* Southern Cooperative Series Bulletin 316, 22–27.

Brazzel, J.R., Newsom, L.D., Roussel, J.S., Lincoln, C., Williams, F.J. and Barnes, G. (1953) Bollworm and tobacco budworm as cotton pests in Louisana and Arkansas. *Louisiana Technical Bulletin* 482, 1–47 (from Neunzig, 1969).

Broadley, R.H. (1980) Preliminary investigations toward establishing economic injury levels of *Heliothis* spp. (Lepidoptera: Noctuidae) larvae in sunflowers. Australian

Sunflower Association, 4th Australian Sunflower Workshop, Shepparton, Victoria, pp. 4–31–36.

Browning, T.O. (1979) Timing of the action of photoperiod and temperature on events leading to diapause and development in pupae of *Heliothis punctigera* (Lepidoptera: Noctuidae). *Journal of Experimental Biology* 83, 261–269.

Burleigh, J.G. and Farmer, J.H. (1978) Dynamics of *Heliothis* spp. larval parasitism in southeast Arkansas. *Environmental Entomology* 7, 692–694.

Butler, G.D., Wilson, L.T. and Henneberry, T.J. (1985). *Heliothis virescens* (Lepidoptera: Noctuidae): Initiation of summer diapause. *Journal of Economic Entomology* 78, 320–324.

Callahan, P.S. (1958) Behaviour of the imago of the corn earworm, *Heliothis zea* (Boddie), with special reference to emergence and reproduction. *Annals of the Entomological Society of America* 51, 271–283.

Callahan, P.S. (1961) Relationship of the crop capacity to the depletion of the fat body and egg development in the corn earworm *Heliothis zea* and fall armyworm, *Laphygma frugiperda*. *Annals of the Entomological Society of America* 54, 819–827.

Callahan, P.S. (1962) Techniques for rearing the corn earworm *Heliothis zea* (Boddie). *Journal of Economic Entomology* 55, 453–457.

Campanhola, C., McCutchen, B.F., Baehreke, E.H. and Plapp, F.W., Jr. (1991) Biological constraints associated with resistance to pyrethroids in the tobacco budworm (Lepidoptera:Noctuidae). *Journal of Economic Entomology* 84, 1404–1411.

Cantelo, W.W. and Jacobson, M. (1979) Corn silk volatiles attract many species of moths. *Journal of Environmental Science and Health* 14, 695–707.

Carner, G.R. and Yearian, W.C. (1989) Development and use of microbial agents for control of *Heliothis* spp. in the U.S.A. In: King, E.G. and Jackson, R.D. (eds) *Proceedings of the Workshop on Biological Control of* Heliothis*: Increasing the Effectiveness of Natural Enemies*, 11–15 November 1985, New Delhi, India, FERRO/USDA, New Delhi, India, 469–481.

Carshatt, E.D. (1964) Factors affecting colour variation of larvae of the tobacco budworm, *Heliothis virescens* (Fab.). MS. Thesis, North Carolina State University, Raleigh, NC, 50 pp. (from Neunzig, 1969).

Coaker, T.H. (1959) Investigations on *Heliothis armigera* in Uganda. *Bulletin of Entomological Research* 50, 87–506.

Cock, M.J.W., van den Berg, H., Odour, G.I. and Osongo, E.K. (1989) *The Population Ecology of* Helicoverpa armigera *in Smallholder Crops in Kenya with Emphasis on its Natural Enemies*. Annual Report 1988–89. CAB International Institute of Biological Control, Kenya Station, Nairobi, 74pp.

Cock, M.J.W., van den Berg, H., Odour, G.I. and Onsongo, E. K. (1991) *The Population Ecology of* Helicoverpa armigera *in Smallholder Crops in Kenya with Emphasis on its Natural Enemies*. Final Report, Phase II: April 1988–March 1991. CAB International Institute of Biological Control, Kenya Station, Nairobi, 179pp.

Colvin, J.T. (1990) Laboratory studies on the regulation of migration of the cotton bollworm, *Heliothis armigera* (Hb.) (Lepidoptera: Noctuidae). PhD Thesis, University College of North Wales, Bangor, 221pp.

Common, I.F.B. (1953) The Australian species of *Heliothis* (Lepidoptera: Noctuidae) and their pest status. *Australian Journal of Zoology* 1, 319–344.

Common, I.F.B. (1985) A new Australian species of *Heliothis* Ochsenheimer (Lepidoptera: Noctuidae). *Journal of the Australian Entomological Society* 24, 129–133.

Commonwealth Institute of Entomology (1951) Distribution Maps of Pests. Commonwealth Institute of Entomology, London (Loose leaf binder).

Coombs, M. (1992) Diel feeding behaviour of *Helicoverpa punctigera* (Wallengren) adults (Lepidoptera: Noctuidae). *Journal of the Australian Entomological Society* 31, 193–197.

Cox, P.G., Marsden, S.G., Brook, K.D., Talpaz, H. and Hearn, A. B. (1991) Economic optimisation of *Heliothis* thresholds on cotton using a pest management model. *Agricultural Systems* 35, 157–171.

Croft, B.A., Adkisson, P.L., Sutherst, R.W. and Simmons, G.A. (1984) Applications of ecology for better pest control. In: Huffaker, C.B. and Rabb, E.L. (eds) *Ecological Entomology*. Wiley, New York, pp. 763–795.

Cullen, J.M. (1969) The Reproduction and Survival of *Heliothis punctigera* Wallengren in South Australia. PhD thesis, University of Adelaide (from Zalucki *et al.*, 1986).

Cullen, J.M. and Browning, T.O. (1978) The influence of photoperiod and temperature on the induction of diapause in pupae of *Heliothis punctigera*. *Journal of Insect Physiology* 24, 595–601.

Dabi, R.K., Puri, M.K., Gupta, H.C. and Sharma, S.K. (1989) Synergistic response of low rate of *Bacillus thuringiensis* Berliner with sub lethal dose of insecticides against *Heliothis armigera* Hübner. *Indian Journal of Entomology* 50, 28–31.

Daly, J.C. (1988) Insecticide resistance in *Heliothis armigera* in Australia. *Pesticide Science* 23, 165–176.

Daly, J.C. and Fitt, G.P. (1990) Resistance frequencies in overwintering pupae and the spring generation of *Helicoverpa armigera* (Hübner) (Lepidoptera: Noctuidae) in northern New South Wales, Australia: selective mortality and gene flow. *Journal of Economic Entomology* 83, 1682–1688.

Daly, J.C. and Gregg, P. (1985) Genetic variation in *Heliothis* in Australia: species identification and gene flow in the two species *H. armigera* (Hübner) and *H. punctigera* (Wallengren) (Lepidoptera: Noctuidae). *Bulletin of Entomological Research* 75, 169–184.

Daly, J.C. and McKenzie J.A. (1986) Resistance management strategies in Australia: the *Heliothis* and 'wormkill' programmes. In: *Proceedings of the British Crop Protection Conference on Pests and Diseases, Brighton, 1986*. BCPC, Farnham, UK, pp. 951–959.

Daly, J.C. and Murray, D.A.H. (1988) Evolution of resistance to pyrethroids in *Heliothis armigera* (Hübner) (Lepidoptera: Noctuidae) in Australia. *Journal of Economic Entomology* 81, 984–988.

Daly J.C., Fisk J.H. and Forrester N.W. (1988) Selective mortality in field trials between strains of *Heliothis armigera* (Hubner) (Lepidoptera: Noctuidae) resistant and susceptible to synthetic pyrethroids: functional dominance of resistance and age-class. *Journal of Economic Entomology* 81, 1000–1007.

Danilevskii, A.S. (1965) *Photoperiodism and Seasonal Development of Insects*. Oliver and Boyd, London, 283pp.

Daoust, R.A. (1974) Weight-related susceptibility of larvae of *Heliothis armigera* to a crude nuclear-polyhedrosis preparation. *Journal of Invertebrate Pathology* 23, 318–324.

Dent, D.R. and Pawar, C.S. (1988) The influence of moonlight and weather on catches of *Helicoverpa armigera* (Hubner) (Lepidoptera: Noctuidae) in light and pheromone traps. *Bulletin of Entomological Research* 78, 365–377.

Dhandapani, N. and Balasubramanian, M. (1980) Effect of different food plants on the development and reproduction of *Heliothis armigera* (Hbn.). *Experientia* 36, 930–931.

Dhingra, S., Phokela, A. and Mehrotra, K.N. (1988) Cypermethrin resistance in the populations of *Heliothis armigera* Hubner. *National Academy of Sciences, India; Science Letters* 11, 123–125.

Dillon, M.L. and Fitt, G.P. (1990) HEAPS: A regional model of *Heliothis* population dynamics. In: *Proceedings of the Fifth Australian Cotton Conference*, 8–9 August 1990. Broadbeach, Queensland. Australian Cotton Growers' Research Association, Brisbane, pp. 337–344.

Doss, S.A. (1979) Effect of host plants on some biological aspects of bollworm, *Heliothis armigera* (Hubner) Lepidoptera, Noctuidae. *Zeitschrift für Angewändte Entomology* 86, 143–147.

Drake, V.A. (1990) Methods for studying adult movement in *Heliothis*. In: Zalucki, M.P. (ed.) *Heliothis: Research Methods and Prospects.* Springer-Verlag, New York, pp. 109–121.

Drake, V.A. and Farrow, R.A. (1988) The influence of atmospheric structure and motions on insect migration. *Annual Review of Entomology* 33, 183–210.

Drake, V.A. and Fitt, G.P. (1990) Studies of *Heliothis* mobility at Narrabri, summer 1989/90. In: *Proceedings of the Fifth Australian Cotton Conference*, 8–9 August 1990, Broadbeach, Queensland. Australian Cotton Growers' Research Association, Brisbane, pp. 295–304.

Drake, V.A., Helm, K.F., Readshaw, J.L. and Reid, D.G. (1981) Insect migration across Bass Strait during spring: a radar study. *Bulletin of Entomological Research* 71, 449–466.

Eger, J.E., Sterling, W.L. and Hartstack, A.W. (1983) Winter survival of *Heliothis virescens* and *Heliothis zea* (Lepidoptera: Noctuidae) in College Station, Texas. *Environmental Entomology* 12, 970–975.

Ellington, J.J. and El-Sokkari, A. (1986) A measure of the fecundity, ovipositional behaviour, and mortality of the bollworm, *Heliothis zea* (Boddie) in the laboratory. *Southwest Entomology* 14, 205–209.

Elzen, G.W., O'Brien, P.J., Snodgrass, G.L. and Powell, J.E. (1987) Susceptibility of the parasitoid *Microplitis croceipes* (Hymenoptera: Braconidae) to field rates of selected cotton insecticides. *Entomophaga* 31, 545–550.

Evans, M.L. (1985) Arthropod species in soybeans in southeast Queensland. *Journal of the Australian Entomological Society* 24, 169–177.

Ewing, K.P., Parancia, C.R. and Ivy, E.E. (1947) Cotton insect control with benzene hexachloride, alone or in mixture with DDT. *Journal of Economic Entomology* 40, 374–381.

Farrar, R.R. and Bradley, J.R. (1985a) Within-plant distribution of *Heliothis* spp. (Lepidoptera: Noctuidae) eggs and larvae on cotton in North Carolina. *Environmental Entomology* 14, 205–209.

Farrar, R.R. and Bradley, J.R. (1985b) Effects of within-plant distribution of *Heliothis zea* (Boddie) (Lepidoptera: Noctuidae) eggs and larvae on larva development and survival on cotton. *Journal of Economic Entomology* 78, 1233–1237.

Farrow, R.A. and Daly, J.C. (1987) Long-range movements as an adaptive strategy in the genus *Heliothis* (Lepidoptera: Noctuidae): a review of its occurrence and detection in four pest species. *Australian Journal of Zoology* 35, 1–24.

Fife, L.C. and Graham, H.M. (1966) Cultural control of overwintering bollworm and tobacco budworm. *Journal of Economic Entomology* 59, 1123–1125.

Firempong, S.K. and Zalucki, M.P. (1990a) Host plant preferences of populations of *Helicoverpa armigera* (Hübner) (Lepidoptera: Noctuidae) from different geographic locations. *Australian Journal of Zoology* 37, 665–673.

Firempong, S.K. and Zalucki, M.P. (1990b) Host plant selection by *Helicoverpa armigera* (Hübner) (Lepidoptera: Noctuidae); role of certain plant attributes. *Australian Journal of Zoology* 37, 675–683.

Firempong, S.K. and Zalucki, M.P. (1991) Host plant selection by *Helicoverpa armigera* (Lepidoptera: Noctuidae): the role of some herbivore attributes. *Australian Journal of Zoology* 39, 343–350.

Fitt, G.P. (1989) The ecology of *Heliothis* in relation to agroecosystems. *Annual Review of Entomology* 34, 17–52.

Fitt, G.P. (1990) Host selection in the Heliothinae. In: Bailey, W.J. and Ridsdill-Smith,

T.J. (eds) *Reproductive Behaviour in Insects – Individuals and Populations.* Chapman & Hall, London, pp. 172–201.

Fitt, G.P. and Daly, J.C. (1990) Abundance of overwintering pupae and the spring generation of *Helicoverpa* spp. (Lepidoptera: Noctuidae) in northern New South Wales, Australia: implications for pest management. *Journal of Economic Entomology* 83, 1827–1836.

Fitt, G.P. and Forrester, N.W. (1988) Overwintering of *Heliothis* – the importance of stubble cultivation. *Australian Cotton Grower* 8, 7–8.

Fitt, G.P. and Pinkerton, A. (1990) A Mark-recapture study of *Heliothis* movement from a source crop in the Namoi Valley. In: *Proceedings of the Fifth Australian Cotton Conference* 8–9 August 1990, Broadbeach, Queensland. Australian Cotton Growers' Research Association, Brisbane, pp. 283–293.

Fitt, G.P., Zalucki, M.P. and Twine, P. (1989) Temporal and spatial patterns in pheromone trap catches of *Helicoverpa* spp. in cotton growing areas of Australia. *Bulletin of Entomological Research* 79, 145–162.

Forrester, N.W. (1989) Six years experience with the strategy. *Australian Cotton Grower* 10, 62–64.

Forrester, N.W. (1990) Designing, implementing and servicing an insecticide resistance management strategy. *Pesticide Science* 28, 167–179.

Forrester, N.W. and Cahill, M. (1987) Management of insecticide resistance in *Heliothis armigera* (Hubner) in Australia. In: Ford, M., Holloman D.W., Khambay, B.P.S. and Sawicki, R.M. (eds) *Combating Resistance in Xenobiotics: Biological and Chemical Approaches.* Ellis Horwood, Chichester, pp. 127–137.

Forrester, N.W., Cahill, M., Bird, L.J. and Layland, J.K. (1993) Management of pyrethroid and endosulfan resistance in *Helicoverpa armigera* (Lepidoptera: Noctuidae) in Australia. *Bulletin of Entomological Research*, Supplement No. 1.

Fox, L.R. and Morrow, P.A. (1981) Specialisation: species property or local phenomenon. *Science* 211, 887–893.

Fye, R.E. and Carranza, R.L. (1973) Cotton pests: overwintering of three lepidopterous species in Arizona. *Journal of Economic Entomology* 66, 657–659.

Garman, H. (1920) Observations on the structure and coloration of the larval corn earworm, the budworm and a few other lepidopterous larvae. *Kentucky Agricultural Experimental Station Bulletin* 227, 55–84 (from Neunzig, 1969).

Ghode, M.K., Nayak, U.K., Ghosh, P.K. and Pawar, A.D. (1988) Avian predation of gram pod borer (*Heliothis armigera*) in Orissa. *Journal of Advanced Zoology* 9, 148.

Gledhill, J.A. 1982. Progress and problems in *Heliothis* management in southwest Asia. In: Reed, W. and Kumble, V. (eds) *Procedings of the International Workshop on* Heliothis *Management*, 15–20 November 1981, ICRISAT Center, Patancheru, A.P., India. ICRISAT, Patancheru, A.P., India, pp. 375–384.

Goodenough, J.L., Gaylor, J.J., Harris, V.E. (*et al.*) (1986) Efficacy of entomophagous arthropods. In: Johnson, S.J., King, E.G. and Bradley, J.R. (eds) *Theory and Tactics of* Heliothis *Population Management: 1- Cultural and Biological Control.* Southern Cooperative Series Bulletin 316, 75–91.

Goodyer, G.J. and Greenup, L.R. (1980) A survey of insecticide resistance in the cotton bollworm *Heliothis armigera* (Lepidoptera: Noctuidae) in New South Wales. *General and Applied Entomology* 12, 37–40.

Gopalakrishnan, C. and Narayanan, K. (1989) Epizootiology of *Nomuraea rileyi* (Farlow) Samson in field populations of *Helicoverpa armigera* Hübner in relation to three host plants. *Journal of Biological Control* 3, 50–52.

Goryshin, N.I. (1958) Ecological analysis of the seasonal cycle of development of the cotton bollworm (*Chloridea obsoleta* F.) in the northern regions of its distribution. *Uchen. Zap. Leningr. gos. Univ.* 240, 3–20.

Graham, H.M., Wolfenbarger, D.A., Nosky, J.B., Hernandez, N.S., Llanes, J.R. and Tamayo, J.A. (1978) Use of rubidium to label corn earworm and fall armyworm for dispersal studies. *Environmental Entomology* 7, 435–438.

Greathead, D.J. (1966) Memorandum on the parasites and possibilities of biological control of East African bollworms. Commonwealth Institute of Biological Control, unpublished report, 21pp.

Greathead, D.J. and Girling, D.J. (1982) Possibilities for natural enemies in *Heliothis* management and the contribution of the Commonwealth Institute of Biological Control. In: Reed, W. and Kumble, V. (eds) *Proceedings of the International Workshop on* Heliothis *Management*, 15–20 November, 1981, ICRISAT Center, Patancheru, A.P., India. ICRISAT, Patancheru, A.P., India, pp. 147–158.

Greathead, D.J. and Girling, D.J. (1989) Distribution and economic importance of *Heliothis* spp. and of their natural enemies and host plants in southern and eastern Africa. In: King, E.G. and Jackson, R.D. (eds) *Proceedings of the Workshop on Biological Control of* Heliothis*: Increasing the Effectiveness of Natural Enemies*, 11–15 November 1985, New Delhi, India. FERRO/USDA, New Delhi, India, pp. 329–345.

Green, M.B. and Lyon, D.J.de B. (eds) (1989) *Pest Management in Cotton.* Ellis Horwood, Chichester, 259pp.

Gregg, P.C., Fitt, G.P., Zalucki, M.P. and Twine, P. (1990) Evidence for spring migration of *Heliothis* spp. from inland Australia to cotton areas. In: *Proceedings of the Fifth Australian Cotton Conference*, 8–10 August 1990, Broadbeach, Queensland. Australian Cotton Growers' Research Association, Brisbane, pp. 327–335.

Gross, H.R. and Young, J.R. (1977) Comparative development and fecundity of corn earworm reared on selected wild and cultivated early-season hosts common to the southeastern U.S. *Annals of the Entomological Society of America* 70, 63–65.

Gunning, R.V. and Easton, C.S. (1989) Pyrethroid resistance in *Heliothis armigera* (Hübner) collected from unsprayed maize crops in New South Wales 1983–1987. *Journal of the Australian Entomological Society* 28, 57–61.

Gunning, R.V., Easton, C.S., Greenup, L.R. and Edge, V.E. (1984) Pyrethroid resistance in *Heliothis armigera* (Hübner) (Lepidoptera: Noctuidae) in Australia. *Journal of Economic Entomology* 77, 1283–1287.

Gunning, R.V., Easton, C.S., Balfe, M.E. and Ferris, I.G. (1991) Pyrethroid resistance mechanisms in Australian *Heliothis armigera. Pesticide Science* 33, 472–490.

Hackett, D.S. (1980) Studies on the biology of *Helicoverpa armigera* in the Sudan Gezira. PhD thesis. University College North Wales, Bangor, Wales.

Hackett, D.S. and Gatehouse, A.G. (1982a) Studies on the biology of *Heliothis* spp. in Sudan. In: Reed, W. and Kumble, V. (eds) *Proceedings of the International Workshop on* Heliothis *Management*, 15–20 November 1981, ICRISAT Center, Patancheru, A.P., India. ICRISAT, Patancheru, A.P., India, pp. 29–38.

Hackett, D.S and Gatehouse, A.G. (1982b) Diapause in *Heliothis armigera* (Hübner) and *H. fletcheri* (Hardwick) (Lepidoptera: Noctuidae) in the Sudan Gezira. *Bulletin of Entomological Research* 72, 409–422.

Haggis, M.J. (1982) Distribution of *Heliothis armigera* eggs on cotton in the Sudan Gezira: Spatial and temporal changes and their possible relation to weather. In: Reed, W. and Kumble, V. (eds) *Proceedings of the International Workshop on* Heliothis *Management*, 15–20 November 1981, ICRISAT Center, Patancheru, A.P., India. ICRISAT, Patancheru, A.P., India, pp. 87–99.

Haile, D.G., Snow, J.W. and Young, J.R. (1975) Movement of adult *Heliothis* released on St Croix to other islands. *Environmental Entomology* 4, 225–226.

Hallman, G.J. (1984) Quantification of the relative importance of 12 uncultivated hosts of *Heliothis virescens* (Fabricius) using field data. *Insect Science and its Application* 5, 465–468.

Hamilton, J.G. and Fitt, G.P. (1990) HEAPS: *Heliothis armigera* and *punctigera* simulation. In: *Proceedings of the Fifth Australian Cotton Conference*, 8–9 August 1990, Broadbeach, Queensland. Australian Cotton Growers' Research Association, Brisbane, pp. 139–145.

Hamm, J.J. (1980) Epizootics of *Entomophthora aulicae* in lepidopterous pests of soybean. *Journal of Invertebrate Pathology* 36, 60–63.

Hampson, G.F. (1903) *Catalogue of the Lepidoptera Phalaenae in the British Museum* 4, 657pp. (from Pearson, 1958).

Hardwick, D.F. (1965) The corn earworm complex. *Memoirs of the Entomological Society of Canada* 40, 247pp.

Harris, V.E. and Phillips, J.R. (1986) Mowing spring host plants as a population management technique for *Heliothis* spp. *Journal of Agricultural Entomology* 3, 125–134.

Hartstack, A.W. (1982) Modeling and forecasting *Heliothis* populations. In: Reed, W. and Kumble, V. (eds) *Proceedings of the International Workshop on* Heliothis *Management*, 15–20 November 1981, ICRISAT Center, Patancheru, A.P., India. ICRISAT, Patancheru, A.P., India, pp. 51–60.

Hartstack, A.W. and Witz, J.A. (1981) Estimating field populations of tobacco budworm moths from pheromone trap catches. *Environmental Entomology* 10, 908–914.

Hartstack, A.W., Witz., J.A., Hollingworth, J.P., Ridgeway, R.L. and Lopez, J.D. (1976) *MOTHZV-2: A Computer Simulation of* Heliothis zea *and* Heliothis virescens *Population Dynamics*. User manual. US Department of Agriculture, Agricultural Research Service S–127, 55pp.

Hartstack, A.W., Hollingworth, J.P., Witz., J.A., Buck, D., R., Lopez, J.D. and Hendricks, D.E. (1978) Relation of tobacco budworm catches in pheromone baited traps to field populations. *Southwestern Entomologist* 3, 43–51.

Hartstack, A.W., Lopez, J.D., Muller, R.A., Sterling, W.L. and King, E.G. (1982) Evidence of long range migration of *Heliothis zea* into Texas and Arkansas. *Southwestern Entomologist* 7, 188–201.

Hartstack, A.W., King, E.G. and Phillips, J.R. (1983) Monitoring and predicting *Heliothis* populations in Southeast Arkansas. In: *Proceedings of the Beltwide Cotton Production Research Conference*, Memphis, Tennessee, pp. 187–190.

Hartstack, A.W., Lopez, J.D., Muller, R.A. and Witz, J.A. (1986) Early season occurrence of *Heliothis* spp. in 1982: Evidence of long-range migration of *Heliothis zea*. In: Sparks, A.N. (ed) *Long-Range Migration of Moths of Agronomic Importance to the United States and Canada: Specific Examples of Occurrence and Synoptic Weather Patterns Conducive to Migration*. US Department of Agriculture, Agricultural Research Services 43, 48–60.

Hassan, S.T.S., Wilson, L.T. and Blood, P.R.B. (1990) Oviposition by *Heliothis armigera* and *H. punctigera* (Lepidoptera: Noctuidae) on okra leaf and smooth-leaf cotton. *Environmental Entomology* 19, 710–716.

Hayes, J.L. and Hopper, K.R. (1987) Trace element labelling of *Heliothis* spp: labelling of individual eggs from moths reared on treated host plants. In: *Proceedings of the Beltwide Cotton Production Research Conference*, Memphis, Tennessee, pp. 311–314.

Hearn, A.B., Ives, P.M., Room, P.M., Thomson, N.J. and Wilson, L.T. (1981) Computer-based cotton pest management in Australia. *Field Crops Research* 4, 321–332.

Hendricks, D.E. and Hartstack, A.W. (1978) Pheromone trapping as an index for initiating control of cotton insects, *Heliothis* spp.: A compendium. In: *Proceedings of the Beltwide Cotton Production Research Conference*, Memphis, Tennessee, pp. 116–120.

Hendricks, D.E., Graham, H.M. and Raulston, J.R. (1973) Dispersal of sterile tobacco budworms from release points in northeastern Mexico and southern Texas. *Environmental Entomology* 2, 1085–1088.

Hendrix, W.H., Mueller, T.F., Phillips, J.R. and Davis, O.K. (1987) Pollen as an indicator of long-distance movement of *Heliothis zea* (Lepidoptera: Noctuidae). *Environmental Entomology* 16, 1148–1151.

Henneberry, T.J. and Clayton, T.E. (1991) Tobacco budworm (Lepidoptera: Noctuidae): temperature effects on mating, oviposition, egg viability and moth longevity. *Journal of Economic Entomology* 84, 1242–1246.

Holt, J., King, A.B.S. and Armes, N.J. (1990) Use of simulation analysis to assess *Helicoverpa armigera* control on pigeonpea in southern India. *Crop Protection* 9, 197–206.

Holzer, T.O., Bradley, J.R. and Rabb, R.L. (1976) Effects of various temperature regimes on the time required for emergence of diapausing *Heliothis zea*. *Annals of the Entomological Society of America* 69, 257–260.

Hopper, K.R. (1989) Conservation and augmentation of *Microplitis croceipes* for controlling *Heliothis* spp. In: Powell, J.E., Bull, D.L. and King, E.G. (eds) *Biological Control of* Heliothis *spp. by* Microplitis croceipes. *Southwestern Entomologist Supplement* No. 12, 95–115.

Hopper, K.R. and Stark, S.B. (1987) A simulation model for making decisions about *Heliothis* control. In: *Proceedings of the Beltwide Cotton Producers Research Conference*, Memphis, Tennessee, pp. 286–291.

Ibraham, A.A. El-Hamid and Fayad, Y.H. (1989) Distribution and economic importance of *Heliothis* spp., their natural enemies, and host plants in Egypt. In: King, E.G. and Jackson, R.D. (eds). *Proceedings of the Workshop on Biological Control of* Heliothis*: Increasing the Effectiveness of Natural Enemies* 11–15 November 1985, New Delhi, India. FERRO/USDA New Delhi, India, pp. 173–176.

ICRISAT (International Crops Research Institute for the Semi-Arid Tropics) (1986) *Annual Report 1985*. ICRISAT, Patancheru, A.P. 502324, India, p. 194.

ICRISAT (International Crops Research Institute for the Semi-Arid Tropics) (1987) *Annual Report 1986*. ICRISAT, Patancheru, A.P. 502324, India, pp. 184–188.

ICRISAT (International Crops Research Institute for the Semi-Arid Tropics) (1988) An overview of cropping systems entomology trials, 1975–1982. ICRISAT, Patancheru, A.P., 502324 India, unpublished report, 82pp.

Ignoffo, C.M. (1983) Development of a viral insecticide: Concept to commercialisation. *Experimental Parasitology* 33, 380–406.

Isenhour, D.J., Wiseman, B.R. and Layton, R.C. (1989) Enhanced predation by *Orius insidiosus* (Hemiptera: Anthocoridae) on larvae of *Heliothis zea* and *Spodoptera frugiperda* (Lepidoptera: Noctuidae) caused by feeding on resistant corn genotypes. *Environmental Entomology* 18, 418–422.

Isley, D. (1935) Relation of hosts to abundance of cotton bollworm. *Arkansas Agricultural Experimental Station Bulletin* 320, 30pp.

Jackson, G.J. (1989) Insecticide resistance – the challenge of the decade. In: Green, M.B. and Lyon, D.J. de B. (eds) *Pest Management in Cotton*. Ellis Horwood, Chichester, pp. 27–30.

Jackson, D.M., Severson, R.F., Johnson, A.F., Chaplin, J.F. and Stephenson, M.G. (1984) Ovipositional response of tobacco budworm moths (Lepidoptera: Noctuidae) to cuticular chemical isolates from green tobacco leaves. *Environmental Entomology* 13, 1023–1030.

Javaid, I., Joshi, J.M., Dadson, R.B. and Nobakht, M. (1991) Leaf feeding resistance in soybean breeding lines to corn earworm (*Heliothis zea* (Boddie)). *Soybean Genetics*

Newsletter, US Department of Agriculture, Agricultural Research Service 18, 271–274.

Jayaraj, S. (1982) Biological and ecological studies of *Heliothis*. In: Reed, W. and Kumble, V. (eds) *Proceedings of the International Workshop on* Heliothis *Management*, 15–20 November 1981, ICRISAT Center, Patancheru, A.P., India. ICRISAT, Patancheru, A.P. India, pp. 17–28.

Jayaraj, S., Rabindra, R.J. and Narayanan, K. (1989) Development and use of microbial control agents for control of *Heliothis* spp. (Lepidoptera: Noctuidae) in India. In: King, E.G. and Jackson, R.D. (eds) *Proceedings of the Workshop on Biological Control of* Heliothis*: Increasing the Effectiveness of Natural Enemies*, 11–15 November 1985, New Delhi, India. FERRO/USDA, New Delhi, India. pp. 483–504.

Jenkins, J.N. (1989) State of the art in host plant resistance in cotton. In: Green, M.B. and Lyon, D.J. de B. (eds) *Pest Management in Cotton*. Ellis Horwood, Chichester, pp. 53–69.

Johnson, C.G. (1969) *Migration and Dispersal of Insects by Flight*. Methuen, London.

Johnson, D.J. (1983) Relationship between tobacco budworm (Lepidoptera: Noctuidae) catches when using pheromone traps and egg counts in cotton. *Journal of Economic Entomology* 76, 182–183.

Johnson, M.W., Stinner, R.E. and Rabb, R.L. (1975) Ovipositional response of *Heliothis zea* to its major hosts in North Carolina. *Environmental Entomology* 4, 291–297.

Johnson, S.J., King, E.G. and Bradley, J.R. (eds) (1986) *Theory and Tactics of* Heliothis *Population Management: 1-Cultural and Biological Control*. Southern Cooperative Series Bulletin 316, 161pp.

Joyce, R.J.V. (1982) A critical review of the role of chemical pesticides in *Heliothis* management. In: Reed, W. and Kumble, V. (eds) *Proceedings of the International Workshop on* Heliothis *Management*, 15–20 November 1981, ICRISAT Center, Patancheru, A.P., India. ICRISAT, Patancheru, A.P., India, pp. 173–188.

Juvic, J.A., Babka, B.A. and Timmermann, E.A. (1988) Influence of trichome exudates from species of *Lycopersicon* on oviposition behaviour of *Heliothis zea* (Boddie). *Journal of Chemical Ecology* 14, 1261–1278.

Kabissa, J.C.B. (1989) Evaluation of damage thresholds for insecticidal control of *Helicoverpa armigera* (Hübner) (Lepidoptera: Noctuidae) on cotton in eastern Tanzania. *Bulletin of Entomological Research* 79, 95–98.

Keerthisinghe, C.I. (1982) Economic thresholds for cotton pest management in Sri Lanka. *Bulletin of Entomological Research* 72, 239–246.

Kehat, M., Gothilf, S., Dunkelblum, E. and Greenburg, S. (1982). Sex pheromone traps as a means of improving control programs for the cotton bollworm. *Environmental Entomology* 11, 727–729.

Kennedy, G.G., Gould, F., Deponti, O.M.B. and Stinner, R.E. (1987) Ecological, agricultural and commercial considerations in the deployment of insect-resistant germplasm. *Environmental Entomology* 16, 327–338.

King, A.B.S., Armes, N.J. and Pedgley, D.E. (1990) A mark-capture study of *Helicoverpa armigera* dispersal from pigeonpea in southern India. *Entomologia Experimentalis et Applicata* 55, 257–266.

King, E.G. (1986) Insecticide use in cotton and the value of predators and parasites for managing *Heliothis*. In: *Proceedings of the Beltwide Cotton Production Research Conference*, Memphis, Tennessee, pp. 155–162.

King, E.G. and Coleman, R.J. (1989) Potential for biological control of *Heliothis* species. *Annual Review of Entomology* 34, 53–75.

King, E.G. and Jackson, R.D. (eds) (1989) *Proceedings of the Workshop on the Biological Control of* Heliothis*: Increasing the Effectiveness of Natural Enemies*. 11–15 November 1985,

New Delhi, India. Far Eastern Regional Research Office, US Department of Agriculture, New Delhi, India, 550pp.

King, E.G., Powell, J.E. and Smith, J.W. (1982) Prospects for utilisation of parasites and predators for management of *Heliothis* spp. In: Reed, W. and Kumble, V. (eds) *Proceedings of the International Workshop on* Heliothis *Management*, 15–20 November 1981, ICRISAT Center, Patancheru, A.P., India, ICRISAT, Patancheru, A.P., India, pp. 103–122.

King, E.G., Powell, J.E. and Coleman, R.J. (1985) A high incidence of parasitism of *Heliothis* spp. (Lepidoptera: Noctuidae) larvae in cotton in southeastern Arkansas. *Entomophaga* 30, 419–426.

King, E.G., Baumhover, A., Bouse, L.F., Greany, P., Hartstack, A.W., Hopper, K.R., Knipling, E.F., Morrison, R.K., Nettles, W.C. and Powell, J.E. (1986) Augmentation of entomophagous arthropods. In: Johnson, S.J., King, E.G. and Bradley, J.R. (eds) *Theory and Tactics of* Heliothis *Population Management: 1- Cultural and Biological Control.* Southern Cooperative Series Bulletin 316, 116–131.

Kirkpatrick, T.H. (1961) Comparative morphological studies of *Heliothis* species (Lepidoptera: Noctuidae) in Queensland. *Queensland Journal of Agricultural Science* 18, 179–194.

Klun, J.A., Plimmer, J.R., Bierl-Leonhardt, B.A., Sparks, A.N. and Chapman, O.L. (1979) Trace chemicals: the essence of sexual communication systems in *Heliothis* species. *Science* 204, 1328–1330.

Klun, J.A., Plimmer, J.R., Bierl-Leonhardt, B.A., Sparks, A.N., Primiani, M., Chapman, O.L., Lee, G.H. and Lepone, G. (1980) Sex pheromone chemistry of female corn earworm moth, *Heliothis zea. Journal of Chemical Ecology* 6, 165–175.

Knipling, E.F. (1960) Use of insects for their own destruction. *Journal of Economic Entomology* 53, 415–420.

Knipling, E.F. (1979) *The Basic Principles of Insect Population Suppression and Management.* US Department of Agriculture, Agricultural Handbook Vol. 512, 632pp.

Knipling, E.F. and Stadelbacher, E.A. (1983) The rationale of area-wide management of *Heliothis* (Lepidoptera: Noctuidae) populations. *Bulletin of the Entomological Society of America* 29, 29–37.

Knutson, L. (1989) Systematics of *Heliothis* species and their natural enemies as a basis for biological control research. In: King, E.G. and Jackson, R.D. (eds) *Proceedings of the Workshop on the Biological Control of* Heliothis*: Increasing the Effectiveness of Natural Enemies*, 11–15 November 1985, New Delhi, India. FERRO/USDA, New Delhi, India, pp. 119–159.

Kogan, M., Helm, C.G., Kogan, J. and Brewer, E. (1989) Distribution and economic importance of *Heliothis virescens* and *Heliothis zea* in North, Central and South America and of their natural enemies and host plants. In: King, E.G. and Jackson, R.D. (eds) *Proceedings of the Workshop on the Biological Control of* Heliothis*: Increasing the Effectiveness of Natural Enemies.* 11–15 November 1985, New Delhi, India. FERRO/USDA, New Delhi, India, pp. 241–298.

Komarova, O.S. (1959) On the conditions determining diapause of hibernating pupae of *Chloridea obsoleta* F. (Lepidoptera: Noctuidae). *Entomological Review, Washington* 38, 318–325.

Kou, R. and Chow, Y. (1987) Calling behaviour of the cotton bollworm, *Heliothis armigera* (Lepidoptera: Noctuidae). *Annals of the Entomological Society of America* 80, 490–493.

Kyi, A., Zalucki, M.P. and Titmarsh, I.J. (1991) An experimental study of early stage survival of *Helicoverpa armigera* (Lepidoptera: Noctuidae) on cotton. *Bulletin of Entomological Research* 81, 263–271.

Lateef, S.S. (1985) Gram pod borer (*Heliothis armigera* (Hub.)) resistance in chickpea. *Agriculture, Ecosystem and Environment* 14, 95–102.

Laster, M.L. (1972) Interspecific hybridisation of *Heliothis virescens* and *H. subflexa*. *Environmental Entomology* 1, 682–687.

Leonard, B.R., Graves, J.B., Sparks, T.C. and Pavloff, A.M. (1988) Variation in resistance of field populations of tobacco budworm and bollworm (Lepidoptera: Noctuidae) to selected insecticides. *Journal of Economic Entomology* 81, 1521–1528.

Leonard, B.R., Graves, J.B., Burris, E., Pavloff, A.M. and Church, G. (1989) *Heliothis* spp. (Lepidoptera: Noctuidae) captures in pheromone traps: Species composition and relationship to oviposition in cotton. *Journal of Economic Entomology* 82, 574–579.

Lingren, P.D., Ridgway, R.L. and Jones, S.L. (1968) Consumption by several common arthropod predators of eggs and larvae of *Heliothis* species that attack cotton. *Annals of the Entomological Society of America* 61, 613–618.

Lingren, P.D., Greene, G.L., Davis, D.R., Baumhover, A.H. and Henneberry, T.J. (1977) Nocturnal behaviour of four lepidopteran pests that attack tobacco and other crops. *Annals of the Entomological Society of America* 61, 161–167.

Lingren, P.D., Sparks, A.N. and Raulston, J.R. (1982) The potential contribution of moth behaviour research to *Heliothis* management. In: Reed, W. and Kumble, V. (eds) *Proceedings of the International Workshop on* Heliothis *Management*, 15–20 November 1981, ICRISAT Center, Patancheru, A.P., India, ICRISAT, Patancheru, A.P., India, pp. 39–47.

Lingren, P.D., Warner, W.B., Raulston, J.R., Kehat, M., Henneberry, T.J., Pair, S.D., Zvirgzdins, A. and Gillespie, J.M. (1988) Observations on the emergence of adults from natural populations of corn earworm, *Heliothis zea* (Boddie) (Lepidoptera: Noctuidae). *Environmental Entomology* 17, 254–258.

Loganathan, M. (1981) Studies on ecology and effect of host plants on the susceptibility of larvae of *Heliothis armigera* (Hübner) to insecticides. MSc (Ag.) thesis. Tamil Nadu Agricultural University, Coimbatore, T. N., India (from Jayaraj, 1982).

Lopez, J.D. and Hartstack, A.W. (1985) Comparison of diapause development in *Heliothis zea* and *H. virescens* (Lepidoptera: Noctuidae). *Annals of the Entomological Society of America* 78, 415–422.

Lopez, J.D., Ridgeway, R.L. and Pinnel, R.E. (1976) Comparative efficacy of four insect predators of the bollworm and tobacco budworm. *Environmental Entomology* 5: 1160–1164.

Lopez, J.D., Hartstack, A.W., Witz, J.A. and Hollingsworth, J.P. (1979) Recovery in blacklight traps of marked bollworms released in a multiple cropped area. *Southwestern Entomologist* 4, 46–52.

Lopez, J.D., Hartstack, A.W. and Beach, R. (1984) Comparative pattern of emergence of *Heliothis zea* and *H. virescens* (Lepidoptera: Noctuidae) from overwintering pupae. *Journal Economic Entomology* 77, 1421–1426.

Lopez, J.D., Shaver, T.N. and Dickerson, W.A. (1990) Population monitoring of *Heliothis* spp. using pheromones. In: Ridgeway, R.L., Silverstein, R.M. and Inscoe, M.N. (eds) *Behaviour-Modifying Chemicals for Insect Management: Applications of Pheromones and Other Attractants*. Marcel Dekker, New York, pp. 473–496.

Lozina-Lozinskii, L.K. (1939) [Expedition for the study of the ecology of the cotton noctuid]. Izv. Azerbaidzh Fil. Akad. Nank. SSSR. 3, 137–138. (from Hardwick, 1965).

Lukefahr, M.J. (1982) A review of the problems, progress and prospects of host plant resistance to *Heliothis* spp. In: Reed, W. and Kumble, V. (eds) *Proceedings of the International Workshop on* Heliothis *Management*, 15–20 November 1981, ICRISAT Center, Patancheru, A.P., India. ICRISAT, Patancheru, A.P., India, pp. 223–231.

Lukefahr, M.J. and Martin, D.F. (1964) The effects of various larval and adult diets on the fecundity and longevity of the bollworm, tobacco budworm and cotton leafworm. *Journal of Economic Entomology* 57, 233–235.

Lukefahr, M.J., Houghtaling, J.E. and Graham, N.M. (1971) Suppression of *Heliothis* populations with glabrous cotton strains. *Journal of Economic Entomology* 64, 486–488.

Lukefahr, M.J., Houghtaling, J.E. and Crumb, D.G. (1975) Suppression of *Heliothis* spp. with cottons containing combinations of resistance characters. *Journal of Economic Entomology* 68, 743–746.

Luttrell, R.G., Yearian, W.C. and Young, S.Y. (1983) Effect of spray adjuvants on *Heliothis zea* (Lepidopotera: Noctuidae) nuclear polyhedrosis virus efficacy. *Journal of Economic Entomology* 76, 162–167.

Luttrell, R.G., Phillips, J.R. and Pfimmer, T.R. (1986) Cultural control in mid-south US cropping systems. In: Johnson, S.J., King, E.G. and Bradley, J.R. (eds) *Theory and Tactics of* Heliothis *Population Management: 1-Cultural and Biological Control.* Southern Cooperative Series Bulletin 316, 28–37.

Luttrell, R.G., Roush, R.T., Ali, A., Mink, J.S., Reid, M.R. and Snodgrass, G.L. (1987) Pyrethroid resistance in field populations of *Heliothis virescens* (Lepidoptera: Noctuidae) in Mississippi in 1986. *Journal of Economic Entomology* 80, 985–989.

McCaffery, A.R., Maruf, G.M., Walker, A.J. and Styles, K. (1988) Resistance to pyrethroids in *Heliothis* spp.: Bioassay methods and incidence in populations from India and Asia. In: *Proceedings, 1988 Brighton Crop Protection Conference – Pests and Diseases.* British Crop Protection Council, UK, pp. 433–438.

McCaffery, A.R., King, A.B.S., Walker, A.J. and El-Nayir, H. (1989) Resistance to synthetic pyrethroids in the bollworm, *Heliothis armigera* from Andhra Pradesh, India. *Pesticide Science* 27, 65–76.

McDaniel, S.G., Sterling, W.L. and Dean, D.A. (1981) Predators of tobacco budworm larvae in Texas cotton. *Southwestern Entomologist* 6, 102–108.

McLaughlin, J.R. and Mitchell, E.R. (1982) Practical development of pheromones in *Heliothis* management. In: Reed, W. and Kumble, V. (eds) *Proceedings of the International Workshop on* Heliothis *Management*, 15–20 November 1981, ICRISAT Center, Patancheru, A.P., India. ICRISAT, Patancheru, A.P., India, pp. 309–318.

McVeigh, L.J., Campion, D.G. and Critchley, B.R. (1990) The use of pheromones for the control of cotton bollworms and *Spodoptera* spp. in Africa and Asia. In: Ridgway, R.L., Silverstein, R.M. and Inscoe, M.N. (eds) *Behaviour-Modifying Chemicals for Insect Management.* Marcel Dekker, New York, pp. 407–415.

Manjunath, T.M., Bhatnagar, V.S., Pawar, C.S. and Sithanantham, S. (1989) Economic importance of *Heliothis* spp. in India and an assessment of their natural enemies and host plants. In: King, E.G. and Jackson, R.D. (eds) *Proceedings of the Workshop on Biological Control of* Heliothis*: Increasing the Effectiveness of Natural Enemies*, 11–15 November 1985, New Delhi, India. FERRO/USDA, New Delhi, India, pp. 179–240.

Martin, P.B., Lingren, P.D. and Greene, G.L. (1976) Relative abundance and host preferences of cabbage looper, soybean looper, tobacco budworm and corn earworm on crops grown in northern Florida. *Environmental Entomology* 5, 878–882.

Masaki, S. (1980) Summer diapause. *Annual Review of Entomology* 25, 1–25.

Ma Shijun and Ding Yanquin (1989). Distribution and economic importance of *Heliothis armigera* and its natural enemies in China. In: King, E.G. and Jackson, R.D. (eds) *Proceedings of the Workshop on Biological Control of* Heliothis*: Increasing the Effectiveness of Natural Enemies*, 11–15 November 1985, New Delhi, India. FERRO/USDA, New Delhi, India, pp. 185–195.

Matthews, G.A. (1989) *Cotton Insect Pests and their Management.* Longman, Harlow, UK, 199pp.

Matthews, G.A. and Tunstall, J.P. (1968) Scouting for pests and the timing of spray applications. *Cotton Growers Review* 45, 115–127.

Matthews, M. (1987) The Classification of Heliothinae (Noctuidae). PhD thesis. British Museum (Natural History) and Kings College, London University, 253pp.

Maxwell, F., Schusker, M.F., Meredith, W.R. and Laster, M.L. (1976) Influence of the nectariless character in cotton on harmful and beneficial insects. In: Jermy, T. (ed.) *The Host Plant in Relation to Insect Behaviour and Reproduction.* Plenum, New York, pp. 157–161.

Meirrose, C., Araujo, J., Perkins, D., Mercardier, G., Poitout, S., Bues, R., Vargas Piqueras, P. and Cabello, T. (1989) Distribution and economic importance of *Heliothis* spp. (Lepidoptera: Noctuidae) and their natural enemies and host plants in Western Europe. In: King, E.G. and Jackson, R.D. (eds) *Proceedings of the Workshop on Biological Control of* Heliothis*: Increasing the Effectiveness of Natural Enemies,* 11–15 November 1985, New Delhi, India. FERRO/USDA, New Delhi, India, pp. 311–327.

Menn, J.J., King, E.G. and Coleman, R.J. (1989) Future control strategies for *Heliothis* in cotton. In: Green, M.B. and Lyon, D.J. de B. (eds) *Pest Management in Cotton.* Ellis Horwood, Chichester, pp. 101–121.

Michael, P.J. (1973) Biological control of *Heliothis* in sorghum. *Journal of Agriculture, Western Australia* 14, 222–224.

Morrison, G., Lewis, W.J. and Nordlund, D.A. (1980) Spatial differences in *Heliothis zea* egg density and the intensity of parasitsm by *Trichogramma* spp.: an experimental analysis. *Environmental Entomology* 9, 79–85.

Morton, M. and Collins, M.D. (1989) Managing the pyrethroid revolution in cotton. In: Green, M.B. and Lyon, D.J. de B. (eds) *Pest Management in Cotton.* Ellis Horwood, Chichester, pp. 155–165.

Morton, N., Smith, R.K., Vigil, O. and Van der Mersch, C. (1981) Evaluation of the 'toxic carpet' spray strategy with cypermethrin in cotton. In: *Proceedings of the British Crop Protection Conference.* British Crop Protection Council, UK, pp. 371–379.

Moyhuddin, A.I. (1989) Distribution and economic importance of *Heliothis* spp. in Pakistan and their natural enemies and host plants. In: King, E.G. and Jackson, R.D. (eds) *Proceedings of the Workshop on the Biological Control of* Heliothis*: Increasing the Effectiveness of Natural Enemies,* 11–15 November 1985, New Delhi, India. FERRO/USDA, New Delhi, India, pp. 229–240.

Mueller, T.F. and Phillips, J.R. (1983) Population dynamics of *Heliothis* spp. in spring weed hosts in southeastern Arkansas: survivorship and stage-specific parasitism. *Environmental Entomology* 12, 1846–1850.

Mueller, T.F., Harris, V.E. and Phillips, J.R. (1984) Theory of *Heliothis* (Lepidoptera: Noctuidae) management through reduction of the first spring generation: A critique. *Environmental Entomology* 13, 625–634.

Muller, R.A. and Tucker, N.L. (1986) Climatic opportunities for the long-range migration of moths. In: Sparks, A.N. (ed.) *Long-Range Migration of Moths of Agronomic Importance to the United States and Canada: Specific Examples of Occurrence and Synoptic Weather Patterns Conducive to Migration.* US Deptartment of Agriculture, Agricultural Research Service 43, pp. 61–83.

Murray, D.A.H. and Zalucki, M.P. (1990a) Survival of *Helicoverpa punctigera* (Wallengren) and *H. armigera* (Hübner) (Lepidoptera: Noctuidae) pupae submerged in water. *Journal of the Australian Entomological Society* 29, 191–192.

Murray, D.A.H. and Zalucki, M.P. (1990b) Effect of soil moisture and simulated rainfall on pupal survival and moth emergence of *Helicoverpa punctigera* (Wallengren) and *H. armigera* (Hübner) (Lepidoptera: Noctuidae). *Journal of the Australian Entomological Society* 29, 193–197.

Nadgauda, D. and Pitre, H. (1983) Development, fecundity and longevity of the tobacco budworm (Lepidoptera: Noctuidae) fed soybean, cotton and artificial diet at three temperatures. *Environmental Entomology* 12, 582–586.

Nanthagopal, R. and Uthamasamy, S. (1989) Life tables for American bollworm, *Heliothis armigera* Hübner on four species of cotton under field conditions. *Insect Science and its Application* 10, 521–530.

Napompeth, B. (1989) Distribution and economic importance of *Heliothis* spp. and their natural enemies and host plants in Southeast Asia. In: King, E.G. and Jackson, R.D. (eds) *Proceedings of the Workshop on the Biological Control of* Heliothis*: Increasing the Effectiveness of Natural Enemies*, 11–15 November 1985, New Delhi, India, FERRO/USDA, New Delhi, India, pp. 299–309.

Nel, J.J.C. (1961) The seasonal history of *H. armigera* (Hüb.) on lupins in the South Western Cape Province. *South African Journal of Agricultural Science* 4, 575–588.

Nesbitt, B.F., Beevor, P.S., Hall, D.R. and Lester, R. (1979) Female sex pheromone components of the cotton bollworm *Heliothis armigera*. *Journal of Insect Physiology* 25, 535–541.

Nesbitt, B.F., Beevor, P.S., Hall, D.R. and Lester, R. (1980) (Z)-9-Hexadecenal: a minor component of the female sex pheromone (Lepidoptera: Noctuidae). *Entomologia Experimentalis et Applicata* 27, 306–308.

Neunzig, H.H. (1964) The eggs and early instar larvae of *Heliothis zea* and *Heliothis virescens* (Lepidoptera: Noctuidae). *Annals of the Entomological Society of America* 57, 98–102.

Neunzig, H.H. (1969) The biology of the tobacco budworm and the corn earworm in North Carolina, with particular reference to tobacco as a host. *North Carolina Agricultural Experimental Station Technical Bulletin* 196, 76pp.

Newsom, L.D. and Brazzel, J.R. (1968) Pests and their control. In: Elliot, F.C., Hoover, M. and Porter, W.K. (eds) *Advances in Production and Utilisation of Quality Cotton: Principles and Practices.* Iowa State University Press, Ames, pp. 367–405.

Nordlund, D.A., Lewis, W.J., Vinson, S.B. and Gross, H.R. (1986) Behavioral manipulation of entomophagous insects. In: Johnson, S.J., King, E.G. and Bradley, J.R. (eds) *Theory and Tactics of* Heliothis *Population Management: 1 – Cultural and Biological Control.* Southern Cooperative Series Bulletin 316, 104–115.

Nyambo, B.T. (1988) Significance of host-plant phenology in the dynamics and pest incidence of the cotton bollworm, *Heliothis armigera* Hübner (Lepidoptera: Noctuidae), in Western Tanzania. *Crop Protection* 7, 161–167.

Nye, I.W.B. (1980) *Heliothis* Ochsenheimer, 1816 (Insecta, Lepidoptera): proposal to designate gender and stem. *Bulletin of Zoological Nomenclature* 37, 186–189.

Nye, I.W.B. (1982) The nomenclature of *Heliothis* and associated taxa (Lepidoptera: Noctuidae): past and present. In: Reed, W. and Kumble, V. (eds) *Proceedings of the International Workshop on* Heliothis *Management,* 15–20 November 1981, ICRISAT Center, Patancheru, A.P., India. ICRISAT, Patancheru, A.P., India, pp. 3–8.

Parsons, F.S. (1940) Investigations on the cotton bollworm, *Heliothis armigera* Hübn. Part 3. Relationships between oviposition and the flowering curves of food plants. *Bulletin of Entomological Research* 31, 147–177.

Parsons, F.S. and Marshall, J. (1939) *Progress Report, Experimental Station, Empire Cotton Growing Corp. 1937–38*, p. 28 (from Pearson, 1958).

Parsons, F.S., Hutchinson, H. and Marshall, J. (1938) *Progress Report, Experimental Station, Empire Cotton Growing Corp. 1936–37*, p. 26 (from Pearson, 1958).

Passlow, T. (1973) Insect pests of grain sorghum. *Queensland Agricultural Journal* 99, 620–628.

Patel, R.C., Patel, R.M., Madhukar, B.V.R. and Patel, R.B. (1974) Oviposition behaviour of *Heliothis armigera* in cotton Hybrid-4. *Current Science* 43, 588–589.

Pawar, C.S., Sithananthan, S., Sharma, H.C., Taneja, S.L., Amin, P.W., Leushner, K. and Reed, W. (1984) Use and development of insect traps at ICRISAT. Paper presented at the National Seminar on the use of Traps in Vector Research and Control, 10–11 March 1984, Kalyani, West Bengal. ICRISAT, Patancheru, A. P., India. Internal Report.

Pawar, C.S., Bhatnagar, V.S. and Jadhav, D.R. (1986a) *Heliothis* species and their natural enemies, with their potential for biological control. *Proceedings of the Indian Academy of Science* (Animal Science) 95, 695–703.

Pawar, C.S., Srivastava, C.P. and Reed, W. (1986b) Some aspects of population dynamics of *Heliothis armigera* at ICRISAT Center. In: *Proceedings of the Oriental Entomology Symposium*, 21–24 February, 1984. Trivandrum, India. pp. 79–85.

Pawar, C.S., Sithananthan, S., Bhatnagar, V.S., Srivastava, C.P. and Reed, W. (1988) The development of sex pheromone trapping of *Heliothis armigera* at ICRISAT. *Tropical Pest Management* 34, 30–43.

Payne, G.T., Blenk, R.G. and Brown, T.M. (1988) Inheritance of permethrin resistance in the tobacco budworm (Lepidoptera: Noctuidae). *Journal of Economic Entomology* 81, 65–73.

Pearson, A.C., Sevacherian, V., Ballmer, G.R., Vail, P.V. and Henneberry, T.J. (1989) Population dynamics of *Heliothis virescens* and *H. zea* (Lepidoptera: Noctuidae) in the Imperial Valley of California. *Environmental Entomology* 18, 970–979.

Pearson, E.O. (1958) *The Insect Pests of Cotton in Tropical Africa.* Commonwealth Institute of Entomology, London, 355pp.

Pedgley, D.E. (1985) Windborne migration of *Heliothis armigera* (Hübner) (Lepidoptera: Noctuidae) to the British Isles. *Entomological Gazette* 36, 15–20.

Pedgley, D.E. (1986) Windborne migration in the Middle East by the moth *Heliothis armigera* (Lepidoptera: Noctuidae). *Ecological Entomology* 11, 467–470.

Pedgley, D.E., Tucker, M.R. and Pawar, C.S. (1987) Windborne migration of *Heliothis armigera* (Hübner) (Lepidoptera: Noctuidae) in India. *Insect Science and its Application* 8, 599–604.

Pedigo, L.P., Pitre, H.N., Whitcomb, W.H. and Young, S.Y. (1983) Assessment of the role of natural enemies in regulation of soybean pest populations. In: Pitre, H.N. (ed.) *Natural Enemies of Arthropod Pests in Soybean.* Southern Cooperative Series Bulletin 285, pp. 39–48.

Persson, B. (1974) Diel distribution of oviposition in *Agrotis ipsilon* (Hüfn.), *Agrotis munda* (Walk.), and *Heliothis armigera* (Hbn.), (Lep. Noctuidae), in relation to temperature and moonlight. *Entomologia Scandinavica* 5, 196–208.

Persson, B. (1976) Influence of weather and nocturnal illumination on the activity and abundance of populations of Noctuids (Lepidoptera) in south coastal Queensland. *Bulletin of Entomological Research* 66, 33–63.

Phillips, J.R. and Newsom, L.D. (1966) Diapause in *Heliothis zea* and *Heliothis virescens* (Lepidoptera: Noctuidae). *Annals of the Entomological Society of America* 59, 154–159.

Phillips, J.R. and Whitcombe, W.H. (1962) Field behaviour of the adult bollworm, *Heliothis zea* (Boddie). *Journal of the Kansas Entomological Society* 35, 242–246.

Piccardi, P., Capizza, A., Cassani, G., Spinelli, P., Arsura, E. and Massardo, P. (1977) A sex pheromone component of the old world bollworm *Heliothis armigera. Journal of Insect Physiology* 23, 1443–1445.

Plapp, F.W. and Bull, D.L. (1989) Modifying chemical control practices to preserve natural enemies. In: King, E.G. and Jackson, R.D. (eds) *Proceedings of the Workshop on Biological Control of* Heliothis*: Increasing the Effectiveness of Natural Enemies*, 11–15 November 1985, New Delhi, India. FERRO/USDA, New Delhi, India, pp. 537–546.

Plapp, F.W. and Vinson, S.B. (1977) Comparative toxicities of some insecticides to the tobacco budworm and its ichneumonid parasite *Campoletis sonorensis*. *Environmental Entomology* 6, 381–384.

Poole, R. (1989) A general synopsis of the systematics of *Heliothis* and *Helicoverpa*. In: King, E.G. and Jackson, R.D. (eds) *Proceedings of the Workshop on the Biological Control of* Heliothis*: Increasing the Effectiveness of Natural Enemies*, 11–15 November 1985. New Delhi, India. FERRO/USDA, New Delhi, India, pp. 161–171.

Powell, J.E., King, E.G. and Jany, C.S. (1986) Toxicity of insecticides to adult *Microplitis croceipes*. *Journal of Economic Entomology* 79, 1343–1346.

Pretorius, L.M. (1976) Laboratory studies on the developmental reproductive performance of *Heliothis armigera* on various food plants. *Journal of the Entomological Society of South Africa* 39, 337–343.

Proshold, F.I., Laster, M.J., Martin, D.F. and King, E.G. (1982). The potential for hybrid sterility in *Heliothis* management. In: Reed, W. and Kumble, V. (eds) *Proceedings of the Workshop on* Heliothis *Management*, 15–20 November 1981, ICRISAT Center, Patancheru, A.P., India: ICRISAT, Pantancheru, A.P., India, pp. 329–339.

Quaintance, A.L. and Brues, C.T. (1905) *The Cotton Bollworm*. US Department of Agriculture, Bureau of Entomology Bulletin 50, 155pp.

Raina, A.K. (1989) Male-induced termination of sex pheromone production and receptivity in mated females of *Heliothis zea*. *Journal of Insect Physiology* 35, 821–826.

Rajagopal, D., and Channa Basavanna, G.P. (1982) Biology of the maize cob caterpillar, *Heliothis armigera* (Hubner) (Lepidoptera: Noctuidae). *Mysore Journal of Agricultural Science* 16, 153–159.

Ramalho, F.S., Parrott, W.L., Jenkins, J.N. and McCarty, J.C. (1984) Effects of leaf trichomes on the mobility of newly hatched tobacco budworms (Lepidoptera: Noctuidae). *Journal of Economic Entomology* 77, 619–621.

Ramaswamy, S.B. (1988) Host finding by moths: sensory modalities and behaviours. *Journal of Insect Physiology* 34, 235–249.

Ramaswamy, S.B. (1990) Periodicity of oviposition, feeding and calling by mated female *Heliothis virescens* in a field cage. *Journal of Insect Behavior* 3, 417–427.

Ramaswamy, S.B., Ma, W.K. and Baker, G.T. (1987) Sensory cues and receptors for oviposition by *Heliothis virescens*. *Entomologia Experimentalis et Applicata* 43, 159–168.

Ramos, V.E. and Morallo-Rejesus, B. (1976) Effects of nutrition on larval coloration of *Helicoverpa armigera* (Hübner). *Philippine Entomology* 3, 201–224.

Raulston, J.R. and Houghtaling, J.E. (1986) Circumstantial ecological evidence for *Heliothis virescens* migration into the Lower Rio Grande Valley of Texas from northeastern Mexico. In: A.N. Sparks (ed) *Long Range Migration of Moths of Agricultural Importance to the United States and Canada: Specific Examples of Occurrence and Synoptic Weather Patterns Conducive to Migration*. US Department of Agriculture, Agricultural Research Services 43, pp. 34–47.

Raulston, J.R., Wolf, W.W., Lingren, P.D. and Sparks, A.N. (1982) Migration as a factor in *Heliothis* management. In: Reed, W. and Kumble, V. (eds) *Proceedings of the International Workshop on* Heliothis *Management*, 15–20 November 1981, ICRISAT Center, Patancheru, A.P., India. ICRISAT, Patancheru, A.P., India, pp. 61–73.

Raulston, J.R., Pair, S.D., Martinez, F.A., Westbrook, J.K., Sparks, A.N. *et al.* (1986) Ecological studies indicating the migration of *Heliothis zea*, *Spodoptera frugiperda* and

Heliothis virescens from northeastern Mexico and Texas. In: Danthanaryana, W. (ed) *Insect Flight, Dispersal and Migration.* Springer-Verlag, Heidelberg, pp. 204–220.

Raulston, J.R., Summy, K.R., Loera, J., Pair, S.D. and Sparks, A.N. (1990) Population dynamics of corn earworm larvae (Lepidoptera: Noctuidae) on corn in the lower Rio Grande Valley. *Environmental Entomology* 19, 274–280.

Reed, W. (1965) *Heliothis armigera* (Hb.) (Noctuidae) in western Tanganyika. 1. Biology, with special reference to the pupal stage. *Bulletin of Entomological Research* 56, 117–125.

Reed, W. and Kumble, V. (1982) (eds) *Proceedings of the International Workshop on* Heliothis *Management,* 15–20 November 1981, ICRISAT Center, Patancheru, A.P., India. ICRISAT (International Crops Research Institute for the Semi-Arid Tropics), Patancheru, A.P., India, 418pp.

Reed, W. and Pawar, C.S. (1982) *Heliothis*: A global problem. In: Reed, W. and Kumble, V. (eds) *Proceedings of the International Workshop on* Heliothis *Management,* 15–20 November 1981, ICRISAT Center, Patancheru, A.P., India. ICRISAT, Patancheru, A.P., India, pp. 9–14.

Reed, W., Cardona, C., Sithanantham, S. and Lateef, S.S. (1989a) Chickpea insect pests and their control. In: Saxena, M.C. and Singh, K.B. (eds) *The Chickpea.* CAB International, Wallingford, UK, pp. 283–318.

Reed, W., Lateef, S.S. and Sithanantham, S. (1989b) Compatibility of host plant resistance and biological control of *Heliothis* spp. (Lepidoptera: Noctuidae). In: King, E.G. and Jackson, R.D. (eds) *Proceedings of the Workshop on the Biological Control of* Heliothis*: Increasing the Effectiveness of Natural Enemies,* 11–15 November 1985, New Delhi, India. FERRO/USDA, New Delhi, India, pp. 529–535.

Ridgway, R.L. and Jones, S.L. (1968) Field-cage releases of *Chrysopa carnea* for suppression of populations of the bollworm and tobacco budworm on cotton. *Journal of Economic Entomology* 61, 892–898.

Ridgway, R.L. and Jones, S.L. (1969) Inundative releases of *Chrysopa carnea* for control of *Heliothis* on cotton. *Journal of Economic Entomology* 62, 177–180.

Ridgway, R.L. and Lingren, P.D. (1972) Predaceous and parasitic arthropods as regulators of *Heliothis* populations. *Southern Cooperative Series Bulletin* 169, 48–56.

Ridgway, R.L., King, E.G. and Carillo, J.L. (1977) Augmentation of natural enemies for control of plant pests in the Western Hemisphere. In: Ridgeway, R.L. and Vinson, S.B. (eds) *Biological Control by Augmentation of Natural Enemies.* Plenum, New York, pp. 379–416.

Riley, J.R., Smith, A.D. and Bettany, B.W. (1990) The use of video equipment to record in three dimensions flight trajectories of *Helicoverpa armigera* and other moths at night. *Physiological Entomology* 15, 73–80.

Riley, J.R., Armes, N.J., Reynolds, D.R. and Smith, A.D. (1992) Nocturnal observations of the emergence and flight behaviour of *Helicoverpa armigera* (Noctuidae) in the post-rainy season in central India. *Bulletin of Entomological Research* 82, 243–256.

Roach, S.H. (1981) Emergence of overwintered *Heliothis* spp. moths from three different tillage systems. *Environmental Entomology* 10, 817–818.

Roach, S.H. and Adkisson, P.L. (1970) Role of photoperiod and temperature in the induction of pupal diapause in the bollworm *Heliothis zea. Journal of Insect Physiology* 16, 1591–1597.

Roach, S.H. and Hopkins, A.R. (1979) *Heliothis* spp.: Behaviour of prepupae and emergence of adults from different soils at different moisture levels. *Environmental Entomology* 8, 388–391.

Robertson, G.A. (1977) Ord River cropping progress. *Journal of Agriculture, Western Australia* 18, 136–140.

Robinson, S.H., Wolfenbarger, D.A. and Dilday, R.H. (1980) Antixenosis of smooth leaf cotton to the ovipositional response of tobacco budworm. *Crop Science* 20, 646–649.

Rogers, D.J., Teakle, R.E. and Brier, H.B. (1983) Evaluation of *Heliothis* nuclear polyhedrosis virus for control of *Heliothis armigera* on navy beans in Queensland, Australia. *General and Applied Entomology* 15, 31–34.

Room, P.M. (1979a) A prototype on-line system for management of cotton pests in the Namoi Valley, New South Wales, Australia. *Protection Ecology* 1, 245–264.

Room, P.M. (1979b) Parasites and predators of *Heliothis* spp. (Lepidoptera: Noctuidae) in cotton in the Namoi Valley, New South Wales. *Journal of the Australian Entomological Society* 18, 223–228.

Room, P.M. (1983) Calculations of temperature-driven development of *Heliothis* spp. (Lepidoptera: Noctuidae)in the Namoi Valley, New South Wales. *Journal of the Australian Entomological Society* 22, 211–215.

Room, P.M., Titmarsh, I.J. and Zalucki, M.P. (1990) Life tables. In: Zalucki, M.P. (ed.) Heliothis: *Research Methods and Prospects.* Springer-Verlag, New York, pp. 69–79.

Roome, R.E. (1975) Activity of adult *Heliothis armigera* (Hb.) (Lepidoptera: Noctuidae) with reference to the flowering of sorghum and maize in Botswana. *Bulletin of Entomological Research* 65, 523–530.

Roome, R.E. (1979) Pupal diapause in *Heliothis armigera* (Hubner) (Lepidoptera: Noctuidae) in Botswana: its regulation by environmental factors. *Bulletin of Entomological Research* 69, 149–160.

Rose, D.J.W., Page, W.W., Dewhurst, C.F., Riley, J.R., Reynolds, D.R., Pedgley, D.E. and Tucker, M.R. (1985) Downwind migration of the African armyworm moth, *Spodoptera exempta*, studied by mark and recapture and by radar. *Ecological Entomology* 10, 299–313.

Rothschild, G.H.L. (1978) Attractants for *Heliothis armigera* and *H. punctigera. Journal of Australian Entomology* 102, 389–390.

Rothschild, G.H.L., Wilson, A.G.L. and Malafant, K.W. (1982) Preliminary studies on the female sex pheromones of *Heliothis* species and their possible use in control programs in Australia. In: Reed, W. and Kumble, V. (eds) *Proceedings of the International Workshop on* Heliothis *Management,* 15–20 November 1981, ICRISAT Center, Patancheru, A.P., India: ICRISAT, Patancheru, A.P. India, pp. 319–327.

Roush, R.T. and Luttrell, R.G. (1989) Expression of resistance to pyrethroid insecticides in adults and larvae of tobacco budworm (Lepidoptera: Noctuidae): Implications for resistance monitoring. *Journal of Economic Entomology* 82, 1305–1310.

Rummel, D.R. and Neece, K.C. (1989) Winter survival of *Heliothis zea* (Boddie) in cultivated soil in the southern Texas High Plains. *Southwestern Entomologist* 14, 117–125.

Rummel, D.R., Neece, K.C. and Arnold, M.D. (1986) Overwintering survival and spring emergence of *Heliothis zea* (Boddie) in the Texas southern High Plains. *Southwestern Entomologist* 11, 1–9.

Samson, P.R. and Blood, P.R. (1980) Voracity and searching ability of *Chrysopa signata* (Neuroptera: Chrysopidae), *Micromus tasmaniae* (Neuroptera: Hemerobiidae), and *Tropiconabis capsiformis* (Hemiptera: Nabidae). *Australian Journal of Zoology* 28, 575–580.

Sawicki, R.M. and Denholm, I. (1987) Management of resistance to pesticides in cotton pests. *Tropical Pest Management* 33, 262–272.

Sawicki, R.M. and Denholm, I. (1989) Insecticide resistance management revisited. In: *Progress and Prospects in Insect Control.* British Crop Protection Council Monograph No. 43, pp. 193–203.

Sawicki, R.M., Denholm, I., Forrester, N.W. and Kershaw, C.D. (1989) Present insecticide-resistance management strategies in cotton. In: Green, M.B. and Lyon, D.J. deB. (eds) *Pest Management in Cotton.* Ellis Horwood, Chichester, pp. 31–43.

Schaefer, G.W. (1976) Radar observations of insect flight. In: Rainey, R.C. (ed.) *Insect Flight.* Symposium Royal Entomological Society London. Blackwell Scientific Publications, Oxford, pp. 157–197.

Schneider, J.C. (1989) Role of movement in evaluation of area-wide insect pest management tactics. *Environmental Entomology* 18, 868–874.

Schneider, J.C., Benedict, J.H., Gould, F., Meredith, W.R., Schuster, M.F. and Zummo, G.R. (1986) Interaction of *Heliothis* with its host plants. In: Johnson, S.J., King, E.G. and Bradley, J.R. (eds) *Theory and Tactics of* Heliothis *Population Management: 1-Cultural and Biological Control.* Southern Cooperative Series Bulletin 316, pp. 3–21.

Sehgal, V.K. and Ujagir, R. (1990) Effect of synthetic pyrethroids, neem extracts and other insecticides for the control of pod damage by *Helicoverpa armigera* (Hübner) on chickpea and pod damage-yield relationship at Pantnagar in northern India. *Crop Protection* 9, 29–32.

Seymour, J.E. and Jones, R.E. (1990) Evaluating natural enemy impact on *Heliothis.* In: Zalucki, M.P. (ed.) Heliothis: *Research Methods and Prospects.* Springer-Verlag, New York, pp. 80–89.

Shaw, A.J. and Colton, R.T. (1987) *Cotton Pesticides Guide 1987–1988.* Department of Agriculture, New South Wales, 40pp.

Shepard, M. and Sterling, W. (1972) Incidence of parasitism of *Heliothis* spp. (Lepidoptera: Noctuidae) in some cotton fields of Texas. *Annals of the Entomological Society of America* 65, 759–760.

Singh, H. and Singh, G. (1975) Biological studies of *Heliothis armigera* in Punjab. *Indian Journal of Entomology* 37, 154–164.

Slosser, J.E., Phillips, J.R., Herzog, G.A. and Reynolds, C.R. (1975) Overwintering survival and spring emergence of the bollworm in Arkansas. *Environmental Entomology* 4, 1015–1024.

Smith, R.F. (1970) Pesticides: Their use and limitations in pest management. In: Rabb, R.L. and Guthrie, F.E. (eds) *Concepts of Pest Management.* North Carolina State University, Raleigh, pp. 103–113.

Smith, D.B. and Hostetter, D.L. (1982) Laboratory and field evaluations of pathogen-adjuvant treatments. *Journal of Economic Entomology* 75, 472–476.

Sparks, T.C. (1981) Development of insecticide resistance in *Heliothis zea* and *Heliothis virescens* in North America. *Bulletin of the Entomological Society of America* 27, 186–192.

Sparks, A.N., Raulston, J.R., Carpenter, J.E. and Lingren, P.D. (1982) The present status and potential for novel uses of pheromone to control *Heliothis.* In: Reed, W. and Kumble, V. (eds) *Proceedings of the International Workshop on* Heliothis *Management,* 15–20 November 1981, ICRISAT, Patancheru, A.P., India, ICRISAT, Patancheru, A.P., India, pp. 291–308.

Srivastava, C.P., Pimbert, M.P. and Reed, W. (1992) Monitoring of *Helicoverpa* (=*Heliothis*) armigera (Hübner) moths with light and its pheromone traps in India. *Insect Science and its Application* 13, 205–210.

Stadelbacher, E.A. (1982) An overview and simulation of tactics for management of *Heliothis* spp. on early season wild host plants. In: *Beltwide Cotton Producers Research Conference,* Memphis, Tennessee, pp. 209–212.

Stadelbacher, E.A. and Martin, D.F. (1980) Fall diapause, winter mortality and spring emergence of the tobacco budworm in the delta of Mississippi. *Environmental Entomology* 9, 553–556.

Stadelbacher, E.A and Pfimmer, T.R. (1972) Winter survival of the bollworm at Stoneville, Mississippi. *Journal of Economic Entomology* 65, 1030–1034.

Stadelbacher, E.A., Powell, J.E. and King. E.G. (1984) Parasitism of *Heliothis zea* and *H. virescens* (Lepidoptera: Noctuidae) larvae in wild and cultivated host plants in the Delta of Mississippi. *Environmental Entomology* 13, 1167–1172.

Stadelbacher, E.A., Graham, H.M., Harris, V.E., Lopez, J.D., Phillips, J.R. and Roach, S.H. (1986) *Heliothis* populations and wild host plants in the southern U.S. In: Johnson S.J., King, E.G. and Bradley, J.R. (eds) *Theory and Tactics of* Heliothis *Population Management: 1 – Cultural and Biological Control.* Southern Cooperative Series Bulletin 316, 54–74.

Sterling, W. (1989) Estimating the abundance and impact of predators and parasites on *Heliothis* populations. In: King, E.G. and Jackson, R.D. (eds) *Proceedings of the Workshop on Biological Control of* Heliothis*: Increasing the Effectiveness of Natural Enemies.* 11–15 November 1985, New Delhi, India. FERRO, USDA, New Delhi, India. pp. 37–56.

Steward, V.B., Yearian, W.C. and Kring, T.J. (1990) Parasitsm of *Heliothis zea* (Lepidoptera: Noctuidae) eggs and larvae and *Celama sorghiella* (Lepidoptera: Noctuidae) eggs on sorghum grain panicles. *Environmental Entomology* 19, 154–161.

Steyskal, G.C. (1971) On the grammar of the name *Heliothis* Ochsenheimer (Noctuidae). *Journal of the Lepidopterists' Society* 25, 264–266.

Stinner, R.A., Rabb, R.L. and Bradley, J.R. (1974) Population dynamics of *Heliothis zea* (Boddie) and *H. virescens* (F.) in North Carolina: a simulation model. *Environmental Entomology* 3, 163–168.

Stinner, R.A., Rabb, R.L. and Bradley, J.R. (1977) Natural factors operating in the population dynamics of *Heliothis zea* in North Carolina. *Proceedings of the XV International Congress of Entomology*, Washington, DC. Entomological Society of America, Washington, DC, pp. 622–642.

Teakle, R.E. (1989) Estimating the abundance and impact of predators and parasites on *Heliothis* populations. In: King, E.G. and Jackson, R.D. (eds) *Proceedings of the Workshop on Biological Control of* Heliothis*: Increasing the Effectiveness of Natural Enemies,* 11–15 November 1985, New Delhi, India. FERRO, USDA, New Delhi, India, pp. 57–67.

Thead, L.G., Pitre, H.N. and Kellogg, T.F. (1987) Predation on eggs and larvae of *Heliothis virescens* (Lep.: Noctuidae) by an adult predator complex in cage studies on cotton. *Entomophaga* 32, 197–207.

Thomson, N.J. and Lee, J.A. (1980) Insect resistance in cotton: A review and prospectus for Australia. *Journal of the Australian Institute of Agricultural Science* 46, 75–86.

Tingle, F.C. and Mitchell, E.R. (1981) Relationship between pheromone catches of male tobacco budworm, larval infestation, and damage levels in tobacco. *Journal of Economic Entomology* 74, 437–440.

Tingle, F.C. and Mitchell, E.R. (1984) Aqueous extracts from indigenous plants as oviposition deterrents for *Heliothis virescens* (F.). *Journal of Chemical Ecology* 10, 101–113.

Titmarsh, I.J. (1985) Population dynamics of *Heliothis* spp. on tobacco in far north Queensland. MSc thesis, James Cook University of North Queensland, Australia (from Zalucki *et al.*, 1986).

Todd, E.L. (1955) The distribution and nomenclature of the corn earworm. *Journal of Economic Entomology* 48, 600–603.

Todd, E.L. (1978) A checklist of species of *Heliothis* Ochsenheimer (Lepidoptera: Noctuidae). *Proceedings of the Entomological Society of Washington* 80, 1–14.

Topper, C.P. (1981) The behaviour and population dynamics of *Heliothis armigera* (Hb) (Lepidoptera: Noctuidae) in the Sudan Gezira. PhD thesis, Cranfield Institute of Technology, Bedford, 245pp.

Topper, C.P. (1987a) Nocturnal behaviour of adults of *Heliothis armigera* (Hübner) (Lepidoptera: Noctuidae) in the Sudan Gezira and pest control implications. *Bulletin of Entomological Research* 77, 541–554.

Topper, C.P. (1987b) The dynamics of the adult population of *Heliothis armigera* (Hübner) (Lepidoptera: Noctuidae) in the Sudan Gezira in relation to cropping pattern and pest control on cotton. *Bulletin of Entomological Research* 77, 525–539.

Tripathi, S.R. and Sharma, S.K. (1984) Biology of *Heliothis armigera* (Hübner) in the Terai Belt of Uttar Pradesh, India (Lepidoptera: Noctuidae). *Giornale Italiano di Entomologia* 2, 215–222.

Twine, P.H. (1971) Cannibalistic behaviour of *Heliothis armigera* (Hübner). *Queensland Journal of Agricultural and Animal Science* 28, 153–157.

Twine, P.H. (1974) Some aspects of the natural control of larval *Heliothis armigera* (Hubn.) in maize. MSc thesis, University of Queensland (from Zalucki *et al.*, 1986).

Twine, P.H. (1989) Distribution and economic importance of *Heliothis* (Lepidoptera: Noctuidae) and of their natural enemies and host plants in Australia. In: King, E.G. and Jackson, R.D. (eds) *Proceedings of the Workshop on Biological Control of* Heliothis*: Increasing the Effectiveness of Natural Enemies*. 11–15 November 1985, New Delhi, India. FERRO, USDA, New Delhi, India, pp. 177–184.

Twine, P.H., Kay, I.R. and Lloyd, R.J. (1983) *Heliothis* in sorghum. *Queensland Agricultural Journal* 109, 185–188.

Ujagir, R. and Khare, B.P. (1987) Preliminary screening of chickpea genotypes for susceptibility to *Heliothis armigera* (Hub.) at Pantnagar, India. *International Chickpea Newsletter* 17, p. 14.

van den Berg, H., Waage, J.K. and Cock, M.J.W. (1988) *Natural Enemies of* Helicoverpa armigera *in Africa: A review*. CAB International Institute of Biological Control, Silwood Park, Ascot, Berks SL5 7PY, 81pp.

van den Berg, H., Nyambo, B.T. and Waage, J.K. (1990) Parasitism of *Helicoverpa armigera* (Lepidoptera: Noctuidae) in Tanzania: Analysis of parasitoid–crop associations. *Environmental Entomology* 19, 1141–1145.

van den Bosch, R. and Hagen, K.S. (1966) Predaceous and parasitic arthropods in California cotton fields. *Californian Agricultural Experimental Station Bulletin* 820, 32pp.

van den Bosch, R., Leigh, T.F. and Gonzalez, D. (1969) Cage studies on predation of the bollworm on cotton. *Journal of Economic Entomology* 62, 1486–1489.

Vanderzant, E.S., Richardson, C.D. and Fort, S.W. (1962) Rearing of the bollworm on artificial diet. *Journal of Economic Entomology* 55, 140.

van Emden, H.F. and Williams, G.F. (1974) Insect stability and diversity in agro-ecosystems. *Annual Review of Entomology* 19, 455–476.

Waite, G.K. 1981. Effect of methomyl on *Heliothis* eggs on cotton in central Queensland, Australia. *Protection Ecology* 3, 265–268.

Wardaugh, K.G., Room, P.M. and Greenup, L.R. (1980) The incidence of *Heliothis armigera* (Hübner) and *H. punctigera* Wallengren (Lepidoptera: Noctuidae) on cotton and other host plants in the Namoi Valley of New South Wales. *Bulletin of Entomological Research* 70, 113–131.

Watson, J.S. (1989) Recent progress in breeding for insect resistance in cotton. In: Green, M.B. and Lyon, D.J.deB. (eds) *Pest Management in Cotton*. Ellis Horwood, Chichester, pp. 44–52.

Wellso, S.G. and Adkisson, P.L. (1966) A long-day short-day effect in the photoperiodic control of the pupal diapause of the bollworm, *Heliothis zea* (Boddie) (Lepidoptera: Noctuidae). *Journal of Insect Physiology* 12, 1455–1465.

West, A.J. and McCaffery, A.R. (1992) Evidence of nerve insensitivity to cypermethrin from Indian strains of *Helicoverpa armigera*. *Proceedings, Brighton Crop Protection Conference 1992: Pests and Diseases.* British Crop Protection Council, UK, pp. 233–238

Westbrook, J.K., Allen, C.T., Plapp, F.W. and Multer, W. (1990) Atmospheric transport and pyrethroid-resistant tobacco budworm, *Heliothis virescens* (Lepidoptera: Noctuidae), in western Texas in 1985. *Journal of Agricultural Entomology* 7, 91–101.

Widmer, M.W. and Schofield, P. (1983) *Heliothis* Dispersal and Migration. TDRI Information Service Annotated Bibliography No. 2. London. Tropical Development Research Institute (now NRI, Natural Resources Institute, Chatham, Kent ME4 4TB), 41pp.

Wilcox, J., Howland, A.F., and Campbell, R.E. (1956) *Investigations of the Tomato Fruitworm, its Seasonal History and Method of Control.* US Department of Agriculture, Technical Bulletin 1147, 47pp.

Willers, J.L., Schneider, J.C. and Ramaswamy, S.B. (1987) Fecundity, longevity and caloric patterns in female *Heliothis virescens*: Changes with age due to flight and supplemental carbohydrate. *Journal of Insect Physiology* 33, 803–808.

Wilson, A.G.L. (1974) Resistance of *Heliothis armigera* to insecticides in the Ord irrigation area, North Western Australia. *Journal of Economic Entomology* 67, 256–258.

Wilson, A.G.L. (1976) Varietal responses of grain sorghum to infestation by *Heliothis armigera*. *Experimental Agriculture* 12, 257–265.

Wilson, A.G.L. (1981) *Heliothis* damage to cotton and concomitant action levels in the Namoi Valley, New South Wales. *Protection Ecology* 3, 311–326.

Wilson, A.G.L. (1982) Past and future *Heliothis* management in Australia. In: Reed, W. and Kumble, V. (eds) *Proceedings of the International Workshop on* Heliothis *Management*, 15–20 November 1981, ICRISAT Center, Patancheru, A.P., India: ICRISAT, Patancheru, A.P., India, pp. 343–354.

Wilson, A.G.L. (1985) Evaluation of pheromone trap design and dispensers for monitoring *Heliothis puntiger* and *H. armiger*. In Bailey, P. and Swincer, D. (eds) *Proceedings of the Fourth Australian Applied Entomological Research Conference*, Adelaide, 24–28 September 1984. South Australia, Government Printer, pp. 74–81.

Wilson, A.G.L. and Greenup, L.R. (1977) The relative injuriousness of insect pests of cotton in the Namoi Valley, New South Wales. *Australian Journal of Ecology* 2, 319–328.

Wilson, A.G.L., Lewis, T. and Cunningham, R.B. (1979) Overwintering and spring emergence of *Heliothis armigera* (Hübner) (Lepidoptera: Noctuidae) in the Namoi Valley, New South Wales. *Bulletin of Entomological Research* 69, 97–109.

Wilson, A.G.L., Desmarchelier, J.M. and Malafant, K. (1983) Persistence on cotton foliage of insecticide residues toxic to *Heliothis* larvae. *Pesticide Science* 14, 623–633.

Wilson, L.T., Gutierrez, A.P. and Leigh, T.F. (1980) Within plant distributions of the immatures of *Heliothis zea* (Boddie) on cotton. *Hilgardia* 48, 12–23.

Wiseman, B.R. (1982) The importance of *Heliothis*–crop interactions in the management of the pest. In: Reed, W. and Kumble, V. (eds) *Proceedings of the International Workshop on* Heliothis *Management*, 15–20 November 1981, ICRISAT Center, Patancheru, A.P., India. ICRISAT, Patancheru, A.P., India, pp. 209–222.

Wiseman, B.R. and Isenhour, D.J. (1990) Impact of resistant corn silks on *Heliothis zea* (Boddie): a laboratory study. *Journal of Economic Entomology* 13, 614–617.

Wiseman, B.R., Widstrom, N.W. and McMillan, W.W. (1977) Ear characteristics and mechanisms of resistance among selected corns to corn earworm. *Florida Entomologist* 60, 97–103.

Wolfenbarger, D.A., Bodegas, V.P.R. and Flores, G.R. (1981) Development of resistance in *Heliothis* spp. in the Americas, Australia, Africa and Asia. *Bulletin of the Entomological Society of America* 27, 181–185.

Yearian, W.C., Luttrell, R.G., Stacey, A.L. and Young, S.Y. (1980) Efficacy of *Bacillus thuringiensis* and *Baculovirus heliothis*–chlordimeform spray mixtures against *Heliothis* spp. on cotton. *Journal of the Georgia Entomological Society* 15, 260–271.

Yearian, W.C., Hamm, J.J. and Carner, G.R. (1986) Efficacy of *Heliothis* pathogens. In: Johnson, S.J., King, E.G. and Bradley, J.R. (eds) *Theory and Tactics of* Heliothis *Management: 1 – Cultural and Biological Control.* Southern Cooperative Series Bulletin 316, 92–103.

Young, J.H. and Price, R.G. (1977) Overwintering of *Heliothis zea* in southwestern Oklahoma. *Environmental Entomology* 6, 627–628.

Zalucki, M.P., Daglish, G., Firempong, S. and Twine, P. (1986) The biology and ecology of *Heliothis armigera* (Hübner) and *H. punctigera* Wallengren (Lepidoptera: Noctuidae) in Australia: what do we know? *Australian Journal of Ecology* 34, 779–814.

4 *Pectinophora* (Lepidoptera: Gelechiidae)

W.R. Ingram

Orchards, Alhampton, Shepton Mallet, Somerset BA4 6PZ, UK

Introduction

The genus *Pectinophora* contains two species which attack cotton. These are the pink bollworm, *P. gossypiella* (Saunders), and the pink-spotted bollworm (Passlow and Sabine, 1963), *P. scutigera* (Holdaway). The pink bollworm is one of the most serious pests of cotton occurring almost throughout the tropics and subtropics, while pink spotted bollworm is restricted to parts of Australasia and the Pacific Islands.

Pink bollworm larvae (Plate III.1) will feed on flowerbuds and flowers but bolls, and the seeds therein, are the main source of food. Under suitable conditions the larvae undergo a lengthy resting stage inside cotton seeds and it is in this state that pink bollworm has been spread by man to most of the cotton growing countries in the world. Its initial spread from India to Egypt, sometime prior to 1910, was unwitting, but most of its subsequent spread has occurred in spite of stringent plant quarantine regulations and phytosanitary precautions.

Pink bollworm damage to buds and flowers only occurs early in season, before there are any green bolls on the cotton plants, and is usually of little consequence. Far more serious is the damage to the developing seeds and the consequent arrest of growth, lock or boll rotting, premature or partial boll opening, reduction of staple length and yarn strength and general lowering of quality through increased trash content in the lint. Yield losses due to pink bollworm vary from country to country and year to year from almost nothing to total crop loss. Schwartz (1983) estimated that yield losses in the United States due to pink bollworm ranged from 9% when controlled to 61% when not controlled.

Taxonomy

The pink bollworm was first described by Saunders (1843) as *Depressaria gossypiella* from specimens collected from American Upland cotton grown at Broach in western India. Other generic assignments are *Gelechia gossypiella* (Saunders), Meyrick (1905), *Pectinophora gossypiella* (Saunders), Busck (1917) and *Platyedra gossypiella* (Saunders), Meyrick (1918). The generic assignment of Busck (1917) was accepted in the United States but in the Old World it was Meyrick's (1918) assignment that was accepted. Common (1958) details the full synonymy and places pink bollworm as *Pectinophora gossypiella* (Saunders). This status is accepted by the British Museum (Natural History) (Dr Klaus Sattler, personal communication).

The pink-spotted bollworm was described by Holdaway (1926) as *Platyedra scutigera* and then placed as *Pectinophora scutigera*, Holdaway (1929). This is accepted by Common (1958) and Dr Klaus Sattler (personal communication). One other species, *P. endema* Common, is known from Australia but it does not attack cotton (Common, 1958).

Distribution and Origins

Pink bollworm is recorded from nearly all the cotton-growing countries of the world (CAB Institute of Entomology, 1990). The only major cotton growing countries from which pink bollworm is still absent appear to be Russia (excluding Uzbekistan), Central America (Belize, Costa Rica, Guatemala, Honduras, Nicaragua and Salvador) and parts of South America (Ecuador, Guyana and Surinam) and Australia (Queensland). Passlow and Sabine (1963) emphasize the need for stringent quarantine measures to prevent pink bollworm gaining entry into Queensland.

The pink-spotted bollworm is found in Australia (Queensland and Northern Territory), Hawaii, Micronesia and Papua New Guinea (CAB Institute of Entomology, 1987). Wilson (1972) queries the record of pink-spotted bollworm from the Northern Territory of Australia but reports it as occurring in part of New South Wales.

Pectinophora gossypiella was first recorded in India in 1842 and Pearson (1958) documents its spread throughout Sri Lanka, Burma and Malaysia (Lefroy, 1906) and China before 1918 (Hunter, 1918). It was first recorded in Australia in 1911 (Wilson, 1972). The first African records were in Tanzania (Vosseler, 1904), Egypt about 1906–1907 (Willcocks, 1916), Sudan in 1914–1915 (Ripper and George, 1965), and it did not reach Malawi until 1939 (Smee, 1940) and Zimbabwe as late as 1959 (Whellan, 1960).

Its spread in the New World started in Hawaii, where it had been imported direct from India in cotton seed (Fullaway, 1909), and from Hawaii to St Croix in 1911 (Hunter, 1918). The spread of pink bollworm in the United States is summarized by Noble (1969). It was imported into Mexico in 1911 in seed from Egypt and thence to Texas in 1917. Early infestations were apparently eliminated by the use of cotton-free zones and extensive clean-up measures, but infestations persisted along the Texas–Mexican border and it gradually spread to most of the cotton-growing areas in Texas, New Mexico, Oklahoma, Arizona, Arkansas and Louisiana during the late 1940s and early 1950s. Pink bollworm was eradicated from cultivated cotton in parts of Florida and Georgia in 1932, but it still exists in wild cotton and backyard cotton in southern Florida. The most recent invasions are to California in 1965 and Nevada in 1966.

Pearson (1958) suggested that the circumstantial evidence pointed to the original home of *P. gossypiella* as having been India, only coming into prominence in 1842 after the introduction of American Upland cottons. Saunders (1843), however, reported that 'the native cotton is sometimes affected by it'. The native cottons *Gossypium arboreum* and *G. herbaceum* developed in the Indus basin, and Pearson (1958) discusses the distribution and spread of *P. gossypiella* in relation to two theories of the origins and spread of these Old World cottons. One theory is that *G. herbaceum* spread westwards to Asia Minor, the Mediterranean, Arabia and Africa, and it is from *G. herbaceum* that *G. africanum* in southern Africa is derived. The other theory is that *G. africanum* was indigenous, spreading along ancient coastal trade routes around the Indian Ocean and to Arabia, Iran and Baluchistan where it evolved into *G. herbaceum*.

Whichever theory is correct *P. gossypiella* certainly did not follow the spread of *G. herbaceum* westwards. It is absent from most of the former USSR and only recently invaded Iran and Iraq. Pearson (1958) considered that pink bollworm is much more closely associated with *G. arboreum*, suggesting that it may have accompanied the original movements of the perennial *arboreum* cotton from India through Southeast Asia and China. The initial infestations of *P. gossypiella* in East Africa and Madagascar antedate its introduction to Egypt, and Pearson (1958) suggested that the pest may have been introduced with *G. arboreum* race *indicum* by traders plying the ancient trade routes of the Indian Ocean.

However, if the area of origin of *P. gossypiella* was western peninsula India then Pearson (1958) suggested that it must have transferred from another host onto *G. arboreum*, as it does not occur naturally on any wild species of *Gossypium* in India. This theory is supported by Holdaway (1926), who suggested that *P. scutigera* had transferred to cultivated cotton from the wild hosts *Hibiscus tiliaceus* and *Thespesia populnea* in Queensland. An Australiasian origin for *P. gossypiella* is further supported by another species, *P. endema* Common, and a complex of some 20 species in closely related genera in Australia, and Common (1958) suggested that pink bollworm originated in an area of the eastern Indian Ocean bounded by northwestern Australia and the Indonesian and Malaysian islands. However, Wilson (1972) stated that *P. gossypiella* had probably been introduced into Australia in cotton seed.

Thus the area of origin of *P. gossypiella* is still not known for certain although the numbers of species of parasites recorded in Pakistan (Cheema *et al.*, 1980a) does tend to support an Indo-Pakistan origin.

Description of Stages

Life History and Feeding Habits

The life history has been studied or reviewed by many workers, amongst whom were Willcocks (1916) and El-Sayed and El-Rahman (1960) in Egypt, Busck (1917) in Hawaii, Taylor (1936) in Uganda, Pearson (1958) worldwide, Ripper and George (1965) in the Sudan, Owen and Calhoun (1932) and Noble (1969) in the United States and Ahmad (1977) in Pakistan and India.

The eggs are pearly-iridescent white when first laid, turning yellowish and, finally, orange-red just prior to eclosion. They are flattened ovals measuring approximately 0.5 mm long by 0.25 mm wide and are sculptured with longitudinal lines. Incubation takes from 3.5–6 days. Hatching takes approximately three hours and usually occurs early in the mornings. Eggs are either laid singly or in small groups of four to five. Early in the season, before any bolls are present, they can be found in sheltered places anywhere on the plant – axils of petioles or peduncles, the underside of young leaves, under older leaves at the junction of the main veins or on buds or flowers. Once there are bolls up to 14 days old these become the favoured oviposition site and eggs are laid either in the sutures near the boll tip or under the bracteoles at the base of the boll. This latter site provides almost complete protection from predators or contact insecticides.

On hatching, the first instar larva is about 1.0 mm long and has a dark brown head and almost colourless body which becomes a glossy white after feeding. The second instar larva is creamy-white, with a conspicuous dark brown head and paler brown thoracic and anal shields. The third instar is about 6 mm long, still a glossy ivory-white in ground colour but with two deep pink transverse dorsal bands, a broad anterior and narrow posterior interrupted by pale median and lateral streaks, per body segment. It should be noted that larvae feeding entirely on flowers or on shed, rotting fruiting forms do not develop the pink colour. The fourth instar is about 11–15 mm long and 2.5 mm round when fully grown; this is normally the final instar but Watson and Johnson (1974) found that about 25% had five larval instars and these took about three days longer to complete development. In equatorial regions the normal larval cycle lasts for 12–20 days and 9–14 days in hotter regions.

The prolegs on the third to sixth abdominal segments carry a single row of 15–20 crotchets in an incomplete circle with the opening facing outwards (Busck, 1917). This arrangement of crochets is useful to distinguish it from other larvae infesting cotton bolls.

The larva will make its way to the nearest bud, flower or boll and attempt to penetrate within two hours of hatching. If attacked buds are less than ten days old when shed the larva dies, whereas when the bud is older the larva can complete its development. Larvae in flowerbuds spin webbing which prevents the flower from opening properly, the so-called rosetted bloom; here the staminal column only may be damaged and normal bolls develop, or the ovary is attacked and shedding results. Butler and Henneberry (1976) found that some 40% fewer rosetted flowers formed bolls than uninfested flowers and that boll weight was unaffected. Larvae attack green bolls from 10–20 days old by boring into the boll wall, often from under the bracteoles, and the entrance hole is extremely small. Most larvae will bore within the carpel wall leaving conspicuous mines before entering the boll, others will enter directly and a callous is formed on the inside. Here the larva will feed on developing seeds; in young bolls it may destroy the entire contents while in older bolls it may complete its development on three–four seeds. The larvae will make neat, round tunnels between seeds or through the loc wall. Several larvae may infest a single boll.

Thus pink bollworm larvae can cause bud shedding, flower damage, the total loss of smaller bolls and partial damage to larger bolls. This partial damage to larger bolls can lead to the complete loss of the undamaged seed cotton from that boll as it may not fluff out sufficiently for a spindle picker to harvest (Fry *et al.*, 1978), and it can also be harder to hand pick (Ingram, 1981). Pink bollworm attack lowers the quantity of both lint and seed and can also lower the quality of the lint, affecting fibre length, fibre-bundle strength and micronaire (Fry *et al.*, 1978). Lint quality is also lowered by staining and the presence of crushed seed after ginning (Ingram, 1981).

The mature larvae are either 'short-cycle' and will go on to pupate, or 'long-cycle' and enter a state of diapause. Short cycle larvae pupating may either cut a round exit hole, about 2 mm in diameter, through the carpel wall and fall to the ground to pupate, or may tunnel to the cuticle, leaving this as a transparent window, and pupate inside. The larval exit holes in the carpel wall predispose the bolls to infection by *Aspergillus flavus* Link and a build up of the potent hepatocarcinogen aflatoxin (Henneberry *et al.*, 1978). Before actually pupating, the larva will spin an elongate, loose-fitting cocoon with a lightly webbed exit at one end. Inside this cocoon the larva undergoes a pre-pupal stage which lasts for 2.5–3.5 days and then it turns into the shiny, brown pupa, measuring 6–8 mm long and about 2.5 mm wide, which is clearly visible inside the cocoon. In equatorial conditions the pupal stage lasts 8–13 days, depending upon temperature.

The long-cycle larvae entering diapause will spin an entirely different cocoon, a tough, thick-walled, closely-woven, spherical cell without an exit hole. This cell is sometimes referred to as an 'hibernaculum'. The larva will remain curled up in its hibernaculum in a head-to-tail position for weeks or months until conditions are right to break diapause and pupate. Most long-term larvae occur towards the end of the cotton season when there are mature bolls present and they often form their hibernaculae inside seeds. The hibernaculum may occupy a single seed in the larger seeded cottons (Willcocks, 1916; Pearson, 1958), or it may consist of two or more tightly bound seeds, the so-called 'double seeds' (Pearson, 1958; Ingram, 1981). Diapause larvae inside *Hibiscus* spp. capsules bind many seeds together (Pearson, 1958). Not all diapause larvae are found inside seeds, they may spin up in the lint of an open boll and, if still active in the ginnery, will spin up on bales of lint, bags of seed or in cracks and crevices. Birds' nests made from seed cotton have been found to contain diapausing larvae (Ingram, 1981).

Squire (1937) reviewed the early theories of the factors inducing the occurrence of long-cycle larvae; some workers thought it was hibernation, others aestivation and still others a combination of aestivation and hibernation. In all cases, however, long-cycle larvae were associated with the end of the cotton crop, be it summer or winter, wet or dry. Squire (1937) theorized that it must be related to the age of the crop, gave experimental evidence from St Vincent to support this and demonstrated that most resting larvae would soon pupate if kept in moist conditions.

Pearson (1958) also reviewed the work on the development of long-cycle larvae and stated that 'not all the larvae in any generation enter a resting state, which is thus a facultative diapause, requiring some external stimulus to bring it about'. Early generations on annual cotton all consist of short-cycle larvae, later generations in some areas have increasing proportions of larvae entering diapause while in other areas no diapause occurs at all. Pearson (1958) then critically examined the proportions of diapause larvae in data from subtropical areas above 25°N, tropical areas between 13 and 22°N and in equatorial areas between 11 and 5°N and areas between 5°N and 7°S. Ripe or dry bolls were thought to be contributory factors but in Texas, when larvae were fed on green bolls throughout the season, more entered diapause late in the season. Diapause in the field could also not be related to atmospheric humidity, nor could it always be associated with low minimum or high maximum temperatures, and so Pearson (1958) postulated that photoperiod interacted with other factors to induce diapause and that the effects of daylength should be investigated.

The termination of diapause was also contrary, in some areas it is at the beginning of the rainy season and in others at the onset of rainless summers. Pearson (1958) also critically examined data on the length of diapause, including Willcocks's (1916) data where some moths emerged 20 months after their collection as larvae. In general, emergences corresponded with the new planting but there seemed to be considerable variation in the duration of the diapause and in the spread of the emergences. Pearson (1958) concluded that diapause was a process of physiological development which was affected by temperature, irrespective of humidity, although moisture appeared to accelerate the later stages of diapause development.

Pearson's conclusions led to a great deal of experimental work on the induction and termination of pink bollworm larval diapause. Experimental work in the United States, sometimes with strains from the West Indies, South America or India, on the effects of photoperiod, moisture, temperature and lipid content of larval food were reported by Bull and Adkisson (1960), Adkisson (1961, 1964), Lukefahr (1961), Adkisson *et al.* (1963), Lukefahr *et al.* (1964), Wellso and Adkisson (1964), Ankersmit and Adkisson (1967), Rice and Reynolds (1971a), Watson *et al.* (1973), Raina and Bell (1974a,b), Gutierrez *et al.* (1977), Butler *et al.* (1978) and Bariola and Henneberry (1980). Elsewhere, diapause was studied in Egypt (El-Sayed and Rustom, 1960a,b; Metwally and Hosny, 1972; and Abul-Nasr *et al.*, 1974a,b), in Barbados (Ingram, 1980), in India (Awaknavar *et al.*, 1982), in China (Zhao and Liu, 1985) and in Brazil (Lukefahr *et al.*, 1985). The general conclusion was that although temperature, humidity and diet all played a role in diapause induction or termination, the critical factor was when daylength went above, or below, about 11.5 hours.

An empirical model of the effects of fluctuating temperatures and decreasing photophases from the data of 216 laboratory experiments on a wild pink bollworm stock from Arizona was developed by Gutierrez *et al.* (1981), and was successfully tested against field data from different sites and years in California and Arizona. Gutierrez *et al.* (1986) used this same model with Brazilian stock and found that it did not differ in its diapause response from the Arizona stock from which the model was first developed. In general, a lengthy diapause is induced by high fluctuating temperatures and short decreasing photoperiods.

When conditions are right for the termination of diapause the larvae pupate. Willcocks (1916) commented that the resting cocoon seemed too small to allow the formation of a straight pupa, however, in trials with double-seeds and bolls some larvae pupated outside the seeds or bolls and others pupated *in situ*, and in cases where infested seed and bolls were buried 80–100 mm in the soil many of the larvae made their way to the surface to pupate. Taylor (1936) reported cases where larvae pupated inside their resting cocoons, El-Sayed and El-Rahman (1960) reported that the majority of resting larvae pupated in the resting cocoon and Ripper and George (1965) also stated that the larvae pupate in the diapause cocoon. Henneberry and Clayton (1983) reported that larvae in bolls buried in the soil tunnelled to, or just below, the soil surface and many of these pupated; moist soil increased this activity.

The adult moth is greyish-brown with blackish bands on the forewings; the hindwings are silvery-grey. The body measures some 8–9 mm long and has a wingspan of 15–20 mm. The moths emerge from the pupae early in the morning or in the evening and are nocturnal, hiding amongst soil debris or soil cracks during the day. They are attracted to mercury-vapour and black-light traps. The adults will feed on dilute honey, sugar syrup or even water and are attracted to a mixture of beer and molasses. The sexes occur in equal numbers.

Mating depends on a light intensity below three foot-candles for at least seven hours, and so begins about this length of time after darkness, peaks between eight and nine hours and continues until about an hour later. About 40% mate on the first night after emergence and 80% by the third night (Lukefahr and Griffin, 1957). Kaae and Shorey (1973) showed that mating nearly always occurred on the upper surfaces of cotton leaves on the outer perimeter of the plant; on calm nights mating took place near the top of the plants, lower down on windy nights and in the shade on moonlit nights. Under cage conditions (Noble, 1969) males mated about four times and females twice, light trapped females had mated up to six times with a mean of 1.1 times. Oviposition begins on the second night, peaking on the third night, with 80% of eggs being laid before midnight (Lukefahr and Griffin, 1957).

The number of eggs laid ranges from 100 to 500, depending on the size and longevity of the female. Caged moths lived for 5–31 days, mean 13 days (Willcocks, 1916). In the United States moths derived from short-cycle larvae lived for about 14 days while in India adults from short-cycle larvae lived 7–9 days but those from long-cycle larvae lived up 20 days (males) and 56 days (females) (Pearson, 1958). El-Sayed and El-Rahman (1960) found that longevity decreased with rising temperatures and that high humidity prolonged adult life.

Pink bollworm moths disperse by both long-range movement (migration) or short-range movement although, interestingly Busck (1917) stated that the moths were 'rather too sluggish for sustained flight'. In long-range movements the moths may use thermals and have been collected up to 1000 m by aircraft over heavily infested fields in Mexico in 1928, and subsequent collections have been made at heights ranging from 30–650 m (Noble, 1969). Observations over six seasons of small plots of cotton planted in an arid area of western Texas, where there were no alternative host plants, showed that they frequently became infested late in the season with pink bollworm, the adults having flown from the nearest cotton 50–100 km away (Noble, 1969). Bariola *et al.* (1973) reviewed earlier work on long-range dispersal and reported work in California where trapped moths must have come from diapause larvae in cotton 50 km away. Stern and Sevacherian (1978) trapped moths in desert areas of California which corresponded with all four generations emerging from cotton fields at least 16 km away, and also showed that moths were continually flying between fields and from one cotton area to another. This short-range dispersal was also investigated by Van Steenwyk *et al.* (1978) catching rubidium marked moths in pheromone traps; males were highly mobile within cotton fields and they exhibited late season dispersion outside the fields. In Egypt, Metwally and Hosny (1974) took samples of buds or bolls at different intervals and in different cardinal directions from villages harbouring overwintering larvae and found that attacks developed on the nearest available cotton and that this generally occurred downwind.

Duration of Life Cycle and Modelling

The duration of the life cycle is dependent on temperature or, more accurately, the number of day-degrees. Willcocks (1916) gave the life cycle from egg to adult for short-cycle larvae as ranging from 27–38 days. Busck (1917) gave 35–50 days from egg to egg, Owen and Calhoun (1932) 21.5–59.5 days egg to adult and Taylor (1936) 29.5–38 days from egg to adult. El-Sayed and El-Rahman (1960) gave egg to adult as 75.4 days at 18°C, 56.1 days at 22°C, 42.2 days at 25°C, 29.9 days at 30.5°C and 24.8 days at 35.5°C. Stone and Gutierrez (1986a) gave the development times per larval instar at seven different constant temperatures and the total ageing process in both buds and bolls.

Because most of the eggs are laid on the bolls and most larvae gain entry into bolls without any possible exposure to contact insecticides, the adults are the only stage which can be treated chemically by conventional insecticides or pheromones. Thus the exact duration of the life cycle of the short-cycle insects is important, and even more important is to forecast when overwintering, long-cycle adults will emerge.

The forecasting of the pink bollworm spring emergence by thermal summation was first proposed by Sevacherian *et al.* (1977) and successfully used in the field in California by Toscano *et al.* (1979). The concept of degree-days, or accumulated heat units, is based on the fact that there are minimum and maximum temperatures, the thresholds, below and above which no development occurs in a given organism. The degree-days are the amount of heat units accumulated within these limits and the total for the completion of a generation is constant for a given species. Sevacherian and El-Zik (1983), using a minimum development threshold of 15.5°C, a spring emergence at about 675 degree-days and the emergence of each generation after each 967 degree-days, provided a table from which to calculate the actual degree-days based on daily maximum and minimum temperatures and a slide rule combining the degree-days for Acala and Deltapine cottons with those of pink bollworm and tobacco budworm.

Another group of workers, Stone and Gutierrez (1986a,b) and Stone *et al.* (1986), reviewed crop modelling and proposed three models for pink bollworm control in southwestern desert cotton in the United States; a crop model, a pest model and a management model using the concept of degree-days. Work on the crop model in California, Arizona, Brazil and the Sudan was reviewed and the model modified by Stone and Gutierrez (1986a). The pink bollworm (PBW) model was also reviewed and modified by Stone and Gutierrez (1986a); and the management model, Stone and Gutierrez (1986b), compared the control of pink bollworm by different strategies – pesticides applied against susceptible eggs and adults or pheromones applied in discrete point sources. Stone and Gutierrez (1986b) concluded that the use of pheromone early in the season, when pink bollworm populations were low, was efficient in suppressing mating, delaying insecticide application and not upsetting the natural enemy complex, but was not effective against high populations late in the season. An economic simulation was then conducted (Stone *et al.*, 1986) in which various pheromone and/or insecticide regimes were followed. Of these, the most economical were insecticides applied alone, especially with two early sprays, and pheromones plus insecticides at the 5% threshold. Stone *et al.* (1986) cautioned that these analyses were purely theoretical and more work was needed, particularly now that a single, season-long twist-tie formulation of pheromone was now available.

Host Plants

A large number of species of plants have been recorded as hosts of pink bollworm worldwide and these have been listed by Noble (1969); this list was based upon those published by Shiller *et al.* (1962) for the United States and northern Mexico and in other parts of the world. The majority of these host plants belong to the Malvales; some are single records which may refer to chance or even to incorrect identifications of other species of *Pectinophora* or other similar genera.

Pearson (1958) stated that the pink bollworm attacks all true cottons, cultivated, commensal and wild, and many species of *Abelmoschus*, particularly the cultivated species *A. esculentus* (lady's finger, okra, ochro, gumbo, bamia), *A. cannabinus* (kenaf) and *A. sabdariffa* (roselle) and also *Corchorus olitorius* (jute). Noble (1969) and Ingram (1981) found that okra is a favoured host in the US and the West Indies, and Ripper and George (1965) stated that okra is a major alternative host plant for pink bollworm and consequently its cultivation is prohibited during the summer months in the Gezira.

Pearson (1958) and Noble (1969) have listed the host plants, but there are additional records in Matthews *et al.* (1965), Sohi (1984) and Gutierrez *et al.* (1986).

Natural Enemies

Most of the parasites and predators of the pink bollworm have been catalogued by Thompson (1946) and Hertig (1975).

Parasitoids

Cheema *et al.* (1980a) compiled a worldwide list of the parasitoids of *P. gossypiella* but there are other records by Pearson (1958), Noble (1969), Legner (1976), Menon and Thangavelu (1979), Sekhon and Varma (1983), Gordh (1984) and Naumann and Sands (1984). Cheema *et al.* (1980a) also carried out a study of the parasitoids of pink bollworm in Pakistan and gave details of the biology and host range of the nine most important species.

The early introductions into Egypt were reviewed by Oatman (1978), these were *Bracon mellitor* Say, which failed to establish, *B. kirkpatricki* (Wlkn.), which was recovered for several seasons before dying out, and *B. lefroyi* (D. & G.), which is said to be established. Oatman (1978) also gives the numbers of several species imported and released in Texas, Mexico and Puerto Rico between 1933 and 1955; some were recovered the season of release but failed to overwinter. The overwintering of introduced parasitoids was examined by Fye and Jackson (1973) and Legner (1979), and although it was established that several species could successfully pass the winter in Arizona and California, none of eight species which reproduced during the year of release was recovered the next summer (Legner and Medved, 1979). Jackson (1980) reviewed early biological control of pink bollworm in the United States and the lines of future research.

In India *Apanteles angalati* Mues. carries over in another host *(Sathbrota simplex* Wlsm.) in stacked cotton sticks (Singh and Sidhu, 1982; Singh *et al.*, 1988), *Chelonus* sp. and *Camptothlipsis* sp. were associated with long-cycle pink bollworm larvae (Sekhom and Varma, 1983) and *C. blackburni* Cam. was reported as overwintering (Pawar *et al.*, 1983). In Egypt (Hekal, 1986), diapausing larvae of pink bollworm were parasitized by *Exeristes roborator* (F.), *Perisierola* sp., *Elasmus platyedrae* Ferr., *Habrocytus* sp. and *Bracon brevicornis* Wesm.

Mass-rearing and release of various parasitoids against pink bollworm has been reported from China, *Bracon nigrorufum* (Cushm.) (Gong *et al.*, 1984) and *Dibrachys cavus* (Wlk.) (Chu, 1978; Cock, 1985), and from India, *Trichogramma brasiliensis* (Ashm.), *Bracon kirkpatricki* and *Chelonus blackburni* (Prasad *et al.*, 1985). Habib and Mohyuddin (1981) discussed native pink bollworm parasitoids and possible introductions in Pakistan.

Predators

Various predators of the pink bollworm have been reported by Noble (1969), Orphanides *et al.* (1971), Irwin *et al.* (1974), Legner (1976), Abul-Nasr *et al.* (1978), Cheema *et al.* (1980b), Ingram (1981), Singh and Singh (1984) and Henneberry and Clayton (1985).

Two species of birds from Brazil and a bird and a lizard from St Vincent are listed by Cheema *et al.* (1980b) who also record the common mongoose and 12 birds feeding on pink bollworm in Pakistan.

The most vulnerable stages of the pink bollworm are the egg and the first instar larvae while they are boring into a bud or boll. Orphanides *et al.* (1971) collected all the possible predators of pink bollworm in Californian cotton fields, carried out laboratory trials to assess their potential against eggs, larval stages and pupae, and confirmed that 12 insects could feed on pink bollworm eggs and that all 12, plus nine spiders, could feed on pink bollworm larvae. Irwin *et al.* (1974) went on to assess the effectiveness of seven predatory species against pink bollworm eggs in field cages in California and *Chrysopa carnea* Stephens, *Geocorus pallens* Stäl. and *Nabis americoferus* Carayon were seen to be the most effective. Further laboratory studies in the US showed that *Collops marginellus* LeConte, *Hippodamia convergens* (Guer.), *Sinea confusa* Caudell and *Orius tristicolor* (White) were also effective egg predators (Henneberry and Clayton, 1985). Two species of *Chrysopa* from Pakistan were introduced into California but failed to establish and *C. spiniscutis*, also from Pakistan, was not released in Arizona (Jackson, 1980).

Cheema *et al.* (1980b) carried out an investigation of the predators of pink bollworm in Pakistan combining field observations with predation of eggs laid in sleeved cages, artificial flower infestation of larvae, ground trash infestation of pupae and diapausing larvae in buried bolls in Pakistan.

Pearson (1958) only recorded a single predator, the mite *Pyemotes ventricosus* (Newport), which he stated attacks pink bollworm in seed stores and the field, and that it also can attack man, causing skin irritation of ginnery workers. Both this species and *P. herfsi* (Oudemans) are widely reported as preying on, and giving good control of, long-cycle larvae in India (Singh and Singh, 1984; Naresh and Balan, 1985), Barbados (Ingram, 1981), Egypt (Abul-Nasr *et al.*, 1978). These mites have also been reported as ruining laboratory experiments on diapause (Ingram, 1980). Noble (1969) reported a predatory mite feeding on pink bollworm eggs and also several other species of predatory mites feeding on the larvae in stored cotton seed in the United States. Noble (1969) reported evidence of spiders preying on moths tagged with P^{+32}.

Entomopathogens

Early work in the US on the use of bacteria for the control of pink bollworm concentrated on *Bacillus thuringiensis* Berliner, and it was shown that control with this and another isolate, *B. thuringiensis* var. *kustaki*, gave rather variable results in field-cage tests (Bell and Henneberry, 1980). In Egypt Abul-Nasr *et al.* (1978) isolated *B. thuringiensis* var *finitimus* and *B. cereus* from among resting pink bollworm larvae and Metwally *et al.* (1986) found that bacteria killed from 53–81% of larvae in the field. Cheema and Muzaffar (1979) isolated *B. cereus* and *Streptococcus* sp. from pink bollworm larvae in Pakistan.

A cytoplasmic polyhedrosis virus was isolated from pink bollworms in a laboratory culture but very high doses were required to obtain an LD_{50}; and a nuclear polyhedrosis virus, isolated from *Autographica californica* (Speyer) but which infected pink bollworm, failed to reduce pink bollworm populations in a field trial in Arizona (Bell and Henneberry, 1980). The addition of feeding stimulants to this virus did reduce the numbers of pink bollworm larvae in bolls early in the season, however the treatment was too costly (Bell and Henneberry, 1980). Metwally *et al.* (1986) reported 19.4% mortality of pink bollworm larvae due to viral infections in Egyptian cotton fields.

Control

A pink bollworm control programme involves both pest management and crop or environment management. The former entails the use of natural enemies, sex pheromones, sterile moth releases or insecticides, and the latter utilizes cultural control and plant resistance.

Discovery and Characterization of the Sex Pheromone

Ouye and Butt (1962) were the first to use a methylene chloride extract of homogenized mating pairs to trap virgin male pink bollworm moths released in cages. Characteristic responses from male moths were then obtained with an extract from the terminal 2–3 abdominal segments of virgin females (Berger *et al.*, 1964). Jones *et al.*(1966) isolated and synthesized 10-propyl-*trans*-5,9-tridecadienyl acetate and named this 'propylure'. Extensive field testing of propylure showed it to be less attractive than crude extracts of virgin female but the addition of the activator 'deet' increased catches of males (Jones and Jacobson, 1968).

Empirical screening of acetate esters of long-chain, unsaturated, aliphatic alcohols by Green *et al.* (1969) led to the discovery that *cis*-7-hexadecen-1-ol-acetate was attractive to male moths and named this compound 'hexalure'. This was extensively tested by Keller *et al.* (1969). Subsequently Hummel *et al.* (1973) showed that the natural sex pheromone of pink bollworm was a mixture of (*Z,Z*)- and (*Z,E*)-7-hexadacadien-1-ol-acetate and named this compound 'gossyplure'. A 1:1 mixture of (*Z,Z*)- and (*Z,E*)-isomers trapped more male moths than either hexalure or 10 virgin females. The chemical structure and the 1:1 ratio of the isomers was confirmed by Bierl *et al.* (1974) in Arizona, although Marks (1976) in Malawi found that the optimum ratio lay between 55 and 60% of the (*Z,Z*)-isomer and 45 and 40% of the (*Z,E*)-isomer. Reed *et al.* (1975), Marks (1976) and Ingram (1980) confirmed that gossyplure was superior to virgin females and hexalure.

Further work on the ratios of the two isomers of gossyplure was carried out by Flint *et al.* (1978), who found that wild males in Arizona showed a statistically significant preference for a bait containing 60% (*Z,Z*)-isomer and 40% (*Z,E*)-isomer early and late in the season but preferred a 50:50 isomer mixture in midseason; irradiated moths showed the same preferences. Flint *et al.* (1979) compared seven ratios of (*Z,Z*)- and (*Z,E*)-isomers of gossyplure in Argentina, Bolivia, Brazil, India, Pakistan, Australia, Egypt and the USA. Traps baited with 50% or 60% of the (*Z,Z*)-isomer showed no significant differences in numbers caught in India, Australia, Pakistan, Argentina and USA; traps with 50% baits were significantly higher in Brazil and 60% baits in Egypt and so it was concluded that baits with either 50% or 60% (*Z,Z*)-isomers would be the best for survey or control work throughout the distribution area of the pink bollworm. Flint and Merkle (1984a) found that high rates of (*Z,Z*)-isomer were as effective as a standard commercial gossyplure flake formulation in suppressing mating in mini-mating stations and in reducing trap catches. This work was continued (Flint and Merkle, 1984b; Flint *et al.*, 1988) and in fields treated with the (*Z,Z*)-isomer 69% of males were caught in traps baited with 9:1 ratio of (*Z,Z*)-:(*Z,E*)-isomers and only 3% in traps baited with gossyplure; representing an induced shift in male response.

The effective life of gossyplure is affected by the substrate. Hummel *et al.* (1973) evaporated the chemicals from stainless steel planchets and Teflon tubes with cross-sectional areas 1/10 or 1/100 of the steel planchets. Bierl *et al.* (1974) used cotton dental roll wicks but found that it lost its activity in 10 days. However, Neumark *et al.* (1972), compared catches of hexalure and gossyplure with and without the antioxidant (*N*-octyl-*N'*-phenyl-*p*-phenylenediamene) on filter paper, and found that the antioxidant increased catches 2.5-fold and remained attractive for at least two months. Marks (1976) used Teflon tubes, 2.5 ml polyethylene vials and polyethylene beem-vials and found that the vials caught moths for at least 44 days, but that the addition of an antioxidant (2,4-di-*tert*-butyl-*p*-cresol) did not extend the life.

Flint *et al.* (1976) used No.1F red rubber tube stoppers impregnated with 1000 μg of gossyplure and Huber *et al.* (1979) used dispensers made from red, gum rubber tubing (ID 6.35 mm), split lengthwise and cut into 12.7 mm lengths and impregnated with 1000 μg gossyplure. Ingram (1980) used the Pherocon PBW rubber caps supplied, already impregnated with gossyplure, by Zoecon Corp. and found that they caught moths for up to 56 days but their trapping efficiency fell below 50% after about 28 days. In India, Dhawan and Sidhu (1987) tested five dispensers – narrow gauge Teflon capilliary tubes, filter paper, gelatine capsules, polyethylene capsules and cotton – and found that the Teflon tubes were still effective after five weeks.

In Australia the pink spotted bollworm was also attracted to gossyplure, a 9:1 mixture of the (*Z,Z*)- and *Z,E*)-isomers being the most attractive to males in Delta traps (Flint and Stone, 1985).

Design of pheromone traps

The design of a pheromone trap can dramatically affect catching efficiency. This can be critical in studies at low population densities or in attempts at mass trapping and thus design has been investigated by many workers. Trapping of pink bollworm male moths began in the US with the work of Ouye and Butt (1962) who suspended their sex lure in modified 225 g paper cups lined with a sticky substance – Tanglefoot – on the inside. Many researchers found that these sticky surfaces – Tanglefoot or Stikem – rapidly became coated with moth scales, required frequent stirring and also became saturated after catching about

50 moths. This led to many workers developing their own pink bollworm traps and experimenting with trapping media.

Guerra and Ouye (1967) used a modified Frick trap consisting of a 1 litre ice cream carton coated with Stikem Special. Sharma *et al.* (1971) used a cylindrical cardboard carton 115 mm long by 105 mm diameter with a 20 mm hole at each end, also coated on the inside with Stikem. This was superseded by an omnidirectional octagonal trap with Stikem inside (Sharma *et al.*, 1973a) and then an omnidirectional cone trap (Sharma *et al.*, 1973b). Bierl *et al.* (1974) modified the Sharma trap, making it from a 14-litre ice cream container to form a closed cylinder 60 mm high and 200 mm diameter with six equally spaced 22 mm holes round the vertical face; it also used Stikem as the trapping medium.

In Israel Neumark *et al.* (1972) used polystyrene cups partly filled with water and a drop of detergent, modifying this with a constant-level liquid device (Neumark and Teich, 1973) and finally (Teich *et al.*, 1977) recommended roofing the cup and using cotton seed oil as trapping medium. Reed *et al.* (1975), in southern India, made a Sharma-type trap from cine-film cans measuring 275 mm diameter and 45 mm high with nine equally spaced 20 mm square holes in the sides; liquid paraffin BP was used as the trapping medium. Marks (1976), in Malawi, compared Sectar (3M) fold-up paper adhesive traps, an omnidirectional Sharma-type trap, a large (600×600×50 mm) water tray trap and a simple collapsible flat plywood disk adhesive trap 600 mm diameter, 2829 cm^2, with a small roof and settled on this last trap as being effective and cheap.

Flint *et al.* (1976), in Arizona, used the Sharma trap and the Huber trap, which consisted of a 300 ml styrofoam cereal bowl with four 22 mm diameter holes cut in the sides; these were used inverted with Stikem on the inside of the lid. Coudriet and Henneberry (1976), also experimented with the Sharma cone trap, the Sharma omnidirectional trap, the Huber trap, the saucer-type trap (Hendricks *et al.*, 1973) and the Pherocon IC trap baited with gossyplure and other cotton insect pest pheromones, looplure and virelure, alone and in all combinations. They found no significant differences between trap catches of pink bollworm and that virelure, but not looplure, could be used in combination with gossyplure without affecting catch of pink bollworm.

Foster *et al.* (1977) compared a modified Frick trap, a Delta trap, a modified Sharma omnidirectional trap and the Huber trap and found that the Delta trap was significantly better than the other three. Further experiments showed that increasing the surface area, covered with Stikem, with the addition of internal baffles improved the efficacy of Delta and Frick traps. Lingren *et al* (1980), using night vision goggles and a low-light level video camera, observed the behaviour of male moths at pheromone traps and designed a funnel live trap (FLT) which proved to be 3–15 times more efficient than other standard designs. This FLT was also the only trap which could detect daily population changes.

Sharma *et al.* (1973b) found that sticky traps were only useful the first night before becoming contaminated with moth scales, insects, dust and debris. Flint *et al.* (1976) found that moths frequently entered and exited their traps because the Stikem was fouled with scales. Huber and Hoffmann (1979) developed the Huber oil trap (Huber *et al.*, 1979), a 240 ml plastic coated paper cup with four 25 mm diameter holes cut in the sides just below the lid and filled with

130–135 ml cotton picker drip oil (Union Oil Co. CP–150) as trapping medium, and compared it with the modified Sharma trap, the Neumark and Teich trap and the Delta trap. This oil trap proved to be vastly superior to the other traps for mass trapping, particularly when moth density was very high (>50/night). Beasley and Henneberry (1988) also found that reservoir traps caught significantly more moths than did sticky Delta traps.

Ingram (1980) found that the Pherocon 1CP sticky trap could only retain about 50 moths a night and developed an omnidirectional water plus detergent trap based on a plastic ice cream tub, lidded and with eight 35 mm holes cut in the sides. Ingram (1980) also found that the catches of the water trap was about ten times that of the sticky trap and that paraffin was superior to water + detergent as the trapping medium, but needed changing after four days. Dhawan and Sidhu (1984a) compared the modified cone trap, the modified Frick trap, the polystyrene cup trap, the modified Sharma trap, the Delta trap and the Pherocon 1CP trap with a locally made trap, the PAU trap, consisting of a 2-kg tin with three 25×50 mm holes cut in the sides, using paraffin or kerosinized water as trapping medium, and found the PAU trap was equal to or better than the others. Megahed *et al.* (1984) found that a water trap was significantly better than a plastic cone trap, the Pherocon 1CP or the Delta trap.

Trap colour

Some investigations have been made of the effect of trap colour on pheromone trap catches of pink bollworm males. Reed *et al.* (1975) compared their cine film cans left silver or painted matt black, and showed no consistent differences. Marks (1976) compared yellow, black, green, blue, red, white and unpainted plywood with no significant differences between catches, but Megahed *et al.* (1984) found that white Pherocon 1CP traps caught more moths than did yellow, green or red traps.

Trap height and placement

Sharma *et al.* (1971), in cotton 1.2 m tall, found that catches of pink bollworm moths in traps at canopy height or 1.8 m were significantly greater than catches in traps at 0, 0.6 or 2.7 m above soil level. Ingram (1980) also found that trap catches rose with height above ground level within the canopy and recommended that trapping should be carried out at just below canopy height. Flint and Merkle (1983) found that traps should be placed within 0.3 m of the tops of cotton beds early in the season and above the canopy later. However, Reed *et al.* (1975) found that the largest catches were at mid-canopy height (0.4–0.8 m), falling off at ground level and above canopy, and Marks (1976) also found peak catches at mid-canopy level (0.5 m).

Trap catches fell off rapidly with distance from cotton fields, but catches were higher in another crop (sorghum) than in bare fallow (Sharma *et al.*, 1971). They also found that trap-catch efficiency fell off with height above ground on bare fallow and hypothesized that pink bollworm males fly close to horizontal surfaces – the cotton canopy or bare ground. Sharma *et al.* (1971) found that traps placed on the edges of cotton fields consistently caught fewer moths than traps in the centre of fields and Ingram (1980) showed that trap catches varied across a field, with catches at the windward end being consistently lower than traps in the centre or the leeward edge of the field. Beasley and Henneberry (1988) also found higher trap catches within cotton field than on the edges.

Mass trapping and communication disruption with sex pheromones

The first attempt to control pink bollworm populations by mass trapping in a cotton field was by Graham *et al.* (1966), who used traps baited with a crude extract from female moths; failure was attributed to the field not being isolated from other infested fields. However, Guerra *et al.* (1969) were successful in suppressing populations by trapping in field cages. Flint *et al.* (1974) released labelled overwintering larvae into a 15 ha cotton field containing 42 gossyplure baited traps and recaptured 43% of the moths. Later studies (Flint *et al.*, 1976) using 12 to 50 traps/ha in a farm containing 18.1 ha alfalfa, 26.1 ha small grains and 35.8 ha cotton trapped *c.* 30,000 males and reduced the acreage requiring treatment at the 10% boll infestation level to 9% compared with 45–100% for the preceding three seasons. Huber *et al.* (1979), using 11 gossyplure-baited traps/ha, in an area-wide, three-year trapping programme in Arizona, reduced pink bollworm larval infestations to 1.9% compared with 4.18% mean infestation for the previous three seasons.

Disruption of communication in pink bollworm was first attempted by McLaughlin *et al.* (1972) using hexalure released from planchet evaporators placed in cotton fields in California. When >10 mg of hexalure per ha was being released per night there was a greater than 90% disruption in pheromone communication, as measured by the numbers of males trapped in Sharma traps baited with virgin females. McLaughlin *et al.* (1972) also tried to prevent females from mating with foliar applications of hexalure to a 7.3 ha cotton field, but were only able to show a reduction in males trapped and in the numbers of females who mated more than once when compared with a 6.5 ha untreated field. Shorey *et al.* (1974) were able to show a reduction in larval boll infestation, comparable to that provided by insecticide applications, by placing 30,000 substrates inoculated with 10 μl of hexalure per week in a 4.8 ha cotton field.

Shorey *et al.* (1976) also used liquid gossyplure in open reservoirs ranging from Teflon capillaries to nylon cloths. Shorey *et al.* (1976) reviewed the work in California using both hexalure and gossyplure and, particularly, that on evaporator release rates, vertical location of evaporators, spacing of evaporators, effectiveness of hexalure versus gossyplure. Gaston *et al.* (1977) then confirmed that gossyplure released into the atmosphere of three cotton fields over the entire growing season led to a reduction in larvae-infested bolls comparable with that in ten insecticide-sprayed cotton fields.

Brooks *et al.* (1979) reported the first registration in the US of a hollow fibre controlled release formulation. In 1977 it had been used on 9300 ha of cotton under an experimental label and use data indicated that pheromone application reduced the need for conventional insecticides by 50–80% with less risk of inducing attacks by late-season secondary pests. In 1978 in an experiment comparing disruption with the hollow fibre formulation, mass trapping and an untreated check, pheromone trap catches and mating table observations confirmed disruption, as did lower larval boll infestation levels between the pheromone treated and untreated checks. Experiments using this hollow fibre formulation in Arizona and Brazil (Brooks *et al.*, 1980) showed that control by mating disruption was an economic and viable alternative to conventional insecticides, and it gave control of pink bollworm throughout the second generation in California (Doane *et al.*, 1983) reducing boll infestations when compared with insecticide treated cotton.

Kydonieus and Beroza (1981), Quisumbling and Kydonieus (1982) and Henneberry *et al.* (1981) also reported work using a plastic laminate flake formulation of gossyplure for the control of pink bollworm by mating disruption. Butler *et al.* (1983) made comparisons of hollow fibre and multilayer plastic laminate formulations of gossyplure in field trials in California and Arizona. Stone and Gutierrez (1986a,b) and Stone *et al.* (1986) have developed simulation models for cotton, pink bollworm, pink bollworm populations as affected by pheromone applications and the economics of different control strategies for pink bollworm.

Work on pink bollworm control using gossyplure was also being carried out elsewhere. Boness *et al.* (1977) reported that the percent boll infestations were reduced by using traps baited with gossyplure in 1975 and 1976 on cotton in Spain. Anon. (1981) described trials in the Indian Punjab where aerial applications of gossyplure in hollow fibres significantly reduced boll infestation and gave a 33.7% increase in yield over untreated cotton. Campion and Nesbitt (1982) found that a pheromone treatment against pink bollworm in Greece and Egypt was comparable with insecticidal control.

Much work has been carried out in Egypt on control of pink bollworm by mating disruption. Johnstone (1982a,b), Critchley *et al.* (1983) and Jutsum *et al.* (1983) reported trials using fixed wing aircraft to apply a microencapsulated gossyplure formulation (Hall *et al.*, 1982) compared with conventional insecticide sprays, and found no differences between percent rosetted blooms, percent boll infestation, gross yield and lint quality; large numbers of beneficials were also recorded in the pheromone treatments. Von Ramm and Krone (1983) described trials involving the hand-application of a laminated flake formulation. McVeigh *et al.* (1983) and Critchley *et al.* (1985) went on to compare hollow fibre, laminated flake and microencapsulated gossyplure formulations in Egypt and found little difference between the pheromone formulations; all three suppressed trap catches, had higher yields and more beneficials than insecticide sprayed cotton. Critchley *et al.* (1984) showed that early season microencapsulated pheromone and late insecticide was an effective combination, causing least harm to beneficials which were destroyed by early insecticide applications.

The microencapsulated gossyplure formulation and its (*Z,Z*)-isomer were also used in Arizona (Flint and Merkle, 1984c), initially applying drops by hand, and then using a hand-held spinning disc applicator; the sprays gave significantly lower trap catches.

The development of yet another slow release gossyplure formulation was reported by Flint *et al.* (1985), this was a polyethylene tube dispenser, which has been called the 'rope' or 'twist-tie' formulation. Hand tied on to cotton in Arizona, a single application at a high rate and three applications at lower rates were compared with untreated and insecticide treated plots. Trap suppression was between 94 and 98% and mating did not occur in any pheromone treated plot. Staten *et al.* (1987) compared this rope dispenser, with insecticide as needed; hollow fibre dispensers, with insecticide as needed; and insecticide as needed in California; and the rope and insecticide comparisons only in Mexico. At both sites there were significant reductions in insecticide usage of 37–42% in the rope treated fields compared with the others, and this reduction in insecticide was considered particularly important as it reduces the likelihood of outbreaks of secondary pests such as whitefly or leaf perforator. Critchley *et al.* (1987) used a single application of the twist-tie dispensers, loaded with both gossyplure and the sex pheromone of spotted and spiny bollworms, to suppress all three pests for the entire season in Pakistan.

Work with pheromone trapping and mating suppression of pink bollworm in China has been described by Chen *et al.* (1984a,b) and Shu *et al.* (1987). Surulivelu (1985) reported trials with hand application of hollow fibre gossyplure dispensers in India, and showed that the pheromone treatment reduced larval infestation, increased seed cotton yields and lowered the number of insecticide applications. Qureshi *et al.* (1985) found that the control of pink bollworm obtained by mating disruption was as good as with fenvalerate. Campion and Murlis (1985) reviewed the work with gossyplure on cotton in Pakistan, Peru and Egypt, and Hosny (1988) reviewed the role of pheromones in the management of pink bollworm in Egypt. Also in Egypt, El-Adl *et al.* (1988) reported large-scale aerial applications of microencapsulated formulation using conventional boom and nozzle, hollow fibre formulation using a specially designed applicator (Funkhouser and Las, 1981) and the laminated flake with another specially designed applicator (Quisumbing and Kydonieus, 1982). Vaissayre (1987) attempted the eradication of pink bollworm in the Ivory Coast with massive and continuous applications of a hollow fibre formulation and reduced boll infestation to 10% compared to 20–70% in untreated areas.

Lure and kill

Another aspect in the use of pheromones for the control of pink bollworm was the development of an 'attracticide' (Lingren, 1982), consisting of gossyplure, an adjuvant and an insecticide, which gave kills of 90% of the males coming into contact with it in small-scale field trials and kept pink bollworm populations below economic levels for three weeks in larger-scale tests. Butler and Las (1983) reported that permethrin added to the sticker used to secure the hollow fibre gossyplure formulation to the foliage was not likely to reduce the population of beneficials in Californian cotton fields. However, Beasley and Henneberry (1984) found that the addition of parathion to the adhesive of the laminated flake caused high mortality of predators. Evans (1984) described the development of a water-based controlled-release pheromone which could be applied together with insecticides, allowing reduced insecticide usage. Qureshi and Ahmed (1987) compared gossyplure, adjuvant and cypermethrin with fenvalerate and found a reduction of 80% in trapped males but no difference in larval infestations in Pakistan.

Haynes *et al.* (1984) studied the potential for the evolution of resistance to gossyplure in the field in California and concluded that although the development of resistance was real, there were current agricultural practices and conditions which would hinder its development.

Critchley *et al.* (1989) reviewed the use of pheromones to control pink bollworm in Egypt, Pakistan and Peru, and Henneberry (1989) in the southwestern United States.

Control Using Sterile Moth Releases

Sterile moth releases have been tried as a pink bollworm management tool for many years and the subject has been reviewed by Bartlett (1978), Henneberry (1980) and Henneberry and Keaveny (1985). Success of this method depends on swamping native populations with ninefold numbers of sterile insects and requires a sophisticated mass-rearing programme. This was developed in the US where the Animal and Plant Health Inspection Service (APHIS) facility at Phoenix reared in excess of a million moths a day (Henneberry, 1980).

Cobalt^{60} gamma radiation of one-, two-, three- and four-day old eggs with doses varying from 1 to 32 krad gave variable results (Bartlett *et al.*, 1973); adult emergence from eggs treated with 2 krad was normal but sterility was not complete. Dosages over 2 krad increased larval development time and the resulting adults had poor competitive ability when compared with untreated adults.

Treatment of last instar larvae with doses of 2 to 32 krad of cobalt^{60} gamma radiation gave varying degrees of chromosomal damage in the larvae and did not have a significant effect on pupation (Bartlett and Lewis, 1973). The numbers of morphologically normal adults was greatly reduced when doses exceeded 4 krad for males and 2 krad for females, but those that did emerge were crossed with all possible combinations of treated and untreated mates and showed reductions in the numbers of progeny. Bartlett (1978) also reported studies of gamma radiation of diapause larvae and obtained normal emergence with doses of 3 and 5 krad.

Ouye *et al.* (1964) showed that the older the pupa at the time of irradiation the less pupal mortality and adult malformation. Irradiation of seven-day-old pupae, i.e. within 24 hours of eclosion, yielded the best results measured by using egg hatch as the criterion for sterility but longevity of the males was reduced. Bartlett (1978) concluded that although the pupal stage was very easy to handle, obtaining sufficient pupae of the right physiological age all at the same time for treatment made it difficult.

Bartlett (1978) reviewed the early work on irradiating pink bollworm adults. Initial studies showed that the sterilizing dose was 40 krad of gamma radiation for males and 30 krad for females. Subsequent experiments with one-day-old adult males treated with varying doses and then mated with untreated females reduced egg-laying by 13% after radiation at 5 krad and 77% at 40 krad. Females were much more sensitive, with egg production and egg hatching severely reduced at

20 krad and virtually eliminated at 30 krad. Flint *et al.* (1973) found no difference in the attractiveness of mass-reared female moths, as compared with native female moths, and also that females irradiated with doses of 10 and 25 krad were as attractive to males as were untreated females.

Experiments evaluating releases of sterile pink bollworm moths were reviewed by Bartlett (1978) and Henneberry (1980) and Bartlett and Butler (1979) used a computer model (WATBUG) to simulate the effects of radiation sterility. Field cage experiments gave rather variable results, generally in direct relationship with the ratio of sterile-to-native releases. Field and large-area experiments were affected by emigration of sterile moths and immigration of native moths which prevented overflooding-ratios being obtained or masked any effects. Irradiated moths were released in the San Joaquin Valley to try to prevent pink bollworm from becoming established there culminating with the release of 400 million moths exposed to 20 krad in 1977. It was concluded that there was little or no reproduction due to overwintering moths but reproduction of immigrants in second, third and fourth generations reduced the overall ratios of sterile-to-native moths. However, the lack of large localized native populations was seen as an indication of the success of the sterility method (Henneberry, 1980).

The factors affecting the efficiency of irradiated moth releases were also reviewed by Henneberry (1980); these included the effects of diet, rearing and handling methods, domestication, irradiation dosages, sperm transfer and numbers of matings.

Henneberry and Keaveny (1985) reported the results of releases of high numbers of sterile moths into cotton plots and their interactions with native moths in St Croix, US Virgin Islands, between late 1980 and March 1982. Adult moths, mass-reared on an artificial diet containing Calco Red oil dye, were dosed with 20 krad of cobalt^{60} gamma radiation at Phoenix and then air-shipped to St Croix where they were released in cotton fields on the afternoon of the day received. Mortality of irradiated moths was 7±5% after 20 to 24 hours in transit, 42±42% after 40 to 56 hours and 100% after more than 56 hours; longevity was not seriously affected nor the percentage of mated females. Between 29/12/80 and 15/04/82 some 22,261,100 dye-marked, sterilized moths were released. The abundance and dispersal of the released moths, the numbers of native male moths, before and after release of the sterilized males, the mating status of released and native females, the ratios of all four possible mating combinations, attractiveness of mass-reared females to both released and native males and the effect of larval infestations after releases are all described.

Crossing experiments with various laboratory strains of irradiated and unirradiated pink bollworms and various different wild strains, including the St Croix strain, showed different effects between males and females in mating frequencies, longevity and sterility of F_1 progeny (Flint and Merkle, 1980; Henneberry *et al.*, 1980b; and Henneberry and Clayton, 1981).

Control by Insecticides

Pearson (1958) reviewed early attempts at the chemical control of the pink bollworm in Egypt and Mexico using arsenicals and fluosilicates but the control obtained was inadequate and never adopted. With the advent of the organic insecticides, chemical control became a practical proposition and DDT was shown to give effective control in Mexico in 1946 (Chapman *et al.*, 1950).

Pearson (1958) described the early work with DDT dusts in Mexico, where 3–11 applications were needed, and Egypt, where two applications sufficed for the control of pink bollworm. Noble (1969) stated that in the United States applications of DDT, azinphos-methyl or carbaryl, beginning when flower counts showed 350 larvae per acre and continued at five-day intervals until the majority of bolls were open, gave good control of pink bollworm. However, resistance to DDT was acquired by pink bollworm in Mexico in 1958–1960 and in Texas by 1962 (Lowry and Berger, 1965) and control with DDT alone was no longer possible, although azinphos-methyl, carbaryl or mixtures of DDT and azinphos-methyl or DDT and HCH were still effective (Noble, 1969). Little insecticide was required for cotton production in the desert areas of Arizona and California but, after the invasion of pink bollworm, as many as 10–15 applications of insecticide were required (Reynolds, 1980).

Currently in the United States insecticidal control is obtained with azinphos-methyl, monocrotophos, carbaryl or synthetic pyrethroids, and Bariola and Lingren (1984) compared the toxicity of 23 insecticides in topical applications against adult pink bollworms. Haynes *et al.* (1986, 1987) have developed a method of pyrethroid resistance monitoring using insecticide-laced, pheromone-baited sticky traps. Hutchison *et al.* (1988) have presented data of the lethal and sublethal effects of monocrotophos, fenvalerate and permethrin on pink bollworm oviposition in the field.

In India, Sidhu and Dhawan (1978) recommended spray schedules of four to five applications of carbaryl, DDT, endosulfan or fenitrothion at 10–14 day intervals, and commencing within two weeks of first flower, for pink bollworm control. Sidhu *et al.* (1979) and Sidhu and Dhawan (1981) tested many organic insecticides and found that only monocrotophos, quinalphos, phosolone and azinphos-methyl were as effective as carbaryl for the control of pink bollworm; triazophos (Saini, 1985) and isofenphos (Sukhija *et al.*, 1985) were also effective. Also in India, Jayaswal and Saini (1981b) found permethrin, cypermethrin, deltamethrin and fenvalerate to be as effective as carbaryl while Aginhothrudu and Gour (1982) and Kurtadikar and Hedgire (1982) found fenvalerate to be the most effective of a wide range of insecticides. Field studies in India with the newer synthetic pyrethroids (Bhamburkar and Kathane, 1984; and Gupta and Katiyar, 1985, 1987) have shown that cyfluthrin, flucythrinate and fenpropathrin are also effective in controlling pink bollworm.

In Egypt many workers have carried out field trials with carbamates, organophosphates and synthetic pyrethroids, used alone or in mixtures, to control pink bollworm. Saleh and El-Din (1982) tried cyfluthrinate, cypermethrin and mephosfolan; Khalil *et al.* (1983) used diflubenzuron in combination with various insecticides; and Watson and Guirguis (1983) tested various granular formulations against pink bollworm. El-Guindy *et al.* (1982) and Kassem *et al.* (1986) compared different insecticides alone or mixed, and also investigated both their synergistic and antagonistic effects against pink bollworm larvae. Watson *et al.* (1988) found that successive sprays of chlorpyriphos, fenpropathrin and then cyanophos gave a 90% reduction in the pink bollworm larval population.

Studies elsewhere have also confirmed the superiority of the synthetic pyrethroids for pink bollworm control. In Sri Lanka, Keerthisinghe (1982) showed that fenvalerate and permethrin were superior to the standard monocrotophos. In Cote d'Ivoire, Angelini *et al.* (1982) showed that deltamethrin, cypermethrin, fenvalerate, cyfluthrin, flucythrinate and fenpirithin were all effective, while Vaissayre (1983) recommended a mixture of cypermethrin and triazophos for pink bollworm control. In China, Wan and Wan (1982) recommended the use of deltamethrin and Luo *et al.* (1986) advocated mixtures of *B. thuringiensis* and fenvalerate for effective pink bollworm control.

The monitoring of a pest population and the establishment of thresholds is a prerequisite of any pest management programme before specific control measures, such as the application of insecicide or pheromone. With pink bollworm, monitoring is essential to ensure the correct timing as the sprays are directed primarily at the adults; eggs and larvae being inaccessible to conventional spraying.

The monitoring of pink bollworm populations in cotton in the United States has been reviewed by Noble (1969) and Toscano and Sevacherian (1980). Early survey methods were based on counts of rosetted flowers at the beginning of the season until 15 days after first flower and then on the percentage of 14- to 21-day-old bolls infested by pink bollworm larvae. Samples of 25 such bolls were taken at random every five paces, starting at least 25 paces from the field edge, from each quadrant of a 10 to 20 hectare field. The 100 bolls were then cracked and carefully examined for the presence of I to IV instar larvae and the percentage infested recorded. Control measures were not warranted unless at least 20% of firm green bolls were attacked.

Insecticide applications based on counts of male moths caught in omnidirectional pheromone traps (Sharma *et al.*, 1973b) baited with gossyplure have reduced treatments by one third to one half in southern California (Toscano *et al.*, 1974). Early to mid-season thresholds for insecticide application are 12 to 15 moths per trap per night, later in the season 3.5 to 4 moths per night is the threshold. Insecticides should be applied within 24 hours and preferably at night. Later work by Henneberry and Clayton (1982) showed that catches of male moths for three to seven days between boll sampling periods were strongly correlated with oviposition on cotton bolls, percentage of infested bolls and numbers of larvae per boll. However, after insecticidal treatments catches of moths remained above the threshold, probably due to immigration and fresh emergence, and so scheduling treatments on the basis of moth catches was only practical for the initial treatment. Later work by Beasley *et al.* (1985) listed recommended trap operation techniques and showed a positive correlation between the numbers of male moths caught three to four days before first bud and the initial flower infestation; the numbers of larvae in bolls were also positively correlated to later flower and boll infestation.

The use of sex pheromone traps to monitor moth populations and to provide spray thresholds has also been recommended in various other parts of the world. In Israel, Melamed and Shoham (1975) recommended spraying when trap catches reached eight moths a night but Teich *et al.* (1977) lowered the threshold to five per night. In Pakistan, Ahmad (1979) monitored year-round pink bollworm occurrence with pheromone traps with a view to ascertaining the optimum time for cotton planting and planning control programmes. In Barbados, Ingram (1980) found a highly significant correlation between moth catches and damaged bolls some ten days later; eight to nine moths per night represented a 10% level of boll damage and spraying was recommended when catches exceeded eight moths per night. In India, Taneja and Jayaswal (1981), working in Haryana, established that sprays should be applied within 24–48 hours after a catch of eight moths per night, but Dhawan and Sidhu (1984b,c), in Punjab, found that spraying at four moths per night gave the best results. Page *et al.* (1984), using gossyplure-baited traps in Queensland, found that mean trap catches of 10–20 moths per night were related to economically damaging attacks of the pink spotted bollworm.

Finally, a word of caution: repeated applications of insecticides for pink bollworm control have led to outbreaks of cotton leaf perforator, tobacco budworm, bollworm and spider mites.

Control by Cultural Methods

Cultural control plays a key role in keeping down the numbers of pink bollworm carrying over between cotton crops. The simplest form of cultural control, which applies mainly to equatorial areas where there are few long-term larvae undergoing diapause, is to ensure that there are no hosts which will allow the continuous breeding of short-term larvae. As cultivated cotton is the pink bollworm's primary host, a close season, where no cotton whatsoever is allowed to be grown and other hosts severely restricted, is almost universally enforced wherever cotton is cultivated.

Growing of stub cotton, ratoon or standover cotton was a common practice in parts of the United States (Noble, 1969), but after the arrival of pink bollworm this practice had to be abandoned as these plants flower much earlier than the newly sown cotton and so allow an extra generation to develop. Bishara (1930) reported on the effects of ratoon cotton and pink bollworm attack in Egypt. In Barbados (Tucker, 1939) the Sea Island cotton used to occupy the land for 11 months or more, but after the advent of pink bollworm the attack was so severe that a cotton close season of over a year was instigated. Subsequently, planting was restricted to a three week period from the beginning of September, uprooting had to be completed by the end of April and a close season proclaimed 1 May to 31 August each year.

Volunteer cotton growing in old fields from seed cotton that has germinated should always be uprooted and the old plants should always be cut below ground level or hand-pulled to prevent sprouting. Taylor (1936) described severe attacks of pink bollworm in Uganda when cotton was not uprooted, and Matthews *et al.*

(1965) also attributed outbreaks in Zimbabwe in 1959 and Malawi in 1962 to ineffective close seasons. The effect of any possible shortening of the close season in Barbados was illustrated diagrammatically by Ingram (1980). Outbreaks have also been reported in the Sudan in areas where ginning seasons have been extended further into the cotton close season (El-Tigani and Khagali, 1978).

Alternative hosts, either wild or cultivated, must also be destroyed. The reasons for this are twofold, because they will permit the breeding of short-term, non-diapause larvae between cotton crops, and because they can provide early hosts for the moths emerging from diapause and thus permit the development of another generation on the cotton. Pearson (1958) stated that there was some carryover on *A. esculentus* (bamia, okra, ochro, gumbo) in equatorial Sudan and that the cultivation of okra is prohibited in the Gezira during the summer months. In Uganda pink bollworm can carryover on *A. esculentus* or *Hibiscus macranthus*; in eastern Tanzania, and possibly Somalia and coastal Kenya, on *Abutilon* spp.; and in southern Ghana and Nigeria on backyard, perennial *G. barbadense* and *G. hirsutum* race *marie-galante* (Pearson, 1958). In Zimbabwe (Matthews *et al.*, 1965) carry over is in *H. dongolensis* and wild cotton, *G. herbaceum* race *africanum*. Wild *G. hirsutum* and okra are major hosts in parts of Florida (Noble, 1969) and wild *G. barbadense* × *hirsutum* hybrid cotton and okra in the West Indies (Ingram, 1981). In southern India, where there are no long-term larvae, carry-over in alternative hosts is considered important (Sohi, 1984).

In areas where the majority of the pink bollworm larvae are long term and undergo diapause the critical factors are the numbers of overwintering larvae and the subsequent emergence of moths to attack the new cotton crop. These long-term larvae can occur in the soil, in crop residues such as unharvested bolls, in seed cotton and in cotton seed, and anything that can be done to reduce their numbers at these times must be beneficial.

Ideally, once the crop has been harvested, the fields should be deep-ploughed to destroy any long-term larvae in the soil and to bury all the crop residues, particularly fallen seed cotton and remaining bolls. This is impractical in much of Africa and Asia where the cotton is cultivated by hand or by animal traction and the soil is rock hard by harvest-time. Noble (1969) reported that pasturing cattle or goats on cotton fields destroyed 90% of overwintering pink bollworm larvae in the United States. Jayaswal and Saini (1981a) reported a 62–65% reduction after grazing by sheep in India and in Pakistan (Anon., 1988) the cotton sticks are usually grazed bare. In Egypt, Turkey, India and Pakistan the cotton sticks after harvest are a valuable source of fuel and are usually kept on homestead roofs or in stacks around villages. Work in India has investigated the effects on diapausing larval mortality of height of stacking the sticks, stacking sticks vertically or horizontally, stacking in the shade or the sun and with the boll bearing portions facing north, east, south or west (Simwat *et al.*, 1982, 1985; Nandal *et al.*, 1984; Nandal and Singh, 1985). The effects of stack height seem to vary somewhat but vertical stacks about 2 m high, in the sun and facing west and south would seem to give the highest larval mortality.

Where the cotton stalks are not required for fuel they are often burnt on the land. Noble (1969) stated that residue destruction by burning killed as many as

90% of pink bollworm larvae but that it was better to use these crop residues for soil improvement. On small farms in the West Indies Ingram (1981) advocated hand-pulling, stacking and burning and recent studies in the United States (Chu and Bariola 1987, 1988) have shown that 44–82% larval mortality was obtained in trash fires.

Henneberry *et al.* (1980a) have reviewed the management of the cotton crop to limit host availability in the western United States where longer growing seasons and mild winters result in heavy pink bollworm attacks. Spring emergence generally begins in March and continues until August, but peaks in May or June, depending on locality and altitude. If the cotton planting here is delayed, so that the bulk of this emergence is suicidal, yields are adversely affected and overwintering larval populations are increased. Diapause larvae begin to appear in early September and develop in bolls that may not contribute much to yield. Methods to manage the late crop and so reduce the numbers of larvae in the soil have included the use of hand stripping, defoliants, desiccants, early irrigation cut-off, plant growth regulators and chemical terminators. Reductions of overwintering larval populations can be achieved but timings and local conditions are critical if yield and quality losses are to be avoided.

Adkisson *et al.* (1962) found that suicidal emergence was very high following delayed plantings and calculated that pink bollworm populations could be reduced two- to fourfold by planting at the optimum time in Texas. In India, Taneja and Dhindwal (1982) and Singh and Sidhu (1983), and in Turkey, Dincer (1984), also found that delayed planting reduced pink bollworm attacks and carry-over.

In Texas, Adkisson *et al.* (1958) found that applying preharvest chemicals for cotton defoliation delayed buildup of pink bollworm infestations and in California Bariola *et al.* (1984) found that early irrigation cut-off or chemical termination could prevent the development of any diapausing larvae. In India, Sundaramurthy *et al.* (1985) found that crop termination with chlorfurenol reduced late-season larval numbers without any reduction in yield.

Fye (1979) reported the effects of various cultural control practices in Arizona between 1971 and 1976. During the first two years no adequate cultural control was carried out but thereafter the practice was for the earliest possible picking, immediate residue destruction and ploughing as soon as possible, and thorough winter irrigation for winter cereal, fallow or cotton. This resulted in reduced catches of males caught in hexalure traps in the subsequent four years, reduced infestations in blooms, reduced larval infestations in bolls and minimal use of insecticides for pink bollworm control.

Winter survival can be reduced by stalk shredding, soil tillage, irrigation and crop rotation and this has been reviewed by Noble (1969) and Watson (1980). In Texas, where most diapausing larvae are found in seed in bolls, stalk shredders were shown to reduce moth emergence, but were not effective in Arizona where the diapausing larvae occur in hibernacula in, or on the soil. Of various tillage operations tried the most practical was shredding, discing and then deep mouldboard ploughing where the soil and crop residues are to a depth of about 150 mm. The depth is critical as most overwintering larvae occur in the top 50 mm (Rice and Reynolds, 1971b). The earlier in the autumn or winter this is carried out the greater the larval mortality. Each spring tillage operation in preparation for planting further reduces moth emergence as does each irrigation. Sowing a cover crop in the winter can delay the formation of pupae, possibly due to a lowering of soil temperatures, and thus the subsequent emergence of moths can be non-suicidal.

Ingram (1980) found that cutting down the standing cotton with a horizontal-rotary slasher, driven across the rows, raking and debris along the rows, stacking and burning it and then disc-ploughing in the cotton stumps across the rows gave almost 100% control of diapausing larvae on an estate scale in Barbados. Moawad (1981) investigated the survival of overwintering pink bollworm larvae in different soil types in Egypt and recommended a winter ploughing.

Control Using Plant Resistance

Natural plant resistance to pink bollworm has been long exploited as a means of reducing boll damage. Work in the US has been reviewed by Noble (1969), Wilson (1980) and Jenkins (1989). The Upland variants which impart resistance to pink bollworm are nectariless, nectariless–glabrous, okra-leaf, super okra-leaf, densely pubescent, high gossypol content, internal-boll antibiosis and earliness. Wilson (1987, 1989) compared different Upland germplasm lines combining the okra-leaf trait and the nectariless trait which have shown reduced percent seed damage and similar yields to standard cultivars when grown without insecticides. One resistant line, WC-12-NL, required 66% fewer insecticide sprays against pink bollworm than Deltapine 61; fibre length and strength, however, were not as good. Long-staple glandless, okra-leaf and pilose Pima cottons also show reduced seed damage from pink bollworm (Wilson, 1980).

Resistance to pink bollworm has also been studied in India. Agarawal *et al.* (1975) exploited earliness, while Singh *et al.* (1984) investigated the effects of super okra-leaf, red frego-bract, nectariless and glandless lines.

Integrated Pest Management

The integrated pest management, or IPM, of the pink bollworm does not occur in name only, it is practised to a greater or lesser degree by all cotton growers. It incorporates very many different aspects which can combine to give a truly viable IPM system where applications of conventional insecticides may not be necessary each season.

Varieties should be tolerant or resistant to pink bollworm, i.e. with internal-boll antibiosis, nectariless, okra-leafed and semi-smoothleafed. They should also be short-term with a determinate growth habit; long-term (Pima, Egyptian, Sea Island) varieties and indeterminate Upland varieties are in the ground long enough to permit the development of an additional generation of pink bollworm.

Cotton seed can be acid delinted, heat treated or fumigated to prevent carry-over of diapausing larvae in the seed. Ideally planting should be timed so that no fruiting bodies will be available when the majority of moths emerge from diapausing larvae in the soil. This can be predicted from the degree-days and is termed 'suicidal emergence'. Planting in any given area – farm, village, parish or whatever – should be carried out in as short a period of time as possible and it should never exceed four weeks, the average duration of a generation of the pink bollworm.

From first bud formation until harvest, pink bollworm eggs or larvae should be monitored at least once a week and, ideally, male moth activity monitored with pheromone traps to provide data on the pink bollworm attack. These data will provide the means of ascertaining if and when pink bollworm controls need to be applied.

During the early stages of the cotton crop sprays of insecticides should be used as little as possible so as to allow natural enemies to establish. Should the pest complex include early-season sucking pests, leafworms, etc., then consideration should be given to the use of soil application of granular pesticides or sprays of very short-persistence products so as to cause the least harm to the natural enemies of pink bollworm. Mating disruption with gossyplure may also be used to control the pink bollworm without any harmful effects to natural enemies or bees.

If insecticidal sprays are needed for the control of an outbreak of pink bollworm then great care should be taken with regard to the choice of active ingredient. Many of the pesticides used for pink bollworm control can lead to outbreaks of other, secondary, pests which in turn may require chemical controls.

Harvesting should be carried out as soon as possible and completed as soon as possible. As long as the plants are carrying any fruiting bodies the pink bollworm will be able to continue multiplying so the crop must be terminated. This may occur naturally with the advent of frosts but in frost-free areas the crop may have to be terminated chemically with defoliants, physically by grazing, ploughing-in or uprooting or by irrigation cut-off.

Crop residues should be destroyed by shredding and burning or ploughing-in, or where the stalks are required for fuel they should be grazed bare of leaves and stripped of unpicked bolls. Wherever it is physically possible the land should also be ploughed during the pink bollworm's larval resting stage so as expose the larvae to fatal surface temperatures or to bury them so deep as to prevent their emergence.

Ginning seasons should be kept as short as possible and never allowed to continue into the next cotton season. Ginneries should be thoroughly cleaned of seed-cotton residues and possibly sprayed with insecticide or fumigated to ensure destruction of resting larvae. Cotton seed for milling should be crushed immediately or fumigated, cotton seed for planting should also be fumigated.

There should also be a legally enforceable cotton close-season during which no cotton may be cultivated or left standing in the field. This should include wild cottons and it may also include other cultivated or wild pink bollworm host plants.

IPM of pink bollworm is covered in the major texts such as the FAO *Guidelines for Integrated Control of Cotton Pests* (Frisbie, 1983) and the University of California *Integrated Pest Management for Cotton in the Western Region of the United States* (University of California Publication 3305, 1984).

Other reviews and studies of IPM and pink bollworm control in the United States are by Henneberry *et al.* (1980c) and Burrows *et al.* (1984). Cotton pest and bollworm IPM techniques have been reviewed by Das and Basu (1981), Jayaswal and Saini (1982) and Agarawal and Gupta (1986) in India; by Hosny *et al.* (1983) and Campion and Hosny (1987) in Egypt; by Jin (1986) and Tao and Jia (1986) in China; and Bleicher *et al.* (1985) in Brazil.

References

Abul-Nasr, S., Awadallah, K.T. and Omar, H.M. (1974a) Effect of aging of cotton bolls on diapause in the pink bollworm *Pectinophora gossypiella* Saunders. *Bulletin of the Entomological Society of Egypt* 58, 303–308.

Abul-Nasr, S., Awadallah, K.T. and Omar, H.M. (1974b) Oil and moisture content of cotton bolls and seeds as factors inducing diapause in the pink bollworm, *Pectinophora gossypiella* Saunders. *Bulletin of the Entomological Society of Egypt* 58, 405–413.

Abul-Nasr, S.E., Tawfik, M.F.S., Ammar, E.D. and Farrag, S.M. (1978) Occurrence and causes of mortality among active and resting larvae of *Pectinophora gossypiella* (Lepidoptera: Gelechiidae) in Giza, Egypt. *Zeitschrift für Angewändte Entomologie* 86, 403–414.

Adkisson, P.L. (1961) Effect of larval diet on the seasonal occurrence of diapause in the pink bollworm. *Journal of Economic Entomology* 54, 1107–1112

Adkisson, P.L. (1964) Action of the photoperiod in controlling insect diapause. *The American Naturalist* 98, 357–374.

Adkisson, P.L., Wilkes, L.H. and Johnson, S.P. (1958) Chemical, cultural,and mechanical control of the pink bollworm. *Texas Agricultural Experiment Station Bulletin* No. 920, 16pp.

Adkisson, P.L., Robertson, O.T. and Fife, L.C. (1962) Planting date as a factor involved in pink bollworm control. In: Martin, D.F. and Lewis, R.D. (eds) *A Summary of Recent Research Basic to the Cultural Control of Pink Bollworm.* Texas Agricultural Experiment Station Miscellaneous Publication No. 579, pp. 16–20.

Adkisson, P.L., Bell, R.A. and Wellso, S.G. (1963) Environmental factors controlling the induction of diapause in the pink bollworm, *Pectinophora gossypiella* (Saunders). *Journal of Insect Physiology* 9, 299–310.

Agarawal, R.A. and Gupta, G.P. (1986) Recent advances in cotton pest management. *Plant Protection Bulletin, India* 38, 51–54.

Agarawal, R.A., Singh, M., Katiyar, K.N. and Singh, V.P. (1975) Resistance to pink bollworm in cotton. *Indian Journal of Agricultural Science* 45, 57–59.

Aginhothrudu, V. and Gour, T.B. (1982) Control of cotton bollworms with fenvalerate in India. *Crop Protection* 1, 231–234.

Ahmad, Z. (1977) A review of the research work done on pink bollworm, *Pectinophora gossypiella* (Saunders) with special reference to Indo-Pakistan sub-continent. *The Pakistan Cottons* 21, 119–130.

Ahmad, Z. (1979) Monitoring the seasonal occurrence of the pink bollworm in Pakistan with sex traps. *Plant Protection Bulletin*, FAO 27, 19–20.

Angelini, A., Trijau, J.P. and Vaissayre, M. (1982) Comparative action of three 'first generation' pyrethroids and of a certain number of new pyrethroids against cotton bollworms. *Coton et Fibres Tropicales* 37, 359–364.

Ankersmit, G.W. and Adkisson, P.L. (1967) Photoperiodic responses of certain geographical strains of *Pectinophora gossypiella* (Lepidoptera). *Journal of Insect Physiology* 13, 553–564.

Anon. (1981) *An Operational Field Trial Project in India for Suppression of the Cotton Pink Bollworm,* Pectinophora gossypiella *(Saunders) (Gelechiidae, Lepidoptera) Employing 'Gossyplure Hollow Fiber' Controlled Release Sex Pheromone Formulation.* Technical Report, Plant Protection Adviser to the Government of India, 44pp.

Anon. (1988) *Cotton Production Plan 1988–89.* Pakistan Central Cotton Committee, Karachi, 49pp.

Awaknavar, J.S., Thontadarya, T.S. and Patil, B.V. (1982) Termination of diapause of the long-cycle larvae of the pink bollworm, *Pectinophora gossypiella* (Saunders), and factors affecting it. *Mysore Journal of Agricultural Sciences* 16, 414–417.

Bariola, L.A. and Henneberry, T.J. (1980) Induction of diapause in field populations of the pink bollworm *Pectinophora gossypiella* in the western USA. *Environmental Entomology* 9, 376–380.

Bariola, L.A. and Lingren, P.D. (1984) Comparative toxicities of selected insecticides against pink bollworm (Lepidoptera: Gelechiidae) moths. *Journal of Economic Entomology* 77, 207–210.

Bariola, L.A., Keller, J.C., Turley D.L. and Farris, J.R. (1973) Migration and population studies of the pink bollworm in the arid west. *Environmental Entomology* 2, 205–208.

Bariola, L.A., Henneberry, T.J., Walhood, V.T. and Brown, C. (1984) Pink bollworm, effects of early maturing, narrow row cotton, insecticides, and chemical termination on seasonal infestations and overwintering larvae. *Southwestern Entomologist* 9, 62–68.

Bartlett, A.C. (1978) *Radiation Induced Sterility in the Pink Bollworm.* US Department of Agriculture, Science & Education Administration, ARM-W-1, 25pp.

Bartlett, A.C. and Butler, C.D. (1979) Pink bollworm: radiation sterility and computer simulation of population growth. *Southwestern Entomologist* 4, 216–233.

Bartlett, A.C. and Lewis, L.J. (1973) Pink bollworm: chromosomal damage and reproduction after gamma radiation of larvae. *Journal of Economic Entomology* 66, 731–733.

Bartlett, A.C., Staten, R.T. and Ridgway, W.O. (1973) Gamma radiation of eggs of the pink bollworm. *Journal of Economic Entomology* 66, 475–477.

Beasley, C.A. and Henneberry, T.J. (1984) Combining gossyplure and insecticides in pink bollworm control. *California Agriculture* 38 (7/8), 22–24.

Beasley, C.A. and Henneberry, T.J. (1988) Effects of trap type and placement on male pink bollworm moth captures. In: *Proceedings of the Beltwide Cotton Production Research Conference,* Memphis, Tennessee, pp. 306–309.

Beasley, C.A., Henneberry, T.J., Adams, C. and Yates, L. (1985) Gossyplure-baited traps as pink bollworm survey, detection, research and management tools in southwestern desert cotton growing areas. *California Agricultural Experiment Station Bulletin* 1915, 15pp.

Bell, M.R. and Henneberry, T.J. (1980) Entomopathogens for pink bollworm control. In: Graham, H.M. (ed.) *Pink Bollworm Control in the Western United States.* US Department of Agriculture, Science and Education Administration, ARM-W-16, pp. 76–81.

Berger, R.S., McGough, J.M., Martin, D.F. and Ball, L.R. (1964) Some properties and the field evaluation of the pink bollworm sex attractant. *Annals of the Entomological Society of America* 57, 606–608.

Bhamburkar, M.W. and Kathane, T.V. (1984) Role of synthetic pyrethroids in the control of bollworms in relation to yield of H4 cotton under rainfed conditions. *Pesticides* 18 (1), 15–18.

Bierl, B.A., Beroza, M., Staten, R.T., Sonnet, P.E. and Adler, V.E. (1974) The pink bollworm sex attractant. *Journal of Economic Entomology* 67, 211–216.

Bishara, I. (1930) Ratoon cotton in relation to insect pests. *Bulletin of the Ministry of Agriculture, Egypt* No. 77, 11pp.

Bleicher, E., Ferraz, C.T. and Lamas, F.M. (1985) Sugestoes para o controle de pragas do algodoeiro no Estado de Mato Grosso do Sul. *Comunicado Tecnico, Empresa de Pesquisa Assistencia Tecnica e Extensao Rural de Mato Grosso do Sol* 4, pp. 15.

Boness, M.B, Eiter, K. and Disselnkotter, H. (1977) Studies on sex attractants of Lepidoptera and their use in crop protection. *Pflanzenschutz-Nachrichten Bayer* 30, 213–236.

Brooks, T.W., Doane, C.C. and Staten, R.T. (1979) Experience with the first commercial pheromone communication disruptive for suppression of an agricultural insect pest. In: Ritter, F.J. (ed.) *Chemical Ecology: Odour Communication in Animals*. Elsevier/North-Holland Biomedical Press, Amsterdam, pp. 375–388.

Brooks, T.W., Doane, C.C. and Haworth, J.K. (1980) Suppression of *Pectinophora gossypiella* with sex pheromone. *Proceedings of the 1979 British Crop Protection Conference – Pests and Diseases*, Vols. 1, 2 and 3. British Crop Protection Council, Croydon, pp. 853–866.

Bull, D.L. and Adkisson, P.L. (1960) Certain factors inducing diapause in the pink bollworm, *Pectinophora gossypiella*. *Journal of Economic Entomology* 53, 793–798.

Burrows, T.M., Sevacherian, V., Moffit, L.J. and Baritelle, J.L. (1984) Economics of pest control alternatives for Imperial Valley cotton. *California Agriculture* 38 (5/6), 15–16.

Busck, A. (1917) The pink bollworm, *Pectinophora gossypiella*. *Journal of Agricultural Research* 9 (10), 343–370.

Butler, G.D. and Henneberry, T.J. (1976) Biology, behaviour, and effects of larvae of pink bollworm in cotton flowers. *Environmental Entomology* 5, 970–972.

Butler, G.D. and Las, A.S. (1983) Predacious insects, effect of adding permethrin to the sticker used in gossyplure applications. *Journal of Economic Entomology* 76, 1448–1451.

Butler, G.D., Hamilton, A.G. and Guttierez, A.P. (1978) Pink bollworm, diapause induction in relation to temperature and photoperiod. *Annals of the Entomological Society of America* 71, 202–204

Butler, G.D., Henneberry, T.J. and Barker, R.J. (1983) *Pink Bollworm, Comparison of Commercial Control with Gossyplure or Insecticides*. US Department of Agriculture, Agricultural Research Service ARM-W-35, iv + 13pp.

CAB International Institute of Entomology (1987) *Distribution Maps of Pests*, Series A (Agricultural), Map no. 14 (revised), June 1987, *Pectinophora scutigera* (Holdaway).

CAB International Institute of Entomology, (1990) *Distribution Maps of Pests*, Series A (Agricultural), Map no. 13 (third revision), June 1990, *Pectinophora gossypiella* (Saund.)

Campion, D.G. and Hosny, M.M. (1987) Biological, cultural and selective methods for control of cotton pests in Egypt. *Insect Science and its Application* 8, 4–6.

Campion, D.G. and Murlis, J. (1985) Sex pheromones for the control of insect pests in developing countries. *Mededelingen van de Faculteit Landbouwwetenschappen, Rijksuniversiteit Gent* 50, 203–209.

Campion, D.G. and Nesbitt, B.R. (1982) Recent advances in the use of pheromones in developing countries with particular reference to mass-trapping for the control of the Egyptian cotton leafworm *Spodoptera littoralis* and mating disruption for the control of pink bollworm *Pectinophora gossypiella*. *Les Mediateurs Chimiques Agissant sur le Comportement des Insectes*. Symposium International, 16–20 November 1981, Versailles, Paris. Institut National de la Recherche Agronomique, Paris, pp. 335–342.

Chapman, A.J., Fife, L.C., Smith, G.L. and Clark, J.C. (1950) DDT for control of pink bollworm in Mexico in 1946. *Journal of Economic Entomology* 43, 491–494.

Cheema A. and Muzaffar, N. (1979) Pathogens associated with the pink bollworm in Pakistan. *Proceedings of the Pakistan Academy of Sciences* 16, 43–44.

Cheema, M.A., Muzaffar, N. and Ghani, M.A. (1980a) Biology, host range and incidence of parasites of *Pectinophora gossypiella* (Saunders) in Pakistan. *The Pakistan Cottons* 24, 37–73.

Cheema, M.A., Muzaffar, N, and Ghani, M.A. (1980b) Investigation on phenology, distribution, host range and evaluation of predators of *Pectinophora gossypiella* (Saunders) in Pakistan. *The Pakistan Cottons* 24, 139–176.

Chen, Y.G., Ge, D.H., He, H.X., Wang, G.H., Zhang, W.L. and Gao, S.Y. (1984a) Hollow fibre formulation of gossyplure and its competing attraction to pink bollworm moths in the cotton field. *Acta Entomologica Sinica* 27, 229–234.

Chen, Y.G., Ge, D.H., Dai, X.J. and Gao, S.Y. (1984b) Investigation of the formulation of insect semiochemical I. Plastic film and laminated plastic capsule formulations of gossyplure and the effect on controlling pink bollworm. *Contributions from Shanghai Institute of Entomology* 4, 31–39.

Chu, C.C. and Bariola, L.A. (1987) Survival of pink bollworm, *Pectinophora gossypiella* (Saunders), in green bolls at high temperatures.*Southwestern Entomologist* 12, 271–277.

Chu, C.C. and Bariola, L.A. (1988) Effect of cotton boll temperatures on larval mortality of pink bollworm, *Pectinophora gossypiella* (Saunders). *Southwestern Entomologist* 13, 185–189.

Chu, H.F. (1978) Strategies and tactics of pest management with special reference to Chinese cotton insects. *Acta Entomologica Sinica* 21, 297–308.

Cock, M.J.W. (1985) The use of parasitoids for augmentative biological control of pests in the People's Republic of China. *Biocontrol News and Information* 6, 213–224.

Common, I.F.B. (1958) A revision of the pink bollworms of cotton (*Pectinophora* Busck (Lepidoptera, Gelechiidae)) and related genera in Australia. *Australian Journal of Zoology* 6, 268–306.

Coudriet, D.L. and Henneberry, T.J. (1976) Capture of male cabbage loopers and pink bollworms, effect of trap design and pheromone. *Journal of Economic Entomology* 69, 603–605.

Critchley, B.R., Campion, D.G., McVeigh, L.J., Hunter-Jones, P., Hall, D.R., Cork, A., Nesbitt, B.F., Marrs, G.J., Jutsum, A.R., Hosny, M.M. and Nasr, El-Sayed A. (1983) Control of pink bollworm, *Pectinophora gossypiella* (Saunders) (Lepidoptera: Gelechiidae), in Egypt by mating disruption using an aerially applied microencapsulated pheromone formulation. *Bulletin of Entomological Research* 73, 289–299.

Critchley, B.R., Campion, D.G., McVeigh, E.M., McVeigh, L.J., Jutsum, A.R., Gordon, R.F.S., Marrs, G.J., Nasr, E.S.A. and Hosny, M.M. (1984) Microencapsulated pheromones in cotton pest management. *Proceedings of British Crop Protection Conference – Pests and Diseases* British Crop Protection Council, Croydon, 1, 241–245.

Critchley, B.R., Campion, D.G., McVeigh, L.J., McVeigh, E.M., Cavanagh, G.G., Hosny, M.M., Nasr, El-Syaed, A., Khidr, A.A. and Naguib, M. (1985) Control of pink bollworm, *Pectinophora gossypiella* (Saunders) (Lepidoptera: Gelechiidae), in Egypt by mating disruption using hollow-fibre, laminated flake and microencapsulated formulations of synthetic pheromone. *Bulletin of Entomological Research* 75, 329–345.

Critchley, B.R., Campion, D.G., Cavanagh, G.G., Chamberlain, D.J. and Attique, M.R. (1987) Control of three major bollworm pests of cotton in Pakistan by a single application of their combined sex pheromones. *Tropical Pest Management* 33, 374.

Critchley, B.R., Campion, D.G. and McVeigh, L.J. (1989) Pheromone control in the integrated pest management of cotton. In: Green, M.B. and Lyon, D.J. de B. (eds), *Pest Management in Cotton.* Ellis Horwood, Chichester, pp. 83–92.

Das, B.B. and Basu, A.K. (1981) Prospects of an integrated control of cotton bollworms. *Indian Journal of Agricultural Research* 15, 79–86.

Dhawan, A.K. and Sidhu, A.S. (1984a) Evaluation of PAU trap to catch males of pink bollworm of cotton. *Indian Journal of Agricultural Sciences* 54, 318–320.

Dhawan, A.K. and Sidhu, A.S. (1984b) Assessment of capture thresholds of pink-bollworm moths for timing insecticidal applications on *Gossypium hirsutum* Linn. *Indian Journal of Agricultural Sciences* 54, 426–433.

Dhawan, A.K. and Sidhu, A.S. (1984c) Timing of sprays against pink bollworm on basis of the moth catch in pheromone baited traps. *Agricultural Science Digest* India, 4, 203–205.

Dhawan, A.K. and Sidhu, A.S. (1987) Field evaluation of different dispensers and trapping media for catches of pink bollworm males. *Indian Journal of Plant Protection* 15, 152–158.

Dincer, J. (1984) Investigations on the possibilities of cotton in the Aegean Region. *Bitki Koruma Bulteni* 24, 15–32.

Doane, C.C., Haworth, J.K. and Dougherty, D.G. (1983) NoMate PBW, a synthetic pheromone formulation for wide area control of the pink bollworm. In: *10th International Congress of Plant Protection* 1, p.265.

El-Adl, M.A., Hosny, M.M., Campion, D.G. (1988) Mating disruption for the control of pink bollworm *Pectinophora gossypiella* (Saunders) in the Delta cotton growing area in Egypt. *Tropical Pest Management* 34, 210–214,243,247.

El-Guindy, M.A., Abdel-Sattar, M.M., Dogheim, S.M.A., Madi, S.M. and Issa, Y.H. (1982) The joint action of certain insecticides on a field strain of the pink bollworm *Pectinophora gossypiella* Saund. *International Pest Control* 24, 154–155.

El-Sayed M.T. and El-Rahman, H.A.A. (1960) On the biology and life history of the pink bolllworm, *Pectinophora gossypiella* (Saunders). *Bulletin of the Entomological Society of Egypt* 44, 71–90.

El-Sayed, M.T. and Rustom, Z.M.F. (1960a) Factors affecting termination of the resting stage of the pink bollworm, *Pectinophora gossypiella* Saunders. *Bulletin of the Entomological Society of Egypt* 44, 265–282.

El-Sayed, M.T. and Rustom, Z.M.F. (1960b) Factors affecting initiation of the resting stage of the pink bollworm, *Pectinophora gossypiella* Saunders. *Bulletin of the Entomological Society of Egypt* 44, 253–264.

El-Tigani, M.E-A. and Khagali, M.A. (1978) Pink bollworm in Barakat Block. In: *Annual Report of the Gezira Research Station and Sub-stations 1970–1971*. Agricultural Research Corporation, Ministry of Agriculture, pp. 98–100

Evans, W.H. (1984) Development of an aqueous-based controlled release pheromone-pesticide system. In: Scher, H.B. (ed.) *Advances in Pesticide Formulation Technology*, American Chemical Society, Washington, DC, pp. 151–162.

Flint, H.M. and Merkle, J.R. (1980) Pink bollworm, irradiation of laboratory and native males. *Journal of Economic Entomology* 73, 764–767.

Flint, H.M. and Merkle, J.R. (1983) Methods for the efficient use of the Delta trap in the capture of pink bollworm moths. *Southwestern Entomologist* 8, 140–144.

Flint, H.M. and Merkle, J.R. (1984a) Pink bollworm, disruption of sexual commmunication by the release of the *(Z,Z)*-isomer of gossyplure. *Southwestern Entomologist* 9, 58–61.

Flint, H.M. and Merkle, J.R. (1984b) The pink bollworm (Lepidoptera: Gelechiidae), alteration of male response to gossyplure by release of its component *(Z,Z)*-isomer. *Journal of Economic Entomology* 77, 1099–1104.

Flint, H.M. and Merkle, J.R. (1984c) Studies on disruption of sexual communication in the pink bollworm, *Pectinophora gossypiella* (Saunders) (Lepidoptera: Gelechiidae), with microencapsulated gossyplure or its component *(Z,Z)*-isomer. *Bulletin of Entomological Research* 74, 25–32.

Flint, H.M. and Stone, M. (1985)*Pectinophora scutigera* (Holdaway) (Lepidoptera: Gelichiidae), monitoring populations and disrupting communications with *Z,Z*- and *Z,E*-7,11–16, Ac in the field.*Journal of the Australian Entomological Society* 24, 281–286.

Flint, H.M., Staten, R.T. and Palmer, D.L. (1973) Gamma-irradiated pink bollworms: attractiveness, mating, and longevity of females. *Environmental Entomology* 2, 97–100.

Flint, H.M., Kuhn, S., Horn, B. and Sallam, H.A. (1974) Early season trapping of pink bollworm with gossyplure. *Journal of Economic Entomology* 67, 238–240.

Flint, H.M., Smith, R.L., Bariola, L.A., Horn, B.R., Forey, D.E. and Kuhn, S.J. (1976) Pink bollworm: trap tests with gossyplure. *Journal of Economic Entomology* 69, 535–538.

Flint, H.M., Smith, R.L., Noble, J.M. and Shaw, D. (1978) Pink bollworm, response of released APHIS strain and native moths to ratios of (*Z,Z*)- and (*Z,E*)-isomers of gossyplure in the field. *Journal of Economic Entomology* 71, 664–666.

Flint, H.H., Balasubramanian, M., Campero, J., Strickland, G.R., Ahmad, Z., Barral, J., Barbosa S. and Khail, A.F. (1979) Pink bollworm, response of native males to ratios of (*Z,Z*)- and (*Z,E*)-isomers of gossyplure in several cotton growing areas of the world. *Journal of Economic Entomology* 72, 758–762.

Flint, H.M., Merkle, J.R. and Yamamoto, A. (1985) Pink bollworm (Lepidoptera: Gelechiidae), field testing a new polyethylene tube dispenser for gossyplure. *Journal of Economic Entomology* 78, 1431–1436.

Flint, H.M., Curtice, N.J. and Yamamoto, A. (1988) Pink bollworm (Lepidoptera: Gelechiidae), further tests with (*Z,Z*)-isomer of gossyplure. *Journal of Economic Entomology* 81, 679–683.

Foster, R.N., Staten, R.T. and Miller, E. (1977) Evaluation of traps for pink bollworm. *Journal of Economic Entomology* 70, 289–291.

Frisbie, R.E. (1983) *Guidelines for Integrated Control of Cotton Pests.* FAO Plant Production and Protection Paper 48. Food and Agriculture Organisation of the United Nations, Rome, xii + 187pp.

Fry, K.E., Kittock, D.L. and Henneberry, T.J. (1978) Effect of numbers of pink bollworm larvae per boll on yield and quality of Pima and Upland cotton. *Journal of Economic Entomology* 71, 499–502.

Fullaway, D.T. (1909) Insects of cotton in Hawaii. *Bulletin Hawaii Agricultural Experiment Station,* No. 18, 27pp.

Funkhouser, W.A. and Las, A.S. (1981) Aerial dissemination of insect pheromones. In: *Proceedings Winter Meeting of the American Society of Agricultural Engineers,* 15–18 December 1981, Chicago.

Fye, R.E. (1979) *Pink Bollworms, a 5-Year Study of Cultural Control in Southern Arizona.* US Department of Agriculture, Science and Education Administration, ARR-W-3, 10pp.

Fye, R.E. and Jackson, C.G. (1973) Overwintering of *Chelonus blackburni* in Arizona. *Journal of Economic Entomology* 66, 807–808.

Gaston, L.K., Kaae, R.S., Shorey, H.H. and Sellers, D. (1977) Controlling the pink bollworm by disrupting sex pheromone communication between adult moths. *Science* 196, 904–905.

Gong, X.W., Meng, G.L., He, D., Zhao, P.Z., Jiang, T.R. and Huang, T.B. (1984) Biology of the braconid wasp, *Bracon nigrorufum* (Cushman) and its use in cotton fields. *Natural Enemies of Insects* 6, 57–61.

Gordh, G. (1984) *Goniozus pakmanus* (Hymenoptera, Bethylidae), a new species imported into California for the biological control of pink bollworm *Pectinophora gossypiella* (Lepidoptera: Gelechiidae) *Entomological News* 95, 207–211.

Graham, H.M., Martin, D.F., Ouye, M.T. and Hardman, R.M. (1966) Control of pink bollworms by male annihilation. *Journal of Economic Entomology* 59, 950–953.

Green, N., Jacobson, M. and Keller, J.C. (1969) Hexalure an insect sex attractant discovered by empirical screening. *Experimentia* 25, 682–683.

Guerra, A.A. and Ouye, M.T. (1967) Catches of male pink bollworms in traps baited with sex attractant. *Journal of Economic Entomology* 60, 1046–1048.

Guerra, A.A., Garcia, R.D. and Leal, M.P. (1969) Suppression of populations of pink bollworm in field cages with traps baited with sex attractant. *Journal of Economic Entomology* 62, 741–742.

Gupta, G.P. and Katiyar, K.N. (1985) Bioefficacy and economics of synthetic pyrethroids for the control of cotton bollworms. *Indian Journal of Entomology* 47, 381–387.

Gupta, G.P. and Katiyar, K.N. (1987) Evaluation of new synthetic pyrethroids and formulations against bollworm complex in cotton. *Pesticides* 21(4), 20–22

Gutierrez, A.P., Butler, G.D., Wang, Y. and Westphal D. (1977) The inter-action of pink bollworm (Lepidoptera: Gelechiidae), cotton and weather: a detailed model. *Canadian Entomologist* 109, 1457–1468.

Gutierrez, A.P., Butler, G.D. and Ellis, C.K. (1981) Pink bollworm: diapause induction and termination in relation to fluctuating temperatures and decreasing photophases. *Environmental Entomology* 10, 936–942.

Gutierrez, A.P., Pizzamiglio, M.A., Dos Santos, W.J., Villacorta, A. and Gallagher, K.D. (1986) Analysis of diapause induction and termination in *Pectinophora gossypiella* in Brazil. *Environmental Entomology* 15, 494–500.

Habib, R. and Mohyuddin, A.I. (1981) Possibilities of biocontrol of some pests of cotton in Pakistan. *Biologica* 27, 107–113.

Hall, D.R., Nesbitt, B.R., Marrs, G.J., Green, A. St J., Campion, D.G. and Critchley, B.R. (1982) Development of microencapsulated pheromone formulations. In: Leonhardt, B.A. and Beroza, M. (eds) *Insect Pheromone Technology: Chemistry and Application.* American Chemical Society Symposium Series No. 190, Washington, DC, pp.131–143.

Haynes, K.F., Gaston, L.K., Pope, M.M. and Baker, T.C. (1984) Potential for evolution of resistance to pheromones, interindividual and interpopulational variation in chemical communication system of pink bollworm moth. *Journal of Chemical Ecology* 10, 1551–1565.

Haynes, K.F., Miller, T.A., Staten, R.T., Li, W.G. and Baker, T.C. (1986) Monitoring insecticide resistance with insect pheromones. *Experientia* 42, 1293–1295.

Haynes, K.F., Miller, T.A., Staten, R.T., Li, W.G. and Baker, T.C. (1987) Pheromone trap for monitoring insecticide resistance in the pink bollworm moth (Lepidoptera: Gelechiidae): new tool for resistance management. *Environmental Entomology* 16, 84–89.

Hekal, A.M. (1986) Percentage of parasitism and sex ratio in different parasites of the diapausing larvae of *Pectinophora gossypiella* Saund. *Annals of Agricultural Science, Moshtohor* 24, 2213–2221.

Hendricks, D.E., Graham, B.M., Guerra, R.J. and Perez, C.T. (1973) Comparison of the numbers of tobacco budworms and bollworms caught in sex pheromone traps vs. blacklight traps in Lower Rio Grande Valley, Texas. *Environmental Entomology* 2, 911–914.

Henneberry, T.J. (1980) Potential of sterile moth releases for pink bollworm management. In: Graham, H.M. (ed.) *Pink Bollworm Control in the Western United States.* US Department of Agriculture Science and Education Administration, ARM-W-16, pp. 52–66.

Henneberry, T.J. (1989) The pink bollworm as a factor in cotton production in the southwestern United States. In: *New Developments on Pest Control and its Impact on Yield and Fiber Quality.* Technical Seminar, 48th Plenary Meeting, International Cotton Advisory Committee, October 1989, Scottsdale, Arizona, pp. 3–11

Henneberry, T.J. and Clayton, T.E. (1981) Effects on reproduction of gamma irradiated laboratory-reared pink bollworms and their F-1 progeny after matings with untreated laboratory reared or native insects. *Journal of Economic Entomology* 74, 19–23.

Henneberry, T.J. and Clayton, T.E. (1982) Pink bollworm of cotton (*Pectinophora gossypiella* (Saunders)): male moth catches in gossyplure-baited traps and relationships to oviposition, boll infestation and moth emergence. *Crop Protection* 1, 497–504.

Henneberry, T.J. and Clayton, T.E. (1983) Pink bollworm (Lepidoptera: Gelechiidae), effects of soil moisture on the behaviour of diapausing larvae and adult emergence from bolls. *Environmental Entomology* 12, 1490–1495

Henneberry, T.J. and Clayton, T.E. (1985) Consumption of pink bollworm (Lepidoptera: Gelechiidae) and tobacco budworm (Lepidoptera: Noctuidae) eggs by some predators commonly found in cotton fields. *Environmental Entomology* 14, 416–419.

Henneberry, T.J. and Keaveny III, D.F. (1985) *Suppression of Pink Bollworm by Sterile Moth Releases.* US Department of Agriculture, Agricultural Research Service, ARS-32, 80pp.

Henneberry, T.J., Bariola, L.A. and Russell, T. (1978) Pink bollworm: chemical control in Arizona and relationship to infestations, lint yield, seed damage, and aflatoxin in cottonseed. *Journal of Economic Entomology* 71, 440–442.

Henneberry, T.J., Bariola, L.A. and Kittock, D.L. (1980a) Pink bollworm control: potential of cotton crop management to selectively limit host availability. In: Graham, H.M. (ed.) *Pink Bollworm Control in the Western United States.* US Department of Agriculture, Science and Education Administration, ARM-W-16, pp. 9–23.

Henneberry, T.J., Clayton, T.E. and Keaveny, D.F. (1980b) Effects of gamma radiation on mating, reproduction, and longevity of laboratory-reared pink bollworms and their F-1 progeny crossed with moths of a laboratory-reared or native St Croix strain. *Southwestern Entomologist* 5, 250–256.

Henneberry, T.J., Bariola, L.A. and Kittock, D.L. (1980c) Integrating methods for control of the pink bollworm and other cotton insects in the southwestern United States. *US Department of Agriculture, Science and Education Administration, Technical Bulletin* No. 1610, 45pp.

Henneberry, T.J., Gillespie, J.M., Bariola, L.A., Flint, H.M., Lingren, P.D. and Kydonieus, A.F. (1981) Gossyplure in laminated plastic formulations for mating disruption and pink bollworm control. *Journal of Economic Entomology* 74, 376–381.

Hertig, B. (1975) *A Catalogue of Parasites and Predators of Terrestrial Arthropods.* Section A, Host or Prey/Enemy. Vol. VI. Lepidoptera, Part I (Microlepidoptera). Commonwealth Agricultural Bureaux, Slough, 218pp.

Holdaway, F.G. (1926) The pink bollworm of Queensland. *Bulletin of Entomological Research* 17, 67–83.

Holdaway, F.G. (1929) Confirmatory evidence of the validity of the species *Pectinophora scutigera* Holdaway (Queensland pink bollworm), from a study of the genitalia. *Bulletin of Entomological Research* 20, 179–185

Hosny, M.M. (1988) The role of pheromones in the management of pink bollworm infestation in Egyptian cotton fields. *Agriculture, Ecosystems and Environment* 21, 67–83.

Hosny, M.M., Saadany, G., Iss-Hak, R., Nasr, E.A., Moawad, G., Naguib, M., Khidr, A.A., Elnagar, S.H., Campion, D.G., Critchley, B.R., Jones, K., McKinlay, D.J., McVeigh, L.J. and Topper, C.P. (1983) Techniques for the control of cotton pests in Egypt to reduce the reliance on chemical pesticides. In: *10th International Congress of Plant Protection* 1, p. 270.

Huber, R.T. and Hoffmann, M.P. (1979) Development and evaluation of an oil trap for use in pink bollworm pheromone mass trapping and monitoring programs. *Journal of Economic Entomology* 72, 695–697.

Huber, R.T., Moore, L. and Hoffmann, M.P. (1979) Feasibility study of area-wide pheromone trapping of male pink bollworm moths in a cotton insect pest management program. *Journal of Economic Entomology* 72, 222–227.

Hummel, H.E., Gaston, L.K., Shorey, H.H., Kaae, R.S., Byrne, K.J. and Silverstein, R.M. (1973) Clarification of the chemical status of the pink bollworm sex pheromone. *Science* 181, 873–875.

Hunter, W.D. (1918) The pink bollworm with special reference to the steps taken by the Department of Agriculture to prevent its establishment in the United States *US Department of Agriculture Bulletin*, No.723, 27pp.

Hutchison, W.D., Henneberry, T.J. and Beasley, C.A. (1988) Efficacy of selected insecticides on pink bollworm oviposition in cotton. In: 1988 *Proceedings of the Beltwide Cotton Production Research Conference*, Memphis, Tennessee, pp. 309–311.

Ingram, W.R. (1980) Studies of the pink bollworm, *Pectinophora gossypiella*, on Sea Island Cotton in Barbados. *Tropical Pest Management* 26, 118–137.

Ingram, W.R. (1981) *Pests of West Indian Sea Island Cotton.* Centre for Overseas Pest Research, London, 35pp + 90 col. plates.

Irwin, M.E., Gill, R.W. and Gonzalez, D. (1974) Field-cage studies of native egg predators of the pink bollworm in southern California cotton. *Journal of Economic Entomology* 67, 193–196.

Jackson, C.G. (1980) Entomophagous insects attacking the pink bollworm. In: Graham, H. M. (ed.) *Pink Bollworm Control in the Western United States.* US Department of Agriculture, Science and Education Administration, ARM-W-16, pp. 71–75.

Jayaswal, A.P. and Saini, R.K. (1981a) Sheep grazing as a method to reduce the population of diapausing larvae of pink bollworm (*Pectinophora gossypiella* Saunders) in cotton. *Cotton Development* 11, 18.

Jayaswal, A.P. and Saini R.K. (1981b) Effect of some synthetic pyrethroids on pink bollworm incidence and yield of cotton. *Pesticides* 15(1), 33–35.

Jayaswal, A.P. and Saini, R.K. (1982) Cotton pest management in Haryana. *Indian Journal of Plant Protection* 9, 29–33.

Jenkins, J.N. (1989) State of the art in host plant resistance in cotton. In: Green, M.B. and Lyon, D.J. de B. (eds) *Pest Management in Cotton.* Ellis Horwood, Chichester, pp. 53–69.

Jin, Z.S. (1986) Integrated control of insect pests on cotton for years. *Natural Enemies of Insects* 8, 25–28.

Johnstone, D.R. (1982a) *Factors Affecting Aerial Application of a Micro Encapsulated Pheromone Formulation for the Control of* Pectinophora gossypiella *(Saunders) by Communication Disruption on Cotton in Egypt.* Miscellaneous Report, Centre for Overseas Pest Research, No. 56, 11pp.

Johnstone, D.R. (1982b) Physical and meteorological factors affecting aerial application of sex pheromone for control of the pink bollworm of cotton in Egypt. *Agricultural Meteorology* 26, 117–126.

Jones, W.A. and Jacobson, M. (1968) Isolation of *N,N*-diethyl-m-toluamide (deet) from female pink bollworm moths. *Science* 159, 99–100.

Jones, W.A., Jacobson, M. and Martin, D.F. (1966) Sex attractant of the pink bollworm moth: isolation, identification, and synthesis. *Science* 152, 1516–1517.

Jutsum, A.R., Marrs, G.J., Gordon, R.F.S., Campion, D.G., Critchley, B.R., McVeigh, L.J., Cork, A., Hall, D.R., Nesbitt, B.F., Hosny, M.M. and Nasr, E.A. (1983) Control of crop pests with microencapsulated insect pheromones. In: *10th International Congress of Plant Protection* 1, pp. 266–267.

Kaae, R.S. and Shorey, H.H. (1973) Sex pheromones of Lepidoptera, 44. Influence of environmental conditions on the location of pheromone communication and mating in *Pectinophora gossypiella. Environmental Entomology* 2, 1081–1084.

Kassem, S.M. I., Aly, M.I., Bakry, N.S. and Zeid, M.I. (1986) Efficacy of methomyl and its mixtures against the Egyptian cotton leafworm and bollworms. *Alexandria Journal of Agricultural Research* 31, 291–300.

Keerthisinghe, C.I. (1982) Synthetic pyrethroids and cotton bollworm control in Sri Lanka. *Tropical Pest Management* 28, 33–36.

Keller, J.C., Sheets, L.W., Green, N. and Jacobson, M. (1969) *Cis*-7-hexadecen-1-ol-acetate (Hexalure), a synthetic sex attractant for pink bollworm males. *Journal of Economic Entomology* 62, 1520–1521.

Khalil, F.A., Watson, W.M. and Guirguis, M.W. (1983) Evaluation of Dimilin and its combinations with different insecticides against some cotton pests in Egypt. *Bulletin of the Entomological Society of Egypt (Economic)* 1978–79, 11, 71–76.

Kurtadikar, J.S. and Hedgire, D.N. (1982) Studies on the efficiacy of new insecticides against cotton bollworms on rainfed H-4 cotton. *Pesticides* 16(9), 33–34.

Kydonieus, A.F. and Beroza, M. (1981) The Hercon dispenser formulation and recent test results. In: Mitchell, E. R. (ed.) *Management of Insect Pests with Semiochemicals: Concepts and Practice*, Plenum, New York, pp. 445–453.

Lefroy, H.M. (1906) *Indian Insect Life.* Superintendent of Government Printing, Calcutta, 318pp.

Legner, E.F. (1976) *Review of Biological Control Efforts Against Pink Bollworm,* Pectinophora gossypiella *(Saunders), in California Compared to Other World Areas.* Division of Biological Control, University of California, Technical Report, US Project 2548, 38pp.

Legner, E.F. (1979) Emergence patterns and dispersal in *Chelonus* spp. near *curvimaculatus* and *Pristomerus hawaiiensis* parasites on *Pectinophora gossypiella. Annals of the Entomological Society of America* 72, 681–686

Legner E.F. and Medved R.A. (1979) Influence of parasitic hymenoptera on the regulation of pink bollworm, *Pectinophora gossypiella*, on cotton in the Lower Colorado Desert. *Environmental Entomology* B8, 922–930.

Lingren, P.D. (1982) Confusing and killing cotton pests. *Agricultural Research, USA* 31, 4–5.

Lingren, P.D., Burton, J., Shelton, W. and Raulston, J.R. (1980) Night vision goggles, for design, evaluating, and comparative efficiency determination of pheromone trap for catching live adult male pink bollworms. *Journal of Economic Entomology* 73, 622–630.

Lowry, W.L. and Berger, R.S. (1965) Investigation of pink bollworm resistance to DDT in Mexico and the United States. *Journal of Economic Entomology* 58, 590.

Lukefahr, M.J. (1961) Factors related to the induction of diapause in the pink bollworm. PhD Dissertation, Texas A & M College.

Lukefahr, M.J. and Griffin, J.A. (1957) Mating and oviposition habits of the pink bollworm moth. *Journal of Economic Entomology* 50, 487–490.

Lukefahr, M.J., Noble, L.W. and Martin, D.F. (1964) Factors inducing diapause in the pink bollworm. *US Department of Agriculture, Technical Bulletin*, No. 1304, 17pp.

Lukefahr, M.J., Braga, S.R. and Vieira, R de M. (1985) Pink bollworm; diapause in the equatorial regions of Brazil. *Southwestern Entomologist* 10, 283–288.

Luo, S.B., Yan, J.P., Chai, C.J., Liang, S.P., Zhang, Y.M., Zhang, Y. and Le, G.K. (1986) Control of pink bollworm, *Pectinophora gossypiella* with *Bacillus thuringiensis* in cotton fields. *Chinese Journal of Biological Control* 2, 167–169.

McLaughlin, J.R., Shorey, H.H., Gaston, L.K., Kaae, R.S. and Stewart, F.D. (1972) Sex pheromones of Lepidoptera. XXXI. Disruption of sex pheromone communication in *Pectinophora gossypiella* with Hexalure. *Environmental Entomology* 1, 645–650.

McVeigh, L.J., Critchley, B.R. and Campion, D.G. (1983) Control of the pink bollworm in Egypt by mating disruption using pheromones. In: *10th International Congress of Plant Protection*, p. 268.

Marks, R.J. (1976) Field evaluation of gossyplure, the synthetic sex pheromone of *Pectinophora gossypiella* (Saund.) (Lepidoptera: Gelechiidae) in Malawi. *Bulletin of Entomological Research* 66, 267–278.

Matthews, G.A., Tunstall, J.P. and McKinley, D.J. (1965) Outbreaks of pink bollworm (*Pectinophora gossypiella* Saund.) in Rhodesia and Malawi. *Empire Cotton Growing Review* 42, 197–208.

Megahed, M.M., Ammar, D., Metwally, A.G. and Rashad A. (1984) Studies on the attraction of males of *Pectinophora gossypiella* Saunders, in Egypt, to the sex pheromone 'gossyplure'. *Agricultural Research Review* (Cairo)(1982 publ. 1984) 60, 181–199.

Melamed, Y. and Shoham, Ch. (1975) *Insect Scouting in Cotton Fields*. Centre for International Agricultural Co-operation, Rehovot, Israel, 17pp.

Menon M.V. and Thangavelu, K. (1979) Survey of beneficial arthropods in the cotton ecosystem at Coimbatore, South India. *Entomon* 4, 281–284.

Metwally, A.G. and Hosny, M.M. (1972) Factors affecting the resting stage duration of the pink bollworm, *Pectinophora gossypiella* (Saunders). *Agricultural Research Review* (*Cairo*) 50, 21–24.

Metwally, A.G. and Hosny, M.M. (1974) The approximate flight range of pink bollworm moths and the rate of infestation at cardinal direction around villages. *Bulletin of the Entomological Society of Egypt* 58, 55–64.

Metwally, A.G., El-Lakwah, F.A., Shalaby, F.F. and El-Gemely, H.M. (1986) Natural role of diseases against the pink bollworm, *Pectinophora gossypiella* (Saund.) in Egyptian cotton fields. *Agricultural Research Review* (*Cairo*) 61, 1–21.

Meyrick, E. (1905) Descriptions of Indian Microlepidoptera. *Journal of the Bombay Natural History Society* 16, 580–619.

Meyrick, E. (1918) *Exotic Microlepidoptera* 2, 136.

Moawad, G.M. (1981) Survival of pink bollworm, *Pectinophora gossypiella* (Saund.), under various soils and climatic conditions. *Agricultural Research Review* (*Cairo*) 59, 99–105.

Nandal, A.S. and Singh, Z. (1985) Effect of different heights of stacks of cotton sticks on the survival of overwintering *Pectinophora gossypiella* (Saunders) larvae. *Haryana Agricultural University Journal of Research* 15, 313–316.

Nandal, A.S., Singh, J.P., Singh, Z. and Lather, B.P.S. (1984) Effect of the direction of stacking cotton sticks on the mortality of the diapausing larvae of pink bollworm. *Indian Journal of Agricultural Sciences* 54, 671–673.

Naresh, J.S. and Balan, J.S. (1985) *Pyemotes ventricosus* (Newport), an ectoparasitic mite on cotton pink bollworm *Pectinophora gossypiella* (Saunders) in Haryana. *Indian Journal of Entomology* 47, 239–240.

Naumann, I.D. and Sands, D.P.A. (1984) Two Australian *Elasmus* spp. (Hymenoptera: Elasmidae), parasitoids of *Pectinophora gossypiella* (Saunders) (Lepidoptera: Gelechiidae): their taxonomy and biology. *Journal of the Australian Entomological Society* 23, 25–32

Neumark, S. and Teich, I. (1973) Pink bollworm, constant-level liquid device for use in trapping moths. *Journal of Economic Entomology* 66, 298.

Neumark, S., Green, N. and Teich, I. (1972) The pink bollworm attractant, Hexalure, improvement by formulation with an antioxidant. *Journal of Economic Entomology* 65, 1709–1711.

Noble, L.W. (1969) *Fifty Years of Research on the Pink Bollworm in the United States*. US Department of Agriculture, Agricultural Research Service, Agricultural Handbook No.357, 62pp.

Oatman, E.R. (1978) Pink bollworm. In: Clausen, C.P. (ed.) *Introduced Parasites and Predators of Arthrodpod Pests and Weeds*. US Department of Agriculture, Agricultural Research Service, Agricultural Handbook No. 480, pp. 186–188.

Orphanides, G.M., Gonzalez, D. and Bartlett, B.R. (1971) Identification and evaluation of pink bollworm predators in southern California. *Journal of Economic Entomology* 64, 421–423

Ouye, M.T. and Butt, B.A. (1962) A natural sex lure extracted from female pink bollworms. *Journal of Economic Entomology* 55, 419–421.

Ouye, M.T., Garcia, R.S. and Martin, D.F. (1964) Determination of the optimum sterilising dosage for pink bollworms treated as pupae with gamma radiation. *Journal of Economic Entomology* 57, 387–390.

Owen, W.L. and Calhoun, S.L. (1932) Biology of the pink bollworm at Presido, Texas. *Journal of Economic Entomology* 25, 746–751.

Page, F.D., Modini, M.P. and Stone, M.E. (1984) Use of pheromone trap catches to predict damage by pinkspotted bollworm larvae in cotton. In: *Pest Control: Recent Advances and Future Prospects.* Proceedings of the Fourth Australian Applied Entomological Research Conference, 24–28 September 1984, Adelaide. South Australia Government Printer, Adelaide, pp. 68–73.

Passlow, T. and Sabine, B.N.E. (1963) Two pink bollworms of cotton. *Queensland Agricultural Journal* 89, 354–356.

Pawar, A.D., Prasad, J., Asre, R. and Singh, R. (1983) Introduction of exotic parasitoid, *Chelonus blackburni* Cameron, in India for the control of cotton bollworms. *Indian Journal of Entomology* 45, 436–439.

Pearson, E.O. (1958) *The Insect Pests of Cotton in Tropical Africa.* Empire Cotton Growing Corporation and Commonwealth Institute of Entomology, London, x + 355pp.

Prasad, J., Pawar, A.D. and Singh, P., (1985) Role of exotic and indigenous parasites *Trichogramma brasiliensis* Ashmead, *T.pretiosum* Riley, *T.acheae* Nagaraja and Nagarkati, *Chelonus blackburni* Cameron and *Bracon kirkpatricki* (Wilkinson) for the control of cotton bollworms in Hissar (Haryana) India. *Proceedings of a National Symposium on Pesticide Residues and Environmental Pollution* 2–4 October 1985, Muzaffarnagar, India. Sanatan Dharm College, Muzaffarnagar, India, pp. 190–194.

Quisumbling, A.R. and Kydonieus, A.F. (1982) Laminated structure dispensers. In: Kydonieus, A. F. and Beroza, M. (eds) *Insect Suppression with Controlled Release Pheromone Systems,* Vol. 1, CRC Press, Boca Raton, Florida, pp. 213–235

Qureshi, Z.A. and Ahmed, N. (1987) Evaluation of attracticide for the control of pink bollworm. *Pakistan Journal of Scientific and Industrial Research* 30, 380–381.

Qureshi, Z.A., Ahmed, N. and Bughio, A.R. (1985) Efficacy of gossyplure for the control of pink bollworm, *Pectinophora gossypiella* (Saund.) (Lepidoptera: Gelechiidae). *Zeitschrift fur Angewandte Entomologie* 100, 476–479.

Raina, A.K. and Bell, R.A. (1974a) Influence of dryness of the larval diet and parental age on diapause in the pink bollworm *Pectinophora gossypiella* (Saunders). *Environmental Entomology* 3, 316–318

Raina, A.K. and Bell, R.A. (1974b) A nondiapausing strain of pink bollworm from southern India. *Annals of the Entomological Society of America* 67, 685–686.

Reed, W., Vedamoorthy, G., Vijaya Raghavan, M. and Rajan, M.P. (1975) Pink bollworm moths, catches in sex attractant traps and nocturnal behaviour in South India. *Cotton Growing Review* 52, 350–359.

Reynolds, H.T. (1980) Insecticides for the control of pink bollworm. In: Graham, H. M. (ed.) *Pink Bollworm Control in the Western United States.* US Department of Agriculture, Science and Education Administration, ARM-W-16, pp. 35–39.

Rice, R.E. and Reynolds, H.T. (1971a) Seasonal emergence and population development of the pink bollworm in California. *Journal of Economic Entomology* 64, 1429–1432.

Rice, R.E. and Reynolds, H.T. (1971b) Distribution of pink bollworm larvae in crop residues and soil in southern California. *Journal of Economic Entomology* 64, 1451–1454.

Ripper, W.E. and George, L. (1965) *Cotton Pests of the Sudan. Their Status and Control.* Blackwell Scientific Publications, Oxford, xv + 345pp.

Saini, R.K. (1985) Triazophos – a promising insecticide for the control of pink bollworm on cotton. *Pesticides* 19(3), 45–46.

Saleh, M.K. and El-Din, A.E.H. (1982) *The Efficiency of some Insecticides against Cotton Bollworms in Relation to their Effect on Fiber Quality of Giza 70 Cultivar.* Research Bulletin, Faculty of Agriculture, Ain Shams University, No. 1919, 17pp.

Saunders, W.W. (1843) Description of a species of moth destructive to the cotton crops in India. *Transaction of the Entomological Society, London* 3, 284–285.

Schwartz, P.H. (1983) Losses of yield in cotton due to insects. In: *Agricultural Handbook, 1983.* US Department of Agriculture, Agricultural Research Service, Beltsville, Maryland, pp. 329–358.

Sekhon, B.S. and Varma, G.C. (1983) Parasitoids of *Pectinophora gossypiella* (Lepidoptera: Gelechiidae) and *Earias* spp. (Lepidoptera: Noctuidae) in the Punjab. *Entomophaga* 28, 45–54.

Sevacherian, V. and El-Zik, K.M. (1983) *A Slide Rule for Cotton Crop and Insect Management.* Cooperative Extension, Division of Agricultural Sciences, University of California, Leaflet 21362, 13pp.

Sevacherian, V., Toscano, N.C., Van Steenwyk, R.A., Sharma, R.K. and Sanders, R.R. (1977) Forecasting pink bollworm emergence by thermal summation. *Environmental Entomology* 6, 545–546.

Sharma, R.K., Rice, R.E., Reynolds, H.T. and Shorey, H.H. (1971) Seasonal influence and effect of trap location on catches of pink bollworm males in sticky traps baited with Hexalure. *Annals of the Entomological Society of America* 64, 102–105.

Sharma, R.K., Rice, R.E., Reynolds, H.T. and Hannibal R.M. (1973a) Effect of trap design and size of Hexalure dispensers on catches of pink bollworm males. *Journal of Economic Entomology* 66, 377–379.

Sharma, R.K., Mueller, A.J., Reynolds, H.T. and Toscano, N.C. (1973b) Techniques for trapping pink bollworm males. *California Agriculture* 27, 14–15.

Shiller, I., Noble, L.W. and Fife, L.C. (1962) Host plants of the pink bollworm. *Journal of Economic Entomology* 55, 67–70.

Shorey, H.H., Kaae, R.S. and Gaston, L.K. (1974) Sex pheromones of Lepidoptera. Development of a method for pheromonal control of *Pectinophora gossypiella* in cotton. *Journal of Economic Entomology* 67, 347–350.

Shorey, H.H., Gaston, L.K. and Kaae, R.S. (1976) Air-permeation with gossyplure for control of the pink bollworm. In: Beroza, M. (ed.) *Pest Management with Insect Sex Attractants.* American Chemical Society, Washington, DC, 1976, pp. 67–74.

Shu, C.N., Gao, C.Y. and Zhang, Y.X. (1987) Control of pink bollworm, by mating disruption with microencapsulated gossyplure. *Chinese Journal of Biological Control* 3, 106–108.

Sidhu, A.S. and Dhawan, A.K. (1978) Note on the evaluation of some new insecticides against the pink bollworm of cotton. *Indian Journal of Agricultural Sciences* 48, 188–190.

Sidhu, A.S. and Dhawan, A.K. (1981) Testing of azinphos-methyl for the control of pink bollworm. *Pesticides* 15(2), 7–9.

Sidhu, A.S., Dhawan A.K. and Singh, K. (1979) Testing of new chemical for control of cotton pests. *Pesticides* 13(11), 7–11.

Simwat, G.S., Sidhu, A.S. and Dhawan, A.K. (1982) Mortality of diapausing larvae of pink bollworm *Pectinophora gossypiella* (Saund.) in cotton stacks during summer in Punjab. *Journal of Research*, Punjab Agricultural University 19, 35–38.

Simwat, G.S., Sidhu, A.S. and Dhawan, A.K. (1985) Effect of height of cotton stacks and some other factors on the survival of diapausing larvae of pink bollworm *Pectinophora gossypiella* (Saund.) during summer in the Punjab. *Indian Journal of Entomology* 47, 328–332.

Singh, J. and Sidhu, A.S. (1982) Carryover sources for *Pectinophora gossypiella* (Saunders) and parasitoid *Apanteles angalati* Muesebeck. *Journal of Research, Punjab Agricultural University* 19, 217–221.

Singh, J. and Sidhu, A.S. (1983) Schematic model of hirsutum cotton phenology and pink bollworm incidence for pest management. *Indian Journal of Ecology* 10, 310–315.

Singh, J., Sandhu, S.S. and Sidhu, A.S. (1988) Adult emergence pattern of parasitoid *Apanteles angaleti* and its known hosts during the off-season in Punjab. *Entomophaga* 33, 309–314.

Singh, P., Nandeshwar, S.B. and Ratan, R. (1984) Marker lines of *Gossypium hirsutum* Linn. in relation to bollworm infestation. *Indian Journal of Agricultural Sciences* 54, 134–136.

Singh, R. and Singh, Z. (1984) A note on the mite, *Pyemotes herfsi* (Oudemans), ectoparasitic on diapausing pink bollworm larvae. *Current Science* 53, 53–54.

Smee, C. (1940) Report of the entomologist 1939. Department of Agriculture, Nyasaland, 11pp, typescript.

Sohi, G.S. (1984) Pests of cotton. In: Pant, N. C. (ed.) *Entomology in India, 1938–1963.* The Entomological Society of India, New Delhi, pp. 111–148.

Squire, F.A. (1937) A theory of diapause in *Platyedra gossypiella* Saund. *Tropical Agriculture* 14, 299–301.

Staten, R.T., Flint, H.M., Weddle, R.C., Quintero, E., Zarate, R.E., Finnell, C.M., Hernandes, M. and Yamamoto, A. (1987) Pink bollworm (Lepidoptera: Gelechidae): large-scale field trials with a high-rate gossyplure formulation. *Journal of Economic Entomology* 80, 1267–1271.

Stern V. and Sevacherian, V. (1978) Long-range dispersal of the pink bollworm into the San Joaquin Valley. *Californian Agriculture* 32(7), 4–5.

Stone, N.D. and Gutierrez, A.P. (1986a) Pink bollworm control in southwestern desert cotton. I. A field-orientated simulation model. *Hilgardia* 54, 1–24.

Stone, N.D. and Gutierrez, A.P. (1986b) Pink bollworm control in southwestern desert cotton. II. A strategic management model. *Hilgardia* 54, 25–41.

Stone, N.D., Gutierrez, A.P., Getz, W.M. and Norgaard, R. (1986) Pink bollworm control in southwestern desert. III. Strategies for control: an economic simulation study. *Hilgardia* 54, 42–56.

Sukhija, H.S., Butter, N.S. and Singh, J. (1985) Efficacy of new organo-phosphatic insecticides against bollworms on hirsutum cotton. *Pesticides* 19(10), 45–46, 52.

Sundaramurthy, V.T., Natarajan, K. and Basu, A.K. (1985) Effect of the crop terminator on the population collapse of late season larvae of *Pectinophora gossypiella* Saunders in the cotton econiche. In: Regupathy, A. and Jayaraj, S. (eds) *Behavioural and Physiological Approaches in Pest Management.* Tamil Nadu Agricultural University, Coimbatore, India, pp. 194–196.

Surulivelu, T. (1985) Effect of gossyplure on field management of pink bollworm *Pectinophora gossypiella* (Saunders) In: Regupathy, A. and Jayaraj, S. (eds) *Behavioural and Physiological Approaches in Pest Management.* Tamil Nadu Agricultural University, Coimbatore, India, pp. 68–72.

Taneja, S.L. and Dhindwal, A.S. (1982) Bollworm incidence as affected by sowing date, nitrogen application and plant population in upland cotton. *Indian Journal of Plant Protection* 10, 1–6.

Taneja, S.L. and Jayaswal, A.P. (1981) Capture thresholds of pink bollworm moths on hirsutum cotton. *Tropical Pest Management* 27, 318–324.

Tao, Z.X. and Jia, P.H. (1986) Advance in research of integrated control of major diseases and insect pests on crops during the sixth five-year plan period. *Plant Protection* 12, 25–26.

Taylor, T.H.C. (1936) Report on a year's investigation of *Platyedra gossypiella* (pink bollworm) in Uganda (March 1935 to April 1936). *Report of the Department of Agriculture, Uganda, 1935–36* Pt. 2, pp. 19–39.

Teich, I., Neumark, S. and Jacobson, M. (1977) The capture threshold of male pink bollworm moths with gossyplure, and its effect on boll infestation and frequency of insecticidal treatment. *Journal of Environmental Science and Health, (A)* 12, 423–430.

Thompson, W.R. (1946) *A Catalogue of the Parasites and Predators of Insect Pests.* Section I. Parasites host catalogue. Part 8. Parasites of Lepidoptera (N–P). Imperial Agricultural Bureaux, London, 523pp.

Toscano, N.C. and Sevacherian, V. (1980) Pink bollworm monitoring methods. In: Graham, H. M. (ed.) *Pink Bollworm Control in the Western United States.* US Department of Agriculture, Science and Education Administration, ARM-W-16, pp. 40–45.

Toscano, N.C., Mueller, A.J., Sevacherian, V., Sharma, R.K., Nilus, T. and Reynolds, H.T. (1974) Insecticide applications based on Hexalure trap catches versus automatic schedule treatments for pink bollworm moth control. *Journal of Economic Entomology* 67, 522–524.

Toscano, N.C., Van Steenwyk, R.A., Sevacherian, V. and Reynolds, H.T. (1979) Predicting population cycles of the pink bollworm by thermal summation. *Journal of Economic Entomology* 72, 144–147.

Tucker, R.W.E. (1939) Cotton growing and cotton pests control. *Agricultural Journal,* Department of Science and Agriculture, Barbados 8, 3–7.

University of California (1984) *Integrated Pest Management for Cotton in the Western Region of the United States.* University of California, Division of Agricultural and Natural Resources Publication 3305, 144pp.

Van Steenwyk, R.A., Ballmer, G.R., Page, A.L., Ganje, T.J. and Reynolds, H.T. (1978) Dispersal of rubidium-marked pink bollworm. *Environmental Entomology* 7, 608–613.

Vaissayre, M. (1983) Pyrethroid–organophosphate combination for the protection of cotton crops: selection of the most effective proportions. *Coton et Fibres Tropicales* 38, 269–273.

Vaissayre, M. (1987) Attempted eradication of the pink bollworm, *Pectinophora gossypiella* (Saunders), by the mating disruption method in the Bouake Station, Ivory Coast. *Coton et Fibres Tropicales* 42, 267–271.

Von Ramm, C. and Krone, W.W. (1983) Hand application of pheromone dispensers for control of agricultural pests. In: *10th International Congress of Plant Protection* 1, p. 290.

Vosseler, J. (1904) Einige Feinde der Baumwollkulturen in Deutsch-Ostafrika. *Mitteilungen aus den Biologisch – Landwirtschaftlichen Institut, Amani* No.18, 4pp.

Wan, S.Y. and Wan, M. (1982) Preliminary report on controlling cotton insects with Decis. *Insect Knowledge* 19, 25–27.

Watson, T.F. (1980) Methods for reducing winter survival of the pink bollworm. In: Graham, H. M. (ed.) *Pink Bollworm Control in the Western United States.* US Department of Agriculture, Science and Education Administration, ARM-W-16, pp. 29–34.

Watson, T.F. and Johnson, P.H. (1974) Larval stages of the pink bollworm, *Pectinophora gossypiella. Annals of the Entomological Society of America* 67, 812–814.

Watson, T.F., Lindsay, M.L. and Slosser, J.E. (1973) Effects of temperature, moisture and photoperiod on termination of diapause in the pink bollworm. *Environmental Entomology* 2, 967–970

Watson, T.F., Crowder, L.A. and Langston, D.T. (1974) Geographical variation of diapause termination of the pink bollworm. *Environmental Entomology* 3, 933–934.

Watson, W.M. and Guirguis, M.W. (1983) Laboratory and field studies on the efficiency of granular insecticides against cotton pests. In: *10th International Congress of Plant Protection* 1, p. 944.

Watson, W.M., Rashad, A.M. and Hussein, N.M. (1988) Potencies of certain insecticides against the pink bollworm, *Pectinophora gossypiella* (Saund.), as influenced by chemical control programmes in Egypt. *Bulletin of the Entomological Society of Egypt (Economic)* 1986 (published 1988) 15, 79–86.

Wellso, S.G. and Adkisson, P.L. (1964) Photoperiod and moisture as factors involved in the termination of diapause in the pink bollworm, *Pectinophora gossypiella*. *Annals of the Entomological Society of America* 57, 170–173.

Whellan, J.A. (1960) Pink bollworm (*Platyedra gossypiella*) in the Federation of Rhodesia and Nyasaland. *FAO Plant Protection Bulletin* 8, 113.

Willcocks, F.C. (1916) *The Insect and Related Pests of Egypt* Vol.1, part 1. Sultanic Agricultural Society, Cairo, 339pp.

Wilson, A.G.L. (1972) Distribution of pink bollworm, *Pectinophora gossypiella* (Saund.) in Australia and its status as a pest in the Ord irrigation area. *Journal of the Australian Institute of Agricultural Science* 38, 95–99.

Wilson, F.D. (1980) Cotton cultivars resistant to the pink bollworm. In: Graham, H. M. (ed.) *Pink Bollworm Control in the Western United States*. US Department of Agriculture, Science and Education Administration, ARM-W-16, pp. 46–51.

Wilson, F.D. (1987) Registration of three cotton germplasm lines. *Crop Science* 27, 820–821.

Wilson, F.D. (1989) Yield, earliness, and fiber properties of cotton carrying combined traits for pink bollworm resistance. *Crop Science* 29, 7–12.

Zhao, D.X. and Liu, B.Z. (1985) The biological characters and models of *Pectinophora gossypiella* (Saunders) population. *Contributions from Shanghai Institute of Entomology* 5, 67–79.

5 *Earias* Spp. (Lepidoptera: Noctuidae)

W. Reed

Waterside, Sherborne Street, Bourton-on-the-Water, Gloucestershire GL54 2BY, UK

The genus *Earias* is widely distributed in the Old World and Australasia, and some are pests of considerable importance in many of the cotton-growing countries of Africa and Asia. In Africa, the common name spiny bollworm is used indiscriminately for two species, *Earias biplaga* Wlk. and *Earias insulana* (Boisd.), and the latter is often also known as the Egyptian bollworm, having been at one time of great importance in Egypt where *biplaga* is unknown. *Earias insulana* is also an important pest of cotton in India, together with *Earias vittella* F., the two being known there collectively as spotted bollworms. A fourth species, *Earias huegeli* Rogenh., is reported to be of some importance on cotton in Australia, where it is sometimes called the rough bollworm, but there are now doubts about the identity of this species. A fifth species, *Earias cupreoviridis* (Wlk.), is said to be serious in China. Two other species have been recorded from cotton, *Earias chlorana* (L.) and *Earias vernana* (Hb.), both of which are virtually restricted to Europe. Neither is of importance as a pest; the former is most commonly found feeding on willow (*Salix viminalis*) and the latter on poplar (*Populus alba*). The major pest species, however, are strongly associated with the order Malvales. On cotton, they feed on the contents of the flower bud and the green boll, and also act as stem borers, penetrating the tip of the main stem or side shoots.

Distribution and Taxonomy

The genus *Earias* (Lepidoptera: Noctuidae) is confined to the Old World and Australasia. There are about 47 accepted species in the genus and others await description. Variation in the coloration of the moths of some of the species has led to considerable confusion in identification. A review of the systematics of the genus appears to be long overdue (J.D. Holloway, personal communication).

Earias insulana has an extremely wide range, covering most of Africa and including Madagascar, Mauritius and the Canary Isles. It extends northwards to the Mediterranean islands and southern Europe, and eastwards through the Near and Middle East, including southern Arabia, to India, China and Southeast Asia. There are typical specimens in the British Museum (Natural History) collection from Japan, Taiwan and the Philippines, and it is represented in Australia by a form (*smaragdina* Butl.) that is now regarded as synonymous with *insulana.*

Earias biplaga is distributed from the Cape Province of South Africa northwards as far as the arid sub-Saharan region. It is also an important pest on cotton and cacao in Madagascar. Although there are references to this species having been recorded in Egypt, it is likely that these are the result of errors in identification and literature abstraction and that *biplaga* is not established north of the Sahara. Most species of *Earias* show some degree of adult polymorphism and this is very pronounced in *biplaga*. The typical colour pattern seems to be restricted to the female, but a series of gradations occurs connecting this with a form originally described as another species, *citrina* Saalm., and in this form it extends as far as Sokotra (Pearson, 1958).

Earias vittella, which was earlier recorded as *Earias fabia* (Stoll.), is widely distributed from the Indus to Australasia. It is particularly important as a pest of cotton in several areas of India, and also damages this crop in China and Thailand. A recent report (Capizzi, 1987) that this species has been discovered in the Sudan feeding on okra (*Abelmoschus esculentus*) requires confirmation.

Earias insulana completely overlaps *biplaga* in Africa and *vittella* in Asia and Australasia, and predominates over these species in the drier parts of their range.

Earias cupreoviridis has an extremely wide, but discontinuous range. It is known from nearly all parts of Africa south of the Sahara, but there are no well-authenticated records of it attacking cotton in that continent. It is absent from north-east Africa and Arabia, but reappears in India, and here again there appear to be no authentic records of it attacking cotton. It appears to be absent from the Burma–Malaya region but is widely distributed in southern China and spreads out to Japan, Taiwan, the Philippines and the Palau Islands. In China, it is recorded as a pest of cotton, and is said to be of some importance (Li and Chou, 1937).

Earias huegeli is reported to be a cotton pest of some importance in Australia. Pearson (1958) noted that this species is obviously very closely related to *vittella*, and occurs as far east as Tahiti and the Marquesas Islands. However, Holloway (1977) pointed out that the type specimen of *huegeli* (in the Natural History Museum, Vienna) appears to be very similar to, and probably synonymous with, *vittella*. He redescribed specimens from Australasia that differ from *vittella* in possessing distinctive forewing markings, including transverse bands, as *Earias perhuegeli*. Thus it appears that there may be some confusion in the records of the species of *Earias* that attack cotton in Australia.

The picture presented by the species of *Earias* that are important pests of cotton is thus of one species, *insulana*, closely associated with *Gossypium* and a wide range of other members of the Malvales, adapted to existence in the arid regions of the Old World and thence spreading out in all directions; and two other species, *biplaga* and *vittella*, which resemble each other in being commoner than *insulana* in the equatorial regions, but which are barred off from each other by the arid areas of the Middle East and North Africa.

Description of the Stages

The *Earias* species that feed on cotton resemble each other in their early stages, and the descriptions that follow refer to all, except where diagnostic differences are mentioned.

Egg

Spherical, slightly under 0.5 mm diameter, light blue-green when laid, decorated with approximately 30 longitudinal ridges, of which alternate ones project upwards to form a crown, thus causing the egg to resemble in miniature the fruit of a poppy or pomegranate.

Larva

Detailed descriptions of the larvae of the species of *Earias* that occur in India have been given by Gardner (1947) and a key to those feeding on cotton in Africa was provided by McKinley (1968). The full-grown larva (Plate III.2) is about 13–18 mm long, somewhat stout and spindle-shaped, bearing a number of long hairs or setae on each segment, and, in addition, on the last two thoracic and all the abdominal segments, two pairs of fleshy tubercles, one of which is dorsal and the other lateral. These tubercles may be long and slender, or short and rounded, according to the species, and are always more prominent on the last two thoracic segments; each bears a hair at its apex. Additional protuberances occur on the last three segments of the abdomen. The ground colour of the larva is light brown, tinged with grey or green, but of variable depth of colour, distinctly paler along the median dorsal line, with dark brown or black spots at the base of the setae, more pronounced on the second and fifth abdominal segments, which tend to be generally suffused. In addition there are yellowish spots at the base of the tubercles, notably those on the thoracic segments. The larvae of *insulana* and *biplaga* can be distinguished by the fact that *insulana* is generally lighter in colour, the pattern being in grey and yellow rather than brown and deep orange, with the pale band on the head broader and continuous, the tubercles are not as long as those of *biplaga*, and the dorsal tubercles on the 8th abdominal segment are white instead of dark brown. In *vittella*, and *cupreoviridis*, the larval tubercles are much less prominent, particularly on the abdomen, than in *biplaga* and *insulana*.

Pupa

Yellow to chocolate brown, head and tip of the abdomen bluntly rounded, surface smooth or slightly rugose, without projections other than three small ones on the terminal segments; about 13 mm long. The pupa is enclosed in a cocoon shaped like an inverted boat and made of tough, felt-like silk of a dirty-white or pale-brown colour, usually attached to the food plant or to a fragment of plant debris on the ground. Exceptionally, pupation may occur in cracks in the soil, the cocoon being attached to a clod of earth. Tripathi and Krishna (1983) found that the cocoon colour of *E. vittella* varied with the substrate on which it was formed. On a white substrate most of the cocoons were white, but on a darker substrate most cocoons were brown.

Adult

The adult moth is usually seen at rest, with its wings flat and snugly folded, overall length about 12 mm, wing expanse 20–22 mm, with a rather dense, soft coating of scales. The abdomen and hindwing are a uniform silvery or creamy-white, while the head, thorax and forewing vary according to the species and form, but share the same ground colour. The principal diagnostic characters of *Earias* species are the patterns on the forewing, but within each species the extent of development of the patterns, and the ground colour, are extremely variable. In *insulana* the ground colour varies from silvery-green to straw-yellow, the distal fringe on the wings being the same colour, and there are three more-or-less distinguishable transverse lines of a darker shade; traces of a fourth transverse line sometimes appear towards the base of the wing. Occasionally there is a faint discoidal spot in the centre of the wing. In *biplaga* the ground colour varies from a metallic green to gold, with a marginal fringe which is invariably a dark, brownish colour, and three more-or-less distinguishable transverse lines and a central discoidal spot. These lines differ from those of *insulana* in being sinuous, almost angled, and when poorly developed merely show up as a number of dark dots, which are the remnants of the enlargements of the lines where they cross the main veins. Superimposed on this basic pattern is a median, chocolate-brown blotch of irregular outline which develops between the discoidal spot and the posterior edge of the wing, occupying the area between the first two transverse lines: this blotch is usually associated with a green background colour and appears to occur only in the female. A similarly placed brownish blotch occurs rarely in *E. insulana*. Sometimes the ground colour is suffused with light brown, spreading out from the centre of the wing to the edges. The two species, *insulana* and *biplaga*, despite the variability of their colouring, can usually be separated without difficulty on the colour of the fringe and the shape of the transverse lines. The green forms in both species occur during the summer, yellow or brown forms becoming commoner towards the end of the season, when it is becoming cooler and (in the case of areas with a monsoon or equatorial climate) drier (Pearson, 1958).

Couilloud (1983a,b) confirmed Pearson's observations that these seasonal differences in both *insulana* and *biplaga* were caused by environmental factors and concluded that relative humidity was the most important factor determining colour form, while temperature played a secondary role. However, recent observations by Klein (1988) have shown that some colour variants of *insulana* are the result of mutations.

Earias cupreoviridis can be readily distinguished from *biplaga* and *insulana* as the forewing is metallic green with a gold patch on the proximal part of the leading edge, a well-marked spot or pair of spots in the centre of the wing, and a broad chocolate margin with three gold spots in its inner edge. In addition, *E. cupreoviridis* moths are generally considerably smaller than the other species.

Earias vittella moths are also quite distinctive: their ground colour is creamy-white or peach, with a central green wedge running from the proximal to the distal edge of the forewing. The confusion concerning *huegeli/perhuegeli* has been mentioned earlier. These are very similar to *vittella*, but distinctive green transverse lines are usually evident in the latter.

Life History and Feeding Habits

The habits of the different stages of the *Earias* species on cotton do not differ in essentials (Pearson, 1958). Eggs are laid singly and on most parts of the plant, but the favoured sites on cotton are the young shoots and, as they become available, the peduncles and bracteoles or flowerbuds and bolls. Eggs are also laid on the surface of flowerbuds and bolls, particularly in the grooves at the apex of the latter. The leaf lamina is little used. After hatching, the larva may move some distance on its host plant before settling down to feed.

Earias is distinguished from other bollworms of cotton by its marked propensity for stem boring. The larva enters the terminal bud of the vegetative shoot and channels downward from the growing point, or it may enter the internode direct. Only soft, growing tissue is attacked. This type of feeding injury seems to be confined to cotton, and it is most noticeable in the early stages of the crop, before any flowerbuds have formed. The main-stem growing-point is usually attacked; if tunnelling is extensive, the top leaves wilt and the whole apex of the main stem collapses, but if only the apical bud is damaged the attack may pass unnoticed until revealed by the 'twinning' of the main stem, due to growth being carried on by axillary monopodial buds.

In feeding on flowerbuds or green fruits most larvae bore right into the part attacked, the tunnel opening being blocked with excrement. However, many *biplaga* larvae are found feeding on the surface, particularly in open flowers. On cotton bolls, the tunnel often enters from below, coming in at a slight angle to the peduncle. The larvae move about considerably, not confining their feeding to a single boll and often not completely eating out the interior of the bud or boll. The damage may thus be disproportionate to the numbers. Only unripe fruits are attacked.

Normally, there are five larval instars, but Megahed *et al.* (1972) reported that some *E. insulana* larvae passed through only four instars. Moulting normally occurs within the tunnel. When ready to pupate, the fully grown larva spins a cocoon either on the plant, frequently between the boll wall and the bracteoles, or attached to a twig or withered leaf, or amongst the surface debris on the soil. In the South Gujerat region of India, however, pupation is said invariably to occur in the soil, the cocoons being attached to clods in cracks and crevices, mostly within the top 10 cm but sometimes down to 30 cm (Deshpande and Nadkarny, 1936).

The moths are normally quiescent through the day, often resting on the leaves, but will fly when disturbed. They are active at night, feeding on nectar.

Mating usually starts in the early morning of the second day after emergence (Meng *et al.*, 1973).

The length of the different stages largely depends on the temperature and varies greatly with the time of year. Food quality also influences the duration of the larval periods of *insulana* (Meisner *et al.*, 1977) and of *vittella* (Mehta and Saxena, 1973). Records from a number of localities for the three main species show that in the summer the incubation period of the egg is about three days, and the duration of larval life and of the pupal stage are each of the order of two weeks. The pre-oviposition period is about the same as the incubation period. Under optimum conditions, then, a generation can be completed in less than five weeks. Megahed *et al.* (1972) were able to rear ten generations of *E. insulana* in a year in the laboratory in Egypt. In the equatorial regions, with only small seasonal differences in temperature, the life cycle does not vary greatly, but in the regions with a monsoon climate the duration of all stages is prolonged in the cool weather, and at the extremities of the range in Africa (South Africa and Egypt) the life cycle in the winter is extremely prolonged owing to the cessation of development.

Continuous exposure to 13°C proves fatal to any stage of *vittella*, although it can withstand short exposures to lower temperatures. Pearson (1958) postulated that although the pupal period is 32 days at the lowest constant temperature (16°C) at which development can be completed, in places such as Egypt, South Africa and the Punjab, with minimum winter temperatures descending to 5°C, the pupal stage of *Earias* species may last for upwards of two months, owing to the fact that a part of each day is spent at temperatures which, while not lethal, permit no effective development.

At the other end of the temperature range, *Earias* shows considerable powers of resistance. A constant temperature of 40°C is fatal to all stages, but eggs can withstand 40°C for 12 hours or 45°C for four hours, and a small proportion survive 50°C for two hours. Larvae and pupae can survive rather longer exposures to such temperatures.

There appears to be no diapause in the three principal species of *Earias* infesting cotton. The hibernation reported to occur in those parts of the range which have a cold winter can be adequately explained as a straightforward retardation by cold, development being resumed with the spring rise in temperature. The only suggestion of a true diapause is in *cupreoviridis*, which in India has been reported to hibernate as a pupa from October to April, a much longer period than can be accounted for by the temperatures it is likely to encounter in those months. None of the considerable numbers of workers who have dealt with the other species of *Earias* in India and Africa have reported the formation of long-term pupae, and studies on pupal behaviour under field conditions and over a range of controlled temperatures carried out in India, in South Gujerat, where the dry season is very prolonged, revealed no suggestion of diapause (Deshpande and Nadkarny, 1936).

The fecundity of *Earias* species has been studied in the laboratory by many workers, with several publications referring to *vittella* and to *insulana*, but few to the other species, including *biplaga*. *Earias* moths are capable of laying several hundred eggs, for example Megahed *et al.* (1972) recorded up to 732 eggs from one female *insulana*, but the average numbers are much lower and are affected by many factors. These workers recorded strong seasonal differences in fecundity, with an average of 422 eggs per female in September–October but only 164 in July–August. Rather surprisingly, changes in temperature did not appear to have a significant effect on the numbers laid. However, Ahmad and Ghulamullah (1939) recorded considerable reductions in the fecundity of *vittella* when the larvae or moths were kept at higher or lower temperatures than the optimum of 25°C.

The number of eggs is greatly affected by the larval food. Moths bred on cotton shoots gave only 88 eggs against the 399 laid by those bred on *Abelmoschus esculentus* pods, whilst a comparison of eggs laid by moths reared on cotton flower buds, cotton bolls and *A. esculentus* pods gave figures of 196, 342 and 451 eggs per female (Pearson, 1958).

Egg-laying is also affected by adult feeding. Adults normally take nectar freely and if deprived of it the number of eggs laid is only a fraction of the normal. This is partly a result of the reduced length of life of unfed moths (Ahmed and Ghulamullah, 1939). Mani and Krishna (1984) found that *vittella* emerging from white pupal cocoons laid more eggs than those from brown ones.

Earias moths tend to be hardy and long lived; for example *insulana* individuals have survived more than 80 days in the laboratory when adequately fed.

Host Plants

The recorded host plants of the species of *Earias* that are pests of cotton in Africa, Asia or Australasia are virtually confined to the Malvales. The only important exception being that infestations of *insulana* can be found on maize in Egypt (Abul-Nasr and Megahed, 1972). Reed (1974) reported a single record of *biplaga* feeding on maize in Tanzania, and Mehta and Saxena (1973) were able to rear *vittella* on the developing seeds of maize in the laboratory in India. There is also a report of *Earias* larvae feeding on sorghum (Bilapate, 1983).

A very considerable range of herbaceous and shrubby plants of the order Malvales have been recorded as breeding *Earias*. The genera recorded as hosts of *Earias* spp. in Africa (Pearson, 1958; Reed, 1974) and in India (Kashyap and Verma, 1987) are listed in Table 5.1. Unfortunately, owing to the difficulty of identifying the species in the larval stage, few of the published records give the species of *Earias* found.

That there is some specificity in the selection of hosts was shown by studies in the Punjab in which regular collections of *Earias* larvae from different hosts were bred out and the adults identified (Khan and Verma, 1946). This work showed that while *insulana* bred on a number of hosts belonging to different genera (*Abutilon*, *Althaea*, *Gossypium*, *Hibiscus*, *Malva*, *Malvastrum* and *Sida*), *vittella* occurred in quantity on *Gossypium* and *Abelmoschus esculentus* but on nothing else, apart from occasional larvae on hollyhock (*Althaea rosea*), while *cupreoviridis* was plentiful only on *Malvastrum* and *Sida*. The Punjab is towards the northern limit of the distribution of *vittella* and it is possible that the absence of this species from some hosts (notably *Abutilon*) might be because they carry flowers and fruit at a

Table 5.1. Malvales genera recorded as host plants of *Earias*.

	In Africa	In India
Bombacaceae	*Ceiba*	
Malvaceae	*Abelmoschus*	*Abelmoschus*
	Abutilon	*Abutilon*
	Althaea	
	Azanza	
	Cienfuegosia	
	Gossypium	*Gossypium*
	Hibiscus	*Hibiscus*
		Malva
		Malvastrum
	Pavonia	
	Sida	*Sida*
	Thespesia	
	Urena	*Urena*
Sterculiaceae	*Dombeya*	
	Melhania	
	Sterculia	
	Theobroma	*Theobroma*
	Waltheria	
Tiliaceae	*Corchorus*	*Corchorus*
	Grewia	
	Triumfetta	

time when climatic conditions may restrict the breeding of *vittella* but observations made in south India are well within the main range of *vitella*; and where this species is, in fact, much commoner than *insulana*, confirm the Punjab evidence (Cherian and Kylasam, 1941). Here, although *vittella* occurs on *Abutilon*, *insulana* is very much commoner; the reverse is true of *Hibiscus vitifolius*, on which the great majority of the *Earias* found were *vittella*.

There is thus some suggestion, amongst the species of *Earias* occurring in India, of varying degrees of specialization: *Cupreoviridis* being confined to the tribes Malvinae and Sidinae in the Malvaceae (*sens. strict.*); *vittella* being predominantly associated with the Hibisceae; while *insulana* feeds on all these and predominates on the Abutilinae (Pearson, 1958).

Studies in Tanzania and Uganda (Reed, 1974) found similar degrees of host preference to those in India. Here, *cupreoviridis* was restricted to *Sida*, but both *biplaga* and *insulana* were reared from larvae collected from most of the Malvaceae. *Earias insulana* were far more common on the *Abutilon* spp. and *biplaga* predominated on the Sterculiaceae and Tiliaceae. It was not uncommon to find the larvae of both species coexisting on a plant.

The relative proportion of *biplaga* and *insulana* collected from any host did not necessarily indicate the relative attraction of that host to the two species, for there was a seasonal variation in the relative abundance of the hosts and of the insects. For example, *Hibiscus cannabinus* flowered in the wet season when *biplaga* was most common, and *H. panduraeformis* flowered in the dry season when *insulana* predominated. In this case it is probable that the predominance of *biplaga* on the former and *insulana* on the latter simply reflected the relative numbers of egg-laying moths of each species that were flying when the plants were flowering, rather than differential host plant attraction.

The differential attraction of some hosts to *biplaga* and *insulana* was shown by the relative proportions of moths of each species derived from *Earias* larvae collected from three host plant species on and around the Ukiriguru research farm in Tanzania from 1963 to 1969 (Fig. 5.1). *Earias biplaga* predominated on *Waltheria indica* and *insulana* on *Abutilon* spp. throughout each year. However, on cotton *biplaga* was most common through the main flowering/fruiting season (February to May), but *insulana* predominated on this host through the dry season (June to November).

Waltheria indica is a shrub bearing very small flowers and fruits on which *Earias* larvae graze. It has been reported (Reed, 1974) that *biplaga* tend to be surface feeders while *insulana* larvae are more often found feeding inside fruits. These habits may partially explain the observed geographical and seasonal

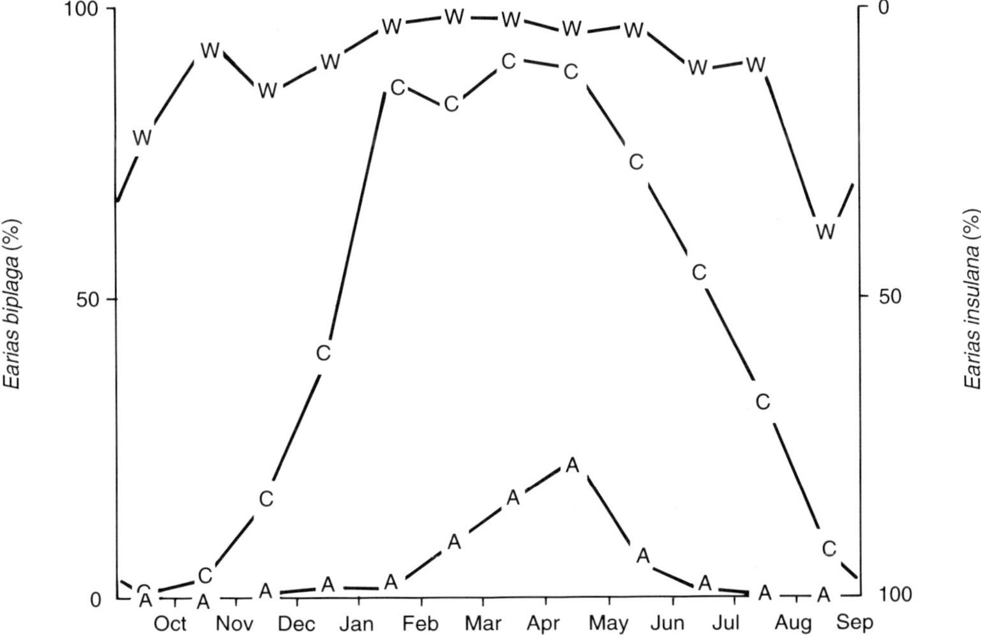

Fig. 5.1. The relative proportions of *E. biplaga* and *E. insulana*, in the total *Earias* moth population, which were recorded from each of three host plants. The moths emerged in the laboratory from larvae collected in each month from Cotton, C—C; *Abutilon* spp., A—A; and *Waltheria indica*, W—W. The data are the means recorded from 1963 to 1969.

distribution differences between the two species, for the surface feeding habit and the long tubercles of *biplaga* make it more vulnerable to desiccation than *insulana*. The long tubercles of *biplaga* provide excellent camouflage when feeding on the small fruits of hosts such as *W. indica* but may be disadvantageous when feeding inside the larger fruited hosts, such as cotton.

Further studies are required to determine why *cupreoviridis* appears to be restricted to *Sida* spp. in Africa and *Sida* and *Malvastrum* spp. in India, but is a major pest of cotton in some areas of China.

Natural Enemies

A large number of species of parasitoids, but relatively few predators have been recorded as feeding on *Earias*. There is little doubt that this imbalance is because it is much easier to observe and record parasitism than predation. It is likely that adequate studies would identify very many species preying on these pests.

Greathead (1966) compiled lists of the insect parasitoids and predators of *Earias* in Africa from published and unpublished records. These lists contained at least 30 species of parasitoids in 24 genera and seven species of predators in five genera. Kashyap and Verma (1987) compiled similar lists for such records in the Indian subcontinent. They listed about 39 species of parasitoids in 20 genera and six species of predators.

Although there have been several species of egg parasitoids (Trichogrammatidae and Mymaridae) recorded from *Earias*, none appear to be very common. The overwhelming majority of the parasitoids recorded attack *Earias* during the larval stage and the only pupal parasitoid of importance appears to be *Goryphus nursei* in India. The lack of pupal parasitism is surprising in view of the fact that the pupa of *Earias* is usually freely exposed, attached to the aerial parts of the host plant (Pearson, 1958).

Most reports indicate that the parasitoids do not appear to discriminate between the species of *Earias*. However, Reed (1970) found that the dipteran parasitoids were more commonly associated with *insulana* and the hymenopteran parasitoids with *biplaga* and *cupreoviridis*. Parasitism rates of more than 30% were common in *Earias* larvae collected from the wild hosts in Tanzania, but lower rates were generally found in the larvae collected from cotton, possibly because the natural enemies were reduced by insecticide use on that crop.

Several pathogens have been reported from *Earias* larvae. Kashyap and Verma (1987) refer to reports that *Serratia marcescens* and *Bacillus cereus* have been isolated from *vittella* in India. Croizier *et al.* (1981) reported that a *Reovirus* caused cytoplasmic polyhedrosis and extensive mortality in both *insulana* and *biplaga*. Reed (1970) recorded that mortality exceeded 60% in *Earias* spp. larvae collected from their major host plants but there was a much lower mortality in those collected from hosts having low infestations. Much of this mortality appeared to be caused by bacterial and viral diseases.

Status of *Earias* as a Pest of Cotton

The importance attached to *Earias* spp. as pests varies considerably from place to place and has in some cases altered greatly with the course of time. In most of the raingrown-cotton areas of Africa it is usually regarded as little more than a nuisance, which only exceptionally causes substantial crop loss. However, in many of the cotton growing areas of North Africa and the Middle East, and in several areas of the Indian subcontinent, it is often the most important component of the pest complex.

The pattern of infestation appears to be similar in most areas where *Earias* occurs. The infestation starts with the shoot-boring phase in the early stages of the crop and is diverted to flowerbuds as they appear. Buds that have been attacked are shed. If they are partly grown the hole made by the larva is conspicuous, but it is likely that some of the losses of very minute buds that are usually ascribed to Mirid damage should really be attributed to very young *Earias* larvae (Pearson, 1958).

Under some circumstances, the attack in the flowerbud stage of the crop may be so severe that the production of bolls is greatly retarded, and a very prolonged, but inefficient, flowering curve is produced. This is notably the case in some parts of India. Finally, as green bolls appear these also are attacked and are either shed or, if beyond the stage at which shedding is possible, remain on the plant as aborted bolls, or as bolls that eventually mature but give partially damaged cotton. The attack often gives the impression of being more severe at the very early and the very late stages of the crop and to ease off in between. This is because in the early, stem-boring stage every larva inflicts clearly visible damage in the shape of a topped plant, whilst later on, when flowerbuds and young bolls are available in quantity, a much larger infestation, feeding in these, will produce effects less readily seen. At an even later stage, when the larvae are feeding in partly grown bolls, their effects are again more conspicuous, and this is accentuated by the secondary invasion of damaged bolls by fungi and bacteria, which often conceal the true cause of the condition. The build-up of the infestation in the crop is essentially a gradual one, there being no sudden influx of egg-laying moths at a particular stage in the crop, as with *Helicoverpa armigera*, nor is there in *Earias* any resting stage such as results, in the case of *Diparopsis* spp., in the bollworm year being started off with a large moth-flight derived from long-term pupae.

Statistics of infestations, where these exist, have chiefly been given in terms of the percentage infestation in green bolls. For example, Bhugio *et al.* (1987) record 23–53% of bolls were damaged by *Earias* in a three-year study in Pakistan. In Egypt, there are occasional reports of 50% or more boll damage. However, a maximum level of 5% damage appears to be more typical (Pearson, 1958). As an indicator of yield loss, counts of damaged bolls suffer from two obvious defects. The first being that it gives no indication of indirect damage due to interference with vegetative structure or the course of flowering. The second being that as the number of green bolls available for attack naturally decreases towards the end of the season, the percentage infestation almost inevitably rises and the highest infestation figures are obtained from the numerically least important part of the crop.

In India, damage by *Earias* has been reported to cause 20% loss of seed yield (Patel, 1949), and 57–80% stained lint (Sidhu and Sandhu, 1977). In most yield-loss studies however, the losses caused by these pests are not distinguished from those caused by the whole of the pest complex. Another problem in quantifying the losses caused by *Earias*, and of other pests, is that most of these studies are conducted in small plots on research farms, rather than in farmers' fields. Pest populations, and the losses they cause, on research stations are often very different to those in farmers' fields. For example, the cropping patterns, alternate host availability and *Earias* populations on a research farm in Tanzania were found to differ markedly from those in the surrounding farmers' fields (Reed, 1974).

Factors Affecting the Status of *Earias*

The principal factors affecting the status of *Earias* as a pest are: its life history; its host plants and their seasonal history; the timing and length of the cotton season; rainfall; and natural enemies. None of these factors works in isolation but it will simplify matters to deal briefly with each in turn and then discuss their interrelations as exemplified by the position of *Earias* in areas where these factors have different weight.

Nature of the life history

The essential feature of the life history of *Earias* is that there is no diapause. The rate of development depends directly on the temperature and if this is favourable the insect must develop, whether other conditions are propitious or not. *Earias* has no means of tiding over periods when food supplies are short. The degree of success of *Earias*, in fact, depends primarily on the goodness of fit between the pattern of the annual cycle of generations imposed on the insect by temperature, and the availability of food supplies.

The supply of food plants is primarily a matter of soil moisture, and the lack of a mechanism in *Earias* for surviving prolonged dry conditions is thus less surprising in the case of *biplaga* and *vittella*, which reach their great prominence in the more equable, equatorial regions of Asia and Africa, in contrast to *insulana*, which is particularly identified with the arid parts of North Africa, North India and the regions connecting them. The prevalence of *insulana* in such regions, however, is associated with irrigated habitats, which provide a potentially continuous supply of food plants. The species has been able to maintain itself successfully in such areas, in fact, purely in association with man. This is in contrast with *Diparopsis*, in which the distribution and the occurrence of a very prolonged pupal resting-stage suggest an ancestral association with rainfed host plants in semi-desert areas.

Host plants

In the Old World, *Earias* larvae are the commonest caterpillars to be found feeding on the flowerbuds and fruits of the genera of Malvales related to

Gossypium. It is natural, therefore, that *Earias* should turn readily to the cultivated cottons, which are bred for the maximum and indeterminate production of relatively large-sized fruits, the spread of which in Africa has provided *Earias* with enormously augmented opportunities for multiplication.

The existence of *Earias* on cotton in any area depends basically upon the existence there of a reservoir of other Malvales, and as the climatic tolerances of cotton and the wild hosts are broadly similar, there are probably no cotton crops in Africa in which *Earias* does not occur. The importance of *Earias* as a pest, however, will depend on the extent to which the cotton crop and the alternative hosts of *Earias* are complementary, or competitive, in respect of the insect.

In most places in Africa cotton, which is normally grown as an annual, and the bulk of the alternative host plants of *Earias*, are fortunately available over about the same period. Most of the important host plants, including several of the *Hibiscus* and *Abutilon* species, seem to be primitively annual in habit, occurring as components of natural or regenerating grasslands in the savannah regions or as riverine and flood-plain species in semi-desert areas, and it is such species that, because of their annual habit, become important weeds of cultivation in nearly all countries. Where irrigation is strictly seasonal, or where rainfall is of the monsoon type, they tend to spring up at the same time as cotton, and their flowering and fruiting periods are much the same as, or start a little later than, those of the crop. Somewhat the same considerations apply where alternate hosts are themselves cultivated plants, notably *Abelmoschus esculentus*, which is particularly favoured as a vegetable crop in Asia and in the arid areas of Africa north of the equator. In all these cases, *Earias* is multiplying on its alternative host plants at the same time as on cotton, and the former do not therefore help it to bridge the period when it is forced to live on something other than cotton.

In the moist equatorial region, with two rainy seasons, conditions may favour the occurrence of a large flora of perennial Malvales. These are shrubby species, occurring as truly wild components of fringing bush surrounding swamps or at the edge of gallery forest, and they provide a supply of food all the year round. Cotton usually occupies only one of the two rainy seasons and it might be expected that it would be heavily infested with the *Earias* population bred up on wild hosts in the other. This must occur to some extent, but as the wild hosts themselves continue to produce flowers and fruit throughout the year, including the time that cotton is available, there is no compulsion for *Earias* to transfer from wild hosts to cotton.

In all circumstances when cotton and other native host plants are fruiting simultaneously the question of host preferences become important. Good evidence is scanty, but on the whole it suggests that cotton is not the first choice of *Earias* (Pearson, 1958).

The main importance of alternative host plants must be sought in the extent to which their fruiting fails to coincide with that of the cotton crop. They may act in one or both of two ways, by providing a food supply on which *Earias* may carry over from one cotton season to the next, or in allowing multiplication early in the season, before cotton is available. Where plant growth is regulated by a strongly seasonal moisture supply, as in areas characterized by a single rainy season or by seasonal flooding, a proportion of the population of annual hosts may survive as biennials or perennials (for example, in tumbledown fallows that have escaped firing, or in intermittently flooded areas). Such plants, having already an established root system, are usually characterized by a flush of flowering well in advance of annual hosts, including cotton. Being, by their nature, in close contact with the cotton crop, such plants provide an early breeding ground for *Earias* which can then transfer readily to the annual crop when this becomes attractive. The alternative host plants, moreover, are not usually uprooted as is the cotton crop, and they may survive in a variety of situations such as fallows, pastures, wasteland, roadsides, river- and canal-banks, and so continue to provide food for *Earias* when cotton is not available.

The timing and length of the cotton season

The ease with which the non-cotton season can be spanned will depend greatly on the time of year in which it occurs. In the savannah regions, cotton is normally grown as a summer, rainfed crop, so that it shares the same growing season as the alternative host plants. But in such regions it is occasionally grown as a dry-season crop on residual moisture, in which case the dry-season cotton crop and the wet-season alternative hosts become complementary to each other and provide *Earias* with an all-year-round food supply. Special interest attaches to regions with an extreme climate where cotton is grown as a summer crop and the winter is cold. This automatically reduces the number of generations through which *Earias* must successfully pass in order to survive the non-cotton season and, if winter irrigation is available or winter rainfall occurs, and alternative hosts survive as weeds or cultivated crops, the close season can readily be spanned.

The length of the cotton season is itself an extremely important factor. In general, the population of *Earias* is at a minimum at the time that cotton first becomes available. Other things being equal, the extent of multiplication in the crop will depend on the number of generations through which the insect passes, and the more protracted the cropping season, the larger will be the size of population of *Earias* eventually reached. *Earias* is thus favoured by varieties of cotton that produce their crop slowly and over a long period, and by pests or diseases that, by interfering with vegetative structure or by causing early flower-bud and boll shedding, tend to prolong the crop. The practice of ratooning constitutes a special case of prolongation of the cotton season, for it provides a particularly favourable host for the build-up of *Earias* before sown cotton becomes available, and is often implicated in cases where *Earias* has become a major pest.

Rainfall

The importance of rainfall as an indirect factor working via the food supply has already been discussed. It seems possible that it also has a direct effect. It has already been remarked that the places where *insulana* has attained real importance as a pest are all arid areas where cotton is grown under irrigation. In regions with a moist climate, *insulana* is replaced by other species, in India by *vittella* and in Africa by *biplaga*. In regions where the species overlap, the proportion of *insulana* in the total *Earias* population on cotton tends to be lowest in the wet weather (Khan, 1945).

Natural enemies

As *Earias* has a large reservoir of plant hosts and lacks a resting stage, thus being available the whole year round, it is usually found to support a considerable parasite population. Although there has been frequent speculation that infestations are largely held in check by the natural enemies, the extent to which *Earias* really is kept in check is not at all well known.

Perhaps the most suggestive evidence regarding the importance of natural enemies comes indirectly. There have been indications from several sources over many years that the use of some insecticides, particularly DDT, to control other pests has resulted in increased damage due to *Earias*, the implication being that natural enemies normally restrain the multiplication of the pest and that insecticides differentially affect these and their host (Reed, 1970).

Interaction of the Factors Affecting the Status of *Earias*

Pearson (1958) provided examples of how he considered the interplay of the factors, discussed in the preceding section, is thought to govern the status of *Earias* in the different regions where it occurs. Although there have been many changes since then, most of his analyses remain relevant and are summarized here.

In Egypt, up to 1913, *E. insulana* was the most serious pest of the national cotton crop. At that time conditions were exactly suited to the pest. Cotton is a summer crop in Egypt and thus *Earias*, entering the crop in May and passing through a first generation developed on buds and stems, was then able to pass through the maximum number of generations under optimum climatic conditions. The strains of cotton commonly cultivated at that time had a prolonged fruiting period and continued to produce their crop well into the autumn and early winter. *Earias* was thus able to continue its multiplication on virtually unlimited food supplies during the summer and autumn months. The Egyptian winter is relatively severe, and the development of *Earias* is retarded to such an extent that the period when food supplies were restricted could be spanned by at most two generations. Agricultural practices up to 1913 were such that even this restriction of food supplies for *Earias* was minimal. Despite the enactment of ordinances regarding the close season for cotton and related crops, plant sanitation was lax and the uprooting of the crop and disposal of crop residues was protracted and inefficient, so that not only were green bolls, containing developing larvae, left on standing plants and in heaps of cut plants, but there was also much growth in the spring from volunteer seedlings and rooted stumps of cotton amongst the succeeding rotation crops. Also, particularly in the northern areas of the Delta, where poor soils prevented good growth of the cotton plant in its first year, the practice of ratooning was prevalent. There was, furthermore, a considerable cultivation of *Abelmoschus esculentus* and *H. cannabinus* as vegetable and fibre crops, and these were not removed at the end of the cotton season.

Thus, in the conditions existing before 1913, the period from October to April inclusive could be spanned by at most three generations, the first matured on the tail end of the cotton crop, the second matured on green bolls in cut or standover cotton, or an alternative hosts, and the third matured on alternative hosts and shoots from standover and ratoon cotton.

However, this situation in Egypt was greatly changed by the advent of the pink bollworm *Pectinophora gossypiella*. This introduced insect was first identified in Egypt in 1910 and by 1913 had become the most important pest in the country. Two effects followed: the first was an immediate one, in that the multiplication of pink bollworm in the cotton crop was so great that the late bolls were largely destroyed, this truncation of the season being subsequently rationalized by the development of earlier-cropping strains such as Sakel. The second, and indirect effect, was that under the threat from pink bollworm more stringent legislation was enacted and enforced, prescribing the uprooting of the cotton crop and prohibiting the ratooning of cotton or the growing of *A. esculentus* and *H. cannabinus* during the close season. The result of these changes was that *Earias*, instead of enjoying a large final multiplication in the late-autumn brood on cotton, followed by at most two winter and spring generations with considerable food available, with the possibility of the first generation of the new year developing on more advanced ratoon plants, now had to survive three winter and spring generations on a greatly diminished food supply, having had its preparatory multiplication much reduced. The effect upon *Earias* was that it sank immediately to the level of a minor pest. More recently, there has been some relaxation in the enforcement of the legislation, and damage caused by *E. insulana* again appears to be of some concern.

A completely different situation occurs in the Gezira region of the Sudan. Here the cotton crop occupies the relatively cooler months of the year, and it is the non-cotton season which is hot, with the result that *Earias* is compelled to pass through about six generations during the close season. This close season is strictly enforced, so that there is no cotton on which *Earias* can breed, whilst alternative hosts are at a minimum in the intensely hot, dry months of April–June, and there is a close season for one of the most important of them, *A. esculentus*, from June–September. As a result, *Earias* is not a major cause for concern in this area.

In the tropical savannah areas north and south of the equatorial zone in Africa, cotton is usually grown in the single rainy season. As many of the principal alternative host plants are annual in habit, they support *Earias* at the same time as does cotton, so diluting the attacks on the crop. Two other facts of importance characterize these areas. The strains of cotton developed in them are essentially early-fruiting, quick cropping types, providing ample food supplies for *Earias* for only a limited period, and a close season has been fairly strictly enforced, at least in some areas. Thus, the multiplication of *Earias* in the cotton crop is restricted and the insect is exposed to the severest conditions during the close season. These conditions are intensified by the fact that although the close season is mainly cool, it becomes extremely hot and dry towards its end, thus providing the greatest compulsion to *Earias* to develop whilst ensuring that food supplies on which it can do so are at a minimum.

In Uganda, in the moist equatorial zone, the situation is more complex. In the wetter areas, in addition to annual and semi-perennial host plants in resting land and rough pasture, there is a reservoir of true perennial Hibisceae at the edge of the bush fringing swamps and watercourses; this, and the fact that rain falls in most months of the year, ensures favourable conditions for the multiplication of *Earias* throughout the year. Cotton might therefore be expected to suffer considerably from *Earias* attack, particularly as the season is relatively long, the crop being sown in June–July and not being uprooted until the following March. This expectation is to some extent fulfilled: stem boring early in the season is sufficiently common to be a serious nuisance on experimental plots and a steady, though never spectacular, toll of flowerbuds and young bolls is taken throughout the season. However, there are evidently factors in Uganda that restrict the invasion of cotton and subsequent multiplication of the pest. There is the likelihood that *Earias* and its parasitoids are in a state of equilibrium on the perennial host-plants, and that the natural enemies hold the insect in much tighter check than in cases where there are wide fluctuations in the insect population during the year. There is also the possibility that rainfall itself plays a part in restricting the development of massive infestations during the greater part of the cotton season, both by washing larvae from the plants and providing high humidity in which some of the insect pathogens thrive.

Status of *Earias* on Cotton in India and Pakistan

Earias is an important pest in nearly all the principal cotton growing areas of the Indian subcontinent. *Earias insulana* is the most abundant species on cotton in the Punjab, except during the rainy season in the less arid regions of the east and southeast of that region, when *vittella* may dominate. In peninsular India, *vittella* is generally the predominant species. Nothing definite is known to account for this difference, but it is possible that *vittella* is unable to tolerate the low minimum temperatures of the northern Punjab winter, since in parts of its range, for example, Indore, Kanpur and Delhi, it experiences as high maximum temperatures and as low relative humidities as occur in the more northern areas where it is rare.

The reasons why *Earias* does so much damage in the Indian subcontinent become clear when conditions there are examined in the light of the evidence from Africa.

In the Punjab, most cotton is grown as a summer crop, under irrigation, being sown in April–May and harvested from October to January. This situation rather resembles that which existed in Egypt before the cotton season there was shortened by pink bollworm. Moreover, temperatures in summer are higher, inducing more rapid development and thus greater multiplication of *Earias*, and the temperatures in winter are rather low, making this period easier for the pest to span. *Earias* survives this period mostly on sprouted cotton and records show that extensive multiplication takes place on such ratoon cotton in April–June (Khan, 1945). This is assisted by breeding on alternative host plants, notably the common perennial weed, *Abutilon indicum*, which flowers and fruits in the spring, from February to April. Conditions in the Punjab have thus been even more favourable to *Earias* than those existing in Egypt at the time when the pest was most serious there. Recently, however, several farmers in some areas of the Punjab have replaced the traditional long duration cotton varieties with a much shorter duration cultivar in response to the very damaging attacks by *Earias* and *Pectinophora* and the increasing costs of controlling these with pesticides.

In much of peninsular India cotton is a raingrown crop, sown at the break of the monsoon, in late June or early July. Long duration cultivars and hybrids are grown in several areas, so the crop is available as a host for *Earias* until February or even later. *Abelmoschus esculentus* provides an important means of carry-over from one cotton season to the next, being grown on a small scale under well-irrigation in the typically hot, dry summers and sown more extensively with the first rains, for it supports large populations of *vittella*. Cotton itself also assists in the carry-over, for plants not removed after harvest continue to produce flowerbuds and bolls, whilst of those which are chopped out, several sprout again and continue to provide food for *Earias*. In addition, in some areas of southern India, an irrigated cotton crop has been grown recently in the summer thus providing hosts throughout each year. This continuous host supply produces build-up conditions similar to those in equatorial Africa.

Control

Cultural Measures

Pearson (1958) considered *Earias* to be a serious pest only in those areas where there is no effective close season between successive cotton crops. He therefore suggested that the control of these pests is primarily a matter of creating such a close season, which will be particularly effective where it coincides with a paucity of alternative hosts.

Legislation in many cotton growing countries requires farmers to uproot and destroy their harvested plants by a designated date to ensure an adequate close season, but this is seldom strictly enforced. Uprooting well-grown plants, particularly from dry or heavy soils, is a strenuous task. Most farmers prefer to simply chop off the stems which can be used as fuel or leave the plants until the next rainy season when they can be more easily uprooted with the aid of a plough. If adequate soil moisture is available the plants and stumps rapidly produce new growth, and so provide food for *Earias* and other pests. Vigorous campaigns are essential, to educate farmers and to convince them of the value of a close season, but in many countries the appropriate resources are not available. Attempts to enforce the legislation with penalties for non-compliance are often ineffective and may be seen to be politically damaging.

Measures to increase the effectiveness of the closed season by persuading farmers to eradicate the alternative host plants have been attempted in some countries and recommended in others. However, many of the alternative hosts are of value, for food, feed or fibre. Also, such eradication may be of doubtful benefit, for it will reduce the pool of natural enemies. In irrigated areas such as the

Sudan Gezira, where a single alternative host, *Abutilon*, appears to be of no value and is known to be the major source of carry-over, host eradication may be feasible, but not in raingrown-cotton areas where there is a plethora of hosts. Farmers in Sukumaland, Tanzania were urged to uproot all *Hibiscus* and *Abutilon* plants (Harris, 1942), but with little success. The futility of such a recommendation was evident when *Waltheria indica* was found to be the predominant natural host for *Earias* in that region (Reed, 1974). As this plant was very common, with an average of over 200 plants per hectare, a recommendation to eradicate it was considered to be impracticable. In recent years, *Earias* infestations on cotton in that same area have noticeably declined. This may be associated with much greater land utilization and overgrazing, which has greatly reduced the numbers of the alternative hosts, particularly in the dry seasons.

Several workers (Kashyap and Verma, 1987) have suggested that farmers should inspect their cotton regularly throughout the vegetative stage and remove all wilting shoots, thus destroying the tip boring *Earias* larvae and consequently reducing the subsequent generations that damage the buds, flowers and bolls. In some areas topping, which is the removal of the terminal shoots of young plants, is widely practised. Some farmers allow their livestock to graze their cotton during the vegetative stage, so producing a similar effect. The benefits, if any, of topping have always been controversial. However, a three-year study in Egypt (Nasr and Azab, 1969) found that topping reduced *E. insulana* infestations and increased yields.

The effects of several other cultural practices on *Earias* have been studied. Close spacing (Abdel-Fatah *et al.*, 1980), early sowing (Bishara, 1969) and deep ploughing (Faseli, 1977) have all been found to decrease damage, but high doses of nitrogenous fertilizers increased the incidence of the pest.

Chemical Control

There are very many reports of research on the use of insecticides against *Earias*, and many chemicals have been found to be effective in the control of this pest (Kashyap and Verma, 1987). Although the larvae are inaccessible to contact insecticides when feeding inside the stems and fruiting bodies, they crawl over the plant surface to seek fresh food sources and so become contaminated. However, some widely used insecticides are ineffective. For example, in a large-scale experiment in the Gash Delta, Sudan the effect of two sprayings with DDT at 1 lb per acre was to increase the infestation of *Earias* in the sprayed fields; this may have been connected with preferential oviposition on the sprayed plants, which developed vigorous secondary growth (Joyce, 1956).

Endosulfan and carbaryl appear to have been the most widely used chemicals for the control of *Earias* in recent years, but the synthetic pyrethroids, including fenvalerate, are particularly effective and are being increasingly used. There has been considerable research interest in the use of neem (*Azadirachta indica*) kernel extract sprays for the control of *Earias* and other bollworm larvae. In Israel, Meisner *et al.* (1978) found that low concentrations of the extract were effective in controlling *E. insulana*, and many workers in India have found it to be effective against several pests. However, there are mixed opinions on the potential for popularizing the use of this natural insecticide.

Resistance

The dangers of inducing resistance to insecticides in *Earias* must be emphasized. In Israel, where endrin was extensively and intensively used for the control of *E. insulana*, strong resistance to this insecticide developed within three years (Ascher and Tahori, 1957).

Population monitoring and thresholds

Economic and environmental factors make it essential that insecticides should only be used against *Earias* when necessary. To ensure this, economic thresholds have to be determined, and some form of monitoring of the *Earias* populations is essential. As the eggs of this pest are very difficult to find, for they are almost the same colour as the plant surfaces on which they are laid, most monitoring relies upon counts of the larvae or of the damaged fruiting bodies. Many factors interact to effect the economic threshold, which will vary across space and time, so it must be determined for each area and be kept under regular review. Thresholds in the region of 50 larvae per 100 plants appear to be commonly recommended. A very large range of damage thresholds has been reported, for example, Mahalle *et al.* (1976) found that *Earias* control was needed when boll damage exceeded 3.4% but Mulchandani *et al.* (1980) found 25% to be the threshold.

Vaissere (1974), working in Madagascar, found that differences in population aggregation within crops rendered sequential sampling methods less useful for *E. insulana* than for *H. armigera.* He found that populations of more than 500 larvae per hectare merited pesticide use.

Light traps

There have been several attempts to use traps, particularly light traps, to monitor *Earias* populations and to use these data to supplement crop scouting data to determine the need for pesticide use. For example, in southern India, light-trap catches were found to be a useful supplement to scouting in determining the timing of pesticide use against *Earias* species and other pests in a large integrated pest management scheme on cotton (Thangavelu, 1982). However, in Tanzania, Reed (1970) found that although there were very good correlations between light-trap catches, when averaged over four-week periods, and counts of larvae on crops, there were very pronounced lunar cycle effects. Catches of *biplaga* in moonless nights were six times greater than those in full moon nights so there was a very weak correlation between an individual night's trap catch and the crop infestation. Consequently, such data were of no use in the daily decision making of whether or not to apply pesticides against this pest. As the lunar cycle effects on *insulana* catches were much smaller (1.7 : 1) a light trap may be a more useful monitoring device for this species. These studies also indicated that each moth caught in a trap, incorporating a 125 W mercury vapour lamp, was equivalent to average populations of 20 *insulana* or 100 *biplaga* moths per hectare. Such low relative catches indicate that light traps are unlikely to be of use in the direct control of these pests through mass trapping.

Timing of pesticide application

As *Earias* tends to be a pest that builds up within the cotton crop through several generations, rather than being an invasionary pest such as *Helicoverpa*, no more than two or three effective pesticide applications will normally be needed to keep the populations of this pest in check. As the generations of *Earias* are seldom synchronized, there are usually mixtures of eggs, larvae and pupae on cotton at any one time, so the timing of pesticide application is far less crucial than in the case of pests such as *Helicoverpa* where pesticides must be applied soon after the egg-laying invasion.

In most cases, farmers are faced not with *Earias* as a sole or key pest, but with a pest complex of which it is but one component. For example, *Earias* is generally found in complexes that include other bollworms, particularly *Helicoverpa*, *Diparopsis* and *Pectinophora.* In such cases the thresholds have to be determined on the combined total pest threat, and the insecticide(s) chosen must be effective against the damaging elements of the complex. There are obvious advantages in narrow spectrum insecticides, particularly because these will not destroy the majority of the fauna including the pests' natural enemies. But most farmers prefer to use broad spectrum products that will control their pest complex, rather than stocking several insecticides, each of which controls a single pest species.

Pheromone Use

Components of the female sex pheromones of *vittella* (Cork *et al.*, 1988) and *insulana* (Hall *et al.*, 1980) have been identified and synthetic formulations have shown considerable attraction to the male moths in the field. The use of such synthetic formulations in the control of these pests through mating disruption has shown promise in trials against both species in Pakistan (Critchley *et al.*, 1987).

Natural Enemies and Biological Control

The large number of natural enemies that feed upon *Earias* has been discussed earlier in this chapter. The populations of some of these can build up to levels that cause high levels of pest mortality. For example, parasitism rates of more than 25% in eggs and 37% in pupae were recorded in the Punjab (Sekhon and Varma, 1983) and about 20% in larvae in Madagascar (Peyrelongue and Bournier, 1974). However, extensive studies over several years in Tanzania (Reed, 1970) recorded that parasitism rates in *Earias* on cotton did not build up to high levels (25%) until very late in each season. There the parasitoids appeared to be much less important than pathogens in regulating larval populations on cotton and other hosts, particularly in the rainy season. Also these studies showed that the average apparent parasitism rate in larvae collected from pesticide treated cotton (14%) was considerably less than in those from pesticide free cotton (23%). However, there are reports that some insecticides, including endosulfan, are considerably less toxic than others to the natural enemies of *Earias*.

There have been several attempts to supplement the natural enemies of *Earias* by releasing laboratory reared egg parasitoids in cotton fields in India. Some of these attempts appear to have been successful. For example, Sangwan *et al.* (1972) reported that releases of *Trichogramma brasiliensis* resulted in parasitism levels of more than 70% and increased cotton yields by 35%. However, several other workers subsequently found that such releases gave much less encouraging results.

In Africa, Greathead (1966) considered that the most promising prospect for biological control of any of the bollworms appeared to be through the introduction of *Earias* parasitoids from India. He suggested that the introduction of the pupal parasitoid *Goryphus nursei* might be the best prospect, and the larval parasitoid *Bracon greeni* might also be useful, since it also attacks *P. gossypiella*. Unfortunately there appears to have been no such introductions. Attempts to establish *T. brasiliensis* on cotton in Uganda and Tanzania were not successful. There appear to be no published records of any specific attempts to augment the predators of these pests.

Several pathogens are known to kill *Earias* larvae, and it might be possible to culture and disseminate one or more of these to augment the natural control elements. Of the known pathogens, a *Reovirus* that causes cytoplasmic polyhedrosis (Croizier *et al.*, 1981) may merit further study. Several bacteria and fungi have been recorded to be pathogenic, but there appears to be little chance of developing any of these as practicable pesticides. Some workers, including Taylor (1974) in Nigeria, have found that sprays containing formulations of *Bacillus thuringiensis* give effective control of *Earias* larvae, but others, including Patti and Carner (1974) in India, found such sprays ineffective against this pest.

Host Plant Resistance

Considerable resistance to *Earias* has been recorded in several wild species of *Gossypium* (Anson *et al.*, 1948), and the widely cultivated *G. hirsutum* has been recorded to be more susceptible than *G. barbadense* (Badawy, 1974) and *G. arboreum* (Butani, 1974). Unfortunately, there has been a paucity of research investment in attempts to exploit such sources of resistance in a sustained plant breeding programme. This is understandable given the economic success of insecticide use and the expensive, long-term nature of host plant resistance breeding. However, there has been a recent surge of interest in screening for resistance to *Earias* and other bollworms, following the realization that the long-term dependence on insecticides may be unwise and uneconomic.

Many workers in India have reported encouraging levels of reduced susceptibility to *Earias* in many cotton genotypes (Kashyap and Verma, 1987). Varieties containing the frego-bract and okra-leaf characters have been found to be less susceptible than commonly used commercial cultivars (Thombre, 1980). In Israel, Zur *et al.* (1980) reported that varieties containing high gossypol levels were less susceptible to *E. insulana* attacks. In the Sudan, Khalifa (1979) reported that varieties combining high gossypol content with the frego-bract or the nectariless character were considerably less attacked than other varieties. Sharma and Agarwal (1984), found considerable differences in stem boring by *E. vittella* between cotton genotypes; those with high levels of tannin and gossypol were the least attacked. Conversely, hirsute varieties (Agarwal and Katiyar, 1974) and glandless varieties (Brader, 1969) have been found to be more susceptible to *Earias* than commercial cultivars.

Unfortunately, some of the characters that have been found to confer reduced susceptibility to *Earias* are not desirable in commercial cultivars. For example, high gossypol reduces the value of the seed, and the frego-bract and okra-leaf characters have been associated with poor lint quality and low yields. Similarly, the hirsute leaf character, which conveys increased susceptibility to *Earias*, is most useful in providing resistance to jassids. There is little doubt that, in spite of such paradoxes, useful levels of resistance to *Earias* could be incorporated in commercial cotton cultivars, given a determined and sustained research effort. However, with emphasis on short-term research in most cotton growing countries, it is unlikely that such host plant resistance projects will be adequately funded in the near future.

However, even in the absence of a resistant cultivar, variety choice can have a substantial effect on the losses caused by *Earias* and other pests. For example, in some areas of northern India the farmers recently abandoned the long duration cultivars which they had traditionally grown in favour of a much shorter duration cultivar that was selected by a local farmer. This change appeared to have been prompted by the fact that the short duration cultivar escaped much of the heavy build up of *Pectinophora* and *Earias* that was so damaging on the traditional cultivars.

References

Abdel-Fattah, M.I., Hosny, M.M. and El-Saadany, G. (1980) The spacing and density of cotton plants as factors affecting populations of the bollworms, *Earias insulana* Boisd. and *Pectinophora gossypiella* (Saund.). *Bulletin de la Societe Entomologique d'Egypte* 60, 85–94.

Abul-Nasr, S. and Megahed, M.M. (1972) Infestation of maize ears by the spiny bollworm, *Earias insulana* Boisd. (Lepidoptera: Agrotidae-Acontiinae). *Bulletin de la Societe Entomologique d'Egypte* 55, 447–454.

Agarwal, R.A. and Katiyar, K.N. (1974) Ovipositional preference and damage by spotted bollworm (*Earias fabia* Stoll) in cotton. *Cotton Development* 4, 28–30.

Ahmad, T. and Ghulamullah, (1939) Ecological studies on the spotted bollworms of cotton and their parasites. *Indian Journal of Entomology* 1, 17–47.

Anson, R.R., Knight, R.L., Evelyn, S.H. and Rose, M.F. (1948) Anglo-Egyptian Sudan, Progress Report of the Cotton Breeding Stations of the Empire Cotton Growing Corporation 1946–47, pp. 49–78.

Ascher, K.R.S. and Tahori, A.S. (1957) Resistance of the spiny bollworm, to endrin in Israel. *Nature* 179, 324.

Badawy, A. (1974) The susceptibility of certain American Upland and Sakel cotton varieties to bollworms infestation (Lepidoptera: Arctiidae and Gelechiidae). *Bulletin de la Societe Entomologique d'Egypte* 58, 261–266.

Bhugio, A.R., Qureshi, Z.A. and Hussain, T. (1987) Influence of boll age, size and moisture contents on pink and spotted bollworms infestation in cotton. *Proceedings of the 5th Pakistan Congress of Zoology*, 8–11 January 1986, University of Karachi. Zoological Society of Pakistan, Karachi, pp. 127–131.

Bilapate, G.G. (1983) Spotted bollworm, *Earias* spp. on cotton in India. A review. *Agriculture Review* 4, 95–107.

Bishara, I. (1969) Effect of agricultural factors on cotton yield and bollworm attack. *Technical Bulletin* No. 2, Plant Protection Department, Ministry of Agriculture, United Arab Republic, 28pp.

Brader, L. (1969) La faune des cottoniers sans glandes dans la partie meridionale du Tchad. II Les chenilles de la capsule. *Coton et Fibres Tropicales* 24, 333–336.

Butani, D.K. (1974) Insect pests of cotton, XVII. Effects of cotton varieties, cultural practices and fertilizers on infestation by pink bollworms. *Coton et Fibres Tropicales* 29, 237–240.

Capizzi, A. (1987) *Earias vittella* (F.) (Lepidoptera: Noctuidae), dannosa al cotone, presente anche in Africa Orientale. *Bolletino di Zoologia Agraria e di Bachicoltura* 19, 199–203.

Cherian, M.C. and Kylasam, M.S. (1941) Preliminary notes on the parasites of the spotted and pink bollworm of cotton in Coimbatore. *Proceedings of the Indian Science Academy* 14(B), 517–528.

Cork, A., Chamberlain, D.J., Beevor, P.S., Hall, D.R., Nesbitt, B.F., Campion, D.G. and Attique, M.R. (1988) Components of female sex pheromone of spotted bollworm, *Earias vittella* F. (Lepidoptera: Noctuidae), identification and field evaluation in Pakistan. *Journal of Chemical Ecology* 14, 929–945.

Couilloud, R. (1983a) Les *Earias* du cotonnier en Cote d'Ivoire: *Earias insulana* (Boisd.); *Earias biplaga* (Wlk.) (Lep. Noctuidae Westermanniinae). Variation de l'importance relative de chacune des especes et variations morphologiques intraspecifiques. *Coton et Fibres Tropicales* 38, 187–200.

Couilloud, R. (1983b) Les *Earias* du cotonnier en Cote d'Ivoire: *Earias insulana* (Boisd.); *Earias biplaga* (Wlk.) (Lep. Noctuidae Westermanniinae). Variation de l'importance relative de chacune des especes et variations morphologiques intraspecifiques (suite et fin). *Coton et Fibres Tropicales* 38, 253–264.

Critchley, B.R., Campion, D.G., Cavanagh, G.G. and Chamberlain, D.J. (1987) Control of three major bollworm pests of cotton in Pakistan by a single application of their combined sex pheromones. *Tropical Pest Management* 33, 374.

Croizier, G., Amargier, A., Jacquemard, P., Couilloud, R. and Croizier, L. (1981) Polyedrose cytoplasmique d'*Earias biplaga* Wlk. Lepidoptere Noctuidae due a un *Reovirus* a tropisme tissulaire large. *Coton et Fibres Tropicales* 36, 127–135.

Deshpande, B.P. and Nadkarny, N.T. (1936) *The Spotted Bollworms of Cotton* (Earias fabia *Stoll. and* Earias insulana *Boisd.) in South Gujerat, Bombay Presidency*. Scientific Monograph of the Imperial Council of Agricultural Research, India, No. 10, 208pp.

Faseli, M.D. (1977) Investigations on the biology and control of *Earias insulana* Boisd. (Noctuidae). *Entomologie et Phytopathologie Appliques* 43, 39–54.

Gardner, J.C.N. (1947) On the larvae of the Noctuidae, III. *Transactions of the Royal Entomological Society of London* 98, 59–90.

Greathead, D.J. (1966) *Memorandum on the Parasites and Possibilities of Biological Control of East African Bollworms.* Commonwealth Institute of Biological Control, Uganda, 21pp.

Hall, D.R., Beevor, P.S., Lester, R. and Nesbitt, B.F. (1980) (E,E)-10,12-Hexadecadienal: A component of the female sex pheromone of the spiny bollworm, *Earias insulana* (Boisd) (Lepidoptera: Noctuidae). *Experientia* 36, 152–153.

Harris, W.V. (1942) *Insects Injurious to Cotton in Tanganyika Territory.* Pamphlet of the Department of Agriculture, Tanganyika No. 29, 11pp.

Holloway, J.D. (1977) *The Lepidoptera of Norfolk Island,* W. Junk, The Hague.

Joyce, R.J.V. (1956) Large scale spraying of cotton in the Gash Delta in Eastern Sudan. *Bulletin of Entomological Research* 47, 399–413.

Kashyap, R.K. and Verma, A.N. (1987) Management of spotted bollworms (*Earias* spp.) in cotton – a review. *International Journal of Tropical Agriculture* 5, 1–27.

Khalifa, A. (1979) Breeding for bollworm resistance in cotton, *Gossypium hirsutum. Coton et Fibres Tropicales* 34, 309–314.

Khan, M.H. (1945) Studies on *Earias* spp. (the spotted bollworms of cotton) in the Punjab. I. The relative abundance of *E. fabia* Stoll. and *E. insulana* Boisd. in various parts in relation to environmental conditions. *Indian Journal of Entomology* 6, 15–27.

Khan, M.H. and Verma, P.M. (1946) Studies on *Earias* spp. (the spotted bollworms of cotton) in the Punjab. III. The biology of the common parasites of *Earias fabia* Stoll., *E. insulana* Boisd. and *E. cupreoviridis* Wlk. *Indian Journal of Entomology* 7, 41–63.

Klein, M. (1988) Color mutations in a laboratory colony of the spiny bollworm (*Earias insulana*). *Phytoparasitica* 16, 355–358.

Li, Feng-Swen and Chou, Shao-Mu (1937) The distribution of important cotton insects recorded in Chinese literature. *Entomology and Pathology* 5, 282–302.

McKinley, D.J. (1968) Key to some larvae of lepidoptera attacking cotton in Central Africa. *Cotton Growing Review* 45, 184–197.

Mahalle, V.P., Ghodka, R.D. and Saxena, J.S. (1976) Economic injury level of cotton bollworms. *Indian Journal of Entomology* 38, 27–32.

Mani, H.C. and Krishna, S.S. (1984) Pupal cocoon colour of mated males and females influencing reproductive potential in *Earias fabia* (Lepidoptera: Noctuidae). *Journal of Advanced Zoology* 5, 52–54.

Megahed, M.M., Tawfik, M.F.S. and El Khateeb, A.A. (1972) Method of rearing a culture with studies on annual generations of *Earias insulana* Boisd. in laboratory (Lepidoptera: Arctiidae). *Bulletin de la Societe Entomologique d'Egypte* 56, 347–360.

Mehta, R.C. and Saxena, K.N. (1973) Growth of the cotton spotted bollworm, *Earias fabia* (Lepidoptera: Noctuidae) in relation to consumption, nutritive value and utilization of food from various plants. *Entomologia Experimentalis et Applicata* 16, 20–30.

Meisner, J., Kehat, M., Zur, M. and Ascher, K.R.S. (1977) The effect of gossypol on the larvae of the spiny bollworm, *Earias insulana. Entomologia Experimentalis et Applicata* 22, 301–303.

Meisner, J., Kehat, M., Zur, M. and Eizick, C. (1978) Response of *Earias insulana* Boisd. larvae to neem (*Azadirachta indica* A. Juss) kernel extract. *Phytoparasitica* 6(2), 85–88.

Meng, H.L., Chang, G.S., Du, S.L. and Ge, J.Y. (1973) On the activity rhythms and interspecific dominance of three species of spotted bollworms (In Chinese). *Acta Entomologica Sinica* 16, 32–38.

Mulchandani, L.H., Shah, A.H. and Mehta, N.P. (1980) Preliminary observations on economic threshold for cotton bollworms in IAN/579/188 cotton (irrigated). *Cotton Development* 10, 11–12.

Nasr, E.A. and Azab, A.K. (1969) Effect of cutting the terminal shoots of cotton plants (topping) on rate of bollworm infestation, cotton yield and fibre quality. *Bulletin de la Societe Entomologique d'Egypte* 53, 325–337.

Patel, R.M. (1949) Control of cotton spotted bollworm (*Earias fabia*) in Baroda. *Indian Cotton Growing Review* 3, 135–144.

Patti, J.H. and Carner, G.R. (1974) *Bacillus thuringiensis* investigations for control of *Heliothis* spp. on cotton. *Journal of Economic Entomology* 67, 415–418.

Pearson, E.O. (1958) *The Insect Pests of Cotton in Tropical Africa.* Empire Cotton Growing Corporation and Commonwealth Institute of Entomology, London, pp. 74–95.

Peyrelongue, J. and Bournier, J.P (1974) *Earias insulana* Boisd. (Lep., Noctuidae) et ses parasites sur *Abutilon asiaticum* L. (Malvaceae) dans la region sud-ouest de Madagascar. *Coton et Fibres Tropicales* 29, 241–245.

Reed, W. (1970) The Ecology and Control of *Earias* spp. (Lepidoptera: Noctuidae), in the Western Cotton Growing Areas of Tanzania. Unpublished PhD thesis, University of Reading.

Reed, W. (1974) Populations and host-plant preferences of *Earias* spp. (Lepidoptera: Noctuidae) in East Africa. *Bulletin of Entomological Research* 64, 33–44.

Sangwan, H.S., Verma, S.N. and Sharma, V.K. (1972) Possibility of integration of exotic parasite, *Trichogramma brasiliensis* Ashmead for the control of cotton bollworms. *Indian Journal of Entomology* 34, 360–361.

Sekhon, A.S. and Varma, G.G. (1983) Parasitoids of *Pectinophora gossypiella* (Lep., Gelechiidae) and *Earias* spp. (Lep., Noctuidae) in the Punjab. *Entomophaga* 28, 45–53.

Sharma, H.C. and Agarwal, R.A. (1984) Factors imparting resistance to stem damage by *Earias vittella* F. (Lepidoptera: Noctuidae) in some cotton phenotypes. *Protection Ecology* 6, 35–42.

Sidhu, A.S. and Sandhu, S.S. (1977) Damage due to the spotted bollworm (*Earias vittella* Fab.) in relation to the age of the bolls of *hirsutum* variety J-34. *Journal of Research, Punjab Agriculture University* 14, 184–187.

Taylor, T.A. (1974) Evaluation of Dipel for control of Lepidopterous pests of okra. *Journal of Economic Entomology* 67, 690–691.

Thangavelu, K. (1982) Some observations on light trap catches of cotton bollworm moths. *Cotton Development* 12, 55–57.

Thombre, M.V. (1980) Association of some morphological characters to confer resistance to bollworm in *Gossypium hirsutum* Cotton. *Andhra Agriculture Journal* 27, 19–20.

Tripathi, A.B. and Kirshna, S.S. (1983) Pupation site and color of the pupal cocoon in *Earias fabia* Stoll. (Lepidoptera: Noctuidae). *Bulletin of the Institute of Zoology, Academia Sinica* 22, 273–275.

Vaissere, M. (1974) Elements pour l'application d'un echantillonnage sequentiel des populations larvaires depredatrices dans le declenchement des interventions sur seuil. *Coton et Fibres Tropicales* 29, 367–370.

Zur, M., Meisner, J., Kabonci, E. and Ascher, K.R.S. (1980) Field evaluation of the response of *Spodoptera littoralis* and *Earias insulana* to cotton strains differing in gossypol content. *Phytoparasitica* 8, 189–194.

6 *Diparopsis* Spp. (Lepidoptera: Noctuidae)

J.P. Tunstall

'Austers', Woodmans Green, Whatlington, Nr Battle, Sussex TN33 0NJ, UK

Introduction

Larvae of moths of the genus *Diparopsis* (Noctuidae), designated generally as the red or Sudan bollworm in the early literature (King, 1908; Jemmett, 1910; Ballard, 1913) are a major pest of cotton in most of the cotton growing regions of Africa, south of the Sahara, that have a single rainy season. The bollworm, in common with the larvae of *Heliothis* sp., *Helicoverpa* spp. and *Earias* spp. feeds upon the buds, flowers and green bolls of cotton and, in the absence of effective insect pest management, can seriously reduce the yield and quality of seed cotton. Larvae of *Helicoverpa armigera*, *Earias* spp. and *Diparopsis* spp., together with larvae of *Pectinophora gossypiella* and *Cryptophlebia leucotreta* constitute the bollworm complex of African cotton. *Diparopsis* has a limited range of alternative host plants confined to a few plant species related to cotton, and the carry-over of the population from one season to the next is generally achieved through the production of diapause or overwintering pupae. A cotton bollworm, closely related to *Diparopsis*, occurs in South America within the genus *Sacadodes*.

Distribution and Taxonomy

Four species of *Diparopsis* have been distinguished by Clements (1951) on the basis of adult morphology, distribution and host-plant, namely, *D. castanea* Hmps, *D. tephragramma* B.-B., *D. watersi* (Roths) and *D. gossypioides* Clements. Cross mating between *D. castanea* (male) and *D. watersi* (female) with production of a few viable eggs has been demonstrated under laboratory conditions (Beevor *et al.*, 1973).

Diparopsis castanea is found in the southern and southeastern region of Africa occurring throughout the cotton areas of the Republic of South Africa, Swaziland, Botswana, Mozambique, Malawi, Zambia and Zimbabwe (Pearson, 1958;

Ingram and Green, 1972; Farrington, 1976; Kabissa and Nyambo, 1989). It also occurs in the extreme south of Tanzania on wild cotton, *Gossypium barbadense*, within the cotton quarantine zone but has not penetrated further northwards into the areas of commercial cotton production (Kabissa and Nyambo, 1989).

Diparopsis tephragramma has only been recorded from Angola where it is widespread throughout the cotton areas (Giraudet, 1968).

Diparopsis watersi, which was included within the species *castanea* prior to the revision of the genus by Clements (1951), is the species found on cotton north of the equator where its distribution extends across the continent from Senegal, Gambia, Ivory Coast, Mali, Ghana, and Nigeria in the west, through Cameroon, Chad and Sudan to Ethiopia and Somalia in the east. It has crossed into the coastal plain of southern Arabia where it occurs in the Abyan Delta of the People's Democratic Republic of Yemen. In the Sudan, the bollworm does not occur in the isolated cotton area of the Tokar Delta or within the Gezira cotton scheme since the introduction of irrigation in the latter. It is found, however, on cotton schemes in Singa district adjoining the Gezira to the south (El-Tigani and El-Amin, 1966). Although it has penetrated northwards along the Nile as far as Dongola in the northern Sudan, it has not reached the cotton areas of Egypt (Pearson, 1958).

Diparopsis gossypioides is identified from a single male moth bred from larvae infesting *Gossypioides kirkii* in southern Tanzania and is not known to infest cotton (Pearson, 1958).

There are no reports of *Diparopsis* having penetrated and established itself in the main cotton areas of the equatorial regions of Zaire, Tanzania, Uganda and Kenya.

Pearson (1958) considered it desirable to restrict the name Sudan bollworm to *D. watersi* and red bollworm to *D. castanea*, with *D. tephragramma* as the Angola red bollworm. However, as the three species are distinctly separated geographically, the retention of the simple common name of red bollworm for all species need not lead to confusion.

Description of the Stages

No detailed comparative descriptions of the egg, larval and pupal stages have been made following the separation of the genus by Clements (1951). The individual stages of *D. tephragramma* have been described in detail by Giraudet (1968). The following descriptions, unless otherwise noted, apply equally to the three species, *watersi*, *castanea* and *tephragramma* attacking cotton.

Adult

The moth is stoutly built with a wing span of 25–35 mm and a body length of about 15 mm; the head and thorax are densely covered with rather long scales of the same shade as the ground colour of the forewings. The general colour of the visible areas of the moth in repose, namely the thorax and forewings, are in various shades of brown that may vary from deep chocolate, through green-brown to light brown and straw coloured. The abdomen and hind wings are silver-cream in colour, the latter slightly infuscate towards the margins which bear a fringe. The forewings are divided by three curved transverse lines into four distinct areas of variable coloration, including a basal segment, a broad central area and two distal bands of similar width beyond which is a fringe. A common pattern is for the central area to be a faded rose colour, with the basal segment and the distal band a darker shade and the penultimate band greyish-brown. Both extremely light and dark chocolate coloured forms may occur (Pearson, 1958).

The four species, in having a similar general colour pattern, are separated on the shape of the frontal process of the head, which is used to break through the hard earthen pupal cell, the male genitalia and the shape of the first transverse line of the forewing which is a smooth curve in *D. watersi*, and sharply angled where it crosses the median vein in *D. castanea* and *D. tephragramma* (Clements, 1951; Pearson, 1954; Giraudet, 1968).

Egg

The egg, spherical in shape with a flattened base and 0.5–0.7 mm in diameter, is light blue in colour changing to grey prior to hatching. The surface is sculptured with horizontal and vertical ribs with small spines arising from the rib junctions (Pearson, 1958; Giraudet, 1968). The size and colour of the egg enable it to be readily seen on the cotton plant with the naked eye, thus facilitating the use of egg counts in scouting procedures for the timing of insecticide sprays. The white shells of hatched eggs may also be seen.

Larva

The larva (Plate III.3) passes through five instars, reaching a length of 25–30 mm in the full grown larva. The first instar larva is pale cream or greyish-white with the head, prothoracic plates, legs, setae, spiracles and anal plates black in colour, while the second instar has a greenish tinge and the previous black markings are lighter in colour (McKinley, 1968). The characteristic rose-red arrowhead markings are pronounced from the third instar, tending to become diffuse in the distended shiny appearance of the fully grown fifth instar. They consist of a medium dorsal mark flanked by an oblique one on either side to form the forward pointing arrowhead on each body segment except the first and last. There is also on each segment a broad lateral rose-red mark above the spiracle and touches of red below it. In later instars, the ground colour tends to be faintly greenish and the head, prothoracic and anal plates and legs become chestnut brown (Pearson, 1958). Maximum head capsule widths for the first to fifth instars, respectively, are given as 0.41, 0.66, 1.07, 1.67 and 2.46 mm for *D. watersi* (El-Tayeb, 1976) and 0.36, 0.58. 0.98, 1.46 and 2.20 mm for *D. tephragramma* (Giraudet, 1968).

Pupa

The pupa, 10-15 mm in length, is stout bodied and distinctly humped. The colour is a uniform yellowish-brown with a greenish tinge when first formed or in the diapause state, but turns to dark brown as morphological development takes place. A distinct knob at the anterior end of the pupa contains the cocoon breaker of the adult moth.

The pupa is enclosed within a well defined oval-shaped and hard-shelled earthen cell, frequently showing a distinct conical protuberance, although the shape may be more irregular in a loose sandy soil.

Internally the cell, with few exceptions, has a silken and silvery lining. The cell wall possesses a particularly porous area associated with the protuberance, if one is present, where the soil particles are less closely held together permitting increased gaseous exchange (Choyce and Reed, 1961). Sweeney (1962) has pointed out that a similar earthen pupal case, in size and texture, is constructed by the larva of the chafer beetle, *Anomala dorsata v. vittipennis* in Malawi but, in soil sampling for *Diparopsis* pupae, the former may be distinguished by the absence of a silken lining and conical hump. On emergence, the moth leaves a circular exit hole at the anterior end of the pupal case.

Life History and Behaviour

Available evidence does not suggest any major differences in the main features of the life history and behaviour of the three species of *Diparopsis* attacking cotton. Little is known about *D. gossypioides*.

The moth, which is strictly nocturnal, is sexually mature on emergence from the pupa and both mating and oviposition may occur on the night of emergence (Tunstall, 1958). During the day, the moth is immobile and hidden, often resting on the underside of a leaf. Galichet (1964) records that adult activity for *D. watersii* commences from one half to one hour after sunset and may continue throughout the night, with individuals interrupting their activity with varying degrees of immobility. In Malawi, the dispersal flight of male moths of *D. castanea* within a cotton crop commenced almost immediately after sunset and before the end of twilight (Marks, 1976b). However, little precise information exists on moth dispersal although, on the basis of the bollworm being found on cotton plots isolated from sources of known infestation by up to 50 km, distances flown by moths may be considerable (Pearson, 1958). In such cases, it is not always possible to rule out the occurrence of intermediary cotton plants or, if plants have been cleared in recent years, the presence of pupae with an extended diapause beyond one year. A mark–recapture technique with *D. castanea* in conjunction with sex pheromone traps has shown that male moths frequently travel distances of over 1.5 km in a single night, a maximum distance of seven kilometres being recorded (Anon. 1969).

In Swaziland, under field conditions at the beginning of the season, moths were shown to be most active soon after dark as indicated by 61% of eggs being laid within the first three hours of darkness (Clarke, 1979b).

Moths are attracted to artificial light, but with light traps showing marked variability in the numbers and proportion of male and female moths caught, dependent upon such factors as the nature of the light source, moon phase, trap site and time of year (King, 1927, 1929; Parsons, 1929; Smith, 1933), their use in measuring moth activity has limitations. A sex ratio of 1:1 is generally accepted (Pearson, 1958; El-Tayeb, 1976).

A sex pheromone is emitted by the virgin female moth to attract males for mating (Rose, 1957) and the isolation, identification and synthesis of a multi-component pheromone has been achieved (Moorhouse *et al.*, 1969; Nesbitt *et al.*, 1973a,b, 1975). The attractive field combination, dicastalure, is shown to be a three component blend of 17% *cis*: 83% *trans*- 9,11 dodecadien-1-yl acetate (IV) and 11-dodecen-1-yl acetate (IIB) in the ratio of 4:1 (Marks, 1976c). A non-attractive component, *trans*-9-dodecen-1-yl acetate (IIA) acts as an inhibitor preventing male response to dicastalure while a fourth compound, dodecen-1-yl acetate (I), is believed to have a modifying effect on IIA (Marks, 1976e). Cross-attraction between *D. castanea* and *D. watersi*, together with similar patterns of electro-antennogram responses to abdominal tip extracts after gas chromatography, make it probable that the sex pheromones of the two species are identical (Beevor *et al.*, 1973). The mating behaviour of male and female moths of *D. castanea* has been extensively studied by Marks (1976a–e, 1977a,b) and Marks *et al.* (1978, 1981) in association with laboratory and field studies on the employment of the sex pheromone in insect pest management.

The eggs are normally laid singly with, very occasionally, two or three together (King, 1926; Smith, 1933; El-Tayeb, 1976). Under laboratory conditions the total number of eggs laid by a female is very variable, with averages of over 200 eggs per female being reported for the three species (Smith, 1933; Pearson, 1958; Tunstall 1968; Giraudet 1968). El-Tayeb (1976) and Tunstall (1968) report that, for *D. castanea*, the majority of eggs are laid within the three to four days following mating; a maximum of 195 eggs laid in one night has been reported for *D. tephragramma* (Giraudet, 1968).

Eggs may be found on most parts of the plant, including leaves, stems, bracts and surfaces of buds, flowers and green bolls. In general, the majority of eggs tend to be found on the younger and more accessible parts of the plant, and moths may show a preference for the more advanced and larger plants in a field (Tunstall, 1958). Variable proportions for the number of eggs laid on different parts of the plant have been given. Seasonal data for Tchad have given proportions for fruiting bodies, leaves and stems of 22, 31 and 47%, respectively, whereas for the Gash Delta, Sudan, proportions of 44 (buds 37), 48 and 9% are reported, with the lower leaf surfaces bearing twice the number of eggs as the upper surfaces (Le Gall, 1952; Tunstall, 1958). In Swaziland, *Diparopsis*, while ovipositing freely on all parts of the plant throughout the season, tended to favour stems (26%), upper leaf surfaces (19%) and petioles (18%) (Clarke, 1979a). Such differences are, no doubt, partly attributable to seasonal and local variability in plant growth and development, for in Tchad the proportion of eggs for fruit and stems of 69 and 19%, respectively, at the beginning of the fruiting period was almost completely reversed later in the season and, in addition, eggs were increasingly found towards the top of the plant as the season advanced

(Couilloud, 1961, 1962). The dependency of oviposition site on crop development is also reported by Beeden (1974) with the preferred egg laying site changing from mainly vegetative parts of the plant to fruiting bodies as the plants become older. Studies have shown that few eggs are laid within the bottom quarter (Beeden, 1974) or bottom third (Clarke, 1979a) of the plant.

Increased egg populations have been recorded from the edges of a cotton field by Choyce and Lyon (1963) for Nigeria and by Beeden (1974) for Malawi. Brader *et al.* (1968), using light traps in Tchad, confirmed that *D. watersi* settles particularly on field borders. Beeden (1974) reports an increase in egg population at the windward edge of a field. This is in contrast to observations in Swaziland where oviposition has been found heaviest on the leeward edges, suggesting that moths may fly upwind on the approach to a cotton field (Clarke, 1979a).

The general distribution of eggs over most of the plants is reflected in egg and larval scouting techniques for this bollworm which recommend that the whole plant (or upper ten nodes on tall cotton above 1.5 m in height) is carefully examined (Anon., 1976, 1985).

Under cage conditions, eggs are readily laid on muslin, paper and other artificial substrates. In contrast to *Earias* and *Helicoverpa*, lack of nectar and water or its equivalent did not exert a significant influence on egg-laying capacity or length of life (Parsons *et al.*, 1938).

Newly hatched larvae will penetrate buds, flowers and green bolls of all ages and in their absence will penetrate the terminal growing point of the plant. The latter site of feeding may be prevalent early in the season, when plants are young and bearing few buds, but it is not one of the generally preferred sites of feeding as in the case of *Earias*. Observations on *D. watersi* in the Gash Delta, Sudan (Tunstall, 1958) and *D. castanea* in Zimbabwe (Anon., 1959, 1960; Tunstall, 1962) showed that the time taken for a larva to reach the site of penetration varied considerably and, to a large extent, was dependent upon the site of egg placement and its proximity to a fruiting point. Larvae of *D. watersi* took from 22–187 minutes in locating a fruiting point with eggs placed on the terminal leaf of a sympodium, and 3–54 minutes for eggs on the terminal growing point of a monopodium. With the former placement site, excessive wandering of the larva apparently resulted from the inability of the larva to locate the leaf petiole and so leave the leaf blade. Dispersal by means of dropping on a thread occurs but was only observed after prolonged periods of wandering. Many larvae confined their wandering to the growing point and adjacent buds, the majority choosing to penetrate one of the four or five buds nearest to the site of egg placement. Little wandering of the larva occurs once a bud or boll has been reached, or if eggs were placed on the bracts of a fruiting point. Similar observations have been made for *D. castanea* where, with eggs laid naturally by moths caged over plants, initial periods of larval wandering averaged 115 minutes with a range of 15–310 minutes. There was no suggestion that buds are preferred to bolls, the selection being a matter of availability, buds being generally infested during the early part of the season and green bolls later in the season, with the more advanced development of the plant.

Initial penetration of the bud or boll by larvae of *D. watersi* was always at a point of contact with a bract, the larva previously joining the two surfaces together with a fine web. Later studies of *D. castanea* showed that most larvae constructed a cell-like structure of silken threads, frass and other debris, sometimes strengthened by a secretion, joining the surfaces of the bract and bud at the point of penetration. The cell may serve to protect the larva during its first moult or early development.

Usually a larva that has entered a flowerbud will remain there feeding until the contents have been eaten. Shedding however may occur before this, when often the bud remains characteristically attached to the plant by silken threads spun across the point of abscission between the peduncle and sympodial stem. Pearson (1958) reports that these threads are made by the larva when entering the bud, but this was not observed in studies conducted in the Gash Delta, Sudan with *D. watersi* and, in Zimbabwe, such webbing has only been noticed during localized larval movement subsequent to the initial penetration. The retention of the shed bud would be expected to assist in survival of the larvae.

The larva, unlike that of *Helicoverpa*, generally feeds with its body totally enclosed within the bud or boll, and will continue to feed on a particular fruiting body until it is shed or its contents consumed. Much of the frass excreted remains inside the boll, tending to plug the entrance hole. Larvae, however, may sometimes make localized movements outside the bud or boll, when it may be seen webbing plant surfaces together or resting within the bracts before re-entering the fruiting body.

The larva may make up to five, but more commonly two or three, transfers from one fruiting body to another during its feeding period when plants are carrying predominantly buds earlier in the season. Later, with the plants providing green bolls as a large single source of food, the majority of larvae make only one transfer, usually from a bud to a boll or boll to boll. Many buds, while infested, will develop into bolls, the feeding of the young larvae keeping pace with bud and boll development for a limited period. There have been a few instances of the larva completing its development entirely within the original infested bud and the subsequent developing boll. Larvae have been observed to regain the plant from infested buds that have fallen to the ground.

In contrast to the somewhat economical feeding habits of *Diparopsis*, larvae of both *Helicoverpa* and *Earias* are, individually, more destructive with recorded transfers from one fruiting body to another of four to 15 (average nine) and seven to 14 (average ten), respectively (Anon., 1961).

When fully developed, the larva vacates the boll and either crawls down the plant to the ground or, more commonly, drops from the plant. On reaching the ground, the larva tends to move away from the plant to find a suitable point, usually a crack or loose soil, from which to burrow into the ground. A small cavity is excavated, often in the side of the crack, within which the hard pupal cell is formed from chewed, moistened fragments of soil. The inner surface is rendered smooth and lined with a tough fine-meshed cocoon. Pupation occurs within the cell after four days.

Egg hatch and initial penetration of a fruiting body takes place from just before dawn and, in Zimbabwe, extended from 04.00–08.00 hours local time. Other major movements of the larva, namely transfers and pupations, took place during the hours of darkness, principally between 18.00 and 24.00 hours.

Transfer movements most commonly took place from the 12th day following egg hatch.

The laboratory rearing of *Diparopsis* presents little difficulty provided a natural diet of buds and/or bolls is provided, but attempts to develop a completely artificial diet have resulted in unsatisfactory larval growth and survival rates (McKinley, 1974). Both a casein–wheatgerm diet with high cellulose and a cotton meat (processed cotton seeds) diet showed promise, but mould problems arise through the sensitivity of the larva to most of the commonly used mould suppressants at effective concentrations.

Pupation may occur to a depth of 150 mm. Observations on *D. castanea* show that the majority of pupae are found within the top 75 mm (Parsons and Ullyett, 1932; McKinstry, 1947; Tunstall 1968) or top 100 mm (Clarke, 1979a), but soil conditions are shown to affect depth of pupation (Smith, 1933) with larvae not penetrating further than 13 mm into hard compacted soil which may particularly occur towards the middle of the inter-row space (Tunstall, 1968). With *D. watersi*, early observations from the Northern Province of the Sudan, suggest that pupation is a little deeper than 75 mm (King, 1927; Giffard, 1929; Bedford 1938), but in the Gash Delta and Nuba Mountain areas of the Sudan, the majority of pupae were found within the top 25 mm and the top 50 mm of soil, respectively, with few pupae lower than 50 mm at both locations (Tunstall, 1958; El-Tayeb 1976). At Tikem, Tchad, 95% of pupae in diapause were within a depth of 80 mm, with over 60% of these at less than 50 mm (Cadou, 1949). Pupae were found throughout the inter-row space for cotton grown on the flat (Zimbabwe) and on the ridge (Malawi), with no pronounced concentration at any particular site which might facilitate mechanical destruction (Tunstall, 1968). Larvae of *D. tephragramma* pupated at a depth of up to 50 mm in the laboratory (Giraudet, 1968). There appears to be no difference in depth of pupation between diapause and non-diapause pupae formed at the same time (Pearson, 1958).

The moth emerging from a non-diapause pupa is likely to be able to follow the path previously made by the larva on entering the soil but, in the case of diapausing pupae which have lain in the soil for a long period, any such path will have been obliterated by rain or land cultivation. Moths are capable of penetrating 75 mm of heavy clay soil when it is moist but cannot penetrate it at all when dry, unless through a crack. They can, however, penetrate 50 mm of dry, lighter clay soil, making a well defined tunnel (Tunstall, 1958).

Length of Life Cycle

Egg

Data reported by Pearson (1958) for *D. castanea* on the relation of incubation period to temperature showing a minimal period of four days at 30°C accord fairly well with other data from South Africa (Smith, 1933), and for *D. watersi* in Tchad (Galichet, 1964) and the Sudan (El-Tayeb, 1976). Egg hatch is greatly reduced below 18°C and above 38°C and is also affected by relative humidities below 60% (Galichet, 1964). Under natural field conditions in Zimbabwe, the period of incubation varied between six and 13 days (mean of nine days), the variation being largely attributed to site of oviposition exposing eggs to different temperature conditions (Tunstall, 1962).

Larva

The minimal development period for *D. castanea* is reported as 11–12 days at 30°C, increasing to 13 days as the temperature is raised to 36°C or lowered to 26°C, below which the rate of development falls off rapidly, taking 23 days at 22.5°C (Pearson, 1958). The larvae did not feed and eventually died at a constant temperature of 20°C and below. Galichet (1964) reported that, at a constant temperature of 30°C and an 11 hour photoperiod, the lengths of the first to the fifth instar periods for *D. watersi* were 48–53, 36–48, 24–48, 48–52 and 72–96 hours, respectively, excluding a prenymphal period of 48–72 hours. Similar data, although with a greater range, is given for the same species under laboratory conditions in the Nuba Mountains area of the Sudan (El-Tayeb, 1976). Under natural field conditions in Zimbabwe, mean larval periods of 38 days (range 22–52 days) and 20 days (range 17–24 days) were recorded for eggs hatching in March in two consecutive seasons (Anon., 1959, 1960). Differences in available food and climatic conditions between the two seasons may have accounted for the variability in length of larval period. Considerable variation in length of larval development periods has previously been reported, namely, 14–42 days for Malawi (King, 1926), 13–38 days and 16–21 days for the northern and western Sudan, respectively (Giffard, 1929; El-Tayeb, 1976) and 18–41 days for South Africa (Smith, 1933).

Pupa

An outstanding character of *D. castanea* and *D. watersi*, which profoundly affects the distribution and status of the pests, is the possession of a facultative diapause in the pupal stage. Generally, pupae formed in the early part of the infestation on cotton mostly give rise to moths within 2–3 weeks, but as the season advances an increasing proportion of pupae formed enter diapause and do not give rise to moths until after much longer periods extending into months or even years, enabling the insect to span the period between seasons when no host plant is available or persist in areas where only the annual commercial crop occurs. Extension of the pupal period into several months is also reported for *D. tephragramma* (Giraudet, 1968). There is considerable diversity in the pattern of moth emergence from diapause pupae both between regions and between seasons within a region, and the extent to which the pattern intercepts the cotton crop at a vulnerable stage to larval attack with subsequent loss or damage of buds and bolls determines the pest status of *Diparopsis*. However, the similarity of the general features of the incidence of the delayed emergence in all regions where *Diparopsis* occurs suggests that the phenomenon is essentially the same for *watersi* and *castanea*.

Factors Affecting the Length of Pupal Period

Early work in Malawi reported by Pearson and Mitchell (1945) and Pearson (1958, 1962) showed that when pupae of *D. castanea* contained within the pupal cells were incubated, from the time of their formation, at a constant temperature of 30°C, the pattern of moth emergence differed markedly from that under field conditions. Following the initial emergence of moths from non-diapausing pupae, the distribution of emergencies from the diapause fraction, instead of the scattered field emergencies tending to a bimodal form extending over a year, showed a unimodal but asymmetrical frequency distribution with a modal value of about 20 weeks and a long tail extending to a year. This pattern of emergence occurred irrespective of date of pupation, except for an earlier emergence from pupae formed in October when temperatures were rising. Similar unimodal patterns of emergence have subsequently been obtained in Zimbabwe (Tunstall, 1968) and for *D. watersi* in Nigeria (Geering and Baillie, 1954). In Zimbabwe, however, the modal value was distinctly earlier at eight to ten weeks. Galichet (1964) records a multimodal distribution for *D. watersi* with a main peak at 14 weeks followed by peaks of decreasing magnitude at 21 and 29 weeks for males and a slightly earlier pattern for females.

Pearson (1958) deduced from the behaviour of pupae kept at a constant temperature of 30°C that the differentiation into diapause and non-diapause pupae is determined before pupal formation and that the ultimate emergence of diapause pupae does not depend primarily upon the application of some external stimulus, but is permitted as a result of the completion of a process of slow development which proceeds, in each individual, at a predetermined rate. He distinguished, following the nomenclature of Andrewartha (1952), between morphogenesis, the process of visible morphological and histological change that constitutes the whole life of the non-diapause pupa and the final phase of the life of the diapause pupa and physiogenesis, or 'diapause development', the process of non-visible, physiologcal change that the diapause pupa has to undergo before morphogenesis is initiated.

Pearson (1958)observed that the development of non-diapause pupae or the process of morphogenesis was minimal between 30 and 36°C at 13–15 days, increasing steeply to nearly eight weeks at 19.7°C, the lowest temperature at which development was completed. The upper limit at which development was completed was 36°C.

The rate of diapause development was most rapid at 28–29°C, below which the rate fell off greatly with a mean period of 38 weeks at 19.7°C. At 16°C and below, the process cannot be completed and pupae become inactivated, although temperatures as low as 8°C are not lethal, development resuming when pupae are returned to a favourable developmental level. As the temperature is increased above the optimum, diapause development takes progressively longer until a point is reached at about 36°C, above which full development is completely prevented. Temperatures higher than 38.5°C lead to desiccation and death if applied continuously. Pearson (1958) concluded that there existed a zone of temperatures above 36°C, the upper limit of which cannot be defined except in terms of temperature and length of exposure, within which diapause development appears suspended. High temperatures alternating diurnally with normal temperatures will continue to arrest development.

In explanation of the emergence pattern of moths from diapause pupae, Pearson (1958) suggests that diapause development involves at least two consecutive physiological processes, the first of which can continue at temperatures of 36°C and above at which the second is inhibited. Applying high temperatures for differing periods suggests that only the terminal one third of the process of physiogenesis is susceptible to heat inhibition. Heat treatment of a population, heterogeneous as regards intrinsic length of diapause, enables the slower individuals to catch up with the faster ones, thus avoiding an otherwise expected sequence of overlapping emergence patterns from pupae formed in successive months.

Pearson (1958) reports that the rate of diapause development is unaffected by the humidity to which pupae are exposed. Removal of the silken lined earthen cocoon however greatly accelerates diapause development, the average length of the pupal period being shortened by approximately one fifth. Pearson suggests that the slower rate of development of enclosed as compared with naked pupae is due to the impedance of gaseous exchange in the former condition resulting in an excess of carbon dioxide and deficiency of oxygen. Cutting off gaseous exchange by submergence of diapause pupae in water further retards development, pupae being able to withstand submergence for as long as four weeks in early pupal life, their ultimate emergence being delayed for a period equal to that of the submergence. Pupae submerged in a late stage of pupal development give rise to a sudden rush of emergences two weeks after the end of submergence which is then followed by a more normal emergence pattern. Heavy pupal mortality occurs as submergence is continued beyond the start of morphogenesis and Pearson (1958) concludes that when diapause development is nearing completion, the effects of submergence are nicely balanced between accelerating the final stages of physiogenesis and suffocating pupae that have started active development. Work by Jacquemard (1976a) demonstrates that, independent of temperature, alternating high night-time (80–90%) and low day-time (40–50%) humidities induce the termination of diapause.

The response of diapause pupae to temperature and other external factors enables an explanation to be given for the variability in moth emergence patterns that are encountered in the different regions where *Diparopsis* occurs.

In the lower Shire Valley of Malawi, Pearson (1958) accounted for the emergence pattern of moths of *D. castanea* from diapause pupae, in which low emergences occurred during September to November followed by a bimodal distribution with a small peak in December and a main peak period covering March–April, by high temperatures initially delaying emergence followed by rain saturation of the surface soil suppressing emergence in January–early February. This generalized picture was subject to variation between and within seasons according to variability in climatic factors. Tunstall (1968) recorded a generally unimodal pattern of moth emergence, using a similar technique of field pupation cages, with no suggestion of a suppression of emergence between late December and late February; while the use of field cages to record emergence from natural field populations of diapause pupae added support to the view that the main periods of moth emergence from diapause pupae in the lower Shire Valley may be very variable, depending upon rainfall and local drainage conditions.

North, in the Lake Shore area of Malawi, with the later onset of the rains, high soil temperatures inhibit pupal development for a longer period and the main period of moth emergences does not begin until January, although scattered emergences occur earlier.

Further south in the middle veld of Zimbabwe, with rains occurring in November and lower temperatures with a higher altitude, the period of heat inhibition is reduced and there is a prolonged period of emergence which may extend from the beginning of October to the middle of April with a main period of emergence from mid-November/early December to mid-February (Tunstall, 1968). Still further south, in the Transvaal and northern Natal of South Africa, the generally lower air temperatures allow the emergence of overwintering pupae to occur throughout September to December and be virtually complete at the end of the period.

Variability in the pattern of moth emergence occurs with *D. watersi* north of the equator. In the Gash Delta of the Sudan, prolonged periods of high inhibiting temperatures may account for the strongly bimodal pattern of emergence and the fact that approximately one third of the pupal population carry over to a second and third season (Tunstall, 1958). In northern Nigeria, however, the strongly bimodal pattern of emergence is accounted for by waterlogging suppressing emergence (Geering and Baillie, 1954) and is an exaggerated version of that reported by Pearson (1958) for the Shire Valley of Malawi.

In the Abyan Delta of the Arabian Peninsular where high and prolonged inhibiting temperatures occur as in the northern Sudan, a tendency to a bimodal distribution of moth emergence results from two periods of lowered temperature brought about initially by the occurrence of low rainfall and later by the practice of flood irrigation (Proctor, 1962). There was some suggestion of the emergence period extending into a second season as in the Gash Delta.

Inducement of Pupal Diapause

A monsoon climate with a warm wet summer and a cool dry winter is found in all areas where *Diparopsis* attacks cotton. Cotton is sown in early summer on the first rains, unless irrigated when sowing may be earlier, and flowerbud production and boll set take place while the weather is warm and wet with bolls maturing as it becomes cooler and drier. Data for *D. castanea* from Malawi (Pearson, 1958; Tunstall, 1968), Zimbabwe (Peat *et al.*, 1942; Tunstall, 1968) and South Africa (Lounsbury 1926; Smith 1933) and for *D. watersi* from the Sudan (Giffard, 1929; Bedford, 1938; Tunstall, 1958), Nigeria (Pomeroy, 1925; Geering, 1951; Geering and Baillie, 1954), Tchad (Galichet, 1964) and Cameroon (Jacquemard, 1976a) show a general picture in which non-diapause pupae are formed predominantly in summer and that, as the weather becomes cooler with the advance of the season an increasing proportion of diapause pupae are produced. In the lower Shire Valley of Malawi, at a time when an extended cotton season permitted breeding to continue past the coolest period, Pearson (1958) showed that the proportion of larvae forming diapause decreased with the onset of warmer weather until the cycle was completed with a return to summer conditions and no diapause.

Pearson established that the nature of the pupal period was determined before the pupa was formed, and that it was necessary to go back at least to the larval stage in search of possible factors inducing diapause. He noted that the formation of diapause pupae appeared to be associated with larval development under conditions where the minimum temperature is below 20°C and the diurnal range exceeds 20°C but was unable to identify any single or multiple factor causing diapause.

Later studies by Galichet (1964) with *D. watersi* in Tchad showed that the onset of diapause is principally dependant upon temperature. Continuous rearing of *Diparopsis* without the interruption of diapause is possible at a constant temperature of 30°C. However, the combination of a low nocturnal temperature (15°C) and a high diurnal one (30°C) over the incubation and larval period, followed by a low (15°C), high (30°C) and very high temperature (38°C) during the pre-pupal stage led to the induction of diapause in 95–100% of the population. The induction of diapause is reduced to 20–60% when alternating low and medium temperatures are not followed by high ones (38°C) after the larva has entered the soil for pupation. Very high temperatures used alone affect up to 20% of the population, mostly males, which may be considered sometimes as a diapause but of a weak intensity. Silk excretion was shown to have an important physiological function in the onset of diapause. In view of the variability of diapause expression in a population of *D. watersi* responding to changes in the environment, it is concluded that the diapause characteristic shows genetic diversity within the species. Further studies by Jacquemard (1976a) in northern Cameroon of the natural incidence of diapause in relation to temperature provide support for the view that alternating high (diurnal) and low (nocturnal) temperatures during the egg and larval stages induce diapause in *D. watersi.*

Host Plants

Diparopsis is narrowly confined in its host range to the genus *Gossypium* and the two closely related genera, *Cienfugosia* and *Gossypioides*. Rare occurrences on other Malvaceous plants have been reported, namely single records of the larva on *Thespesia* in Malawi (Anon., 1927) and *Abutilon* in Mozambique (Peat, 1930) and observation of egg laying and larval feeding on *Hibiscus calycinus* (Parsons, 1928) in South Africa, although later work showed that although eggs were occasionally laid, larvae failed to mature (Marshall *et al.*, 1937). North of the equator, larvae of *D. watersi* have been recorded on *Hibiscus cannabinus* and *H. sabdariffa* (Monteil, 1934; Bedford, 1938). Galichet (1964) advises caution in accepting such reports, unless the adult moth has been identified with certainty, as more recent reports of *Diparopsis* larvae on *H. cannabinus* in Tchad (Le Gall, 1950) have been shown to be confused with the very similar larvae of *Ectolopha viridescens* (Agrotinae).

Diparopsis watersi is found on all the introduced cottons present within its area of distribution, namely, *G. herbaceum* race *acerifolium*, *G. arboreum* race *soudanense*, *G.*

hirsutum in its three forms, *latifolium*, *punctatum* and *marie-galante* and *G. barbadense* and its variety, *brasiliense* (Pearson, 1958). It has not been found occurring naturally on wild species of *Gossypium* but it will infest *G. anomalum* and, to a less extent, *G. somalense* when grown experimentally.

The only known host plants of *D. castanea* south of the equator, apart from the cultivated cottons, *G. hirsutum* race *latifolium* and a small area of *G. barbadense* in Zimbabwe, are *G. herbaceum* race *africanum*, the wild cotton of the southern African low veld, *G. barbadense* variety *brasiliense*, the house-yard kidney cotton bush of northern Malawi and Zambia and the indigenous shrub, *Cienfugosia hildebrandtii*. *Cienfugosia gerrardii* has been reported to carry larvae (Parsons, 1936) but these do not mature and it is not regarded as a host plant (Marshall *et al.*, 1937).

Later studies of *C. hilderbrandtii* in Swaziland (Clarke, 1979a) have confirmed its importance as a host plant which is present either in small scattered patches or in colonies reaching 10,000 plants over 2.8 km^2. Such plants have been found heavily infested with eggs and larvae in all stages of development.

In Angola, *D. tephragramma* is found infesting the wild cotton species, *G. anomalum*, localized in the semi- desert region of the southwest and is a major pest in all the areas of cultivated cotton (Giraudet, 1968).

Diparopsis gossypioides is known only from *Gossypioides kirkii* in southern Tanzania. *Diparopsis castanea* has not been found to occur naturally on *G. kirkii*, although on one occasion it has been found to lay eggs on cultivated plants and a resultant larva was reared to maturity on the fruits (MacDonald *et al.*, 1946).

Natural Enemies

The overall survival rate of *Diparopsis* on cotton untreated with insecticides appears to be generally low, with reported high mortality rates occurring in the egg, larval and pupal stages on occasions. Overall mortality seems rarely to be less than 80% as judged from comparisons of the number of eggs laid and the size of the population of larvae, especially those of the fifth instar, made in Malawi (Pearson, 1958), Nigeria (Geering and Baillie, 1954) and the Gash Delta, Sudan (Tunstall, 1955). Full mortality from the time of initial larval penetration of the plant until moth emergence reached 98% in Zimbabwe, ants accounting for 69% of the mortality (Tunstall, 1962). There is little or no information on mortality of the adult moth stage. The number of recorded parasites and predators is not large but certain of them have particular influence. Table 6.1 lists natural enemies.

Low rates of egg parasitism are reported for *D. watersi* in northern Nigeria (Geering and Baillie, 1954), in the Sudan (Tunstall, 1958) and Tchad (Galichet, 1964). However, variable and sometimes high losses of eggs, amounting to 80–100%, have been recorded for this species in Tchad and for *D. castanea* in Zimbabwe (Anon., 1959, 1960). A high proportion of this loss is attributable to predators, especially ants. In Zimbabwe, the ant *Pheidole megacephala* F has been observed removing eggs from the plants but substantial egg losses also occur for no apparent reason. The fact that there is a high loss of eggs following rain suggests that a loss of adhesion to the plant is influenced by the wetting of eggs.

The Tachinid, *Carcelia evolans* Wied. is a common and important larval parasite of *D. watersi* in northern Nigeria (Geering and Baillie, 1954; Reed and Choyce, 1961), in Tchad (Galichet, 1964) and in northern Cameroon (Jacquemard, 1976a,b), with parasitism rates reaching 25–30%; no comparable parasitism by Tachinids appears to occur with *D. castanea* (Pearson, 1958). Reed and Choyce (1961) showed that the life cycle of *C. evolans* is closely adapted to that of its host with a synchronization of non-diapausing and diapausing generations. The relationship has been studied further by Jacquemard (1976a,b) who concludes that hormonal secretions of the host in diapause induce quiescence in the first larval stage of the parasite and, later, terminate it on resumption of pupal development, suggesting a certain specificity of *Carcelia* to *Diparopsis*. *Apanteles* sp. (*ultar* Reinh. group) is recorded as a common parasite of early instars of *D. watersi* in the Gash Delta, Sudan, while *Bracon* sp., probably *breviconis*, commonly attacked later instars, reaching 20% parasitism of fifth instar larvae (Tunstall, 1958). The liberation of *B. brevicornis* in the Sudan against *Diparopsis* did not affect field rates of parasitism (Pearson, 1958). Several species of ants have been reported as attacking larvae in Zimbabwe with recorded predation of 31% principally by *Pheidole megacephala* (Tunstall, 1962).

Ants, particularly *Pheidole spp.* are recorded as being important predators of pupae, the ants having penetrated the pupal cell wall leaving small holes (Parsons and Pearson, 1942; Pearson, 1943; Mitchell 1943, 1944; Tunstall, 1962, 1968; Clarke, 1979a,b).

A nematode (Mermithidae) is recorded as being of some importance in Tchad (Galichet, 1964) and a high natural mortality of late instar larvae from a virus infection is reported from the Ivory Coast (Anon., 1963). Subsequent studies of viral (Angelini and Vandamme, 1964; Atger, 1969; Croisier *et al.*, 1980) and bacterial (Jacquemard, 1965) infections have been reported.

Other general predators, such as spiders, have been recorded. Larvae of *Helicoverpa armigera* will attack and destroy larvae of *Diparopsis* if encountered in the course of feeding on buds or bolls (Anon., 1960) and Pearson (1958) records rats as significant predators of pupae in Malawi.

Origin and Spread of *Diparopsis*

A detailed consideration of the origin and subsequent spread of *Diparopsis* is given by Pearson (1958), a summary of which is given here with the additional information that a survey of *Gossypium anomalum* in the arid region of southwest Angola has confirmed this species of wild cotton as a natural host of *D. tephragramma* (Giraudet, 1968).

The known facts of *Diparopsis* point to an ancestral association with perennial host plants in areas of low and strongly seasonal rainfall, and that these must be sought within the genera *Gossypioides*, *Cienfugosia* and *Gossypium*. It is unlikely that *Gossypioides* with its associated distinct species of *D. gossypioides* occupies a basic position in the evolution of *Diparopsis*. Although the coastal distribution of *Gossypioides kirkii* from Kenya southwards to Natal penetrates well into the range

Table 6.1. List of natural enemies.

Species	Origin	Reference	Stage under attack
Parasites			
Trichogrammatidae			
Trichogramma luteum (Gir)	South Africa*, Zimbabwe	Peat *et al.* (1942)	Egg
Unidentified	South Africa	Smith (1933), Parsons and Pearson (1942)	Egg
Braconidae			
Apanteles diparopsidis Lyle	South Africa*, Malawi	Smee 1941	Larva
A. earterus Wlkn.	Sudan*, Malawi	Smee (1944)	Larva
A. sp. (*ultor* group)	Sudan	Tunstall (1958)	Larva
Bracon brevicornis Wesm	South Africa*, Sudan	Tunstall (1958)	Larva
Elasmidae			
Elasmus johnstoni Ferriere	Sudan	Cameron *et al.* (1946)	Larva
Tachinidae			
Carcelia evolans Wied.	N. Nigeria	Geering and Baillie (1954)	
	Chad	Galichet (1964)	Larva
Nemoraea capensis (R-D)	Zimbabwe*		Larva
Sturmia inconspicua (Mg)	South Africa	Parsons and Pearson (1942)	Larva
S. imberbis (Wied)	Sudan	Tunstall (1958)	Larva
Thelaira nigripes F.	Malawi	Smee (1940)	Larva
Palexorista quadrizonula (Thompson)	South Africa	Crosskey (1970)	Larva
NEMATODES			
Mermithidae			
Unident. sp.	Chad	Galichet (1961)	Larva
Steinernematidae			
Neoaphlectana sp.	Malawi	Campion (1968)	Larva
Predators			
Pentatomidae			
Macrorhaphis acuta Dall.	Malawi	King (1926), Smee (1942)	
Glypsus conspicuus (Westw)	South Africa	Anon. (1928)	
Agonoscelis versicolor F.	Zimbabwe	Anon. (1959)	
Formicidae			
Anoplolepis braunsi var. *parsoni* Santo.	Swaziland	Parsons and Pearson (1942)	(on *Cienfugosia hildebrandtii*)

Table 6.1. List of natural enemies.

Species	Origin	Reference	Stage under attack
Myrmicaria natalensis var. *eumenoides* (Gerst)	Malawi	Pearson (1943)	
Pheidole capensis Mayr.	South Africa	Parsons and Pearson (1942)	
P. megacephala (F.)	South Africa	Parsons and Pearson (1942)	
Dorylus conradtri Emery	Zimbabwe	Tunstall (1968)	
Chrysopidae			
Chrysopa boninensis Okamoto	Zimbabwe	Brettell (1979)	

* Listed by Thompson, W.R. (ed.) (1943) *A Catalogue of the Parasites and Predators of Insect Pests.* Commonwealth Bureau of Biological Control.

of *D. castanea*, *G. kirkii* is not a host plant of this species and, furthermore, the species is not associated with *G. brevilunatum* in Madagascar.

Cienfugosia hildebrandtii has a similar distribution to *G. kirkii* but in greater depth, penetrating up the Zambesi Valley and flourishing in specialized localities of southern Mozambique, Swaziland and Zululand where it supports the only populations of *D. castanea* known to exist independently of the true cottons. *Cienfugosia hilderbrandtii* appears to be particularly adapted to *Diparopsis*, in that the main period of cropping is during the last three months of the year coinciding with the main period of the carry-over moth flight derived from diapause pupae. In this it differs from the other wild host plants of the area, *G. herbaceum* race *africanum*, for example tends to fruit during the first three months of the year, although both host plants may have flashes of flower buds at other times during the summer depending upon situation and rainfall. *Diparopsis castanea* has not been recorded from *C. hilderbrandtii* in Kenya or Tanzania where the plant is very scarce. *Cienfugosia hilderbrandtii* does not provide any clue to the distribution of *D. tephragramma* and *D. watersi*, but *C. digitata* may do so as this has a discontinuous distribution, occurring north and south of the equator from Senegal to the Sudan, and from the Limpopo Valley westwards to Angola. *Diparopsis castanea* has, however, never been recorded from this host. It is significant that a South American species of a genus closely related to *Diparopsis*, *Sacadodes pyralis* Dyar, also a pest of cotton, is found on *C. affinis*, the only naturally occurring wild host plant, the range of which closely coincides with *S. pyralis*. It would thus seem possible that *Sacododes* and *Diparopsis* have a common derivation in association with *Cienfugosia*.

It is possible that *Diparopsis* originated in association with *Gossypium* itself: including firstly, *G. herbaceum* race *africanum* which is, on botanical and historical evidence considered to be a truly wild cotton of the savannah areas of southern Mozambique, the northeastern Transvaal and adjacent parts of Zimbabwe, extending down to Swaziland and northern Natal; and secondly, three xerophytic perennial shrubs adapted to semi-desert conditions, namely, *G. anomalum*, distributed throughout the sub-Saharan region from western Africa to Eritrea and, also, in southern Angola and Namibia, *G. triphyllum* also in southern Angola and Namibia and *G. somalense* in Somalia and the Sudan. *Diparopsis castanea* is readily found on *G. herbaceum* race *africanum* and *D. tephragramma* is recorded from *G. anomalum* but not from *G. triphyllum*. No naturally occurring wild host for *D. watersi* has been recorded, although it would seem most probably that on the basis of the Angolan evidence and the fact that the species will feed on *G. anomalum* if offered it, *G. anomalum* may well be a host plant for *D. watersi*.

Pearson (1958) states that, having regard to the close relationship between the genera of host plants to which *Diparopsis* is confined, to the close relationship between the species into which *Diparopsis* has been split and to the present day distribution and biology of these species and the host plant genera, one concludes that *Diparopsis* arose in association with an ancestral form of either *Cienfugosia* or *Gossypium* during an inter-pluvial period when central Africa was dry. As conditions became moister in the equatorial regions xerophytic plants and their associated insects were pushed out centrifugally. *Diparopsis* split up into four groups which eventually differentiated into the four species with defined areas of the present day.

Evidence of the occurrence or otherwise of *D. watersi* on a wild species of *Gossypium* or of this species and *D. tephragramma* on *Cienfugosia* is necessary to help resolve the respective ancestral roles of the two plant genera.

The subsequent history of the spread of *Diparopsis* may be explained in terms of the introduction and dissemination of cultivated lint bearing cottons (see Chapter 1), initially the Asiatic cottons, *G. arboreum* race *soudanense*, *G. herbaceum* race *acerifolium* and *G. arboreum* race *indicum*, and later by *G. hirsutum* race *punctatum* and *G. barbadense* including *G. b.* var. *brasiliense*, intercepting the indigenous wild populations of *Diparopsis*.

Diparopsis is absent from the equatorial regions and Pearson (1958) suggests that there may be a physiological barrier to its establishment in that minimum temperatures are sufficiently high and the daily range of temperature sufficiently low to prevent the production of diapause pupae and, thus, the ability of *Diparopsis* to carry-over from one season to the next. This supposition is strengthened by further understanding of the factors inducing diapause (Galichet, 1964).

There is, however, the possibility of *Diparopsis* surviving in the equatorial regions provided there is an adequate continuous food supply from perennial or stand-over cotton. Two possible access points exist to the otherwise isolated equatorial region, namely through southern Tanzania bordering on Zambia, Malawi and Mozambique, and from Ubangi-Shari in the Central African Republic into the Ubangi Province of Zaire.

There are early records of *D. castanea* being recorded on cotton in Tanzania north of the Ruvuma River near Songea, and Kabissa and Nyambo (1989) report more recent occurrences of red bollworm larvae on wild cotton in the region. In order to restrict the northward movement of *Diparopsis*, a quarantine zone of 100–250 km in depth in which no cotton may be grown, was established in 1946, along the length of the Tanzania border with Zambia, Malawi and Mozambique.

Diparopsis castanea has not yet been found in commercial cotton areas north of the quarantine zone, but Kabissa and Nyambo (1989) mention an increase of cotton production within the quarantine zone near Mbeya, less than 90 km from infested cotton areas of Zambia, which could be at risk and emphasize the danger that this poses to the main cotton growing areas.

Regarding the northern access point, Pearson (1958) reports that cotton is now grown in sufficient continuity in the Central African Republic for there to be no physical barrier to the spread of *D. watersi* south-eastwards into Zaire.

Status and Course of Infestation in Cotton

The status of *Diparopsis* will depend upon: the absolute size and duration of the carry-over moth flight, and the extent to which this is intercepted by the crop at a vulnerable stage of development and reinforced by previous breeding on wild host plants, stand-over cotton or exceptionally early sowings; the effect of weather and natural enemies in limiting the build-up of the infestation; and the relation between the incidence of the pest population and formation of the crop during the season.

Diparopsis has in the past been generally identified as an important pest of cotton where it occurs in Africa, with particularly serious damage being caused to the rainfed crops of Malawi and Mozambique (Pearson, 1958). Serious infestations of *D. castanea* are also known from Zimbabwe, where it has vied for importance with *Helicoverpa armigera*, and from Swaziland and South Africa. *Diparopsis watersi*, although of significance as a pest of raingrown cotton of sub-Saharan savannah regions, has been generally accorded less importance than *castanea*, south of the equator. *Diparopsis tephragramma* is known to be of importance in Angola. *Diparopsis* can be a serious pest of irrigated cotton, although *D. watersi* is absent from the irrigated areas of the Gezira and Tokar Delta in the Sudan.

With the advent of the successful use of insecticides against *Diparopsis* in the early 1960s, this bollworm has become progressively of less immediate importance, especially in relation to *Helicoverpa* which has become increasingly damaging in many areas. *Diparopsis castanea*, however, continues to maintain its importance in areas such as Malawi, where there are appreciable areas unsprayed.

The course of infestation of *Diparopsis* in the crop has been described and discussed in detail by Pearson (1958) for *D. castanea* and by Galichet (1964) for *D. watersi*, with special reference to Malawi and Tchad respectively. A characteristic course of infestation by *Diparopsis* in rainfed-cotton crops is for moths to be emerging already from diapause pupae, carried over from the previous season, at the time the cotton is sown, shortly after the break of the rains. Effective egg laying occurs from the commencement of flowerbud formation with larvae feeding upon and destroying the developing flowerbuds. The first generation of larvae give rise largely, if not entirely, to the short-term pupae producing a second generation of moths some five to six weeks after the start of an infestation. Successive generations of moths derived from short-term pupae follow at intervals which, with falling tempertures, are increasingly prolonged. Because emergences from the previous season's diapause pupae may continue over a long period, the carry-over moth flight can overlap that due to the first generation of pupae bred in the crop and, with variations in rate of individual development, this prevents the appearance of a succession of marked peaks of egg laying corresponding to successive generations. As the plants develop, the green bolls come increasingly under attack with larvae continuing to feed within full size bolls prior to boll opening, thus causing a direct loss of seed cotton.

Control Measures

Measures to reduce the damage caused by *Diparopsis* have been based on limiting the effectiveness of the carry-over from one season to the next, promoting the early establishment of a crop and applying insecticides. More recently, the role of the components of the sex pheromone in pest management programmes has been examined. In the case of a pest like *Diparopsis*, in which the whole of the infestation is derived from diapause pupae formed the previous season, the importance of attacking the carry-over population is obvious and most of the work carried out on control, prior to the advent of modern insecticides, has been with this object in mind.

The effectiveness of any control measure, including that of insecticide application, in raising yields will be dependent upon sound cultural practices promoting the maximum early production of fruiting points and subsequent setting of green bolls. Under such circumstances plants are more able to produce a crop in the face of a *Diparopsis* attack despite some loss of fruiting points. Poorly grown plants, on the other hand, are unable to develop sufficient undamaged bolls to cause senescence and will, instead, continue to produce a low level of buds and bolls enabling a long and late build-up of *Diparopsis*. The whole object of the agriculturist and farmer must be, in circumstances where heavy bollworm infestations may occur, to obtain a crop from the earliest formed fruiting points.

Limitation of the Carry-over

The carry-over of *Diparopsis* from one season to the next may be limited either by minimizing the production of diapause pupae or by preventing the effective survival of those produced. The former can only be achieved, in the absence of insecticides, by a change in cropping period to one in which temperature conditions would be less favourable to the induction of diapause. A change of sufficient magnitude to be effective would not be possible for cotton dependent upon rainfall and even under irrigation, unsuitable climatic conditions may preclude the growing of the crop at the required time. Pearson (1958) considered that earlier sowing and uprooting dates for cotton in the lower Shire Valley of Malawi might achieve a marked reduction in diapause pupae by bringing boll formation forward to a time when larvae are forming predominantly non-diapause pupae. An earlier growing regime was put into operation but, while this has been maintained, *Diparopsis* continues to be a major pest in the region.

Unfortunately, data do not exist to indicate whether or not there has been some, although inadequate, reduction in diapause pupal population resulting from the change in regime.

While there is little that can be presently done, other than by the application of insecticides, to reduce the carry-over pupal population produced in the crop itself for most rainfed areas, it is most important to prevent its augmentation by continued breeding on ratoon and stand-over cotton. In 1990 areas of abandoned stand-over cotton in the Transvaal were heavily infested with *D. castanea* (Matthews, personal communication). In most cotton producing areas the danger of such a situation is fully recognized and ratooning of cotton is prohibited, with an enforced close season requiring cotton to be uprooted or otherwise destroyed by a certain date.

The manner in which the period of interception of the new crop by the carry-over moth flight might be minimized and/or the mortality of pupae increased to reduce the incidence of *Diparopsis* has been examined. Evasion of a significant portion of the carry-over moth flight through an alteration of sowing date is impracticable for rainfed cotton and it is only under a system of irrigation that there could be sufficient flexibility in time of sowing to avoid, substantially, the normally prolonged period of moth emergence from diapause pupae. No successful instance of affecting control of *Diparopsis* by a change of sowing date has been reported.

It has been shown that, under certain circumstances, it is possible to bring forward the period of moth emergence, and to alter a bimodal pattern of emergence with its often damaging second peak, through field practices which effectively remove the inhibiting high temperature factor in pupal development. Proctor (1962), utilizing the effect of lowered soil temperature through watering in promoting diapause pupal development, proposed a pest management strategy for cotton in the Abyan Delta under which as much as possible of the previous cotton land is flooded in order to advance the second period of the bimodal emergence pattern, thus enabling the cotton crop to evade a significant portion of the carry-over moth flight. He further suggested that the possibility of utilizing the initial early floods to achieve a much greater forward shift in emergence pattern which, in the presence of a close season, would increase the effectiveness of control.

In Zimbabwe, Tunstall (1968) showed that the unimodal emergence pattern of moth emergence could be advanced significantly by burying pupae to a depth of 20 cm where they would avoid heat inhibition. In this way, the majority of moths emerge before the cotton crop is able to support an infestation. Attempts however to achieve a similar effect on a natural field population by double hand-trenching were unsuccessful, but further investigation of soil inversion techniques may be merited. Avoiding high soil temperatures with heavy grass mulches was shown to inhibit pupal development, possibly through too great a lowering of temperature and increased wetness of the soil.

Cultural practices, such as ploughing followed by heavy rolling or dragging have shown some reduction in pupal numbers in the Sudan (Bedford, 1937) and Zimbabwe (McKinstry, 1949). Disc-harrowing in the Gash Delta, Sudan (Tunstall *et al.*, 1956) and mould-board ploughing in the Abyan Delta (Brettell, 1966) are reported to have reduced pupal populations by about half, but such practices alone are probably unlikely to affect materially subsequent levels of crop infestation.

Where irrigation is practised, the destruction of diapause pupae by prolonged waterlogging is considerable, especially where water is applied by flood irrigation and this may have been a major factor in the elimination of *Diparopsis* from the Gezira irrigation scheme in the Sudan. The minor importance of *Diparopsis* in the Shire River region of Malawi in the past, when cotton was grown as a winter crop following the recession of the annual floods, may also have been partialy due to destruction of pupae by flooding (Pearson, 1958).

Pupal mortality from predators was increased both in cultivated soils and under mulches during the hot dry season (Pearson, 1958).

Finally, attention has been given in the past to diverting the carry-over moth flight to trap crops such as ratoon cotton (Peat, 1935, 1936), but with a fuller realization of the nature of the moth flight and of the danger of breeding an early generation of bollworm, the idea was abandoned. The possibility of using *Cienfugosia*, with its strong attraction to ovipositing moths, as a trap crop when grown concurrently with cotton has also been examined in Swaziland with some encouraging results (Clarke, 1979b).

Application of Insecticides

Direct action against the larvae of *Diparopsis* with the application of modern insecticides commenced in the 1950s. Early work on *D. castanea* in Zimbabwe (McKinstry, 1948; McKinstry and James, 1952; Staples, 1956), Republic of South Africa (Faure, 1953), Swaziland (McKinlay, 1957) and Mozambique (Carvalho and Barbosa, 1953; Barbosa and Carvalho, 1958) and on *D. watersi* in Nigeria (Choyce, 1957), Sudan (Ripper and George, 1965) and the Republic of Yemen (Proctor, 1962), which included the insecticides DDT, HCH, toxaphene, parathion and endrin, gave inconsistent yield increases and no general recommendations could be made. In Zimbabwe, this early work was followed by an intensive programme of laboratory and field toxicological studies to assess a range of insecticides against the bollworm pest complex with special reference to *D. castanea*, which, at the time, was considered the major limiting factor to increased yields (Matthews, 1966a,b,c). A major breakthrough came with the advent of carbaryl which, with its particularly high level of toxicity to *Diparopsis* larvae gave excellent control of this bollworm in the field together with a marked reduction in the population of diapause pupae. Its relatively low level of mammalian toxicity made it an acceptable insecticide for use in peasant agriculture. Effective control of *Diparopsis* highlighted the importance of controlling other insects of the pest complex, particularly *Helicoverpa* which may have become increasingly attracted to the well developed plants protected from *Diparopsis* and, possibly, *Empoasca*, the latter having a debilitating effect on the plant. The pest management strategy developed following the achievement of adequate *Diparopsis* control and adopted, with local modifications, in most of the cotton areas of southern Africa where this bollworm is a major pest, is described in Chapter 27.

In respect of *Diparopsis* and, also of *Helicoverpa*, the strategy involved targeting the first instar larvae from the time of egg hatch to penetration of a bud or boll requiring, because of the varied distribution of egg placement and the short period of larval exposure, a precise timing of spray application together with a high level of spray penetration and coverage of the plant. These requirements were met by timing spray applications on the basis of twice weekly egg counts and achieving effective deposition of the insecticide with a multi-directional vertical spray boom ('tailboom'). Other methods of spray application, notably spinning disc sprayers, have since proved effective and are widely used.

In Zimbabwe (Anon., 1985), Malawi (Anon., 1976) and in other areas where *D. castanea* occurs, carbaryl continues to be recommended although thiodicarb and the pyrethroids (with restrictions on their period of use) are listed as alternative insecticides.

It is, perhaps, noteworthy that in Zimbabwe where the entire cotton crop has been treated with insecticide for *Diparopsis* control since 1960 there has been no reported incidence of resistance of *Diparopsis* to carbaryl or other insecticides. Generally, in Zimbabwe, *Diparopsis* has become of less immediate importance perhaps as a direct result of the later sprays of the season effectively reducing the carry-over population of diapause pupae below a critical threshold. However *Diparopsis* remains a most serious potential threat and any breakdown in control could mean a return to the devastating attacks before the introduction of carbaryl.

Within the francophone cotton areas north of the equator, *D. watersi* does not appear to have attained quite the level of importance as *D. castanea* in the south. Effective control of the bollworm is obtained with the insecticide combination of a pyrethroid and an organo-phosphorus component for sucking insect control applied with a spinning disc sprayer. The insecticide applications are timed according to a fixed schedule. Carbaryl and/or pyrethroid insecticides with spinning disc sprayers are used for *Diparopsis* control in the Gambia (Tunstall and King, 1979).

It is important that the relative ease with which *Diparopsis* can be currently controlled with insecticides does not divert attention from cultural and biological factors that reduce the magnitude and effectiveness of the carry-over of the bollworm from one season to the next and minimize the dependency upon insecticides.

Sex Pheromone

The employment of the sex pheromone within a pest management strategy for *Diparopsis* has yet to be defined and developed in a practical manner. Marks (1976c) reports significant communication disruption, as measured by reduced trap catches of male *Diparopsis*, when female-baited traps in 100 m^2 cotton plots were surrounded by point sources of either dicastalure or the inhibitory IIA acetate, and economic factors associated with a pheromone control programme for *Diparopsis* have been considered by Farrington (1976). With the reported success of employing a pheromone in the control of *Pectinophora gossypiella* (see

Chapter 4), further work with *Diparopsis* may well be justified. A possible further use of the pheromone is as an additional tool for the monitoring of *Diparopsis* infestations.

References

Andrewartha, H.G. (1952) Diapause in relation to the ecology of insects. *Biological Review* 27, 50–107.

Angelini, A. and Vandamme, P. (1964) An intestinal virus disease of *Diparopsis watersi. Coton et Fibres Tropicales* 19, 265–270.

Anon. (1927) *The Sudan Bollworm* (Diparopsis castanea). Proceedings of the South and East African Agricultural Conference, 1926, Nairobi, pp. 177–178, 234–235.

Anon. (1928) Entomological notes No. 39. Bollworm parasites and predators. *Farming in South Africa* July 1928, report 2 pp.

Anon. (1959) *Cotton Pest Research Scheme.* Annual Report of the Department of Agriculture, Part II, Nyasaland Government for 1958/59. Government Printer, Zomba.

Anon. (1960) *Cotton Pest Research Scheme.* Annual Report of the Department of Agriculture, Nyasaland Government for 1959/60. Government Printer, Zomba.

Anon. (1961) *Cotton Pest Research Scheme.* Annual Report of the Department of Agriculture, Nyasaland Government for 1960/61. Government Printer, Zomba.

Anon. (1963) Rapport Annuel, 1962–63, Institut de Recherches du Coton et des Textiles exotiques. Station de Bouake, Cote d'Ivoire.

Anon. (1969) *Annual Report of the Agricultural Research Council of Malawi 1969.* ARC Malawi, Zomba.

Anon. (1976) *Cotton Handbook of Malawi.* Ministry of Agriculture and Natural Resources, Malawi.

Anon. (1985) *Cotton Handbook.* Commercial Cotton Growers' Association, Zimbabwe.

Atger, P. (1969) A virus disease with nuclear localisation in *Diparopsis watersi. Coton et Fibres Tropicales* 24(2), 205–206.

Ballard, E. (1913) Some cotton and tobacco pests of Nyasaland. *Nyasaland Government Gazette*, Supplement 30, April 1913, Zomba.

Barbosa, A.J. da Silva and Carvalho, M. de (1958) Resultados dos ensaios de insecticidas realizados pelo C.I.C.A. durante as companhas de 1953–54, 1954–55, 1955–56 and 1956–57. *Memorias e trabalhos Centro de Investigação cientifica algodedeira* No. 29, 18 pp.

Bedford, H.W. (1937) Entomological section, Report of the Agricultural Research Service, Sudan, 1936, p. 50.

Bedford, H.W. (1938) Entomological section, Report of the Agricultural Research Service, Sudan, 1938, p. 50.

Beeden, P. (1974) Bollworm oviposition on cotton in Malawi. *Cotton Growers Review* 51, 52–61.

Beevor, P.S., Campion, D.G., Moorhouse, J.E. and Nesbitt, B.F. (1973) Cross-attractancy and cross-mating between the red bollworm *Diparopsis castanea* Hmps and the Sudan bollworm *Diparopsis watersi* (Roths) (Lep., Noctuidae). *Bulletin of Entomological Research* 62, 439–442.

Brader, L.M., Brader, L., Delalande, P. and Atger, P. (1968) Quatre annees d'observations aux pieges lumineux en culture cotonniere au Tchad. *Coton et Fibres Tropicales* 23, 469–475.

Brettell, J.H. (1966) Eleven years work in Abyan (South Arabia) by entomologists of the Empire Cotton Growing Corporation. *Empire Cotton Growing Review* 43, 286–295.

Brettell, J.H. (1979) Green lacewings (Neuroptera: Chrysopidae) of cotton fields in

Central Rhodesia I Biology of *Chrysopa boninensis* Okamoto and toxicity of certain insecticides to the larva. *Rhodesian Journal of Agricultural Research* 17, 141–149.

Cadou, J. (1949) Rapport de campagne 1948–1949, Tikem (Tchad) (Unpublished) L'Institut de Recherches du Coton et des Textiles (in Galichet, 1964).

Cameron, W.P.L., Cowland, J.W. and Maxwell Darling, R.C. (1946) Regional contribution, Sudan (Unpublished).

Campion, D. (1968) A nematode parasite of the red bollworm *Diparopsis castanea* Hmps and the implications of such a parasite as a possible biological control agent. *Tropical Science* 10, 155–159.

Carvalho, M. de and Barbosa, A.J. da Silva (1953) Resultados dos ensaios de insecticidas realizados pelo C.I.C.A. durante as companhas de 1949–50, 1950–51, 1951–52 and 1952–53. *Memorias e trabalhos Centro de Investigação cientifica algodedeira* No. 13, 27pp.

Choyce, M.A. (1957) Insecticide trials on cotton in Northern Nigeria. *Report of the West African Cotton Research Conference*, 1957, pp. 118–130.

Choyce, M.A. and Lyon, D.J. de B. (1963) Entomology. Progress Report from Experimental Stations in N. Nigeria, 1962–63 Season. Cotton Research Corporation, London.

Choyce, M.A. and Reed, W. (1961) The pupal cell of the red bollworm. *Empire Cotton Growing Review* 38, 182–188.

Clarke, R.O.S. (1979a) Cotton Entomology Research Unit, Annual Report of the Agricultural Research Division 1976–77. University of Botswana and Swaziland.

Clarke, R.O.S. (1979b) Cotton Entomology Research Unit, Annual Report of the Agricultural Research Division 1977–78. Ministry of Agriculture and Cooperatives, Swaziland Government.

Clements, A.N. (1951) A revision of *Diparopsis* Hmps (Lepidoptera: Agrotidae). *Bulletin of Entomological Research* 42, 491–497.

Couilloud, R. (1961) Rapport de campagne (1960–1961), Section d'Entomologie Station de Bebedjia, Tchad (Unpublished) L'Institut de Recherches du Coton et des Textiles Exotique, 1962 (in Galichet, 1964).

Couilloud, R. (1962) Rapport de campagne (1961–1962), Section d'Entomologie Station de Bebedjia, Tchad. L'Institut de Recherches du Coton et des Textiles Exotique, 1963 (in Galichet, 1964).

Croisier, G., Armargier, A., Godse, D.B., Jacquemard, P. and Duthoir, J.L. (1980) A nuclear polyhedrosis virus discovered in the Noctuid moth, *D. watersi* (Roths) – a new variant of the baculovirus of *Autographa californica* (Speyer). *Coton et Fibres Tropicales* 35, 415–423.

Crosskey, R.W. (1970) The identity of *Palexorista quadrizonula* (Thompson) (Diptera), a Tachinid parasite of Lepidopterous pests in Africa. *Bulletin of Entomological Research* 59, 579–583.

El-Tayeb, Y.M. (1976) Biology and control of *Diparopsis watersi* in the Nuba mountains area of Western Sudan. *Journal of Economic Entomology* 70, 553–556.

El-Tigani, M. and El-Amin (1966) Preliminary bollworm survey on cotton in the Singa District, Sudan. *Wissenschaftliche Zeitschrift der Universität Rostock Mathematik und Naturwissenschaft 15* math-nat Reihi pt 2, 319–326.

Farrington, J. (1976) *Economic Factors Associated with the Large Scale Control by Sex Pheromones of Red Bollworm* (Diparopsis castanea *Hmps) on cotton in Malawi.* Report No. CVR/76/6, Centre for Overseas Pest Research, Ministry of Overseas Development, London.

Faure, J.C. (1953) Field experiments with insecticides against cotton bollworms 1951–52. *Entomology Memoirs Department of Agriculture Union of South Africa*, No. 2, 503–522.

Galichet, P.F. (1961) Parasitisme multiple chez *Diparopsis watersi* Roths (Lep. Agrotidae). *Entomophaga* 6, 203–205.

Galichet, P.F. (1964) *Diparopsis watersi* Rothschild (Lepidoptera: Noctuidae), Ravageur du cotonnier en Afrique Centrale, Monographie-Ecologie des populations – Etude experimentale de la diapause. *Coton et Fibres Tropicales* 19, 437–518.

Geering, Q.A. (1951) Progress Report of the Experiment Stations of the Empire Cotton Growing Corporation, 1949–50, p. 100.

Geering, Q.A. and Baillie, A.F.H. (1954) The biology of red bollworm, *Diparopsis watersi* Roths, in Northern Nigeria. *Bulletin of Entomological Research* 45, 661–681.

Giffard, W.E. (1929) The Sudan bollworm, *Diparopsis castanea* Hamp in the Sudan. *Bulletin of the Wellcome Tropical Research Laboratory, Entomology Section* No. 27, 17pp.

Giraudet, L.C. (1968) Etude de *Diparopsis tephragramma* B-B (Lepidoptera: Noctuidae), ravageur des cotonniers en Angola. *Boletin do Institute de investigação cientifica de Angola* 5, 5–28.

Ingram, W.R. and Green, S.M. (1972) Sequential sampling of bollworms on raingrown cotton in Botswana. *Cotton Growers Review* 49, 265–275.

Jacquemard, P. (1965) Bacterial disease of *D. watersi. Coton et Fibres Tropicales* 20, 283–286.

Jacquemard, P. (1976a) La diapause de *Diparopsis watersi* (Roths) (Lepidoptera: Noctuidae) dans le nord du Cameroun. *Coton et Fibres Tropicales* 31, 297–311.

Jacquemard, P. (1976b) Relations entre la diapause de *Diparopsis watersi* (Roths) (Lep. Noct.) et la diapause de son parasite *Eucarcelia* sp. [? evolans (Wied)] (Dipt. Tachin.) dans le nord du Cameroun. *Coton et Fibres Tropicales* 31, 313–321.

Jemmett, C.W. (1910) Annual report on entomological work, S. Nigeria (in Pearson, 1958).

Kabissa, J.C.B. and Nyambo, B.T. (1989) The red bollworm, *Diparopsis castanea* Hmps (Lepidoptera: Noctuidae) and cotton production in Tanzania. *Tropical Pest Management* 35, 190–192.

King, C.B.R. (1926) Some notes on the red (Sudan) bollworm (*Diparopsis castanea* Hampson) in Nyasaland. *Empire Cotton Growing Review* 3, 352–364.

King, C.B.R. (1929) Progress Report of the Experiment Stations of the Empire Cotton Growing Corporation 1927–28, p. 250.

King, H.H. (1908) Report on economic entomology, *Third Report.* Wellcome Research Laboratory, pp. 201–248 (in Pearson, 1958).

King, H.H. (1927) Report of the Government Entomologist for the year 1926. *Bulletin of the Welcome Tropical Research Laboratory, Entomology Section.* No. 24, 6pp.

Le Gall, J. (1950) Rapport de campagne (Tchad) (Unpublished). Institut de Recherches du Coton et des Textiles Exotiques (in Galichet, 1964).

Le Gall, J. (1952) Rapport de campagne (Tchad) 1950–51 (Unpublished). Institut de Recherches du Coton et des Textiles Exotiques (in Galichet, 1964).

Lounsbury, C.P. (1926) Report of the Chief, Division of Entomology, 1925–26. *Farming in South Africa* 1, 334–338 (in Pearson, 1958).

Macdonald, D., Fielding, W.L. Ruston, D.F. and King, H.E. (1946) Progress Report of the Experiment Stations of the Empire Cotton Growing Corporation 1944–45, p. 22.

McKinlay, K.S. (1957) A preliminary note on the control of the red bollworm *Diparopsis castanea*, with insecticides. *Empire Cotton Growing Review* 34, 253–257.

McKinley, D.J. (1968) Key to some larvae of Lepidoptera attacking cotton in Central Africa. *Cotton Growers Review* 45, 184–197.

McKinley, D.J. (1974) *An Attempt to Develop a Completely Artificial Diet for Rearing the Red Bollworm,* Diparopsis castanea. Centre for Overseas Pest Research, Miscellaneous Report No. 16. Foreign and Commonwealth Office, London.

McKinstry, A.H. (1947) Progress Report of the Experiment Stations of the Empire Cotton Growing Corporation 1945–46, p. 42.

McKinstry, A.H. (1948) Progress Report of the Experiment Stations of the Empire Cotton Growing Corporation 1946–47, pp. 43–47.

McKinstry, A.H. (1949) Progress Report of the Experiment Stations of the Empire Cotton Growing Corporation 1947–48, p. 42.

McKinstry, A.H. and James, R.W. (1952) Progress Report of the Experiment Stations of the Empire Cotton Growing Corporation 1950–51, p. 98.

Marks, R.J. (1976a) Mating behaviour and fecundity of the red bollworm *Diparopsis castanea* Hmps. (Lepidoptera, Noctuidae). *Bulletin of Entomological Research* 66, 145–158.

Marks, R.J. (1976b) Female sex pheromone release and timing of male flight in the red bollworm *Diparopsis castanea* Hmps. (Lepidoptera, Noctuidae), measured by pheromone traps. *Bulletin of Entomological Research* 66, 219–241.

Marks, R.J. (1976c) Field studies with the synthetic sex pheromone and inhibitor of the red bollworm *Diparopsis castanea* Hmps. (Lepidoptera, Noctuidae) in Malawi. *Bulletin of Entomological Research* 66, 243–265.

Marks, R.J. (1976d) The influence of behaviour modifying chemicals on mating success of the red bollworm *Diparopsis castanea* Hmps (Lepidoptera, Noctuidae) in Malawi. *Bulletin of Entomological Research* 66, 279–300.

Marks, R.J. (1976e) Laboratory evaluation of the sex pheromone and mating inhibitor of the red bollworm *Diparopsis castanea* Hampson (Lepidotpera, Noctuidae) *Bulletin of Entomological Research* 66, 427–435.

Marks, R.J. (1977a) The influence of climatic factors on catches of the red bollworm *Diparopsis castanea* Hampson (Lepidoptera, Noctuidae) in sex pheromone traps. *Bulletin of Entomological Research* 67, 243–248.

Marks, R.J. (1977b) Assessment of the use of sex pheromone traps to time chemical control of red bollworm, *Diparopsis castanea* Hampson (Lepidoptera: Noctuidae) in Malawi. *Bulletin of Entomological Research* 67, 575–587.

Marks, R.J., Nesbitt, B.F., Hall, D.R. and Lester, R. (1978) Mating disruption of the red bollworm of cotton *Diparopsis castanea* Hampson (Lepidoptera: Noctuidae) by ultra-low-volume spraying with a micro-encapsulated inhibitor of mating. *Bulletin of Entomological Research* 68, 11–29.

Marks, R.J. Hall. D.R., Lester, R., Nesbitt, B.F. and Lambert, M.R.K. (1981) Further studies on mating disruption of the red bollworm, *Diparopsis castanea*, Hampson (Lepidoptera:Noctuidae) with a microencapsulated mating inhibitor. *Bulletin of Entomological Research* 71, 403–418.

Marshall, J., Parsons, F.S. and Hutchinson, H. (1937) Studies on the red bollworm of cotton, *Diparopsis castanea* Hampson, Pt. I. The distribution and ecology of two natural food plants, *Cienfugosia hildebrandtii* Gurke and *Gossypium herbaceum* var. *africana* Watt. *Bulletin of Entomological Research* 28, 621–632.

Matthews, G.A. (1966a) Investigations of the chemical control of insect pests of cotton in Central Africa. I. Laboratory rearing methods and tests of insecticides by application to bollworm eggs. *Bulletin of Entomological Research* 57, 69–76.

Matthews, G.A. (1966b) Investigations of the chemical control of insect pests of cotton in Central Africa. II Tests of insecticides with larvae and adults. *Bulletin of Entomological Research* 57, 77–91.

Matthews, G.A. (1966c) Investigations of the chemical control of insect pests of cotton in Central Africa. III Field trials. *Bulletin of Entomological Research* 57, 193–197.

Mitchell, B.L. (1943) Progress Reort of the Experiment Stations of the Empire Cotton Growing Corporation 1941–42, p. 149.

Mitchell, B.L. (1944) Progress Report of the Experiment Stations of the Empire Cotton Growing Corporation 1942–43, p. 156.

Monteil, L. (1934) Les insectes nuisibles au cotonnier en Afrique equatoriale francaise. *Agronomie coloniale* No. 193, 11–18 (in Pearson, 1958).

Moorhouse, J.E. Yeadon, R., Beevor, P.S. and Nesbitt, B.E. (1969) Method for use in studies of insect chemical communication. *Nature*, London 223, 1174–1175.

Nesbitt, B.F., Beevor, P.S., Cole, R.A., Lester, R. and Poppi, R. G. (1973a) Sex pheromones of two noctuid moths. *Nature New Biology* 244, 208–209.

Nesbitt, B.F., Beevor, P.S., Cole, R.A., Lester, R. and Poppi, R.G. (1973b) Synthesis of both geometric isomers of the major sex pheromone of the red bollworm moth. *Tetrahedron Letters* No. 47, 4669–4670.

Nesbitt, B.F., Beevor, P.S., Cole, R.A., Lester, R. and Poppi, R.G. (1975) The isolation and identification of the female sex pheromones of the red bollworm moth, *Diparopsis castanea*. *Journal of Insect Physiology* 12, 1091–1096.

Parsons, F.S. (1928) Progress Report of the Experiment Stations of the Empire Cotton Growing Corporation 1926–27, p. 66.

Parsons, F.S. (1929) Progress Report of the Experiment Stations of the Empire Cotton Growing Corporation 1927–28, p. 55.

Parsons, F.S. (1936) Progress Report of the Experiment Stations of the Empire Cotton Growing Corporation 1934–35, p. 24.

Parsons, F.S. and Pearson, E.O. (1942) Regional contribution S. Africa (Unpublished) (in Pearson, 1958).

Parsons, F.S. and Ullyett, G.C. (1932) Progress Report of the Experiment Stations of the Empire Cotton Growing Corporation 1930–31, p. 14.

Parsons, F.S., Hutchinson, H. and Marshall, J. (1938) Progress Report of the Experiment Stations of the Empire Cotton Growing Corporation 1936–37, p. 26.

Pearson, E.O. (1943) Progress Report of the Experiment Stations of the Empire Cotton Growing Corporation 1941–42, p. 149.

Pearson, E.O. (1954) The relationship between the African and South American red bollworms of cotton, *Diparopsis* and *Sacadodes*. *Empire Cotton Growing Review* 31, 171–177.

Pearson, E.O. (1958) *The Insect Pests of Cotton in Tropical Africa*. Empire Cotton Growing Corporation and Commonwealth Institute of Entomology, London, 355pp.

Pearson, E.O. (1962) Diapause as a factor determining the status of *Diparopsis* as a pest of cotton. *Annals of Applied Biology* 50, 604–606.

Pearson, E.O. and Mitchell, B.L. (1945). *A Report on the Status and Control of Insect Pests of Cotton in the Lower River Districts of Nyasaland*. Government Printer, Zomba, 48pp.

Peat, J.E. (1930) Progress Report of the Experiment Stations of the Empire Cotton Growing Corporation 1929–30, p. 121.

Peat, J.E. (1935) Progress Report of the Experiment Stations of the Empire Cotton Growing Corporation 1933–34, p. 52.

Peat, J.E. (1936) Progress Report of the Experiment Stations of the Empire Cotton Growing Corporation 1934–35, p. 50.

Peat, J.E., McKinstry, A.H. and Prentice, A.N. (1942) Regional contribution S. Rhodesia (Unpublished) (in Pearson, 1958).

Pomeroy, A.W.J. (1925) The cotton bollworms of southern Nigeria. *Fourth Annual Bulletin*, Department of Agriculture, Nigeria, pp. 89–108.

Proctor, J.H. (1962) The biology and control of the Sudan bollworm, *Diparopsis watersi* (Roths), in the Abyan Delta, West Aden Protectorate. *Bulletin of Entomological Research* 53, 311–335.

Reed, W. and Choyce, M.A. (1961) Observations on *Carcelia evolans*, a parasite of *Diparopsis watersi*, in Northern Nigeria. *Bulletin of Entomological Research* 52, 785–793.

Ripper, W.E., and George, L. (1965) *Cotton Pests of the Sudan.* Blackwell Scientific Publications, Oxford, 345pp.

Rose, D.W. (1957) *Annual Report Gatooma Research Station, Rhodesia 1957.*

Smee, C. (1940) Report of the Entomologist. Department of Agriculture, Nyasaland 1939, 11 pp. typescript.

Smee, C. (1941) Report of the Entomologist. Department of Agriculture, Nyasaland 1940, 7pp. typescript.

Smee, C. (1942) Regional contribution, Nyasaland (Unpublished) (in Pearson, 1958).

Smee, C. (1944) Report of the Entomologist. Department of Agriculture, Nyasaland 1943, 11pp. typescript (in Pearson, 1958).

Smith, A.J. (1933) Report on cotton insect and disease investigations. Pt 3. Notes on the red bollworm (*Diparopsis castanea* Hampson) on cotton in South Africa. *Science Bulletin, Department of Agriculture, South Africa* No. 114, 29pp.

Staples, R.R. (1956) Gatooma Research Station Rep. Sec. Fed. Minist. Agric. Fed. Rhod. Nyasaland 1954–55, 70–73.

Sweeney, R.C.H. (1962) *Insect Pests of Cotton in Nyasaland* II *Coleoptera (Beetles).* Bulletin No. 19, Government Printer, Nyasaland.

Thompson, W.R. (ed.) (1943-) *A Catalogue of the Parasites and Predators of Insect Pests.* Imperial Parasite Service and (later) Commonwealth Bureau of Biological Control, Belleville, Ont. (in Pearson, 1958).

Tunstall, J.P. (1955) Gash Research Project, Annual Report 1953–54 (Unpublished). Research Division, Ministry of Agriculture, Sudan.

Tunstall, J.P. (1958) The biology of the Sudan bollworm *Diparopsis watersi* (Roths) in the Gash Delta, Sudan. *Bulletin of Entomological Research* 49, 1–23.

Tunstall, J.P. (1962) The biology of cotton bollworms in the Federation of Rhodesia and Nyasaland. *Proceedings of the First Federal Science Congress*, Rhodesia.

Tunstall, J.P (1968) Pupal development and moth emergence of the red bollworm (*Diparopsis castanea* Hmps) in Malawi and Rhodesia. *Bulletin of Entomological Research* 58(2), 233–254.

Tunstall, J.P. and King, W.J. (1979) *The Gambia Cotton Handbook.* Natural Resources Institute, Overseas Development Administration, London.

Tunstall, J.P., Patterson, C.A. and Martin, E.O. (1956) Gash Research Project, Entomological Report for the 1954–55 season (Mimeographed). Research Division, Ministry of Agriculture, Sudan.

7

Cryptophlebia leucotreta (Meyrick) (Lepidoptera: Tortricidae)

R. COUILLOUD

Département du Centre de Cooperation Internationale en Recherche Agronomique pour le Développement, BP 5035, 34032 Montpellier Cedex, France

INTRODUCTION

This pest of citrus was originally referred to as *Carpocapsa* sp. in Natal by Fuller in 1900 and in the Transvaal by Simpson in 1905. In the years that followed there was confusion with a neighbouring species *Enarmonia batrachopa* (Meyrick). The species *leucotreta*, described by Meyrick in 1913 under the generic name *Argyroploce* was used for this pest (Gunn, 1921; Pomeroy, 1925; Ford, 1934); it was then transferred to the genus *Cryptophlebia* (Bradley, 1952; Clarke, 1958).

Cryptophlebia leucotreta (Meyrick) gets its common name, false codling moth, from the close resemblance of its habits to those of the codling moth *Cydia pomonella* (L.), the worldwide pest of deciduous fruits, which belongs to the same family.

GEOGRAPHICAL DISTRIBUTION

The origin of *C. leucotreta* is not known, but it may have originated in South Africa or the Ethiopian region. Its area of distribution includes central and southern Africa (Hodgson, 1966; Catling and Aschenborn, 1974) and the islands in the Indian Ocean. It has spread steadily in Africa, first to the coastal areas of West Africa and then to inland areas. Spread has been enhanced by the intensification of maize growing and an increasing number of fruit plantations. The northern boundary of the distribution is in the countries south of the Sahara at latitude 15°N (Couilloud, 1988).

Description of Stages (Couilloud, 1988)

Egg (Plate III.4a)

Oval, lenticular and flat, adhering strongly, 0.94–0.98 mm in the longest direction, translucent, whitish with iridescent highlights.

Larva (Plate III.4b)

At hatching: 1.2 mm long, creamy-white, black cephalic capsule and brown first thoracic segment, dark pinacula giving the dorsal region a dotted appearance. At the end of the larval stage: 12–18 mm long orangey-pink becoming paler along the sides and yellow in the ventral region; the head and the first thoracic segment are brown; the prolegs bear retractile crochets of varying lengths, curved towards the outside and arranged in a full circle; the last segment has an 'anal comb' with two to seven spines (Stofberg, 1948).

Several other species may display similarities to *C. leucotreta* at the larval stage: *Pectinophora gossypiella* (Saunders), *Mussidia nigrivenella* (Ragonot), *Mometa zemiodes* Durrant, *Pyroderces simplex* Walsingham, *Crocidosema plebeiana* Zeller and *Cryptoblabes guidiella* (Milliere) (Le Gall, 1966; McKinley, 1968; Delattre, 1972; Staeubli, 1977).

Pupa (Plate III.4c)

Length 8–10 mm in a very lightly woven cocoon. It is brown and darker in the dorsal area; the abdominal segments bear two transverse rows of spines set in small tubercles; the extremity of the hind segment is rounded and set with spines.

Adult (Plate III.4d)

Body length 6–8 mm, wing span: female 17–20 mm, male 15–18 mm. General colour brown with pale and blackish mottling. The head is covered with erect black and brown scales. The antennae are setiform with distinct segments. At rest, the wings are laid back in a ridge shape.

Biology

The moths, in the most shaded parts of plants, are motionless during the day and are difficult to see. Most activity is nocturnal, with oviposition preferably on green bolls. The sex ratio is 1 male : 2 females and adults live for six to ten days. The eggs are laid individually or in small groups overlapping like tiles during the early evening. Females can mate several times and lay 100–400 eggs. The eggs hatch after three to six days depending on temperature.

After hatching the first instar larvae move about for a while before penetrating the bolls; a twisted viscous deposit characteristic of the species is ejected through the perforation. There are five instars lasting a total of 10–20 days depending on temperature. Fully grown larvae leave the bolls and pupate on the ground after weaving a loose cocoon. Pupation lasts 10–15 days and the moths hatch in the early morning.

The length of the life cycle varies according to climatic differences between regions. Thus it can extend to over 200 days in the cold season in South Africa, while being only 30–45 days in equatorial or sub-equatorial Africa, and demonstrates the ecological flexibility of the species (Gunn, 1921; Bredo, 1933; Ripley *et al.*, 1939; Ghesquiere, 1940; Omer-Cooper, 1940; Alibert, 1946; Myburgh, 1965; Myburgh and Bass, 1969; Angelini and Labonne, 1970; Staeubli, 1976; Couilloud, 1988).

Host Plants

Cryptophlebia leucotreta is highly polyphagous and attacks over 70 species within 40 families of both tropical and temperate (introduced) herbaceous shrub and woodland plants. The larva feeds preferably on the pulp of fleshy fruits or various other growing parts of plants, such as pods, seeds and bolls while they are still soft. Citrus fruits, maize and cotton can sustain considerable economic damage (Pearson, 1958; Staeubli, 1976; Nonveiller, 1984; Couilloud, 1988).

Damage to Cotton Plants

Infestations of cotton crops usually occur only when the plants are old enough to bear bolls with soft, part-grown tissues on which the larva prefers to feed. Young larvae penetrate the carpel, tunnel beneath the epiderm in the pericarp and reach a loc. The tissues of immature boll locs are watery and rich in sugars; the larvae eat the seeds while still soft and the fibres attached to them. Intercarpel walls are also penetrated. Larval activity allows the introduction of microorganisms which cause rotting and complete loss of the boll. Attack of young bolls is less frequent and results in shedding. Damage to flower buds is rare, while damage to the stem is very rare (Angelini and Houiller, 1955).

Control

Chemical Control

For logistic reasons, most chemical control programmes in sub-Saharan Africa are aimed at overall control of the different pest species, which attack flowerbuds and bolls, including *C. leucotreta.* Choice of insecticide is important because of the larvae feeding inside bolls; several pyrethroids have been very effective (Angelini and Couilloud, 1974, 1976; Nyiira, 1974; Angelini *et al.*, 1982).

Other Control Methods

Agronomic control by using a single sowing date, cultivation techniques including ploughing after harvest and destroying crop residues does limit populations especially in areas with a maize–cotton cropping cycle. In areas of citrus, removal of fallen fruit and other phytosanitary procedures are important (Angelini and Labonne, 1970; Staeubli, 1977).

Numerous parasitoids of the egg and larval stages have been listed and their effectiveness studied under laboratory conditions (Searle, 1964; Broodryk, 1969; Delattre, 1973; Catling and Aschenborn, 1974; Reed, 1974; Anon., 1984; Leclant, 1988). Egg and larval parasitoids have been released in Mauritius on litchis, in South Africa on citrus and in Togo on maize and cotton, but it is difficult to assess their impact (Schwartz, 1980; Marais, 1982; Bournier, 1986).

In South Africa, attempts were made to control populations in citrus by the release of sterile males (Schwartz, 1979). Granulosis viruses have been studied (Angelini *et al.*, 1965; Amargier *et al.*, 1968; Jacquemard, 1983). The advantages of using these viruses and pathogenic bacteria (*Bacillus thuringiensis*) have been shown in Cote d'Ivoire and Togo (Staeubli, 1976; Bournier 1986).

Much research has been carried with pheromones and traps used to monitor populations. There have also been attempts at evaluating direct control by mass trapping of males, but experiments have yet to be carried out using the confusion technique (Read *et al.*, 1968; Angelini and Labonne, 1970; Angelini and Couilloud, 1972a,b; Angelini *et al.*, 1976, 1980; Staeubli, 1977; Zagatti, 1979, 1981, 1985; Zagatti *et al.*, 1983).

References

Alibert, H. (1946) Note sur quelques insectes déprédateurs des plantes cultivées ou spontanées en Côte-d'Ivoire. *Agronomie Tropicale* 1 (7–8), 388–399.

Amargier, A., Angelini, A., Vandamme, P. and Vago, C. (1968) Un complexe de viroses: granulose – plyédrie cytoplasmique chez le Lépidoptère *Argyroploce leucotreta* Meyrick. *Coton et Fibres Tropicales* 23, 413–416.

Angelini, A. and Couilloud, R. (1972a) Observations sur le piégeage sexuel chez *Cryptophlebia* (=*Argyroploce*) *leucotreta* (Meyr.). *Coton et Fibres Tropicales* 27, 273–281.

Angelini, A. and Couilloud, R. (1972b) Les moyens de lutte biologique contre certains ravageurs du cotonnier et une perspective sur la lutte intégrée en Côte-d'Ivoire. *Coton et Fibres Tropicales* 27, 283–289.

Angelini, A. and Couilloud, R. (1974) Résultats de l'expérimentation insecticide de 1972–1973 contre les principales chenilles des capsules du cotonnier en Côte-d'Ivoire. *Coton et Fibres Tropicales* 29, 199–206.

Angelini, A. and Couilloud, R. (1976) Premiers résultats obtenus en Côte-d-Ivoire avec les pyréthrinoïdes dans la lutte contre les ravageurs du cotonnier. *Coton et Fibres Tropicales* 31, 323–326.

Angelini, A. and Houiller, M. (1955) Sur une forme de parasitisme d'*Argyroploce leucotreta* observée pour la première fois en AOF. *Coton et Fibres Tropicales* 10, 49–53.

Angelini, A. and Labonne, V. (1970) Mise au point sur l'étude de *Cryptophlebia* (*Argyroploce*) *leucotreta* (Meyr.) en Côte-d'Ivoire. *Coton et Fibres Tropicales* 25, 497–500.

Angelini, A. and Le Rumeur, C. (1962) Sur une maladie à virus d'*Argyroploce leucotreta* découverte en Côte-d'Ivoire. *Coton et Fibres Tropicales* 17, 291–296.

Angelini, A., Amargier, A., Vandamme, P. Duthoit, J.L. (1965) Une virose à granules chez le Lépidoptère *Argyroploce leucotreta. Coton et Fibres Tropicales* 20, 277–282.

Angelini, A., Couilloud, R., Delabarre, M. and Lhoste J. (1976) Effet attractif des isomères de l'acétate de 8 dodécényl pour les mâles de *Cryptophlebia leucotreta* (Meyr.) (Lepidoptera). *Coton et Fibres Tropicales* 31, 373–374.

Angelini, A., Descoins, C., Le Rumeur, C. and Lhoste, J. (1980) Nouveaux résultats obtenus avec un attractif sexuel de *Cryptophlebia leucotreta* (Meyr.) (Lepidoptera). *Coton et Fibres Tropicales* 35, 277–281.

Angelini, A., Trijau, J.-P. and Vaissayre, M. (1982) Activité comparée de trois pyrethri-noïdes de "première génération" et d'un certain nombre de pyréthrinoides nouveaux contre les chenilles de la capsule. *Coton et Fibres Tropicales* 37, 359–364.

Anon. (1984) Possibilities for the biological control of the false codling moth, *Cryptophlebia leucotreta* (Lep., Tortricidae). *Biocontrol News and Information,* Institute of Biological Control 5, 217–220.

Bournier, J.-P (1986) Programme de lutte intégrée dans le système de culture maïs-coton, au Togo. IRCT-CIRAD, Division phytosanitaire, Montpellier, document ronéotypé (non publié), 19pp.

Bradley, J.D. (1952) Some important species of the genus *Cryptophlebia* Walsingham, 1899, with descriptions of three new species (Lepidoptera: Olethreutidae). *Bulletin of Entomological Research* 43, 679–689.

Bredo, H.J. (1933) Note sur *Argyroploce leucotreta* (Meyr.). *Bulletin Agronomique du Congo belge* 24, 150–156.

Broodryk, S.W. (1969) The biology of *Chelonus (Microchelonus) curvimaculatus* Cameron (Hymenoptera: Braconidae). *Journal of the Entomological Society of South Africa* 32, 169–189.

Catling, H.D. and Aschenborn, H. (1974) Population studies of the false codling moth *Cryptophlebia leucotreta* (Meyr.), on *Citrus* in the Transvaal. *Phytophylactica* 6, 31–38.

Clarke, J.F.G. (1958) *Catalogue of the Type Specimens of Microlepidoptera in the British Museum (N.H.) Described by Edward Meyrick*, vols 1–8. British Museum, London, 600pp.

Couilloud, R. (1988) *Cryptophlebia (=Argyroploce) leucotreta* (Meyrick). Lepidoptera, Tortricidae, Olethreutinae. *Coton et Fibres Tropicales* 43, 319–351.

Delattre, R. (1972) Eléments de base pour une lutte intégrée dans les cultures cotonnières d'Afrique. IRCT-Paris, Division Phytosanitaire, document ronéotypé (non publié), Bébedjia-Bamako, 24pp.

Delattre, R. (1973) *Parasites et Maladies en Culture Cotonnière.* Manuel phytosanitaire. IRCT – Paris, 146pp.

Ford, W.K. (1934) Some observations on the bionomics of the false codling moth *Argyroploce leucotreta* (Meyr.) (Family Eucosmidae) in Southern Rhodesia. *Publications of British South African Commonwealth* 3, 9–34.

Ghesquiere, J. (1940) Catalogues raisonnés de la Faune Entomologique du Congo Belge. Lépidoptères, Microlépidoptères (première partie). *Annales Musée Congo Belge, Tervuren,* C. Zoologie, Série III (II), 7 (1), 100–104.

Gunn, D. (1921) The false codling moth (*Argyroploce leucotreta* Meyr.). *Union of South African Pretoria, Department of Agriculture Science Bulletin* 21, 1–28.

Hodgson, C.J. (1966) The problem of the false codling moth. *Rhodesian Agricultural Journal* 63, 3–5.

Jacquemard, P. (1983) Activités: Laboratoire d'études sur les entomopathogènes. *Coton et Fibres Tropicales* 38, 27–28.

Leclant, S. (1988) Etude morphologique, biologique et éthologique de *Tetrastichus israeli*

(Mani et Kurian, 1952) (Hym.: Eulophidae) endoparasitoïde nymphal. Thèse de Doctorat, Université des Sciences et Techniques du Languedoc, Montpellier, 162pp.

Le Gall, J. (1966) *Les 'Platyedra', dans Traité d'Entomologie appliquée à l'Agriculture*, Tome II, Vol. 1, Masson, Paris, pp. 399–442.

McKinley, D.J. (1968) Key to some larvae of Lepidoptera attacking cotton in Central Africa. *Cotton Growers Review* 45, 184–197.

Marais, A.J. (1982) Valskodlingmot in die citrusdal omgewing. *Citrus and Subtropical Fruit Journal* 576, 22–23.

Myburgh, A.C. (1965) Low temperature sterilization of false codling moth *Argyroploce leucotreta* (Meyr.) in export Citrus. *Journal of the Entomological Society of South Africa* 28, 277–285.

Myburgh, A.C. and Bass, M.W. (1969) Effect of low temperature storage on pupae of false codling moth *Cryptophlebia leucotreta* (Meyr.). *Phytophylactica* 1, 115–116.

Nonveiller, G. (1984) Catalogue des insectes du Cameroun d'intérêt agricole. *Institut pour la Protection des Plantes*, Beograd, Teodora Drajzera of Yugoslavia, Mémoire 15, 210pp.

Nyiira, Z.M. (1974) Insecticide trials for the control of the false codling moth, *Cryptophlebia leucotreta* (Meyr.) (Lepidoptera: Tortricidae; subfamily Olethreutinae). *Pesticide Science* 5, 1–5.

Omer-Cooper, J. (1940) Remarks on the false codling moth. Multigraphed, Rhodes University, Grahamstown, S. Africa, 12 + 8pp. *Reviews in Applied Entomology*, Series A, 29, 227–228.

Pearson, E.O. (1958) *The Insect Pests of Cotton in Tropical Africa.* Empire Cotton Growing Corporation and Commonwealth Institute of Entomology, London, 356pp.

Pomeroy, A.W.J. (1925) The cotton bollworms of Southern Nigeria. *Fourth Annual Bulletin*, Department of Agriculture, Nigeria, pp. 89–108.

Read, J.S., Warren, F.L. and Hewitt, P.H. (1968) Identification of the sex pheromone of the false codling moth *Argyroploce leucotreta. Chemicals Communications* 792–793.

Read, J.S., Hewitt, P.H., Warren, F.L. and Myberg, A.C. (1974) Isolation of the sex pheromone of the moth *Argyroploce leucotreta. Journal of Insect Physiology* 20, 441–450.

Reed, W. (1974) The false codling moth, *Cryptophlebia leucotreta* (Meyr.) (Lepidoptera: Olethreutidae) as a pest of cotton in Uganda. *Cotton Growers Review* 51, 213–225.

Ripley, L.B., Hepburn, G.A. and Dick, J. (1939) Mass breeding of false codling moth *Argyroploce leucotreta* (Meyr.) in artificial media. *Union of South Africa Pretoria, Department of Agriculture and Forestry, Science Bulletin* 207, 1–18.

Schwartz, A. (1979) Ondersoek na die steriele-mannetjies tegniek as moont like beheermaatreel vir valskodlingmot by sitrus: vrylating van sterile motte. *Citrus and Subtropical Fruit Journal* 553, 10–12.

Schwartz, A. (1980) Eier-parasiet van valskodlingmot: evaluasie van'n teel-en vrylaatprogram. *Citrus and Subtropical Fruit Journal* 554, 6–8.

Searle, C.M. St. L. (1964) A list of insect enemies of *Citrus* pests in Southern Africa. *Technical Communication Plant Protection Research Institute, Pretoria* 30, 18pp.

Staeubli, A. (1976) Contribution à l'étude biologique et ecologique de *Cryptophlebia leucotreta* (Meyr.) (Lep. Tortricidae) en culture cotonnière au Dahomey. Thèse de Docteur ès Sciences Techniques, Ecole Polytechnique Fédérale, Zurich, 85pp.

Staeubli, A. (1977) Contribution à l'étude de *Cryptophlebia leucotreta* (Meyr.) particulièrement au Bénin. *Coton et Fibres Tropicales* 32, 325–349.

Stofberg, F.J. (1948) Larval structure as a basis for certain identification of false codling moth (*Argyroploce leucotreta*, Meyr.) larvae. *Journal of the Entomological Society of South Africa* XI, 68–75.

Zagatti, P. (1979) Etude de la sécrétion phéromonale chez *Cryptophlebia leucotreta* Meyr.

(Lepidoptera: Tortricidae) et chez l'écaille fileuse *Hyphantria cunea* Drury (Lepidoptera: Arctiidae) défoliateur forestier nouvellement introduit en France. *Mémoires DEA Entomologie, Université Pierre et Marie Curie, Paris VI*, 55pp.

Zagatti, P. (1981) Microcomportements induits par les phéromones sexuelles chez quelques Lépidoptères ravageurs des cultures en milieu sahélien. Diplôme Doctorat 3e cycle, Université Pierre et Marie Curie, Paris VI, 161pp.

Zagatti, P. (1985) Approche évolutive du comportement précopulatoire chez les Lépidoptères Ditrysiens. Thèse de Docteur ès Science, Université Pierre et Marie Curie, Paris VI, 191pp.

Zagatti, P., Lalanne-Cassou, B., Descoins C. and Gallois, M. (1983) Données nouvelles sur la sécrétion phéromonale de *Cryptophlebia leucotreta* (Meyr.) (Lepidoptera: Tortricidae). *Agronomie* 3, 75–80.

8 Other Lepidoptera

G.A. Matthews

International Pesticide Application Research Centre, Imperial College at Silwood Park, Buckhurst Road, Sunninghill, Ascot, Berkshire SL5 7PY, UK

Boll Feeding

Sacadodes pyralis Dyar (Lepidoptera: Noctuidae)

The South American red bollworm, also known as the Colombian pink bollworm has been recorded in Guyana, Venezuela, Colombia, Panama, Nicaragua and El Salvador. A record of it occurring in Paraguay has not been substantiated in recent studies there. It has a very similar appearance, life history and host plant range to *Diparopsis* in Africa (Withycombe, 1926). The larvae remain inside bolls and rarely migrate to another boll unless food is no longer available. The pupae are protected inside an earthen cocoon, some of which can remain for up to five months, suggesting the presence of a diapause (Szumkowski, 1953). It is parasitized by *Apanteles thurberiae* Mues. (Murillo, 1937). At present it is considered to be a relatively minor pest on cotton treated with insecticides against other pests.

Mometa zemoides Durrant (Lepidoptera: Gelechiidae)

The larva, commonly known as the false pink bollworm, closely resembles that of *Pectinophora gossypiella*, but is smaller when fully grown (8–9 mm long). The pink bands on each segment are more interrupted, with the colour more strongly developed than in pink bollworm. The plates at the base of the body setae are minute. The pupae are also similar to pink bollworm, but smaller (5–6 mm long) with a straight cremaster and usually without a cocoon. They are formed inside a seed or amongst the lint of the open cotton boll. The adult moth has a wing span of about 15 mm. The forewing has a rich, lustrous brownish-black colour with three distinctly defined creamy-white marks, consisting of an anterior transverse band, a median oblique blotch (forming a posteriorly directed V when the wings are folded) and a distal spot on the outer margin; the body, except for the creamy head and palps and the hindwings are smoky-brown.

The larvae are often found in old or diseased bolls and rarely attack entirely undamaged, unripe bolls. It may be common locally but it is doubtful that it has any economic importance. The Chalcidid, *Brachymeria olethria* (Wtstn.) is recorded as a parasite in southern Nigeria. Other species of *Mometa* are recorded on *Hibiscus*.

Sathrobota spp. (Lepidoptera: Cosmopterigidae)

The larvae of these small moths, commonly known as pink scavenger worms, are also associated with diseased and damaged open bolls usually feeding on the seeds. The species are polyphagous. *Sathrobota (Pyroderces) simplex* Walsingham is distributed throughout Africa and was reported recently in Pakistan (Chamberlain, 1993), while *P. rileyi* Wlsm. occurs in the New World and *P. coriacella* (Sn.) in Asia.

The egg is sub-oval, white, 0.25 mm long and longitudinally striated. The larva is 7–8 mm long when fully grown; yellowish- or greyish-white with a narrow, transverse pink band on the anterior and posterior margin of each segment, sometimes with a dark, purplish suffusion over the whole body. It is distinguished from *Pectinophora gossypiella* by its smaller size and the complete circle of crochets on the prolegs. Pupa is slender, 4–6 mm long, yellowish to chestnut-brown, pilose and without terminal hooks, usually without a cocoon. The adult moth has a wingspan of 9–11 mm, with very narrow fore- and hindwings, each deeply fringed; ground colour pale chocolate-brown, antennae long and slender, with alternate, narrow dark and light rings, forewing with zigzag darker markings, each with a light edging, and a terminal dark spot.

The Chalcidid, *Brachymeria olethria* (Wtstn.) (Lambourn, 1914) and the Braconid *Apanteles sagax* Wlkn. have been recorded as parasites (Pearson, 1958). It is considered to be a very minor pest.

Cutworms and Armyworms (Lepidoptera: Noctuidae)

The name 'cutworm' is applied to the larvae of several species of Noctuid moths which have in common a greasy-grey appearance and the habit of biting through the stems of seedlings at ground level, as well as that of eating leaves or entire plants. All cutworms have a wide range of food plants, both wild and cultivated, amongst which cotton is not specially preferred, and their attacks on cotton are almost invariably associated with, and frequently actually derived from, infestations developed on these other hosts. Severe attacks tend to be sporadic, but tend to be worse in areas of intense cultivation, abundant soil moisture and weeds.

Agrotis spp. (Lepidoptera: Noctuidae)

Agrotis ipsilon (Hufnagel), the common cutworm has a worldwide distribution. The smooth-skinned larvae are at first pale yellowish-green with a black head and cervical shield, and studded with black tubercles. In later stages the larvae are slate-grey with slightly darker longitudinal lines. It is thought that the pest cannot tolerate high temperatures as in the summer in Egypt, so that populations are derived from moths migrating from elsewhere. Several parasitoids have been recorded. Control can be obtained by good weeding and flooding as the larvae are unable to tolerate a succession of heavy irrigations. Similar damage is done by other widely distributed *Agrotis* species, notably *A. segetum* and *A. longidentifera.*

Spodoptera exigua (Hübner)

The lesser or beet armyworm has a worldwide distribution, but has only been recorded as a pest on cotton in irrigation areas of northern Sudan, when small areas were sown and then heavily irrigated. There were then areas susceptible over a long period so populations could increase locally. Reverting to sowing large areas synchronously with lighter irrigations led to less damage.

Spodoptera exempta (Walker), the African armyworm, virtually never attacks cotton, but has been recorded in the Sudan Gezira and in Zimbabwe (Pearson, 1958), and in Malawi (Sweeney, 1962). Other *Spodoptera* species, including in the New World *S. frugiperda* (J.E. Smith), the black or fall armyworm, *S. praefica* Grote, the western yellow-striped armyworm and *S. ornithogalli* (Guenee), the yellow-striped armyworm are all of relatively minor importance.

Foliage Feeders

A number of insects may feed on cotton leaves. Generally they are of little importance, but certain species may be pests in a particular cotton growing area or be extremely serious on a more localized scale.

Bucculatrix thurberiella Busck (Lepidoptera: Lyonetiidae)

The cotton leaf perforator occurs throughout the New World. The young larvae hatch from minute brownish-white eggs in about four days and spend about five days mining in the leaves. Larvae then feed from the undersurface causing windows covered by the upper epidermis. The moult between instars takes place curled up in a protective web, the so-called ‘horseshoe’ stage. Large populations lead to skeletonizing of the leaves. The six to seven day pupal stage is in a tough white cocoon on the petiole, stem or trash. The moth is silvery-white with brownish wingtips.

Alabama argillacea (Hübner) (Lepidoptera: Noctuidae)

This cotton leafworm occurs throughout South and most of Central America. Moths migrate north into Mexico and the USA in May–June. The female can lay up to 600 eggs singly on the undersides of leaves. The greenish-blue eggs are flattened and ridged, and turn grey before hatching in three days. The slender greenish larvae are semi-loopers and have two broad velvet-black dorsal stripes separated by a thin white line, and more black and white stripes towards the sides. Numerous round black spots are scattered over the body, with four of them forming a square on the back of each segment. The last pair of the five prolegs protrudes behind the body. There are usually six to seven larval instars and three to seven generations in a year. They can defoliate young plants leaving the main ribs to give the plants a skeletonized appearance (Plate IV.2), but populations are usually regulated by natural enemies, principally by *Trichogramma*, but larvae are also attacked by Tachinids. If an insecticide is applied it needs to be selective, for example *Bacillus thuringiensis*, to avoid triggering outbreaks of *Helicoverpa* and *Heliothis*.

It has also been recorded on *Cienfugosia* spp., *Hampea* spp., *Thespesia populnea, Urena lobata* and *Malachra rotundifolia.*

Spodoptera littoralis (Boisduval) (Lepidoptera: Noctuidae) (Plate IV.1)

This widely distributed insect seldom causes serious damage, except in Egypt, where cotton is infested during May–July following migration from the previous host plant berseem clover (*Trifolium alexandrinum*). There are many other host plants, including most common annual crops. Another species, *S. ornithogalli* (Guenee), is prevalent in the New World.

The egg, 0.4 mm diameter, is spheroidal, somewhat flattened and sculptured with approximately 40 longitudinal ribs. They change from pearly-green to black before hatching, and are laid on the undersurface of leaves in clusters of several hundred eggs covered by brown, hair-like scales from the body of the female. The young larvae are greenish with a black head and conspicuous black tubercles bearing a long hair. They are gregarious at first but later spread over the plant and become brown to grey-brown or black with irregular spots and lines. Pupation occurs just below the soil surface. The moths have brown forewings with paler lines along the veins, but with variable intensity of marking. Hindwings are pearly-white.

In Egypt much attention has been given to using a baculovirus (see Chapter 25) and monitoring moth populations using pheromone traps (see Chapter 25). The traditional method of controlling the pest on cotton has been the employment of teams of children to remove egg masses (Salama, 1983), but this may also affect the significant control by predators and parasitoids. Use of insecticides led to rapid selection of resistant populations, so where an insecticide is required, preference is given to an insect growth regulator, such a diflubenzuron, to conserve the natural enemies. Unfortunately this has usually been mixed with an organophosphate or carbamate to obtain a more rapid effect.

Syllepte derogata (Fabricius) (Lepidoptera: Crambidae)

The cotton leaf roller which is distributed throughout the rainfed-cotton growing areas of Africa and Asia, rolls up leaves to protect the greenish-white, semi-translucent larvae while feeding. Eight straight spines with hooked tips at their extremity are found on the brown pupae. The moth is a light cream colour, but the wings are traversed by numerous brown or black wavy lines and a black border with greyish fringe. The head and thorax are also dotted black and the abdomen has brown rings. Infestations frequently occur in shady and weedy conditions, and are generally more serious where broad-spectrum insecticides have prevented control by parasitoids, notably species of *Apanteles* (Odebiyi, 1982). However, locally severe attacks can occur and Sweeney (1962) records the occurrence of severe widespread damage in Kordofan, Sudan. A detailed account of *Syllepte* is given by Silvie (1990).

Anomis flava (Fabricius) (Lepidoptera: Noctuidae)

Outbreaks of the semi-looper, previously referred to as *Cosmophila*, are often sporadic, the larvae feeding on the leaves. Significant loss of leaf area is only likely to occur when young plants are attacked. It occurs in Africa, Asia and Australia, whereas the closely related species *A. erosa* is restricted to the New World. The pale yellowish-green larvae have five fine white lines on the dorsal surface.

Acrocercops difasciata (Washington) (Lepidoptera: Gracillariidae)

Formerly referred to as *Acrocercops bifasciata*, this leaf miner is confined to Africa. Slender reddish-coloured larvae feed on the leaf tissue, forming serpentine mines which coalesce into conspicuous patches below the leaf cuticle, which remains as a silvery-white membrane. It is commonly found on cotton which is shaded.

Xanthodes graellsii (Feisthamel) (Lepidoptera: Noctuidae)

Pale yellow spherical, ribbed eggs are laid singly on the underside of leaves where the semi-looper feeds. The larva is darkish green, with a broad median yellow stripe and narrower lateral stripes, between which are three to five prominent black tubercles. These tubercles on the first eight abdominal segments are linked by a prominent U-shaped black mark. A black transverse bar occurs on the first four abdominal segments and a red one on the tenth. The larva pupates in a cocoon up to 4 cm below the soil surface. Diapause pupae have been reported (Tunstall, 1958).

Tarache spp. (Lepidoptera: Noctuidae)

Several species of the genus *Tarache* attack cotton in India and Africa, including *T. nitidula* F. and *T. leucotrigona* Hmps., but are of minor importance. The stoutly built moths have creamy coloured forewings with leaden-grey markings which give an appearance, at rest, resembling a bird-dropping. The biology is similar to *Xanthodes graelsii.*

Thysanoplusia orichalcea (Fabricius) (Lepidoptera: Noctuidae)

This polyphagous species is widely distributed in the Old World, and has been recorded on cotton. A similar species *Chrysodexis chalcites* (Esper) has also occurred on cotton, while in the New World *Trichoplusia ni* is sometimes important on cotton. All are usually parasitized by Ichneumonids and Tachinids.

Spilosoma investigatorum (Karsch) (Lepidoptera: Arctiidae)

A stout-bodied, silvery-white moth; forewings with a creamy tinge and numerous black spots, hindwings with fewer larger spots and an abdomen with transverse black bars. The larvae are densely clothed in long hairs. Other related species have also been recorded on cotton, but all have a wide host plant range.

Amsacta flavicosta Hampson (Lepidoptera: Arctiidae)

This has a very hairy larva which has been recorded on cotton, but the attack may be due to migration of larvae from weeds, on which the eggs had been laid.

Euproctis rubricosta Fawcett (Lepidoptera: Lymantriidae)

The brightly coloured larvae have strongly developed tufts of urticaceous hairs. Other Lymantriids have also been reported on cotton, but they attack a very wide range of host plants including trees. Defoliation can occur.

Estigmene acraea (Drury) (Lepidoptera: Arctiidae)

The salt-marsh caterpillar has been recorded on cotton in Central America and southern parts of the USA.

References

Chamberlain, D.J. (1993) *Sathrobota (Pyroderces) simplex* Walsingham (Lepidoptera: Cosmopterygidae): a secondary pest of cotton in Pakistan. *International Journal of Pest Management* 39, 19–22.

Lambourn, W.A. (1914) The agricultural pests of the Southern Provinces, Nigeria. *Bulletin of Entomological Research* 5, 197–214.

Murillo, L.M. (1937) Sentido de una lucha biologica. *Revista de la Academia Colombiana de Ciencias exactas, fisicas y naturales*, Bogota 1, 376–410.

Odebiyi, J.A. (1982) Parasites of the cotton leafroller *Sylepta derogata* (F.) (Lepidoptera: Pyralidae), in south-western Nigeria. *Bulletin of Entomological Research* 72, 329–333.

Pearson, E.O. (1958) *The Insect Pests of Cotton in Tropical Africa.* Commonwealth Institute of Entomology, London.

Salama, H.S. (1983) Cotton-pest management in Egypt. *Crop Protection* 2, 183–191.

Silvie, P. (1990) *Syllepte derogata* (Fabricius, 1775) (Lepidoptera, Pyraloidea, Crambidae, Spilomelinae). *Coton et Fibres Tropicales* 45, 199–227.

Sweeney, R.C.H. (1962) Insect pests of cotton in Nyasaland. III. Lepidoptera (butterflies and moths). Department of Agriculture Bulletin No. 20.

Szumkowski, W. (1953) Nota preliminar sobre el gusano rosado grande del algodenero *Sacadodes pyralis* (Dyar.) (Lepidoptera: Noctuidae) en Venezuela. *Agronomie Tropicale, Maracay* (Rev. Inst. nac. Agr. Venezuela) 2, 267–273.

Tunstall, J.P. (1958) The biology of the Sudan bollworm *Diparopsis watersi* (Roths.) in the Gash delta, Sudan. *Bulletin of Entomological Research* 49, 1–23.

Withycombe, C.L. (1926) The South American boll-worm of cotton (*Sacadodes pyralis* Dyar.) *Bulletin of Entomological Research* 17, 623–624.

9 *Anthonomus* (Coleoptera: Curculionidae)

J.W. SMITH[1] AND F.A. HARRIS[2]

[1] *United States Department of Agriculture, Boll Weevil Research Unit, PO Box 5367, Mississippi 39762, USA;* [2] *Mississippi State University, Delta Research and Extension Center, Stoneville, Mississippi 38776, USA*

The boll weevil (*Anthonomus grandis* Boheman) (see Plate IV.3) is the most costly insect pest of US cotton, causing estimated yield losses of 8% annually over *c.* 3,200,000 hectares. In addition, it is the target for approximately 30% of all insecticides used in US agriculture (National Cotton Research Task Force, 1973). It has been credited with the loss of $15 billion to the United States economy since 1892, with current losses ranging from $150 million to $300 million annually, depending on the severity of the infestation, the acreage, and the price of cotton. The cost of control efforts each year is estimated to average $75 million. Biological control is generally ineffective in suppressing the boll weevil below economically damaging levels; so insecticides must be applied to reduce damage. The cost of chemical control, environmental contamination, ecological disruption, intensification of secondary pest problems, insecticide resistance and continued yield losses has led to an aggressive effort to alleviate the effects of the boll weevil. Although it is a key pest in only half of the US cotton area, its impact on cotton and the surrounding environment amounts to a 'key problem' confronting the American people (Bottrell, 1974).

EARLY HISTORY

Carl Boheman first described the boll weevil in 1843 when he received a specimen from Vera Cruz, Mexico. Edward Suffrian also made a similar collection in Cuba in 1871. Dr Edward Palmer later sent specimens to Washington from Monclova district of northern Mexico where the insect had stopped cotton production (Hunter and Hinds, 1912). These early reports were forgotten until 1894 when a new pest was recorded in Texas. The boll weevil had apparently crossed the Rio Grande into the US around 1892. C.H. Townsend who was sent to Texas in 1895 to look into the boll weevil problem, recognized the tremendous damage that boll weevils were doing and suggested: the abandonment of cotton growing over the infested region; and the permanent maintenance of a wide cotton free zone along the Rio Grande River bordering Mexico. Several other entomologists studied and suggested ways to overcome the boll weevil. An act, passed by the Texas legislature in 1903 offered $50,000 as a cash reward for a practical remedy for controlling the boll weevil, but it produced no effective remedy. Officials, who otherwise may have made useful contributions, were occupied full time investigating numerous useless inventions and ideas. Hand picking and destruction of infested flowerbuds, to kill developing larvae, became a common practice by cotton farmers, but this practice resulted in little success against the boll weevil (Sanderson, 1904).

When it became evident that other states east of Texas were threatened with boll weevil infestation, the US Congress appropriated money in 1901 for the first boll weevil investigations to discover means of preventing its spread. W.D. Hunter inaugurated cooperative work on eight farms to demonstrate cultural control of the boll weevil recommended by the USDA Division of Entomology. These demonstrations developed into the Farmers' Cooperative Demonstration Work of the Bureau of Plant Industry and, later, into the present Cooperative Extension Service of the Department (Parencia, 1978).

Estimates of damage caused by the boll weevil during this period varied greatly, depending on the source. The most reliable estimates placed the losses in newly infested areas at about 50% of the crop. About one third of the area planted to cotton in Texas was then infested and the loss to the 1903 crop was conservatively estimated by the cotton industry at $15 million.

All cotton-growing states enacted quarantines against the introduction of infested material, but the boll weevil advanced eastward across Texas, Louisiana and Arkansas, and crossed the Mississippi River into Mississippi in 1907. The spread, believed to have been almost entirely by natural dispersal, was of paramount interest to the people in its line of advance. The most severe damage was usually caused during the first few years before the infestation became stabilized and growers learned the best methods of growing cotton in the presence of weevils. Land values plummeted and were slow to recover as cotton production did not soon return to pre-weevil levels. Cotton was the only cash crop of most farmers, but after the arrival of the boll weevil some areas went out of production while the crop was introduced into other areas (Helms, 1977).

The eastward spread continued unabated and by 1922 the boll weevil had spread over North Carolina and into Virginia, thus covering over 1,500,000 square kilometres of territory since its entry into the United States 30 years previously. About 85% of the cotton in the United States was produced in the infested territory, and the rapid expansion of cotton production to the West was partly due to the absence of the boll weevil in that area.

Weather conditions were the primary factor in limiting or enhancing the destructiveness of the boll weevil. Less important were the use of poisons and the utilization of various methods of cultivation. A mild, wet summer ensured a large population by the beginning of autumn. A mild, dry winter was most favourable to a high spring survival rate.

Since its entry into the United States, the boll weevil has affected the economic and social welfare of more Americans than any other insect (Loftin,

1946). Smith (1989) recounts the importance of the boll weevil in the development of entomology as a discipline in the southern United States. Several bibliographies have been published, namely by Bishopp (1911), Dunn (1964) Mitlin and Mitlin (1968) and Parencia *et al.* (1985).

BIOLOGY

Description

Boll weevils are small, hard-shelled snout beetles (Curculionidae) averaging about 6 mm in length. They are yellowish, greyish or brownish colour, becoming nearly black with age. The life cycle and developmental periods vary depending on location and climatic conditions.

Boll weevils have four stages: egg, larva, pupa and adult. Under favourable conditions, it completes the cycle in two and a half to three weeks. High temperatures and humidity result in faster development, while low temperatures slow it down. As many as seven generations may develop in a year in the extreme southern part of the US cotton belt.

Distribution

The question of the species' original home and original plant hosts has been possibly resolved. Burke *et al.* (1986) has convincing evidence that the boll weevil was indigenous to Central America and only recently moved to *Gossypium* from a related wild host *Hampea.*

Burke (1968) segregated the species into three taxonomic forms, each characterized by morphological and behavioural traits and geographic distribution. Cross (1975) discussed the geographic distribution of the three forms. The form that attacks cotton in the US is also found in Hispaniola and parts of Colombia, Venezuela and recently in Brazil.

Host Plants

Upland or American cotton (*Gossypium hirsutum* Linnaeus) is the most significant host of the boll weevil. Varieties of this species account for 99% of the cultivated cotton in the US. *Gossypium barbadense* Linnaeus, grown originally in Central and South America and now chiefly grown in northern Africa, the Middle East and the southwestern United States, is also a host if grown within the boll weevil's range.

Cross *et al.* (1975) noted that the principal plant hosts of the boll weevil are in four closely related genera of the Malvaceae: *Gossypium*, *Cienfuegosia*, *Thespesia* and *Hampea*. Outside the genus *Gossypium*, the most important host in the United States, is *Cienfuegosia drummondii* (Gray) Lewt. in southern Texas. However, the complexity of hosts increases in tropical areas. Other plant species reported by

Cross *et al.* (1975) as hosts are *Gossypium harknessii* Brandg., *G. lobatum* Gentry, *G. laxum* Phill., *Cienfugosia rosei* Fryx., *Hampea rovirosae* Standle. and *Pseudabutilon lozani* (Rose) R.E. Fries.

Mitchell and Cross (1969, 1971) reported on boll weevil reproduction in cotton fields. Starting in the spring, females normally lay eggs singly in flower-buds and these hatch in three to five days. Inside buds or bolls larvae feed for seven to 12 days and the pupal stage lasts three to five days. The adults cut their way out and after feeding for three to seven days, mating occurs. Females deposit almost one egg per hour at fairly uniform intervals throughout the day starting at about 08.00 hours. Slight peaks in oviposition activity occur between 09.00 and 10.00 hours and between 12.00 and 13.00 hours. The rate of oviposition is reduced in the overwintered generation.

When the boll weevil population is high and there are few buds, two or more eggs may be laid on one bud. Late in the season, eggs are laid both on buds and young bolls. Males and females make feeding punctures but the eggs are deposited in deeper punctures, which are sealed with a mucous secretion and frass.

Both kinds of punctures cause damage. The bracts around buds with an egg puncture will flare and turn yellow. After a few days punctured buds and young bolls will drop to the ground, but large bolls are retained and are also damaged by invasion of microorganisms as well as the larval feeding. Weevil-infested locs produce little, if any, cotton and the quality of that produced is inferior.

Migration and Dispersal

Migration of the boll weevil usually occurs late in the season when populations often develop in extremely large numbers and new sources of food and oviposition sites are needed. Cross (1974) reported that many observations by various entomologists since 1894 indicated that the boll weevil has the capacity to disperse and migrate as an airborne adult. The extreme example in 1909 was an advance of over 190 km into previously uninfested areas of Mississippi apparently due to cyclonic storms. On average an annual gain of 80 km was the rule during the boll weevil's 30-year march across the US cotton belt. Mechanical dispersal by man was probably minimal compared to this natural dissemination. The flight ability of the boll weevil has been substantiated by sustained flying to 18 km on a flight mill and by the capture of adults to 600 m altitude.

With pheromone (Grandlure) baited traps to attract female weevils, more accurate data were obtained on the dispersal distances of the weevil. In 1968 boll weevils were captured in male-baited wing traps, along the highway from Juarez to Chihuahua in Mexico, though the traps were located 40–70 km from cotton. Although the dispersal by overwintered boll weevils is predominantly short range, it may cover long distances especially when cotton is not replanted near the hibernation site. The most distant migration occurs in the last reproducing generation, especially by the females.

In one study Guerra (1988) recovered six boll weevils near Brownsville, Texas, that were marked and released six to seven weeks earlier near Jimines,

Tamps, Mexico. The distance travelled was 272 km. One weevil travelled 48 km in a single day. Rummel *et al.* (1977) reported on seasonal variation in the height of boll weevil flight. Pheromone traps placed at heights ranging from ground level to 9.1 m were used to study boll weevil flight in the vicinity of an infested cotton field during 1972–1973. A definite seasonal variation in the height of boll weevil flight was noted. During the spring, weevil flight from hibernation areas to cotton appeared to be confined mostly to low levels. Over 90% of the overwintered weevils were from ground level to 4.6 m, with *c.* 70–80% captured from ground level to 1.1 m. This low-level flight appears to be correlated with relatively short-range movement by most overwintered weevils. During the late summer and autumn migratory period, a significant increase in the height of boll weevil flight was recorded.

McKibben *et al.* (1991) stated that weather is a major factor influencing boll weevil dispersal with both wind and temperature playing primary roles in displacement. In the study, they used a stochastic dispersal model to simulate dispersal of the boll weevil from cotton fields.

Hibernation

Adult boll weevils hibernate in surface trash in woodlands and along ditch banks near cotton fields, and in trash and litter around gins and farm buildings. In the spring they return to the cotton fields. A study of the distance that the boll weevil moves into woods adjoining a cotton field revealed that 90% hibernate in the first 55 m, with most being within the first nine to 14 m (Fye *et al.*, 1959). In North Carolina, most weevils were collected from deciduous leaf trash; pine straw was not a preferred hibernation site. More weevils were in slightly moist trash rather than excessively wet or dry material. Gaines (1959) states that boll weevils may become active during the hibernation period and fly when temperatures exceed 16.7°C.

Average life of caged boll weevils after emergence from hibernation was 20–22 days but varied from one to 141 days (Fye *et al.*, 1959). Average winter survival at Tullulah, Louisiana, over a 15-year study was 1.2%. (Gaines, 1935). However, Davis *et al.* (1975) found that survival of boll weevils in woodland surface trash at Waco, Texas, from 1960–1973 ranged from 14.4% in 1968 to 100% in 1965. Those surviving in field cages from 1940 to 1958 ranged from 0.02% in 1951 to 21.3% in 1941. These studies show that winter weather conditions and quality of hibernation sites play an important role in winter survival.

Cotton growing areas of the Mississippi Delta have historically fewer boll weevils than other areas of similar climate, probably due to water saturation or flooding in winter of most of the preferred hibernation sites. Clearing most of the hardwood forests in this area has further reduced favourable hibernation sites.

Several factors may influence the emergence of overwintering weevils.

Moisture and temperature definitely have a great effect (Davis *et al.*, 1967). Reinhard (1943) states that after emergence from hibernation has started, rainfall has an important influence on the rate at which boll weevils leave the overwintering shelter.

Grossman (1931), with experiments in Florida, showed that most weevil departure from hibernating quarters occurs during the day, but emergence at night has also been observed. Weevils emerge from the beginning of March to mid-July when physiologically ready, regardless of rainfall, increasing temperature or daylight. Gaines (1935) reported that weevils normally emerge prior to mid-June at College Station, Texas, and migrate to the nearest cotton field to feed and lay eggs. During an 18-year period prior to 1963 at College Station, Reinhard (1943) stated that 50% of hibernating weevils had emerged by mid-May. At Waco, Texas, Davis *et al.* (1967) reported that 97% of weevils had emerged by mid-June (18-year study from 1940 to 1958)

Mitchell (1971) believed temperature was most important in determining when weevils start emerging early in the spring. The average night temperature must approach 20–21°C before weevils begin emerging. Peaks of weevil emergence almost always followed rainfall after the average temperature was above the 20–21°C threshold. Overwintered weevil emergence continues over several months with the peak emergence being from about 20 May–10 June. When weevils emerged in the spring, they had consumed most of their body fat. This suggests that fat consumption may strongly influence the time that weevils emerge from hibernation, although other research does not confirm this (Mitchell and Taft, 1966).

Over 50% of emerging females contain sperm in their spermathecae, indicating that they were ready to begin oogenesis and commence oviposition. The gradual increase of females containing sperm indicated that they were either moving around in trash and mating before emergence or mating shortly after emergence with males that had already emerged, fed, and had begun spermatogenesis.

Mitchell *et al.* (1972), in tests with traps baited with male boll weevils, indicated that early-spring temperature influenced emergence of overwintered weevils, but that the influence of rainfall increased as temperatures increased. Since higher proportions of males than females were captured early in the spring and the reverse became true later, the pheromone may serve initially as an aggregating factor and then subsequently as a true sex attractant. Most weevils emerged in the spring with little fat reserve, but the variation in the amount of fat reserve among individual weevils may account for the emergence over an extended period.

Hopkins *et al.* (1971) reported that when movement of weevils into and out of cotton fields was monitored with screens, weevils entered and left throughout the season, with the greatest numbers in August and September. Peak numbers coincided with emergence of generations. In some cases, movement may have been across the study area from another heavily infested field to a favourable hibernation site. The flight screens were more efficient at monitoring weevils entering a field than leaving it.

Diapause

Sixty-four years after the first practical application of the principle of boll weevil control by the interruption of diapause, Brazzel and Newsom (1959) reported diapause as the physiological condition in which the boll weevil overwinters. They found diapause characterized by the cessation of gametogenesis, atrophy of gonads, increase in fat content, decrease in water content and decrease in respiratory rate Lambremont (1961) confirmed the last two characters. An additional factor, increase in glycogen content, was elucidated as characteristic of boll weevil diapause by Nettles and Betz (1965).

Photoperiod, temperature and food have been implicated as diapause inducing stimuli in the boll weevil (Earle and Newsom, 1964; Lloyd *et al.*, 1967). Sensitivity to diapause inducing stimuli commonly is limited to one for a few stages or instars, and the sensitive and responsive stages may frequently be different (De Wilde, 1962). The boll weevil is sensitive to photoperiod in the immature stages and responsive in the adult stage (Earle and Newsom, 1964; Lloyd *et al.*, 1967).

Studies by Carter and Phillips (1973) in Arkansas suggest that boll weevil diapause is related to changes in fruiting activity of the cotton plant. They believe that the boll weevil not only responds to short photoperiods that are characteristic during the autumn in the temperate zone, but also may respond throughout the season to changes in fruiting activity of the cotton plant.

In the laboratory, a heterogeneous species, such as the boll weevil, shows a diapause response in some individuals even under the most intense diapause suppressing conditions and exhibits a non-diapause response in some individuals even under the most intense diapause inducing conditions (Earle and Newsom, 1964; Lloyd *et al.*, 1967).

Interest has since developed in the interruption of the seasonal history of the boll weevil as it enters diapause. Controlling diapause boll weevils with the application of insecticides in the autumn has been extensively researched (Brazzel, 1959; Brazzel *et al.*, 1961; Lloyd *et al.*, 1964, 1966). The control of diapausing boll weevils in the autumn is a major tool in most eradication schemes (Davich, 1959; Brazzel *et al.*, 1961; Lloyd *et al.*, 1966).

McGovern *et al.* (1976) reported that overwintered male boll weevils produced only very small quantities of pheromone. The ratio of the four components in the pheromone of these weevils differed from that of the standard formulation of Grandlure and was less attractive.

Mitchell *et al.* (1973) found that there was a correlation between the time of entrance into hibernation and the time of emergence the following spring and summer. The late emerging weevils apparently went into diapause early and the early emerging weevils were those that were in the fields late the previous year.

Jones and Sterling (1979) found a strong relationship between temperature and emergence of boll weevils from overwintering sites. The threshold temperature above which overwintering weevils accumulated heat units was 10.85±1.3°C. Early autumn cohorts required a higher number of accumulated heat units than did late autumn cohorts. First-emerging weevils of early cohorts had a lower tendency to subsequently initiate flight activity than did first-emerging weevils of later cohorts. The temperature threshold for flight of 95–98% of newly emerging weevils was 20°C.

Boll Weevil Pheromone (Grandlure)

It has been known since 1962 that male boll weevils emit a wind-borne pheromone that attracts females. Keller *et al.* (1964) confirmed Cross and Mitchell's (1966) earlier 1964 field observations that a substance produced by male boll weevils attracted females for mating purposes. Air continuously drawn over males in a cage was passed through charcoal which was then extracted with chloroform. Removal of the solvent from the extract yielded a substance to which females quickly responded. Cross and Mitchell (1966) reported observations of mating behaviour of the boll weevil in the field. Males were not observed to respond to females over distances greater than a few centimetres; however, females often sought males from distances of more than nine metres, especially from downwind positions. They were of the opinion that this response was due to a wind-borne pheromone released by the males.

Hardee *et al.* (1967) reported laboratory tests using a newly developed olfactometer that gave results that substantiated the field observations of Cross and Mitchell (1966). In both these studies five- to seven-day-old males attracted females consistently, whereas, males were not attracted to females.

Utilizing the laboratory bioassay procedure developed by Hardee *et al.*(1967), Tumlinson *et al.* (1968) reported the isolation of the pheromone in a steam-distilled dichloromethane extract of faecal material of male and mixed-sex boll weevils. Later, Tumlinson *et al.* (1969, 1970, 1971) identified two terpene alcohols and two terpene aldehydes from male boll weevils and their faecal material. The biochemical production of the pheromone and its relationship to the fat bodies has recently been researched (Wiygul *et al.*, 1990). The identification of the component chemicals made possible the synthesis of a synthetic male pheromone called Grandlure. The best ratio at which to combine the four chemical components of Grandlure was determined by the attractancy of several different ratios in field tests (Hardee *et al.*, 1974). Traps baited with Grandlure wicks having higher concentrations of the two alcohols and lower concentrations of the aldehydes captured more weevils. The ratio presently used by percentage is 30:40:30 for alcohol I, alcohol II, and aldehydes III and IV.

Mitchell and Hardee (1976) found that Grandlure (3 mg) wicks attracted weevils to traps at distances as great as 150 to 180 m. Mitchell and Hardee (1974b) reported that a high percentage (over 50%) of the trap-captured weevils in his study were males. The response to the male pheromone in the spring is apparently not limited to females, indicating that an aggregation response is a strong and dominant factor in the attraction of boll weevils in cotton fields. As the season progressed the proportion of males decreased and the pheromone acted more as a true sex attractant.

The external sensory receptors of the boll weevils were identified by Jordan (1977). Dickens (1990) identified the function of several pheromone receptors that had not been specified.

Control Methods

Prior to the invasion of the US by the boll weevil in 1892, there was relatively little concern about the damage to cotton by insects. This may have been due to fewer pest problems in the more diverse agroecosystem of the time, but was more probably due to tolerance to pest damage and acceptance of low yields by cotton producers. The bollworm (*Helicoverpa zea*), cotton leafworm (*Alabama argillacea* Hubner), and the cotton aphid (*Aphis gossypii* Glover) infrequently appeared in sufficient numbers to cause alarm among growers. Most growers did not attempt to control these pests. This changed when the boll weevil entered the United States and inflicted great losses. At that time an intensive search began for more effective insecticides and improved application machinery.

Methods for controlling boll weevil populations fall into three broad categories: cultural, biological and chemical. These methods may be used individually or may be combined in a variety of ways for use in a functional suppression or eradication programme.

Cultural Control

Cultural control measures such as stalk destruction, burning cotton fields and adjacent areas, and deep ploughing of fields in the autumn and winter were recommended by Townsend (1895), Howard (1897), Sanderson (1905) and Hunter and Coad (1922). Cultural control involves modifying the crop environment in ways that make it less favourable for pest reproduction and survival (DeBach, 1974).

Host plant resistance

Manipulation of the cotton plant by management and by genetic means are ways to provide varying degrees of protection from insect attack. In the early days of the boll weevil when effective insecticides were not available, emphasis was given to cultural controls and early maturing cotton varieties. This made it possible to produce a crop before the boll weevil could build up extremely high populations. With the advent of better insecticides and the need for higher yields, varieties of indeterminate growth were developed for the rainbelt and the irrigated areas of the West. Such varieties and production practices favoured the boll weevil, pink bollworm and other pests. Though good crop yields were produced with the intensive use of insecticides, large populations diapaused in the autumn, enhancing survival for infesting the subsequent crops and necessitating repetition of the control cycle year after year.

When resistance to the organochlorine insecticides developed in the boll weevil in the mid-1950s, research turned to alternative methods of controlling the boll weevil and other cotton insects. Though considerable effort was expended, progress in the development of varieties resistant to the boll weevil was slow. In the 1970s Frego bract cottons reduced boll weevil oviposition by 50% or more when compared with normal bract varieties. Frego bract cottons have distorted bracts that do not envelop the bud as is done with normal bract cottons. The exposed bud condition in Frego bract cotton appears to agitate the female weevil causing her to spend more time searching for an oviposition site, consequently resulting in a preference for normal bract cotton when she has a choice. However, the reduction in oviposition has been considerably less than when the variety was grown under 'no-choice' situations. Frego bract varieties, with acceptable agronomic characters, might have a place in boll weevil control when field plantings are interspersed with trap crop plantings of normal bract cottons to attract overwintered weevils which could then be killed with insecticides. Unfortunately Frego bract cottons are more susceptible to lygus bugs. Jenkins *et al.* (1973) reported on the role of the boll weevil resistant cotton strain, Frego bract, in the management of cotton pests. In 22 fields in Yalobusha County, Mississippi, in 1971, the use of the Frego cotton suppressed the boll weevil population 69% and 79%, respectively, with and without a diapause programme. In addition, the beginning of weekly applications of insecticides for boll weevil control was delayed four weeks longer in the Frego fields than in non-Frego fields.

Walker (1980), reporting on earliness in cotton to escape the boll weevil, showed that where a build-up to injurious numbers of boll weevils was delayed until the second generation, rapid-fruiting cotton escapes a significant amount of injury and can produce a substantial increase in lint production compared to later-maturing cotton. A delayed infestation allows larger proportions of buds to reach the boll stage, which is much less preferred than buds for feeding and oviposition by the boll weevil. As a boll ages its susceptibility to boll weevil attack decreases. Therefore, the more rapid bolls are set and accumulated, the greater the yield possibilities. Certain genotypes and narrow-row planting schemes can hasten the progression from predominantly flowerbuds to mostly bolls. One commercial variety, TAMCOT SP-37, has demonstrated that it suffers less boll damage than later maturing cottons.

A wide range of cotton cultivars that are resistant to insect attack are available (Niles *et al.*, 1974). Although many promising resistant characters are known, movement of these characters into suitable commercial cultivars has been limited. A very complete bibliography of literature related to host plant resistance to the boll weevil has been compiled (Benedict and George, 1979).

Short-season cotton

Modern short-season cotton varieties set fruit more rapidly than older cotton varieties. Such varieties increase the likelihood that an acceptable crop can be produced at an early stage in the growing season, before boll weevil numbers increase to levels capable of causing significant reductions in yield.

Some short-season determinate cotton varieties are now grown extensively in Texas and show promise in reducing late-season insect damage, especially by the boll weevil. They may also help alleviate some damage caused by bollworms by maturing before the late-season population peak occurs. In most areas in which this technique is practised, early and uniform planting is essential, together with regular field observations, based on scouting or pheromone trapping, to detect boll weevil infestations that appear likely to reach damaging levels during the critical 30-day flowering period. By restricting insecticide treatments to fields in which damaging boll weevil numbers appear most likely to occur, and eliminating the need for many midseason and late-season treatments, this short-season technique has been shown to provide effective boll weevil control while reducing insecticide inputs required for successful cotton production (Bottrell and Adkisson 1977; Frisbie *et al.*, 1983).

A second short-season technique, developed and widely used in the Rolling Plains Region of Texas, relies on a uniform delay in planting date. Growers have to ensure that cotton plants are in a very early stage of development, which is unsuitable for boll weevil reproduction, during the period when the majority of adult weevils emerge from winter diapause. This results in a high degree of 'suicidal emergence', in which adult weevils emerging from diapause are unable to feed or reproduce (Bottrell and Adkisson, 1977; Frisbie *et al.*, 1983). Following this suppression of the boll weevil population during spring, numbers remain low during the 30-day flowering period that is required for successful crop production.

Cotton-free periods

Mandatory post-harvest stalk destruction and prohibitions against the cultivation of perennial cotton are designed to provide a 60-day period between growing seasons in which no cotton is available for boll weevil feeding or breeding. The combination of rapid crop maturation, prompt harvesting and the destruction of cotton stalks following harvest interferes with preparation for diapause by adult weevils and thus contributes to a reduction in the size of overwintering populations of the pest (Frisbie *et al.*, 1983).

Trap crops

'Trap cropping' is the deliberate planting of strips or small blocks of cotton that are particularly attractive to adult boll weevils, in order to concentrate adults in small areas where they can be killed by localized insecticide treatments. As early as 1901, Malley suggested the use of an early planted trap crop to control the boll weevil, as the early planted fields were infested with weevils well in advance of the adjoining later planted fields. He suggested planting a few rows of cotton earlier than the normal planting date and using an early maturing variety to try to concentrate the emerging weevils on a few rows. He suggested that the work and cost of controlling the pest on a few early planted rows was considerably less than for the entire crop. He suggested planting about two rows across the middle of every 20 acres of cotton and along hibernation sites as well.

Isley (1950) reported that trap crops of cotton 3–30 rows wide were successful in concentrating overwintering boll weevils which were subsequently killed with insecticides.

In a cultivation system based on uniform planting of a short-season cotton variety, for example, the 'trap' crop is normally a strip or block that is planted two to three weeks earlier than the remainder of the crop. The 'trap' area thus contains larger plants, which bear more mature and attractive fruit than the remainder of the crop, during the period when adult weevils are emerging from winter diapause and searching for feeding and reproduction sites. In other cultivation systems, trap areas may contain a cotton variety that is highly attractive to weevils while non-trap areas contain a different variety such as a Frego bract type that is known to be less attractive or more resistant to weevil attack. By limiting the need for insecticide treatments to a relatively small area in which adult boll weevils have been concentrated, the trap-crop method can assist in providing boll weevil control while reducing the use of insecticides within the crop as a whole (Frisbie *et al.*, 1983).

Crop rotation

Annual crop rotation is a widely-used agricultural technique that helps to prevent the uninterrupted increases in pest populations that can occur in fields in which only a single crop is grown. Because the boll weevil is a highly specialized insect that requires cotton plants and fruits for successful feeding and reproduction, removal of cotton through crop rotation can be an effective tool for short-term suppression of local populations.

Prohibitions against the planting of cotton in environmentally sensitive areas or in areas that are physically inaccessible to chemical treatment, could be helpful in ensuring that all boll weevil outbreaks occurring in a region can be treated with insecticides if such action becomes necessary.

Biological Control

Biological methods of boll weevil suppression fall into three categories:

- Reliance on natural enemies (parasites, predators or pathogens)
- Mass trapping using pheromone traps
- Mass release of sterilized insects.

Parasites, predators and pathogens

Cotton fields contain a substantial entomophagous fauna. Reynolds *et al.* (1975) states that it would not be economical to produce cotton if parasitic and predaceous species did not suppress the pest complex. The surveys of predatory arthropods in Arkansas cotton made by Whitcomb and Bell (1964) revealed about 600 species associated with the crop, not including parasites. Predators are free living throughout their life cycle and are capable of feeding on almost any prey they can capture. Also, some predators have both phytophagous and entomophagous habits. Over 50 species of both parasites and predators attack the boll weevil but few, if any, manage to give adequate control. This situation is consistent with the generally accepted theory of biological control whereby an introduced pest (such as the boll weevil), which is adapted to survive in a location to which it is not indigenous, will escape natural controls in the new location (Ables *et al.*, 1983). An example to the contrary, Sterling (1978) reported that the red imported fire ant (*Solenopsis invicta* Buren), an introduced insect that is generally considered a pest, has proven capable of providing some biological control of local populations of boll weevils.

Because of the possibility of introduction or augmentation of parasites for boll weevil control, this area has received considerable attention. These arthropods aid control of the weevil in the cotton field by feeding internally and externally on its immature stages. All the major orders of insects contain species that parasitize other insects, but only representatives of the Hymenoptera, Diptera and Coleoptera contain species known to parasitize the boll weevil. Including several introduced by man, there are as many as 36 species of Hymenoptera, five species of Diptera, one species of Coleoptera and three species of Acarina that have been definitely implicated as boll weevil parasites. It is not known whether any parasites were introduced with the boll weevil into the United States but, during the first three years of its existence in Texas, it was attacked by three important species.

The first large scale survey to determine the status of boll weevil parasites across the Cotton Belt was conducted in 1908 (Pierce *et al.*, 1912). At this time, 29 species were found to attack the weevil.

Probably none of the known boll weevil parasites exhibits a singular host preference and most also attack other species of weevils and other insects. Hunter (1917) refers to two fields with a high mortality of weevils due to parasites. One field near Robson, Louisiana, showed 77% control by parasites, while during the same period in a field near Victoria, Texas, parasites killed 61% of the boll weevil immature stages

Chesnut and Cross (1971) reported on a 1965 survey made across the Cotton Belt from South Carolina through Texas to determine the current status of parasitism of the boll weevil. Samples were taken from 42 cotton fields. Also, results of similar surveys made in 1934, 1935 and 1936 were summarized and compared with those of the 1927, 1930 and 1965 surveys. In 1965, 12.77% of the boll weevils were parasitized compared with 4.05% in 1934. More parasites were found in weevil larvae in hanging buds than in larvae in fallen buds, and parasitism was higher in weevils collected from cotton grown in the hill region than in weevils collected from cotton grown in the coastal plains. Some developing resistance to insecticides in the parasites was suggested by records of heavy parasitism in treated fields. *Bracon mellitor* Say was much more important in natural parasitism than all others combined and accounted for 74.5% of the parasites recovered in 1965. *Aliolus curculionis* (Fitch) and *Eurytoma gossypii* Bugbee ranked second and third, and accounted for 12.3% and 5.3% of the parasites, respectively.

Cross and Mitchell (1968) reviewed reports of species of arthropods which had been found parasitizing the boll weevil in its Mexican range. Of these, about 14 species are now known to occur naturally in Mexico. Most records are of a qualitative nature and for a limited number of localities. A tarsonemid mite, *Pyemotes ventricosus* (Newport), has been reported attacking the boll weevil in areas of northern Mexico. All others found were in the insect order Hymenoptera.

Considerable attention has been given to pathogens as potential alternatives to conventional control of cotton insects with insecticides. Evidence of the occurrence of pathogens has been known for many years. At the Boll Weevil Research Laboratory results of biological control tests showed that the fungus, *Beauveria bassiana*, can infect larvae, pupae and adult boll weevils in the laboratory, but it cannot be cultured for practical use in the field. Laboratory cultures of the boll weevil became infected with a new species of neogregarine that subsequently was named *Mattesia grandis* McLaughlin. The ensuing epizootic resulted in destruction of the boll weevil colony. To be effective, the neogregarine had to be ingested by the adult boll weevils. A cottonseed oil bait was developed to enhance ingestion of the pathogen. Production and application problems have prevented the pathogen from becoming a practical tool in the control of the boll weevil.

Pheromone traps

The first pheromone traps used extensively to capture boll weevils were plywood wing traps and an oblique funnel live trap (Cross and Hardee, 1968). The wing traps were coated with a sticky substance to capture weevils. The oblique funnel trap was made from clear Plexiglas with screen windows near the top to allow for air circulation. Each side of the trap had a screen wire funnel. The funnels converged upward and opened into the trap interior where weevils that climbed the funnels were captured. Live male weevils caged in small boxes were used as an attracting pheromone source.

The wavelength sensitivity of the weevil's eye was first determined by Hollingsworth *et al.* (1964). They found the compound eye to have maximum sensitivity in the blue-green (490 to 515 nm) region of the spectrum. Cross *et al.* (1976) tested the responses of weevils to traps painted with nonfluorescent vs. daylight fluorescent and 'visible' vs. 'invisible' paints in the field and their response to coloured lights in the laboratory. They confirmed that the 500 to 525 nm region of the spectrum has the greatest attraction for weevils. Red was again found to be the colour to which weevils were the least responsive (originally reported by Taft *et al.*, 1969). Weevils were more responsive to daylight fluorescent paints with pigment in the blue-green (500 to 525 nm) spectral region. The increased attraction of daylight fluorescent paints was thought to be due to the increased intensity of light reflected. They also concluded that boll weevils 'see' traps painted daylight fluorescent Saturn Yellow as blue-green. The colour appears yellow to the human eye because of our greater sensitivity to yellow and decreased sensitivity to blue in the spectrum.

The development of Grandlure eliminated the use of live males as the pheromone source in weevil traps. The Leggett trap (Leggett and Cross, 1971) was the first trap to efficiently use Grandlure. The Leggett trap is made from a pressed-paper conical floral liner painted daylight fluorescent Saturn Yellow and topped with a screen wire funnel leading upward into a plastic capture box. Hardee *et al.* (1972) showed that under field conditions the boll weevil pheromone was 80% as effective in attracting overwintered weevils as caged live males fed flowerbuds in traps once or twice a week.

The efficiency of both sticky traps and the Leggett trap declines during the fruiting of cotton in midsummer, due to competition from male weevils in the field feeding on flowerbuds and producing natural pheromone. In addition, these traps interfere with farming practices if placed in the cotton field during the growing season. The in-field trap (Mitchell and Hardee, 1974a) solved both of these problems since it is placed on cotton plants in the field where it is effective in capturing weevils. The original in-field trap differed from the Leggett trap in that the floral liner of the Leggett trap was replaced by a smaller plastic cup. In addition, the bottom of the cup (the cup is inverted to make the trap) had 10 to 12 triangular holes. These holes allow weevils that crawl up the inside surface of the cup to be easily captured. At present, manufactured traps are available from several sources and are about the same size and shape as the in-field trap but there are some differences in construction.

Currently, in pest management and area-wide trapping programmes, pheromone traps (Mitchell and Hardee, 1974a; Dickerson *et al.*, 1981) are placed around field borders and used for detection and prediction (Rummel *et al.*, 1980; Johnson and Gilreath, 1981; Leggett *et al.*, 1981; Hamer *et al.*, 1983; Benedict *et al.*, 1985). Traps now being used to capture adult weevils in survey and control programmes contain a small quantity (10 mg) of the Grandlure, along with a small quantity of an insecticide such as propoxur or chlorpyrifos to ensure that captured weevils do not escape the trap.

Numerous field studies have shown that pheromone traps are effective tools for the detection of boll weevil populations, including the very small populations that may be present during early spring or following insecticide treatment (Lloyd *et al.*, 1983). Roach *et al.* (1972) found that the positioning of baited traps in relation to overwintering sites had an influence on the numbers of emerging overwintering boll weevils captured. The weevils apparently overwintered in greater numbers as close to the cotton fields as suitable overwintering sites could be found.

During a three-year study, Hollingsworth *et al.* (1977) reported the number of emerging overwintered boll weevils captured in Leggett traps was positively correlated throughout the emergence period with the number of boll weevils found previously in samples of surface trash in woodlands. Thus, the traps can detect a potential early field population on individual farms.

Mitchell and Hardee (1974a) found that Grandlure-baited traps placed in the field captured boll weevils when manual surveys failed to detect signs of infestations. An aggregating response was obtained from both sexes in early and late season as in previous trap studies, but significant numbers of boll weevils, primarily females, were captured in midseason, thus indicating a true sex pheromone at this time of year. Previous trap studies failed to obtain this midseason response.

Mitchell and Hardee (1974b) reported that boll weevils captured in traps were not always representative of field populations with respect to sex ratio and status of diapause. From June to midwinter, the ratios of captured females to males are higher than they are in field populations. In the autumn, the proportion of boll weevils in the field that are in diapause is higher than that among those captured in traps.

Rummel *et al.* (1980) reported on a pheromone trap index system for predicting need for overwintered boll weevil control. A positive relationship was established between the mean number of overwintered weevils captured per trap, per field just prior to the appearance of buds suitable for oviposition and the proportion of oviposition damaged buds during the early one-third-grown buds period. A trap index threshold was developed and applied to actual field situations.

Suppression of boll weevil populations may be achieved with pheromone traps when the traps are used at sufficiently high densities. Lloyd *et al.* (1983) showed effective suppression of boll weevil with pheromone traps when populations have been reduced to low levels by other factors. For population suppression, traps may be concentrated in brushy areas near field margins in sites where overwintering weevils are known or suspected to occur, or may be placed in-field to capture weevils in low-density population situations. Research on the concept of trapping overwintering weevils as they emerged from the hibernation sites was initiated by Cross and Hardee (1968) and Cross *et al.* (1969).

Pheromone traps used in concentrations near hibernation sites and/or in fields provide a promising approach to survey and monitor efforts, to evaluate an elimination programme, and to actually manage and suppress the boll weevil.

Sterile insect technique

The most widely publicized and successful proposal for the use of genetic techniques in insect control is the Sterile Insect Release Method (SIRM, or Sterile Insect Technique, SIT) first conceived by E.F. Knipling in 1938 (Lindquist, 1955).

Initial attempts to irradiate boll weevils used large dosages that caused unacceptably high mortality (Davich and Lindquist, 1962), and significantly reduced longevity in both the field and laboratory. From results of a field cage test, Davich *et al.* (1965) estimated the mating competitiveness of irradiated males to be roughly 20% that of normal males.

Chemosterilization was tried as an alternative method, but it also reduced vigour and sterility and was not permanent (Haynes, 1963; Lindquist *et al.*, 1964; Gassner *et al.*, 1974; Earle and Leopold, 1975; Haynes *et al.*, 1975; McHaffey and Borkovec, 1976; Borkovec *et al.*, 1978). However, chemosterilization with busulfan and hempa was chosen as the best method for sterilizing the weevils released in the Pilot Boll Weevil Eradication Experiment conducted in south Mississippi, Louisiana, and Alabama from 1971–1973.

The use of fractionated irradiation on boll weevil pupae was begun in the mid 1970s. Males emerging from pupae subjected to a series of 25 treatments of about 250 rad per treatment (Haynes *et al.*, 1977) were 23% as competitive as normal males (Villavaso *et al.*, 1979). In comparison, adult males allowed to remain on the surface of the larval media for three to four days after emergence and then treated with a single dosage (acute irradiation) of 7 krad followed by a five second dip in a 0.02% solution of diflubenzuron in acetone (Earle *et al.*, 1978) were 36% as competitive as normal males. Although it worked relatively well, pupal fractionation was dropped because of its inherent operational complexity and inefficiency, and its failure to produce males any more competitive than those treated with acute irradiation.

Diflubenzuron had been found to be an effective means of preventing hatch of eggs laid by irradiated females mated to fertile males without causing increased mortality (Moore and Taft, 1975; Moore *et al.*, 1978), but administration of diflubenzuron to males not yet hardened after emergence severely reduced their ability to inseminate females (Earle *et al.*, 1979). The mating ability of males allowed to age four or more days before treatment with diflubenzuron was not affected. However, diflubenzuron was applied as an acetone dip, and acetone was found to severely impair the flight ability of treated weevils (Earle and Simmons, 1979; Haynes *et al.*, 1981).

Pheromone production for both pupal fractionation and acute irradiation was approximately equal. Even though the pupal fractionation group was newly emerged, pheromone production averaged 2.0 micrograms per male, per day for days one to three after emergence; this rose to 4.5 μg per male, per day for days four to six. The weevils that received the single dosage of 7 krad had been allowed to feed on the surface of the larval media for three to four days before treatment; however, their level of pheromone production was not significantly higher than that of the pupal fractionation group indicating that diet might be as important as age in the onset of pheromone production by males (Villavaso *et al.*, 1979).

The laboratory work of Earle *et al.* (1978) had stimulated renewed interest in acute irradiation as a method to sterilize the boll weevil. Acceptable levels of field competitiveness in male weevils sterilized by acute irradiation re-established the feasibility of using this treatment in mass-release programmes (Villavaso *et al.*, 1979). However, the use of acute irradiation would not have come about without the advent of the following three factors: mass-rearing of boll weevils relatively free of pathogenic bacteria (Sikorowski *et al.*, 1977; Sikorowski, 1984); use of diflubenzuron to bring about complete sterility of treated females (Moore *et al.*, 1978); and, perhaps most importantly, the lowering of the formerly acceptable standard of 50–70% survival of treated males for three weeks after treatment to a more realistic one. A sterilizing treatment is now considered to be acceptable if males are able to attract and inseminate females for at least seven days after treatment (Villavaso *et al.*, 1980).

The first field tests designed explicitly to establish estimates of the competitiveness of irradiated males were conducted in 1977 (Villavaso *et al.*, 1979). Using basically the same procedures established in 1977, small plot tests were conducted simultaneously in Louisiana and North Carolina (Villavaso *et al.*, 1980) to determine the competitiveness of males sterilized by three methods.

In 1979 sterile males were released as part of the Boll Weevil Eradication Trial on about 7500 hectares of cotton in Virginia and North Carolina. An autumn diapause programme in which all the cotton area had been treated with organophosphate insecticides significantly reduced the number of weevils entering diapause. It was followed by spring applications of sterile insects, pheromone trapping, aerial applications of organophosphates and the insect growth regulator diflubenzuron. Though the boll weevil was eradicated from the trial area by this combination of techniques, the effect of each technique could not be measured separately. Only seven native weevils were captured in the trial area prior to the release of 11.2 million sterile weevils; thus, the role played by the sterile insects in eradication remains unclear. The treatment selected to sterilize the weevils released in the trial consisted of feeding weevils on slabs of diet containing 0.01% diflubenzuron for the first five days after they had emerged followed by 10 krad of gamma-irradiation (Wright *et al.*, 1980).

In 1979, 1980 and 1981 weevils treated by the same method used in the previous trial were tested for competitiveness in the field. Competitiveness of the sterile males versus untreated laboratory reared males for laboratory reared females averaged 10.6% for the first seven days following release. Competitiveness of the sterile versus native males for the native females averaged only 6%. In general, the treated weevils were only effective during the first four days of the seven-day period.

In the early 1980s, a method of sterilization was developed that resulted in the highest competitiveness value that had been obtained for sterile boll weevils in small plot field tests. Males fed an ecdysteroid rather than diflubenzuron for five days prior to irradiation were 43.7% as competitive as untreated laboratory-reared males for laboratory-reared females, as compared with 12.5% for the diflubenzuron-fed, irradiated males (Villavaso and Thompson, 1984). Weevils treated by the ecdysteroid plus irradiation technique were 50.4% as competitive as native males that naturally infested three small field plots (Villavaso *et al.*, 1986).

In 1983, mass-reared and sterilized weevils were released into the Mississippi cotton fields by two new methods. Dropping weevils onto freshly cultivated or hot soils (>46°C) in early 1983 resulted in very low numbers of weevils reaching the cotton plants (Roberson and Villavaso, unpublished). The loose soil prevented the weevils from leaving the ground where they had fallen and, if soil temperatures reached lethal levels as they often did during the release periods, the weevils died on the ground before reaching the plants. The importance of developing a method of release that resulted in a large portion of the weevils reaching the cotton plants was clearly seen, and a method by which released weevils would be containerized for mass-release was subsequently developed.

In 1987 and 1988, a test of the effectiveness of mass-reared, sterilized (irradiation plus aqueous dip in diflubenzuron, Haynes and Smith, 1989a), containerized and aerially-dropped weevils was conducted on about 2000 and 1200 hectares, respectively, in Fayette County, Alabama (Smith *et al.*, 1988). In 1987 the test area had native populations, higher than desirable for sterile weevil efficacy. However, even with these high populations, the fertility of the native females was reduced by about 39%. The LT_{50} of samples of sterile males held in individually screened containers on cotton plants averaged 9.1 days, which was the highest figure ever obtained for a mass-reared, mass-sterilized group of weevils exceeding the previous high for a test of this type by 15%. The 1988 populations were smaller than those of the previous year, and intensive sampling of selective fields showed the effectively reduced reproduction.

Mass-reared and sterilized boll weevils have typically exhibited mean post-irradiation survival times of five to seven days or lower (Villavaso, 1982; Haynes and Smith, 1989a). Improvements in rearing and sterilization techniques have increased survival time in mass-reared, mass-sterilized weevils to as long as 9.1 days (Smith *et al.*, 1988). Genetic selection increased post-irradiation survival in a strain of boll weevils held under laboratory conditions after treatment (Enfield *et al.*, 1981). By the 20th generation, 14-day post-irradiation survival was 90% in the selected strain compared to 35% in the control (Enfield *et al.*, 1983). Some advantages in using the selected strains were observed in recent field studies (Villavaso *et al.*, 1993).

Accumulated genetic sterility in progeny of boll weevils treated with apholate was reported by Haynes and Smith (1989b).

Chemical Control

Use of chemical insecticides has been the basis for effective control of the boll weevil throughout the history of its infestation of US cotton. Cultural and biological control methods have been important components of boll weevil management systems, but have been inadequate to suppress weevils to levels at which the pest is not a limiting factor for cotton production in infected areas. Certain undesirable side effects are inherently associated with reliance on chemical insecticides, namely insecticide resistance, secondary pest outbreaks, and both direct and indirect undesirable consequences of chemical residues in non-target areas. Nevertheless, the most modern systems developed for boll weevil management and eradication continue to be dependent upon effective chemical insecticides.

The following discussion of chemical control will deal first with conventional use of chemical insecticides with some emphasis on the history of insecticidal control of boll weevil. Then four specific adaptations of chemical control methods for boll weevil management will be discussed: (i) insect growth regulators; (ii) pinhead bud applications; (iii) reproduction–diapause boll weevil control; (iv) use of attractive baits combined with insecticides.

Conventional insecticides

Insecticides have been used as dusts or sprays to prevent or reduce damage to buds and young bolls by boll weevil feeding and oviposition. Other common 'conventional' methods of application to cotton include systemic insecticides as seed treatments, in-furrow granules or sprays at planting and side-dress granules or sprays, but these have been of limited use for boll weevil control.

The first insecticide recommended for control of the boll weevil was arsenic (Paris green, London purple or lead arsenate) applied as a spray with molasses (Townsend, 1895; Malley, 1901), but control was unsatisfactory because of poor formulations and inadequate machinery for applying the sprays.

The first aerial application experiments using calcium arsenate dusts on cotton at the USDA Federal Delta Laboratory at Tallulah, Louisiana, obtained effective control of the cotton leafworm (Coad *et al.*, 1924), and in 1923, they demonstrated that the boll weevil could also be controlled. The success of this method was confirmed in Georgia, Mississippi and Texas during the period 1925–1927 (Post, 1924; Thomas *et al.*, 1929). Dusting continued until the early 1950s when low volume sprays were developed. When cotton aphid became a serious pest following widespread use of calcium arsenate which destroyed its natural arthropod enemies, nicotine sulfate was added to the arsenical dust. The bollworm also became a much more serious pest during this period for apparently the same reason. However, calcium arsenate was fairly effective against very young larval bollworms and provided satisfactory control when applied to the crop regularly at short intervals.

The era of synthetic organic insecticides began at the end of World War II with the introduction of DDT, HCH, and toxaphene. These were followed by other synthetic organic chlorinated hydrocarbon insecticides including aldrin, dieldrin, endrin, heptachlor, Strobane (polychloroterpenes) and TDE. The new insecticides possessed two qualities of great importance, namely the high initial toxicity to the cotton insect pests and sufficient persistence to control newly emerging insects migrating from untreated to treated areas.

The chlorinated hydrocarbon insecticides had a great impact on cotton production. For the first time, cotton producers were able to achieve highly effective control of all arthropod pests of the crop. It then became profitable for producers to increase the use of fertilizer, irrigation and incorporate long-growing indeterminate varieties. Spectacular yield increases were obtained at higher profit levels for the next 20 years.

The apparent victory over insect pests of cotton was not lasting. By the mid-1950s, the boll weevil in Louisiana and Mississippi had developed resistance to these chemicals. The resistant strain spread rapidly across the southern and southwestern states and all infested areas were reporting chlorinated-hydrocarbon-resistant weevils by 1960 (Roussel and Clower, 1957; Brazzel *et al.*, 1961).

This problem was solved by a switch to the organophosphorus insecticides, mainly methyl parathion, azinophosmethyl, ethyl parathion and malathion. The organophosphorus compounds were highly toxic to the boll weevil at relatively low concentration rates, but higher rates were needed for the bollworm and tobacco budworm.

Beginning in 1973, another change in the pesticide usage pattern on cotton occurred when the Environmental Protection Agency banned the use of DDT. DDT combined with toxaphene had provided satisfactory control of the boll weevil, bollworm, cotton fleahopper and plant bugs in cotton producing states east of Texas. (Methyl parathion was frequently added at a low rate for added weevil control.) The banning of DDT forced southern cotton producers to shift to high concentrations of organophosphate insecticides for pest insect control. Thus, the banning of DDT increased selection pressure for the development of organophosphate-resistant bollworm and tobacco budworm strains. Inevitably, the cotton producers of the South suffered the chemical pest control problems of decreasing effectiveness and increasing cost of insecticides which were once confined to Texas and Mexico.

The use of insecticides for boll weevil control has created numerous problems. One of the problems is public opposition to toxicity hazards to man and wildlife. Persistent residues cause concern because of possible pollution of the environment. Another problem is the adaptive ability of the boll weevil to develop resistance to insecticides. Strong resistance to chlorinated hydrocarbons has been encountered, giving good reason to suspect that resistance may develop to organophosphates and carbamate insecticides. A more subtle type of problem is the destruction of many predacious and parasitic insects by the insecticides applied for boll weevil control. The destruction of the beneficial insects results in an increase in other pests. For example, bollworms, spider mite and cotton aphid which are now serious pests of cotton might seldom develop to economically damaging levels of infestation in the presence of natural enemies.

The systemic insecticide aldicarb was developed in the 1960s for use as an in-furrow granule application at sowing or as a side-dress granule application to cotton at first buds. When applied in the seed furrow at sowing, aldicarb killed overwintered boll weevil and sucking insects, such as fleahoppers for up to eight weeks after germination. It kills boll weevils developing in cotton flowerbuds when applied to plants as a side dressing on young cotton plants. The application of this compound, especially as a side dressing to young fruiting cotton, often results in increased subsequent infestations of bollworms. Research in the early 1970s showed that aldicarb may have potential in boll weevil eradication or suppression programmes when about 2% of each field is sown as a trap crop at least two weeks before the rest of the field is sown and treated with it. However, cost, application convenience and efficacy of alternative methods, as well as the secondary bollworm problem, have greatly limited the use of aldicarb for boll weevil control.

Chemical controls have in the past been associated primarily with conventional and systemic insecticides. However, with the advances that have been made in the identification of naturally occurring compounds such as pheromones, hormones and deterrents, and the synthesis of these natural products, the adoption of a broader view of chemical control is appropriate.

Insect growth regulators

Of the selective chemical pesticides that interfere in different ways with the normal development of certain life stages of the treated arthropods, diflubenzuron is being utilized on a limited basis to control the boll weevil in area-wide management programmes as it appears to have little adverse effect on populations of beneficial species associated with cotton (Bull *et al.*, 1983). Laboratory data indicate that there is synergism in diflubenzuron–malathion mixtures (Haynes, 1991), so this combination may have important potential in boll weevil control and eradication, not only because of improved boll weevil control, but also because the diflubenzuron suppresses certain secondary pests, such as beet armyworm.

Pinhead-bud treatment

The stage of cotton crop development when tiny flowerbuds can be observed in the terminal bud area is often called the 'pinhead-square' stage. Events affecting cotton around the time when the flowerbuds and subsequent bolls are small are important since the crop's future development may be significantly affected by injury to young fruiting forms and the terminal at this time. Therefore, decisions about controlling early insect infestations such as *Lygus* and occasionally bollworms are important at this stage.

Boll weevil infestations at the pinhead stage of crop development rarely reach densities sufficient to cause injury because of winter mortality. However, the pinhead stage is an important weak link in boll weevil life history because weevils in the crop at this stage and earlier cannot reproduce. Female boll weevils entering cotton may feed, become physiologically reproductive and mate, but they cannot successfully oviposit until cotton flowerbuds with bud diameter of 3 mm or greater are available for oviposition. Therefore, an effective boll weevil insecticide application at the time of first flowerbuds may prevent a high proportion of potential reproduction. Only those weevils that survive the insecticide treatments and those that emerge later from hibernation can contribute to mid and late season population growth.

Successful early flowerbud treatment may require only one insecticide application, or sometimes two applications with a five to seven day interval between applications and occasionally, under high infestation density situations, three applications with similar intervals between applications. Timing and number of applications may be routinely scheduled, and based on stage of crop development and knowledge of favourable hibernation sites in the area and a history of high infestation potential. However, monitoring of boll weevil emergence from hibernation with pheromone traps or some other detection method, such as scouting by visual observation of weevils in the crop, makes the timing of this treatment more effective and efficient as a boll weevil control tactic. The pheromone trap method of monitoring boll weevils is probably the best tool for deciding which fields need treatments, when to apply and how many applications should be applied at this stage for boll weevil control.

Reproduction–diapause control

The late summer–early autumn period of cotton development is important for boll weevil survival. It is during this time – the last reproductive generation of weevils in cotton fields and the time of initiation of diapause – that the overwintering portion of the population becomes prepared for hibernation and seeks suitable hibernation sites. Insecticidal control of the last reproductive generation of weevils and of diapausing weevils before they enter hibernation sites has proven to be highly effective in reducing boll weevil infestations the following year (Lloyd *et al.*, 1966). The tactic of attacking boll weevil population development with insecticides to prevent significant numbers of weevils in firm diapause from entering hibernation has become known as 'reproduction–diapause' boll weevil control.

Reproduction–diapause control is enhanced by lengthened developmental time for egg hatch, larval growth, pupal development and by lengthened time required for adults to enter diapause, when temperatures, especially at night time, become cooler than optimum. Harris *et al.* (1966) studied boll weevil development under a simulated September and October 1964 temperature regime and conservatively estimated that a boll weevil egg laid after the beginning of October could not develop into a diapausing adult in time to survive winter.

The number of insecticide applications required for successful reproduction–diapause boll weevil control varies depending on the objective, the population being controlled and subsequent weather conditions. A single insecticide application mixed with a chemical defoliant may be effective if moderate suppression is the goal and winter mortality is relatively high. On the other hand, when eradication is the goal, several applications are usually required.

Lure and kill baits

Baits used with insecticide were investigated for controlling the boll weevil soon after its entry into this country (Hunter and Hinds, 1905), but their effectiveness was limited. The discovery and synthesis of the pheromone produced by the male boll weevil (Tumlinson *et al.*, 1969), and the discovery of potent feeding stimulants for the boll weevil (Daum *et al.*, 1967; McKibben *et al.*, 1985) prompted more recent work with baits. Application of liquid baits, primarily cottonseed oil emulsions, to cotton plants and cotton field borders was attempted with only limited success. Daum *et al.* (1967) reported on a field-cage experiment, in which almost 60% of the adult weevil population was attracted to and fed on a paste-like bait formulation containing 16% crude cottonseed oil that was applied to vigorously growing cotton plants at a rate of 50–120 litres per hectare. Davich *et al.* (1968) noted that the effectiveness of the bait has been measured with incorporated pathogens, a chemosterilant and a dye. The pathogens, and the chemosterilant, greatly suppressed population build-up when compared with an untreated check. Approximately 70–80% of a field population became visibly marked when the bait contained a dye.

Mitchell *et al.* (1976) researched in-field traps and insecticides for suppression and elimination of populations of boll weevils. Grandlure-baited in-field traps, used at the rate of 25 per hectare, captured 76% of a population of overwintered boll weevils, estimated to number about 25 per acre, that emerged between the time cotton was planted and the time the plants produced pinhead flowerbuds. A combination of in-field traps and insecticides, captured or killed all of the emerging overwintered weevils during the 23 days during the early bud stage of plant growth. The traps alone captured about 96% of a late emerging population of about 2.5 boll weevils per hectare during 6–31 July. The problem of clumping of F_1 and F_2 progeny proved to be an important factor in the efficiency of the traps.

Leggett *et al.* (1981) reported on the detection and suppression of low population levels of F_1 generation boll weevils with in-field traps. In other tests Lloyd *et al.* (1980) described detection and suppression of F_1 generation boll weevils with several trapping densities in fields where populations of known size were evaluated. They reported that the increased efficiency of pheromone traps was related to very low overwintered boll weevil populations. The presence of competing males appeared to reduce the efficiency of pheromone traps for suppression.

McKibben *et al.* (1971) formulated the synthetic pheromone in polyethylene glycol and dispensed it in cotton dental rolls and later cigarette filters. In 1972, a commercial gel formulation of the boll weevil pheromone was prepared by Zoecon Corporation (Hardee *et al.*, 1974). Bull (1976) slowed the release rate by placing the cigarette filter dispenser in a glass vial which functioned as a physical barrier. McKibben *et al.* (1980) used a commercially available polyester-wrapped cigarette filter in place of the 'filter-in-a-vial' method.

McKibben *et al.* (1985) identified feeding stimulants of the boll weevil. This led to a new device that is now in development that shows promise as a boll weevil suppression tool. The bait devices, in the form of a coated stake with a PVC cap containing Grandlure, use only a fraction of the normal insecticide per unit area as compared with conventional application. As with any pheromone system, the devices only attract the target species. Researchers at the Boll Weevil Research Unit in Mississippi State, had the idea to combine the pheromone, a feeding stimulant and a toxicant into a plastic bait. Alternatively, formulations can be prepared containing various natural feeding stimulants and attractants, such as crude cottonseed oil.

Laboratory and field bioassays have shown that boll weevils are attracted to and are killed by the new bait sticks. Cyfluthrin is currently being incorporated into the plastic cap and stake coating to kill the boll weevils that ingest it, or stay in contact with it for a short time.

A large field test in Mississippi during 1990 evaluated the bait stick as a suppression tool in a boll weevil management/eradication programme. In 80 hectare blocks the bait sticks at 2.5 per hectare during early season reduced trap catch significantly. One area that had 2.5 bait sticks early and 10 per hectare starting 29 June, had a 70% reduction in trap catch from a nearby untreated area of equal size.

The boll weevil bait stick may be an economical, effective, environmentally sound approach that can be applied over a large area as needed without the use of highly skilled personnel. The deployment of these bait sticks against the weevil at its overwintering site should prevent a high percentage of the overwintering population from ever reproducing in cotton.

Using the bait sticks against late season weevil population might also be an effective strategy. There is some question as to whether a significant number of weevils going into diapause can be attracted and killed. Villavaso and Earle (1974) showed in field and laboratory tests that males destined to enter diapause were significantly less attractive to females than reproductive males.

Unpublished data from the Boll Weevil Research Unit at Mississippi State indicated that diapause and other late-season populations of boll weevils produced only very limited quantities of pheromone. This condition could cause the new bait sticks to be even more effective late-season.

Review of Boll Weevil Eradication

In 1958, the National Cotton Council called for increased research and development to provide technology for the eradication of the boll weevil from the United States. At the request of the House and Senate Agricultural Committees of the US Congress, a working group subsequently appointed by USDA to review existing boll weevil research programmes, recommended the establishment of the Boll Weevil Research Laboratory on the Mississippi State University campus and to augment funds at other USDA and State Experiment Stations.

The Boll Weevil Research Laboratory was dedicated in 1962 with the stated goal of ultimately eradicating the boll weevil. By 1969, a special study committee concluded that adequate technology had been developed to justify large-scale field testing and recommended that a pilot Boll Weevil Eradication Experiment be conducted in south Mississippi and adjacent areas of Alabama and Louisiana in 1970. The objective of the experiment was to assess the technical and operational feasibility of boll weevil eradication.

The pilot experiment was located in 30 counties in south Mississippi, five parishes in Louisiana, and two counties in Alabama. An area of 8000–10,000 hectares was divided into zones with an outer buffer zone of about 80 km wide to reduce immigration to the inner zone where evaluation was done.

Boll weevil reproduction was suppressed below detectable levels in 203 of 236 fields in the eradication zone. All of the infested fields were located in the northern one third of the eradication zone and less than 40 km from substantial populations further north. In the southern two thirds of the eradication zone, no reproduction could be detected in any of the 170 fields.

Following careful critical analysis of the trial results, the cotton industry asked that USDA conduct another eradication experiment in view of the fact that they had not 'conclusively' demonstrated the feasibility of eradication. They also believed that utilizing other research findings developed since the previous trial, particularly the pheromone trap, was important. They suggested locating the evaluation area a sufficient distance from uncontrolled cotton to prevent migration from confounding the results.

This second eradication trial was located in northeast North Carolina and southern Virginia at the eastern extremity of the cotton belt, and cotton located outside the eradication zone was over a 100 km to the southwest. The area included 6500 hectares the first year and increased to 13,500 hectares by the third and last year of the trial. About 20% of the cotton was located within the buffer zone between the evaluation zone and outside non-programme cotton.

The components of the eradication trial consisted of:

- Late season insecticide treatments which began when diapause was detected and continued until cotton was destroyed.
- Pheromone traps to monitor populations and determine if in-season treatments were needed.
- Diflubenzuron applied to pinhead-square cotton as needed.
- Sterile weevil releases in early fruiting period.
- Defoliant applied to destroy food and breeding sites prior to stalk destruction.
- Stalk destruction as soon as possible after harvest.
- Monitoring for insects other than boll weevil (primarily bollworms) and treating as needed.

Data indicated that eradication had been achieved by the second year of the 3-year trial, and that the improved pheromone trap with a controlled release formulation would trap out very low populations of boll weevil in early spring.

Continuing studies (Carlson and Suguiyama 1982) show the economic returns growers can expect following eradication of boll weevil. Using 4-year averages before and after eradication, pesticide costs to produce a crop decreased from \$126 to \$42 per hectare. Further, there was about a $5\,g\,kg^{-1}\,ha^{-1}$ yield increase following eradication. While difficult to quantify, there must have been some environmental benefits derived from the dramatic reductions in pesticide use in the area. In fact, this reduction in pesticide use on cotton was to some extent mirrored in pesticide use on other crops in the area.

The eradication trial was successfully completed in 1980, and a containment programme was conducted during 1981–1982 in the buffer zone to prevent reinfestation of the eradication area during the extensive evaluation process.

The programme that was initiated on 1 July 1983, included all of the cotton area in North and South Carolina which was infested with boll weevils. This amounted to about 40,000 hectares, and included the buffer zone and southern portions of the original eradication trial area where immigration had occurred.

In 1984–1985 the same procedures were employed and the programme was successfully completed. Eradication had been achieved, with the exception of a few scattered fields in the area and the buffer zone between South Carolina and Georgia outside the programme area. These areas were addressed during the holding period until the next increment of the programme could be initiated.

In 1985–1986 the programme expanded to include western Arizona, southern California and northwestern Mexico where heavy infestations had developed. Eradication has been completed in these areas and the programme was further expanded in 1987 into parts of Georgia, Alabama, all of Florida and in 1988 to the remainder of the southwestern infestation in Arizona.

During 1986–1987 the cotton industry worked with growers in Georgia, Florida, and portions of Alabama to expand the programme. The necessary referendums were passed by the growers and the programme began with the initiation of the late season treatments in early September 1987, and continued on the same area into December, depending on the condition of cotton. All cotton which contained fruiting forms suitable for food for diapause development was treated in this phase of the programme, since boll weevils were present in all fields. The programme area in this phase was approximately 162,000 hectares with an average of slightly over eight treatments being applied.

The first crop year (1988) was divided into three periods based upon the strategy to be employed. These were: the pinhead square stage in June; the midseason containment stage in July and August; and the late season stages extending to the end of the crop year when food and breeding sites are destroyed. These periods are approximate and may vary from area to area.

All treatments during these three phases were based upon the trap data from individual fields. No blanket or automatic treatments were made, except in the buffer zone adjacent to cotton outside the programme area during the late season phase.

Treatments were based upon the numbers of boll weevils caught in traps around each field. A field was designated as up to 16 hectares in size. Approximately 2.5 traps per hectare were used, with traps arrayed around field borders with more used near good hibernation sites. The number of boll weevils trapped to trigger treatment varied with the phase of the programme.

Pinhead flowerbud phase

Success at the last opportunity to destroy the overwintering population was dependent upon the trapping effort. A trap catch (all traps around a field) of two to three boll weevils triggered treatment. Two treatments were made at seven-day intervals beginning at the eight-leaf stage of cotton development, but continued until catches were below the threshold.

Midseason containment phase

Treatments made in this phase (July and August) were designed to prevent boll weevil spread from isolated established populations to adjacent uninfested fields, and to prevent population build-up in midseason which would cause economic loss to growers. The trap catch per field to trigger treatment was five boll weevils per field. Fields were treated at seven-day intervals. If trap catches began to increase, they shortened the interval between treatments to five days and in a few cases to three-day intervals.

Late season phase

During this time of the season, boll weevil migration had begun from the earlier fields which were nearing harvest. There were flushes of boll weevils from fields when they were defoliated, when harvested and again when stalks were shredded. Migrating populations were not of serious consequences, particularly in November and later. Major diapause populations developed in late September and October and primarily in those fields where populations developed during midseason. During this phase, the trigger for treatment was raised to ten weevils per field. Also, the treatment interval gradually increased as the season progressed to up to 14–21 days.

During the 1988 season there was a programme area of nearly 200,000 hectares of cotton. These fields were treated on a field-by-field basis according to the trap catch trigger cited above. Average number of applications ranged from 3.8 in the South Carolina buffer to 11.2 in the Eufaula, Alabama area. For the programme as a whole, an average of 8.6 applications were made until the end of October.

Both 1988 and 1989 were difficult years for the programme. Although progress was made, the programme was behind schedule. Several factors contributed to this delay including underestimating boll weevil populations and reproductive potential in the expanded eradication area; overestimating natural mortality, specifically winter kill; starting the initial diapause programme with insufficient time to prepare; involvement with environmental concerns and a lawsuit that precluded normal operations and diluted staff time; and finally inadequate funding for the situation.

Low temperatures over the entire eradication area during December of 1989 set the tone for an extraordinarily successful suppression effort during 1990. Most boll weevil experts predicted that the unusually low winter temperatures would drastically reduce the spring emerging weevil population. Early trap captures throughout the programme area bore this out. The 1990 cotton crop had the best start in three years. Weevil catches were very low. In early June fields with zero weevils ranged from 60% for Albany, Georgia, to 94% for Jay, Florida. Spring spray operations began on 13 May 1990, and peaked during the period 15 June, at about 48,000 hectares. Although the 1990 programme area in the eradication zone was about 31% above 1989 levels, only 76% of the estimated projection was sprayed. The programme was summarized at the 19 September 1990, Southeastern Boll Weevil Eradication Meeting in Atlanta, Georgia, as 'super good'. Overall, weevils are low and all indications are that eradication is close at hand.

References

Ables, J.R., Jr., Goodenough, J.L., Hartstack, A.W. and Ridgway, R.L. (1983) Entomophagous Arthropods. In: Ridgway, R.L., Lloyd, E.P. and Cross, W.H. (eds), *Cotton Insect Management With Special Reference to the Boll Weevil.* United States Department of Agriculture, Agricultural Research Service, Agriculture Handbook No. 589, pp. 103–127.

Benedict, J.H. and George, D.M. (1979) A bibliography of host plant resistance literature for the boll weevil, *Anthonomus grandis. Bulletin of Entomological Society of America* 25, 19–23.

Benedict, J.H., Urban, T.C., George, D.M., Segers, J.C., Anderson, D.J., McWhorter, G.M. and Zummo, G.R. (1985) Pheromone trap thresholds for management of overwintered boll weevils (Coleoptera: Curculionidae). *Journal of Economic Entomology* 78, 169–171.

Bishopp, F.C. (1911) *An Annotated Bibliography of the Mexican Cotton Boll Weevil.* United States Department of Agriculture, Bureau of Entomology Circular 140, 30pp.

Borkovec, A.B, Woods, C.W., and Terry, P.H. (1978) Boll weevil: Chemosterilization by fumigation and dipping. *Journal of Economic Entomology* 17, 862–866.

Bottrell, D.G. (1974) The boll weevil as a key pest. In: *Proceedings of a USDA–ARS Conference on Boll Weevil Suppression, Management, and Elimination Technology*, Memphis, Tennessee, pp. 5–8.

Bottrell, D. G. and Adkisson, P. L. (1977) Cotton insect pest management. *Annual Review of Entomology* 22, 451–481.

Brazzel, J.R. (1959) The effect of late-season applications of insecticides on diapausing boll weevils. *Journal of Economic Entomology* 52, 1042–1045.

Brazzel, J.R. and Newsom, L.D. (1959) Diapause in *Anthonomus grandis* Boh. *Journal of Economic Entomology* 52(4), 603–611.

Brazzel, J.R., Davich, T.B. and Harris, L.D. (1961) A new approach to boll weevil control. *Journal of Economic Entomology* 54, 603–611.

Bull, D.L. (1976) Formulations of grandlure. In: *Detection and Management of the Boll Weevil with Pheromone.* Texas Agricultural Experiment Station Research Monograph 8, pp. 1–9.

Bull, D.L., Ables, J.R. and Lloyd, E.P. (1983) Insect growth regulators with emphasis on the use of benzoylphenyl ureas. In: Ridgway, R.L., Lloyd, E.P. and Cross, W.H. (eds), *Cotton Insect Management With Special Reference to the Boll Weevil.* United States Department of Agriculture, Agricultural Research Service, Agriculture Handbook No. 589, pp. 207–235.

Burke, H.R. (1968) Geographic variation and taxonomy of *Anthonomus grandis* Boheman. Department of Entomology, Texas A&M University, ERD Contract Number 12.14–100–7733 (33), 152pp.

Burke, H.R., Clarke, W.E., Cate,J.R. and Fryxell, P.A. (1986) Origin and dispersal of the boll weevil. *Bulletin of the Entomological Society of America* 32, 228–238.

Carlson, G.A. and Suguiyama, L.F. (1982) Economic evaluation of the boll weevil eradication trial in North Carolina, 1978–80. In: Ridgway, R.L., Lloyd, E.P. and Cross, W.H. (eds), *Cotton Insect Management With Special Reference to the Boll Weevil.* United States Department of Agriculture, Agricultural Research Service, Agriculture Handbook No. 589, pp. 497–517.

Carter, F.L., and Phillips, J.R. (1973) Diapause in the boll weevil, *Anthonomus grandis* Boheman, as related to fruiting activity in the cotton plant. *Arkansas Academy of Sciences Proceedings* XXVII, 16–20.

Chesnut, T.L. and Cross, W.H. (1971) Arthropod parasites of the boll weevil, *Anthonomus grandis* 2. Comparisons of their importance in the United States over a period of thirty-eight years. *Annals of the Entomological Society of America* 64, 549–557.

Coad, B.R., Johnson, E. and McNeil, G.L. (1924) Dusting cotton from airplanes. *United States Department of Agriculture Bulletin* 1204, 40.

Cross, W.H. (1974) Importance of dispersal and migration of the boll weevil to an eradication program. *Proceedings of Beltwide Cotton Production Research Conferences*, Memphis, Tennessee, p. 130.

Cross, W.H. (1975) Biology, control and eradication of the boll weevil. *Annual Review of Entomology* 18, 17–46.

Cross, W.H. and Hardee, D.D. (1968) Traps for survey of overwintered boll weevil populations. *Cooperative Economic Insect Report* 8(20), 430.

Cross, W.H. and Mitchell, H.C. (1966) Mating behavior of the female boll weevil. *Journal of Economic Entomology* 59, 1503–1507.

Cross, W.H. and Mitchell, H.C. (1968) Parasites of the boll weevil in Mexico. VI National Entomological Congress Organized for the Society of Mexican Entomology, 23–26 October 1967. *Folia Entomologica Mexicana* 18–19, 24.

Cross, W.H., Hardee, D.D., Nichols, F., Mitchell, H.C., Mitchell, E.B., Huddleston, P.M. and Tumlinson, J.H. (1969) Attraction of female boll weevils to traps baited with males or extracts of males. *Journal of Economical Entomology* 62, 154–161.

Cross, W.H., Lukefahr, M.J., Fryxell, P.A. and Burke, H.R. (1975) Host plants of the boll weevil. *Environmental Entomology* 4, 19–26.

Cross, W.H., Mitchell, H.C., and Hardee, D.D. (1976) Boll weevils: Response to light sources and colors on traps. *Environmental Entomology* 5, 565–71.

Daum. R.J., McLaughlin, R.E. and Hardee, D.D. (1967) Development of the bait principle for boll weevil control: Cottonseed oil, a source of attractants and feeding stimulants for the boll weevil. *Journal of Economic Entomology* 60, 321–325.

Davich, T.B. (1959) New approaches to cotton insects. *Farm Chemicals* 122, 62,64,65.

Davich, T.B. and Lindquist, D.A. (1962) Exploratory studies on gamma radiation for the sterilization of the boll weevil. *Journal of Economic Entomology* 55, 164–167.

Davich, T.B., Keller, J.C., Mitchell, E.B., Huddleston, P.A., Hill, R., Lindquist, D.A., McKibben, G.H. and Cross, W.H. (1965) Preliminary field experiments with sterile males for eradication of the boll weevil. *Journal of Economic Entomology* 58, 127–131.

Davich, T.B., Daum R.J. and McLaughlin, R.E. (1968) Development of a bait for boll weevil control and ecological studies. *Folia Entomologica Mexicana* 18, 19–25.

Davis, J.W., Cowan, C.B.,Jr. and Parencia, C.R., Jr. (1967) Emergence of overwintered boll weevils from hibernation sites near Waco, Texas. *Journal of Economic Entomology* 60, 1102–1104.

Davis, J.W., Cowan, C.B., Jr. and Parencia, C.R., Jr. (1975) Boll weevil: Survival in hibernation cages and in surface woods trash in central Texas. *Journal of Economic Entomology* 68, 797–799.

DeBach, P. (1974) *Biological Control by Natural Enemies.* Cambridge University Press, Cambridge.

De Wilde, J. (1962) Photoperiodism in insects and mites. *Annual Review of Entomology* 7, 1–26.

Dickens, J.C. (1990) Specialized receptor neurons for pheromones and host plant odors in the boll weevil, *Anthonomus grandis* Boh. (Coleoptera: Curculionidae). *Chemical Senses* 15, 311–331.

Dickerson, W.A., McKibben, G.H., Lloyd, E.P., Kearney, J.F., Lann, J.J., Jr. and Cross, W.H. (1981) Field evaluation of a modified in-field boll weevil trap. *Journal of Economic Entomology* 74, 280–282.

Dunn, H.A. (1964) *Cotton Boll Weevil* (Anthonomus grandis *Boh.). Abstracts of Research Publications 1843–1960.* United States Department of Agriculture Miscellaneous Publication No. 985, 194pp.

Earle, N.W. and Leopold, R.A. (1975) Sterilization of the boll weevil: Vacuum fumigation with hempa combined with feeding busulfan-treated diet *Journal of Economic Entomology* 68, 283–286.

Earle, N.W. and Newsom, L.D. (1964) Initiation of diapause in the boll weevil. *Journal of Insect Physiology* 10, 131–139.

Earle, N.W. and Simmons, L.A. (1979) Ability to fly affected by acetone, irradiation, and diflubenzuron. *Journal of Economic Entomology* 72, 573–575.

Earle, N.W., Simmons, L.A., Nilakhe, S.S., Villavaso, E.J., McKibben, G.H. and Sikorowski, P.P. (1978) Pheromone production and sterility in boll weevils: Effect of acute and fractionated gamma irradiation. *Journal of Economic Entomology* 71, 591–595.

Earle, N.W., Nilakhe, S.S. and Simmons, L.A. (1979) Mating ability of irradiated male boll weevils treated with diflubenzuron or penfluron. *Journal of Economic Entomology* 72, 334–336.

Enfield, F.D., North, D.T. and Erickson, R. (1981) Response to selection for resistance to gamma radiation in the cotton boll weevil. *Annals of the Entomological Society of America* 74, 422–424.

Enfield, F.D., North, D.T., Erickson, R. and Rotering, L. (1983) A selection plateau for radiation resistance in the cotton boll weevil. *Theoretical and Applied Genetics* 65, 277–281.

Frisbie, R.E., Phillips, J.R., Lambert, W.R.A. and Jackson, H.B. (1983) Opportunities for improving cotton insect management programs and some constraints on beltwide implementation. In: Ridgway, R.L., Lloyd, E.P. and Cross, W.H. (eds), *Cotton Insect Management with Special Reference to the Boll Weevil.* United States Department of Agriculture, Agricultural Research Service, Agricultural Handbook No. 589, pp. 521–527.

Fye, R.E., Hopkins, A.R., McMillian, W.W., and Walker, R.L., (1959) The distance into woods along a cotton field at which the boll weevil hibernates. *Journal of Economic Entomology* 52, 310–312.

Gaines, R.C. (1935) Cotton boll weevil survival and emergence in hibernation cages in Louisiana. *United States Department of Agriculture Technical Bulletin* No. 486, p. 28.

Gaines, R.C. (1959) Ecological investigations of the boll weevil. Tallulah, Louisiana, 1915–1958. *United States Department of Agriculture Technical Bulletin* No. 1208, p. 20.

Gassner, G.D., Childress, D., Pomonis, G. and Eaton, J. (1974) Boll weevil chemosterilization by hypobarometric distillation. *Journal of Economic Entomology* 67, 278–280.

Grossman, E.F. (1931) Insect enemies of the cotton boll weevil. *Florida Entomologist* 15, 8–10.

Guerra, A.A. (1988) Seasonal boll weevil movement between Northeastern Mexico and the Rio Grande Valley of Texas, USA. *Southwestern Entomologist* 13, 261–271.

Hamer, J.L., Andrews, G.L., Seward, R.W., Young, D.F., Jr. and Head, R.B. (1983) Optimum pest management trial in Mississippi. In: R.L. Ridgway, E.P. Lloyd and W.H. Cross (eds), *Cotton Insect Management with Special Reference to the Boll Weevil.* United States Department of Agriculture, Agricultural Research Service, Agricultural Handbook 589, pp. 385–407.

Hardee, D.D., Huddleston, P.M. and Mitchell, E.B. (1967) Procedure for bioassaying the sex attractant of the boll weevil. *Journal of Economic Entomology* 60, 169–171.

Hardee, D.D., McKibben, G.H., Gueldner, R.C., Mitchell, E.B., Tumlinson, J.H. and Cross, W.H. (1972) Boll weevils in nature respond to grandlure, a synthetic pheromone. *Journal of Economic Entomology* 65, 97–100.

Hardee, D.D., Graves, T.M., McKibben, G.H., Johnson, W.L., Gueldner, R.C. and Olsen, C.M. (1974) A slow-release formulation of Grandlure, the synthetic pheromone of the boll weevil. *Journal of Economic Entomology* 67, 44–46.

Harris, F.A., Lloyd, E.P and Baker, D.N. (1966) Effects of the fall environment on the boll weevil in northeast Mississippi. *Journal of Economic Entomology* 59, 1327–1330.

Haynes, J.W. (1963) Chemical sterility agents as they affect the boll weevil, *Anthonomus grandis* Boheman. M.S. Thesis, Mississippi State University, Mississippi State, Mississippi.

Haynes, J.W. (1991) Laboratory results of boll weevil insecticide synergism. *Proceedings of the 1991 Mississippi Insect Control Conference.* Cooperative Extension Center, Mississippi State University, Mississippi.

Haynes, J.W. and Smith, J.W. (1989a) Evaluation of a new method for sterilizing boll weevils (Coleoptera: Curculionidae) by dipping in a diflubenzuron suspension followed by irradiation. *Journal of Economic Entomology* 82, 64–68.

Haynes, J.W. and Smith, J.W. (1989b) Accumulated genetic sterility in progeny of boll weevils treated with apholate. *Journal of Agricultural Entomology* 6, 147–152.

Haynes, J.W., Mitlin, N., Davich, T.B., Nail, B.J. and Dawson, J.R. (1975) Mating and sterility of male boll weevils treated with busulfan plus hempa. *Environmental Entomology* 4, 315–318.

Haynes, J.W., Mitlin, N., Davich, T.B., Dawson, J.R., McGovern, W.L. and McKibben, G.H. (1977) Sterilization of boll weevil pupae with fractionated doses of gamma irradiation. *Entomologia Experimentalis et Applicata* 21, 57–62.

Haynes, J.W., McGovern, W.L. and Wright, J.E. (1981) Diflubenzuron (solvent-water suspension) dip for boll weevils: effects measured by flight, sterility, and sperm transfer. *Environmental Entomology* 10, 492–495.

Helms, J.D. (1977) *Just Lookin' For a Home: The Cotton Boll Weevil and the South.* PhD Dissertation, Florida State University.

Hollingsworth, J.P., Wright, R.L. and Lindquist, D.A. (1964) Spectral response characteristics of the boll weevil. *Journal of Economic Entomology* 57, 38–41

Hollingsworth, J.P., Taft, H.M. and Roach, S.H. (1977) Leggett traps as a substitute for woods trash examinations as an indicator of potential field populations. *Journal of Economic Entomology* 70, 445–446.

Hopkins, A.R., Taft, H.M. and Agee, H.R. (1971) Movement of the boll weevil into and out of a cotton field as determined by flight screens. *Annals of the Entomological Society of America* 64, 254–257.

Howard, L.O. (1897) The Mexican cotton boll weevil (*Anthonomus grandis* Boheman) *United States Department of Agriculture, Division of Entomology Circular* 18.

Hunter, W.D. (1917) The boll weevil problem, with special reference to means of reducing damage. *United States Department of Agriculture Farmers Bulletin* 848, 1–40.

Hunter, W.D. and Coad, B.R. (1922) The boll weevil problem. *United States Department of Agriculture Farmers Bulletin* 1261, 19–20.

Hunter, W.D. and Hinds, W.E. (1905) The Mexican cotton boll weevil. *United States Department of Agriculture Bureau of Entomology Bulletin* 51, 181pp.

Hunter, W.D. and Hinds, W.E. (1912) Mexican cotton-boll weevil: a summary of the results of the investigations of this insect up to December 31, 1911. US Department of Agriculture, *Bureau of Entomology Bulletin* No. 114, S. Document 3. *Entomological Society of America* 64, 549–557.

Isley, D. (1950) Trapping weevils in spots with early cotton. *Arkansas Agricultural Experiment Station Bulletin* 496, 35–36.

Jenkins, J.N., Parrott, W.L. and McCarty, J.C. (1973) The role of a boll weevil resistant cotton in pest management research. *Journal of Environmental Quality* 2, 337–340.

Johnson, D.R. and Gilreath, M.E. (1981) Boll weevil, *Anthonomus grandis* Boheman, pheromone trapping as an index of population trends. *Journal of Georgia Entomological Society* 17, 429–433.

Jones, D. and Sterling, W.L. (1979) Temperature thresholds for spring emergence and flight of the boll weevil. *Environmental Entomology* 8, 1118–1122.

Jordan, P. (1977) Fine structure and ultrastructure of boll weevil sensory receptors with special emphasis on the antennae. Master's Thesis, Auburn University, Auburn, Alabama.

Keller, J.C., Davich, T.B., McKibben, G. and Mitchell, E.B. (1964) A sex attractant for female boll weevils from males. *Journal of Economic Entomology* 57, 609–610.

Lambremont, E.N. (1961) Homogenate respiration of diapausing and nondiapausing boll weevils. *Annals of the Entomological Society of America* 54(2), 313–316.

Lees, A.D. (1955) *The Physiology of Diapause in Arthropods.* Cambridge University Press, Cambridge, 151pp.

Leggett, J.E. and Cross, W.H. (1971) A new trap for capturing boll weevils. *Cooperative Economic Insect Report* 21 (45–48), 773–774.

Leggett, J.E., Lloyd, E.P. and Witz, J.A. (1981) Efficiency of infield traps in detecting and suppressing low population levels of boll weevils. *Environmental Entomology* 10(1), 125–130.

Lindquist, A.W. (1955) The use of gamma irradiations for control or eradication of the screw-worm. *Journal of Economic Entomology* 48, 467–469.

Lindquist, D.A., Gorzycki, L.J., Mayer, M.S., Scales, A.L. and Davich, T.B. (1964) Laboratory studies on sterilization of the boll weevil with apholate. *Journal of Economic Entomology* 57, 745–750.

Lloyd, E.P., Master, M.L. and Merkl, M.E. (1964) A field study of diapause, diapause control, and population dynamics of the boll weevil. *Journal of Economic Entomology* 57, 433–436.

Lloyd, E.P., Tingle, F.C., McCoy, J. and Davich, T.B. (1966) The reproduction-diapause approach to population control of the boll weevil. *Journal of Economic Entomology* 59, 813–816.

Lloyd, E.P., Tingle, F.C., Merkl, M.E., Burt, E.C., Smith, D.B. and Davich, T.B. (1967) Comparison of three rates of application of ultra-low-volume azinphosmethyl in a reproduction-diapause control program against the boll weevil. *Journal of Economic Entomology* 60, 1696–1699.

Lloyd, E.P., McKibben, G.H., Knipling, E.F., Witz, J.A., Hartstack, A.W., Leggett, J.E. and Lockwood, D.F. (1980) Mass trapping for detection, suppression, and integration with other suppression measures against the boll weevil. In: *Proceedings, International Colloquium on the Management of Insect Pests with Semio-chemicals*, Gainesville, Florida.

Lloyd, E.P., McKibben, G.H., Leggett, J.E. and Hartstack, A.W. (1983) Pheromones for survey, detection, and control. In: Ridgway, R. L., Lloyd, E. P. and Cross, W. H.

(eds), *Cotton Insect Management with Special Reference to the Boll Weevil.* United States Department of Agriculture, Agricultural Research Service, Agriculture Handbook No. 589, pp. 179–205.

Loftin, U.C. (1946) Living with the boll weevil for fifty years. *United States Department of Agriculture Publication* 3827, 273–291.

McGovern, W.L., Cross, W.H., Leggett, J.E., McKibben, G.H., Johnson, W.L., McCoy, J.R. and Haynes, J.W. (1976) Boll weevils: field competitiveness among several laboratory strains, chemosterilized weevils, and field weevils. *Environmental Entomology* 5, 354–356.

McHaffey, D.G. and Borkovec, A.B. (1976) Vacuum dipping: A new method of administering chemosterilants to the boll weevil. *Journal of Economic Entomology* 69, 139–143.

McKibben, G.H., Hardee, D.D., Davich, T.B., Gueldner, R.C. and Hedin, P.A. (1971) Slow-release formulations of Grandlure, the synthetic pheromone of the boll weevil. *Journal of Economic Entomology* 64, 317–319.

McKibben, G.H., Johnson, W.L., Edwards, R., Kotter, E., Kearny, J.F., Davich, T.B., Lloyd, E.P. and Ganyard, M.C. (1980) A polyester-wrapped cigarette filter for dispensing Grandlure. *Journal of Economic Entomology* 73, 250–251.

McKibben, G.H., Thompson, M.J., Parrott, W.L., Thompson, A.C. and Lusby, W.R. (1985) Identification of feeding stimulants for boll weevils from cotton buds and anthers. *Journal of Chemical Ecology* 9, 1229–1237.

McKibben, G.H., Smith, J.W., Willers, J. and Wagner, T., A. (1991) Stochastic Model for Studying Boll Weevil Dispersal. *Environmental Entomology* 20(5), 1327–1332.

Malley, F.W. (1901) Arsenate of lead against cotton insects. *Texas Farm Congress Proceedings* 4, 103.

Mitchell, E.B. (1971) Manipulation and Reduction of Boll Weevil Field Populations with Plant and Sex Attractants. Ph.D. Dissertation. Mississippi State University.

Mitchell, E.B. and Hardee, D.D. (1974a) In-field traps: A new concept in survey and suppression of low populations of boll weevils. *Journal of Economic Entomology* 67, 506–508.

Mitchell, E.B. and Hardee, D.D. (1974b) Seasonal determination of sex ratios and condition of diapause of boll weevils in traps and in the field. *Environmental Entomology* 3, 386–388.

Mitchell, E.B. and Hardee, D.D. (1976) Boll weevils: Attractancy to pheromone in relation to distance and wind direction. *Georgia Entomological Science* 11, 113–117.

Mitchell, E.B. and Taft, H.M. (1966) Starvation method for obtaining diapausing boll weevils able to survive the winter in hibernation. *Journal of Economic Entomology* 59, 55–57.

Mitchell, E.B., Hardee, D.D., Cross, W.H., Huddleston, P.M. and Mitchell, H.C. (1972) Influence of rainfall, sex ratio, and physiological condition of boll weevils on their response to pheromone traps. *Environmental Entomology* 1, 438–440.

Mitchell, E.B., Huddleston, P.M., Wilson, N.M. and Hardee, D.D. (1973) Boll weevils: Relationship between time of entry into diapause and time of emergence from overwintering. *Journal of Economic Entomology* 66, 1230–1231.

Mitchell, E.B., Lloyd, E.P., Hardee, D.D., Cross, W.H. and Davich, T.B. (1976) In-field traps and insecticides for suppression and elimination of populations of boll weevils. *Journal of Economic Entomology* 69, 83–88.

Mitchell, H.C. and Cross, W.H. (1969) Ovipositioned by the boll weevil in the field. *Journal of Economic Entomology* 62, 604–605.

Mitchell, H.C. and Cross, W.H. (1971) Mating of boll weevils in the field. *Journal of Economic Entomology* 64, 73–74

Mitlin, L.L. and Mitlin, N. (1968) *Boll Weevil* (Anthonomus grandis *Boheman*). *Abstracts of Research Publications 1961–65.* United States Department of Agriculture Miscellaneous Publication No. 1092, 32pp.

Moore, R.F. and Taft, H.M. (1975) Boll weevil: Chemosterilization of both sexes with busulfan plus Thompson-Hayward TH-6040. *Journal of Economic Entomology* 68, 96–98.

Moore, R.F, Leopold, R.A. and Taft, H.M. (1978) Boll weevils: Mechanism of transfer of diflubenzuron from male to female. *Journal of Economic Entomology* 71, 587–590.

National Cotton Research Task Force (1973) *NCR Task Force Report*, 153pp.

Nettles, W.C., Jr. and Betz, N.L. (1965) Glycogen in the boll weevils with respect to diapause, age and diet. *Annals of the Entomological Society America* 58, 721–726.

Niles, G.A., Walker, J.K. and Gannaway, J.R. (1974) Breeding for insect resistance. *Proceedings of the Beltwide Cotton Production Research Conference*, Memphis, Tennessee, pp. 84–86.

Parencia, C.R., Jr. (1978) *One Hundred and Twenty Years of Research on Cotton Insects in the United States.* United States Department Agriculture Handbook 515, 75pp.

Parencia, C.R., Scot, W.P. and Smith, J.W. (1985) *Boll Weevil* (Anthonomus grandis *Boheman). Abstracts of Research Publications 1966–1980.* United States Department of Agriculture, Agricultural Research Service Miscellaneous Publications (reproduced by National Technical Information Service P385–14920), 531pp.

Pierce, W.D., Cushman, R.A. and Hood, C.E. (1912) The insect enemies of the boll weevil. *United States Department of Agriculture, Bureau of Entomology Bulletin* 100, 99pp.

Post, G.B. (1924) Boll weevil control by airplane. Georgia State College of Agriculture, *Extension Division Bulletin* 301, Vol. 13, p. 304.

Reinhard, H.J. (1943) Hibernation of the boll weevil. *Texas Agricultural Experiment Station Bulletin* No. 638, p.45.

Reynolds, H.T., Adkisson, P.L. and Smith, Ray F. (1975) Cotton insect pest management. In: Metcalf, R.L. and Luckmann, W.H. (eds) *Introduction to Insect Pest Management*, John Wiley & Sons, New York.

Roach, S.H., Agee, H.R. and Ray. L. (1972) Influence of position and color of male-baited traps on capture of boll weevil. *Environmental Entomology* 1, 530–532.

Roussel, J.S. and Clower, D.F. (1957) Resistance to the chlorinated hydrocarbon insecticides in the boll weevil. *Journal of Economic Entomology* 50, 463–468.

Rummel, D.R., Jordan, L.B., White, J.R. and Wade, L.J. (1977) Seasonal variation in the height of boll weevil flight. *Environmental Entomology* 6, 673–678.

Rummel, D.R., White, J.R., Carroll, S.C. and Pruitt, G.R. (1980) Pheromone trap index system for predicting need for overwintering boll weevil control. *Journal of Economic Entomology* 73, 806–810.

Sanderson, E.D. (1904) The cotton boll weevil in Texas. *Society for Promotion of Agricultural Science Proceedings* 25, 157–170.

Sanderson, E.D. (1905) Some observations on the cotton boll weevil. *United States Department of Agriculture, Bureau of Entomology Bulletin* 52, 29–42.

Sikorowski, P.P. (1984) Pathogens and microbial contaminants: Their occurrence and control. In: Sikorowski, P.P., Griffin, J.G., Roberson, J.L. and Lindig, O.H. (eds), *Boll Weevil Mass Rearing Technology.* University Press of Mississippi, Jackson, Mississippi, pp. 115–169.

Sikorowski, P.P., Wyatt, J.M. and Lindig, O.H. (1977) Methods of surface sterilization of boll weevil eggs. *Southwestern Entomologist* 2, 32–36.

Smith, J.W. (1989) History of the Southeastern Branch Entomological Society of America. Centennial Issue of the Entomological Society of America Bulletin, Vol. 35, No. 3, pp. 84–91.

Smith, J.W., Villavaso, E.J., McGovern, W.L. and Brazzel, J.R. (1988) Sterile boll weevil releases as part of a boll weevil eradication program. *Proceedings of Beltwide Cotton Production Conference*, Memphis, Tennessee pp. 286–287.

Sterling, W.L. (1978a) Fortuitous biological suppression of the boll weevil by the red imported fire ant. *Environmental Entomology* 7, 564–568.

Sterling, W.L. (1978b) Imported fire ant . . . may wear a gray hat. *Texas Agricultural Progress* 867(24), 19–20.

Taft, H.M., Hopkins, A.R. and Agee, H.R. (1969) Response of overwintered boll weevils to reflected light, odor, and electromagnetic radiation. *Journal of Economic Entomology* 62, 419–424.

Thomas, F.L., Owen, W.L., Jr., Gaines, J.C. and Sherman, F., III. (1929) Boll weevil control by airplane dusting. *Texas Agricultural Experiment Station Bulletin* No. 394, 40pp.

Townsend, C.H.T. (1895) Report on the Mexican cotton boll weevil in Texas (*Anthonomus grandis* Boh.). *Insect Life* 7, 295–309.

Tumlinson, J.H., Hardee, D.D., Minyard, J.P., Thompson, A.C., Gast, R.T. and Hedin, P.A. (1968) Boll weevil sex attractant: Isolation studies. *Journal of Economic Entomology* 61, 470–474.

Tumlinson, J.H., Hardee, D.D., Gueldner, R.C., Thompson, A.C., Hedin, P.A. and Minyard J.P. (1969) Sex pheromones produced by male boll weevils: Isolation, identification, and synthesis. *Science* 166, 1010–1012.

Tumlinson, J.H., Gueldner, R.C., Hardee, D.D., Thompson, A.C., Hedin, P.A. and Minyard, J.P. (1970) The boll weevil sex attractant. In: Beroza, M. (ed.), *Chemicals Controlling Insect Behavior.* Academic Press, New York, pp. 41–59.

Tumlinson, J.H., Gueldner, R.C., Hardee, D.D., Thompson, A.C., Hedin, P.A. and Minyard, J.P. (1971) Identification and synthesis of the four compounds comprising the boll weevil sex attractant. *Journal of Organic Chemistry* 36, 2616–2621.

Villavaso, E.J. (1982) Boll weevil: Field competitiveness of diflubenzuron-fed irradiated males 1980–1981. *Journal of Economic Entomology* 75, 662–664.

Villavaso, E.J., and Earle, N.W. (1974) Attraction of female boll weevils to diapausing and reproducing males. *Journal of Economic Entomology* 67, 171–172.

Villavaso, E.J. and Thompson, M.J. (1984) Field competitiveness of boll weevils (Coleoptera: Curculionidae) sterilized by the feeding of chemosterilants followed by irradiation or fumigation. *Journal of Economic Entomology* 77, 583–587.

Villavaso, E.J., Nilakhe, S.S. and McGovern, W.L. (1979) Field competitiveness of sterile male boll weevils. *Environmental Entomology* 5, 279–280.

Villavaso, E.J., Lloyd, E.P., Lue, P.S. and Wright, J.E. (1980) Boll weevils: Competitiveness of sterile male boll weevils in isolated field plots. *Journal of Economic Entomology* 73, 213–217.

Villavaso, E.J., Roberson, J.L., Sikorowski, P.P. and Thompson, M.J. (1986) Competitiveness of sterile boll weevils (Coleoptera: Curculionidae) relative to a native population in small field plots. *Journal of Economic Entomology* 79, 76–78.

Villavaso, E.J., North, D.T., McGovern, W.L., Smith, J.W. and Enfield, F.D. (1993) Field cage, greenhouse, and small field plot evaluation of a radiation-resistant strain of boll weevil (Coleoptera: Curculionidae). *Journal of Economic Entomology* 86(6), 1693–1699.

Walker, J.K., Jr. (1980) Earliness in cotton and escape from the boll weevil. In: *Biology and Breeding for Resistance to Arthropods and Pathogens in Agricultural Plants.* Texas Agricultural Experiment Station Miscellaneous Publication MP-1451, pp. 113–123.

Whitcomb, W.H. and Bell, K. (1964) Predaceous insects, aphids, and mites of Arkansas cotton fields. *Arkansas Agricultural Experiment Station Bulletin* 690.

Wiygul, G., Dickens, J.C. and Smith, J.W. (1990) Effect of juvenile hormone III and beta-bisabolol on pheromone production in fat bodies from male boll weevils, *Anthonomus grandis* Boheman (Coleoptera: Curculionidae). *Comparative Biochemistry and Physiology* 95B, 489–491.

Wright, J.E., Roberson, J.L. and Dawson, J.R. (1980) Boll weevil: Effects of diflubenzuron on sperm transfer, mortality, and sterility. *Journal of Economic Entomology* 73, 803–805.

10 Other Coleoptera

S.W. Broodryk[1] and G.A. Matthews[2]

[1]South African Development Trust Corporation Ltd, PO Box 213, Pretoria 0001, South Africa; [2]International Pesticide Application Research Centre, Imperial College at Silwood Park, Buckhurst Road, Sunninghill, Ascot, Berkshire SL5 7PY, UK

Syagrus spp. (Coleoptera: Eumolpidae)

The five *Syagrus* species which feed on cotton in Africa are *S. calcaratus* (F.) from West Africa to Uganda (Peacock, 1913; Hancock, 1926); *S. morio* Har. in East and South Africa (Mason, 1917), *S. rugiceps* Lef. in Somalia (Anon., 1928), *S. rosae* Bryant in Uganda (Bryant, 1936) and *S. rugifrons* Baly in Tanzania (Ritchie, 1925) and the eastern lowveld of Transvaal, Swaziland and northern Natal (Broodryk, 1965b).

Syagrus rugifrons devastated cotton crops in northern Natal from 1924 to 1930 when planting ceased for economic reasons. Cotton growing was resumed in 1947, and then the Syagrus beetle started appearing in crops in the eastern Transvaal, Swaziland and northern Natal and became a major pest. The following information is based on a study of *S. rugifrons* (Broodryk, 1965a).

Eggs are laid in groups of eight to ten on or near the cotton stem at soil level. They are cylindrical, about 1 mm × 0.5 mm and are covered with a secretion that binds soil particles together to form an egg packet which enables the developing larvae to survive inside the chorion for months if necessary. Eggs normally hatch within ten days. Newly-hatched larvae enter the soil and feed on rootlets and later attack the taproot which is eventually encircled by feeding channels. These are debilitating enough to affect plants which then show symptoms of wilting and serious moisture stress. Larvae are creamy-white with a brown head and pronotum and six well-developed legs. When fully developed, they are about 9 mm long and 3 mm wide. Fully grown larvae pupate in an earthen cell at a depth of 5 to 35 cm and the adult emerges ten days later. Teneral adults are creamy-white with pale brown appendages.

Adults (Fig. 10.1) are about 8 mm long, black and the head and thorax are rugose. The elytra are striated and broader than the thorax. They feed on young foliage, hence the name cotton leaf beetle. During overcast weather adults hide in the growing tip foliage or bracts of squares. They often hide under clods and

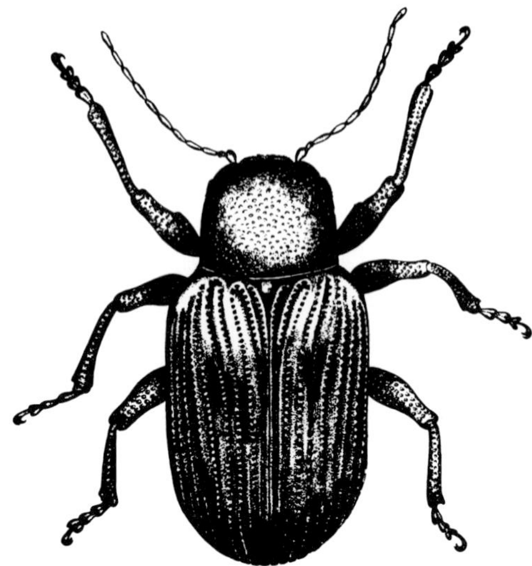

Fig. 10.1. *Syagrus rugifrons* adult.

debris on the soil, preferring to feed at night. Feeding holes are usually elongated and seldom reach the leaf edge. The beetles feign death when disturbed and fall to the ground.

Adults overwinter in sheltered positions under debris on the soil and start feeding when temperatures rise and rains enable the first seedlings to grow. If these are heavily attacked replanting may become necessary. In situations where the cotton plants are not ploughed in, burnt, or are only partially destroyed after harvest, a population of larvae survives on the roots and the resulting adults emerge in spring to join in the extensive feeding activity on the seedlings. Farmers may have to replant up to four times in these circumstances. Adults eventually mate and oviposit on plants in the growing crop. Larval damage only becomes noticeable when the midsummer dry spells cause moisture stress. At this time damaged plants, often carrying a heavy boll load, start wilting and gradually die off. One or two generations are completed per year.

Host plants of the genus *Syagrus* appear to be restricted to the Malvaceae. Pearson and Maxwell Darling (1958) list the following:

S. calcaratus	*Hibiscus esculentus* (now *Abelmoschus esculentus*)
	H. sabdariffa
	H. rosa-sinensis
	Urena lobata
	Sida carpinifolia
S. morio	*H. esculentus*
	Azanza
S. rugifrons	
	Cienfuegosia hildebrandtii
	C. gerrardii

Gossypium herbaceum race *africanum*
Abutilon sonneratianum

Very few natural enemies appear to prey on *S. rugifrons*. Apart from guinea fowl, the only other predator encountered was the reduviid, *Coranus sordidus* Miller. It is doubtful whether any of these exert meaningful control on *Syagrus* populations.

The proximity of abundant hosts affects the pest status of *S. rugifrons*, but failure to destroy plants and crop residues after harvest is a major factor. It has extended its distribution to the Limpopo River and central Transvaal as stand-over cotton provided larvae with food on which to overwinter. Crop rotation is a practice worth considering to limit larval survival.

The cotton leaf beetle is susceptible to most chemicals used to protect cotton, including carbaryl, endosulfan and the organophosphates. When damage is noticed, a dust or spray should be applied to seedlings, ensuring that good

Fig. 10.2. Gall-like swellings caused by *Apion*.

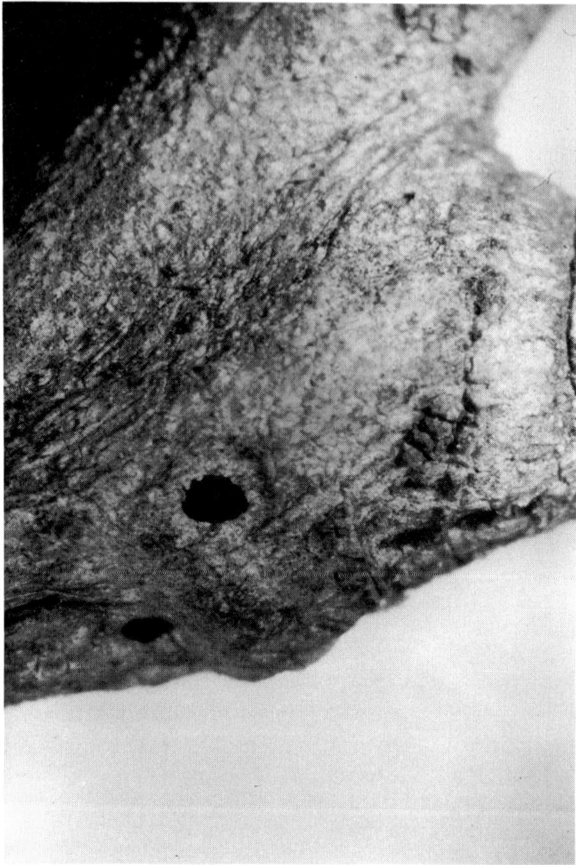

Fig. 10.3. Exit holes of adult of *Apion* after pupation.

cover is achieved. Treatment may have to be repeated after rain. It is important that populations are controlled before they oviposit in the crop.

Apion soleatum Wagner (Coleoptera: Apionidae)

Apion soleatum Wagner (see Plate IV.4), the cotton stem weevil, is the most important of a number of species of *Apion* attacking cotton. Its distribution has expanded from Tanzania, Mozambique and Malawi (Pearson and Maxwell Darling, 1958) to include the eastern Transvaal, Swaziland and northern Natal (Broodryk and Mansfield, 1985).

The beetle is about 2 mm long with the head shiny black and the thorax and elytra dull black with white hairs. Eggs are minute and are laid in the bark near leaf axils or in the stem-end of older bolls. The larvae are creamy-white, legless with a brownish head. They bore into the woody stem tissue and their feeding causes gall-like swellings (Fig. 10.2) near the branch axils. This feeding weakens the branches to the point where they collapse as soon as the boll load becomes sufficiently large.

Larvae pupate in the burrows and pupae are initially white and darken towards completion of the stage. Adults drill exit holes (Fig. 10.3), often near the axils or in the swollen nodes. They are strong fliers and disperse rapidly. Cotton is usually attacked late in the growing season and stand-over or ratoon cotton is crucially important to bridge the dry season. *Apion* has been consistently recorded in southern Africa in association with stand-over cotton, and legislation to ensure a closed season was formulated as a result of its destructive capabilities (Niles and Greeff, 1989).

Its importance in Tanzania was probably due to an abundance of hosts (Pearson and Maxwell Darling, 1958). The natural hosts are not known. An Eulophid parasitoid, *Entedon apionidis* Ferrière, parasitizes the larvae of the stem beetle throughout its distribution range.

Alcidodes spp. (Coleoptera: Curculionidae)

These weevils damage cotton by stem-girdling and boring, which leaves a characteristic ring of frayed fibres. An egg is laid on an enlarged part of the girdle and the larva is thought to feed downwards in the pith. The upper part of the plant dies and often breaks off at the girdle. Attacks are rarely concentrated, but in some areas a large number of plants may be lost. *Alcidodes brevirostris* (Boh.) and the smaller *A. gossypii* (Hust.) are recorded in East and West Africa respectively. The adult is a typical weevil with a black thorax and head, and rusty-red elytra with several longitudinal rows of deep-set punctures and obscurely patterned with golden scales.

Eutinobothrus gossypii Pierce and *E. braziliensis* Hamb. (Coleoptera: Curculionidae)

These weevils damage the stem of young cotton plants in South America.

Myllocerus maculosus Desb. (Coleoptera: Curculionidae)

The grey weevil can damage cotton in India, Pakistan and Nepal by feeding on leaves and bracts, and larvae feeding on roots. Root damage can kill young seedlings and cause wilting, but the adults do not cause significant yield losses (Taneja and Jayaswal, 1984).

Sphenoptera spp. (Coleoptera: Buprestidae)

A number of species of *Sphenoptera* stem borers, for example *S. gossypii* Cotes and *S. neglecta* occur on cotton, chiefly in the drier savannah or semi-desert areas.

Sphenoptera gossypii is found in India, Pakistan and to a lesser extent in the Sudan, while *S. neglecta* is widely distributed in Africa. The eggs are laid singly in fissures in the bark at the base of the stem, and the larvae feed on the wood just below the bark. Uprooting and destroying of crop residues keeps this pest controlled.

Podagrica spp. (Coleoptera: Chrysomelidae)

Flea beetles have specially adapted hind-legs for jumping when disturbed. Several species attack crops, but the majority associated with cotton belong to the genus *Podagrica*. Two species *P. puncticollis* Weise and *P. pallida* (Jacoby) can attain pest status in the Sudan, if populations exceed 20 flea beetles per 100 leaves. Infestations occur more on early sown cotton or when the rains are late, but irrigated plants suffer little damage when the plants are more than 14 days old. Adults feeding on leaves cause a shot-hole effect.

The adult beetle is small (3–4 mm long), elliptical in outline, dorsally strongly convex, with filiform antennae and swollen hind femora; body surface smooth, minutely punctate; eyes black, rest of body in *puncticollis* is light brown and shining, in *pallida* is greyish-ivory in tint. The eggs are laid in the soil and the larvae feed on rootlets. The adults aestivate in the soil, except when alternative hosts remain green through the dry season, and emerge with the first heavy rains.

Other host plants include *Adansonia digitata*, *Corchorus olitorius* and *C. fascicularis, Abutilon pannosum, Hibiscus cannabinus*, *H. esculentus*, *H. sabdariffa* and *Sida* spp. *P. pallida* is also recorded on *Cienfuegosia digitata* and *Grewia villosa* as well as some legumes, namely *Dolichos lablab, Phaseolus vulgaris* and *Cajanus cajan*.

Another Chrysomeloid beetle that caused some damage to cotton in Zimbabwe was *Monolepta gossypiperda* Bryant (Bryant, 1948).

References

Anon. (1928) Actes du conseil international scientifique agricole. Premiere session, 1927. Institute of International Agriculture.

Broodryk, S.W. (1965a) The biology of *Syagrus rugifrons* Baly. *South African Journal of Agricultural Science* 8, 147–170.

Broodryk, S.W. (1965b) The morphology of *Syagrus rugifrons* Baly (Coleoptera: Eumolpidae). *South African Journal of Agricultural of Science* 8, 455–468.

Broodryk, S.W. and Mansfield, J. (1985) A new cotton pest in Transvaal. *Die OTKaner*, Bethal, Republic of South Africa 29, 15.

Bryant, G.E. (1936) Some new injurious Phytophaga from British East Africa (Coleoptera). *Proceedings of the Royal Entomological Society*, London B(5), 12pp.

Bryant, G.E. (1948) A new species of *Monolepta* (Galerucidae: Coleoptera) from South Africa feeding on cotton. *Annals and Magazine of Natural History, London* 14, 584–586.

Hancock, G.L.R. (1926) Annual report of the Assistant Entomologist. Report of the Department of Agriculture of Uganda.

Mason, C. (1917) Report of the Government Entomologist, Nyasaland Protectorate. Annual Report of the Department of Agriculture of Nyasaland.

Niles, G.A. and Greeff, M.S. (1989) *Ratoon Cotton in Perspective.* Technical Communication No. 217. Department of Agriculture and Water Supply, Republic of South Africa.

Peacock, A.D. (1913) Entomological pests and problems of Southern Nigeria. *Bulletin of Entomological Research* 4, 191–220.

Pearson, E.O. and Maxwell Darling, R.C. (1958) *The Insect Pests of Cotton in Tropical Africa.* Commonwealth Institute of Entomology, London.

Ritchie, A.H. (1925) Report of the Entomologist. Report of the Department of Agriculture of Tanganyika 1924–25.

Taneja, A.L. and Jayaswal, A.P (1984) Seasonal activity of grey weevil (*Myllocerus maculosus* Desb.) on different varieties of cotton in Haryana. *Indian Journal of Agricultural Research* 18, 148–150.

11 *Dysdercus* (Hemiptera: Pyrrhocoridae) and Other Heteroptera

S.W. Broodryk[1] and G.A. Matthews[2]

[1]*South African Development Trust Corporation Ltd, PO Box 213, Pretoria 0001, South Africa;* [2]*International Pesticide Application Research Centre, Imperial College at Silwood Park, Buckhurst Road, Sunninghill, Ascot, Berkshire SL5 7PY, UK*

Dysdercus (Hemiptera: Pyrrhocoridae)

Introduction

Cotton stainers of the genus *Dysdercus* occur throughout the tropics of each continent from where their distribution extends mainly into the southern subtropics. Their natural host plants are usually of the order Malvales and all stages feed on developing or mature seed. Penetration of developing cotton bolls occasions transmission of fungi which develop on the immature lint and seed, rendering the latter inviable and staining the lint to the typical yellow colour of stained cotton. Feeding, as such, on immature as well as ripe cotton seed seriously affects the crop mass, oil content and hence its total value. The severity of stainer attacks usually depends on extraneous factors related to the weather and natural hosts in the surrounding veld, as well as crop age and time left to maturity.

Taxonomy and Distribution

Although stainers are brightly coloured in red, yellow and orange the variability and cumbersome nature of colour descriptions for identification have been avoided in taxonomic classification.

According to Freeman (1947) structural and genital characters offer a good basis for grouping and separating most of the species of *Dysdercus* that he studied. Most important of these are the parameres, vertical processes of the ninth segment, the aedeagus and general shape of the male ninth segment and the spermatheca. The following species have been recorded from the regions listed:

Africa

fasciatus Signoret — Equatorial zone to southern Africa

nigrofasciatus Stål

Species	Distribution
superstitiosus (Fabricius) syn. *voelkeri* Schmidt	
melanoderes Karsch	West African forest and Zaïre
haemorrhoidalis Signoret	West Africa, Lake Victoria, Uganda
cardinalis Gerstaecker	Northeastern region of South Yemen to Tanzania
intermedius Distant	Southeastern region of Natal, Republic of South Africa to Tanzania
pretiosus Distant	Zaïre and eastern Africa
orientalis Schouteden	Kenya, Uganda, Tanzania
ortus Distant	Seychelle Islands
flavidus Signoret	Malagasy Republic, Mauritius and Rodriguez
Asia, Australasia and Pacific Islands	
koenigii (Fabricius)	India, Pakistan, Sri Lanka
similis Freeman	India, Sri Lanka
cingulatus (Fabricius)	India, Thailand, Vietnam, China, Queensland (Australia)
sidae Montrouzier	Australia, Fiji, Samoa, Papua
America	
andreae (L.)	USA (Florida), Cuba, Jamaica, Hispaniola, Puerto Rico, Lesser Antilles
bimaculatus (Stål)	USA (Florida, Arizona), Mexico, Central America, Colombia, Venezuela
chaquensis Freiberg	Brazil, Argentina
collaris Blöte	Colombia, Venezuela
concinnus Stål	USA (Texas), Mexico, Central America, Colombia, Venezuela
fernaldi Ballou	Venezuela, Suriname
flavolimbatus Stål	Mexico, Guatemala, Panama
fulvoniger (De Geer)	Venezuela, Trinidad, Br. Guyana, Suriname, Brazil, Lesser Antilles
honestus Blöte	Peru, Colombia, Venezuela, Br. Guyana, Suriname, Brazil, Paraguay, Bolivia
immarginatus Blöte	Argentina, Trinidad
jamaicensis Walker	Jamaica
maurus Distant	Brazil, Trinidad, Venezuela, Suriname
mimulus Hussey	USA (Texas, Florida), Mexico, Central America, Bahamas, Jamaica, Cuba, Hispaniola

mimus (Say)	Mexico, Central America, Venezuela, Brazil, Ecuador, Bolivia, Suriname
obliquus (Herrich-Schäffer)	Mexico, Guatemala, Central America
obscuratus Distant	Mexico, Central America, Colombia, Venezuela, Ecuador, Peru, Bolivia
ocreatus (Say)	Hispaniola, Lesser Antilles
peruvianus Guérin Méneville	Peru, Ecuador, Colombia, Venezuela, Brazil, Paraguay, Argentina, Bolivia
ruficollis (L.)	Br. Guyana, Suriname, Brazil, Argentina, Paraguay
rusticus Stål	Colombia, Peru, Brazil, Bolivia
suturellus (Herrich-Schäffer)	USA (Florida), Bahamas, Jamaica, Mexico, Guatemala, Nicaragua

The species listed for Asia, Australasia and America are not complete and only represent those mentioned in current scientific literature as occurring on cotton. New World species are listed according to a revision by Van Doesburg, 1968.

Description of Life Stages

The eggs are ovoid, 1.5 mm × 0.9 mm, with size varying according to species. The chorion is smooth and initially creamy-white, changing to orange as the embryo develops. Five nymphal instars occur, the first being yellow or bright red, without markings and about 2 mm long. The length at subsequent ecdyses increases to about 3, 5, 6, 9 and 12 mm. The adult colour pattern gradually appears on the thorax and abdomen and the antennae and legs gradually darken. In the third instar the rudimentary wings become visible as black bars on each side on the hind part of the thorax. In the fourth instar they are larger and square-shaped and in the fifth instar extend over the anterior abdomen.

Adults (Plate IV.5) are elongate, about 15 mm long × 4.5 mm broad, the size varying according to species. The head is red or reddish-orange with a white prothoracic collar. The membrane of the hemelytron is dark and on the broadest part of the forewing there is usually a black mark in the form of a spot, ellipse, band or line. The antennae are more than two thirds of the body length and the rostrum is folded beneath the body to reach to or beyond the second abdominal segment.

Life History

Eggs are laid in a shallow depression in the soil or under debris in batches of about 100. They hatch within five to 13 days depending on temperature and species. The first instar nymphs usually remain together at the oviposition site until the first moult after which they disperse in search of food. The gregarious tendency persists and they congregate around suitable ripe or decaying seeds of the Malvales for feeding. The stylets are not yet long enough to utilize unopened fruit but quite hard and dry seeds can be penetrated if adequate free water is available. Water in the form of dew, rainwater and nectar from flowers and extra-floral nectaries is important for all stages. The third and subsequent instars can successfully penetrate unopened fruit and reach the developing seeds for feeding.

Nymphs (Plate IV.6) tend to congregate in sheltered situations before each moult but after the final moult the young adults disperse. They colonize suitable hosts, and move on when these become nutritionally unsuitable or disperse for other reasons. Flights of up to 15 km have been recorded (Parsons, 1928).

Adults mate about two days after the final moult and remain *in copula* until oviposition occurs three to eight days later. Egg production differs widely between species, averaging from 300 for *D. superstitiosus* to 900 for *D. nigrofasciatus*. Fecundity is affected by availability and quality of food. Stainers remain sexually active throughout the year except when low temperatures or lack of food inhibits breeding. No resting stage occurs in winter and their survival depends on adequate food and moisture being available.

The rate of development of the egg and five nymphal instars is dependent on temperature and nutrition in the broad sense, which includes water supply. Seasonal variation in temperature influences the length of development in *D. fasciatus* and *D. intermedius*, since in Malawi the nymphal development (instars II to IV) is extended from 22 days in summer to 48 in the former and 24 to 47 days in the latter. This effect appears to be broadly similar for all *Dysdercus* species studied. *Dysdercus superstitiosus* at 30°C requires four days for egg development and 18 days to complete the five nymphal stages in the presence of adequate moisture. At 25°C the respective times are extended to 6.5 and 33.5 days.

The optimal temperature level for development appears to be about 30°C. The eggs and first instar nymphs are susceptible to temperature extremes and low humidity, while the actively feeding nymphs are comparatively resistant provided that they have access to moisture. Adults tolerate the widest range of conditions, surviving exposure to subzero temperatures for five hours in the case of *D. koenigii* and seven days in the case of *D. sidae*. The lethal maximum for *D. sidae* is 40°C and for *D. koenigii* 41–42°C. Copulation and oviposition are severely affected by extreme temperatures.

Temperature and humidity appear to determine the prevalence of *Dysdercus* species, operating especially through the egg and first instar nymphal stage. The success of members of the genus depends largely on the selection by adults of suitable sites for survival of their offspring.

Feeding Habits

Dysdercus nymphs normally feed on seeds of plants belonging to the suborder Malvales. The state of development of the seed is important in that nymphal development is the more successful the riper the seed. Nymphs in the first instar cannot penetrate hard seeds and depend for nutrition on water and plant exudates. In the second instar they can utilize seeds, given access to water. The subsequent instars are able to penetrate fruit sufficiently to utilize the developing seed material. Those hosts on which development is most rapid give rise to the most fecund adults.

The seeds of Malvales are considered to be essential for normal development of stainers. *Dysdercus superstitiosus* is an exception, since it is able to develop normally and produce fertile adults on developing seeds of grain sorghum (*Sorghum vulgare*) and bulrush millet (*Pennisetum typhoides*). This species also accepts cowpea (*Vigna unguiculata*), but egg production on this food source is lower than on cotton. *Dysdercus cingulatus* is similarly capable of utilizing sorghum and *Pennisetum typhoides* and producing fertile offspring on these hosts in India (Ahmad, 1979a). Lower egg production is achieved than on cotton seed. Feeding on young bolls in cotton may occur primarily in search of water and sugar, whereas the older bolls with seeds further developed are attacked in search of the proteins necessary for reproduction.

Adults can survive for several weeks on water alone at the end of the cotton-growing season. In this situation fat reserves are utilized and longevity depends on their exhaustion. On a diet of water and nectar, adult survival is equal to survival on water and cotton seed, since the adults are able to synthesize fat on the former diet. Interspecific differences occur in synthesis and rate of utilization and may explain why some species survive the dry season. *Dysdercus nigrofasciatus* may survive on water alone until nectar becomes available from early flowering trees, like *Dombeya rotundifolia*, allowing survival until suitable seeds are produced by the preferred hosts.

Damage to Plants

Cotton is the major cultivated host plant of stainers. Other cultivated hosts include *Abelmoschus esculentus* or okra, the *Hibiscus* species grown for plant fibre and *Ceiba pentandra* or kapok. *Dysdercus superstitiosus* and *D. cingulatus* will attack grain sorghum and bulrush millet, and also the jute substitute *Urena lobata* (Harris, 1983).

Damage to the cotton crop may be purely mechanical as a result of feeding or may be pathological when disease organisms are introduced into the unopened boll causing internal boll disease to develop. The extent of damage in both categories above depends on the age of the boll, and hence its stage of development, at the time of the attack. In the following discussion of damage, the normal maturation period from flowering to boll-split is taken to be about ten weeks.

1. Effects of mechanical feeding. Feeding on unopened bolls leaves no external signs and the extent of damage only becomes apparent at boll-split. If young bolls are inspected, only the seeds penetrated by the stylets show signs of damage. The inside of the boll wall shows whitish spongy pustules of proliferating tissue surrounding the puncture, discernible as a dark dot. Stylet tracks can be distinguished as brownish threads. Bolls of up to two weeks old may be shed after a

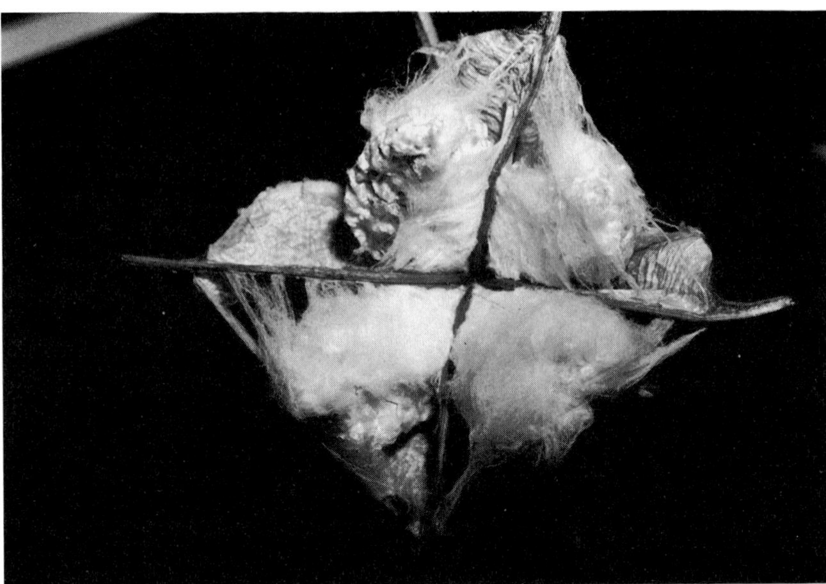

Fig. 11.1. Damaged lint.

severe attack. If retained, they may open at the normal time with unevenly developed locks and immature fibre. In older bolls the feeding mark on the inside boll wall shows as a slightly raised circular surface with a glassy, water-soaked appearance. Once the embryo is fully formed, the seed seems to withstand non-infected punctures.

2. Effect of *Ashbya*, formerly known as *Nematospora*, infection on boll development. Bolls infected within the first two weeks of development are shed, since the embryos are destroyed. The lint is usually reduced to a dark brown membrane, closely enveloping each seed. Bolls infected when three or four weeks old, are slightly smaller than healthy ones. The lint is stained brown or yellow, remains compressed and often adheres to the carpels which are only partly reflexed and extended. This causes characteristic webbing where strands of disintegrating lint are tightly stretched from the carpels to the parent seeds (Fig. 11.1). Bolls infected at five to seven weeks of age develop to the normal size and open normally when they are mature. The lint is, however, stained shades of yellow with the oldest infection returning the darkest shade of colour. The stained cotton of commerce originates from this type of boll. Bolls infected at eight weeks or older show virtually no effects.

Levels of boll loss due to feeding by uninfected insects are related to the abundance of actively feeding insects during boll formation, especially early in the growing season. The prevalence of staining, on the other hand, is related to the number of infected individuals present during later phases of boll development. Both the feeding damage and spread of the fungal pathogen may also be caused by the Pentatomids, *Calidea* and *Nezara*, but their numbers in cotton fields are usually negligible.

3. The third type of damage which occurs is the result of feeding by stainers on cotton seed which is exposed by normal opening of the mature boll. Feeding at this stage causes seeds to lose viability and mass. Since seed contributes about two thirds of the mass of seed cotton, the loss of mass directly affects the return to the grower. Poor seed quality also affects the success of oil extraction. The Lygaeid, *Oxycarenus hyalinipennis* also attacks ripe seed in the field. Since young stainer nymphs and *Oxycarenus* of all stages congregate in open bolls, both are frequently squashed during picking, resulting in yellow body fluid contaminating the cotton. The yellow lint is often mistakenly taken to be due to the fungus and may cause downgrading of the crop.

The Fungal Pathogen, *Ashbya gossypii* (Ashby and Nowell) Guillerm

The origin of fungi that stain cotton is unclear. Of the three fungus species that have been isolated from cotton bolls in Africa, *Ashbya gossypii* is by far the most prevalent. *Ashbya coryli* has a wide host range in addition to beans and coffee, and *A. phaseoli* attacks beans. Both these pathogens are also able to attack cotton, causing less severe staining than *A. gossypii.* The latter has a wider host range than cotton, also attacking various legumes, coffee, citrus and tomatoes (Pridham and Raper, 1950).

Dysdercus nigrofasciatus adults have been shown to transmit *A. gossypii* from wild cotton, *Gossypium herbaceum* var. *africanum* and *Hibiscus vitifolius.* Seeds of *Abelmoschus esculentus* and *Abutilon* sp. in Uganda were found to carry the pathogen as did those of *Azanza garckeana* in Malawi. The evidence therefore suggests that stainers become infected while feeding on their natural hosts and carry the pathogen into the cotton crop as they migrate to a more suitable food source. Transmission of the organism is mechanical as the stainers move from plant to plant. The spores or mycelium are carried as external contaminants on the mouth-parts or in the stylet pouches (Pridham and Raper, 1950). Newly moulted individuals appear to be lightly infected since the pathogen does not appear to persist through the moulting process.

The level of damage to the crop depends on the ratio of stainers to bolls. This relationship is complex, also being influenced by the level of infected stainers in the population and the age of the crop at the time of influx of stainers into the field.

Host Plants

Cotton is the major cultivated host of stainers and where it is grown as an annual crop followed by a cotton-free period, natural hosts are important for the continued survival of stainer populations. Pearson and Maxwell Darling (1958) gives a detailed description of the interrelationship between natural hosts and stainer populations in Africa. Some elements of his findings are given below, to serve as basic guidelines that may be applicable to the behaviour of stainers in other countries.

As a general rule stainers breed successfully only if they have seeds of plants belonging to the order Malvales as food. *Dysdercus superstitiosus* and *D. cingulatus* are the exceptions since they can breed successfully on sorghum and *Pennisetum typhoides* (Egwuatu, 1980). Other orders are listed as hosts of stainers in Brazil, e.g. the Rosales, Geraniales, Cucurbitales and Rubiflorae, but observations did not extend beyond presence and feeding (De Almeida *et al.*, 1981), and their role as true hosts is unclear.

The arboreal hosts *Adansonia* (baobab) and *Azanza* occur in the monsoon savannah, represented by regions with a single rainy season in summer and vegetation typefied by a grass cover and upper storey of trees. Species of the genus *Sterculia* occur from the central Sudan to South Africa as small trees associated with rocky hills and outcrops or well-drained sandy soils. The taller species occur on heavier, often alluvial soils at lower altitudes in Malawi and other parts of southeastern Africa.

Perennial herbaceous hosts are restricted to areas of low rainfall and terrain too broken to produce a dense grass cover, because they cannot survive veld fires, or to the edges of swamps and fringes of forests. They are the most important host plants in the Transvaal, Swaziland, and also in Uganda. *Dysdercus nigrofasciatus* is the predominant stainer species in these areas.

Annual herbaceous hosts are important in the savannah areas. Various *Hibiscus* species are included in this group of plants which aggressively exploit the pioneer situations offered by roadsides, cultivated fields and other sites of disturbed soil.

The arboreal hosts usually produce a single crop of fruit which, in the case of trees with dehiscent fruits, provide an abundant food supply for a restricted period after which deterioration sets in and healthy seeds are removed by other animals. This happens in the case of *Sterculia, Bombax, Azanza* and others. The fruit of *Adansonia*, which are indehiscent, fall over a longer period and are only accessible to stainers once they have been opened by other animals. The food supply in this case is available for a longer period.

The herbaceous perennial hosts have an indeterminate growth habit and will continue to produce flowers and fruit as long as adequate moisture is available and temperatures are not too low. Like stand-over cotton, they are able to produce fruit even before the onset of rain, making use of the established root system and residual moisture present in the soil. Herbaceous annuals can only respond to the first rains to germinate and develop the necessary framework for flowering and fruiting, and therefore have a much more restricted fruiting period than the perennial hosts.

Under natural conditions the wild hosts are invaded by flying adults which remain on the plants if sufficient food, in the form of ripening fruit, is available. They mate and oviposit at the base of the plants. The ensuing nymphs complete their development if their requirements of sufficient moisture and accessible seed material are met, and the newly moulted adults disperse in search of a new food source. Even the perennial shrubs which flower over an extended period produce fruit in a succession of ripening crops. Conditions rarely allow more than one generation to mature successfully at one site and as a consequence there is an almost continual movement of winged adults in search of new hosts. This behaviour pattern does not necessarily occur in response to deterioration of host plant quality.

In the moist equatorial regions, the occurrence of two rainy seasons is likely to ensure an uninterrupted succession of hosts in a suitable developmental stage throughout the year. In the monsoon regions, the long dry season, often with low temperatures, threatens the continued existence of stainers. Two factors aid their survival: first is the fact that most of the natural hosts are at peak fruit production at the start of the dry season; and secondly the lower metabolic rate induced by winter temperatures lengthens the development period of this generation to a level where only a single generation can be completed. Once temperatures increase towards the end of winter, some of the savannah trees begin to flower and this situation enables *D. nigrofasciatus* to produce a generation on *Dombeya rotundifolia* in South Africa and *D. superstitiosus* on *Triumfetta* in Zambia. True aestivation to bridge the dry season does not seem to occur in *Dysdercus.*

The cotton crop is usually invaded by adults that fly into the fields at about the time when the oldest bolls approach maturity. This invasion may, on rare occasions, be large enough and sufficiently early to destroy all the green bolls, as often happens with untimely destruction of stand-over cotton when the annual crop is vulnerable. The initial influx does not usually cause serious damage, but on some occasions a small invasion can cause significant stainer damage. Once the bolls start opening the adults start copulating and ovipositing. The ensuing multiplication within the crop is usually the cause of serious economic damage. Seeds in open bolls are then available to the young nymphs, developing bolls are penetrated by older nymphs and apart from damage to the embyonic seeds, the introduction of *Ashbya gossypii* results in destruction and staining of the immature lint.

The size of the infestation reaching a new cotton crop is often dependent on the management of the previous season's crop. If stainer populations have been kept at low levels by spraying against bollworms throughout the vulnerable period and the old crop is properly destroyed in a timely manner, the source of infestation is usually limited to the natural host plants. Where stand-over cotton is present it far outweighs the importance of natural hosts, since it can produce huge number of stainers.

List of host plant genera

Africa

Bombacaceae
- *Adansonia*
- *Bombax*
- *Ceiba*

Malvaceae
- *Abelmoschus*
- *Abutilon*
- *Cienfuegosia*

Sterculiaceae
- *Dombeya*
- *Melhania*
- *Sterculia*
- *Waltheria*

Tiliaceae
- *Triumfetta*

Gossypium
Hibiscus
Malvastrum
Pavonia
Sida
Azanza
Urena
Wissadula

Rhamnaceae
Zizyphus

Gramineae
Pennisetum
Sorghum

Records of plants listed above imply that development can be completed on seeds of certain species belonging to these genera. The following records indicate that stainers were observed on the plants.

Central and South America

Bombacaceae
Pseudobombax
Ceiba

Malvaceae
Cienfuegosia
Gossypium
Hibiscus
Malvastrum
Sida
Sidastrum
Thespesia
Urena

Sterculiaceae
Sterculia

Asia and Australia

Bombacaceae
Ceiba

Malvaceae
Abutilon
Hibiscus
Gossypium

Gramineae
Pennisetum
Sorghum

Natural Enemies

Pearson and Maxwell Darling (1958) includes the following species which are all African except *A. peruviana* and *H. chilensis* from Peru. A comprehensive list of recorded natural enemies of *Dysdercus* in Africa is given by Pearson and Maxwell Darling (1958).

Parasitoids

Sarcophagidae	*Sarcophaga dispersa* Villen. *S. dysderci* Villen.
Tachinidae	*Actia comitata* Villen. *Alophora multisetosa* Villen. *A. nasalis* (Bezzi) *Bogosia helva* Wied. *B. minor* (Villen.) *Bogosiella fasciata* (F.) *Acaulona peruviana* Tns. *Hyalomyia chilensis* Macq.

Van Doesburg (1968) lists a number of New World parasitoids in addition to the above:

Tachinidae	*Acaulona brasiliana* Townsend *Alophoropsis brasiliensis* Townsend *Euomogenia dysderci* Townsend *Euophorantha dysderci* Townsend *Eutrichopoda abdominalis* Townsend *Hyalomyodes brasiliensis* Townsend *Paraphasiana dysderci* Townsend *Paraphorantha brasiliana* Townsend *P. dimidiata* Townsend *P. politana* Townsend *Phoranthella mendesi* Townsend

Predators

Reduviidae	*Cosmolestus pictus* (Klug) *Phonoctonus fasciatus* (P. de B.) *P. lutescens* (G. & P.) *P. nigrofasciatus* Stål *P. poultoni* Schout. *P. picturatus* Fairm. *P. principalis* Gerst. *P. subimpictus* Stål *Rhinocoris bicolor* (F.) *R. segmentarius* Germ. *R. tibialis* (Stål)

Pearson and Maxwell Darling (1958) records *Bogosiella fasciata* (=*pomeroyi* Villen.) and *Alophora nasalis* (=*nigeriensis* Villen.) as the most common with distribution through tropical and southern Africa. Single eggs are laid externally on the adults or fifth instar nymphs. On hatching, the larva penetrates into the body tissues where it feeds and eventually leaves through the anal aperture to pupate. Parasitized fifth instar hosts may moult successfully to adults and parasitized females may lay eggs although in reduced numbers.

On wild host plants in the moister areas of Uganda 40–100% parasitism by *Bogosiella* of *D. nigrofasciatus* and *D. superstitiosus* has been recorded. In the drier areas, parasitism decreases to as low as 2%. Higher parasitism is recorded also for *A. nasalis* and *Bogosia epineura* associated with natural host plants. It therefore appears that the known parasitoids may exert some degree of control of stainers on wild hosts, but are unlikely to control them significantly in the cotton crop.

The Reduviid predators of stainers (Plate IV.7) tend to resemble their prey in appearance. They are never sufficiently abundant in cotton to exert meaningful control on *Dysdercus*. On the natural hosts, especially the trees *Adansonia, Ceiba* and *Sterculia* they may become more abundant, especially in the presence of other Pyrrhocorid bugs like *Odontopus* on which *Phonoctonus* feeds. The Reduviids may contribute indirectly to the control of stainers.

Domestic fowls and guinea fowl feed on stainers and Hill (1975) states that stainers may be controlled by caging chickens in cotton plots provided that they are not treated chemically.

Status as a Pest

In situations where natural hosts abound or harvested crop plants are not destroyed as prescribed or where insecticides are not applied, due to economic reasons, stainers have the potential to cause tremendous crop losses.

Where cotton is grown as an annual crop, a closed season is observed and insecticides are used against other pests, stainers do not appear to be important. In spraying trials the proportion of stained cotton and damaged seed is significantly greater on the untreated plots (Matthews *et al.*, 1968).

Few records of resistance to insecticides under field conditions appear in current scientific literature although stainers do react to selection under laboratory conditions (Nyamasyo and Karel, 1982).

Control Measures

Current trends towards integrated pest management require that all available methods be used in order to reduce pest numbers to levels where economic damage does not occur. In the case of cotton stainers, cultural and chemical control methods are often combined to achieve the desired result.

Planting date

Cotton is grown as an annual crop in most countries

Early sowing, as soon as the rains or temperatures allow, will often enable the bolls to mature before significant stainer invasions occur. Natural hosts offer sufficient food, especially in the monsoon savannah regions, to sustain stainer populations for a while and migration to cotton may occur late enough to minimize damage to developing bolls.

Crop destruction and a cotton-free period

Since cotton is perennial it will produce a crop annually for many seasons without replanting. The practice of cutting the main stem back to about 20 cm above soil level is called ratooning.

If the crop is not pruned back it is called stand-over cotton. The practice of ratooning or leaving the crop to regrow has been tried in most countries and has subsequently been abandoned, due to an increase in pests which eventually become uncontrollable (Niles and Greeff, 1989).

In the case of *Dysdercus*, stand-over or ratoon cotton bridges the gap between the old and the new plantings at a time when stainers are able to multiply in the presence of adequate food and rising temperatures. This extra generation at the start of the season allows stainer populations to attain huge numbers to the detriment of the new crop (Rainey and Smit, 1950).

Observing the proper crop destruction dates and techniques forces stainer populations to depend only on natural host plants, which may be scarce and unsuitable at the crucial time. Also of vital importance is the need to clean residues of seed from the ginneries, where large populations of stainers can survive.

Chemical control

Stainers appear to be susceptible to the insecticides generally used on cotton. Laboratory screening trials by Matthews (1966) showed that stainers were killed by contact with chemicals in the carbamate and organochlorine groups. Trial work by Brettell and Madende (1977) in Zimbabwe indicated that effective control of stainers could be achieved by application of the pyrethroid cypermethrin, widely used for bollworm control, as well as dimethoate and carbaryl. Singh (1979) demonstrated the residual effectiveness of propoxur against *D. cingulatus*. It is possible that immigrant stainers might be attracted to a bait impregnated with insecticide, but this has not been studied. Stadler and Schang (1989a,b) have reported on the effect of cyfluthrin on *D. albofasciatus* in Argentina and side effects on a natural enemy *Acaulona brasiliana*.

Resistance to insecticides has been demonstrated in *Dysdercus cardinalis, D. fasciatus, D. nigrofasciatus* and *D. superstitiosus* by Nyamasyo and Karel (1982). They found that field strains of all four species showed slight resistance to lindane and methidathion. Laboratory and field strains of *D. fasciatus* readily developed resistance to carbaryl after selection and a laboratory strain of the same species developed resistance to lindane accompanied by cross-resistance to carbaryl. No significant increase in resistance to carbaryl was observed in *D. nigrofasciatus* after selection.

Chemosterilization of *Dysdercus* has been considered (Campion and Critchley, 1971), but is not a practical control tactic. Resistance to the chemosterilant HMAC occurred in *D. cingulatus* after surviving adults of four successive generations were treated at 1% concentration that initially induced 91% sterility (Ahmad, 1979b). The percentage sterility in the F_4 generation had decreased to 43%.

If spray applications specifically to control stainers are considered, they should be applied in response to the presence of damaging populations recorded during accurate and regular scouting of the fields (Matthews and Tunstall, 1968). *Dysdercus* is a mobile pest in the crop and the possibility exists that a relatively low number of immigrant adults, especially if carrying the spores of the fungus *Ashbya*, can cause considerable damage. Scouting methods and economic thresholds require careful examination bearing in mind that stainers tend to hide during the heat of the day and are thus not easily observed.

Attempts have been made to control populations on the arboreal hosts by spraying a persistent insecticide around the base of trees, such as baobab trees, where the stainers feed on fallen seeds. Attempts to remove large numbers of trees from an area were also unsuccessful due to the flight range of the adults. Neither of these ideas is considered to be economic or environmentally acceptable due to the large numbers of alternative host plants.

Biological control

Various natural enemies attack stainers (Banerjee and Datta, 1980; Fadare, 1980; Fagoonee and Umrit, 1980; De Almeida *et al.*, 1981; Lamas, 1981; Cardona *et al.*, 1982; Zaidi, 1985; Schaefer and Ahmad, 1987), but proof is lacking that they are capable of causing significant mortality among stainers in the cotton. The most common Tachinids, *Bogosiella* and *Alophora* are sufficiently well adapted to allow limited survival and even egg-laying by their hosts.

Both the parasitoids and predators apparently exert their main pressure on stainer populations feeding on natural hosts. Sweeney (1961) considered that *Phonoctonus* was an important predator on arboreal hosts, but it tended to migrate to cotton long after the stainers had done so. Despite high percentages of stainers destroyed at times, the role of natural enemies appears to be insignificant in controlling stainer numbers on cotton.

Other Heteroptera

Calidea spp. (Heteroptera: Pentatomidae)

Over thirty species of shield bugs have been recorded on cotton, with virtually all attacking green bolls, but most are of no importance. However, species of two genera, *Calidea* and *Nezara*, interfere with the maturation of the bolls by feeding on the developing seeds. They are also implicated in the transmission of the fungus *Ashbya gossypii*, the cause of internal boll disease mainly transmitted by *Dysdercus* spp. Invasions tend to be very transitory, with large populations of up to 100,000 adults per hectare arriving and then disappearing as suddenly for no apparent reason, and subsequently green bolls are found to be heavily punctured and stained, with no insect agent then visible. They will also attack sorghum and sunflowers and have appeared on cotton mainly in areas associated with *Brachystegia-Isoberlinia* or similar woodland, especially in Tanzania and Zambia.

Adults of the genus *Calidea* are strikingly coloured (Plate IV.8) the undersurface being usually orange or reddish and the upper surface a brilliant iridescent blue or green, often with a bronze tinge, with a bold pattern or dark spots and stripes. The scutellum is greatly extended, so that it covers the whole dorsal surface of the abdomen and, together with the thorax, is strongly convex, the insect thus having a distinctly obovate form. The size ranges from about 8.5–17 mm long × 4–8.5 mm broad, according to the species.

The genus is restricted to the Ethiopian region, including Madagascar and Arabia. The main species is *Calidea dregii* Germ., which has been recorded in Tanzania, Zambia, Malawi, Mozambique, Cote d'Ivoire, Zaire and South Africa. Other species recorded on cotton are: *C. bohemani* (Stal) in Zaire, Mozambique and Tanzania; *C. duodecimpunctata* (F.) and *C. natalensis* in the Sudan; and *C. humeralis* in Malawi.

The eggs of *C. dregii* are spherical, about 1 mm in diameter and laid in batches of up to 40 arranged in a closed spiral round a stalk, or dried leaf. They are white at first but turn red before hatching. The nymphs are oval, at first somewhat flattened, and orange or reddish in colour, with a pattern of dark, metallic blue, which becomes more extensive as the nymph develops (Duerden and Evans, 1954). The period from egg to adult can take eight weeks in Tanzania, but *C. duodecimpunctata* in the southern Sudan took much less (23–28 days) (Pollard, 1954). Adults can enter diapause to survive dry periods. The genus is polyphagous.

Insecticide sprays against *Helicoverpa* have apparently been effective, but early reports suggested that *Calidea* was not very susceptible to chemical control (Duerden and Evans, 1954).

Nezara viridula (L.) (Heteroptera: Pentatomidae)

The only member of the genus *Nezara* that is of importance on cotton is *N. viridula* (L.), which is distributed throughout Africa, southern Europe, most of tropical and subtropical Asia, the West Indies and southeastern United States, restricted parts of Central and South America, and parts of Australia, New Zealand and several oceanic islands. This polyphagous bug has a host range of over 30 families of dicotyledonous plants and some monocotyledonous plants (Todd, 1989), and is essentially a pest of fruit and vegetable crops. It is of some importance on cotton in the West Indies, where it can transmit *Ashbya coryli*, but is normally a minor pest of cotton in Africa. *Nezara robusta* Dist. is a yellowish species with dark green longitudinal markings recorded on cotton in Malawi. In Egypt it has been shown to transmit the fungus *Rhizopus nigricans* which can cause rotting of bolls.

The eggs are barrel shaped, 1.2 mm long × 0.75 mm diameter, yellowish-white when laid and developing pink marks prior to hatching. There are five nymphal instars; the first stage is golden brown with mottlings that become darker; the next two stages are dark brown to black, with yellow margins to the thorax and two rows of white dots on each side of the upper surface of the abdomen, one marginal and one median; in the later stages, the head, thorax, wing pads and ground colour of the abdomen turn green, the edges of the abdomen become pinkish and the legs and antennae pale brown. The adults, about 15 mm long × 8 mm broad, occur in three colour forms. One form, var. *smaragdula*, is a uniform apple green above and a rather lighter shade below, while another form, var *torquata*, is similar, except that the head in front of the eyes and the anterior margin of the pronotum are yellow. A rarer form has a straw-yellow dorsal surface decorated with a number of green dots. In all forms the green may be replaced by a reddish-brown.

A scelionid egg-parasite *Microphanurus basalis* (Woll.), is believed to control *N. viridula* effectively in many areas. A Tachinid, *Trichopoda pennipes* (F.) is effective against nymphs in the West Indies and in the USA.

Oxycarenus hyalinipennis (Costa) (Heteroptera: Lygaeidae)

While some Lygaeids such as *Geocoris* are predatory, the genus *Oxycarenus* is particularly associated with ripe seeds of cotton and other Malvaceous plants, the various species being known collectively as cotton seed bugs. *Oxycarenus hyalinpennis* (Costa) is recorded throughout continental Africa, except Algeria and Morocco where *O. lavaterae* occurs and extends eastwards through Egypt, the Middle East, to India, Indo-China and the Philippines. It has also been recorded in Brazil, where it was presumably accidentally introduced. Several other species have been recorded in Africa, but the genus is absent from North America and the West Indies.

The adults are rather small, elongate bugs with somewhat pointed heads, about 4 mm long and 1.5 mm broad, dull black or very dark brown except for the abdomen, which may have a reddish tinge and the hemelytra, which have a translucent, dusky or slightly pearly appearance. The egg is creamy, oval, about 1 mm × 1.25 mm, longitudinally striated and with six projections at the anterior end. The nymphs resemble the adults, but lack wings and ocelli, and are lighter and browner in tint. All stages are characterized by a powerful smell when they are crushed. They feed on the seeds and in large numbers can reduce the weight of seeds as well as their viability. The bugs can discolour the lint if crushed, and so are sometimes referred to as dusky cotton stainers, although they are not involved in the transmission of any internal boll disease. Any damage is minimized by early harvesting. A detailed account of *O. hyalipennis* in Egypt was given by Kirkpatrick (1923).

References

Ahmad, I. (1979a) Effects of different plant food on fecundity, fertility and development of *Dysdercus cingulatus* (Fabr.). *Indian Journal of Zoology* 7, 13–15.

Ahmad, I. (1979b) Studies on the development of resistance to HMAC: 1, 6-hexamethylene bis (1-agiridine carboxamide) in *Dysdercus cingulatus* (Fabr.). *Journal of Pesticide Science* 4, 515–516.

Banerjee, P. and Datta, S. (1980) Biological control of red cotton bug, *Dysdercus koenigii* Fabricius by (a) mite *Hemipteroseius indicus* (Krantz & Khot). *Indian Journal of Entomology* 42, 265–267.

Brettell, J.H. and Madende, M. (1977) Stainer control trial. In: Annual Report of the Cotton Research Institute for 1977/78. Gatooma, Rhodesia.

Campion, D.G. and Critchley, B.R. (1971) Interference with growth and reproduction as a means of controlling insect pests of cotton. In: *Proceedings of the Cotton Pest Control Conference*, March 1971, Malawi, pp. 239–246.

Cardona, C., Pacheco, L.C. and Rendon, F. (1982) Populations of pest and beneficial insects on ratoon cotton on the Atlantic Coast. *Revista Colombiana de Entomologia* 5, 3–12.

Couilloud, R. (1989) Heteroptera pests on cotton in Africa and in Madagascar. *Coton et Fibres Tropicales* 44, 185–227.

De Almeida, J.R., Mizuguchi, Y., De Xerez, R. and Da Silva, G.M. (1981) Parasitism in cotton stainers, *Dysdercus* spp. (Hemiptera: Pyrrhocoridae). *Revista Brasileira de Entomologia* 25, 55–60.

Derr, J.A. (1981) Light-trap catches of populations of *Dysdercus bimaculatus* Stål (Heteroptera: Pyrrhocoridae) in relation to weather and the fruiting cycle of its host-plants. *Bulletin of Entomological Research* 71, 47–56.

Duerden, J.C. and Evans, A.C. (1954) A note on the biology of *Calidea dregei* germ. *East African Agriculture Journal* 19, 188–192.

Egwuatu, R.I. (1980) The role of food plants in the development and survival of *Dysdercus superstitiosus* F. (Hemiptera: Pyrrhocoridae). *Revue de Zoologie Africaine* 94, 780–790.

Fadare, T.A. (1980) Efficiency of *Phonoctonus* spp. (Hemiptera: Reduviidae) as regulators of populations of *Dysdercus* spp. (Hemiptera: Pyrrhocoridae). *Nigerian Journal of Entomology* 3, 45–48.

Fagoonee, I. and Umrit, G. (1980) Biology of *Dysdercus flavidus* Sign. and its control by *Ageratum conyzoides*. *Revue Agricole et Sucriecre de l'Ile Maurice* 59, 122–128.

Freeman, P. (1947) A revision of the genus *Dysdercus* Boisduval (Hemiptera: Pyrrhocoridae), excluding the American species. *Transactions of the Royal Entomological Society, London* 98, 373–424.

Harris, P.J.C. (1983) Effect of cotton stainers (*Dysdercus* spp.) on *Urena lobata* seed quality. *Tenth International Congress of Plant Protection 1983*. British Crop Protection Council, Croydon, vol. 1, p. 107.

Hill, D.S. (1975) *Agricultural Insect Pests of the Tropics and their Control*. Cambridge University Press, Cambridge, xi + 516pp.

Kirkpatrick, T.W. (1923) The Egyptian cotton seed bug (*Oxycarenus hyalinipennis* Costa): its bionomics, damage and suggestions for remedial measures. *Bulletin of Ministry of Agriculture Egypt* 35, 107pp.

Lamas, C.J.M. (1981) Control of insect pests of cotton crops in Peru. Outline of the planning of a campaign of integrated control and the problems involved. *Revista Peruana de Entomologia* 23, 1–6.

Leston, D. (1980) The natural history of some West African insects. Part 16. The identity and distribution of cotton stainers, *Dysdercus* spp. (Hemiptera: Pyrrhocoridae) in Ghana. *Entomologist's Monthly Magazine* 115, 113–116.

Matthews, G.A. (1966) Investigations of the chemical control of insect pests of cotton in Central Africa Part II. *Bulletin of Entomological Research* 57, 77–91.

Matthews, G.A. and Tunstall, J.P. (1968) Scouting for pests and the timing of spray applications. *Cotton Growing Review* 45, 115–127.

Matthews, G.A., McKinley, D.J. and Tunstall, J.P. (1968) Seed and lint quality of sprayed and unsprayed cotton in Malawi. *Cotton Growing Review* 45, 27–35.

Niles, G.A. and Greeff, M.S. (1989) *Ratoon Cotton in Perspective.* Technical Communication No. 217, Department of Agriculture and Water Supply, Republic of South Africa.

Nyamasyo, G.H.N. and Karel, A.K. (1982) Studies on insecticide resistance in cotton stainers, *Dysdercus* spp. (Hemiptera: Pyrrhocoridae), in Kenya. *Bulletin of Entomological Research* 72, 461–465.

Parsons, F.S. (1928) Progress Report of the Experiment Stations of the Empire Cotton Growing Corporation 1926–27, p. 66.

Pearson, E.O. and Maxwell Darling, R.C. (1958) *The Insect Pests of Cotton in Tropical Africa.* Commonwealth Institute of Entomology, London, pp. 256–297.

Pollard, D.G. (1954) A note on the biology of *Calidea duodeciumpunctata* (Fabr.) in the Sudan. *Bulletin Society Fouad Ier Entomology* 38, 291–295.

Pridham, T.G. and Raper, K.B. (1950) *Ashbya gossypii* – its significance in nature and in the laboratory. *Mycologia* 42, 603–623.

Rainey, R.C. and Smit, B. (1950) Cotton and its pests in South Africa. Ratooning as a threat to the revival of cotton growing in the Union. *Scientific Bulletin of the Department of Agriculture of South Africa* No. 308, 17pp.

Schaefer, C.W. and Ahmad, I. (1987) Parasites and predators of Pyrrhocoroidea (Hemiptera), and possible control of cotton stainers by *Phonoctonus* spp. (Hemiptera: Reduviidae). *Entomophaga* 32, 269–275.

Singh, D.R. (1979) Residual effectiveness of Baygon (propoxur) against *Dysdercus cingulatus* Fabr. (Hemiptera: Pyrrhocoridae). *Zeitschrift für Angewändte Zoologie* 66, 139–142.

Stadler, T. and Schang, M.M. (1989a) Comparative toxicities of four insecticides and the diasteroisomers of cyfluthrin to the cotton stainer *Dysdercus albofasciatus* (Berg, 1878) (Heteroptera: Pyrrhocoridae). *German Journal of Applied Zoology* 76, 349–356.

Stadler, T. and Schang, M.M. (1989b) Side effects of two pesticides on a natural enemy of the cotton stainer *Dysdercus albofasciatus* (Berg, 1878) correlated to the biology and behaviour of both species. A Model for Cotton-IPM in Argentina. *Proc. Int. DLG-Symp.*, pp. 783–787.

Stahle, P.P. (1980) The immature stages of the harlequin bug, *Dindymus versicolor* (Herrich-Schaeffer) (Hemiptera: Pyrrhocoridae). *Journal of the Australian Entomological Society* 19, 271–276.

Sweeney, R.C.H. (1961) *Insect Pests of Cotton in Nyasaland I. Hemiptera (Bugs).* Bulletin No. 18, Department of Agriculture, Zomba, Nyasaland.

Taylor, D.E. (1982) The cotton stainer. *Zimbabwe Agricultural Journal* 79(6), 203–204.

Todd, J.W. (1989) Ecology and behavior of *Nezara viridula. Annual Review of Entomology* 34, 273–292.

Van Doesburg, P.H. (1968) A revision of the New World species of *Dysdercus* Guérin Méneville (Heteroptera, Pyrrhocoridae). *Zoologische Verhandelingen* 97, 215 (16 plates).

Wadnerkar, D.W., Gaikawad, B.B. and Thombre, U.T. (1979) New record of an alternate host plant of red cotton bug, *Dysdercus koenigii* (Fab.). *Indian Journal of Entomology* 41, 185.

Zaidi, Z.S. (1985) Proteolytic activity in the gut of the red cotton bug, *Dysdercus cingulatus* Fabr. and its predator, *Antilochus cocquebertii* (Fabr.) (Heteroptera: Pyrrhocoridae). *Current Science*, India 54, 252–253.

12 Aphids (Hemiptera: Aphididae)

F. LECLANT[1] AND J.P. DEGUINE[2]

[1]*Ecole Nationale Supérieure Agronomique, 2 Place Pierre Viala, 34060 Montpellier Cedex 1, France;* [2]*Centre Coopération Internationale en Recherche Agronomique pour le Développement, Institut de la Recherche Agronomique, BP 22, Mároua, Cameroon*

Only two of the aphid species on cotton mentioned in the literature (Table 12.1) are of real importance in agriculture. These are *Acyrthosiphon gossypii* Mordvilko and especially *Aphis gossypii* Glover.

ACYRTHOSIPHON GOSSYPII MORDVILKO, 1914

Described on the basis of specimens collected on cotton in the Amou Daria delta in western Asia, this species is easily distinguished by its large size (apterae 2.5 to 3.8 mm, alatae 2.1 to 3.5 mm) and its long, slender cornicles (3 times as long as the cauda and 0.4 to 0.5 times body length in apterous virginipara compared to 0.3 to 0.4 times in the alatae). *Acyrthosiphon gossypii* is green and covered with whitish wax giving both larvae and adults a greyish appearance. This character enables easy distinction from *Acyrthosiphon pisum* Harris when the two species occur simultaneously on legumes. Only the larvae of *A. pisum* display wax, and cornicles are not more than a quarter of body length.

The species is widespread from India and Central Asia to the Mediterranean

Table 12.1. Species of aphids found on cotton.

Species
Acyrthosiphon gossypii Mordvilko
Aphis craccivora Koch
Aphis fabae Scopoli
Aphis gossypii Glover
Aphis maidiradicis Forbes
Macrosiphum euphorbiae Thomas
Myzus persicae Sulzer
Smynthurodes betae Westwood

and the Atlantic through southern Ukraine and the Middle East (Iran, Iraq, Turkey and Israel). It is also found in North Africa (Algeria, Tunisia, Morocco and Egypt) and sub-Saharan Africa (Sudan, Kenya, Cameroon, Burkina Faso). It is observed in Sicily.

Acyrthosiphon gossypii is considered as a serious cotton pest in Central Asia (Shuravleva, 1956; Narzikulov and Umarov, 1969). Major attacks cause a decrease in plant sugar contents and premature fruit shedding. The insect is found on cotton during most of the season in Anatolia where, together with *Thrips tabaci* Lindeman and *Aphis gossypii*, it is one of the main pests at the start of the cycle. In contrast, in the Sudan there is no infestation by this species which may however pullulate on *Vicia faba* where it can be confused with *Acyrthosiphon pisum*.

According to Nevsky (1929), the most important host plants for *A. gossypii* are legumes (*Dolichos*, *Phaseolus*, *Vigna*, *Medicago sativa*, *Pisum sativum*, *Trifolium alexandrinum* and *Alhagi camelorum*) and also Malvaceae. The same author reported outbreaks fairly late in the season on stems and leaf undersides of *Gossypium hirsutum* and *G. herbaceum*, but considered these plant species to be occasional hosts. In contrast, legumes – especially perennial species – are the hosts of the bisexual generation and enable the species to overwinter. This is the case for *Alhagi camelorum* on which *A. gossypii* displays sexual forms in Central Asia. *Acyrthosiphon gossypii* is in fact a typical species on legumes and is found on over thirty species of plants in steppe regions including – in addition to those already mentioned – Brassicaceae, Convolvulaceae, Asteraceae and Rosaceae (Shuravleva, 1956) when legumes are not available (or receptive). This behaviour of *A. gossypii* should be compared to that of *Brachyunguis harmalae* Das, a typically monoecious holocyclic species (with wingless males) on *Pegamum harmala*. The species is found, in the middle of the dry season in steppe regions of Mali and Senegal, on *Calotropis procera* (Asclepiadaceae), *Anacardium* (closely related to Terebinthaceae) and even *Citrus*.

Acyrthosiphon sesbaniae, described on a Papilionnaceae by Kanaraj (1956), is thought to be a distinct species of *A. gossypii* by Daoud and El-Haidari (1968). However, Szelegiewicz (1963), Eastop (1971) and Blackman and Eastop (1984) consider them to be synonyms, together with *A. skrjabani* Mordvilko, which lives on *Malva neglecta* in Kirghizia.

The results of transfer experiments performed by both Kanaraj (1956) and Müller and El Tigani (1986), and the contradictory observations reported by various authors concerning the *A. gossypii* host range according to region, might lead us to think that there are several biological races which differ in their particular food preferences.

Some authors consider that the Sudan race is a distinct subspecies, although no morphological differences have been detected in material collected on *Vicia faba* and *Pisum sativum* in the Sudan and specimens of homologous generations and morphs collected on cotton (*Gossypium hirsutum*) at Tashkent in Central Asia (Müller, 1975; Müller and El Tigani, 1986). This is contested by other authors (Remaudière, personal communications) and considered as absurd.

Table 12.2. Synonyms of *Aphis gossypii* Glover, 1877 (from Eastop and Hille Ris Lambers, 1976).*

gossypii Glover, 1877 is a member of a group of interfertile species and subspecies not yet fully understood but including:
capsellae Kaltenbach, 1843
frangulae Kaltenbach, 1845
  ? mamontovae Davletshina, 1964
  rhamni Kaltenbach, 1843 nec Boyer de Fonscolombe, 1841
frangulae subsp. beccabungae Koch, 1855
frangulae subsp. testacea Thomas, 1968
gossypii Glover, 1877
  affinis var. *gardeniae* del Guercio, 1913
  aurantii var. *limonii* (del Guercio, 1917)
  bauhiniae Theobald, 1918
  bryophyllae Shinji, 1922
  chloroides Nevsky, 1929
  circezandis Fitch, 1870
  ? citri Ashmead, 1887
  citrulli Ashmead, 1882
  colocasiae Matsumura, 1917
  commelinae Shinji, 1922
  commelinae Shinji, 1924
? convolvulicola Ferrari, 1872
  cucumeris Forbes, 1882
  cucurbiti Buckton, 1879
  ? ficus Theobald, 1918
  flava Nevsky, 1929
  gossypii var. *callicarpae* Takahashi, 1921
  gossypii var. *viridula* Nevsky, 1929
  hederella Theobald, 1915
  helianthi del Guercio, 1916
  ? heliotropii Macchiati, 1885
  hibiscifoliae Shinji, 1922
  inugomae Shinji, 1922
  leonuri Takahashi, 1921
  ligustriella Theobald, 1914
  ? lilicola Williams, 1911
  malvacearum van der Goot ex Das, 1918
  malvoides Das, 1918 nec van der Goot, 1917
  ? minuta Wilson, 1911
  monardae Oestlund, 1887
  ? oxalis Macchiati, 1884
  parvus Theobald, 1915
  ? perillae Shinji, 1922
  pomonella Theobald, 1916
  pruniella Theobald, 1918
  shirakii Takahashi, 1921
  solanina Passerini, 1863
  tectonae van der Goot, 1917
  tridacis Theobald, 1929
  vitifoliae Shinji, 1922
A. gossypii var. callicarpae Takahashi, 1921 = gossypii Glover, 1877
gossypii var. lutea Nevsky, 1929
gossypii var. obscura Nevsky, 1929
gossypii var. viridula Nevsky, 1929 = gossypii Glover, 1877

* Synonyms are in italic; valid names are in roman.

Aphis gossypii Glover, 1877

Introduction

Aphis gossypii is a cosmopolitan species widely distributed in tropical, subtropical and warm temperate regions. The aphid has now colonized all zones with mild winters and remains in regions with colder winters thanks to greenhouses, in which it is a major pest. Ability to overwinter in the egg form has also been reported by several authors. *Aphis gossypii* is thus present in all cotton-growing areas of the world and in temperate zones, principally as a pest on vegetable crops.

It was long considered as a minor pest on cotton and only when plant growth conditions were unfavourable, but now appears to cause serious problems in many production regions.

Description – Diagnosis

This aphid is extremely variable in colour (Wall, 1933) and colonies are almost always made up of individuals of different colours: dirty green, greyish green to dark green and sometimes very blackish brown (Pergande, 1895; Patch, 1925). Yellowish, brown with green marbling, orangey yellow, lemon yellow, dirty yellow and very small whitish specimens have also been observed. Size also varies considerably (apterae 0.9 to 1.8 mm, alatae 1.1 to 1.8 mm). This explains why the species has been described under different names. Some forty synonyms have been counted (Eastop and Hille Ris Lambers, 1976) (Table 12.2).

Apterae (Plate V.1) The antennae generally have six segments (five in small specimens) and are pale except for basal segments I and II and the two terminal segments. Tibias are pale and darker brown at the extremity. Tarsi are dark. Cornicles are black and twice as long as the paler cauda which has five to seven hairs. According to Wall (1933), the size of specimens is well correlated with colour. Pale, yellowish and whitish specimens are generally small (< 1 mm) with short appendices: cornicles are pale, brown only at the tips and in this case not longer than 1.5 times that of the cauda, which is also pale. Such individuals are observed under hot conditions in large populations.

The dorsal cuticle of prepared specimens (cleared and mounted) is membranous and not pigmented. Only the head is dusky. Cornicles are dark and always more so than the cauda. Antenna length is slightly greater than half the body length. The terminal process of the last article is two to three times longer than the base. Cornicles are conical in the basal half and become distinctly cylindrical towards the apex.

Alatae (Plate V.2) Head and thorax are dusky, as are the antennae. The latter always consist of six segments, enabling distinction between *A. gossypii* and *A. kachkouli* Remaudière. Cornicles are black and 1.5 to twice as long as the smoky ochraceous cauda; the latter is black in large specimens. The abdomen bears

Table 12.3. Complex formed by *frangulae* and subspecies: biological and colour characteristics (after Müller, 1986).

Correct name (taxon)	Generation sequence	Dominant colour in summer	Gynoparae	Males	Host suitability: Potato	Host suitability: Cucumber	Host suitability: Cotton
Aphis frangulae Kaltenbach, 1845 s.str.	Facultative host alternation on *Frangula alnus*	Dark bluish green to nearly black	Alate	Alate	None	None	None
A. frangulae beccabungae Koch, 1855		Brownish yellow to greenish yellow	Alate	Alate	Good	None	None
A. frangulae testacea Thomas, 1968	Mainly monoecious on *Frangula alnus*	Yellowish brown	Alate	Alate	None	None	None
A. frangulae capsellae Kaltenbach, 1843	Without host alternation refuses *Frangula alnus*	Green to rather dark bluish green	Alate	Alate	Good	None	None
A. frangulae gossypii Glover, 1877*	Anholocyclic	Very variable: whitish, light yellow to very dark green	n.a	n.a	Good	Maximum	Maximum
	Holocyclic (monoecious or heteroecious)		Alate	Alate	Variable	Variable	Variable

* Some clones originated from cotton do not establish on Cucurbitaceous plants and vice versa.
n.a., Not applicable.

well-developed marginal and post-siphuncular sclerites, a well-pigmented narrow transverse stripe on segment VIII and small scattered sclerites consisting of a few pigmented cells. Antennae are two thirds to three quarters of body length with a terminal process 2.5 to three times as long as its base; segment III bears three to nine secondary sensoria in a row along the whole length of the segment.

In addition to morphological variability, studied in particular by Batchelder (1927), Wall (1933) and Behura and Acharya (1983), *A. gossypii* displays biological variability, causing difficulties for taxonomists.

Taxonomic Status

The least that can be said concerning *Aphis gossypii* at the world scale is that its taxonomic status is particularly complex, confused and problematic. Interpretation of the various biological data in the literature is hence very difficult.

For European authors, the 'true' *Aphis gossypii* Glover is a polyphagous, anholocyclic species closely related to several European species (*A. capsellae* Kaltenbach, *A. beccabungae* Koch) in the *frangulae* Kaltenbach group and considered to be a subspecies of the latter. In contrast, *A. frangulae* s.str., *A. beccabungae* and *A. capsellae* are heteroecious holocyclic species which accept *Frangula alnus* as primary host, indicating palearctic origin. *Aphis testacea* Thomas, another subspecies of the group, is monoecious holocyclic on *F. alnus* in Germany (Thomas, 1968; Stroyan, 1984; Müller, 1986). Table 12.3 can be used to separate these five entities according to biological criteria.

However, in America, Japan, China and Korea, *Aphis gossypii* is considered as a completely holocyclic heteroecious species (Eguchi, 1937; Moritsu, 1948; Shibata, 1955; Kring, 1955, 1959; Sorin, 1975; Inaizumi, 1980, 1981; Takada, 1988). In Connecticut for example, Kring (1959), ruling out the confusion made by Patch (1925) with *A. sedi* Kaltenbach in Maine, reported holocyclic development in what he termed *A. gossypii*, with wintering as eggs on *Catalpa bignonioides* and *Hibiscus syriacus*. According to Stroyan (1984), this species may be what Mamontova (1953) in the former USSR described under the name of *A. catalpae* and which differs from *gossypii* by the presence of two to four secondary sensoria on article IV of the antenna on the alatae.

Although the species is generally anholocyclic in the southern United States, very small numbers of oviparous females have been observed on cotton regrowth at the end of November in Mississippi and Louisiana (O'Brien *et al.*, 1990, 1991), thus confirming *A. gossypii*'s ability to overwinter in the sexual form on cotton in the southern states, as is known for more northern latitudes (Sartor *et al.*, 1976).

In Japan, Inaizumi (1981) drew a distinction between four *Aphis gossypii* biotypes:

- two heteroecious holocyclic types:
  one with *Hibiscus syriacus* as primary host and various herbaceous plants, including potato, as secondary hosts;

the other whose bisexual generation is on *Rhamnus* and *Celestria* (as primary host) and whose summer generations accept numerous and very varied plants;

- a monoecious anholocyclic form with apterous males found throughout the year on *Rubia*;
- finally, an anholocyclic form with continuous development of parthenogenetic females on potato and various other plants without sexupara and sexuals in the autumn.

Takada (1988) confirmed Inaizumi's observations and added an androcyclic form, as is known in *Myzus persicae* Sulzer (Blackman, 1971, 1972), whereas Komazaki (1982) reported possible overwintering as eggs on *Citrus*.

Zhang and Zhong (1990) showed that three major categories of cycle could be observed in China. Basing their position on a great number of experiments on the transfer of different clones kept on different host plants and under different microclimatic conditions, they questioned Mordvilko's (1928) theory according to which, occurrence of anholocycly is probably mainly caused by macroclimate. They considered that prolonged maintenance on the same host-plants under microclimatic conditions, rather than the paleoclimate, enabled parthenogenetic development leading to a change in the biological cycle and enabling anholocycly. Zhang and Zhong thus listed:

- an anholocyclic form mainly using Cucurbitaceae as host-plants. The form has two variants:

  one can only develop on Cucurbitaceae (*Cucumis sativus* and *Cucurbita pepo*) and does not survive on cotton. It consists mainly of apterous specimens, most of which are green;

  the other consists mainly of yellow specimens; it develops on Cucurbitaceae but also on cotton – with a certain degree of difficulty. Change to cotton is easier for populations from *C. pepo*.
- a heteroecious holocyclic form with *Zanthoxylum simulans* (Rutaceae), *Rhamnus* spp. and *Punica granatum* as primary hosts. Cotton, *C. pepo* and a few other plants are secondary hosts.
- finally, a monoecious holocyclic form developing on *Hibiscus syriacus*, *Gossypium*, etc. Parthenogenetic reproduction and the bisexual generation are on the same hosts.

Zhang and Zhong consider that *Rhamnus*, whose spring growth is later than that of *Zanthoxylum*, is probably a secondary host for *A. gossypii* and must have gradually achieved primary host function during evolution. The same authors also showed that interspecific crosses between *Aphis glycines* Matsumura and *A. gossypii* could easily occur in the laboratory and also outdoors, but nevertheless with a low success rate in the latter case. *Aphis glycines* is a heteroecious holocyclic species in China and very different from *A. gossypii*. However, they share the same primary host, bushes of the genus *Rhamnus*. The secondary host is *Glycine max* (Wang *et al.*, 1962). Hybrid eggs hatch into fundatrices whatever the cross direction, but they and their progeny can only develop on the host of the oviparous female which laid the eggs. In other words, the hybrid descendants of an oviparous 'cotton' female can only survive on cotton. The same principle applies to hybrid offspring with an *A. glycines* female parent.

These contradictory observations lead to the following alternative: either different nearctic and palearctic species have been confused under the same name; or, more likely, *A. gossypii* reacquired a holocycle in North America and the Far East using the new primary hosts available (Blackman and Eastop, 1984). This is similar in some respects to the behaviour of *Therioaphis trifolii f. maculata* Buckton, which was anholocyclic when introduced in the southern United States (1953) and subsequently capable of producing viable sexual forms, and completely anholocyclic when it spread further north in the 1960s (Manglitz *et al.*, 1966; Blackman, 1986).

In short, *Aphis gossypii* is a *cosmopolitan* insect with such a *broad range of hosts* that it can be considered to be omnivorous. These two characteristics, related to the prodigious adaptability of the group, cause extraordinary variety at several levels and considerably complicate taxonomy. In the *frangulae s.l.* group, the *gossypii* subspecies itself is a mosaic of subunits which have not been thoroughly listed. In Europe, Africa and generally in the southern US states, the species displays anholocyclic behaviour with parthenogenetic reproduction by viviparous females throughout the year. However, holocyclic populations overwintering on Rhamnaceae are probably of European origin. Those overwintering on other hosts are possibly of oriental or North American origin.

Biology

Host plants

Aphis gossypii, commonly referred to as the cotton aphid or the melon aphid, has a large range of hosts, as stated above, and is practically omnivorous. Essig (1947) and Leonard *et al.* (1971) in the United States recorded it on 350 and 200 host plants respectively. Remaudière and Autrique (1985) observed it in Burundi on 83 species in 35 different families including 23 Compositae, 5 Polygonaceae and 5 Malvaceae; it attacked 64% of families and appeared in 16% of the plant/aphid combinations listed in the country (Remaudière *et al.*, 1985). Deguine (1992) collected it in Cameroon on over 100 plants of different species during the inter-season, and Eastop (1958) mentions it on some 15 families in East Africa and on some 60 species in West Africa (1961). In Asia, Roy and Behura (1983a) found it in India on 200 species belonging to 46 families (30 Solanaceae, 17 Malvaceae, 19 Cucurbitaceae and 22 Asteraceae). Finally, Higuchi and Miyazaki (1969) in Japan listed it on plants belonging to about 100 families and in Australia Cottier (1953) reported it on some 20 plant families. Over 900 hosts have thus been listed worldwide (Inaizumi, 1981).

The most frequently mentioned target crops are cotton and cucurbits (melon, courgette, gumbo) followed by *Citrus*, coffee, cocoa, pimento, eggplant, potato, okra, cowpea and peanut and numerous ornamental plants such as *Lantana*, *Hibiscus*, *Lagerstroemia*, *Bougainvillea*, *Chrysanthemum*, etc.

Although *A. gossypii* is strongly polyphagous, not all host plants have the same nutritional value. Some may cause greater fecundity than others (Table

Table 12.4. Fecundity of *Aphis gossypii* depending on host plant (from Ekukolé, 1990).

Host plant	Family	Mean fecundity in 24 h*	Amount of infestation
Okra *Abelmoschus esculentus*	Malvaceae	5.16a†	+++
Cotton *Gossypium hirsutum*	Malvaceae	4.76a	+++
Kenaf *Hibiscus cannabinus*	Malvaceae	3.30b	+++
Melon *Citrullus vulgaris*	Cucurbitaceae	2.88bc	++
Roselle *Hibiscus sabdarifa*	Malvaceae	2.66bc	++
Urena sp.	Malvaceae	2.20cd	+
Sida sp.	Malvaceae	1.44de	+
Groundnut *Arachis hypogea*	Leguminosae	0.64e	+

* Temperature 25 ± 1°C.
† Means with the same letter are not significantly different.

12.4). The existence of numerous anholocyclic clones with special associations with certain host plants has also been reported in the literature. Eastop (1958) reported in East Africa a dark green and black form on legumes and Compositae that reproduced slower than other forms when transferred to *Hibiscus* and did not produce small pale specimens. This biotype also has somewhat shorter cornicles than specimens of the same size living on other plants. Furk *et al.* (1980) gave another example: the aphids in chrysanthemum greenhouses in Great Britain do not colonize Cucurbitaceae and vice versa. In addition, these two biotypes cannot develop on cotton. The 'chrysanthemum' form also developed resistance

Table 12.5. Development time for the cotton aphid, newborn nymph to adult, on cotton seedlings (from Akey and Butler, 1989).

Temperature (°C)	No. of aphids	Days ± SEM[a]
10.0	97	24.6 ± 0.25
12.5	90	21.1 ± 0.19
15.0	70	13.8 ± 0.22
17.5	83	10.8 ± 0.14
20.0	158	8.1 ± 0.09
22.5	34	6.9 ± 0.22
25.0	165	5.7 ± 0.08
27.5[b]	79	5.0 ± 0.07
30.0	30	5.4 ± 0.09
32.5	92	5.9 ± 0.09

[a] Standard error of mean.
[b] Optimal temperature.

Table 12.6. Fecundity of the cotton aphid on cotton seedlings (from Akey and Butler, 1989).

Temperature (°C)	No. of aphids	$\bar{x}$ Nymphs/days ± SEM[a]
15.0	22	0.39 ± 0.12
17.5	15	0.50 ± 0.15
20.0	16	0.75 ± 0.24
22.5	19	1.83 ± 0.32
25.0[b]	20	2.85 ± 0.43
27.5	19	2.61 ± 0.26
30.0	24	2.71 ± 0.34
32.5	15	1.05 ± 0.20

[a] Standard error of mean.
[b] Optimal temperature.

to carbamate and organophosphorus compounds whereas the 'Cucurbitaceae' form did not display this feature. Host incompatibility between Cucurbitaceae and cotton had been reported by Paddock (1919) and by Isely (1946), and many clones collected recently on cotton by Deguine (1992) in Cameroon and Chad do not survive on melon or eggplant.

The role of abiotic factors in development

Under optimal temperature conditions (25–30°C), development lasts for some 4–6 days and average daily fecundity is 2.3–3 larvae (Tables 12.5 and 12.6). According to Liu and Perng (1987), the optimum temperature for larval growth is 27°C. At this temperature, the intrinsic growth rate is 0.541 and the final rate 1.718; a generation develops in an average of 7–9 days. A female is considered to be able to produce up to 80 larvae during the reproduction period lasting an average of some two weeks (Paddock, 1919). The average life span is 28 days. Paddock counted up to 57 generations during one cotton season. Goff and Tissot (1932) obtained 51 generations in one year on young melon plants kept in an insectarium in Florida, with temperatures varying from 25 to 29.5°C from May to September and 15 to 24°C from October to April.

Duration of development and reproduction rate reported in the literature

Table 12.7. Comparison of optimum development and fecundity from three sources.

	Isely (1946)	Komazaki (1982)	Akey and Butler (1989)
Host plant	Cotton	Citrus	Cotton
Optimum temperature of immature stage	28°C	29.7°C	27.5°C
Days for development	5.18	6.17	5
Optimum temperature for reproduction	20°C	19.8°C	25°C
Corresponding period for fecundity	2.69	1.01	2.85

nevertheless vary (Table 12.7) and the results should be considered with caution, especially when data may be used to model epidemics. Indeed, clones of different origin may react to rearing conditions in different ways because of the selection pressure exerted on the insects in their original environment. For example, optimum temperature may differ according to origin. In addition, as in other aphid species, development rate, longevity and reproduction rate may vary between apterae and alatae under the same temperature conditions; performance of the latter is generally poorer. Isely (1946) evaluated maximum potential reproduction of *A. gossypii* apterae to be 63.73 ± 1.77 larvae at 19°C; this is the temperature of the longest reproduction period giving a daily reproduction rate of 2.45 ± 0.1 larvae. Comparison of these figures, with those obtained by Nassar (1962) and Khalifa and Sharaf El-Din (1964) in Egypt and those of Vaissayre (1970) in the Central African Republic (also on apterae), shows that the optimal temperature of the African 'races' is higher than Isely's American 'race', but similar to that of the American 'race' described by Akey and Butler (1989) collected in the arid environment of the southwestern United States.

Influence of light on the development of *A. gossypii* was studied in the laboratory by Auclair (1967b). Reared on artificial medium, aphids given the choice seek fairly low lighting (50–500 lux) and strong lighting (6000 lux) reduces longevity. Rearing in yellow-orange light (λ = 525–595 nm) enhances reproduction and survival whereas green, blue and violet should not be used.

Heavy rainfall reduces populations directly by washing them away. Rainfall of 20 mm falling in four hours at the height of pullulation may reduce a population by 20–40%. Farmers consider that 'a good rainfall is as good as spraying'. In contrast, high relative humidity enhances the appearance of Entomopathogenic fungi (*Neozygites fresenii* Nowakowski and *Cladosporium* sp.).

The water conditions (irrigation) to which the plants are subjected have a direct influence on their physiological state (sap density, turgor). Any modification in water conditions causes changes in the quantity and quality of aphid food. These factors thus have a distinct effect on duration of insect growth, longevity and fertility and the appearance of alatae, as has been shown by other authors (Lamb, 1955; Kennedy *et al.*, 1958; Weismann and Vallo, 1963).

Over and above nutritional factors, Weismann *et al.* (1970, 1971) report that abiotic factors (particularly the temperature and moisture in cotton plantations) may play an essential role in population dynamics and pullulation of aphids. These factors directly affect vegetative development (growth and branching) and phenology of cotton plants. Favourable conditions are prevalent from flowering onwards, especially following an irrigation period affecting plant physiology. This application of water causes vegetative development and increased shade within the plant. The largest populations are then found below the leaves of the lower third of the plants where they are partially protected from high temperatures.

The role of food factors

Much work has been performed on aphid nutritional requirements (Auclair, 1963). Although *Acyrthosiphon pisum*, *Myzus persicae* and *Aphis fabae* Scopoli have received the most attention, some work has been carried out on *Aphis gossypii*, mainly concerning nitrogen, amino acids, sugars and pH; the results have not revealed any special features in the species.

High nitrogen fertilization enhances infestation (El-Fattah, 1975). Reduction in nitrogen application lowers daily reproduction rate and total descendants by 12.5% and 16.5% respectively according to Isely (1946). Similarly, Ratchford *et al.* (1989) observed significantly higher population levels in cotton plots with sowing row placement of NPK fertilizer than in controls. Similar results were reported by Vaissayre (1970) in the Central African Republic. Reduction in nitrogen and potassium fertilization causes population decrease whereas reducing potassium fertilizer alone causes increased fertility. Vaissayre (1970) accounts for maximum siting of aphids at apices during flowering by the nitrogen richness of this part of the plant.

The sulphur amino acids cysteine and methionine are indispensable to both *A. gossypii* (Turner, 1971) and *M. persicae* (Massonnié, 1971). Sucrose is a necessary phagostimulant for the success of aphid-rearing on synthetic medium; it must be available in sufficient quantities (Auclair, 1967a) and in certain proportions in relation to maltose for a high survival rate (Auclair, 1967b).

An alkaline medium pH (7–7.8) enhances reproduction (Auclair, 1967a). The phloem pH is also clearly alkaline (Mittler, 1988). These features are found in cotton leaves, but various physiological processes may change levels and hence affect the population. Some phenolics (flavanols and flavonols) may change the behaviour of insects, and especially of aphids (Todd *et al.*, 1971). The same applies to compounds that are more specific to cotton such as gossypol or chrysanthemin (Bell, 1986) whose role in resistance to certain Lepidoptera and Coleoptera is known.

Finally, various chemical stimulation methods have been identified which can also affect feeding, reproduction and dispersion of arthropods. Indirect stimulation of aphid and mite populations by certain pesticides (trophobiosis) is well known (Dunnam and Clark, 1941; Chaboussou, 1969; Risch, 1987) and should not be forgotten. Various aspects have been described at length by Slosser *et al.* (1989).

The role of biotic factors

The information available on the natural enemies of *Aphis gossypii* in the various cotton-growing regions consists mainly of lists and inventories accompanied by a few approximate assessments (Mao and Xia, 1983; Belikova and Kosaev, 1985; Michel and Prudent, 1987; Selim *et al.*, 1987; Nozato and Toshio, 1988). Few serious studies of their true impact on aphid populations are available. It has nevertheless been known for a long time (Folsom and Bondy, 1930) that fresh infestation may occur after insecticide applications to control other pests. This is an indirect demonstration of the effectiveness of beneficial insects (Anon., 1960). For example, 300 to 600 parasite and predator species were recorded on cotton growing in the USA according to climatic region, excluding entomopathogens and vertebrates (mainly birds) (Whitcomb and Bell, 1964; Van Den Bosch and Hagen, 1966). All the authors nevertheless agree that predators are more frequent and more effective than parasitoids. Observations on entomopathogens are rare.

Predators

Coccinellidae (Plates V. 3a–d) are most frequently mentioned in the literature. Their essential role in population regulation is recognized (Agarwala and Ghosh, 1988). They are followed by Hemerobiideae and Chrysopidae, which have much the same importance as Syrphidae, and then the Anthocoridae. A common characteristic is the lag between the gradation of prey and predator populations (Bishop and Blood, 1978). These are 'density-dependent' beneficial insects whose numbers increase when aphid colonies are attractive and already large. This means that it is frequently too late and chemical control is often required to reduce populations to acceptable levels.

Parasitoids

These are not usually effective enough in the USA and Asia and are practically non-existent in the whole Ethiopian region. In Africa, most aphid species and hence their parasites cannot withstand excessively high temperatures or prolonged drought affecting the vitality of their host plants. In the Sahelian and Sudan zones with a very long dry season, aphids practically disappear for part of the year when the vegetation wilts, when there are bush fires and trees and shrubs go through a resting period. These severe conditions probably prevent the survival of Aphidiides and Aphelinides parasitoids except perhaps around dwellings, in villages or in marshy areas (Plate VI.1). Shuja-Uddin (1977) nevertheless showed that *Lipolexis* sp. mummies in diapause state could withstand the severe conditions in certain parts of India. Aphid populations and their natural enemies are only maintained permanently in central, eastern and southern Africa where highland temperatures approach temperate levels and the dry season is short. It is only under such conditions that *Aphidius colemani* Viereck, a widely distributed polyphagous species (Mediterranean region, Central Asia, tropical zones in India, Africa, South America and Australia), was collected on *Gossypium* at altitudes of 800 to 2100 m in Burundi (Stary *et al.*, 1985). In the United States, *Lysiphlebus testaceipes* Cresson may, in some years, help to limit populations locally (Kerns and Gaylor, 1991a,b).

Several cases of parasitism by Aphelinides Hymenoptera (*Aphelinus albipodus* Hayat, *A. gossypii* Timberlake) have also been reported (Vaissayre, 1970; Zhi-Yi and Guo-Pei, 1986; Traoré-Célini, personal communication), but these species have practically no impact as they arrive too late and their numbers probably remain too small for the reasons given above; Vaissayre (1970) reported for example that the first mummies appeared towards the 40th day after sowing, i.e. after the first infestation. The limiting effect of the parasitoid is thus only exerted on residual populations.

It should be noted that in addition to their parasite effect, Aphelinides are also predators and destroy numerous aphids by feeding punctures (Jervis and Kidd, 1986). This behaviour has been described and studied in particular in *Aphelinus asychis* Walker, a parasitoid of *Schizaphis graminum* Rondani (Cate *et al.*, 1973, 1974), and in *Aphelinus gossypii* on *Aphis gossypii* by Traoré-Célini (personal

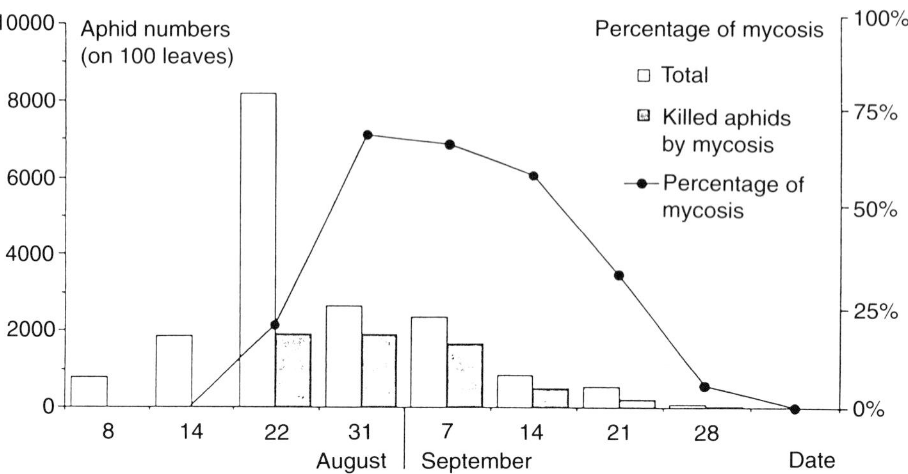

Fig. 12.1. Population dynamics of *Aphis gossypii* in a cotton field at Bebedjia (Chad) in August–September 1987 (count on 100 leaves collected at random from a total of 6400 plants). Variations in total number, specimens killed by *Neozygites fresenii* and the percentage of mycosis. The observations also revealed that aphids were parasitized by Hymenoptera, but the rates never exceeded 5% and are not shown here (from Silvie and Papierok, 1991).

communication) in the Central African Republic. The limiting effect of this parasitoid/predator is nevertheless a rare case in Africa. However, observations in China by Zhi-Yi and Wei-Nian (1982), showed that parasitism may reach 30%, which reduces population levels, but such parasitism, caused by the combined effect of an Aphidiid (*Trioxys* sp.) and an Aphelinid (*Aphelinus* sp.) does not prevent initial population growth.

Microbiological agents

Under certain conditions, entomopathogens especially fungi (Entomophthorales) form a particularly interesting group of beneficial organisms (Papierok, 1987) as their presence does not, in general, depend on host density. Five main genera have been found on aphids (*Conidiobolus*, *Entomophthora*, *Erynia*, *Neozygites* and *Zoophthora*); the rare accurate observations made on *A. gossypii* in cotton growing concern *Neozygites fresenii* Nowakowski and are mainly for Chad (Silvie and Papierok, 1991) and the United States (Steinkraus *et al.*, 1991). *Neozygites fresenii* has also been reported on aphid in Burundi, Côte d'Ivoire, Uganda, Tanzania and Togo.

In nature, infection begins by penetration of aphid cuticle by a germinative tube originating in a conidium (thin-walled spore) on the cuticle. The mycelium develops and invades main cavity of the host, leading to death after a few days. If conditions are favourable (moisture is necessary), fungal conidiophores develop, forming a dusty whitish coating on the body, and form conidia. These primary conidia are projected to a distance of a few centimetres and infect other hitherto healthy aphids. A new 'sporulation–infection–incubation–death–sporulation' cycle takes place, leading to an epizooty. In unfavourable conditions, thick-walled resting spores are formed which can survive in the soil for several months until conditions are suitable for germination and evolution into conidiophores producing conidia capable of infecting insects. Silvie and Papierok (1991) reported parasitism rates of up to 70% in Chad, coinciding precisely with the sharp fall in aphid populations in August, September and October (Fig. 12.1); the results agree perfectly with those of Steinkraus *et al.* (1991) for the United States.

Myrmecophily

Association of ants and aphids is well known (Sudd, 1987). Ants feed on aphid honeydew (Way, 1963); they build special sheaths or soil ridges at the necks of plants, keep aphids in their nests and transport them. Some species cannot remain without the help of ants and others possess true anatomical adaptation of their anal plate for myrmecophilic purposes (*Metopeurum fuscoviride* Stroyan or some Fordinae, which are radicolous on their secondary hosts for example). Finally, ants protect aphids from their enemies, which can considerably reduce parasite or predator effectiveness and compromise biological control operations. This is the case for the red imported fire ant *Solenopsis invicta* Buren (Vinson and Scarborough, 1989).

It has also been suggested that ants may directly modify aphid biology and population dynamics (Banks, 1958). Schmutterer (1969) observed in South Africa that the most seriously infected young cotton plants were those in the immediate vicinity of nests of *Myrmicaria natalensis* F. Smith. A special over-wintering mode involving two ant species (*Formica nigricans* Emery and *F. cinerea* Mayr) was described by Radev (1965) in Bulgaria. In a series of experiments conducted over five years, consisting of infecting different perennial grass species in the autumn to observe the overwintering of parthenogenetic females of *A. gossypii*, Radev noted that aphids did not survive low winter temperatures. However, it could be observed in end-of-cycle cotton plants that numerous individuals were grouped at the base of the plants near the ants' nests, into which the ants carried them when cold weather started. Careful examination of the contents of the ants' nests revealed *A. gossypii* at all larval and adult stages mixed with *Smynthurodes betae* Westwood (Fordinae) and other aphid species. The aphids were taken back to the surface in the spring and were able to walk to the nearest host plants. On this subject, it is noted that in work on apterous aphid movement, Thygesen (1968) recorded *A. gossypii* marked with ^{32}P 1.8 m from the site at which they were released.

Relations with other biocenosis elements

There are many known examples of competition between several organisms, especially for an ecological niche or for food. The aphid *Rhopalosiphum padi* L. progressively eliminates all other aphid species (*Schizaphis graminum*, *Sitobion avenae* F. and *Metopolophium dirhodum* Walker) when the different species are reared as a mixture on cereal. Hermoso de Mendoza and Moreno (1989) reported that *A. gossypii* generally dominates *A. spiraecola* Patch on *Citrus*. Likewise, strong infestation of cotton by the whitefly *B. tabaci* may prevent the development of *A. gossypii* colonies, as both insects are in the same ecological niche (Hector and Hodkinson, 1989).

Rakotofiringa (1989) reported the difficulties of *Spodoptera littoralis* Boisduval larvae in occupying cotton plants strongly infected by *A. gossypii*, whereas, inversely, leaves rolled by *Syllepte derogata* F. hinder the aphid. Finally, honeydew appears to attract *Helicoverpa zea* Bodie as a food source. The usual predators of young *Heliothis* caterpillars prefer aphids (Isely, 1946; Ables *et al.*, 1978), which tends to reduce their effectiveness in control of Lepidoptera populations.

However, these different types of behaviour are fairly anecdotal and do not cause serious changes to aphid population dynamics.

Population dynamics – dispersion and distribution on the plant

Methodological aspects

Population dynamics studies require counting and sampling under standard conditions. These are indispensable for defining tolerance thresholds and application of integrated control. There are many techniques according to aphid species, type of host plant and the aim of the work.

Trapping alatae (Zettler *et al.*, 1967; Irwin, 1980; Robert *et al.*, 1987; Labonne *et al.*, 1989) provides information on the composition of aphid fauna and the average seasonal rhythm of flying activity of each species. It also makes it possible under certain conditions to monitor one or more species in time. Some traps give an absolute estimate of specific aerial density or the landing rate of aphids on plants.

Various other techniques have been put forward for estimating and monitoring plant aphid populations and in particular:

- estimation by infestation category: numbers are divided into classes using a logarithmic (Delattre, 1973), linear (Cauquil *et al.*, 1982) or geometrical scale (Leclant and Remaudière, 1970);
- accurate counting of all individuals (O'Brien *et al.*, 1991);
- estimation of the percentage of plant organs attacked.

The number of samples to be observed or collected, the number of plants to be sampled and the sector or portion of plant to be observed or sampled may vary considerably. Cauquil *et al.* (1983), using the results of Denechere (1981), estimated the number of leaves infested by at least one aphid, considering only the five well-developed terminal leaves of 20 cotton plants. The technique proposed by Denechere was developed to compare the effectiveness of aphicide active ingredients and is based on very detailed statistical considerations. Choosing leaves in the upper part of cotton plants might be criticized; this doubtless makes observation faster but introduces systematic error as *A. gossypii* is mainly sited in the lower part of the plant, as observed by O'Brien *et al.* (1991). In fact, from a practical point of view, the existence of a highly significant correlation between the calculated average population of the six terminal leaves and of the entire plant justifies this sampling method for the objective sought. Excellent correlation between the logarithm of the population and the percentage of leaves infested is also observed. Other authors thus use the percentage of plants infested by observing the whole plant or only part of it. It is noted that according to the purpose of study, counting can be performed in the field or subsequently in the laboratory using a binocular microscope or after treatment of the leaves by various techniques (e.g. according to Heathcote 1972; Rabasse and Bouchery, 1977; Burris *et al.*, 1990).

It is particularly difficult to model infestation because of numerous interacting factors. Two attempts can nevertheless be reported. That of Xia and Sterling (1987) in China is a dynamic determinist model aimed at forecasting evolution of field populations of *A. gossypii*. The variables are population growth rate, temperature, natural enemies and plant phenology; the model satisfies its authors in spite of insufficient data on predators and aphid alates. The other model – also from China – is by Zhang *et al.* (1987), who presented a 'dynamic simulation model for analysing the control of aphids by ladybirds and for determining optimum ladybird/aphid ratio'.

Practical aspects

In Central Asia, *A. gossypii* overwinters as parthenogenetic females on various weeds and spontaneous plants. It can survive temperatures as low as −9°C (Kozhaeva, 1965). Alates appear in April and disperse among all available host plants – including cotton and various Cucurbitaceae colonized and infested in the second half of May. Kozhaeva reports population decrease in July and August with a fresh increase in the autumn. There are thus some twenty successive generations during the whole of the cotton season. One generation develops in 5 to 12 days depending on suitability of the weather. Average reproduction rate is 40 to 50 larvae per female.

Aphis gossypii can be observed throughout the year in Egypt, but it is more abundant in spring and early summer. Attacks are earlier in the Cairo region than further north in the delta. The first sowing of cotton is attacked between March and June and then, after a respite in June and July, there are strong infestations in August and September, especially in the north of the country.

Under conditions in Upper Egypt where irrigated cotton is grown, Weismann *et al.* (1971) identified four phases in the cotton vegetative cycle with changes in the number of plants attacked and in the structure and density of aphid populations and their distribution on the plants.

Phase 1 All stages of development are observed at the start of plant growth at fairly low temperatures with high relative humidity.

Phase 2 Temperatures rise during the driest period when plants grow and start to ramify. The volume of vegetation is still small, the soil is not heavily shaded and leaves occupied by aphids are subjected to fairly high temperatures and high dryness. The aphids lose a great deal of moisture by transpiration and as leaf turgor pressure is low cannot compensate by feeding. Exogenous alates are observed together with a few young larvae who can only develop if the crop is irrigated.

Phase 3 After the first two phases of only about a month each, the plants have ramified and the number of leaves and the proportion of shaded soil increase. The aphid population grows. Irrigation enables the development of apterae accompanying numerous larvae at all stages with the exception of larvae with pterothecae (wing-cases). The phase lasts until flowering. The number of plants attacked and population density are closely related to waterings, as is the shape of the gradation curve. It is noted that these results appear to be in contradiction with those of El-Fatah (1975). According to the latter, the smaller and more frequent the waterings, the higher the population. Vaissayre (1970) considered that water deficiency causes an increase in sap nutrients, enhancing aphid multiplication, but also observed an antagonistic effect caused by increase in osmotic pressure, as has been reported by Weismann *et al.* (1970, 1971). The result of a combination of the two effects is thought to be a decrease in the multiplication potential of the insects after three days (Vaissayre, 1970).

Phase 4 Finally, in the fourth phase the plants in neighbouring rows overlap and 40% to 95% of the soil surface is shaded. The aphid population and the number of infested plants increase rapidly, especially with watering. At this stage, 30% to 90% of the leaves on each plant may bear aphids, with numerous larvae with pterothecae which spread to other cotton plants when they reach adult stage.

In a series of observations in Louisiana and Mississippi, O'Brien *et al.* (1991) identified three phases in population evolution. Numbers were fairly small before the first flowers opened. This was followed by a strong, sharp population increase in July at flowering soon followed by a decrease. Moderate increase was observed in the second half of August and the population then returned to the initial level at the end of September. The same authors also observed that vertical distribution of aphids in the different plant strata varied during the season and also according to crop management, variety and plant height. They also stressed the danger of applying a single sampling plan, the unsuitability of sampling only the top of the plant and the pointlessness of early sprayings of small populations, confirming the results of Parker and Huffman (1991).

In West and Central Africa, where cotton is dry-farmed in the rainy season, aphid colonization begins practically at emergence. Numerous alatae are already available from the many host plants which enabled aphids to survive the dry season. Research is in progress to assess the role which may be played by allochthonous populations following the northward movement of the Inter-Tropical Convergence Zone (ITCZ) in infestation of apparently azoic cotton zones at the end of the dry season. In fact, we have shown that over 300 different plant species could serve as hosts for *A. gossypii* in the inter-season, often only in very small numbers. Larvae with pterothecae appear on the plants around the fields shortly before the trapping and detection on cotton plants of the first alatae responsible for primary infestation. Only the cotyledons of seedlings are visible against the bare soil (Kennedy *et al.*, 1961; Duviard and Mercadier, 1973); they are sufficiently attractive to be located and colonized by alate aphids during local flights (Robert, 1988). Similar observations of the same species, but under totally different ecological conditions, were made on melon in southeast France.

After heterogeneous primary infestation, new plants are colonized by apterous aphids. These are generally dark-hued and large and are the progeny of the exogenous alates. The first two generations which develop on cotton consist mainly of apterae which spread by infesting neighbouring plants. This type of infestation is the most important and effective type at field level. According to Thygesen (1968), *A. gossypii* apterae are frequently found 1.8 m from their release site. It is noted that there may also be vertical movements on the same plant, often enhancing complete colonization. Such movements vary according to the time of day and could affect sampling and counting quality. Finally, about a fortnight after the arrival of exogenous alatae on the crop, the proportion of larvae with pterothecae increases considerably (third and fourth generations) followed by the appearance of alates which cause secondary infestation and fresh outbreaks.

These three types of colonization (exogenous alatae, apterae and indigenous alates) render attacks homogeneous and can result in 90% infestation of cotton plants in the third week after the arrival of the first alates, causing direct damage (crumpled leaves) and indirect damage (honeydew) and slowed plant growth. Infestation decreases slightly and then stabilizes from physiological flowering onwards because of the balance attained between plants, aphids and natural enemies. There is a spectacular decrease in aphid populations 50 to 60 days after the start of primary colonization as a result of Entomophthorales infection, mainly caused by *Neozygites fresenii* (Fig. 12.1), and predators such as the ladybirds *Cheilomenes sulphurea* Olivier and *Exochomus* spp. (Duviard and Mercadier, 1973). During fruiting and boll maturation, the largest populations are found beneath the oldest leaves, at the base of the plant and near the stem (Cauquil *et al.*, 1982; Banerjee and Raychaudhuri, 1985) (Fig. 12.2).

At the end of the cycle, when protection against various boll pests has ceased and the crop is safe, there is generally fresh infestation caused by residual populations under the leaves in the lower or mid stages of the plants whose nutrient quality is decreasing. The insects 'migrate' towards the extremities where

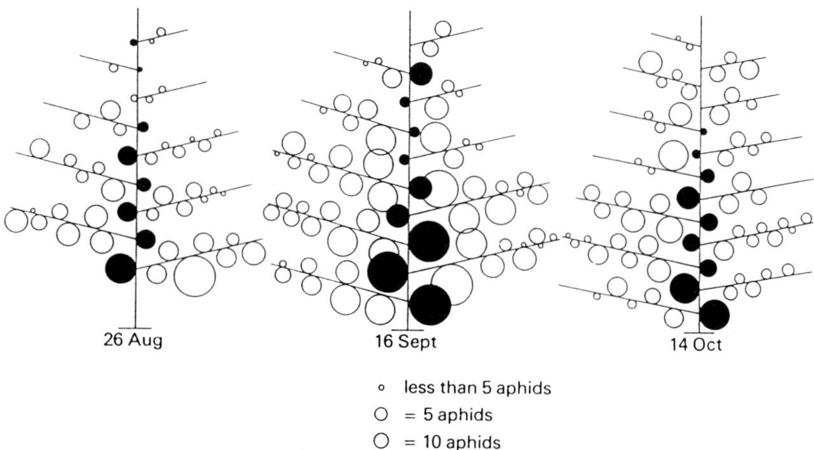

Fig. 12.2. Position of aphids on plants (from Cauquil *et al.*, 1982).

the leaves still have suitable appetency for sap-feeding insects. The honeydew causes serious damage to fibre.

Damage – economic importance

Damage

Plant-sucking insects in general and aphids in particular are responsible for two types of damage to the plants that they colonize or that they sometimes visit without colonizing or feeding:

- direct damage, resulting from search for food which induces plant deformation;
- indirect damage, caused either by the sugary exudate (honeydew) which soils plants, or by the transmission and spread of various phytopathogens (namely viruses in the case of aphids) causing serious diseases with specific symptomatology.

Aphis gossypii is thus a particularly harmful species, causing not only quantitative decrease in the harvest but also decreased fibre quality.

Direct damage The feeding mechanism of *A. gossypii* on cotton leaf was studied in great detail by Pollard (1958, 1973), who described the way in which the stylets penetrate the different tissues: the phloem, where actual feeding takes place, is attained by an essentially intercellular route through the epiderm and the mesophyll. *Aphis gossypii* does not cause any apparent mechanical damage to tissues by laceration or biochemical damage by spread of saliva or plasmolysis as with other aphids or mirids and jassids (Johnson, 1934; Brzezina *et al*., 1986). Only large populations which have been feeding for some time cause the leaf crumpling and curling characteristic of aphid attacks, generally on very hairy cotton varieties (Dunnam, 1936; Dunnam and Clark, 1938).

Indirect damage – honeydew, sticky cotton There are several causes of sticky cotton. One is physiological, resulting from cellulose precursors and extrafloral nectar secretion, but the main cause is entomological (80–90% of cases) in the form of sugary excreta (honeydew) from various sucking insects during infestation at the end of the cycle. The main species involved is *Bemisia tabaci*, but *A. gossypii* is also involved, and to a much more limited degree Pseudococcinae scale insects (*Ferrisia virgata* Cockerell and *Phenacoccus solenopsis* Tinsley). Phloem sap, which generally has a very high sugar content and a low amino acid content, is taken up in large quantities by the insects to meet essential amino acid requirements. Excreta therefore contain a concentrated solution of the excess sugars ingested (sucrose) or partially converted (melezitose) and projected in fine drops on to the underlying vegetation.

Honeydew is an excellent medium for the various saprophyte fungi which cause sooty moulds hindering light absorption by chlorophyll and affect plant respiration. In addition, they soil the seed cotton in open bolls (Figs 12.3 and 12.4), harming quality, but above all the little drops of honeydew – often crystallized – are not eliminated during ginning. They cause sticking, and con-

Fig. 12.3. Honeydew and sooty moulds on lint in open boll.

siderably complicate various fibre treatment processes (rolling and drawing) from carding to spinning; these difficulties have increased considerably in the

Fig. 12.4. Droplets of honeydew on lint.

past decade with the evolution of weaving techniques and the use of high capacity machines with high rotation speeds (Gutknecht, 1988). The work of Hector and Hodkinson (1989) contains an interesting review of the sticky cotton problem from the entomological causes to the technological consequences.

In addition to the use of chemical control which is often found to be ineffective today, various measures, cultural precautions in particular, can be used to reduce or prevent stickiness: no late fertilizer applications, sow early and harvest early in several operations, sow okra-leaf varieties and use chemical or manual defoliation (Ekukolé, 1992a,b). These are promising methods and deserve to be developed, especially in Sahelian Africa, by informing supervisors at village level and by financial encouragement for growers through the purchase price for seed cotton.

Indirect damage – spread of viruses diseases Although these are less serious than the diseases spread by whitefly, many virus diseases spread by aphids have been reported throughout the world, especially in francophone Africa and in South America (Cauquil and Follin, 1983). *Aphis gossypii* is the vector of some sixty virus diseases, mainly of the non-persistent type (Eastop, 1983), in a very large range of plants. Together with *Myzus persicae*, it is the most frequently mentioned aphid vector in the literature, mainly because it is polyphagous and ubiquitous.

In Central Africa, *Aphis gossypii* is the vector of Blue disease in cotton, a non-persistent virus infection characterized by plant bushiness caused by slowed growth and shortening of internodes (Cauquil and Vaissayre, 1971). Epinasty may be observed with crawling habit and leaf drop or even total withering of leaves. The leaves are generally thick, brittle and darker than normal; laminae are revolute, flowering sparse and bolls are shed before maturity. Yield is small. As with all virus diseases, the earlier they occur the more serious they are (Cauquil, 1977). This disease is found in particular in the Central African Republic where it was observed for the first time in 1949. It has spread to Zaïre, Chad, Cameroon, Benin and Côte d'Ivoire. Early chemical control improves the situation and is profitable in high yield crops when production potential is high enough (Cauquil *et al.*, 1978). Tolerant varieties have also been developed (Mahama and Cauquil, 1976).

Similar symptoms to Blue disease, also spread by *A. gossypii*, have been reported in the Philippines (Nour, 1960; Escober *et al.*, 1963) and Brazil (Costa and Carvalho, 1965). *Aphis gossypii*, together with *A. craccivora* Koch and *M. persicae* is considered to be responsible for the spread of a disease with leaf curl type symptoms and shiny, thick, brittle leaves (Tarr, 1964). The epidemiological facies of another disease observed in Argentina (Follin and Campagnac, 1981), called 'Flétrissement pourpre' (purple wilt) or 'Murcha vermelha', and to which the variety Reba P 279 bred in Paraguay by the Institut de Recherches du Coton et des Textiles exotiques (IRCT) is particularly sensitive, suggest that *A. gossypii* is the vector. The symptoms are sudden wilting of the plant and then reddening of the foliage, and are similar to typical Anthocyanosis ('Vermelhao') described in Brazil in 1956 by Costa and which is also spread by *A. gossypii* (Costa and Sauer, 1954). The leaf symptoms are characteristic and appear when the plant has two or four true leaves. Lamina chlorosis is followed by the appearance of dark red patches bounded by the main veins and covering the whole leaf except for a narrow strip along the veins. It is economically the most serious virus disease in Brazil where average losses of 10% have been recorded.

Economic importance

Aphis gossypii used to be considered as a minor pest on cotton, but over the past decade has emerged as one of the main problems in numerous cotton growing regions in the world. This is the case in Iraq (Khalid and Al-Zarari, 1983), in Israel, where it is considered to be the most serious pest and alone justifies 80% of insecticide treatments, in Turkey, Syria (Broza, 1986) and Romania (Ullah and Paul, 1985). In Turkey, *Aphis gossypii* is accompanied by *Acyrthosiphon gossypii* and *Macrosiphum euphorbiae* during the principal cotton growth phases but most of the impact on yield is in the company of other pests during boll maturation. In China (Nan *et al.*, 1987), the most severe damage caused by *A. gossypii* is at the beginning and in the middle of the cycle during boll formation; the problem is mainly centred in northern China along the Yellow River.

In Central Asia (Kozhaeva, 1965), damage occurs shortly after emergence (May) and causes deformation in very young seedlings. The leaves crumple and branched plants with two stems are formed. Attacks on older plants reduce overall vigour and growth. Boll formation and flowering are reduced, maturation affected and bolls are shed. The degree of damage depends on the period of the attack and the size of aphid populations. An attack at the very beginning of the cycle, with aphids on the plants for about thirty days, can cause losses of 25–50%. Fibre quality is also affected in hard-hit plants. Populations decrease in July and August and then increase again in the autumn. Seed cotton production is reduced and fibre is considerably soiled by honeydew, in years in which populations are high at the end of the cycle.

Aphis gossypii, together with *Bemisia tabaci*, is a serious pest in India (Roy and Behura, 1983a), Madagascar (Rakotofiringa, 1989), Sudan and Egypt and in West Africa (Onu, 1989; Toe, 1992), Central Africa (Deguine, 1992) and Zambia (Javaid *et al.*, 1987), whereas Couilloud in 1965 reported small *A. gossypii* populations in Chad. According to Cauquil and Vincens (1982), *A. gossypii* is probably responsible for seed cotton losses in Africa of up to 6.5%. Losses in late-sown cotton may exceed 20% in cases of early pullulation, thus putting *A. gossypii* on a par with bollworms as the main cotton pests in Cameroon. Early leaf crumpling causes considerable lag in plant growth resulting in smaller plants with smaller leaf areas and a decrease in girth and weight of stems and roots (Deguine, 1992). It is also a factor enhancing sticky cotton and late fibre maturity.

In South America, Vendramin and Nakano (1981) in Brazil considered *A. gossypii* to be responsible for the decreased plant growth, reduced seed cotton yield and late maturing. Likewise, Michel and Prudent (1985) found that *A. gossypii* was one of the main causes of yield losses in Paraguay.

Until the 1940s, *A. gossypii* was one of the serious cotton pests in most of the cotton growing regions in the United States, where it caused yield losses of 276 kg ha^{-1} (Ewing, 1943). This status was often ascribed in the past to the use of calcium arsenate in control of *Anthonomus grandis* Boheman and the destruction of beneficial insects (Paddock, 1919; Folsom and Bondy, 1930; Isely, 1946). The introduction of organophosphorus insecticides in the 1950s made it possible to prevent serious pullulation following use of DDT to control *Heliothis* (Bottrell and Adkisson, 1977), and the aphid problem decreased considerably (Anon., 1960; Hardee and O'Brien, 1990). The situation is changing today. According to Sterling *et al.* (1989), the arthropods generally considered as 'key' pests and which dominate control programmes vary according to production region: *Lygus* spp. and Tetranychidae in California; pink bollworm (*Pectinophora gossypiella* Saunders) and the *Heliothis* spp. complex in arid zones in the southwest; *A. grandis* from Texas to the east coast; and dominance of the mirid *Pseudalomoscelis serriatus* Reuter (Fleahopper) in Texas and many parts of Oklahoma and Louisiana. The *Heliothis* complex is dominant wherever cotton is grown intensively (irrigation, fertilization, much spraying and long season plant growth strategies). It is thus surprising to note that hardly any mention is made of *A. gossypii* in the work by Frisbie *et al.* (1989), where it is considered as a secondary pest, as is *B. tabaci*. In addition, none of the extension services in producer states suggests a tolerance threshold for the species (Benedict *et al.*, 1989). In contrast, the reports by King *et al.* presented at the Beltwide Cotton Conference in 1986, 1987 and 1988, mention presence of *A. gossypii* in 12 cotton states; serious damage has been caused, especially in 1986, and four to seven treatments are necessary annually. Price *et al.* (1983) reported possible losses of up to 100 kg ha^{-1}. According to Bagwell *et al.* (1991), strongly infested plants suffer considerable stress; they are small and less branched than lightly attacked plants and yield losses are caused above all by considerable fruit shedding. In addition to the appearance of resistance and the elimination of beneficial insects, simultaneous, convergent change in certain factors in the immediate environment of the insect (especially bioclimatic and nutritional factors) may also result in new conditions more favourable to the aphids which may account for the new situation (King and Phillips, 1989; Slosser *et al.*, 1989).

On a global scale, *A. gossypii* is found throughout almost the whole year on cotton, but numbers vary according to phenological stage and from one region to another. The resulting damage also varies considerably according to production region.

Control Measures

Until recent years, the main aim was to protect cotton fruit organs from the pests which formed the greatest threat to seed cotton production – mainly bollworms and the boll weevil. Although aphids were observed and reported at the start of the cycle in many countries, there was generally no special treatment. The same applied to the period running from boll opening to harvesting. Treatments were thus performed with non-aphicide products with a broad spectrum of activity such as pyrethroids, often applied using ULV in many countries.

Pullulations observed in many regions in the early 1980s led specialists to seek the causes and envisage specific aphid treatments. The destruction of beneficial insects mentioned above has long been suggested as one of the causes.

Today, increasing numbers of cases of resistance, especially to carbamates, organophosphorus compounds and also pyrethroids are reported in *A. gossypii* on cotton in Africa (Gubran, 1991; Gubran *et al.*, 1992), the USA (Grafton-Cardwell, 1991; Mitchell *et al.*, 1991; Kerns and Gaylor, 1992; O'Brien *et al.*, 1992), Israel (Ishaaya and Mendelson, 1987), China (Jianguo *et al.*, 1987; Tang *et al.*, 1988) and Japan (Saito, 1991).

In regions like francophone Africa, where cotton is dry-farmed and there are relatively few insecticide treatments, resistance has yet to be observed in *A. gossypii* and the effectiveness of aphicides does not need to be questioned. However, ultra-low volume (ULV) spraying was found to be totally unsuitable both at the beginning of the cycle, when coverage is poor because of the small size of cotton plants, and during the season when aphid populations are grouped in the lower half of the plant (Cauquil *et al.*, 1982; Cauquil, 1990). Use of varieties with laciniate leaves ('okra' varieties) enhancing spray penetration and preference for very low volume (VLV) techniques, which are more economical (Gaudard, 1992), improve protection of cotton from aphids and also from the whitefly, *Bemisia tabaci.*

Three main parts of the cotton seasons can be considered (Matthews, 1989) for crop protection purposes:

At the very start of the season – from sowing to the first flower – plants must become well-established and vigorous and the aim is to control early infestation, which often affects late sown crops less than one month old, by using substances with little effect on beneficial insects. In addition to leaf treatment of young crops using plant oils, which has given interesting results (Butler *et al.*, 1988a), coating or dusting appropriately delinted seed with systemic chemicals can prove satisfactory; several active ingredients in various categories can be suggested, and especially carbamates (e.g. furathiocarb, carbosulfan and thiodicarb) (Elmsheuser and McCracken, 1989). New products (imidacloprid from the nitroguanidin family) and improvement of techniques (pelleting) are promising for situations with a risk of serious damage. However, conflicting results have been published regarding the advantages of aldicarb; unlike Carlson and Mohamed (1986) in the Sudan, Parker and Huffman (1991) reported no significant difference in yield, maturity or fibre quality after application at sowing in Texas, even though they observed a population decrease in the fields treated, whereas Butler *et al.* (1988b) in Israel reported that a single application on 15 July provided effective control of aphids but not whitefly.

During the vegetative phase (flowering and boll formation) following the juvenile period, the insecticides generally sprayed on leaves are aimed at other pests (Heteroptera, Lepidoptera and Coleoptera) and generally keep *A. gossypii* populations down to a tolerable level when there is no resistance. For example, in sub-Saharan francophone Africa, the control programme, generally extended ('lutte étagée ciblée' – targeted staggered control), consists of adapting leaf application to the insects present by:

- choice of active ingredients according to the pests;
- choice of dose according to pest pressure and to conserve beneficial insects as much as possible;

Table 12.8. Potential host plant resistance traits in cotton for sucking pests (Aphids, Jassids and Aleurodids) (from El-Zik and Thaxton, 1989).

Resistance traits	Aphids	Jassids	Aleurodids
Glabrous	S	S	R
Pilose (gene H2) (high density hairs)	S	R	S
Hirsute (gene H1)	S/?	R	N
Nectariless	?	?	N
Okra-shaped leaf	?	N	R
Frego bract	?	N	N
Red plant colour	R	N	N
High gossypol	?	R	S
High tannin	?	N	?

S, susceptible.
R, resistant.
N, no effect.
?, conflicting evidence or not verified.

- taking into account the phenological stage of the crop, as in Cameroon (Gaudard, 1992).

Spraying is performed from the 45th day after emergence once a fortnight (VLV) with 10 litres of mix per hectare. Observations for decision-making consist of examining the five terminal leaves of 25 cotton plants chosen diagonally in 0.25 ha per ha of crop; the percentage of infested leaves (cf. Denechère, 1981) is evaluated and related to a tolerance threshold which is in fact the spraying threshold. There are numerous active ingredients: organophosphorus compounds (monocrotophos, omethoate, dimethoate, ethyl or methyl chlorpyriphos), carbamates (carbosulfan, butocarboxime, benfuracarb), biphenthrine and esfenvalerate (pyrethroids) or imidacloprid (nitroguanidine) (Cauquil *et al.*, 1978, 1982; Cauquil, 1985; Pinchard and Cauquil, 1992). However, the least toxic substances should be preferred, e.g. those listed as WHO Class III (1989), and those which conserve beneficial insects. Use of combinations makes it possible to play on ratios and delay the start of resistance.

Boll pest pressure tends to decrease at the end of the season whereas the complex of sucking insects, including *A. gossypii*, resurges. Spraying foliage with aphicides at the end of the cycle is generally ineffective and preference should be given to cultural and preventive methods. Early picking in several stages is the most effective technique. Manual defoliation, consisting of topping the plants at the end of the season when sucking insects concentrate on parts which still retain appetency, is practised in China but is a demanding and time-consuming operation. Both could be encouraged financially through the seed cotton price. Chemical defoliation (Mikhidtdinov, 1985; Meyerdirk *et al.*, 1986; Zalom and Natwick, 1987), by desiccating defoliants or growth regulators inhibiting growth of axillary buds, is also used to control sap-feeding insects at the end of the season. This requires large amounts of water and there are risks for users. It may also damage various cotton products and by-products (fibre, seeds, oil and oil cake).

Protection of cotton from aphids and other sucking insects must be performed logically from sowing onwards. Sowing date is often an important feature (El-Fattah, 1975; Dhawan *et al.*, 1987; Deguine, 1992). For example, in addition to the previously mentioned advantages of early sowing (avoidance of early infestation and its consequences), it generally results in faster defoliation than in late-sown plots, on condition that the soil type is not a limiting factor through high water retention. Likewise, for agronomic reasons it is strongly recommended not to practise late fertilization. Irrigation should be stopped as early as possible to prevent continued vegetation or regrowth of plant tops (Von Arx *et al.*, 1983). These are two important factors which enhance sucking insects at the end of the cycle and hence sticky fibre. Use of resistant varieties is an attractive solution. In fact, little research has been carried out on cotton resistance to *A. gossypii*. Red varieties are always attacked less severely, whereas the hairy, jassid-resistant varieties are sensitive to aphids and whitefly (El-Zik and Thaxton, 1989), although some results have been intermediate (Khan and Agarwal, 1990) or contradictory (Table 12.8). Laboratory tests have also shown the high toxicity of gossypol for aphids (Niles, 1980).

In Central Asia, Kozhaeva (1965) recommends deep ploughing in of weeds

Table 12.9. Desirable characteristics for a cotton crop to reduce the probability of occurrence of sticky lint (from Hector and Hodkinson, 1989).

Varietal characteristics

1. tall, open canopy
2. okra or super okra leaf form
3. glabrous (smooth) leaves
4. absence of extrafloral nectaries
5. a high gossypol content

Good agronomic practices

1. wide row width
2. restricted use of nitrogenous fertilizers
3. adequate weeding use of appropriate herbicides
4. removal of plant stands after harvest
5. cotton should be grown apart from other potential alternative hosts of whitefly and aphid
6. termination of irrigation when bolls open
7. early crop termination by defoliation
8. tractor or ground spraying
9. use of appropriate insecticides and defoliation

Other factors

1. regular picking rather than 'once over'
2. hand or spindle picking
3. avoid cotton crops subject to restricted growing season, low night temperatures, drought and disease.

used by *A. gossypii* in the autumn and elimination of all host plants in headland, around crops and along irrigation canals. According to the same author, cultivation of cotton next to cucurbits should be avoided. However, intercropping cotton with other plants which are not *A. gossypii* hosts (wheat and oil crops as in China, and maize and *Vigna* elsewhere) (Vieira *et al.*, 1983) is an interesting technique that is also widespread in India.

Conclusion

Aphis gossypii is an ubiquitous, remarkably flexible species. Its cycles vary considerably according to climatic region: strict monoecy with or without sexual reproduction on various 'secondary' hosts; anholocycly, which may be accompanied by androcycly; heteroecy, also with various primary hosts and a broad variety of secondary hosts. It is practically omnivorous, although specialized biotypes are adapted to a host plant or a particular botanical family. It is polymorphic in size and colour.

Although it is possible to control infestation effectively early in the season, protection of crops against late aphid pullulation at the end of the cycle poses serious problems. *Aphis gossypii* has been a major cotton pest in numerous cotton growing regions for a number of years, because of its contribution (with the whitefly *Bemisia tabaci*) to sticky cotton resulting from deposits of honeydew on open bolls. These sugary excreta are a problem during spinning operations and bring seed cotton prices down considerably.

Several reasons have been put forward to account for the increase in stickiness caused by sucking insects. It is difficult to identify the most important ones and ranking doubtless varies according to region. The following list is an example:

- elimination of natural enemies and resistance following over-use or inappropriate use of insecticides (organophosphorus compounds and pyrethroids);
- planting higher-yielding cotton varieties that are more sensitive;
- a high level of nitrogen fertilization and pursuance of irrigation until a late date;
- a favourable climatic sequence;
- changes in cropping techniques during cultivation (weeds), harvest (early or not) and after the harvest (crop residues left in the fields).

More detailed research on the biology of these two honeydew-producing species should therefore be undertaken or continued in relation with the sticky cotton problem. The aim should be: 'How can economically profitable control methods, which respect the agroecosystem, be used by growers to control pest populations to conserve both good fibre quality and production levels?'

The only hope for solving this crucial textile industry problem at its source lies in application of integrated control strategies, with particular emphasis on varietal characteristics and good agricultural practices (Table 12.9).

References

Ables, J.R., Jones, S.L. and McCommas, D.W., Jr. (1978) Response of selected predator species to different densities of *Aphis gossypii* and *Heliothis virescens. Environmental Entomology* 7, 402–404.

Agarwala, B.K. and Ghosh, A.K. (1988) Key records of aphidophagous Coccinellids in India. A review of bibliography. *Tropical Pest Management* 34, 1–14.

Akey, D.H. and Butler, G.D., Jr. (1989) Developmental rates and fecundity of apterous *Aphis gossypii* in seedlings of *Gossypium hirsutum. Southwestern Entomologist* 14, 295–299.

Anon. (1960) *The Cotton Aphid: How to Control It.* US Department of Agriculture Leaflet No. 467, 8pp.

Auclair, J.L. (1963) Aphid feeding and nutrition. *Annual Review of Entomology* 8, 439–440.

Auclair, J.L. (1967a) Effects of pH and sucrose on rearing the cotton aphid, *Aphis gossypii*, on a germ-free and holidic diet. *Journal of Insect Physiology* 13, 431–446.

Auclair, J.L. (1967b) Effects of light and sugar on rearing the cotton aphid, on a germ-free and holidic diet. *Journal of Insect Physiology* 13, 1247–1248.

Bagwell, R.D., Tugwell, N.P. and Wall, M.L. (1991) Cotton aphid: insecticide efficacy and an assessment of its damage to the cotton plant. In: *Proceedings of the Beltwide Cotton Production Research Conference*, Memphis, Tennessee, pp. 693–695.

Banerjee, T.K. and Raychaudhuri, D. (1985) Behavioural response (feeding preference and dispersal posture) of *Aphis gossypii* Glover on brinjal crop. *Proceedings of the Indian Academy of Science (Animal Science)* 94, 295–301.

Banks, C.J. (1958) Effects of the ant, *Lasius niger* (L.), on the behavior and reproduction of the bean aphid *Aphis fabae* Scopoli. *Bulletin of Entomological Research* 49, 701–714.

Batchelder, C.H. (1927) The variability of *Aphis gossypii. Annals of the Entomological Society of America* 20, 263–278.

Behura, B.K. and Acharya, M. (1983) Biometrical studies of the cotton aphid *Aphis gossypii* Glover with regard to three different host plants. *Pranikee* 4, 65–74.

Belikova, E.V. and Kosaev, E.M. (1985) The biology of the most important species of Coccinellidae and their role in controlling aphids in a cotton–lucerne rotation. *Izvestiya Akademii Nauk turkmenskoj SSR, Seriya biologicheskikh Nauk* 5, 61–63.

Bell, S.J. (1986) Physiology of secondary products. In: Mauney, J.R. and Stewart, J.M. (eds) *Cotton Physiology.* The Cotton Foundation, Memphis, Tennessee, pp. 597–621.

Benedict, J.H., El-Zik, K.M., Oliver, L.R., Roberts, P.A. and Wilson, L.T. (1989) Economic injury levels and thresholds for pests of cotton. In: Frisbie, R.E., El-Zik, K.M. and Wilson, L.T. (eds) *Integrated Pest Management Systems and Cotton Production.* John Wiley & Sons, New York, pp. 121–153.

Bishop, A.L. and Blood, P.R.B. (1978) Temporal distribution and abundance of coccinellid complex as related to aphid populations on cotton in south-east Queensland. *Australian Journal of Zoology* 26, 153–158.

Blackman, R.L. (1971) Variation in the photoperiodic response within natural populations of *Myzus persicae* (Sulzer). *Bulletin of Entomological Research* 60, 533–546.

Blackman, R.L. (1972) The inheritance of life-cycle difference in *Myzus persicae* (Sulz.) (Hem., Aphididae). *Bulletin of Entomological Research* 62, 281–294.

Blackman, R.L. (1986) Species, sex and parthenogensis in aphids. In: Forey, P.L. (ed.) *The Evolving Biosphere.* Cambridge University Press, Cambridge, pp. 75–85.

Blackman, R.L. and Eastop, V.F. (1984) *Aphids on the World's Crops. An Identification Guide.* John Wiley & Sons, Chichester, 466pp.

Bottrell, D.G. and Adkisson, P.L. (1977) Cotton insect pest management. *Annual Review of Entomology* 22, 451–481.

Broza, M. (1986) An aphid outbreak in cotton fields in Israel. *Phytoparasitica* 14, 81–85.

Brzezina, A.S., Spiller, N.J. and Llewellyn, M. (1986) Mesophyll cell damage of wheat plants caused by probing of the aphid, *Metopolophium dirhodum. Entomologia Experimentalis et Applicata* 42, 195–200.

Burris, E., Pavloff, A.M., Leonard, B.R., Graves, J.B. and Church, G. (1990) Evaluation of two procedures for monitoring populations of early season insect pests (Thysanoptera: Thripidae and Homoptera: Aphididae) in cotton under selected management strategies. *Journal of Economic Entomology* 83, 1064–1068.

Butler, G.D., Jr. Coudriet, D.L. and Henneberry, T.J. (1988a) Toxicity and repellancy of soybean and cotton seed oils to the sweet-potato whitefly and the cotton aphid on cotton in greenhouse studies. *Southwestern Entomologist* 13, 81–86.

Butler, G.D., Jr. Hutchinson, W.D. and Broza, M. (1988b) Effect of aldicarb treatments to cotton on *Bemisia tabaci* and *Aphis gossypii* populations in Israel. *Southwestern Entomologist* 13, 87–93.

Carlson, G.A., Mohamed, A. (1986) Economic analysis of cotton-insect control in the Sudan Gezira. *Crop Protection* 5, 348–354.

Cate, R.H., Archer, T.L., Eikenbary, R.D., Starks, K.J. and Morrison, R.D. (1973) Parasitization of the greenbug by *Aphelinus asychis* and the effect of feeding by the parasitoid on aphid mortality. *Environmental Entomology* 2, 549–553.

Cate, R.H., Sauer, J.R. and Eikenbary, R.D. (1974) Demonstration of host feeding by the parasitoïd *Aphelinus asychis* (Hymenoptera: Eulophidae). *Entomophaga* 19, 479–482.

Cauquil, J. (1977) Etude sur une maladie d'origine virale du cotonnier, la maladie bleue. *Coton et Fibres Tropicales* 32, 259–278.

Cauquil, J. (1985) La protection des cotonniers contre les ravageurs en Afrique francophone au Sud du Sahara: principe et évolution des techniques. *Coton et Fibres Tropicales* 40, 187–191.

Cauquil, J. (1990) New developments in cotton crop protection against pests in French-speaking sub-Saharan Africa. *Coton et Fibres Tropicales* 45, 45–51.

Cauquil, J. and Follin, J.C. (1983) Les maladies du cotonnier attribuées à des virus ou à des mycoplasmes en Afrique au sud du Sahara et dans le reste du Monde. *Coton et Fibres Tropicales* 38, 293–308.

Cauquil, J. and Vaissayre, M. (1971) La 'maladie bleue' du cotonnier en Afrique: transmission de cotonnier à cotonnier par *Aphis gossypii* Glover. *Coton et Fibres Tropicales* 26, 463–466.

Cauquil, J. and Vincens, P. (1982) Maladies et ravageurs du cotonnier en Centrafrique: expression des dégâts et moyens de lutte. *Coton et Fibres Tropicales*, Supplement 1, 32pp.

Cauquil, J., Guillaumont, M. and Jouve, G. (1978) Premiers résultats obtenus en Empire Centrafricain sur la lutte chimique contre *Aphis gossypii* Glover, vecteur d'une 'virose' du cotonnier: la maladie bleue. *Coton et Fibres Tropicales* 33, 335–351.

Cauquil, J., Vincens, P., Denechere, M. and Mianze, T. (1982) Nouvelle contribution sur la lutte chimique contre *Aphis gossypii* Glover, ravageur du cotonnier en Centrafrique. *Coton et Fibres Tropicales* 37, 333–350.

Cauquil, J. Vincens, P. and Girardot, N. (1983) La lutte contre le puceron du cotonnier. (*Aphis gossypii* Glover) en Centrafrique. *Mededelingen van de Faculteit Landbouwwetenschappen Rijksuniversiteit Gent* 48, 341–347.

Cauquil, J., Couilloud, R., Girardot, B., Gozé, E., Jouve, G. and Vaissayre, M. (1989) *Méthodologie de l'Expérimentation Phytosanitaire en Culture Cotonnière*. CIRAD-IRCT, Montpellier, 63pp.

Chaboussou, F. (1969) Les déséquilibres biologiques provoqués par les traitements pesticides de la plante. *Publication de OEPP* Série A, 52, 33–44.

Costa, A.S. (1956) Anthocyanosis, a virus disease of cotton in Brazil. *Phytopathologische Zeitschrift* 42, 113–138.

Costa, A.S. and Carvalho, A.M.B. (1965) *Molestias de virus in cultura e adubaçao do algodeiro.* Instituto brasileiro de la Potassa, Experimentaçoes e Pesquisas, San Paulo, 567pp.

Costa, A.S. and Sauer, H.F.G. (1954) Vermelhao do algodeiro. *Bragantia* 13, 237–246.

Cottier, W. (1953) *Aphids of New Zealand.* N.Z. Department of Scientific and Industrial Research, Wellington, Bulletin 106, 382pp.

Couilloud, R. (1965) Observations sur la faune du cotonnier dans le bassin du Logone, Tchad (exception faite des chenilles de la capsule). *Coton et Fibres Tropicales* 20, 517–530.

Daoud, A.A.K. and El-Haidari, H. (1968) Recorded aphids from Iraq. *Iraq Natural History Museum Publications* 24, 37pp.

Deguine, J.P. (1992) Considérations pour une lutte intégrée vis-à-vis d'*Aphis gossypii* Glover sur culture cotonnière en Afrique Centrale. *Revue scientifique du Tchad* 2, 74–92.

Delattre, R. (1973) *Parasites et Maladies en culture Cotonnière.* Institut de Recherches du Coton et des Textiles exotiques, Paris, 146pp.

Denechère, M. (1981) Note sur la distribution et l'évaluation des populations d'*Aphis gossypii* Glov. (Homoptère, Aphididae) sur cotonnier en République Centrafricaine. *Coton et Fibres Tropicales* 36, 3, 271–280.

Dhawan, A.K., Simwat, G.S. and Sidhu, A.S. (1987) Effect of sowing dates on incidence of sucking pests and bollworms in *arboreum* cotton. *Journal of Research, Punjab Agricultural University* 24, 75–85.

Dunnam, E.W. (1936) Pilosity of the cotton plant in relation to adherence of dusted calcium arsenate. *Journal of Economic Entomology* 29, 1086–1087.

Dunnam, E.W. and Clark, J.C. (1938) The cotton aphid in relation to pilosity of cotton leaves. *Journal of Economic Entomology* 31, 663–666.

Dunnam, E.W. and Clark, J.C. (1941) Cotton aphid multiplication following treatment with calcium arsenate. *Journal of Economic Entomology* 34, 587–588.

Duviard, D. and Mercadier, G. (1973) Les invasions saisonnières de pucerons en culture cotonnière: origine et mécanismes. *Coton et Fibres Tropicales* 28, 483–491.

Eastop, V.F. (1958) *A Study of the Aphididae (Homoptera) of East Africa.* HMSO, Colonial Office, London, 126pp.

Eastop, V.F. (1961) *A Study of the Aphididae (Homoptera) of West Africa.* British Museum (Natural History) London, 93pp.

Eastop, V.F. (1971) Keys for the identification of *Acyrthosiphon* (Homoptera: Aphididae). *Bulletin of the British Museum (National History) (Entomology)* 26, 1–115.

Eastop, V.F. (1983) The biology of the principal virus vectors. In: Plumb, R.T. and Thresh, J.M. (eds) *Plant Virus Epidemiology.* Blackwell Scientific Publications, Oxford, 115–132.

Eastop, V.F. and Hille Ris Lambers, D. (1976) *Survey of the World's Aphids*, W. Junk, The Hague, 533pp.

Eguchi, M. (1937) Results of examinations and surveys on *Aphis gossypii* and its control. *Bulletin of the Agricultural Experiment Station of the Government general of Chosen* 9, 379–416 (in Japanese).

Ekukolé, G. (1990) Effects of selected plants on the fecundity of *Aphis gossypii* Glover under laboratory conditions. *Coton et Fibres Tropicales* 45, 263–266.

Ekukolé, G. (1992a) Preliminary results on the effect of pruning cotton plants in *Aphis gossypii* Glover population in Maroua, North Cameroon. *Coton et Fibres Tropicales* 47, 135–138.

Ekukolé, G. (1992b) Effect of some agronomic and chemical control practices in *Aphis gossypii* populations and stickiness in cotton. *Coton et Fibres Tropicales* 47, 139–143.

El-Fattah, M.I.A. (1975) Effect of certain cultural practices on the infestation of cotton by *Aphis gossypii* Glover (Homoptera: Aphididae). II. Sowing dates and planting spaces. *Journal of Applied Ecology* 78, 301–305.

Elmsheuser, H. and McCracken, A. (1989) Treatment of cotton seeds by plant protection agents. In: Green, M.B. and de B. Lyon, D.J. (eds) *Pest Management in Cotton.* Ellis Horwood, Chichester, pp. 142–152.

El-Zik, K.M. and Thaxton, P.M. (1989) Genetic improvement for resistance to pests and stresses in cotton. In: Frisbie, R.E., El-Zik, K.M. and Wilson, L.T. (eds) *Integrated Pest Management Systems and Cotton Production.* John Wiley & Sons, New York, pp. 191–224.

Escober, J.T., Agati, J.A. and Bergonia, H.T. (1963) Une nouvelle virose du Cotonnier aux Philippines. *Bulletin phytosanitaire de la FAO* 11, 76–81.

Essig, E.D. (1947) Aphids feeding on violaceous plants in California. *Hilgardia* 17, 597–617.

Ewing, K.P. (1943) Cotton aphid damage and control in Texas. *Journal of Economic Entomology* 36, 598–601.

Follin, J.C. and Campagnac, N. (1981) Une maladie nouvelle du cotonnier, présumée d'origine virale observée en Argentine. *Coton et Fibres Tropicales* 36, 313–317.

Folsom, J.W. and Bondy, F.F. (1930) Calcium arsenate as a cause of aphid infestation. *US Department of Agriculture, Circular* No. 116, 11pp.

Frisbie, R.E., El-Zik, K.M. and Wilson, L.T. (1989) *Integrated Pest Management Systems and Cotton Production.* John Wiley & Sons, New York, 437pp.

Furk, C., Powell, D.F. and Heyd, S. (1980) Pirimicarb resistance in the melon and cotton aphid, *Aphis gossypii* Glover. *Plant Pathology* 29, 191–196.

Gaudard, L. (1992) Les traitements insecticides au Nord-Cameroun. *Revue scientifique du Tchad* 2, 104–106.

Goff, C.C. and Tissot, A.N. (1932) The melon aphid, *Aphis gossypii* Glover. *Florida Agricultural Experiment Station Bulletin* No. 252, 23pp.

Grafton-Cardwell, E.E. (1991) Geographical and temporal variation in response to insecticides in various life stages of *Aphis gossypii* (Homoptera: Aphididae) infesting cotton in California. *Journal of Economic Entomology* 84, 741–749.

Gubran, E.E. (1991) *Etude de la résistance chez le Puceron du Coton* Aphis gossypii *Glover (Homoptère: Aphididae) à divers produits insecticides au Soudan.* Thèse de Doctorat en Sciences, Université de Paris VI, 181pp.

Gubran, E.E., Delorme, R., Augé, D. and Moreau, J.P. (1992) Insecticide resistance in cotton aphid (*Aphis gossypii* Glover) in the Sudan Gezira. *Pesticide Science* 35, 101–107.

Gutknecht, J. (1988) Assessing cotton stickiness in the mill. *Textile Monthly* 53–57.

Hall, R.A. and Papierok, B. (1982) Fungi as tropical control agents of arthropods of agricultural and medical importance. *Parasitology* 84, 205–240.

Hardee, D.D. and O'Brien, P.J. (1990) Cotton aphids: current status and future trends in management. In: *Proceedings of the Beltwide Cotton Production Research Conference*, Memphis, Tennessee, pp. 169–171.

Heathcote, G.D. (1972) Evaluating aphid populations on plants. In: Van Emden, H.F. (ed.) *Aphid Technology.* Academic Press, London, pp. 105–145.

Hector, D.J. and Hodkinson, I.D. (1989) *Stickiness in Cotton.* CAB International, Wallingford, UK, 43pp.

Hermoso de Mendoza, A. and Moreno, P. (1989) Cambios cuantitativos en la fauna afidica de los citros valencianos. *Boletin de Sanidad vegetal de Plagas* 15, 139–142.

Higuchi, H. and Miyazaki, M. (1969) A tentative catalogue of host plants of Aphidoidea in Japan. *Insecta Matsumurana*, suppl. 5, 66pp.

Inaizumi, M. (1980) Studies on the life-cycle and polymorphism of *Aphis gossypii* Glover

(Homoptera: Aphididae). *Special Bulletin of the College Agriculture, Utsunamiya University* No. 37, 132pp (in Japanese with English summary).

Inaizumi, M. (1981) Life-cycle of *Aphis gossypii* Glover (Homoptera: Aphididae) with special reference to biotype differenciation on various host plants. *Kontyû* 49, 219–240 (in Japanese with English summary).

Irwin, M.E. (1980) Sampling aphids in soybean fields. In: Kogan, M. and Herzog, D.C. (eds) *Sampling Methods in Soybean Entomology*, Springer Verlag, New York, pp. 239–259.

Isely, D. (1946) The cotton aphid. *Arkansas Agricultural Experiment Station Bulletin* No. 462, 29pp.

Ishaaya, I. and Mendelson, Z. (1987) The susceptibility of the melon aphid, *Aphis gossypii* to insecticides during the cotton growing season. *Hassadeh* 67, 1772–1773 (in Hebrew).

Javaid, I., Zulu, J.N., Matthews, G.A., Norton, G.A. (1987) Cotton insect pest management on small scale farms in Zambia. – I. Farmer's perceptions. *Insect Science and its Applications* 8, 1001–1006.

Jervis, M.A. and Kidd, N.A.C. (1986) Host-feeding strategies in hymenopteran parasitoids. *Biological Review* 61, 395–434.

Jianguo, T., Fujie, T. and Ziping, Y. (1987) Preliminary studies on the insecticide – resistance of cotton aphids in North China. *Journal of Nanjing Agricultural University* 12, 13–21.

Johnson, D.R. and Studebaker, G.E. (1991) Cotton aphid control and management in Arkansas. In: *Proceedings of the Beltwide Cotton Production Research Conference*, Memphis, Tennessee, pp. 689–690.

Johnson, H.N. (1934) Nature of injury to forage legumes by the potato leafhopper. *Journal of Agricultural Research* 49, 379–406.

Kanaraj, D.S. (1956) Notes on south Indian aphids. 1. Description of four new species. *Indian Journal of Entomology* 18, 1–9.

Kennedy, J.S., Lamb, K.P. and Booth, C.O. (1958) Responses of *Aphis fabae* Scop. to water shortage in host plants in pots. *Entomologia Experimentalis et Applicata* 1, 274–291.

Kennedy, J.S., Booth, C.O. and Kershaw, W.J.S. (1961) Host finding by aphids in the field. III. Visual attraction. *Annals of Applied Biology* 49, 1–21.

Kerns, D.L. and Gaylor, M.J. (1991a) Insecticide resistance in field populations of cotton aphids and relative susceptibility of its parasitoid *Lysiphlebus testaceipes*. In: *Proceedings of the Beltwide Cotton Production Research Conference*, Memphis, Tennessee, pp. 682–685.

Kerns, D.L. and Gaylor, M.J. (1991b) Induction of cotton aphid outbreaks by the insecticide sulprofos. In: *Proceedings of the Beltwide Cotton Production Research Conference*, Memphis, Tennessee, pp. 699–701.

Kerns, D.L. and Gaylor, M.J. (1992) Insecticide resistance in field populations of the cotton aphid (*Homoptera: Aphididae*). *Journal of Economic Entomology* 85, 1–8.

Khalid, R.A. and Al-Zarari, A.J. (1983) Estimation of the economic threshold of infestation for cotton aphid, *Aphis gossypii* in cotton in Mosul, Iraq. *Mesopotamia Journal of Agriculture* 17, 71–78.

Khalifa, A. and Sharaf El-Din, M. (1964) Biological and ecological study of *Aphis gossypii* Glover. *Bulletin de la Société Entomologique d'Egypte, Le Caire* 57, 131–153.

Khan, Z.R. and Agarwal, R.A. (1990) Mechanism of resistance to aphid (*Aphis gossypii*) in cotton. *Indian Journal of Entomology* 52, 236–240.

King, E.G. and Phillips, J.R. (1989) 42nd annual conference report on cotton insect research and control. *Proceedings of the Beltwide Cotton Production Research Conference*, Memphis, Tennessee, pp. 180–191.

King, E.G., Philips, J.R. and Head, R.B. (1986) 39th annual conference report on cotton insect research and control. *Proceedings of the Beltwide Cotton Production Research Conference*, Memphis, Tennessee, pp. 126–135.

King, E.G., Philips, J.R. and Head, R.B. (1987) 40th annual conference report on cotton insect research and control. *Proceedings of the Beltwide Cotton Production Research Conference*, Memphis, Tennessee, pp. 170–192.

King, E.G., Philips, J.R. and Head, R.B. (1988) 41st annual conference report on cotton insect research and control. *Proceedings of the Beltwide Cotton Production Research Conference*, Memphis, Tennessee, pp. 188–202.

Komazaki, S. (1982) Effects of constant temperatures on population growth of three aphid species, *Toxoptera citricidus* (Kirkaldy), *Aphis citricola* Van der Goot, and *Aphis gossypii* Glover (Homoptera: Aphididae) on Citrus. *Applied Entomology and Zoology* 17, 75–81.

Kozhaeva, K. (1965) The melon aphid on cotton. *Zaschita Rastenij of Vreditelej i Boleznej* 1965, 36–37 (in Russian).

Kring, J.B. (1955) Biological separation of *Aphis gossypii* Glover and *Aphis sedi* Kaltenbach. *Annals of the Entomological Society of America* 48, 442–444.

Kring, J.B. (1959) The life cycle of the melon aphid, *Aphis gossypii* Glover, an example of facultative migration. *Annals of the Entomological Society of America* 52, 284–286.

Labonne, G., Lauriaut, F. and Quiot, J.B. (1989) Comparaison de trois types de pièges pour l'èchantillonnage des pucerons ailés. *Agronomie* 9, 547–557.

Lamb, K.P. (1955) Physiological Relations between Aphids and their Host Plants with Special References to Wilting. Ph.D. thesis, University of Cambridge, 139pp.

Leclant, F. and Remaudière, G. (1970) Prise en considération des Aphides dans la lutte intégrée en vergers de pêchers. *Entomophaga* 15, 53–81.

Leonard, M.D., Walker, H.G. and Enari, L. (1971) Host plants of *Aphis gossypii* at Los Angeles State and country arboretum, Arcadia, California (U.S.A.). *Proceedings of the Entomological Society of Washington* 74, 9–16.

Liu, Y.C. and Perng, J.J. (1987) Populations growth and temperature-dependent effect of cotton aphid, *Aphis gossypii* Glover. *Chinese Journal of Entomology* 7, 95–111 (in Chinese).

Mahama, A. and Cauquil, J. (1976) La sélection de variétés résistantes à la maladie bleue du cotonnier dans l'Empire Centrafricain. *Coton et Fibres Tropicales* 31, 439–446.

Mamontova, V.A. (1953) *Les pucerons des cultures dans les régions de steppes entrecoupées de forêts de Russie.* Zoological Institute, Academy of Sciences, Ukrainian SSR, 72pp. (in Russian).

Manglitz, G.R., Calkins, C.O., Walstrom, R.J., Hintz, S.D., Kinder, S.D. and Peters, L.L. (1966) Holocyclic strain of the spotted alfalfa aphid in Nebraska and adjacent states. *Journal of Economic Entomology* 59, 636–639.

Mao, G.H. and Xia, Z.C. (1983) Observations on the population dynamics of the natural enemies of the cotton aphid on cotton. *Insect Knowledge* 20, 217–219.

Massonnie, G. (1971) L'élevage des aphides sur milieu synthétique. *Annals de Zoologie – Ecologie animale* 3, 103–123.

Matthews, G.A. (1989) *Cotton Insect Pests and Their Management.* Longman Scientific & Technical, Harlow, Essex, 199pp.

Meyerdirk, D.E., Coudriet, D.L. and Prabhaker, N. (1986) Population dynamics and control strategy for *Bemisia tabaci* in the Imperial Valley, California. *Agricultural Ecosystems and Environment* 17, 61–67.

Michel, B. and Prudent, P. (1985) Acariens et insectes déprédateurs du cotonnier (*Gossypium hirsutum* L.) au Paraguay. *Coton et Fibres Tropicales* 40, 219–230.

Michel, B. and Prudent, P. (1987) Prédateurs et parasitoïdes des ravageurs du cotonnier au Paraguay. *Coton et Fibres Tropicales* 42, 165–172.

Mikhidtdinov, M. (1985) The role of defoliation in the reduction of abundance of the main sucking pests in the cotton agrobiocoenosis. *Izvestiya Academii Nauk Tadzhikskoj SSR, Seriya biologicheskikh Nauk*, 3, 64–65.

Mitchell, H.R., Hatfield, L.D. and Flatt, T.I. (1991) Response of cotton aphids to pyrethroids and organophosphate insecticide applications. In: *Proceedings of the Beltwide Cotton Production Research Conference*, Memphis, Tennessee, pp. 696–698.

Mittler, T.E. (1988) Applications of artificial feeding techniques for aphids. In: Minks, A.K. and Harrewijn, P. (eds) *Aphids: Their Biology, Natural Enemies, and Control*, vol. 2 B. Elsevier Science Publishers, Amsterdam, pp. 145–170.

Mordvilko, A. (1928) The evolution of cycles and the origin of heteroecy (migrations) in plant-lice. *Annals and Magazine of Natural History* 10(2), 570–582.

Moritsu, M. (1948) Injurious aphids in Japan. 1. Aphids of potatoes. *Mushi* 18, 67–75 (in Japanese).

Müller, F.P. (1975) Incidence of the aphid *Acyrthosiphon gossypii* Mordvilko on legumes and on cotton (Homoptera: Aphididae). *Beiträge zur Entomologie* 25, 257–260.

Müller, F.P. (1986) The role of subspecies in aphids for affairs of applied entomology. *Zeitschrift für Angewändte Entomologie* 101, 295–303.

Müller, F.P., El Tigani, Mohamed El Amin (1986) Further notes on Sudanese aphids including the description of a new *Brachyunguis*. *Deutsche Entomologische Zeitschrift* 33, 1–9.

Nan, L.Z., Wang, D.A., Sun, X., Tian, X.G., Zhang, W.H., Wang, G.X., Han, Y.G. and Zhang, J.Q. (1987) Study of protection and application of natural enemies of cotton insect pests. *Natural Enemies of Insects* 9, 125–129.

Narzikulov, M.N. and Umarov, S.A. (1969) *Aphids (Homoptera, Aphidinea) in Tajikistan and Adjacent Regions of Central Asia*. Fauna of the Tajikstan, SSR, tome IX, part II, Dushambe, 229pp. (in Russian).

Nassar, S.A. (1962) Some studies on the biology of the cotton aphid *Aphis gossypii* Glover. *Alexandria Journal of Agricultural Research* 10, 3–22.

Nevsky, V.P. (1929) The plant lice of Central Asia. *Uzbekistan Plant Protection Experimental Station*, Tashkent 16, 425pp. (in Russian).

Niles, G.A. (1980) Breeding cotton for resistance to insect pests. In: Maxwell, F.G. and Jennings, P.R. *Breeding Plants Resistant to Insects*. John Wiley & Sons, New York, pp. 337–369.

Nour, M.A. (1960) On leaf curl of cotton in the Philippines. *FAO Plant Protection Bulletin* 8, 55–56.

Nozato, K. and Toshio, A. (1988) Effects of predation of *Coccinella septempunctata bruckii* (Coleoptera: Coccinellidae) and temperature on seasonal changes of the survival rate of *Aphis gossypii* Glover (Homoptera: Aphididae) population in the warmer region of Japan. *Japanese Journal of Applied Entomology and Zoology* 52, 198–204.

O'Brien, P.J., Stoetzel, M.B. and Hardee, D.D. (1990) Verification of the presence of male and oviparous morphs of the cotton aphid in Midsouth cotton (*Gossypium hirsutum* L.). *Journal of Entomological Science* 25, 73–74.

O'Brien, P.J., Leonard, B.R. and Graves, J.B. (1991) Population dynamics of *Aphis gossypii* Glover. In: *Proceedings of the Beltwide Cotton Production Research Conference*, Memphis, Tennessee, pp. 686–688.

O'Brien, P.J., Abdel-Aal, Y.A., Ottea, J.A. and Graves, J.B. (1992) Relationship of insecticide resistance to carboxylesterases in *Aphis gossypii* (Homoptera: Aphididae) from Midsouth cotton. *Journal of Economic Entomology* 85, 651–657.

Onu, I. (1989) Current status of cotton pests management in Nigeria. In: Berger, M. and Frydrych, D. (eds.) *Actes de la 1ère Conférence de la Recherche cotonnière africaine*, Lomé, Tome II, CIRAD-IRCT, Montpellier, pp. 147–156.

Paddock, F.B. (1919) The cotton or melon louse. *Texas Agricultural Experiment Station Bulletin* No. 257, 54pp.

Papierok, B. (1987) Importance des champignons Entomophthorales dans la régulation naturelle des populations d'insectes d'intérêt économique en zone tropicale. *Mededelingen van de Faculteit Landbouwwetenschappen Rijksuniversiteit Gent* 52, 165–169.

Parker, R.D. and Huffman, R.L. (1991) Effects of early season aphid infestations on cotton yield and quality under dryland conditions in the Texas Coastal Bend. In: *Proceedings of the Beltwide Cotton Production Research Conference*, Memphis, Tennessee, pp. 702–704.

Patch, E. (1925) The melon aphid. Life history studies. *Maine Agricultural Experiment Station Bulletin* No. 326, pp. 185–196.

Patch, E. (1938) Food-plant catalogue of the aphids of the world including the Phylloxeridae. *Maine Agricultural Experiment Station Bulletin* No. 393, 431pp.

Pergande, T. (1895) The cotton or melon plant louse (*Aphis gossypii* Glover). *Insect Life* 7, 309–315.

Pinchard, V. and Cauquil, J. (1992) La protection chimique du cotonnier en Afrique francophone: bilan critique et évolution possible. *Revue scientifique du Tchad* 2, 93–96.

Pollard, D.G. (1958) Feeding of the cotton aphid (*Aphis gossypii* Glover). *Empire Cotton Growing Review* 35, 244–253.

Pollard, D.G. (1973) Plant penetration by feeding aphid (Hemiptera, Aphidoidea): A review. *Bulletin of Entomological Research* 62, 631–714.

Price, J.R., Slosoer, J.E. and Puterka, G.J. (1983) Cotton aphid control, Chillicothe, TX, 1981. *Insecticide and Acaricide Tests* 8, 197–198.

Rabasse, J.M. and Bouchery, Y. (1977) Nouvelles données sur les méthodes d'évaluation des populations de pucerons (Homoptères, Aphididae): séparation des insectes de leur plante-hôte et dénombrement. *Annales de Zoologie – Ecologie animale* 9, 407–423.

Radev, R. (1965) On the overwintering of the cotton leaf aphid. *Rastitelna Zashchita* 13, 7–8 (in Bulgarian).

Rakotofiringa, H. (1989) *Spodoptera littoralis* en culture cotonnière Malagasy. In: Berger, M., Frydrych, D. (eds) *Actes de la 1ère Conference de la Recherche cotonnière africaine*, Lomé, Tome II, CIRAD-IRCT, Montpellier, pp. 165–172.

Ratchford, K.J., Burris, E., Leonard, B.R. and Graves, J.B. (1989) Evaluation of infurrow fungicide, insecticide, and starter fertilizer treatments for effects on early season cotton pests and yields. In: *Proceedings of the Beltwide Cotton Production Conference*, Memphis, Tennessee, pp. 293–295.

Remaudière, G. and Autrique, A. (1985) Ecologie des aphides du Burundi. In: Remaudiere, G. and Autrique, A. (eds) *Contribution à l'Écologie des Aphides Africains*. FAO, Rome, pp. 11–75.

Remaudière, G., Aymonin, G. and Autrique, A. (1985) Les plantes hôtes des pucerons africains. In: Remaudiere, G., Autrique, A. (eds) *Contribution à l'Ecologie des Aphides Africains*. FAO, Rome, pp. 103–139.

Risch, S.J. (1987) Agricultural ecology and insect outbreaks. In: Barbosa, P. and Schultz, J.C. (eds) *Insect Outbreaks*. Academic Press, San Diego, pp. 217–238.

Robert, Y. (1988) Particularités éthologiques des aphides. Cycles. Comportement de vol. *Annales de l'Association Nationale de Protection des Plantes (ANPP)* 2 (1/1), 61–70.

Robert, Y., Dedryver, C.A. and Pierre, J.S. (1987) Sampling techniques. In: Minks, A.K. and Harrewijn, P. (eds) *Aphids: Their Biology, Natural Enemies, and Control*, Vol. 2B. Elsevier Science Publishers, Amsterdam, pp. 1–20.

Roy, D.K. and Behura, B.K. (1983a) Notes on host-plants, feeding behaviour, infestation and ant attendance of cotton aphids *Aphis gossypii* Glov. *Journal of the Bombay Natural History Society* 80, 654–656.

Roy, D.K. and Behura, B.K. (1983b) On the infestation of *Aphis gossypii* Glov., on cotton (*Gossypium herbaceum*). *Pranikee* 4, 438–440.

Saito, T. (1991) Insecticide resistance of the cotton aphid, *Aphis gossypii* Glover (Homoptera: Aphididae). V. Relationship between host preference and organophosphorus resistance. *Japanese Journal of Applied Entomology and Zoology* 35, 145–152.

Sartor, C., Young, D., Hamer, J. and Mitchell, H.C. (1976) Cotton Scounting manual. *Mississipppi State Extension Service Publication* No. 988, 20pp.

Sawicki, R.M., Denholm, I., Forrester, N.W. and Kershaw, C.D. (1989) Present insecticide – resistance management in cotton. In: Green, M.B. and de B. Lyon, D.J. (eds) *Pest Management in Cotton.* Ellis Horwood, Chichester, pp. 31–43.

Schmutterer, H. (1969) *Pests of Crops in Northeast and Central Africa with Particular Reference to the Sudan.* Gustav Fischer Verlag, Stuttgart, 296pp.

Selim, A.A., El-Rafai, S.A. and El-Gantiry, A. (1987) Seasonal fluctuations in the population of *Aphis craccivora* Koch, *Myzus persicae* (Sulz.), *Aphis gossypii* (Glov.) and their parasites. *Annals of Agricultural Science* 32, 1837–1848.

Shibata, B. (1955) Ecological studies of plant lice (9) with special reference to the ecological cycle. *Bulletin of the College of Agriculture, Utsunomiya University* 3, 1–8. (in Japanese with English summary).

Shuja-Uddin, (1977) Observations on normal and diapausing cocoons of the genus *Lipolexis* Foerster (Homoptera: Aphidiidae) from India. *Bolletino del Laboratori di Entomologia agraria 'Fillippo Silvestri'* 34, 51–54.

Shuravleva, I.A. (1956) Biology and noxiousness of the big cotton aphid (*Acyrthosiphon gossypii* Mordv.) in Usbekistan. *Trudy Instituta Zoologii Parasitologii Akademii Nauk Uzbekistanskoj, SSR*, Tashkent 7, 31–48 (in Russian).

Silvie, P. and Papierok, B. (1991) Les ennemis naturels des insectes du cotonnier au Tchad. Premières données sur les champignons de l'ordre des Entomophthorales. *Coton et Fibres Tropicales* 46, 293–303.

Slosser, J.E., Pinchak, W.E. and Rummel, D.E. (1989) A review of known and potential factors affecting the population dynamics of the cotton aphid. *Southwestern Entomologist* 14, 302–313.

Sorin, M. (1975) Aphids of fruit trees in Japan (1.). *Rostria Transactions of the Hemiptera Society of Japan* 25, 167–169 (in Japanese).

Stary, P., Remaudiere, G. and Autrique, A. (1985) Les Aphidiides parasites de pucerons en région éthiopienne. In: Remaudiere, G. and Autrique, A. (eds) *Contribution à l'Écologie des Aphides Africains.* FAO, Rome, pp. 95–102.

Steinkraus, D.C., Kring, T.J. and Tugwell, N.P. (1991) *Neozygites fresenii* in *Aphis gossypii* on cotton. *Southwestern Entomologist* 16, 118–122.

Sterling, W.L., Wilson, L.T., Gutierrez, A.P., Rummel, D.R., Phillips, J.R., Stone, N.D. and Benedict, J.H. (1989) Strategies and tactics for managing insects and mites. In: Frisbie, R.E., El-Zik, K.M. and Wilson, L.T. (eds) *Integrated Pest Management Systems and Cotton Production*, John Wiley & Sons, New York, pp. 267–325.

Stroyan, H.L.G. (1984) Aphids – Pterocommatinae and Aphidinae (Aphidini). Homoptera: Aphididae. In: Fitton, M.G. (ed.) *Handbooks for the Identification of British Insects*, Vol. 2, Part 6. Royal Entomological Society, London, 232pp.

Sudd, J.H. (1987) Ant–aphid mutualism. In: Minks, A.K. and Harrewijn, P. (eds) *Aphids: Their Biology, Natural Enemies, and Control*, Vol. 2 A. Elsevier Science Publishers, Amsterdam, pp. 355–365.

Szelegbiewicz, H. (1963) Aphididae (Homoptera) aus Vorderasien, samt Beschreibung einer neuen Untergattung und Art. *Annales Zoologici* 21, 53–60.

Takada, H. (1988) Interclonal variation in the photoperiodic response for sexual morph production of Japanese *Aphis gossypii* Glover (Hom., Aphididae). *Journal of Applied Entomology* 106, 188–197.

Tang, Z.H., Gong, K.Y. and You, Z.P. (1988) Present status and countermeasures of insecticide resistance to agricultural pests in China. *Pesticide Science* 23, 189–198.

Tarr, S.A.J. (1964) Virus diseases of cotton. *Miscellaneous Publication* No. 18 Commonwealth Mycological Institute, Kew, 23pp.

Thomas, K.H. (1968) Die Blattlaäuse aus der engeren Verwandschaft von *Aphis gossypii* Glover und *Aphis frangulae* Kaltenbach unter besonderer Berücksichtigung ihres Vorkommens an Kartoffel. *Entomologische Abhandlungen, Straatliches Museum für Tierkunde in Dresden* 35, 337–389.

Thygesen, T. (1968) The dispersal of apterous *Aphis gossypii* Glov. marked with ^{32}P. *Acta Agriculturae Scandinavica* 18, 196–198.

Todd, G.W., Getahun, A. and Cress, D.C. (1971) Resistance in barley to the greenbug, *Schizaphis graminum*. 1. Toxicity of the phenolic and flavonoid compounds and related substances. *Annals of the Entomological Society of America* 64, 718–722.

Toe, A. (1992) Synthèse des résultats phytosanitaires des états d'Afrique de l'Ouest. *Revue scientifique du Tchad* 2, 15–19.

Turner, R.B. (1971) Dietary amino acid requirement of cotton aphid *Aphis gossypii*: the sulphur containing amino acid. *Journal of Insect Physiology* 17, 2451–2456.

Ullah, K. and Paul, P. (1985) Chemical control of cotton aphid. *Pakistan Journal of Agricultural Research* 6, 213–217.

Vaissayre, M. (1970) *Biologie du Puceron du Cotonnier, Aphis gossypii Glover, en Conditions Naturelles*. Rapport ORSTOM, CIRAD-IRCT, Montpellier, 53pp.

Van Den Bosch, R. and Hagen, K.S. (1966) Predaceous and parasitic arthropods in California cotton fields. *California Agricultural Experiment Station Bulletin* No. 820, 32pp.

Vendramin, J.D. and Nakano, O. (1981) Damage estimation of *Aphis gossypii* (Homoptera: Aphididae) in the cotton cultivar Lac 17. *Anais de Sociedade entomologica do Brasil* 10, 40–49 (in Portuguese).

Vieira, F.V., Santos, J.H.R. and Oliveira, F.J. (1983) Influence of intercropping on the incidence of the aphid *Aphis gossypii* Glover on semi-perennial cotton. *Fitosaninade* 5, 26–30.

Vinson, S.B. and Scarborough, T.A. (1989) Impact of the imported fire ant on laboratory populations of cotton aphid (*Aphis gossypii*) predators. *Florida Entomologist* 8, 307–344.

Von Arx, R., Baumgartner, J. and Delucchi, V. (1983) A model to simulate the population dynamics of *Bemisia tabaci* Genn. (Stern. Aleyrodidae) on cotton in Sudan Gezira. *Zeitschrift für Angewändte Entomologie* 96, 341–363.

Wall, R.E. (1933) A study of color and color-variation in *Aphis gossypii* Glover. *Annals of the Entomological Society of America* 26, 425–463.

Wang, C.L., Xiang, L.Y., Zhang, G.X. and Chu, H.F. (1962) Studies on the soybean aphid, *Aphis glycines* Matsumura. *Acta Entomologica Sinica* 11, 31–44 (in Chinese with English summary).

Way, M.J. (1963) The mutualism between ants and honeydew producing Homoptera. *Annual Review of Entomology* 8, 307–344.

Weismann, L. and Vallo, V. (1963) *Die schwarze Rübenblattlaus* (Aphis fabae *Scop.)*. Vydavatelštvo Slovenskej Akadémie Vied, Bratislava, 301pp. (in Czechoslovak).

Weismann, L., Ashur, N., Afify, A.M. and Zaki, F.N. (1970) Reaktion der Blattlaus *Aphis gossypii* Glover auf die Veränderung des physiologischen Zustandes der Baumwollpflanze. *Biologia* 25, 527–535.

Weismann, L., Jasic, J., Afify, M. and Zafi, F.N. (1971) Populations dynamik der

Baumwollblattlaus *Aphis gossypii* Glover in Baumwollpflanzungen unter der ariden Bedingungen Oberägyptens. *Biologia* 26, 601–610.

Whitcomb, W.H. and Bell, K. (1964) Predaceous insects, spiders, and mites of Arkansas cotton fields. *Arkansas Agricultural Experiment Station Bulletin* No. 690, 84pp.

World Health Organization (1989) *The WHO Recommended Classification of Pesticides by Hazard and Guidelines to Classification 1988–1989.* World Health Organization, WHO/VBC/88.953, Geneva, 39pp.

Xia, X.Y. and Sterling, W.L. (1987) Computer simulation of cotton aphid population dynamics. *Acta Phytophylactica Sinica* 14, 151–156 (in Chinese).

Zalom, F.G. and Natwick, E.T. (1987) Developmental time of sweet-potato whitefly (Homoptera: Aleyrodidae) in small field cages on cotton plants. *Florida Entomologist* 70, 427–431.

Zettler, F.W., Louie, R. and Olson, A.M. (1967) Collections of winged aphids from sticky traps compared with collections from bean leaves and water pan traps. *Journal of Economic Entomology* 60, 242–244.

Zhang, G.X. and Zhong, T.S. (1990) Experimental studies in some aphid life-cycle patterns and hybridization of the sibling species. In: Campbell, R.K. and Eikenbary, R.S. (eds) *Aphid Plant Genotype Interactions*, Elsevier Science Publishers, Amsterdam, pp. 37–50.

Zhang, L.X., Niu, X.T. and Wu, M. (1987) A dynamic simulation model of the biological control system of cotton aphids. In: *Proceedings of the 1st International Conference on Agricultural Systems Engineering*, 11–14 August 1987, Changchun, China. China Machine Press, Jilin University of Technology, Changchun, China, pp. 456–464.

Zhi-Yi, L. and Guo-Pei, G. (1986) Population dynamics of cotton aphids on cotton during square-boll stage and the relation between population age structure and parasitism. *Acta Entomologica Sinica* 29, 56–61.

Zhi-Yi, L. and Wei-Nian, Z. (1982) The fluctuation of hymenopterous of cotton aphid and the effect of insecticide application in cotton field. *Contribution of the Shangai Institute of Entomology* 18, 247–256.

13 *Bemisia* and *Trialeurodes* (Hemiptera: Aleyrodidae)

G.D. Butler, Jr. and T.J. Henneberry

United States Department of Agriculture, Agricultural Research Service, Western Cotton Research Laboratory, 4135 E Broadway, Phoenix, Arizona 85040, USA

Introduction

Several species of the insect family Aleyrodidae, known as whiteflies (Plate VI.2), infest cotton, *Gossypium* spp. (Russell, 1948, 1957; Pearson and Maxwell Darling, 1958; Gerling, 1967). Most whiteflies occur in tropical and subtropical regions roughly corresponding to the areas between the 30th parallels. The sweetpotato whitefly, *Bemisia tabaci* Gennadius, has been reported most frequently as an economic pest on a worldwide basis. In a recent survey, 16 of 27 cotton-producing countries throughout the world reported *B. tabaci* as a major pest in mid to late periods of the cotton growing season (Anon., 1989). The bandedwing whitefly, *Trialeurodes abutilonea* (Haldeman), and greenhouse whitefly, *T. vaporariorum* (Haldeman), occur as pests of economic importance sporadically in the United States (Clower *et al.*, 1971; Clower and Watve, 1973; Jones *et al.*, 1975, Flint *et al.*, 1984). Also, Gerling (1967) reported cotton in California as a new host record for the iris whitefly, *Aleyrodes spiraeoides* Quaintence, but the species has not been reported to cause damage of economic significance. *Trialeurodes lubia* (El-Khidir and Khalifa), has been associated with cotton as well as other crops in the Sudan (El-Khidir and Khalifa, 1962; Khalifa and El-Khidir, 1964).

Distribution, Synonyms and Description of Sweetpotato Whitefly Stages

The geographical distribution and role of *B. tabaci* as a damaging economic pest of cultivated crops in Southern Europe, the United States, Central America, West Indies, South America, Africa and Asia has been reported by a number of authors (Hutchison *et al.*, 1950; Silberschmidt and Tommasi, 1955; Russell, 1957; Varma, 1963; Marchoux *et al.*, 1970; Maramorosch, 1975).

The insect has been referred to in the literature as the cassava whitefly, tobacco whitefly and cotton whitefly (Mound, 1963), as well as *B. tabaci*. The taxonomic relationships and the synonomy of the species have been reported by Russell (1957). Additionally, Gameel (1972) and Mound and Halsey (1978) reviewed the synonymy of *B. tabaci* and Gameel (1972) listed the synonyms as determined by a number of authors as follows:

Aleurodes tabaci Gennadius
Aleurodes inconspicua Quaintance
Bemisia inconspicua (Quaintance) Quaintance and Baker
Bemisia costa – limai Bondar
Bemisia signata Bondar
Bemisia bahiana Bondar
Bemisia gossypiperda Misra and Lamba
Bemisia hibisci Takahashi
Bemisia longispina Priesder and Hosny
Bemisia goldingi Corbett
Bemisia nigeriensis Corbett
Bemisia tabaci (Gennadius) Takahashi
Bemisia rhodesiansis Corbett

The taxonomic confusion has arisen because of the variation in numbers and distribution of dorsal setae in several stages of the insect's life cycle, its association with many different host plants and the effect of host plants on morphological characteristics (Gameel, 1972). All stages of the insect were redescribed in detail and the reader is referred to that work for additional information, although an abbreviated description of the stages from Hill (1969), Gameel (1972), and others is included in the following sections.

Sweetpotato whitefly race and/or biotype designation have also often been recorded in past literature in relation to host affinities and virus transmission vector interactions (Brown and Bird, 1992). The possibility of biotype occurrence has received new interest with the increasing importance of sweetpotato whitefly in crop production on a worldwide basis. The detection of differences in electrophoretic isozyme patterns, biology and extended host range have provided evidence for the existence of different sweetpotato whitefly biotypes as compared to sweetpotato whitefly previously encountered in the desert southwestern United States crop production areas (Brown *et al.*, 1989; Costa and Brown, 1990, 1991). The biotype has been referred to in the literature as the sweetpotato whitefly strain B. Based on this information and expanded studies using polymerase chain-reaction-based DNA differentiation test, allozymic frequency analysis, crossing experiments, and mating behaviour observations, Perring *et al.* (1993) suggested that the sweetpotato whitefly strain B is a distinct species. The authors also named the new species the 'silverleaf whitefly'.

Host Plants

Bemisia tabaci has been reported from 315 host plants worldwide (Mound and Halsey, 1978; Anon., 1981), including 172 in Egypt (Azab *et al.*, 1970), 52 in Israel

(Avidov and Harpaz, 1969), 115 in the Sudan (Gameel, 1972), and 87 in Taiwan (Ghong, 1969). Additional hosts have been reported in the United States (Coudriet *et al.*, 1985, 1986).

Life History and Biology

Adult *B. tabaci* are small, the female being 1.1–1.2 mm in length, and the male slightly smaller. The bodies are pale yellow in colour with two pairs of white wings. Antennae of females are longer than those of males. Genitalia of females, consisting of outer and inner vulvulae, are rounded in appearance, whereas parameres of males are extended, narrow and pointed in appearance. Adults emerge dorsally from the pupae through a T-shaped fissure in the integument. Most emergence occurs between 08.00 and 12.00 hours (Husain and Trehan, 1933; Azab *et al.*, 1971; Butler *et al.*, 1983, 1986b). Parthenogenetic reproduction occurs with unfertilized eggs producing males only (Husain and Trehan, 1933) and male to female ratios of emerging adults range from close to 1:1 to 1:6 (Azab *et al.*, 1971; Butler *et al.*, 1983). The adults frequently remain on or near the pupal case for 10–20 minutes following emergence to spread and dry their wings (Avidov, 1956); they begin feeding soon thereafter. Copulation occurs within a few hours to 1–2 days after emergence (Avidov, 1956; Khalifa and El-Khidir, 1964; Azab *et al.*, 1971; Gameel, 1974). Pairing generally occurs after 08.00 in the morning and may continue to occur as late as 16.00 in the afternoon, and males actively seek female mates (Butler *et al.*, 1986b). Males exhibit aggressive behaviour, approaching the female with antennal stroking and wing fluttering that may last as long as 20 minutes, with the actual time *in copula* being no longer than 4 minutes (Gameel, 1974).

Under laboratory conditions preoviposition periods range from 3.3 to 4.9 days at 28°C and 14°C, respectively. Under field and insectary conditions the preoviposition period ranges from one to eight days at average temperatures of 21.7–26.5°C, to as long as 22 days at temperatures of 13.8°C (Avidov, 1956; Avidov and Harpaz, 1969; Azab *et al.*, 1971). The temperature threshold for oviposition may be about 14°C (Avidov, 1956). However, Husain and Trehan (1933) observed under field conditions that no eggs were laid at temperatures below 22.8°C. Females insert their ovipositor through the leaf epidermal layer into mesophyll tissue so eggs are firmly embedded in the leaf tissue (Avidov, 1956), and vertically oriented to the leaf surface. Most eggs are laid on the underside of the leaves. Eggs are eliptically shaped, narrow at the apical end, broadly rounded at the base, with a pedicel or stalk. They are yellowish-white, turning brownish prior to hatch, and clearly exhibiting red eyespots. Eggs retain their shape after hatching (Hill, 1969). Numbers of eggs per female may range from 28 to 43 (Husain and Trehan, 1933), and from 72 to 81, at 26.7°C and 32.2°C, respectively (Butler *et al.*, 1983). Other reports suggest fecundity of 50 eggs per female (Avidov and Harpaz, 1969), and 161 eggs per female (Azab *et al.*, 1971). A single female was reported to lay 300 eggs (Avidov, 1956). Maximum oviposition occurs within the first week of adult life, declining thereafter (Gameel, 1974).

A number of factors apparently affect oviposition and oviposition site preference. Adults are highly mobile within and between plants, moving to preferred oviposition sites, and adults may be distributed on leaves throughout the cotton plant (Mabbett *et al.*, 1980). Generally, females prefer younger upper leaves of the plant for oviposition (Avidov, 1956). Eggs are generally found in the top, larvae in the middle, and pupae on the bottom portions of cotton plants (Gerling *et al.*, 1980). Increased numbers of *B. tabaci* have been associated with increased leaf-nitrogen content of cotton plants (Joyce, 1958). Some observations have suggested that *B. tabaci* prefer leaves with the highest pH values (Husain *et al.*, 1936b). Under laboratory conditions, using a synthetic diet and membrane-feeding technique, adults preferred food with a pH ranging from 6.0 to 7.5 (Magdal *et al.* 1982). Also, the leaf pH from 120-day-old cotton plants averaged pH 6.8, while that of leaves from 60-day-old cotton plants averaged pH 5.9, suggesting a partial explanation for rapid rates of *B. tabaci* increase late in the season. Similar results have been reported by other authors (Berlinger *et al.*, 1983) who also showed that a definite diet preference exists for 15% sucrose solutions compared with 0.5, 10, 20 and 25% concentrations. Mor *et al.* (1982) found that in sprinkler-, drip-, and non-irrigated plots, higher populations were found on plants exposed to water stress; peak numbers were reached where water potential in leaves at noon was -24 and -26 bars (-2400 and -2600 kPa). Additionally, Mor (1983) reported an increase in the total number of larvae per cm^2 of leaf, as well as per plant, in water-deficient cotton plants. Thus, the stage of plant growth, cultural practices such as irrigation and fertilization (Jackson *et al.*, 1973), and weather (Seif, 1981), and the physiology of the plant probably all interact to affect the population dynamics of the species.

Relative humidity is apparently not a factor in egg viability since eggs are inserted into the leaf tissue of host plants (Avidov, 1956). Temperature, however, does affect egg hatch. Published information indicates that at 15°C and 20°C, 62% and 81%, respectively, and at 25°C to 35°C from 92% to 98% of the eggs hatch, but egg viability is reduced to 59% at 40°C (Gameel, 1978). Butler *et al.* (1983) found that egg hatch, under fluctuating temperatures with a mean of 26.7°C or 32.2°C, was 68% and 75%, respectively. Eggs develop under laboratory conditions of constant temperatures in 22.5, 11.5, 9.9, 7.6, 6.1, 5.4 and 5.0 days at 16.7, 20.0, 22.5, 25.0, 27.5, 30.0 and 32.5°C, respectively. No hatch occurred at 36.0°C (Butler *et al.*, 1983). Similar results have been reported by El-Helaly *et al.* (1971) and Gameel (1978), but the egg development times of Von Arx *et al.* (1983a) agree up to 28°C but deviate considerably from some of these reports at temperatures above 28°C. Under field conditions in California, eggs required 27.0 degree-day heat units (10°C and 32°C thresholds) for development (Zalom and Natwick, 1987). The reports of number of days to egg hatch in the field and insectary appear to agree well with laboratory data (Avidov, 1956; Khalifa and El-Khidir, 1964; El-Khidir, 1965).

The immature stages, other than the egg, consist of three larval instars and pupae (Azab *et al.*, 1969). Although immature stages undergo incomplete metamorphosis and should probably be referred to as nymphs, the more common larval term is retained to conform to literature citations. From Gameel (1972), first instar larvae are yellowish, proceeding to greenish-yellow in the second and third instars. Dorsal surfaces are slightly convex to convex with margins smooth in the first instar to weakly and slightly crenulated in the second and third instars, with apparent segmentation. Ventral surface exhibits antennae and segmented appendages. All larval instars have a colourless waxy coating. Pupae are yellowish to light brown, with crenulated margins and covered dorsally with a layer of colourless wax. Development of the first, second and third larval instars and the pupae at 25.4°C and 31.0°C occurs in 3.0, 2.0, 1.9 and 4.7 days and 4.5, 2.7, 2.6 and 6.2 days, respectively, under laboratory conditions (El-Helaly *et al.*, 1971). At 14°C and 25°C, development of the four stages occurs in 9.0, 7.0, 9.5 and 27.5 days and 2.8, 2.4, 3.0 and 4.7 days, respectively (Sharaf and Batta, 1985). Similar rates of development occur under field conditions (Khalifa and El-Khidir, 1964). Zalom and Natwick (1987) found that development of the larval instars required an average of 19.0, 140.5, 55.0 and 54.0 degree days, respectively, in field cages in California using developmental thresholds of 10°C and 32°C. Total development from egg to adult has been found to occur in 65.1, 48.7, 34.7, 27.8, 23.6, 17.8 and 16.6 days at 14.9, 16.7, 20.0, 22.5, 25.0, 27.5 and 30.0°C, respectively (Butler *et al.*, 1983). Total developmental times from egg to adult reported by Sharaf and Batta (1985), as 69.5 and 19.3 days at 14°C and 25°C, respectively, are similar. Also, Von Arx *et al.* (1983a) obtained similar development rates at temperatures of 20–27°C, but slower developmental rates at lower or higher temperatures. The differences that occur may be explained because development time from egg to adult appears to be influenced by the cultivated or wild host inhabited (Coudriet *et al.*, 1985, 1986). Additionally, egg, larval and the first half of the pupal stage have been found to develop faster under long-day photoperiods as opposed to short days at the same temperatures (El-Helaly *et al.*, 1977). In any event, the insect can develop 11–12 generations per year in many tropical and subtropical areas of the world (Avidov, 1956; Azab *et al.*, 1971; Gameel, 1978).

Adult females are longer lived than males (Avidov, 1956; El-Khidir, 1965; Azab *et al.*, 1971; Butler *et al.*, 1983). Adult longevity is dependent on temperature, and under laboratory conditions males may live 8.2–11.7 days and females 14.8–19.2 days at 14–28°C (Sharaf and Batta, 1985). Males and females lived, respectively, for 7.6 and 8.0 days at 26.7°C and for 10.4 and 11.7 days at 32.2°C (Butler *et al.*, 1983). Under field and insectary conditions the range of adult longevity is considerably wider, depending on the time of year. For example, male longevities ranging from 6.4 to 34.0 days and female longevities from 14.5 to 55.3 days with temperatures ranging from 12.7°C to 26.5°C have occurred under outdoor conditions at various times of the year (Avidov, 1956).

Bemisia tabaci infestations in cotton fields are initiated during May or earlier in the year when adults move from overwintering hosts and/or other cultivated hosts (Gerling 1967; Butler and Henneberry, 1985) (Figs 13.1 and 13.2). Populations develop slowly through June and early July, expand rapidly in late July and August, reaching peak numbers in September, declining thereafter until the crop is harvested (Figs 13.3 and 13.4). The seasonal distribution may vary somewhat depending on the cotton-growing region but population growth rates are similar. Life table analysis of *B. tabaci* population development on cotton in Israel showed higher survival rates from egg to adult in early and midseason generations as

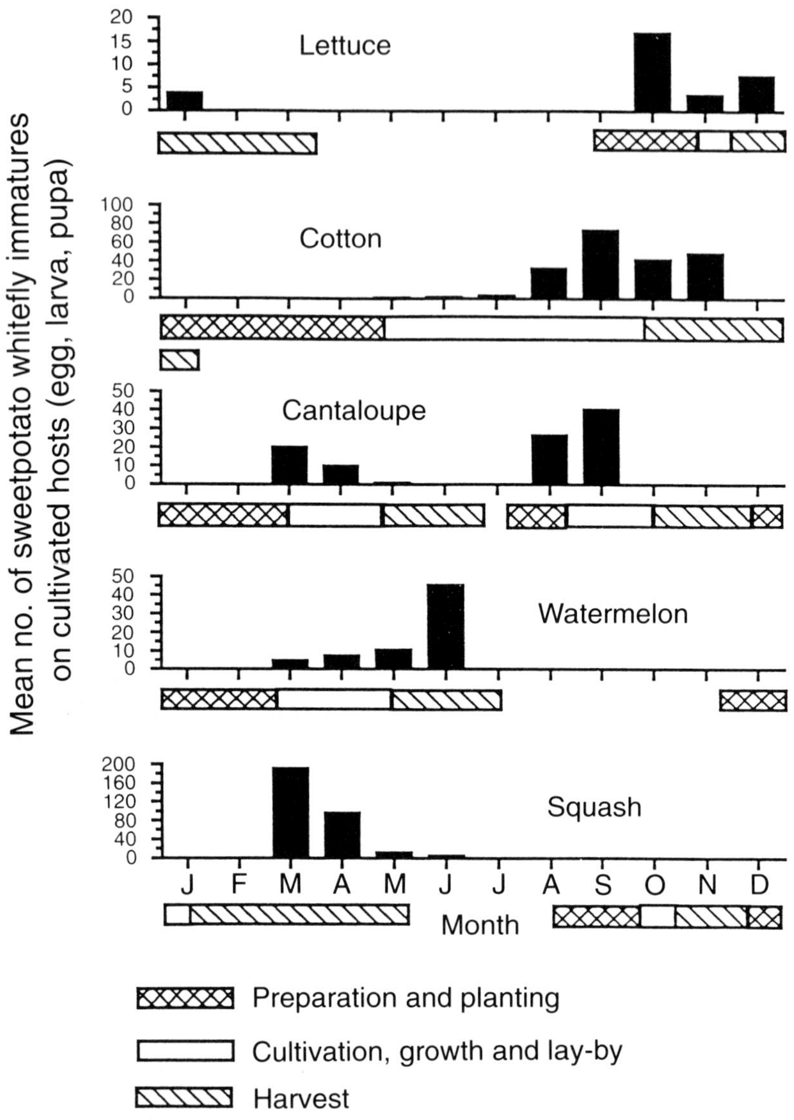

Fig. 13.1. Selected cultivated crop planting and harvest dates and seasonal distribution of immature sweetpotato whiteflies per leaf.

compared to late-season generations (Horowitz *et al.*, 1984). The authors proposed that favourable temperatures and host nutritional status contributed to early and midseason survival, and plant senescense and lower temperatures contributed to increased mortality of late-season generations. The majority of *B. tabaci* mortality occurs during the first-instar larval stages (Horowitz *et al.*, 1984; Horowitz, 1986) and first-instar larval mortality contributes most to the total generation mortality. First instar mortality has been attributed primarily to climatic factors (e.g. low humidity, high temperatures) and to predation by lacewings, spiders and *Orius* spp., and possibly host-plant effects. Moderate mortality occurring in the two remaining larval instars and pupae has been attributed to weather and to parasitism by *Eretmocerus mundus* Merc. and *Encarsia lutea* (Masi).

During the winter months *B. tabaci* populations are generally at their lowest point. The decline in the relative population density may be due to an increase in the availability of leaf substrate as a result of dispersal to many hosts, a decline in reproduction or an increase in mortality (Ohnesorge *et al.*, 1980, 1981). In some areas, such as the Jordan Valley, Israel, the amount of leaf area available for

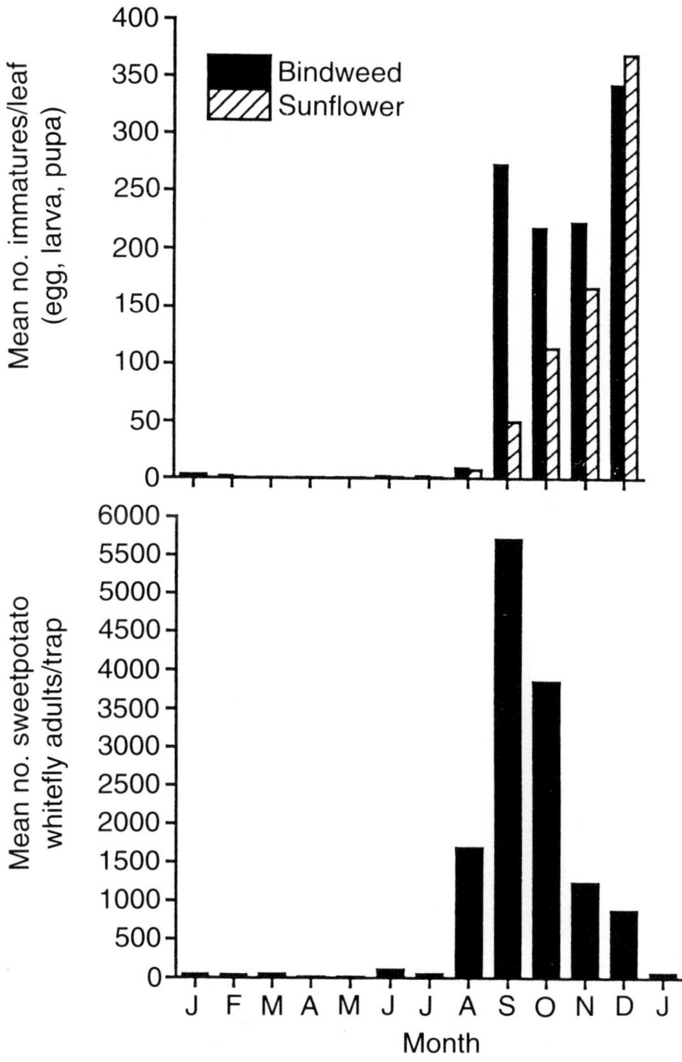

Fig. 13.2. Seasonal distribution of sweetpotato whitefly adults per sticky trap and immatures per leaf in field bindweed and wild sunflower habitats.

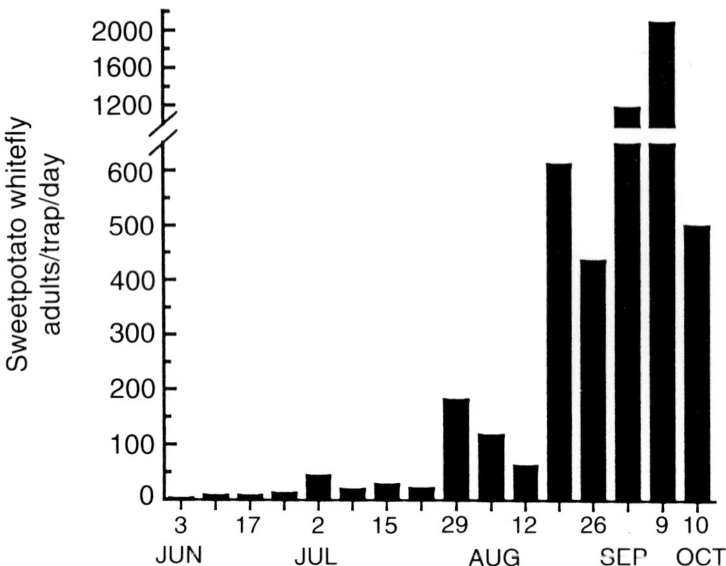

Fig. 13.3. Seasonal distribution of adult sweetpotato whiteflies on yellow sticky traps in cotton.

infestation is greatly multiplied by the planting and production of numerous host plants such as tomatoes (*Lycopersicon* spp.), squash (*Cucurbita* spp.), potatoes (*Solanum tuberosum* L.) and cauliflower (*Brassica oleracea* var. *botrytis* L.) (Ohnesorge *et al.*, 1981). Among the cultivated plants in the Punjab upon which overwintering takes place are rape (*Brassica napul* L.), cauliflower, turnips (*Brassica rapa* L.) and

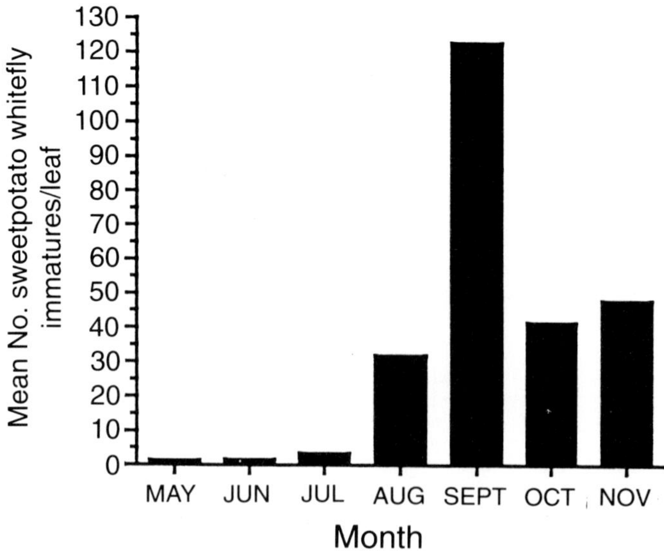

Fig. 13.4. Seasonal distribution of immature sweetpotato whiteflies per cotton leaf.

potatoes. The more important weeds include *Sonchus* spp. *Euphorbia* spp., and *Convolvulus arvensis* L. (Husain *et al*., 1936a). In Israel, Gerling (1984) conducted a survey which revealed 19 plant species that were hosts during the winter. This list includes: *Abutilon grandifolium* L., *Lantana camara* L., *Chrysanthemum indicum* L., *Malva parviflora* L., *Gerbera* spp., *Solanum vilosum* L., *Withania somnifera* (L.) Dunal, *Celtis sastralis* L., *Lonicera etrusca* L., *Verbena* spp., *Cercis siliquastrum*, *Convolvulus arvensis* L., *Plumbago europaea, Alcea setosa* L., *Tropaeolum majus* L., *Calendula* spp. and *Sonchus oleraceus* L. Particularly high population levels occurred on *L. camera*, *A. grandifolium*, and *Impomoea batatas* L. Lam.

In the Sudan, *Ipomoea cordofana* L. is an important winter host (Gameel, 1978). In southern California, Gerling (1967) found all developmental stages of *B. tabaci* on *Malva parviflora* L. This plant is a common weed and germinates during September and October and blooms during February and March. It is frost resistant and thus the only broad-leaved weed species that survives frost. Immature stages, adults and parasites are found on this host plant. Butler and Henneberry (1984b) observed that *Helianthus annuus* L., *Convolvulus arvensis* L. and *Lactuca serriola* L. were also important overwintering hosts in the Imperial Valley. Coudriet *et al.* (1985) found that a number of crops grown in winter months in southern California were hosts upon which *B. tabaci* can complete development. These plants include carrot *(Daucus carota* L.), broccoli (*Brassica oleracea* L.), squash (*Curcurbita* spp.), eggplant (*Solanum malongena* L.), guar (*Cyamopsis tetragonoloba* (L.) Taub.), guayule (*Parthenium argentatus* A. Gray), alfalfa (*Medicago sativa* L.) and lettuce (*Lactuca sativa* L.).

Oviposition is greatly reduced at low winter temperatures. Only 18 offspring were produced per parent pair during studies in Israel (Avidov and Harpaz, 1969). In the Jordan Valley, oviposition took place throughout the winter although it was greatly reduced (Ohnesorge *et al*., 1981). In the Punjab, adults emerge about the end of January and multiply on cultivated and weed host plants (Husain *et al*., 1936a). At least one generation and a partial second was completed on lettuce in southern California during the winter. An estimate of the number of empty pupal cases in the late autumn indicated that one adult *B. tabaci* was produced for every 25 mature lettuce leaves or about 5000 adults per hectare of lettuce. The egg laying from these adults should be sufficient to maintain the species until warmer weather and appearance of other host plants. Development during the winter occurred on a number of crop hosts such as lettuce, carrots, broccoli, squash, eggplant and alfalfa (Coudriet *et al*., 1985).

One of the major effects of low winter temperatures is the asynchrony of *B. tabaci* development and leaf ageing. Low temperatures cause the development of eggs and larvae to slow down or cease while leaf ageing and destruction by microorganisms continues: leaf ageing is thus considered to be a key mortality factor during the winter months. Plants vary greatly in the retention of their leaves: some hosts such as cauliflower and shrubs such as *Lantana* spp. retain leaves, while on other plants, such as squash and tomatoes, older leaves die before the completion of *B. tabaci* development when temperatures are low. The population density of *B. tabaci* during December in Jordan indicated that low temperatures in combination with heavy rains increased the mortality of the immature stages (Ohnesorge *et al*., 1981). Although from December onwards the number of adults decreased considerably in the Punjab, egg and nymphs remained on alternate hosts throughout the winter (Husain *et al.*, 1936a).

Damage and Economic Losses

Economic losses in cotton can result from direct feeding damage and reduced yield (Husain and Trehan, 1933; Mound, 1965b; Larios, 1979), lint contamination with honeydew and associated fungi (Gerling *et al.*, 1980, Hector and Hodkinson, 1989). *Bemisia tabaci* is also considered the most important aleyrodid vector of agents causing plant diseases, and in cotton transmits the aetiological organism causing cotton leaf crumple and cotton leaf curl. Physiological changes in the cotton plant may also occur as a result of infestations that could partially account for reduced yields. Fowler (1956) found that the nitrogen index was higher in lightly infested cotton plants than in heavily infested plants until about 65 days after planting, but was higher in heavily infested plants thereafter; total sugars were generally lower in heavily infested plants and the phosphorus index was higher. The author proposed that under heavy infestations these changes may have significant effects on boll and lint development.

Honeydew produced by *B. tabaci* accumulates on the leaves and exposed cotton lint fibres of open bolls (Nuessly *et al.*, 1989). *Bemisia tabaci* feed by inserting stylets between epidermal and parenchyma cells to reach phloem tissues. Adults and all the larval instars excrete copious amounts of honeydew. These excretions make the fibre sticky and also provide a medium for the growth of fibre-staining fungi that reduce cotton quality. Heavy fungal growth on honeydew-covered leaves may lead to premature leaf drop and increased exposure of open bolls to potential contamination from honeydew. The resulting sticky cotton is a serious problem in the cotton industry. Cotton ginning output may be reduced and more frequent replacement of gin blades required when sticky cotton is encountered (Khalifa, 1980; Khalifa and Gameel, 1982). At the textile mill honeydew-contaminated lint interferes in the cotton fibre processing. Sugar residue deposits may interfere with smooth machinery operation during carding, drawing, roving or spinning (Perkins, 1971; Hector and Hodkinson, 1989). This situation may have been aggravated in the spinning activity in recent years because of new advances in high-speed spinning technology (Gutnecht and Fournier, 1988). Sticky cotton has become a limiting factor in some cotton-producing countries and a highly important quality consideration in the textile industry (Khalifa and Gameel, 1982; Perkins, 1983).

Diagnosis of the problem is complicated by the need for different techniques necessary to identify: *natural physiological sugars* (mostly reducing sugars) produced by the cotton plant that do not appear to be a major cause of cotton lint stickiness as opposed to; *sugars of honeydew-producing insects* (reducing sugars, non-reducing sugars and polyols) that do cause stickiness; and *the metabolic products of these sugars* after enzymatic degradation by various biological agents (Gutnecht and Fournier, 1988). Studies by these authors suggest that effective control of honeydew-producing insects can reduce the cotton lint stickiness problem. The importance of physiological sugars in the sticky cotton problem has not been firmly established, but cotton stickiness has most often occurred because of insect honeydew (Hector and Hodkinson, 1989).

The infectious agent of cotton leaf crumple disease is transmitted by *B. tabaci* in the United States (Dickson *et al.*, 1954). Infected plants exhibit floral hypertrophy, severe foliar malformation and yield reduction. The disease was first described in California in 1954 and in Arizona in 1960 (Dickson *et al.*, 1954; Allen *et al.*, 1960). Studies have been conducted on transmission (Laird and Dickson, 1959; Duffus and Flock, 1982), host range (Duffus and Flock, 1982; Brown and Nelson, 1986), symptomology (Dickson *et al.*, 1954; Erwin and Meyer, 1961) and economic impact (Allen *et al.*, 1960; Van Schaik *et al.*, 1962; Russell, 1982). The virus-like particles of the organism found in partially purified preparations of infected bean plants have been identified and described (Brown and Nelson, 1984, 1986). As a result of an extensive host-range study, over 20 plant species within the Leguminosae and Malvaceae, in addition to *Gossypium* species, are now recognized as hosts of the cotton leaf crumple virus (Brown and Nelson, 1986). Viruliferous *B. tabaci* have been collected from weeds such as cheeseweed (*Malva parviflora* L.), field bindweed (*Convolvulus arvensis* L.), globe-mallow (*Sphaeralcea coulteri* (Wats.) Gray), prickly lettuce (*Lactuca serriola* L.) and sweetpotato (*Ipomea batatas* L. Lam.), in addition to commercial cotton (Butler and Brown, 1985). Cotton plants infected early in the season develop severe foliar symptoms and yield reduction (Butler *et al.*, 1986a). Significant yield reductions may also occur when plants are infected later in the season, despite only mild foliar symptoms (Brown *et al.*, 1985).

Leaf curl diseases transmitted by *B. tabaci* are characterized by strong curling of the leaves without typical mosaic symptoms, plants may be stunted or show leaves with diffuse chlorosis and vein clearing or thickening (Costa, 1976; Bird and Maramorosch, 1978). Cotton leaf curl occurs primarily in Africa (Bird and Maramorosch, 1978), and was originally reported in 1912, found to be virus-induced in 1926, and vectored by *B. tabaci* in 1930 (Tarr, 1949). Combinations of symptoms may occur that include prominent curling of leaf margins, crinkled leaf appearance with vein thickening and chlorotic spots or streaks (Tarr, 1949). Cotton plants infected at an early age may suffer severely reduced yield as opposed to infections occurring in older plants past the peak fruiting cycle. Effective control methods have not been developed, but resistant varieties, vector control, elimination of virus reservoirs and adoption of agronomic practices to minimize exposure of the plant to the vector and/or crop rotation appear to be the best approaches to minimizing the incidence of the disease. These methods apply equally well to cotton leaf crumple in the United States.

Factors Affecting Pest Status

The pattern of *B. tabaci* population dynamics in cotton is typically characterized by exponential growth (Gameel, 1969; Gerling *et al.*, 1980; Von Arx *et al.*, 1983b; Butler *et al.*, 1985b; Zalom *et al.*, 1985; Natwick *et al.*, 1984). The exponential growth pattern observed is not surprising in view of the insect's short generation developmental time, relatively long reproductive period, and high reproductive rate (Butler *et al.*, 1983; Von Arx *et al.*, 1983a; Zalom *et al.*, 1985; Youngman *et al.*, 1986).

Conventional chemical control is difficult to achieve because of the distribution of the immature forms solely on the underside of leaves, with older larvae and pupae located lower in the plant canopy (Gerling *et al.*, 1980), the development of insecticide resistance (Prabhaker *et al.*, 1985) and the waxy coating of immature stages (Gameel, 1972).

Bellows and Arakawa (1988) found that more than half of the immature *B. tabaci* population was found in the middle third of the cotton plant. The *B. tabaci* problem in the cotton-growing region of the Sudan Gezira appears to have occurred soon after large-scale spraying with DDT (Joyce, 1955; Joyce and Roberts, 1959; van der Laan, 1961) for control of other insect pests, followed by increased numbers of sprays as well as mixtures of insecticides (Eveleens, 1983). Several explanations have been suggested, including reduced natural enemy populations (Joyce, 1955), less competition for *B. tabaci* as a result of the effective control of the jassid, *Empoasca lybica* de Berg (Pearson and Maxwell Darling, 1958; Joyce, 1959), and an effect of DDT on cotton-plant physiology that resulted in increased *B. tabaci* fecundity (van der Laan, 1961). Eveleens (1983) concluded that the increasing problem probably resulted because of the development of resistance and reduced impact of natural mortality factors, principally the effectiveness of certain hymenopterous parasites, particularly in the late season. Dittrich *et al.* (1985) suggested that *B. tabaci* fertility stimulation by DDT residues and the development of resistance to organophosphate insecticides were the major factors contributing to the emergence of the insect as a primary pest of cotton in the Sudan. The authors considered that the reduced role of parasitoids (Gameel, 1969; Hassan, 1970) in insecticide-treated fields did not significantly contribute to the increasing problem. Gutknecht and Fournier (1988) proposed a combination of factors that include change from organophosphorus to pyrethroid insecticide for cotton insect control, reduced rainfall, increased crop fertilization and lengthening the growing season as possible causes for the increasing *B. tabaci* problems in the Sudan Gezira cotton-growing areas of Africa. Prabhaker *et al.* (1985) proposed several other possibilities with regard to the increasing problem in the Imperial Valley of California: specifically these included the availability of overwintering hosts, abundance of summer hosts, moderate winter temperatures and rainfall, low populations of natural enemies and insecticide resistance.

A number of other factors contribute to intensifying the pest status of *B. tabaci*. The diversity of the cultivated and weed host plants attacked increases the complexity of the problem and the seasonal distribution and population dynamics of the insect. For example, in India, *B. tabaci* emerge in mid-January from weeds, such as *Convolvulus arvensis* L., *Euphorbia* spp. and cultivated *Brassica* spp., spreading to cotton, cucurbits, and *Hibiscus* spp. by early April and returning to *Brassica* spp., *Solanum* spp. and various weed species at the end of the cotton season (Husain and Trehan, 1933; Husain *et al.*, 1936a). Similar host sequences have been reported in southern California (Johnson *et al.*, 1982; Butler and Henneberry, 1985), Israel (Melamed-Madjar *et al.*, 1979), Thailand (Nachapong and Mabbett, 1979; Mabbett *et al.*, 1980) and Iran (Habibi, 1975). Thus, the many crops involved and the extreme mobility of *B. tabaci* are important factors affecting pest status of the insect. Widespread dispersal occurs even though *B. tabaci* has been characterized as a very poor flier (Byrne *et al.*, 1988). The role of inter- and intra-crop movement of *B. tabaci* is not well understood, but some insight into the magnitude of the problem is becoming evident. Short-distance movement within and between cultivated crops, weed host plants, or both is well documented. Long-range dispersal of *B. tabaci* adults has been reported after collections in aircraft-mounted nets at high altitudes. Moreover, the density of the airborne insects may increase several fold, to reach economic levels during redistribution of the insects as a result of the concentrating effect of vertical convection currents and horizontal air movement associated with weather fronts (Joyce, 1983). The importance of this intercrop movement within the agricultural community cannot be overstated. For example, consider the effect of *B. tabaci* dispersal from heavily infested cotton and melon fields in Arizona to cultivated lettuce fields 1.4 to 4.8 km distant (Butler and Henneberry, 1989). *Bemisia tabaci* were first trapped during 08.30 to 09.00 hours. Peak activity occurred between 10.30 and 11.00 hours, when an average of 384 adults per 7.5 × 10 cm yellow sticky trap were caught during a 30-minute sampling period (Fig. 13.5). Similar periods of *B. tabaci* diel activity patterns were reported by Bellows *et al.* (1988). One trap captured 770 adults in 30 minutes. The rate of daily increase in adults in lettuce was b = 0.1444 (r = 0.99), demonstrating the extreme population pressure that occurs in the lettuce crop as a result of the mobile, migrating *B. tabaci* from cotton fields. This dispersal activity is of particular importance when insect vectors are involved that transmit infectious agents of plant diseases. In the

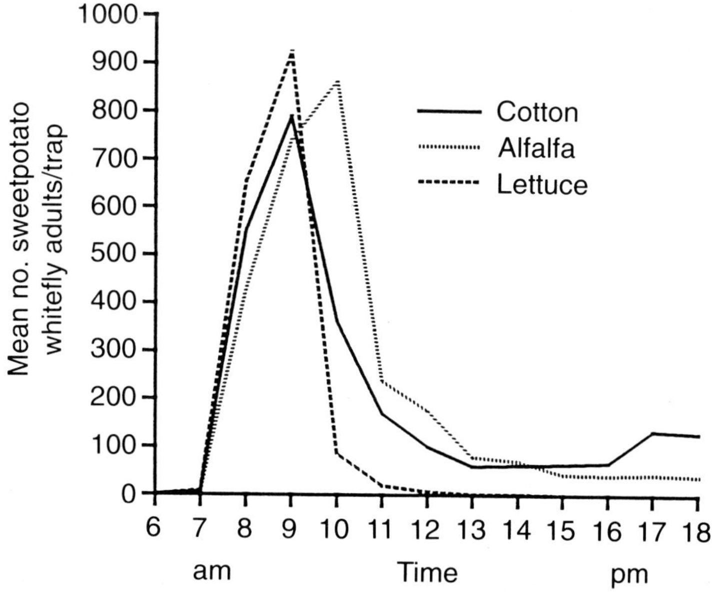

Fig. 13.5. Daily activity of sweetpotato whitefly adults as measured by sticky trap catches in cotton, lettuce and alfalfa.

current example, *B. tabaci* collected in early September were capable of transmitting lettuce-infectious yellows virus to 23–25% of susceptible plants and those collected late in September, 23–55% of susceptible plants. The high numbers of viruliferous adults infect susceptible plants in the field, which in turn serve as secondary sources of virus.

Natural Enemies and Insecticide Interactions

The role of natural enemies in regulating *B. tabaci* populations is not very well understood. However, worldwide outbreaks have appeared to intensify with the use of synthetic organic insecticides (Anon., 1981), which suggests that natural enemies may have a beneficial impact in limiting population development. However, reports of natural-enemy population regulation and suppression have been variable (Gerling, 1986). Characteristically low levels of parasitism occur early in the season increasing later in the season in conjunction with increasing *B. tabaci* populations (Sharaf, 1982; Bellows and Arakawa, 1988).

A recent review of published natural-enemy records shows that all reported parasites, except two, are Aphelinidae in the genera *Eretmocerus*, *Encarsia*, *Prospaltella*, and *Pteroptrix* (Anon., 1981). Chrysopid, coccinellid, and several phytoseiid mite species are reported to be predators, and a single fungal pathogen is known. The assessment of parasite activity is complex and highly variable, depending on the plant substrate. Most natural-enemy records pertain to cotton and information is scarce on other cultivated and weed hosts. Possibly, more effective parasites may be found if the origin of *B. tabaci*, which is not clear, can be established. Gerling (1986) suggests that Pakistan is in the region of *B. tabaci* origin.

Three species of aphelinid parasites of *B. tabaci* were reported in southern California (Gerling, 1967): *Encarsia formosa* Gahan, *E. meritoria* Gahan and *Eretmocerus haldemani* Howard. The complex was dominated by *E. haldemani* in both the Coachella and the Imperial Valleys, and *E. meritoria* was more abundant than *E. formosa*. Percentage parasitism was not distinguished for *B. tabaci* alone, as two other whitefly species were included in the study. However, the percentage parasitism by the total parasite complex, primarily on *B. tabaci* and bandedwing whitefly, reached 74–87% by October in both untreated and insecticide-treated fields. More typical seasonal parasitism rates ranged from 20% to 40% where increases in the percentage parasitism generally followed the exponential increases in *B. tabaci* population density. Insecticide applications greatly reduced whitefly species and parasite populations, but both hosts and parasites generally returned to pretreatment levels in two weeks. Recently *Encarsia formosa* n. spp. from southern California was described as a new species of *B. tabaci* parasite (Gerling and Rivnay, 1984). The insect resembles *E. formosa* (Gahan), but has a light brown head and thorax, is smaller and biparental. In Israel *B. tabaci* populations in cotton planted in May begin to increase in July, with peak populations occurring in September and decreasing thereafter (Gerling *et al.*, 1980). The two parasite species encountered were: *Encarsia lutea* (Masi), which reached peak populations in early September; and *Eretmocerus mundus* Mercet, approximately one week later. *B. tabaci* and the parasites *E. lutea* and *E. mundus* were found on a number of weed and cultivated hosts throughout the winter in Israel (Gerling, 1984). Similarly, Coudriet *et al.* (1986) found *B. tabaci*, *Eretmocerus* spp. and *Encarsia* spp. parasites throughout the year in the Imperial Valley on wild lettuce, *Lactuca serriola* L. and wild sunflower, *Helianthus annus* L. Seasonal trends of parasite populations were similar, but *Eretmocerus* parasitism was always greater than that of *Encarsia*. The results of the studies suggested to the authors that parasite populations did not exert a significant impact on early season *B. tabaci* populations. Further, the parasites were dependent on host density for population establishment and increase.

In the Sudan, a two-year study of *B. tabaci* population dynamics revealed that the parasite complex was dominated by *E. lutea* (66%) with *E. mundus* comprising the remaining 34% (Gameel, 1969). No hyperparasites were recovered. Percentage parasitism also increased with increasing whitefly abundance. The maximum parasitism by the total complex was 44%, with maximum parasitism by *E. lutea* and *E. mundus* averaging 30% and 15%, respectively.

Hafez *et al.* (1979), in Egypt, reported that during two cropping seasons, the average percentages of *B. tabaci* parasitized by *E. mundi* on insecticide-treated cotton in various parts of the country were from 44.4% to 73.0%, from 34.0% to 55.4% on insecticide-treated cabbage (*Brassica oleracea* L.), and from 78.6% to 80.8% on untreated *Lantana camara* L. The authors concluded that *E. mundi* had a major role in regulating *B. tabaci* populations.

Much less information is available concerning the interaction of *B. tabaci* and predator populations. Meyerdirk and Coudriet (1985) found that *Euseuis hibisci* (Chant), a phytoseiid mite, fed and developed to the adult stage on a mixed diet of *B. tabaci* eggs and first- and second-instar larvae. Adult female *E. hibisci* consumed an average of 4.5 eggs per day, suggesting that *E. hibisci* may have potential as a biological agent for control of *B. tabaci*. In other areas of the world, several additional phytoseiid mite species have been reported to prey on *B. tabaci* (Teich, 1966; Elbadry, 1967, 1968; Swirski and Dorzia, 1968, 1969; Swirski *et al.*, 1970; Gameel, 1971) and appeared to reduce populations in the field (Teich, 1966; Gameel, 1971).

The common green lacewing has also been observed in relation to *B. tabaci* populations on cotton in Israel (Or and Gerling, 1985). The author suggested that adult lacewings were not attracted to whitefly-infested fields and that lacewing population build-up was slowed by low oviposition and slow immature developmental rates. Nevertheless, lacewing larvae of all stages voraciously attack the various whitefly stages on cotton leaves under laboratory conditions, and reduced adult whitefly visitations and oviposition activity occur on plants inhabited by lacewing larvae (Butler and Henneberry, 1988).

The potential of natural enemies for *B. tabaci* control has not been well researched, and the results reported are variable. The life table studies of Horowitz *et al.* (1984), and a subsequent report (Horowitz, 1986) suggest that *B. tabaci* parasitization was not a key factor affecting population regulation in Israel. Classical biological control approaches have considerable merit and, in view of the increasing economic importance of *B. tabaci*, should be a principal focus of attention. Particular attention should be given to quantifying natural enemy impact, defining their role in *B. tabaci* integrated management programmes, and explaining their apparent lack of effectiveness under current crop production systems.

Control Technology

The increasing worldwide concern regarding the adverse impact of *B. tabaci* in crop-production systems emphasizes the need to develop control strategies based on the biology, ecology and population dynamics of the species in relation to cultivated and wild host ecosystems. The problem, as previously discussed, is complex because a great many hosts are involved and short distance movement within and between cultivated crops, weed host plants or both, as well as long-range movement occurs (Husain and Trehan, 1933; Husain *et al.*, 1936a; Mabbett, 1978; Melamed-Madjar *et al.*, 1979, Nachapong and Mabbett, 1979; Johnson *et al.*, 1982; Joyce, 1983).

In the desert agricultural areas of southern California and Arizona, cotton appears to be a major factor in population development of the insect (Meyerdirk *et al.*, 1986). Numbers increase exponentially and have been observed to double in six days (Natwick *et al.*, 1984; Butler and Henneberry, 1985). Movement of *B. tabaci* from cotton into autumn vegetable crops causes a severe disease complex (Duffus and Flock, 1982). A continuing sequence of cultivated crops and weed hosts that serve as *B. tabaci* reproductive hosts (Butler and Henneberry, 1985; Coudriet *et al.*, 1985), as well as virus reservoirs (Duffus and Flock, 1982), make it difficult to consider elimination of these weed hosts or a crop rotation system to eliminate the source of the insect or the viruses.

Conventional chemical-control technology has not been satisfactory because of the distribution of the insect within the plant canopy on the underside of leaves, the high reproductive potential of the insect and development of resistant *B. tabaci* strains. Possibly chemical application-technology research, to improve canopy penetration and selectively provide better coverage on the underside of leaves, would result in increased chemical control efficiency. Some success has been reported with systemic insecticides, but additional studies are needed to determine efficacy in management systems. In any event, total reliance on chemical control must be considered to be a temporary measure until more satisfactory approaches can be developed.

Incorporation of plant characters resistant to the vector may have potential for control: for example, smooth-leaved cottons support lower populations of *B. tabaci* than do hirsute cottons (Mound, 1965a; Butler and Henneberry, 1984a; Butler and Wilson, 1984). Also, reduced populations of *B. tabaci* occur in okra-leaf cottons (Butler and Wilson, 1984). Incorporation of these characters into desirable agronomic cotton types may help in reducing *B. tabaci* populations (Ozgur and Sekerolyu, 1984). In addition to inherent resistant mechanisms, it has been determined that insecticide spray penetration and percentage of leaf area covered were increased 64% and 26% respectively in okra-leaf versus normal-leaf cotton (James and Jones, 1985): this increase in penetration could result in better *B. tabaci* control.

Bemisia tabaci build-up was significantly reduced in the Sudan on okra-leaf cottons. Plants with this character tend to have an open plant canopy which differs from the shady, humid and warm conditions usually favoured by the species. The reduced leaf area may affect abundance (Khalifa and Gameel, 1983). In Arizona, okra-leaf selections also had fewer *B. tabaci* adults than normal-leaf cottons (Butler and Wilson, 1984).

The rapid expansion of *B. tabaci* populations late in the cotton-growing season may provide a focus for reducing the extent of the problem. Short-season cotton cultivars and short-season cotton-production systems have been of benefit in reducing other late-season insect populations (Heilman *et al.*, 1977, 1979; Walhood *et al.*, 1981) and may be effective in managing *B. tabaci.* Cultural practices such as water management (Watson, 1980) and the use of plant growth regulators (Bariola *et al.*, 1976), can also be used to reduce late-season cotton growth and have been used effectively to shorten the growing season and to reduce overwintering populations of other cotton insects. These approaches have not been considered in *B. tabaci* management systems, but should be investigated.

Bandedwinged Whitefly

The bandedwinged whitefly, *Trialeurodes abutilonea* (Haldeman), occurs in 33 of the 50 United States and parts of Mexico and the West Indies. Russell (1948) considered *T. erigerontis* Maskill, *T. wolfsii* (Quaintance), *T. fitchi*, and *Aleyrodes fitchi* Quaintance, as synonymous with *T. abutilonea* and recognized considerable variation in size and certain morphological characteristics associated with hosts and geographical areas. The host range is wide and involves about 140 plant species (Russell, 1963). The bandedwinged whitefly has been known for a number of years to occur on cotton in the United States (Russell, 1963). At constant temperatures of 23.9, 29.4 and 35.0°C, the duration of the egg stage is 7.8, 5.2 and 5.1 days, respectively. Eggs develop to pupae and adults at the same temperatures in 29.9, 20.8 and 17.2 days and 30.7, 25.8 and 20.7 days, respectively. Under fluctuating temperatures and under greenhouse conditions, the developmental times are slightly longer. Generally, populations in cotton are regulated by natural enemies (Stoner and Butler, 1965; Butler, 1967), but outbreaks have occurred causing serious problems (Clower *et al.*, 1971). Six parasite species, *Eretmocerus haldimani* Howard, *Encarsia formosa* Gahan, *E. meritoria* Gahan, *E. coquilleti* Howard, *Encarsia* spp. and *Prospaltella* spp. were reported to parasitize bandedwinged whitefly in California (Gerling, 1967) and *Encarsia lutea* (Masi) was reported as a bandedwinged whitefly parasite in Arizona (Stoner and Butler, 1965). The role of the natural enemy complex in regulating bandedwinged whitefly populations has not been quantified as a group or as individuals, however, increasing bandedwinged whitefly populations have been recorded with increasing insecticide use. Epidemic infestations in some cotton growing areas have been associated with development of resistance to insecticides (Clower *et al.*, 1971). The species has been reported to vector an infectious agent of a plant disease (Hildebrand, 1960). Bandedwinged whiteflies appear to occur earlier in the year on cotton (early spring to July) than *B. tabaci* and populations decrease when *B. tabaci* is increasing late in the season (Gerling, 1967; Byrne *et al.*, 1986). Higher populations generally occur in hirsute vs. glabrous cotton types (Butler and Muramoto, 1967; Lambert *et al.*, 1982). When epidemic infestations occur, conventional chemical control efforts are subject to the same limitations that exist for *B. tabaci.*

Greenhouse Whitefly

The greenhouse whitefly, *Trialeurodes vaporariorum* (Haldeman), apparently originated in Central America, but was described in England by Westwood (1856). It was first reported as a greenhouse pest of tomatoes in 1905 (Morrill, 1905), and has since become a major pest of greenhouse crops throughout the world, with more than 200 reported hosts (Russell, 1948, 1963, 1977). The complete synonymy of the species was traced by Russell (1948). The species has been referred to in the literature under a number of synonyms that include *Asterochitan lecanioides* Maskcl, *A. sonchi* (Kotinsky), *Trialeuroides mossopi* Corbett, *T. natalensis* Corbett, *Aleurodes nicotianae* (Maskell), *A. papillifer* Maskell, *T. sesbaniae* Corbett (Russell, 1948). The author also suggested that several other species may be synonymous with *T. vaporarium*, but due to lack of available information, was unable to confirm her findings. Russell (1948) noted that from these known representative species of *Trialeurodes* that *vaporariorium* appeared most closely related to species occurring in the western and southwestern parts of North America.

Under greenhouse conditions, Lloyd (1922) reported that the numbers of eggs laid averaged 130 per female, Burnett (1949) 210 at 21°C and Curry and Pimental (1971) 92–250 at 21°C, depending on the tomato variety host. At 26.1°C, females lay approximately 20 eggs per female (Collman and All, 1980). Eggs are laid on the underside of leaves and visible embryonic development occurs within 24 hours at 26.7°C, and eggs begin to hatch within six days (Collman and All, 1980), and about 15 and ten days at 18.0°C and 22.5°C, respectively (Madueke and Coaker, 1984). The authors found the egg stage to be 7.6 days at 27°C. Egg viability may exceed 90% (Curry and Pimental, 1971). Eggs are elliptical in shape, yellow-white turning to dark brownish-black within a few days of oviposition (Hill, 1969). The egg shell flattens laterally after hatching.

Davis (1912), Speyer (1927) and Milliron (1940) found development from oviposition to adult emergence to range from 28 to 47 days. At 21°C, Curry and Pimental reported about 48 days as the time of development from egg to adult emergence, Collman and All (1980) 21 days at 26.0°C, and Madueke and Coaker (1984) as 44.9, 30.4 and 22.4 days at 18.0, 22.5 and 27.0°C. The latter authors found the thresholds of development for eggs and pupae to be 8.5°C and 10°C, respectively, and that for the various larval stages to range from 7°C to 11°C.

Damage caused by the species is similar to that reported for other whiteflies. Feeding weakens plant growth and the production of honeydew and associated fungal growth reduces commodity quality. The greenhouse whitefly is also a vector of infectious agents causing disease in ornamentals (O'Reilly, 1974) and vegetable crops (Duffus, 1965).

The greenhouse whitefly has at sporadic intervals been a secondary pest of cotton in San Joaquin Valley, California. Outbreaks have been associated with excessive use of insecticides applied for the control of other insect pests (Flint *et al.*, 1984). Thus, management recommendations stress avoidance of practices that reduce natural enemy populations.

The impact of parasitoids on greenhouse whitefly populations under field conditions, including cotton culture, is not well known. However, biological control under greenhouse conditions is well documented (Vet *et al.*, 1980). The various factors affecting the biological control efficiency of *Encarsia formosa* Gahan under these conditions are well known, and use of the parasitoid has proven to be an effective control method when temperatures can be maintained to efficiently maintain its reproduction rate. Other greenhouse whitefly parasitoid species, in addition to *E. formosa*, have been reported (Vet *et al.*, 1980). Some biological information is available on these parasitoids, but their potential as biological control agents in the greenhouse or under field conditions is largely unknown (Vet *et al.*, 1980). Several parasite species have been reported attacking greenhouse whitefly on cotton but their potential for conservation, augmentation and/or introduction to manage populations has not been established, but should be considered, since conventional control methods are not adequate under field conditions. In fact, chemical control approaches may be a contributing factor to inducing greenhouse whitefly problems on cotton in the areas where it becomes an economic pest.

Summary

The whitefly complex continues to be a serious problem in tropical and subtropical areas of the world and research is urgently needed to continue to develop efficient, reproducible sampling techniques that incorporate both the within-plant and within-field distribution of the insect. In addition, economic thresholds and information on the effects on yield need to be developed. This is a particularly difficult task in the case of a virus vector within a multicrop system. The biology, ecology, behaviour and population dynamics of the species need to be further defined, and data incorporated into models that effectively interface with crop and weed host systems which include natural control factors that will allow us to simulate and predict population outbreaks. One model (Von Arx *et al.*, 1983b), incorporates life-table information in relation to a cotton plant model. It has been useful in identifying *B. tabaci* cotton interactions and has potential for use in developing alternative *B. tabaci* cotton management systems.

A simple degree-day exponential population growth-rate equation can be used to provide overall population projections using historical temperature data and initial *B. tabaci* population densities (Zalom *et al.*, 1985). The model has been effectively used (Natwick and Zalom, 1985), but the authors stress that effective use and accuracy of the model is sensitive to initial *B. tabaci* population-density estimates, and because parasitism or host-plant effects are not included, the impact of these variables cannot be accounted for or simulated. These modelling efforts can provide a better understanding of *B. tabaci* population dynamics, and can form the basis for identifying and evaluating population-suppression technology which together can build an effective and efficient *B. tabaci* crop-management system.

Bemisia tabaci is becoming of increasing worldwide concern regarding its adverse impact on crop-production systems. The many hosts involved, movement within and between cultivated crops and weed host plants complicate the problem. Cotton, in some areas, appears to be a major factor in population development because of the extended growing period and its excellent status as a reproductive host. Movement from cotton into autumn vegetable crops causes a severe disease complex, and a continuing sequence of cultivated crops and weed hosts serve as reproductive hosts, as well as virus reservoirs.

Conventional chemical-control technology has not been satisfactory as discussed in the text. Varietal resistance may have potential for control. Also, the seasonal development of populations showing exponential growth late in the cotton-growing season suggests that short-season cotton cultivars and short-season cotton-production systems provide a key to development of population management systems.

References

Allen, R.M., Tucker, H. and Nelson, R.A. (1960) Leaf crumple disease of cotton in Arizona. *Plant Disease Reporter* 44, 246–250.

Anon. (1981) Possibilities for the use of biotic agents in the control of the whitefly, *Bemisia tabaci. Biocontrol News and Information, Commonwealth Institute of Biological Control* 2, 1–7.

Anon. (1989) Survey on cotton production practices. Report Prepared by the Secretariat for the 48th Plenary Meeting, Scottsdale, Arizona. International Cotton Advisory Committee, Washington, DC.

Avidov, Z. (1956) Bionomics of the tobacco whitefly *Bemisia tabaci* (Genn.) in Israel. *Israel Ktavn* 7, 25–41.

Avidov, Z. and Harpaz, I. (1969) *Plant Pests of Israel.* University Press, Jerusalem. 549pp.

Azab, A. K., Megahed, M. M. and El-Mirsawi, H. D. (1969) Parasitism of *Bemisia tabaci* (Genn.) in the UAR. *Bulletin of the Entomological Society of Egypt* 53, 439–441.

Azab, A.K., Megahed, M.M. and El-Mirsawi, H.D. (1970) On the range of host-plants of *Bemisia tabaci* (Genn.). *Bulletin of Entomological Society of Egypt* 54, 319–326.

Azab, A.K., Megahed, M.M. and El-Mirsawi, D.E. (1971) On the biology of *Bemisia tabaci* (Genn.). *Bulletin of Entomological Society of Egypt* 55, 305–315.

Bariola, L.A., Kittock, D.L., Arle, H.F., Vail, P.V. and Henneberry, T.J. (1976) Controlling pink bollworm: Effects of chemical termination of cotton fruiting on populations of diapausing larvae. *Journal of Economic Entomology* 69, 633–636.

Bellows, T.S. and Arakawa, K. (1988) Dynamics of premaginal populations of *Bemisia tabaci* (Homoptera: Aleyrodidae) and *Eretmocerus* sp. (Hymenoptera: Aphelinidae) in southern California. *Environmental Entomology* 17, 483–487.

Bellows, T.S., Jr., Perring, T.M., Arakawa, K. and Farrar, C.A. (1988) Patterns in diel flight activity of *Bemisia tabaci* (Homoptera: Aleyrodidae) in cropping systems in southern California. *Environmental Entomology* 17, 225–228.

Berlinger, M.J., Magal, Z. and Benzioni, A. (1983) The importance of pH in food selection by tobacco whitefly, *Bemisia tabaci. Phytoparasitica* 11, 151–160.

Bird, J. and Maramorosch, K. (1978) Viruses and virus diseases associated with whiteflies. *Advances in Virus Research* 22, 55–110.

Brown, J.K. and Brid, J. (1992) Whitefly-transmitted geminiviruses and associated disorders in the Americas and the Caribbean Basin. *Plant Disease* 76, 220–225.

Brown, J.K. and Nelson, M.R. (1984) Geminate particles associated with cotton leaf crumple disease in Arizona. *Phytopathology* 74, 989–990.

Brown, J.K. and Nelson, M.R. (1986) Host range study of the cotton leaf crumple virus. *Arizona Agricultural Experiment Station Bulletin* P-63, 171–176.

Brown, J.K., Mihail, J.D. and Nelson, M.R. (1985) The effects of cotton leaf crumple on cotton inoculated at different growth stages. *Arizona Agricultural Experiment Station Bulletin* P-63, 152–155.

Burnett, T. (1949) The effect of temperature on an insect-host parasite population. *Ecology* 30, 113–134.

Butler, G.D., Jr. (1967) Development of the bandedwing whitefly at different temperatures. *Journal of Economic Entomology* 60, 877–878.

Butler, G.D., Jr. and Brown, J.K. (1985) Sweetpotato whitefly infection of cotton leaf crumple from weed hosts in 1984. *Arizona Agricultural Experiment Station Bulletin* P-63, 149–151.

Butler, G.D., Jr. and Henneberry, T.J. (1984a) *Bemisia tabaci* effect of cotton leaf pubescence on abundance. *Southwestern Entomologist* 9, 91–94.

Butler, G.D., Jr. and Henneberry, T.J. (1984b) *Bemisia tabaci* as a cotton pest in the desert cotton-growing areas of the Southwestern United States. In: Brown, J. (ed.) *Proceedings of Beltwide Cotton Production Research Conference*, Memphis, Tennessee, pp. 195–197.

Butler, G.D., Jr. and Henneberry, T.J. (1985) *Bemisia tabaci* (Genn.), a pest of cotton in the southwestern United States. *US Department of Agriculture, Agricultural Research Service Technical Bulletin No. 1701*, 19pp.

Butler, G.D., Jr. and Henneberry, T.J. (1988) Laboratory studies of *Chrysoperla carnea* predation on *Bemisia tabaci. Southwestern Entomologist* 13, 165–170.

Butler, G.D., Jr. and Henneberry, T.J. (1989) Sweetpotato whitefly migration, population increase and control on lettuce with cottonseed oil sprays. *Southwestern Entomologist* 14, 287–293.

Butler, G.D., Jr. and Muramoto, H. (1967) Banded-winged whitefly abundance and cotton leaf pubescence in Arizona. *Journal of Economic Entomology* 60, 1176–1177.

Butler, G.D., Jr. and Wilson, F.D. (1984) Activity of adult whiteflies (Homoptera: Aleyrodidae) within plantings of different cotton strains and cultivars as determined by sticky trap catches. *Journal of Economic Entomology* 77, 1137–1140.

Butler, G.D., Jr., Henneberry, T.J. and Clayton, T.E. (1983) *Bemisia tabaci* (Homoptera: Aleyrodidae) development, oviposition, and longevity in relation to temperature. *Annals of Entomological Society of America* 76, 310–313.

Butler, G.D., Jr., Henneberry, T.J. and Brown, J.K. (1985a) Cotton leaf crumple disease of Pima cotton. *Arizona Agricultural Experiment Station Bulletin* P-63, 158–159.

Butler, G.D., Jr., Henneberry, T.J. and Natwick, E.T. (1985b) *Bemisia tabaci* 1982 and 1983 populations in Arizona and California cotton fields. *Southwestern Entomologist* 10, 20–25.

Butler, G.D., Jr., Wilson, F.D. and Henneberry, T.J. (1985c) Cotton leaf crumple disease in okra-leaf and normal-leaf cottons. *Journal of Economic Entomology* 78, 1500–1502.

Butler, G.D. Jr., Brown, J.K. and Henneberry, T.J. (1986a) Effect of cotton seedling infection by cotton-leaf crumple virus on subsequent growth and yield. *Journal of Economic Entomology* 79, 208–211.

Butler, G.D., Jr., Henneberry, T.J. and Wilson, F.D. (1986b) *Bemisia tabaci* (Homoptera:

Aleyrodidae) adult activity and cultivar oviposition preference. *Journal of Economic Entomology* 79, 350–354.

Byrne, D.N., Von Bretzel, P.N. and Hoffman, C.J. (1986) Impact of trap design and placement when monitoring the banded-winged and sweetpotato whiteflies (Homoptera: Aleyrodidae). *Environmental Entomology* 15, 300–304.

Byrne, D.N., Buchmann, S.L. and Spangler, H.G. (1988) Relationship between wing loading, wing beat frequency and body mass in homopteran insects. *Journal of Experimental Biology* 133, 135–140.

Clower, D.F. and Watve, C.M. (1973) The bandedwing whitefly as a pest of cotton. In: *Proceedings of the Beltwide Cotton Production and Research Conference*, Memphis, Tennessee, pp. 90–91.

Clower, D.F., Watve, C.M., Melville, D.R. and Gaves, J. B. (1971) Whiteflies – a new insect problem in cotton. *Louisiana Agriculture* 14, 8–9.

Collman, G.L. and All, J.N. (1980) Quantification of the greenhouse whitefly life cycle in a controlled environment. *Journal of Georgia Entomological Society* 15, 433–438.

Costa, H.S. (1976) Whitefly-transmitted plant diseases. *Annual Review of Phytopathology* 14, 429–449.

Costa, H.S. and Brown, J.K. (1990) Variability in biological characteristics: isozyme paterns and virus transmission among populations of *Bemisia tabaci* Genn. in Arizona. *Phytopathology* 80, 888.

Costa, H.S. and Brown, J.K. (1991) Variation in biological characteristics and in esterase patterns among populations of *Bemisia tabaci* Genn. and the association of one population with silverleaf symptom development. *Entomologia Experimentia et Applicata* 61, 211–219.

Coudriet, D.L., Prabhaker, N., Kishaba, A.N. and Meyerdirk, D.E. (1985) Variation in development rate on different hosts and overwintering of the sweetpotato whitefly, *Bemisia tabaci* (Homoptera: Aleyrodidae). *Environmental Entomology* 14, 516–519.

Coudriet, D.L., Meyerdirk, D.E., Prabhaker, N. and Kishaba, A.N. (1986) The bionomics of the sweetpotato whitefly (Homoptera: Aleyrodidae) on wild hosts in the Imperial Valley, California. *Environmental Entomology* 15, 1179–1183.

Curry, J.P. and Pimental, D. (1971) Life cycle of the greenhouse whitefly, *Trialeurodes vaporariorum*, and population trends of the whitefly and its parasites. *Annals of the Entomological Society of America* 64, 1188–1190.

Davis, J.S. (1912) The greenhouse whitefly. *27th Report of Illinois State Entomological Society*, 13pp.

Dickson, R.C., Johnson, M. McD. and Laird, E.F., Jr. (1954) Leaf crumple, a virus disease of cotton *Phytopathology* 44, 479–480.

Dittrich, V., Hassan, S.O. and Ernst, G.H. (1985) Sudanese cotton and the whitefly: a case study of the emergence of a new primary pest. *Crop Protection* 4, 161–176.

Duffus, J.E. (1965) Beet pseudo-yellows virus transmitted by the greenhouse whitefly (*Trialurodes vaporariorum*). *Phytopathology* 55, 450–453.

Duffus, J. E. and Flock, R. A. (1982) Whitefly-transmitted disease complex of the desert southwest *California Agriculture* 36 (11–12), 4–6.

Elbadry, E.A. (1967) Three new species of phytoseiid mites preying on the cotton whitefly, *Bemisia tabaci*, in the Sudan (Acarina: Phytoseiidae). *Entomophaga* 13, 323–329.

Elbadry, E.A. (1968) Biological studies on *Amblyseius aleyrodis*, a predator of the cotton whitefly (Acarina: Phytoseiidae). *Entomophaga* 14, 273–279.

El-Helaly, M.S., El-Shazli, A.Y. and El-Gayar, F.H. (1971) Biological studies on *Bemisia tabaci* Genn. (Homoptera: Aleyrodidae) in Egypt. *Zeitschrift für Angewändte Entomologie* 69, 48–55.

El-Helaly, M.S., Ibrahim, E.G. and Rawash, I.A. (1977) Photoperiodism of the whitefly, *Bemisia tabaci* Gennadius (Homoptera: Aleyrodidae). *Zeitschrift für Angewändte Entomologie* 83, 393–397.

El-Khidir, E. (1965) Bionomics of the cotton whitefly (*Bemisia tabaci* Genn.), in the Sudan and the effects of irrigation on population density of whiteflies. *Sudan Agricultural Journal* 1, 8–22.

El-Khidir, E. and Khalifa, A. (1962) A new aleyrodid from the Sudan. *Proceedings of Royal Entomological Society, London* 31, 47–51.

Erwin, D. C. and Meyer, R. (1961) Symptomatology of the leaf crumple disease in several species and varieties of *Gossypium* and variation of the causal virus. *Phytopathology* 51, 472–477.

Eveleens, K. G. (1983) Cotton-insect control in the Sudan Gezira: analysis of a crisis *Crop Protection* 2, 273–287.

Flint, M.L., Rude, P.A. and Clark, J.K. (1984) *Integrated Pest Management for Cotton in the Western Region of the United States.* University of California Division of Agriculture and Natural Resources, Oakland, California.

Fowler, H.D. (1956) Some physiological effects of attack by whitefly (*Bemisia gossypiperda*) and spraying parathion on cotton in the Sudan Gezira. *Empire Cotton Growing Review* 33, 288–299.

Gameel, O.I. (1969) Studies on whitefly parasites *Encarsia lutea* and *Eretmocerus mundus* Mercet (Hymenoptera: Aphelinidae). *Revue de Zoologie et Botanie Africaine* 79, 65–77.

Gameel, O.I. (1971) The whitefly eggs and first larval stages as prey for certain phytoseiid mites *Revue de Zoologie et Botanie Africaine* 84, 79–82.

Gameel, O.I. (1972) A new description, distribution and hosts of the cotton whitefly, *Bemisia tabaci* (Gennadius), (Homoptera: Aleyrodidae). *Revue de Zoologie et Botanie Africaine* 84, 50–64.

Gameel, O.I. (1974) Some aspects of the mating and oviposition behaviour of the cotton whitefly, *Bemisia tabaci* (Genn.). *Revue de Zoologie et Botanie Africaine* 88, 784–788.

Gameel, O.I. (1978) The cotton whitefly, *Bemisia tabaci* (Genn.) in the Sudan Gezira. In: *Ciba-Geigy, Third Seminar on the Strategy for Cotton Pest Control in the Sudan.* 8–10 May 1978 Basle, Switzerland, Giba-Geigy, pp. 111–131.

Gerling, D. (1967) Bionomics of the whitefly–parasite complex associated with cotton in southern California (Homoptera: Aleurodidae; Hymenoptera: Aphelinidae). *Annals of the Entomological Society of America* 60, 1306–1321.

Gerling, D. (1984) The overwintering mode of *Bemisia tabaci* and its parasitoids in Israel. *Phytoparasitica* 12, 109–118.

Gerling, D. (1986) Natural enemies of *Bemisia tabaci*: biological characteristics and potential as biological control agents: A review. *Agricultural Ecosystems and Environment* 17, 99–110.

Gerling, D. and Rivnay, T. (1984) A new species of *Encarsia* (Hymenoptera: Aphelinidae) parasitizing *Bemisia tabaci* (Homoptera: Aleyrodidae). *Entomophaga* 29, 439–444.

Gerling, D., Motro, U. and Horowitz, R. (1980) Dynamics of *Bemisia tabaci* (Gennadius) (Homoptera: Aleyrodidae) attacking cotton in the coastal plain of Israel. *Bulletin of Entomological Research* 70, 213–219.

Ghong, Y. (1969) Host plant and morphological variation of *Bemisia tabaci* (Gennadius) (Homoptera: Aleyrodidae) in Taiwan. *Plant Protection Bulletin, Taiwan* 2, 23–32.

Gutnecht, J. and Fournier, J. (1988) Major research carried out by IRCT on the origin and detection of sticky cotton. In: *Proceedings of the International Cotton Advisory Committee Meeting on Cotton Test Methods*, Bremen, Germany, CAB International, Wallingford, UK, pp. 19–31.

Habibi, J. (1975) The cotton whitefly *Bemisia tabaci* Gen., bioecology and methods of control. *Entomologie et Phytopathologie Appliquées* 38, 3–4.

Hafez, M., Awadallah, K.T., Tawfik, M.F.S. and Sarhan, A. A. (1979) Impact of the parasite *Eretmocerus mundus* Mercat on population of the cotton whitefly, *Bemisia tabaci* (Genn.) in Egypt. *Bulletin of the Entomological Society of Egypt* 62, 23–32.

Hassan, H.M. (1970) Progress in chemical control of pests in the Gezira. In: Siddig, M.A. and Hughes, L.C. (eds) *Cotton Growth in the Gezira Environment*, Ministry of Agriculture, Sudan, pp. 232–246.

Hector, D.J. and Hodkinson, I.D. (1989) Stickiness on cotton. *ICAC Review Article on Cotton Production Research* No. 2, CAB International, Wallingford, UK, 43pp.

Heilman, M.D., Lukefahr, M.J., Namken, L.N. and Norman, J.W. (1977) Field evaluation on a short season production system in the lower Rio Grande Valley of Texas. In: *Proceedings of the Beltwide Cotton Production and Research Conference*, Memphis, Tennessee, pp. 80–83.

Heilman, M.D., Homaker, L.N., Norman, J.W. and Lukefahr, M.J. (1979) Evaluation of an integrated short-season management production system for cotton. *Journal of Economic Entomology* 72, 895–899.

Hill, B.G. (1969) A morphological comparison between two species of whitefly, *Trialeurodes vaporariorum* (Westw.) and *Bemisia tabaci* (Homoptera: Aleyrodidae) which occur on tobacco in the Transvaal. *Phytopathlactica* 1, 127–146.

Hildebrand, E. M. (1960) The feathery mottle virus complex of sweetpotato. *Phytopathology* 50, 751–759.

Horowitz, A.R. (1986) Population dynamics of *Bemisia tabaci* (Gennadius): with special emphasis on cotton fields. *Agricultural Ecosystems and Environment* 17, 37–47.

Horowitz, A.R., Podoler, H. and Gerling, D. (1984) Life table analysis of the tobacco whitefly, *Bemisia tabaci* (Gennadius) in cotton fields in Israel. *Acta Oecologica/Oecologia applicata* 5, 221–233.

Husain, M.A. and Trehan, K.N. (1933) The life-history, bionomics and control of the white-fly of cotton (*Bemisia gossypiperda* M. & L.). *Indian Journal of Agricultural Science* 3, 701–753.

Husain, M.A., Trehan, K.N. and Verma, P.A. (1936a) Studies on *Bemisia gossypiperda* M. & L. No. 3: Seasonal activities of *Bemisia gossypiperda* M. & L. (the whitefly of cotton) in the Punjab. *Indian Journal of Agricultural Science* 6, 893–903.

Husain, M.A., Puri, A.N. and Trehan, K.N. (1936b) Cell sap acidity and the incidence of whitefly *Bemisia gossypiperda* on cotton. *Current Science* 41, 486–487.

Hutchison, J.B., Knight, R.L. and Pearson, E.O. (1950) Response of cotton to leaf-curl disease. *Journal of Genetics* 50, 100–111.

Jackson, J.E., Burham, H.O. and Hassan, H.M. (1973) Effects of season, sowing date, nitrogenous fertilizer and insecticide spraying on the incidence of insect pests on cotton in the Sudan Gezira. *Journal of Agricultural Science* 81, 491–505.

James, D. and Jones, J.E. (1985) Effects of leaf and bract isolines on spray penetration and insecticidal efficiency. In: *Proceedings of the Beltwide Cotton Production Research Conference*, Memphis, Tennessee, pp. 395–396.

Johnson, M.W., Toscano, N.C., Reynolds, H.T., Sylvester, E.S., Kido, K. and Natwick, E.T. (1982) Whiteflies cause problems for southern California growers. *California Agriculture* 36, 24–26.

Jones, J.E., Clower, D.F., Milam, M.R., Caldwell, W.D. and Melvill, D.R. (1975) Resistance in upland cotton to the bandedwinged whitefly, *Trialeurodes abutilonea* (Haldeman). In: *Proceedings of the Beltwide Cotton Production Research Conference*, Memphis, Tennessee, pp. 98–99.

Joyce, R.J.V. (1955) Cotton spraying in the Sudan Gezira. II. Entomological problems arising from spraying. *FAO Plant Protection Bulletin* 3, 97–103.

Joyce, R.J.V. (1958) Effect of the cotton plant in the Sudan Gezira on certain leaf-feeding insect pests. *Nature* 182, 1463–1464.

Joyce, R.J.V. (1959) Recent program in entomological research in the Sudan Gezira. *Empire Cotton Growing Review* 36, 179–186.

Joyce, R.J.V. (1983) Aerial transport of pests and pest outbreaks. *European and Mediterranean Plant Protection Organization Bulletin* 13, 111–119.

Joyce, R.J.V. and Roberts, P. (1959) The determination of the size of plant surface for cotton spraying experiments in the Sudan Gezira. *Annals of Applied Biology* 47, 287–305.

Khalifa, A. (1980) Cotton stickiness. *International Textile Bulletin, World Edition, Spinning* 2, p. 80.

Khalifa, A. and El-Khidir, E. (1964) Biological study on *Trialeurodes lubia* and *Bemisia tabaci* (Aleyrodidae). *Bulletin of the Entomological Society of Egypt* 48, 115–129.

Khalifa, A. and Gameel, O. I. (1982) Control of cotton stickiness through breeding cultivars resistant to whitefly (*Bemisia tabaci* (Genn.)) infestation. In: *Proceedings of Improvement of Oilseed Industrial Crops by Induced Mutations*, International Atomic Energy Agency, Vienna, pp. 181–186.

Khalifa, H. and Gameel, O.I. (1983) Breeding cotton cultivars resistant to whitefly (*Bemisia tabaci* (Genn.)). *NATO Advanced Studies Institute Series A. Life Sciences* 55, 231–236.

Laird, E.F., Jr. and Dickson, R.C. (1959) Insect transmission of the leaf-crumple virus of cotton. *Phytopathology* 49, 366–376.

Lambert, L., Jenkins, J.N., Parrott, W.L. and McCarty, J.C. (1982) Greenhouse technique for evaluating resistance to the bandedwinged whitefly (Homoptera: Aleyrodidae) used to evaluate thirty-five foreign cotton cultivars. *Journal of Economic Entomology* 76, 1166–1168.

Larios, J.F. (1979) Niveles criticos de insectos quetrasmi ten fitopatogenos el caso de mosca blanca (*Bemisia tabaci* Genn.). *Turrialba* 29, 237–241.

Lloyd, P. (1922) The control of the greenhouse whitefly (*Asterichiton vaporariorum*) with notes on its biology. *Annals of Applied Biology* 9, 1–32.

Mabbett, T. (1978) A revision of the economic insect pests of cotton in Thailand. Description, infestation and control. I Cotton whitefly (*Bemisia tabaci* Gennadius). In: *Cotton Pest Management Project*. Department of Agriculture, Bangkok, Report No. 13, pp. 1–12.

Mabbett, T., Nachapong, M. and Mekdaeng, J. (1980) The within-canopy distribution of adult cotton whitefly (*Bemisia tabaci* Gennadius) incorporating economic thresholds and meteorological conditions. *Thai Journal of Agricultural Science* 13, 98–108.

Madueke, E.D. and Coaker, T.J. (1984) Temperature requirements of the whitefly *Trialeurodes vaporariorum* (Homoptera: Aleyrodidae) and its parasitoid *Encarsia formosa* (Hymenoptera: Aphelinidae). *Entomological General* 9, 149–154.

Magdal, Z., Berlinger, M.J. and Benzioni, A. (1982) influence of pH and sucrose content on the attraction of *Bemisia tabaci* in vivo and in vitro. *Phytoparasitica* 10, 294–295.

Maramorosch, K. (1975) Etiology of whitefly-borne diseases. In: Bird, J. and Maramorosch, K. (eds) *Tropical Diseases of Legumes*. Academic Press, New York, pp. 171–221.

Marchoux, G., Ledant, F. and Mathai, P.J. (1970) Maladies de type jaunisse at maladies voisines affectant principalement les solanocies et transmises par des insects. *Annals of Phytopathology* 2, 735–773.

Melamed-Madjar, V., Cohen, S., Chen, M., Tam, S. and Rosilio, D. (1979) Observations on populations of *Bemisia tabaci* Gennadius (Homoptera: Aleyrodidae) on cotton adjacent to sunflower and potato in Israel. *Israel Journal of Entomology* 18, 71–78.

Meyerdirk, D.E. and Coudriet, D.L. (1985) Predation and developmental studies of *Euseius hibisci* (Chant) (Acarine: Phytoseiidae) feeding on *Bemisia tabaci* (Gennadius) (Homoptera: Aleyrodidae). *Environmental Entomology* 14, 24–27.

Meyerdirk, D.E., Coudriet, D.L. and Prabhaker, N. (1986) Population dynamics and control strategy for *Bemisia tabaci* in the Imperial Valley, California. *Agricultural Ecosystems and Environment* 17, 61–67.

Milliron, H.E. (1940) A study of some factors affecting the efficiency of *Encarsia formosa* Gahan, an aphelenid parasite of the greenhouse whitefly, *Trialeurodes vaporariorum* (Wester.). *Michigan Experiment Station Technical Bulletin* 173, 23pp.

Mor, U. (1983) Cotton yields and quality as affected by *Bemisia tabaci* under different regimes of irrigation and pest control. *Phytoparasitica* 11, 65.

Mor, U., Marani, A. and Applebaum, S.W. (1982) The relationship between *Bemisia tabaci* populations and cotton physiological conditions. *Phytoparasitica* 10, 295.

Morrill, A.N. (1905) The greenhouse whitefly *Aleurodes vaporariorum*, Westwood. *US Bureau of Entomology Circular* S7, 9pp.

Mound, L.A. (1963) Host-correlated variation of *Bemisia tabaci* (Gennadius) (Homoptera: Aleyrodidae). *Proceedings of Royal Entomological Society of London* 38, 171–180.

Mound, L.A. (1965a) Effect of leaf hair on cotton whitefly populations in the Sudan Gezira. *Empire Cotton Growing Review* 42, 33–40.

Mound, L.A. (1965b) Effect of whitefly (*Bemisia tabaci*) on cotton in the Sudan Gezira. *Empire Cotton Growing Review* 42, 290–294.

Mound, L.A. and Halsey, S.H. (1978) *Whitefly of the World.* John Wiley & Sons, Chichester, 340pp.

Nachapong, M. and Mabbett, T. (1979) A survey of some wild hosts of *Bemisia tabaci* Genn. around cotton fields in Thailand. *Thai Journal of Agricultural Science* 12, 217–222.

Natwick, E.T. and Zalom, F.G. (1985) Verification of the cotton whitefly population model, *Bemisia tabaci* Gennadius (Homoptera: Aleyrodidae). In: *Proceedings of the Beltwide Cotton Production Research Conference*, Memphis, Tennessee, pp. 174–177.

Natwick, E.T., Zalom, F.G., Toscano, N.C. and Kido, E. (1984) Monitoring of the cotton whitefly *Bemisia tabaci* (Gennadius) studies on the insect's development and control in cotton. In: *Proceedings of the Beltwide Cotton Production Research Conference*, Memphis, Tennessee, pp. 197–202.

Nuessly, G.S., Henneberry, T.J. and Perkins, H.H., Jr. (1989) Effect of sweetpotato whitefly population density on cotton fiber stickiness and reducing sugars. Book 1. In: *Proceedings of the Beltwide Cotton Production Research Conference*, Memphis, Tennessee, pp. 281–284..

Ohnesorge, B., Sharaf, N. and Allawi, T. (1980) Population studies on the tobacco whitefly *Bemisia tabaci* Genn. (Homoptera: Aleyrodidae) during the winter season. I. The spacial distribution on some host plants. *Zeitschrift für Angewändte Entomologie* 91, 226–232.

Ohnesorge, B., Sharaf, N. and Allawi, T. (1981) Population studies on the tobacco whitefly *Bemisia tabaci* Genn. (Homoptera: Aleyrodidae) during the winter season. II. Some mortality factors of the immature stages. *Zeitschrift für Angewändte Entomologie* 92, 127–136.

Or, R. and Gerling, D. (1985) The green lacewing, *Chrysoperla carnea*, as a predator of *Bemisia tabaci. Phytoparasitica* 13, 1.

O'Reilly, C. J. (1974) Investigations on the Biology and Biological Control of the glasshouse whitefly, *Trialeurodes vaporariorum* (Westwood). Thesis, University of Dublin, Dublin, Ireland.

Ozgur, A.F. and Sekeroglyu, E. (1984) Population developments of *Bemisia tabaci* (Genn.) (Homoptera: Aleyrodidae) on various cotton varieties in Cukurova, Turkey. *XVII International Congress of Entomology*. Abstract Vol., p. 568.

Pearson, E.O. and Maxwell Darling, R.C. (1958) *Bemisia tabaci* (Gennadius) (Aleyrodidae). In: *Insect Pests of Cotton in Tropical Africa*. Empire Cotton Growing Corp. and Commonwealth Institute of Entomology, London, pp. 232–236.

Perkins, H.H., Jr. (1971) Some observations on sticky cottons. *Textile Industries* 135, 49–64.

Perkins, H.H. (1983) Effects of whitefly contamination on lint quality of US cottons. In: *Proceedings of Beltwide Cotton Production Research Conference*, Memphis, Tennessee, pp. 102–103.

Perring, T.M., Cooper, A.D., Rodriques, R.J., Fanar, C.A. and Bellows, T.S., Jr. (1993) Identification of a whitefly species by genomic and behavioural studies. *Science* 259, 74–77.

Prabhaker, N., Coudriet, D.L. and Meyerdirk, D.E. (1985) Insecticide resistance in the sweetpotato whitefly, *Bemisia tabaci* (Homoptera: Aleyrodidae). *Journal of Economic Entomology* 78, 387–409.

Russell, L.M. (1948) The North American species of the genus *Trialeurodes. United States Department of Agriculture Miscellaneous Publication* No. 635.

Russell, L. M. (1957) Synonyms of *Bemisia tabaci* (Gennadius) (Homoptera: Aleyrodidae). *Bulletin of the Brooklyn Entomological Society* 52, 122–123.

Russell, L.M. (1963) Host and distribution of five species of *Trialeurodes* (Homoptera: Aleyrodidae). *Annals of the Entomological Society of America* 56, 149–153.

Russell, L.M. (1977) Hosts and distribution of the greenhouse whitefly, *Trialeurodes vaporariorum* (Westwood) (Hemiptera: Homoptera: Aleyrodidae). *United States Department of Agriculture Cooperative Plant Pest Report* 2, 449–458.

Russell, T.E. (1982) Effect of cotton leaf crumple (CLC) diseases on stub and planted cotton. *Arizona Agricultural Experiment Station* P-56, 43–47.

Seif, A.A. (1981) Seasonal fluctuations of adult populations of the whitefly, *Bemisia tabaci*, on cassava. *Insect Scientific Application* 1, 363–364.

Sharaf, N. (1982) Parasitization of the tobacco whitefly, *Bemisia tabaci* Genn. (Homoptera: Aleyrodidae) on *Lantana camara* L. in the Jordan Valley. *Zeitschrift für Angewändte Entomologie* 94, 263–271.

Sharaf, N. and Batta, Y. (1985) Effect of some factors on the relationship between the whitefly *Bemisia tabaci* Genn. (Homoptera: Aleyrodidae) and the parasitoid *Eretmocerus mundus* Mercet (Hymenoptera: Aphelinidae). *Zeitschrift für Angewändte Entomologie* 99, 267–276.

Silberschmidt, K. and Tommasi, L.R. (1955) Observacoes e estudos sobre especies de plantas susetiveis a clorose infecciosa das malvaceas. *Annals Academy Brasil Ciene* 27, 195–214.

Speyer, E.R. (1927) An important parasite of the greenhouse whitefly. *Bulletin of Entomological Research* 17, 301–308.

Stoner, A. and Butler, G.D., Jr. (1965) *Encarsia lutea* as an egg parasite of bollworm and cabbage looper in Arizona cotton. *Journal of Economic Entomology* 58, 1148–1150.

Swirski, E. and Dorzia, N. (1968) Studies on the feeding, development and oviposition of the predaceous mite *Amblyseius limonicus* Garman and McGragor (Acarina: Phytoseiidae) on various kinds of food substances. *Israel Journal of Agricultural Research* 18, 71–75.

Swirski, E. and Dorzia, N. (1969) Laboratory studies on the feeding, development and fecundity of the predaceous mite, *Typhlodromus occidentalis* Nesbitt (Acarina: Phytoseiidae) on various kinds of food substances. *Israel Journal of Agricultural Research* 19, 143–145.

Swirski, E., Amitai, S. and Dorzia, N. (1970) Laboratory studies on the feeding habits, post-embryonic survival and oviposition of the predaceous mites, *Amblyseius chilenensis* Dosse and *Amblyseius hibisci* Chant (Acarina: Phytoseiidae) on various kinds of food substances. *Entomophaga* 15, 93–106.

Tarr, S.A.J. (1949) *Leaf Curl Disease of Cotton.* Commonwealth Mycological Institute, Kew, Surrey, 55pp.

Teich, Y. (1966) Mites of the family Phytoseiidae as predators of the tobacco whitefly, *Bemisia tabaci* Gennadius. *Israel Journal of Agricultural Research* 16, 141–142.

van der Laan. (1961) Stimulating effect of DDT treatment of cotton on whiteflies (*Bemisia tabaci* Genn. Aleyrodidae) in the Sudan Gezira. *Entomologia Experimentalis et Applicata* 4, 47–53.

Van Schaik, P.H., Erwin, D.C. and Garber, M.J. (1962) Effects of time of symptom expression of the leaf-crumple virus on yield and quality of fiber of cotton. *Crop Science* 2, 275–277.

Varma, P.M. (1963) Transmission of plant viruses by whiteflies. *National Institute of Science of India Bulletin* 24, 11–23.

Vet, L.E.M., Vanlenteren, J.C. and Worts, J. (1980) The parasite–host relationship between *Encarsia formosa* (Hymenoptera: Aphelinidae) and *Trialeurodes vaporariorium* (Homoptera: Aleyrodidae). *Zeitschrift für Angewändte Entomologie* 90, 26–51.

Von Arx, R.V., Baumgärtner, J. and Delucchi, V. (1983a) Developmental biology of *Bemisia tabaci* (Genn.) (Sternorrhyncha: Aleyrodidae) on cotton at constant temperatures. *Bulletin Société Entomologique Suisse* 56, 389–399.

Von Arx, R.V., Baumgärtner, J. and Delucchi, V. (1983b) A model to simulate the population dynamics of *Bemisia tabaci* (Aleyrodidae) on cotton in the Sudan Gezira. *Zeitschrift für Angewändte Entomologie* 96, 341–363.

Walhood, V.T., Henneberry, T.J., Bariola, L.A., Kittock, D.L. and Brown, C.M. (1981) Effect of short-season cotton on overwintering pink bollworm larvae and spring moth emergence. *Journal of Economic Entomology* 74, 297–302.

Watson, T. F. (1980) Methods for reducing survival of the pink bollworm. In: Graham, H. (ed.) *Pink Bollworm Control in the Western United States.* US Department of Agriculture, Science & Education Administration, ARM-W-16, pp. 24–34.

Westwood, J.D. (1856) The new *Aleyrodes* of the greenhouse. *Gardener's Chronicle*, p. 853.

Youngman, R.R., Toscano, N.C., Jones, V.P., Kido, K. and Natwick, E.T. (1986) Correlations of seasonal trap counts of *Bemisia tabaci* (Homoptera: Aleyrodidae) in southeastern California. *Journal of Economic Entomology* 79, 67–70.

Zalom, F.G. and Natwick, E.T. (1987) Developmental time of sweetpotato whitefly (Homoptera: Aleyrodidae) in small field cages on cotton plants. *Florida Entomologist* 70, 427–431.

Zalom, F.G., Natwick, E.T. and Toscano, N.C. (1985) Temperature regulation of *Bemisia tabaci* (Homoptera: Aleyrodidae) populations in Imperial Valley cotton. *Journal of Economic Entomology* 78, 61–64.

14 Jassids (Hemiptera: Cicadellidae)

G.A. Matthews

International Pesticide Application Research Centre, Imperial College at Silwood Park, Buckhurst Road, Sunninghill, Ascot, Berkshire SL5 7PY, UK

Introduction

Several species of leafhopper feed on cotton as well as a wide range of other host plants. They apparently introduce a toxin that impairs photosynthesis in proportion to the amount of feeding, and this causes the edges of leaves to curl downwards, the leaf to become yellowish and then redden, before drying out and shedding. Severe hopperburn can severely stunt young plants and reduce yields.

Amrasca devastans (Distant) is the Indian cotton jassid, previously referred to as *A. biguttula* (Ishida) or *Empoasca devastans* Distant. In Africa *Jacobiasca lybica* (de Bergevin) occurs predominantly in the north, but in East Africa this species overlaps with *Jacobiella facialis* (Jacobi) (Plate VII.1), which is common in southern, equatorial and West Africa. Both these species were previously in the genus *Empoasca. Empoasca decipiens* Paoli is of minor importance in Egypt. *Empoasca distinguenda* Paoli occurs in Zaire and South Africa as well as *Empoasca dolichi* Paoli, which has also been recorded in Somalia. *E. terrae reginae* (Paoli) is in Queensland, Australia (Ghauri, 1963). *Erythroneura lubiae* China is abundant on legumes, but may also be found on American Upland cottons in the irrigated areas of the Sudan.

Description of *Jacobiasca fasciatus*

Egg

Curved, 0.7–0.9 × 0.15–0.2 mm, curved, greenish, deeply embedded in the midrib or a large vein on either surface of the leaf, or in the petiole or young stem, but never in the lamina.

Nymph

Somewhat frog-shaped and flattened; pale yellowish-green; length from about 0.5 mm in the first instar to 2 mm. in the fifth and final instar. The nymphs are very shy, with a characteristic rapid, crab-like, sideways movement when disturbed. They are confined to the undersurface of leaves during the daytime, but can be found anywhere on the leaves at night.

Adult

Small elongate, wedge-shaped; about 2.5 mm. long in *J. lybica* but rather larger in *J. fasciatus*; body pale green with semi-transparent, shimmering wings; very active, having a sideways walk like the nymphs, but quick to hop and fly when disturbed.

Eggs usually hatch in about 6–10 days. A generation takes 3–4 weeks in the summer. *Amrasca devastans* is estimated to have 11 generations a year in India.

Host Plants

Apart from feeding on cotton, jassids have a very wide range of host plants, including herbaceous cultivated plants and weeds, chiefly amongst the Malvales, Leguminosae and Solanaceae.

Effect on Cotton

The importance of jassids depends very much on which cotton varieties are sown, those with hirsute leaves being less affected than glabrous types (Plate VII.2). Detailed studies were carried out in the Sudan (Joyce, 1961), where it was recognized that large plots were essential (Joyce and Roberts, 1959) to reduce the interplot effects due to spray drift. Jackson *et al.* (1965) reported yield losses of approximately 350 kg lint per hectare and a deterioration in lint quality, and in most of Africa and India hairy leaved cultivars are grown, as glabrous cultivars would require an extended insecticide spray programme to survive. In India, yield loss from jassids could be reduced from 25% to 12% by growing a hairy variety but these losses were small compared to that due to bollworms (Bhat *et al.*, 1984). Research in Africa assessed the hairiness of leaves in terms of the number of hairs that attained specified lengths measured at right angles to the undersurface of the leaf, or radially from the surface of the midrib.

Research by Parnell *et al.* (1949) with *J. facialis* concluded that:

- Variation is continuous between full susceptibility and very high resistance. The relative susceptibility of strains is independent of the level of infestation.
- Relative susceptibility, visually assessed, corresponds closely with relative infestation, determined by counts of nymphs.
- Relative susceptibility and relative infestation are correlated with both density

and length of hairs on the underside of the leaf lamina, but length is of prime importance. High density without length is ineffective. Very high inverse correlations ($r > 0.95$) between infestation or susceptibility and hairiness are obtained when the latter are measured as the density of hairs that exceed 0.3 mm height above the leaf surface, and even greater refinements in distinguishing between the most resistant strains can be obtained by using densities at heights exceeding 0.5 mm.

- Hairiness of the midrib is usually correlated with that of the lamina, but examination of hybrid material in which the two characters were segregating suggested that both influence resistance, although high midrib-hairiness is not essential if the lamina is hairy. Hairs on the stem and petiole are of little importance.
- Hairiness develops as the plants grow, the first few leaves being virtually glabrous. In consequence, even plants that are very resistant when older are susceptible as seedlings. Each leaf, however, has its full complement of hairs before it unfolds, so the resistance of individual leaves will tend to fall as it expands and its hair density diminishes.
- The relationship between hairiness and resistance is consistent over a very wide range of material of diverse origins.

Studies in India have also shown the importance of hair length on the undersurface of the leaf in reducing oviposition, especially on cultivars on which hair length exceeded the length of the ovipositor (Khan and Agarwal, 1984). Some semiglabrous *G. barbadense* cultivars tolerate higher jassid populations without showing severe symptoms of infestation in the Sudan. Pearson (1958) postulated that this could be related to the life-span of the leaves, which may be shed before symptoms appear; or an irrigated crop may be more tolerant than a rainfed crop. In India it has also been suggested that there could be chemical differences in the leaves which can affect the jassid populations (e.g. Sharma and Agarwal, 1983). The high density of gossypol containing glands present on *G. barbadense* may also be a factor.

Jassids can attack cotton leaves at all stages of development, but they cause the most significant damage to resistant varieties during the first two–three weeks after germination, before the young leaves fully develop their hairiness. Pearson (1958) recorded peak populations in southern Africa as late as April on plants four months old, after which there was a rapid decline as plants senesced.

Natural Enemies

Natural enemies are not considered to have a significant effect on populations of jassids, although a number of egg parasites have been recorded, for example *Anagrus empoascae* Dozier, *Stethynium empoascae* Subba Rao, and *Arescon enocki* (Subba Rao & Kaur) (Subba Rao *et al.*, 1968). The population normally declines during the dry season but jassids can survive on small patches of weeds, especially along drainage ditches and irrigation canals. In some areas, jassids do not enter cotton fields until later in the season after building up on weeds during the early rains. In the Sudan, weeds in fallows were an important source of jassids, but some of the infestation was thought to come by migration from evergreen trees, *Balanites aegyptiaca*, which are abundant to the South of the Gezira scheme. Jassids in this area have been regarded as less serious than other pests in recent years, probably due to changes in the cropping programme and the increased use of insecticides.

Insecticide Control

Insecticidal control measures are commonly considered necessary when the jassid population reaches two nymphs per leaf although with glabrous varieties the threshold may be lowered to one per leaf (Anon., 1985). The same threshold was adopted by Mabbett *et al.* (1984) in Thailand. In some cases numbers of adults invading a crop may be sufficiently high to justify control before a build-up of nymphal stages (Anon., 1976). An early insecticide spray, e.g. carbaryl or dimethoate, was recommended in Central Africa (Tunstall and Matthews, 1961), while later the sprays against bollworms will be effective. Early sprays against jassids, however, should be avoided, if at all possible, to conserve natural enemies of the total insect pest complex. Where resistant varieties are not sown, insecticide sprays can reduce jassid populations, but as reinfestation can occur and plant growth causes rapid dilution of any spray deposits, so repeated sprays are likely to adversely affect the natural enemies of other insect pests.

In order that jassid resistance should continue to be recognized and maintained wherever possible, it is necessary for breeding programmes to include an assessment of cultivars in the absence of insecticide application. This is standard practice in some countries and is especially important in peasant agriculture, where costs of insecticides must be kept to a minimum. It is also essential to minimize sprays to reduce selection for resistance by any of the cotton pests.

In Mali and parts of Burkina Faso and Cote d'Ivoire another leafhopper *Orosius cellulosus* is the vector of a mycoplasma disease called 'virescence' or 'phyllody'. Flowers are transformed into a rosette of green leaf-like petals and cause sterility of the plant (Cauquil, 1988).

References

Anon. (1976) *Cotton Handbook of Malawi.* Ministry of Agriculture, Lilongwe, Malawi.

Anon. (1985) *Cotton Handbook of Zimbabwe.* Commercial Cotton Growers Association, Harare, Zimbabwe.

Bhat, M.G., Joshi, A.B. and Singh, M. (1984) Relative loss of seed cotton yield by jassid and bollworms in some cotton genotypes (*Gossypium hirsutum* L.). *Indian Journal of Entomology* 46, 169–173.

Cauquil, J. (1988) Cotton pests and diseases in Africa south of the Sahara. Supplement to *Coton et Fibres Tropicales.* Institut Recherche de Coton et Textiles, Paris, 92pp.

Ghauri, M.S.K. (1963) Distinctive features and geographical distribution of two closely similar pests of cotton *Empoasca devastans* Dist. and *E. terraereginae* Paoli (Homoptera: Cicadellidae). *Bulletin of Entomological Research* 53, 653–656.

Jackson, J.E., Schultz, L.R. and Faulkner, R.C. (1965) Effect of jassid attack on cotton yield and quality in the Sudan Gezira. *Empire Cotton Growing Review* 42, 295–299.

Joyce, R.J.V. (1961) Some factors affecting numbers of *Empoasca libyca* (de Berg.) infesting cotton in the Sudan. *Bulletin of Entomological Research* 52, 191–230.

Joyce, R.J.V. and Roberts, P. (1959) The determination of the size of plot suitable for cotton spraying experiments in the Sudan Gezira. *Annals of Applied Biology* 47, 287–305.

Khan, Z.R. and Agarwal, R.A. (1984) Ovipositional preference of jassid, *Amrasca biguttula biguttula* Ishida on cotton. *Journal of Entomological Research* 8, 78–80.

Mabbett, T.H., Nachapong, M., Monglaku, K. and Mekdaeng, J. (1984) Distribution of *Amrasca devastans* and *Ayyaria chaetophora* in relation to pest scouting techniques for Thailand. *Tropical Pest Management* 30, 133–141.

Parnell, F.R., King, H.E. and Ruston, D.F. (1949) Insect resistance and hairiness of the cotton plant. *Bulletin of Entomological Research* 39, 539–575.

Pearson, E.O. (1958) *The Insect Pests of Cotton in Tropical Africa.* Commonwealth Institute of Entomology, London.

Sharma, H.C. and Agarwal, R.A. (1983) Role of some chemical components and leaf hairs in varietal resistance in cotton to jassid, *Amrasca biguttula biguttula* Ishida. *Journal of Entomological Research* 7, 145–149.

Subba Rao, B.R., Parshad, B., Ram A., Singh, R.P. and Srivastava, M.L. (1968) Distribution of *Empoasca devastans* and its egg parasites in the Indian Union. *Entomologia Experimentalis et Applicata* 11, 250–254.

Tunstall, J.P. and Matthews, G.A. (1961) Cotton insect control recommendations for 1961–62 in the Federation of Rhodesia and Nyasaland. *Rhodesia Agricultural Journal* 58, 233–253.

15 *Paurocephala gossypii* Russell (Homoptera: Psyllidae)

G.K.C. Nyirenda

University of Malawi, Bunda College of Agriculture, Lilongwe, Malawi

Distribution

Psyllidae are widely distributed (Imms, 1970; Hodkinson, 1978; Hodkinson and White, 1979) with a number of species which are pests of crops of economic importance.

Relatively little has been published on the cotton psyllid, *Paurocephala gossypii* Russell, or on the extent of its occurrence in relation to the incidence of the plant disorder, psyllose. The insect occurs in Zaire (Pearson, 1958), Central African Republic (Cauquil and Follin, 1983), Mozambique (McKinley, 1965; Cauquil and Follin, 1983; Evaristo, personal communication), Zimbabwe (Gledhill and Brettell, personal communication) and possibly in Zambia. *Paurocephala gossypii* is present in Kenya and Tanzania (Pearson, 1958; McKinley, 1965) and the author has observed some on *Azanza garckeana* in southern Tanzania. In Malawi the insect is distributed throughout the country in cotton and non-cotton growing areas, where it is found on its alternative host, *Azanza garckeana*. McKinley (1965) did not record the insect below 519 m above sea level, but it has been found on *Azanza garckeana* in the Shire Valley 50–100 m above sea level, albeit at much lower populations than those found in upland areas. The cotton psyllid may not be present in South Africa as a check list (Capener, 1970) includes only three species of the genus *Paurocephala*, and excludes *Paurocephala gossypii.*

Description and Biology

Egg

The egg is oval, about 0.2 mm long by 0.12 mm broad, with the apex drawn into a thin filament (Fig. 15.1a). The base is rounded with one side drawn into a peduncle which is used to anchor the egg to the main veins, veinlets and the base

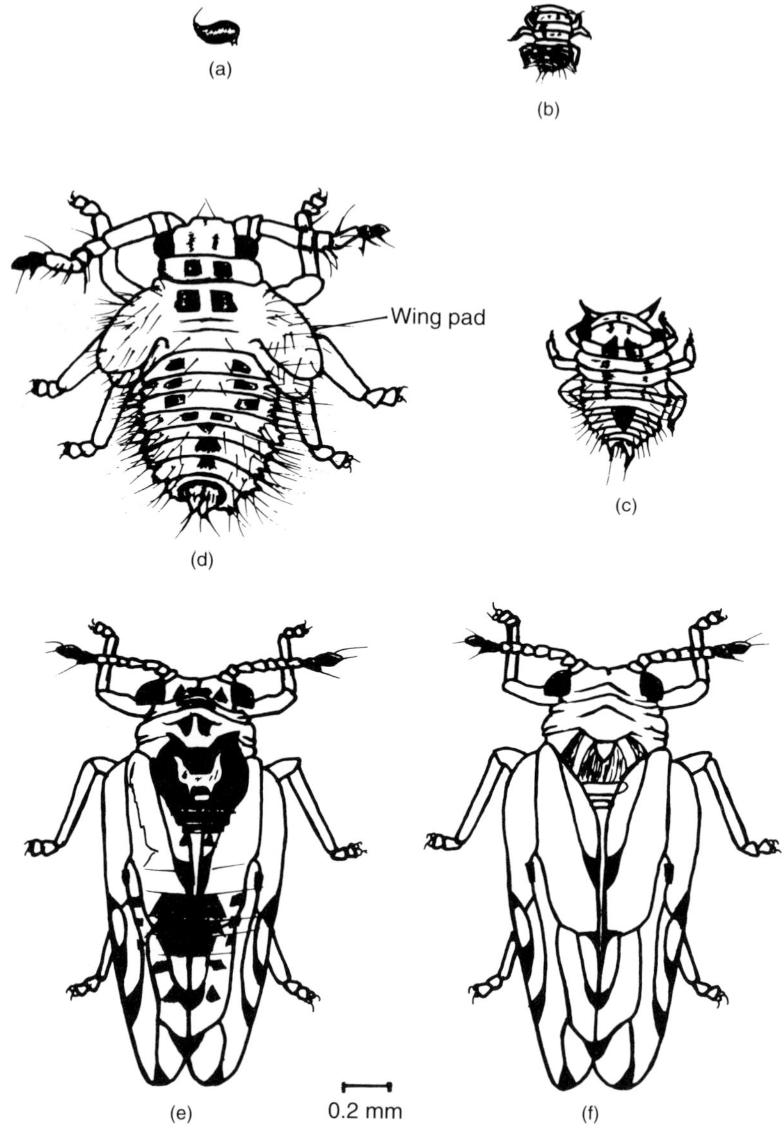

Fig. 15.1. Life stages of the cotton psyllid, *Paurocephala gossypii* (Russell), dorsal view: (a) egg; (b) 1st instar nymph; (c) 2nd instar nymph; (d) 5th (last) instar nymph; (e) male adult; (f) female adult.

of hairs on the underleaf surface of leaves. Eggs are laid singly or in groups. When freshly laid the egg is yellowish-white or opalescent, changing to brown when it is about to hatch. Just before hatching, eyes of the future nymph appear as two red spots. The egg shell is left on the leaf after hatching. The incubation period of the egg depends on the temperature at which eggs are kept. In Malawi, eggs reared at 16°C and 23°C and 70% RH, took an average of 20 days (range 15–25) and 12 days (range 8–15) to hatch.

Nymph

The freshly emerged nymph is whitish or translucent and moves sluggishly soon after hatching. Future wings are barely seen as slight swellings just behind the head (Fig. 15.1b,c). The colour changes to pale yellow or greyish in later instars with some brown markings. The mean body length from head to the last abdominal segment increases from 0.34 mm at hatching to 1.5 mm in the fifth instar. Wing pads become more distinct in later instars (Fig. 15.1d). Nymphs are dorsoventrally flattened, thus distinguishing them from aphids. In light to moderate population densities, nymphs settle and only move again to change feeding positions, or when they are disturbed. They attach their mouthparts to main leaf veins and veinlets or vein junctions, where they are clustered on the underleaf surfaces. In dense populations, nymphs, particularly those in later instars, redistribute to all parts of the plant including stems, petioles, flowers and buds, but do not feed.

The development period of nymphs ranges between 20 and 58 days (Fig. 15.2). At temperatures > 25°C during the whole life cycle, egg to adult takes 26 days (Pearson, 1958). This may increase to 70 days in cold temperatures below 16°C.

Adult

The average adult is about 2 mm long from head to the last abdominal segment. On emergence, wings are folded and pale yellow or translucent but within a few hours wings are expanded and become transparent. Gradually the body and the wing edges develop a number of pigments which show as dark brown spots, distinguishing adults from aphids and jassids. The female is larger than the male (Fig. 15.1f) which has deeper yellow wings and more heavily pigmented brown spots (Fig. 15.1e). The adult is a weak flyer and walks limited distances on the underleaf surfaces, but moves away by flying or jumping when disturbed.

Mating takes place 2–3 days after emerging and oviposition usually two days later. The longevity of the adult is unknown but is believed to be about ten days. Pearson (1958) recorded 12–21 eggs were laid per day.

Natural Enemies

Little is known on the impact of natural enemies on *Paurocephala gossypii* but a number of arthropods feed on both nymphs and adult psyllids. Certain predators have been seen on cotton and *Azanza*. Of the coccinellids, *Ostalia ochracea* Weise probably can significantly reduce psyllid populations, but *Platynaspis capcila* Crotch and *Scymnus morelliti* Mulsant, a general feeder, may be important. A fungus also attacks the nymphs.

Populations of lacewings *Semidalis* spp. often rise when high psyllid populations occur but whether it has an impact on psyllids is uncertain. Larvae of *Chrysopa* spp. have also been found but their impact is probably low, and *Ceratochrysa antica* Walker is only found in low numbers.

The parasitoid *Tetrasticus* (Tamarxia) sp., estimated to cause 5% parasitism, is common at high densities of psyllid, particularly on *Azanza*. A number of spiders have been seen to feed on psyllids. One species, *Heliophanus*, is quite abundant and, although it is a general feeder, its impact on psyllid populations may be significant. A number of ants, on several occasions, have been observed on cotton to carry both nymphs and adult psyllids, but their impact as predators is unknown.

Host Plants

Paurocephala gossypii has only been recorded in Malawi on cotton and, especially in the closed season for cotton, on *Azanza garckeana*. Even in non-cotton growing areas, very high numbers have been found on the latter host plant and subsequently cause sooty moulds to develop. On cotton oviposition is mostly between April and September, whereas on *Azanza* it increases from July, declines temporarily in October and rises again in November to December (Fig. 15.3).

Damage and Pest Status

Damage

High psyllid populations stress plants and this may have a considerable direct effect on yield of seed cotton. Young cotton plants are stunted. Both nymphs and adults exude copious amounts of honeydew which covers leaf surfaces and the ground. This sticky honeydew encourages sooty mould which inhibits photosynthesis and downgrades the lint.

Psyllids transmit the disease ‘psyllose’, caused by a mycoplasma (Cauquil and Follin, 1983). In the early stages young leaves become very pale yellow at the edges, while older, mature leaves turn dark greenish to reddish in colour and are shed. Leaves formed later are reduced in size to scalelike structures, become olive-green in colour and are shed. The internodal distance is much reduced so the plant is stunted with a large number of leaves near the top. The whole plant turns purple or deep red with green tinges. If psyllose occurs before flowering, no bolls and flowers are formed and the plant dies or rarely grows above 0.5 m tall. Immature and nearly mature bolls become flaccid and fail to open, but mature bolls may open with poor quality lint, the locs remaining hard. Buds and flowers are reduced in size, may become purple or deep red, and are shed. In Malawi the disease usually spreads from the edge of the field, but often nymphs and adults are rarely found on the plant by the time symptoms of the disease are fully developed.

Isolation and identification of the mycoplasma is difficult due to its instability (Stead, personal communication). Transmission has not been studied, but the few experiments carried out imply that the disease is carried by a small population of infective psyllids. In every year a few plants can develop psyllose even in well-sprayed cotton. The disease is more evident if some cotton has been left throughout the closed season. Whether *Azanza* is a reservoir of psyllose has not been investigated, but plants can show some symptoms such as smaller leaves becoming pale yellow; the purple or red colour symptoms do not show in *Azanza*. McKinley (1965) associated the disease with wet overcast weather, even though this may be compounded with the presence of stand-over cotton plants throughout the year. It is also not clear why psyllose is only prevalent in specific areas but not others, while the vector and its hosts are present in both. Psyllose does not occur at Kadoma, Zimbabwe, in spite of having high psyllid populations on *Azanza* (Gledhill and Brettell, personal communication). It is therefore possible that a number of factors interact before the symptoms of psyllose disease can develop.

Loss in Yield

Seriously affected plants eventually die, but yield loss will depend on the stage at which the crop was attacked. In upland areas of southern Malawi yields below 100 kg ha^{-1} are common when there is a serious psyllose attack on farmers' fields. Cotton planted adjacent to either ratoon or stand-over cotton infested with psyllids, whether sprayed with different insecticides including carbaryl, DDT and dimethoate, or left unsprayed, gave an average yield of 19 kg ha^{-1} of seed cotton (range 0–66 kg ha^{-1}) (Nyirenda, unpublished). However, sprayed yields on plots

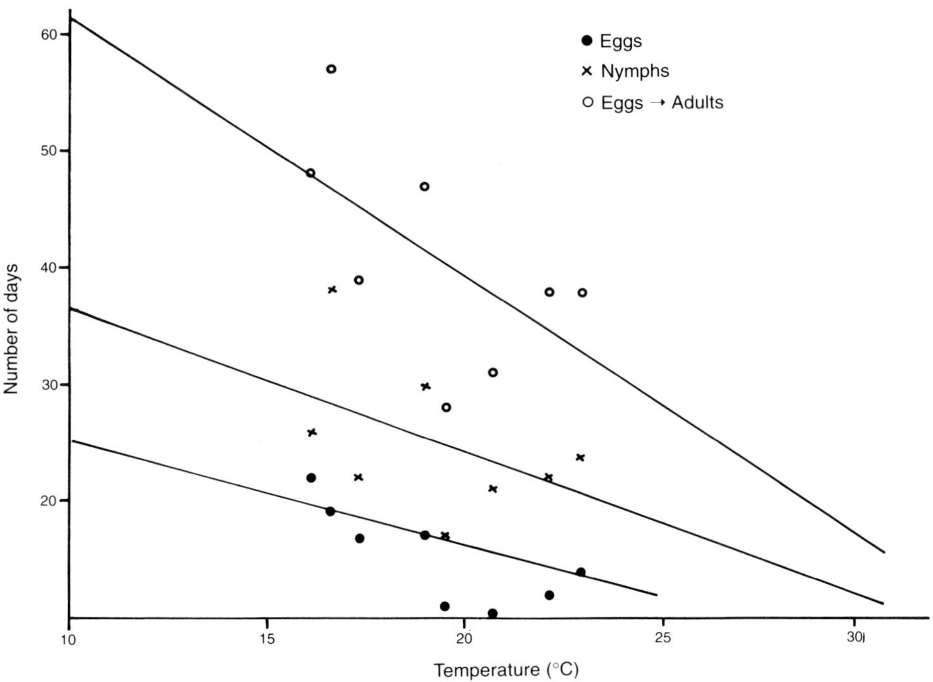

Fig. 15.2. Developmental periods of eggs and nymphs in days at different temperatures.

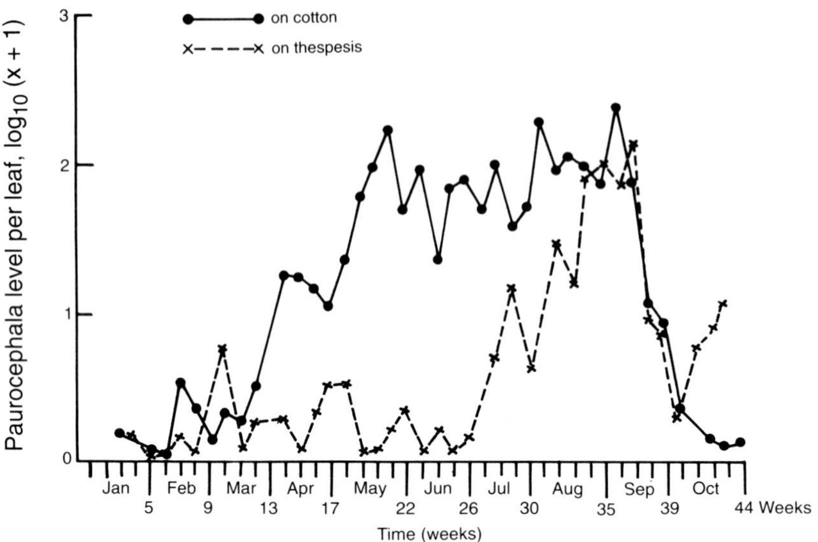

Fig. 15.3. Oviposition of *Paurocephala* on cotton and *Azanza garckeana* at Makoka.

sited about 500 m from the same source of infestation averaged more than 1500 kg ha^{-1} of seed cotton.

Control of psyllose and *P. gossypii* should, to a certain extent, be separated. The insect is easily controlled by sprays applied to control bollworms or sucking insects (Nyirenda, 1986), using insecticides such as synthetic pyrethroids, carbaryl, thiodicarb, endosulfan, dimethoate or DDT. Early sprays between four to eight weeks after germination of cotton are important to control the low-level psyllid populations and limit development of the psyllose. Irregular sprays allow a few psyllids to transmit the disease. In psyllose prone areas sprays should be applied after February when psyllid populations increase and continue until 50% of the bolls are mature on 75% of the crop. Once the mycoplasma is in plants, insecticides cannot reverse the symptoms or disease development. Similarly, once 30–40% of plants in a field have developed symptoms of the disease, spraying should cease to avoid wastage of insecticides, as eventually the whole field will be affected. Early and regular prophylactic sprays may be necessary, particularly in wet weather conditions to maintain control.

Pest Status

Paurocephala gossypii may be more widespread than has been reported, but psyllose appears to be only locally important in a few countries in Africa, including Zaire, Mozambique, Malawi and the Central African Republic (Pearson, 1958; McKinley, 1965; Cauquil and Follin, 1983). No loss figures are available from these countries but Pearson (1958) concluded that the nature of the disease was a threat that required careful watching. In Malawi the disease is so important locally and seasonally, especially in the upland areas of the southern and central regions, that some farmers give up growing cotton the following season. In these areas it is, therefore, of more economic importance than aphids and red spider mites. There is no doubt that cotton left undestroyed after harvest plays an important role in the development of psyllose in the following season.

References

Capener, A.L. (1970) Southern Africa Psyllidae (Homoptera) – I: A check list of species recorded from South Africa, with notes on the Pettey collection. *Journal of the Entomological Society of Southern Africa* 33, 195–226.

Cauquil, J. and Follin, J.C. (1983) Presumed virus and mycoplasma-like organism diseases of cotton in subsaharan Africa and in the rest of the world. *Coton et Fibres Tropicales* 38, 293–308.

Hodkinson, I.D. (1978) The psyllid (Homoptera: Psyllidae) of Alaska. *Systematic Entomology* 3, 333–360.

Hodkinson, I.D. and White, I.M. (1979) New psyllids from France with redescriptions of the type species of Flora Low and Amblyrihina Low (Homoptera: Psyllidae), *Entomologica Scandinavica* 10, 55–63.

Imms, A.D. (1970) *A General Textbook of Entomology*. Ninth edition, revised by Richards, O.W. and Davies, R.G. Chapman and Hall, London, 886pp.

McKinley, D.J. (1965) Occurrence of *Paurocephala gossypii* Russell (Psyllidae) on cotton in Malawi. *Empire Cotton Growing Review* 42, 209–210.

Nyirenda, G.K.C. (1986) Studies of the Effects of Insecticide Application on Cotton in Malawi. Unpublished PhD thesis, University of London, 374pp.

Pearson, O.E. (1958) *The Insect Pests of Cotton in Tropical Africa*. Commonwealth Institute of Entomology, London, 355pp.

16 *Lygus* (Hemiptera: Miridae) and Other Hemiptera

T.F. Leigh (deceased)[1] and G.A. Matthews[2]

[1] *University of California, Shafter Research Station, 17053 Shafter Avenue, Shafter, California 93263, USA;* [2] *International Pesticide Application Research Centre, Imperial College at Silwood Park, Buckhurst Road, Sunninghill, Ascot, Berkshire SL5 7PY, UK*

Introduction

Plant bugs in the genus *Lygus* are pests of cotton in some areas of Africa, the United States of America (USA) and China. Graham *et al.* (1984) have provided a literature review of this genus for 1900–1980, and Hedlund and Graham (1987) have reviewed the economic importance and biological control of *Lygus* and *Adelphocoris* in North America. Detailed taxonomic reviews of these plant bugs are provided in the the cited references (Kelton, 1971, 1975) for all species except *Taylorilygus vosseleri*, which was described by Taylor (1947). The four principal pest species are *Lygus hesperus* Knight, *L. lineolaris* (Palisot de Beauvois), *Taylorilygus vosseleri* (Poppham) and *L. lucorum* Meyer-Dur. *Lygus hesperus* is a major pest of cotton in California, while the tarnished plant bug (*L. lineolaris*) occasionally causes significant crop loss in other areas. Other species are occasionally reported to be pests of USA cotton, particularly in Arizona and New Mexico. *Taylorilygus vosseleri* is widely distributed over Africa and is recorded from Madagascar, but is a major pest of cotton only in regions of the moist equatorial zone. Other species have been reported on African cotton, but have not been confirmed to be pests. *Lygus lucorum* is reported to be a significant pest of transplanted cotton in a coastal cotton growing area near the Yangtze River in China.

Lygus bugs apparently prefer to feed on the apical meristematic tissues, developing floralbuds and developing seeds of major crops including alfalfa, bean, beet, clovers, cotton and cruciferous crops. Feeding damage can cause floralbud and young fruit destruction, resulting in severe reductions in seed set and quality. *Taylorilygus vosseleri* in Africa and *L. lucorum* in China, Europe and eastern Canada are very destructive to meristematic tissues in terminals of plants (Dale and Coaker, 1958; Wang, 1984; Zhang *et al.*, 1986) although boll damage by *T. vosseleri* is reported by Hancock (1935). Meristem feeding results in abortion of plant terminals. Secondary vegetative growth is stimulated and produces plants with many vegetative branches.

Distribution

Lygus hesperus is commonly found on cotton throughout all states of the western United States and has also been recovered in Nebraska, in New Mexico and in western Texas (Kelton, 1975). The tarnished plant bug occurs throughout the United States, parts of Canada and Mexico (Kelton, 1975), and is most commonly reported to be a crop pest east of the continental divide. *Taylorilygus vosseleri* is reported to occur throughout Africa and in Madagascar, but is reported as a pest of cotton only in the moist equatorial zone causing 'leaf tattering' (Plate VII.3) (Taylor, 1945). *Lygus lucorum* is reported from eastern Canada (Kelton, 1971), Great Britain, Europe, Asia (Butler, 1923) and the Yangtze River region of China where it is a pest of cotton (Zhang *et al.*, 1986).

Description of Stages

For detailed descriptions of the lygus bug species cited see: Kelton (1975) for *Lygus hesperus* and *L. lineolaris*, Kelton (1971) for *L. lucorum*, and Taylor (1947) for *T. vosseleri.*

Eggs are a typical Capsid shape, tubular, slightly curved, rounded posteriorly and truncated at the cap. They are about 1 mm long. Usually only the cap is visible, since the remainder of the egg is inserted into plant tissue.

Nymphs of all species are ovoid and 1–4 mm long, fragile in appearance, smooth and slightly pubescent body, with relatively long and slender antennae and legs. The first and second instar nymphs are pale green and may be mistaken for aphids, but differ in moving about more rapidly and having reddish tips on their antennae. The legs of all stages have brownish mottlings. Older nymphs have five characteristic black spots on their backs: two spots on the first segment of the thorax just behind the head, two more on the next segment, and one spot in the centre of the abdomen. Older nymphs of *L. hesperus* and *L. lineolaris* may be pale to medium brown in colour (Kelton, 1975; Anon., 1984).

Adult female lygus bugs range between approximately 4 and 6.5 mm in length and 1.6 mm to 2.8 mm in width. Males are usually slightly smaller than females. The American species tend to be in the larger size range, while *T. vosseleri* is smaller. Several species are pale green, straw-yellow, or reddish-brown in colour with a conspicuous triangle, the scutellum, in the centre of the back. This scutellum is yellow or pale green on summer forms of *L. hesperus* and yellow-brown on the tarnished plant bug. Darker forms occur in the autumn, winter and spring. All species have long slender and usually reddish-brown antennae.

Life History and Phenology

Lygus hesperus overwinters as adults in a reproductive diapause (Beards and Strong, 1966), usually in the crowns of low-growing host plants but may be seen sunning themselves when daytime temperatures are sufficiently high. They become reproductively active in late November (Leigh, 1966), when the rainy season begins, after which this species will oviposit in available crop, native and weed hosts. As a result of cool winter temperatures, nymphs of the first brood are not usually found until March or April (Strong *et al.*, 1970). Development during midsummer can be completed in as little as 21 days. More typically egg incubation and nymphal development will require nine and 18 days respectively (Shull, 1933; Cave and Gutierrez, 1983). Leigh (1963) reported a six to nine day preoviposition period at constant 26.7°C, while Cave and Gutierrez (1983) indicate a ten to 12 day period under field conditions, with a longer preoviposition period on cotton than on alfalfa. Fecundity was highly variable under field conditions (Cave and Gutierrez, 1983) and ranged between 15 and 114 eggs, mean 50 eggs, with up to eight eggs per reproductive day; whereas Leigh (1963) and Beards and Strong (1966) reported 202 and 178 eggs per female respectively, reared in the laboratory on green beans, *Phaseolus vulgaris* L.

The life history of the tarnished plant bug is very similar to that of *L. hesperus*, and the detailed report of Geering (1953) describes a similar life history for *T. vosseleri*. The life history of *L. lucorum* is not well studied.

There are one or two generations of *L. hesperus* and tarnished plant bugs on spring hosts and at least three generations may occur on cotton. Wang (1984) reports movement of second generation *L. lucorum* into cotton where three additional generations occur. While massive flights into cotton may occur at any time during the growing season, this is dependent on abundance of natural and crop hosts and their termination through natural maturation, drying or harvest (Pearson, 1958; Sevacherian and Stern, 1975; Fleischer *et al.*, 1988). Populations also gradually increase in numbers within cotton fields as the season progresses.

Behaviour

Flights of lygus bugs are strongly crepuscular (Mueller and Stern, 1973). These pests are usually evident in the terminals of plants near sunrise and sunset and can be seen actively flying during that period of the day. Male *L. hesperus* and tarnished plant bugs are responsive to the pheromone call of females (Scales, 1968; Strong *et al.*, 1970) and may be seen mating during the pre-dawn and early post-dawn period (Aldrich *et al.*, 1988).

Infestations may appear in cotton during the early fruiting development of plants, or at any time that source hosts mature or are destroyed by cultivation or harvest (Stern *et al.*, 1967; Fleischer *et al.*, 1988). Migrations to cotton may continue over a protracted period of time as source hosts mature and dry (Mueller and Stern, 1973, 1974; Sevacherian and Stern, 1975). *Lygus hesperus* exhibits a strong preference for alfalfa and some other leguminous hosts.

Where preferred hosts occur, infestations on cotton may appear and disappear in as little as two or three days, doing little damage (Anon., 1984). Persistent infestations may develop from eggs deposited on cotton and devastate the crop.

On cotton, lygus bugs demonstrate a strong tendency to feed and oviposit on the upper one third of the plant (Benedict *et al.*, 1981a;, Wilson *et al.*, 1984; Leigh *et al.*, 1988). With the aid of a black light, *L. hesperus* can be found in the terminals of plants at night (Stern and Mueller, 1968), apparently resting and feeding. When disturbed in the daytime adults will readily fly, usually to a nearby plant.

Adult female bugs in this group apparently prefer to deposit their eggs in the newer and presumably more tender plant growth (Graham and Jackson, 1982). Benedict *et al.* (1981a) reported egg laying on cotton was most common in the upper one third of the plant and then most frequently in the leaf pulvinus, although they were also found in other plant parts. Feng and Zhang (1987) report most egg laying by *L. lucorum* is in the leaf veins and petioles. Benedict *et al.*(1983) reported greater numbers of eggs deposited on the more pilose genotypes, but nymphal survival was reduced on these plants.

Host Plants

An annotated host list for *L. hesperus* has been prepared by Scott (1977). Host lists for *L. lineolaris* and citations of host contributions to regional populations have been prepared by Snodgrass *et al.* (1984a,b) for the Arkansas, Louisiana and Mississippi Delta, by Womack and Schuster (1987) for part of Texas, and by Fleischer and Gaylor (1987) for Alabama. Taylor (1945) cites hosts of *T. vosseleri*. The hosts of *L. lucorum* include carrot, mulberry, nettles (*Urtica* sp.), mugwort (*Artemisia* sp.), tansy (*Tanasetum* sp.), hemp (*Eupatorium cannabinum*), agrimony and other plants (Butler, 1923; Southwood and Leston, 1959; Kelton, 1971). Host lists for the first two species are very extensive including many plant families, and the host list for *T. vosseleri* includes 15 families and more than 28 species.

Infestations of plant bugs are commonly associated with nearby native weed, and crop hosts (Smith, 1942; Pearson, 1958; Sevacherian and Stern, 1975). Among the most notorious crop sources of *L. hesperus* are alfalfa (Stern *et al.*, 1964) (whether grown for hay or for seed), safflower (Mueller and Stern, 1974), beets grown for seed, weed hosts in these crops, an array of Compositae,such as *Erigeron annuus* (Pers.) and *E. canadensis* (Fleischer and Gaylor, 1987). In arid areas *L. hesperus* is a more consistent pest near riparian outflows from mountains.

Severity of infestation, often expressed as crop damage, is reported to be greatest in areas of fields that lie adjacent to an infestation source. Schowalter and Stein (1987), Stern *et al.* (1964, 1967) and Sevacherian and Stern (1975) have studied local movement of *L. hesperus* and report field to field movement. Stride (1968) presented a similar relationship between *T. vosseleri* and its native and crop hosts in Uganda, while Hancock (1935) reported common bean, chickpea, sorghum and millet to be hosts. When a crop host is harvested simultaneously over a large area, there is a sudden exodus of the pest and a corresponding invasion of the cotton crop. Fleischer *et al.* (1988) report movement of *L. lineolaris* from weed hosts to cotton to be largely a diffusion process. There may be little movement of lygus bugs from alternate hosts to cotton when these other plants are succulent and growing rapidly. Major emigration occurs as hosts mature, dry, or are harvested or destroyed, and infestations of *L. hesperus* commonly develop in fields in desert areas isolated by several miles from known sources of infestation.

Natural Enemies

Numbers of immature lygus bugs may be greatly supressed by natural enemies, and increase greatly if their predators are destroyed by insecticides. The principal natural enemies of the Miridae, or plant bugs, are several generalist predators and parasitic wasps. The predators include several species of spiders (Araneida), predacious members of the Lygaeidae (*Geocoris* spp.), Miridae (*Deraeocoris* spp.), Nabidae (*Nabis* spp.), and pirate (*Orius* spp.) bugs (Taylor, 1945; Whitcomb *et al.*, 1963; Cao, 1986; Zhang *et al.*, 1986; Young, 1989). Species of these natural enemies are found widely distributed in native vegetation, weed and crop growth areas. Their distribution appears to include the plant hosts frequented by lygus bugs (Fleischer and Gaylor, 1987; Young, 1989). We presume their movement into cotton is by ballooning in the case of spiders and by localized diffusion flights of the true bugs. Lygus bug infestation increases are often associated with destruction of these natural enemies by broad spectrum insecticides (Leigh *et al.*, 1966).

Parasites of lygus bugs in the Braconidae, Tachinidae, Nematoda and Mymaridae have been reported by Clancy (1968), Scales (1973) and Graham *et al.* (1986). In the western USA the most common parasitoids appear to be *Anaphes iole* (*ovijentatis*) (Crosby and Leonard) and *Leiophron uniformis* (Gahan). Taylor (1945) reported hymenopterous parasites in the above groups for *T. vosseleri* in Africa, indicating that they were common to a number of Capsids. Graham *et al.* (1986), in their intensive studies of *A. iole* (*ovijentatis*) and *L. uniformis* in Arizona, suggest that parasitoids are localized in areas of plant bug distribution and are not highly migratory since they are most abundant in cotton adjacent to perennial crop or native vegetation hosts of lygus bugs.

Pest Status

Greatest damage to cotton by the tarnished plant bug and *L. hesperus* is to the floralbuds (squares) (Leigh *et al.*, 1988), many of which shed from the plants. There is also destruction of meristematic tissues in the growing points. Boll feeding will occur when numbers are high. Loss of fruit and apical growing points will result in secondary vegetative growth, producing many branched plants with little fruit. These lygus bugs demonstrate preference for feeding on small squares and light infestations may only remove the squares. This loss of fruit stimulates vegetative growth which results in tall, whip-like plants. *Taylorilygus vosseleri* is reported to feed preferentially on the plant terminals (Taylor, 1945), causing leaf tattering due to the destruction of developing tissue. Taylor also reports square shedding and boll damage. Internodes are often shortened on plants on which lygus bugs have fed (Hancock, 1935; Smith, 1942; Taylor, 1945) and nodes may appear swollen (Ewing, 1929).

Infestations of plant bugs on cotton are usually derived from populations that develop on alternative host plants. Infestations of *L. hesperus*, the pale legume bug (*L. elisus* Van Duzee) and *L. desertinus* frequently move to cotton from alfalfa, clovers, safflower and other seed crops (Stern *et al.*, 1967; Mueller and Stern,

1973). Taylor (1945) and others report similar outbreaks of *T. vosseleri* on cotton from those sources. These infestations may be assessed by sweeping plants with a net (Anon., 1984), shaking plants over a drop cloth, or observation of their presence in plant terminals (Tugwell *et al.*, 1976; Fleischer *et al.*, 1985). Damage to cotton can be assessed by observing flowers for evidence of anther damage (Pack and Tugwell, 1976) and by cutting open squares to evaluate anther damage (Williams *et al.*, 1987; Leigh *et al.*, 1988). In Africa damage has been assessed mainly through evaluation of leaf damage (Taylor, 1945).

In California, the treatment thresholds recommended consider the fruiting stage of cotton plants and the probability that cotton squares will develop to provide harvestable fruit (Anon., 1984). The recommendations are based on twice weekly surveys of cotton fields during the fruiting season to determine numbers of bugs using a sweep net and the numbers of squares per 4.0 m^2. An infestation is considered treatable if the average number of bugs (all stages) per 50 sweeps exceeds three per 100 flowerbuds per 4 m^2 (1/1000 acre).

Control Measures

Several insecticides applied as ground or aerial sprays provide effective control of lygus bugs, although pesticide resistance has become a common problem in the USA (Leigh and Jackson, 1968; Leigh *et al.*, 1977). Where insecticides are used, the systemic and more persistent materials may provide control for up to three weeks, or longer. Less persistent insecticides may control only the active stages, and an infestation of nymphs can develop again in a few days from eggs deposited in the plants. Cotton fields may also be reinvaded. If infestations persist within the field, or migrations from other hosts continue, three to five applications of insecticide may be necessary during the flowerbud stage to assure early fruit set (Mueller and Stern, 1974).

Source management

In California (USA) the threat of movement by *L. hesperus* into cotton from safflower is so great that farmers are advised to treat the safflower crop with an insecticide to protect their cotton from damage. Mueller and Stern (1974) developed guidelines for scheduling the safflower treatments. Some farmers practise trap cropping and management of a more attractive host such as alfalfa (lucerne) (Stern *et al.*, 1964, 1967; Stern, 1969; van den Bosch and Stern, 1969). These plantings, as well as nearby forage alfalfa fields are managed to attract *L. hesperus* from cotton. There are no published references to other suggested trap crops such as sesame (*Sesamum* sp.) and cowpea (*Vigna sinensis*). In Alabama, Fleischer and Gaylor (1987) have recommended management of the weed host *Erigeron canadensis* to attract the tarnished plant bug from cotton fields. Taylor reported sorghum acts as a trap crop for *T. vosseleri*, although it is not utilized for that purpose. As an alternative to trap cropping, some growers attempt to isolate production of lygus source crops from cotton plantings. Farmers who produce both alfalfa and cotton may harvest only portions of their alfalfa crop at a time to assure the presence, at all times, of some of this crop in a growth stage attractive to lygus bugs.

Genetic modification of plants may be used to reduce the threat of lygus bug attack. As lygus bugs feed on nectar, utilization of nectariless cotton cultivars may provide infestation suppression (Benedict *et al.*, 1981b; Bailey *et al.*, 1984; Snodgrass *et al.*, 1984b; Scott *et al.*, 1988), however, Scott *et al.* (1988) reported a significant reduction of some lygus bug predators. Wilson and George (1986) cited greater loss of flowerbuds to *L. hesperus* from smooth leaf cotton than from hirsute and semi-smoothleaf lines. Genetic lines of cotton that carry resistance to lygus bugs have been identified (Shepherd *et al.*, 1986; Kitten *et al.*, 1987; Jenkins *et al.*, 1988) and are currently available for utilization in development of resistant cultivars.

Agronomic practices may be modified to reduce the threat of lygus bugs to cotton. Leigh *et al.* (1969) correlated high lygus bug numbers with excess use of nitrogen fertilizer and irrigation, and recommended limiting use of these agronomic strategies to make plants less attractive to lygus bugs. Taylor (1945) recommended avoiding cultural practices that make plants succulent and thereby more attractive to *T. vosseleri*.

Other Hemiptera: Stem Sucking Bugs

Anoplocnemis curvipes (Fabricius) (Hemiptera: Coreidae)

Anoplocnemis curvipes has been recorded as a very minor pest on cotton in Zaire, the Central African Republic, Cote d'Ivoire, Madagascar, Nigeria, Malawi, Zambia, Tanzania and South Africa. The adult bugs feed on bolls, stems and shoots, but its main damage is inflicted when it attacks the apical shoot and the tips of side branches, which usually die beyond the point of attack. Populations are generally very low and a few plants may be attacked at random.

Helopeltis spp. (Hemiptera: Miridae)

The genus *Helopeltis* occurs throughout the moist tropics of the Old World, and is polyphagous, principally feeding on crops other than cotton. *Helopeltis* is essentially restricted to the forested regions and their outskirts, including the high-grass savannah that replaces the equatorial forest where this is destroyed. Rainfall is usually in excess of 1000 mm with two well-defined peaks. *Helopeltis schoutedeni* occurs in Africa on cotton (Plate VII.4) and a range of other crops including mango, guava, castor, pigeon pea and sweet potato. It has been noted in Côte d'Ivoire, Ghana, Togo, southern Nigeria and in parts of the Central African Republic, Zaire, Uganda, the Equatoria Province of the Sudan and in the wooded upland areas of Mozambique, Malawi and Zambia. It is absent or of little importance in the drier savannah areas of the Sudan, Chad, Zimbabwe and South Africa. *Helopeltis poppius* is also found on cotton in Africa. *Helopeltis theivora* and *H. antonii* occur in Asia but have not been recorded from cotton. In areas where

Helopeltis occurs, attacks are generally irregular from one season to the next and with very pronounced irregularity in the distribution of damage within any one area in one season. Control against *Helopeltis* has not been recommended on cotton as insecticide programmes against other pests appear to provide adequate control. Several natural enemies have been recorded but little is known of their importance.

Damage is due to the insect sucking sap from the young green developing tissue, and, while doing so, injecting saliva that cause severe breakdown of the tissue and the formation of cankers, thus adversely affecting further growth. Feeding on the leaf lamina can cause small discrete pale brown spots, which are not to be confused with the dark brown spots due to bacterial blight, *Xanthomonas malvacearum*, and referred to as angular leaf spot. If lesions are formed on the main veins of a leaf before the lamina is fully expanded, the latter will be puckered like a piece of material gathered on a thread drawn through it. A field of young cotton severely attacked by *Helopeltis* looks as if it had been scorched by fire. Later in the season, feeding on young green bolls stimulates intense meristematic activity below the puncture which lifts up the circular scab and extrudes a worm-like spiral of soft tissue. Later feeding on green bolls can cause so many lesions that the boll splits prematurely and allows entry of bacteria and fungi that rot the contents.

Pearson (1958) considered that the common species on cotton is *H. schoutedeni* Reuter. References to other species in Africa are probably confused by the presence of a red and yellow form of the female.

Description of **Helopeltis schoutedeni**

Egg Elongate, slightly curved, 1.5–1.8 mm × 0.2–0.3 mm, white, with two filaments of unequal length at one end.

Nymph Slender, fragile insects with long legs and antennae; creamy-yellow with brick-red decorations on the body and, as hoops, on the femora. There are five instars. The characteristic pin-like process which projects vertically from the scutellum appears in the second instar and is about 2 mm long in the adult.

Adult Similar to the nymph in build, body 7–10 mm long × 1.5–2.0 mm broad, legs about the same length as the body and antennae nearly twice as long. The head, antennae, feet and membrane of the hemelytra are black, with a lozenge-shaped smoky area lengthwise in the centre of the hemelytra; the rest of the body is yellow in the males and some of the females, the remainder of which are blood-red. Carayon and Delattre (1948) considered that the blood-red colour was acquired by the females on reaching sexual maturity.

Life history

The eggs are buried individually so that only the end with the filaments attached is visible, in slits that the female makes with her ovipositor, sometimes in the bark of the stem or branches, but more usually in the leaf veins or peduncle and most frequently in the petiole. About 30–60 eggs per female have been recorded and these hatch in just under two weeks. The nymphal stage is about three weeks, except in cooler weather. Adults can survive six–ten weeks, with oviposition starting when the female is several days old. *Helopeltis* is polyphagous; host plants include several leguminous crops, such as pigeon pea (*Cajanus cajan*), sweet potato (*Ipomoea batatas*), certain semicultivated fruit trees, such as mango (*Mangifera indica*), guava (*Psidium guajava*) and cashew (*Anacardium occidentale*), as well as wild plants and weeds belonging to the Solanaceae and Euphorbiaceae. In Zaire *Bersama pachytyrsa* (Melianthaceae), *Tetracera alnifolia* (Dilleniaceae) and *Jussiaea abyssinica* (Onagraceae) are also considered to be important hosts (Schmitz, 1952), while in equatorial Sudan *Polygonum limbatum* (Polygonaceae) is another important dry season host plant (McDermid, 1956).

Coccoidea

Scale insects are not normally considered to be of any importance on cotton (Plate VII.5), but there have been infestations on the stems of plants allowed to stand over from one season to the next. Mealybugs (Pseudococcidae) are often found on poorly growing cotton plants or plants growing in shade, but outbreaks have occurred on cotton following the use of broad spectrum insecticides .

References

Aldrich, J.R., Lusby, W.R., Kochansky, J.P., Hoffmann, M.P., Wilson, L.T. and Zalom, F.G. (1988) Lygus bug pheromones *vis-à-vis* Stink bugs. In: *Proceedings of the Beltwide Cotton Production Conferences*, Memphis, Tennessee, pp. 213–215.

Anon. (1984) *Integrated Pest Management for Cotton in the Western Region of the United States.* University of California Publications No. 3305, 145pp.

Bailey, J.C. (1986) Infesting cotton with tarnished plant bug (Heteroptera: Miridae) nymphs reared by improved laboratory rearing methods. *Journal of Economic Entomology* 79, 1410–1412.

Bailey, J.C., Scales, A.L. and Meredith, W.R., Jr. (1984) Tarnished plant bug (Heteroptera: Miridae) nymph numbers decreased on caged nectariless cottons. *Journal of Economic Entomology* 77, 68–69.

Beards, G.W. and Leigh, T.F. (1960) A laboratory rearing method for *Lygus hesperus* Knight. *Journal of Economic Entomology* 53, 327–328.

Beards, G.W. and Strong, F.E. (1966) Photoperiod in relation to diapause in *Lygus hesperus* Knight. *Hilgardia* 37, 345–362.

Benedict, J.H., Leigh, T.F., Frazier, J.L. and Hyer, A.H. (1981a) Ovipositional behavior of *Lygus hesperus* on two cotton genotypes. *Annals of the Entomological Society of America* 74, 392–394.

Benedict, J.H., Leigh, T.F., Hyer, A.H. and Wynholds, P.F. (1981b) Nectariless cotton: Effect on growth, survival and fecundity of lygus bugs. *Crop Science* 21, 28–30.

Benedict, J.H., Leigh, T.F. and Hyer, A.H. (1983) *Lygus hesperus* (Heteroptera: Miridae) oviposition behavior, growth and survival in relation to cotton trichome density. *Environmental Entomology* 12, 331–335.

Bottger, G.T. (1966) Lygus bugs. In: Smith, C.N. (ed.) *Insect Colonization and Mass Production.* Academic Press, New York, pp. 425–427.

Butler, E.A. (1923) *A Biology of the British Hemiptera–Heteroptera.* Witherby, London, pp. 416–417.

Cao, R.L. (1986) Field survey of the predators of cotton plant bugs and their effects. *Chinese Journal of Biological Control* 2, p. 129 (in Chinese).

Carayon, J. and Delattre, R. (1948) Les *Helopeltis* (Hem. Heteroptera) nuisibles de Cote d'Ivoire. *Revue de pathologie végétale et d'entomologie agricole de France* 27, 185–194.

Cave, R.D. and Gutierrez, A.P. (1983) *Lygus hesperus* field life table studies in cotton and alfalfa (Heteroptera: Miridae). *Canadian Entomologist* 115, 649–654.

Clancy, D.W. (1968) Distribution and parasitization of some *Lygus* spp. in western United States and central Mexico. *Journal of Economic Entomology* 61, 443–445.

Dale, J.E. and Coaker, T.H. (1958) Some effects of feeding by *Lygus vosseleri* Popp. (Heteroptera: Miridae) on the stem apex of the cotton plant. *Annals of Applied Biology* 46, 423–429.

Ewing, K.P. (1929) Effects on the cotton plant of the feeding of certain Hemiptera in the family Miridae. *Journal of Economic Entomology* 22, 761–765.

Feng, C.Y. and Zhang, G.W. (1987) Investigation on oviposition habits of cotton plant bugs. *Insect Knowledge* 24, 213–215 (in Chinese: from *Review of Applied Entomology*).

Fleischer, S.J. and Gaylor, M.J. (1987) Seasonal abundance of *Lygus lineolaris* (Hemiptera: Miridae) and selected predators in early season uncultivated hosts: Implications for managing movement into cotton. *Environmental Entomology* 16, 379–389.

Fleischer, S.J., Gaylor, M.J. and Edelson, J.V. (1985) Estimating absolute density from relative sampling of *Lygus lineolaris* (Heteroptera: Miridae) and selected predators in early to mid-season cotton. *Environmental Entomology* 14, 709–717.

Fleischer, S.J., Gaylor, M.J. and Hue, N.V. (1988) Dispersal of *Lygus lineolaris* (Heteroptera: Miridae) adults through cotton following nursery host destruction. *Environmental Entomology* 17, 533–541

Geering, Q.A. (1953) Studies of *Lygus vosseleri* Popp. (Heteroptera: Miridae), a pest of cotton in East and Central Africa. I. A method for breeding continuous supplies in the laboratory. *Bulletin of Entomological Research* 44, 351–362.

Graham, H.M. and Jackson, C.G. (1982) Distribution of eggs and parasites of *Lygus* spp. (Hemiptera: Miridae), *Nabis* spp. (Hemiptera: Nabidae), and *Spissistilus festinus* (Say) (Homoptera: Membracidae) on plant stems. *Annals of the Entomological Society of America* 75, 56–60.

Graham, H.M., Negm, A.A. and Ertle, L.R. (1984) *Worldwide Literature of the* Lygus *Complex (Hemiptera: Miridae), 1900–1980.* US Department of Agriculture, Agricultural Research Service, Bibliographies and Literature of Agriculture No. 30, 205pp.

Graham, H.M., Jackson, C.G. and Debolt, J.W. (1986) *Lygus* spp. (Hemiptera: Miridae) and their parasites in agricultural areas of southern Arizona. *Environmental Entomology* 15, 132–142

Hancock, G.L.R. (1935) Notes on *Lygus simonyi* Reut. (Capsidae), a cotton pest in Uganda. *Bulletin of Entomological Research* 26, 429–438.

Hedlund, R. C. and Graham, H. M. (1987) *Economic Importance and Biological Control of* Lygus *and* Adelphocoris *in North America.* US Department of Agriculture, Agricultural Research Service, ARS-64.

Jenkins, J.N., Parrott, W.L., McCarty, J.C., Jr. and Shepherd, R.L. (1988) Registration of three noncommercial germplasm lines of upland cotton tolerant to tobacco budworm and tarnished plant bug. *Crop Science* 28, 869–870.

Kelton. L. A. (1971) Review of lygocoris species found in Canada and Alaska (Heteroptera: Miridae). *Memoirs of the Entomological Society of Canada* 83, 6–8.

Kelton, L.A. (1975) The lygus bugs (genus *Lygus* Hahn) of North America (Heteroptera: Miridae). *Memoirs of the Entomological Society of Canada* 95, 101pp.

Kitten, W.F., Bridge, R.R. and Laster, M.L. (1987). Response of advanced breeding lines of cotton to early season tarnished plant bug injury. *Research Report, Mississippi Agricultural and Forestry Experiment Station* 12(2), 5pp.

Leigh, T.F. (1963) Life history of *Lygus hesperus* (Heteroptera: Miridae) in the laboratory. *Annals of the Entomological Society of America* 56, 865–867

Leigh, T.F. (1966) A reproductive diapause in *Lygus hesperus* Knight. *Journal of Economic Entomology* 59, 1280–1281.

Leigh, T.F. and Jackson, C.E. (1968) Topical toxicity of several chlorinated hydrocarbon, organophosphorus, and carbamate insecticides to *Lygus hesperus. Journal of Economic Entomology* 61, 328–330.

Leigh, T.F., Black, J.H., Jackson, C.E. and Burton, V.E. (1966) Insecticides and beneficial insects in cotton fields. *California Agriculture* 20(7), 4–6.

Leigh, T.F., Grimes, D.W., Yamada, H., Stockton, J.R. and Bassett, D. (1969) Arthropod abundance in cotton in relation to some cultural management variables. In: *Proceedings of the Tall Timbers Conference on Ecological Animal Control by Habitat Management.* Tall Timbers Research Station, Tallahassee, Florida 1, 71–83.

Leigh, T.F., Jackson, C.E., Wynholds, P.F. and Cota, J.A. (1977) Toxicity of selected insecticides to *Lygus hesperus. Journal of Economic Entomology* 70, 42–44.

Leigh, T.F., Kerby, T.A. and Wynholds, P.F. (1988) Cotton square damage by the plant bug, *Lygus hesperus* (Hemiptera: Heteroptera: Miridae),and abscission rates. *Journal of Economic Entomology* 81, 1328–1337.

Lockley, T.C. and Young, O.P. (1988) Prey of the striped lynx spider *Oxyopes salticus* (Araneae: Oxyopidae), on cotton in the delta area of Mississippi. *Journal of Arachnology*. 14, 395–397.

McDermid, E.M. (1956) Insect pests of cotton in Zande district of Equatoria Province, Sudan. Pt. II. *Empire Cotton Growing Review* 33, 44–66.

Mueller, A.J. and Stern, V.M. (1973) *Lygus* flight and dispersal behavior. *Environmental Entomology* 2, 361–364.

Mueller, A.J. and Stern, V.M. (1974) Timing pesticide treatments on safflower to prevent *Lygus* from dispersing to cotton. *Journal of Economic Entomology* 67, 77–80.

Pack, T.M. and Tugwell, P. (1976) Clouded and tarnished plant bugs on cotton: A comparison of injury symptoms and damage on fruit. *Arkansas Agricultural Experiment Station Report Series* 226.

Pearson, E.O. (1958) *The Insect Pests of Cotton in Tropical Africa.* Empire Cotton Growing Corporation and Commonwealth Institute of Entomology, London, pp. 236–247.

Scales, A. L. (1968) Female tarnished plant bugs attract males. *Journal of Economic Entomology* 61, 1466–1467.

Scales, A. L. (1973) Parasites of the tarnished plant bug in the Mississippi Delta. *Environmental Entomology* 2, 304–305

Schmitz, G. (1952) Comment limiter les degats de l'*Helopeltis* du cotonnier dans l'Ubangi-Ulele? Bulletin d'information Institute national pour l'etude agronomique du Congo belge 1, 191–204.

Schowalter, T.D. and Stein, J.D. (1987) Influence of Douglas-fir seedling provenance and proximity to insect population sources on susceptibility to *Lygus hesperus* (Heteroptera: Miridae) in a forest nursery in western Oregon. *Environmental Entomology* 16, 984–988.

Scott, D.R. (1977) An annotated listing of host plants for *Lygus hesperus* Knight. *Bulletin of the Entomological Society of America* 23, 19–22.

Scott, W.P., Snodgrass, G.L. and Smith, J.W. (1988) Tarnished plant bug (Hemiptera: Miridae) and predacious arthropod populations in commercially produced selected nectariless cultivars of cotton. *Journal of Entomological Science* 23, 280–286.

Sevacherian, V. and Stern, V. M. (1975) Movements of lygus bugs between alfalfa and cotton. *Environmental Entomology* 4, 163–165.

Shepherd, R.L., Jenkins, J.N., Parrott, W.L. and McCarty, J.C. (1986) Registration of eight nectariless-frego bract cotton germplasm lines. *Crop Science* 26, 1260.

Shull, W.E. (1933) An investigation of the *Lygus* species which are pests of beans (Hemiptera: Miridae). *Idaho Agricultural Experiment Station Bulletin* 11, 42pp.

Smith, G.L. (1942) California cotton insects. *University of California Agricultural Experiment Station Bulletin* 660, pp. 3–10.

Snodgrass, G.L., Scott, W.P. and Smith, J.W. (1984a) An annotated list of the host plants of *Lygus lineolaris* (Hemiptera: Miridae) in the Arkansas, Louisiana and Mississippi Delta. *Journal of the Georgia Entomological Society* 19, 93–101.

Snodgrass, G.L., Scott, W.P. and Smith, J.W. (1984b) Host plants and seasonal distribution of the tarnished plant bug (Hemiptera: Miridae) in the delta of Arkansas, Louisiana, and Mississippi. *Environmental Entomology* 13, 110–116.

Southwood, T.R.E. and Leston, D. (1959) *Land and Water Bugs of the British Isles.* Frederick Waite & Co. London, p. 283.

Stern, V.M. (1969) Interplanting alfalfa in cotton to control lygus bugs and other insect pests. In: *Proceedings of the Tall Timbers Conference on Ecological Animal Control by Habitat Management.* Tall Timbers Research Station, Tallahassee, Florida 1, 55–69.

Stern, V.M. and Mueller, A. (1968) Techniques of marking insects with micronized fluorescent dust with especial emphasis on marking millions of *Lygus hesperus* for dispersal studies. *Journal of Economic Entomology* 58, 1232–1237.

Stern, V.M., van den Bosch, R. and Leigh, T.F. (1964) Strip cutting alfalfa for lygus bug control. *California Agriculture* 18(4), 4–6.

Stern, V.M., van den Bosch, R., Leigh, T.F., McCutcheon, O.D., Sallee, W.R., Houston, C.E. and Garber, M.J. (1967) *Lygus* control by strip cutting alfalfa. *University of California Agricultural Extension Service AXT-241*, 13pp.

Stride, G.O. (1968) On the biology and ecology of *Lygus vosseleri* (Heteroptera: Miridae) with special reference to its host plant relationships. *Journal of the Entomological Society of South Africa* 31, 17–59.

Strong, F.E., Sheldahl, J.A., Hughes, P.R. and Hussein, E.M.K. (1970) Reproductive biology of *Lygus hesperus* Knight. *Hilgardia* 40, 105–147.

Taylor, T.H.C. (1945) *Lygus simonyi* Reut., as a cotton pest in Uganda. *Bulletin of Entomological Research* 36, 121–148.

Taylor, T.H.C. (1947) On the identity of the cotton capsid of Uganda. *Bulletin of Entomological Research* 37, pp 503–505.

Tugwell, P., Young, S.C., Jr., Dumas, B.A. and Phillips, J.R. (1976) Plant bugs in cotton: Importance of infestation time, types of cotton injury, and significance of wild hosts near cotton. *Arkansas Agricultural Experiment Station Report Series* 227, 24pp.

van den Bosch, R. and Stern, V.M. (1969) The effect of harvesting practices on insect populations in alfalfa. In: *Proceedings of the Tall Timbers Conference on Ecological Animal Control by Habitat Management*, Vol. 1. Tall Timber Research Station, Tallahassee, Florida 1, 47–54.

Wang, K. (1984) The occurring pattern of *Lygus lucorum* in coastal cotton growing areas and the control strategy. *China Cotton* 3, 47–48 (in Chinese: original article not seen).

Whitcomb, W.H., Exline, H. and Hunter, R.C. (1963) Spiders of the Arkansas cotton field. *Annals of the Entomological Society of America* 56, 53–660.

Williams, L., III, Phillips, J.R. and Tugwell, N.P. (1987) Field technique for identifying causes of pinhead square shed in cotton. *Journal of Economic Entomology* 80, 27–531.

Wilson, F.D. and George, B.W. (1986) Smoothleaf and hirsute cottons: response to insect pests and yield in Arizona. *Journal of Economic Entomology* 79, 229–232.

Wilson, L.T., Leigh, T.F., Gonzalez, D. and Forestiere, C. (1984) Distribution of *Lygus hesperus* (Knight) (Miridae: Hemiptera) on cotton. *Journal of Economic Entomology* 77, 1313–1319.

Womack, C.L. and Schuster, M.F. (1987) Host plants of the tarnished plant bug (Heteroptera: Miridae) in the Northern Blackland Prairies of Texas. *Environmental Entomology* 16, 1266–1272.

Young, Orrey P. (1989) Predation by *Pisaurina mira* (Aranea: Pisauridae) on *Lygus lineolaris* (Heteroptera: Miridae) and other arthropods. *Journal of Arachnology* 17, 43–48.

Zhang, Y.X., Cao, Y.P., Bai, L.X. and Cao, C.Y. (1986) Plant bug damage on cotton in different growing stages and the threshold for control. *Acta Phytophylactica Sinica* 13, 73–78 (in Chinese: original article not seen).

17 Thysanoptera

J.P. Bournier

Institut Recherches du Coton et Textiles Exotiques, Département du Centre de Coopération Internationale en Recherche Agronomique pour le Développement, BP 5035, 34032 Montpellier Cedex, France

Introduction

Thysanoptera or thrips are an order of insects in the Condylognata group, i.e. whose mandibles have become stylets. The order consists of two large suborders:

- Terebrantia, in which the females possess an ovipositor;
- Tubulifera, which do not possess ovipositors and whose 10th abdominal segment is in the form of a tube.

Numerous species of Thysanoptera have been reported in different cotton growing zones in the world, but only a few cause substantial damage.

The following five species are polyphagous and often widely distributed.

- *Thrips tabaci*
- *Thrips palmi*
- *Frankliniella schultzei*
- *Frankliniella occidentalis*
- *Caliothrips impurus*

Thrips are generally one of the main early season cotton pests. Direct damage due to deformation and decrease in plant density can be considerable, but the economic impact is usually caused by delayed vegetative development leading to a late harvest.

Description

Adult (Fig. 17.1)

Adults have an elongated shape and are 1.0–2.5 mm long. Speed of movement on the surface of plant organs varies according to species; some may jump and fly for

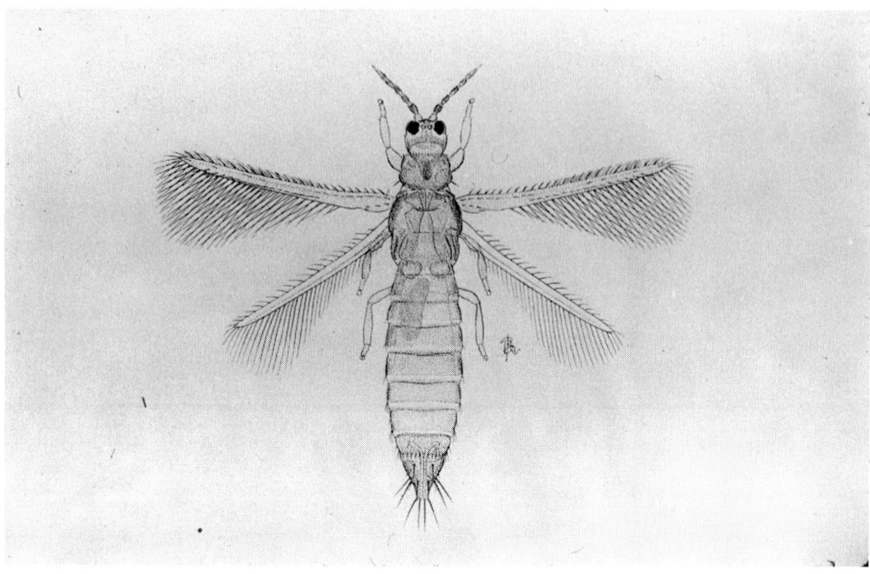

Fig. 17.1. Adult thrips, female.

varying distances by means of their two pairs of wings. These are narrow with fringes of long cilia (whence their name: 'thysano' meaning fringes) which increase the bearing surface during flight.

The head bears two eyes consisting of ommatidia, three ocelli in a triangular layout and two antennae with six to nine segments depending on the species. The mouth cone, which guides the stylet during feeding, is on the underside. There are three stylets: one mandible stylet on the left mandible; the right mandible is totally atrophied (unique in entomology); two maxillary stylets corresponding to the maxillary lacinia; they have a U-shaped cross-section and are coadapted to form a sucking tube. The latter is connected directly to the pharyngeal pump and a saliva pump in the hypopharynx. Each maxillary lobe bears a maxillary palp with several segments; there are also two labial palps on the labium.

The prothorax is approximately rectangular, depending on the species, and bears several pairs of large setae whose dimensions and position are important in taxonomy. The pterothorax bears two pairs of wings fringed with long cilia. Some species are apterous and others, which are usually macropterous, display brachypterous generations. The legs are usually short. The anterior femurs and tibias are sometimes denticulated. In Terebrantia, the posterior tibias bear a series of spines used to comb the wing cilia before flight.

The abdomen is elongated and consists of ten segments. The 11th is a tiny sclerite. In female Terebrantia, the eighth and ninth sternites each have two gonapophyses which form an ovipositor. In the male, the tenth tergite is almost completely set in the ninth. The median abdominal sternites have an ovoid sensorial area in some species. The exsertile penis has a central apophysis (aedeagus) and two lateral apophyses (parameres). The two phallobase sclerites are at the base of the latter. In the Tubulifera, abdominal tergites II to VII bear a pair of sigmoid setae on each side which retain the wings against the body when the insect is not in flight. The tenth segment is tubular. There is no ovipositor.

Egg

Eggs are fairly large in relation to adult size: 200 to 300 μm long and 100 to 150 μm broadest width. They are usually reniform in Terebrantia and perfectly ellipsoid in Tubulifera.

Larvae

These are approximately the same shape as adults with similar mouthparts, but have no wings. The tegument is soft and usually white to pale yellow. They can feed heavily.

Nymphs

There are two nymphal stages in Terebrantia (pronymph and nymph) and three (pronymph, nymph I and nymph II) in Tubuliferae. They differ from the larval stages in the absence of a functional mouth and the appearance of wing buds. The latter are longer in the nymph and the antenna buds are folded against the dorsal part of the head and prothorax.

Biology

In Terebrantia, eggs are laid singly and placed beneath plant epidermis by means of the ovipositor. A single female can lay two to five eggs a day, reaching a total of some 60 to 100. In Tubulifera, groups of three to five eggs are deposited on the surface of the plant and fixed by the mucilaginous substance which covers the chorion. Incubation varies from a few days to several weeks according to climatic conditions (three to 20 days).

After hatching, the stage I larva begins to feed on plant tissue. Unlike most sucking insects, Thysanoptera do not feed on sap; most of them empty plant cells. They inject their saliva, which causes lysis of cell contents, and then use their pharyngeal pump to suck in the resulting substance. They moult after several days (three or four, depending on climatic conditions), and the stage II larvae behave similarly. After five to 12 days pupation may occur at the growth site or in a cavity on the plant in which a cocoon is spun (Aeolothripidae) or by falling to the ground and burrowing for 2–25 cm, depending on the soil. This is the case, among others, of *Thrips palmi* and *Frankliniella occidentalis*. The pronymph becomes a nymph after one to three days and then after a similar period becomes an adult (Terebrantia) or a nymph II (Tubulifera). Like the stage I and II larvae, the adults feed on the contents of plant parenchyma cells.

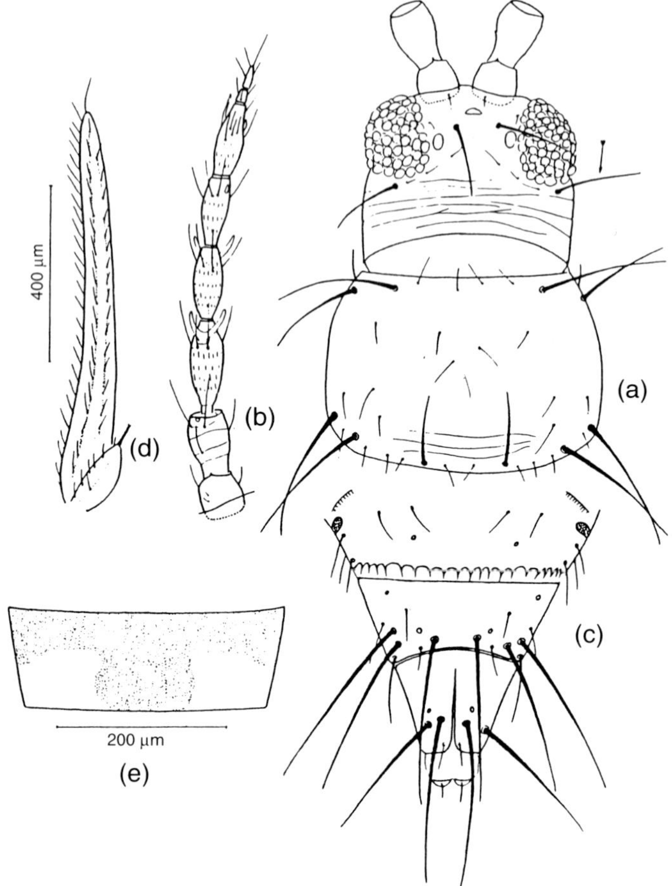

Fig. 17.2. *Frankliniella occidentalis* (Pergande): a, head and pronotum; b, antenna (right); c, tergite VIII–X; d, right wing (anterior and posterior fringes are not represented); e, abdominal tergite coloration of aestival forms.

Thrips live from eight to 25 days. Some species, such as *Thrips tabaci*, can go through the cold season as adults on the ground. There can, therefore, be two to three generations per year for a species in temperate climate and 15 to 18 in hot countries.

Damage

Very different types of damage can be caused by Thysanoptera according to the part of the plant attacked, its age and the extent of the necrotic zone. Most damage occurs during the vegetative phase when it is easier for the insects to find young, soft epidermal tissues. Several parts can be affected in cotton.

Seedlings at Emergence

Cotyledons can be pricked by different thrips immediately after emergence. The damage is in the form of small necrotic patches which vary in shape and generally measure no more than 3–5 mm. Recent damage can be recognized by its pale colour; the area is more or less silvery since the cell contents are sucked out and replaced by air. The damaged cells tend to darken when damage is less recent. The patches are usually pale brown with a varying degree of suberification. Cell multiplication cannot take place in these necrotic patches when the cotyledons develop; this results in deformation. In serious attacks the leaf becomes thickened and blistered (Plate VII.6).

Axial Buds and Young Plants

Since they are small, thrips can reach the heart of the axial buds of plantlets. Pricking by stylets destroys areas of follicle tissue and apical meristem. The results of such damage (and the larva/larvae responsible) are not seen immediately. The various deformed parts are only seen when the plantlet or plant reaches the 2- or 4-leaf stage. The young leaves have incomplete laminae which are thickened and curl downwards at the edges. Leaves are crumpled and torn when a large area of cells has been destroyed.

When the apical meristem has been seriously damaged, the axial bud withers and is shed. Growth of the young plant is halted or at least considerably slowed until secondary branches develop. Growth may be stopped completely in cases of serious wounding, and the plant withers and dies. However, plants often continue to grow and progressively display further deformations. The most common is the formation of forked plants, in which growth of the main stem has been halted and growth continues with two or sometimes three vegetative on fruit-bearing branches. This type of 'architecture' results in branch breakage at the fork during boll formation. Twisted shoots and shortened, swollen internodes may also be observed. Plant development and boll formation are delayed in all cases.

Stems

The stems of young plants may also be deformed as a result of pricking by larvae or adult thrips, or by wounding caused by the ovipositors of egg-laying females. These different types of damage are frequently seen in many cotton zones and especially in countries where irrigation or flood recession farming is practised. Climatic conditions with low relative humidity are particularly favourable for outbreaks of Thysanoptera at the beginning of the cotton growing season.

Leaves

Damage may also be caused to older leaves. Almost all the leaves are affected when there is a large thrips population, with damage usually more serious near the main veins as a result of a thigmotaxis preference in the insects. Strongly infested cotton fields display a characteristic silvery colour. The necrotic areas then brown very rapidly and if the whole thickness of the lamina has been destroyed the leaf withers and falls. This occurs, for example, with *Thrips palmi* in the Philippines (Bournier, 1983).

Flowers

Different species of Thysanoptera can be observed on flowers where they cause various types of damage. *Megalurothrips sjostedti* Trybom in Benin causes necrotic patches which brown very rapidly and deform the petals, which then become rounded and sometimes hinder flower opening. *Frankliniella schultzei*, an extremely polyphagous species, is frequently found in flowers; it can feed on pollen like most flower-dwelling species but may also cause damage to stigmas, stamen filaments and ovaries. The formation of young fruits is endangered in the latter case.

The main Thysanoptera damage of economic importance is usually the result of attacks at a very early stage (seedlings and young plants). Leaf damage caused later is only serious if the pest population is very large.

Thrips tabaci, Lindeman, 1889

The most polyphagous, harmful and widespread of the Thysanoptera species is found in practically all the cotton growing areas in the world.

The colour of macropterous adults can vary from pale yellow to dark brown, the latter particularly in colder climates. Their antennae have seven segments, with the first segment always paler than segment II which is usually dark throughout. Segments III and IV are pale in the proximal half only; V, VI and VII are entirely dark. The wings, cilia and body setae are usually pale. The pair of ocellar setae III are in the ocellar triangle. Forewings have four or five distal setae (six in rare cases) on the first vein. The anterior edge of the abdominal tergites is generally marked by a brown band. There are three setae on the side of abdominal tergite II. The posterior edge of tergite VIII bears a comb with long, thin teeth. There is a single pair of pores (anterior pair absent) on tergite IX. The sculptured lines on the pleurotergites bear ciliate microtrichia.

Reproduction of the species is mainly parthenogenetic, with the comparatively rare male forming 1/60 to 1/3000 of the populations. The life cycle duration can vary between 18 to 22 days at 20°C and 10 to 11 days at 30°C. Pupation takes place in the soil. In temperate climates *T. tabaci* usually overwinters as adults in cracks in the soil or in plant debris.

Thrips tabaci damage to young cotton seedlings, flowers and stems is particularly serious at the beginning of the season in regions with low relative humidity. In Egypt, Ghabn (1948) reported that in some years *T. tabaci* was responsible for the loss of 50% of young cotton plants. Bournier and Couilloud (1969) observed the same symptoms in Iran, particularly in the east in Khorassan where there were serious outbreaks. Passlow (1971) also reported serious damage caused by *T. tabaci* in Australia. It has also been reported as being the species – or one of the species – which causes most damage to cotton crops in Pakistan (Baloch and Soomro, 1980), India (Dahiya and Ram Singh, 1982), the USA (Klein *et al.*, 1986), China (Pan *et al.*, 1987) and the USSR (Shukurilloev, 1979).

Thrips palmi Karny, 1925

This species originated in Indonesia and began to be reported in different countries in Southeast Asia in the 1980s; it is particularly harmful in the Philippines (Bournier, 1983). It is extremely polyphagous and spreading on a worldwide basis appears to be unavoidable It reached Martinique and Guadeloupe in 1985, causing extremely serious damage to vegetable crops and in horticulture, and has since spread through the Caribbean. In the Dominican Republic, it was reported on cotton in 1989 for the first time.

Thrips palmi has not yet reached cotton growing areas of the Middle East, Mediterranean, Africa and mainland America, but there is a risk of it spreading to these regions in the near future. It has been reported recently in Florida (South, 1991).

The adult is pale yellow and is easily confused with the pale forms of *T. tabaci* and *T. flavus*. The antennae have seven segments. Segments I and II are always pale; III to VII are increasingly dark, with a pale part at the base of III and IV in lightly coloured specimens. The wings are slightly 'smoky' and bear brown cilia the same colour as the body. The pair of ocellar setae III is outside the ocellar triangle. The forewings bear two or three distal setae on the first vein. There are four setae on the side of abdominal tergite II (in comparison with three in *T. tabaci*).

The *T. palmi* cycle lasts approximately 11 days at 30°C and 24 days at 20°C. Although the sex distribution is always weighted towards females, males are much more common than in *T. tabaci* at 10–40%. The fertility of the females (over 60 eggs), their longevity and the number of eggs laid per day cause very serious infestations on certain plants. *Thrips palmi* attacks the undersides and surfaces of cotton leaves, especially along the main ribs. Damage is sustained by both young and old plants, leading to characteristic necrosis and deformation.

Frankliniella schultzei (Trybom, 1910) (Cotton Bud Thrips)

This species has numerous synonyms, the most common being *F. dampfi* Priesner. It is distributed very widely throughout the world, from Australia to South America and is particularly common in Africa and Madagascar. It is polyphagous and can spread the tomato spotted wilt virus in vegetable crops. The cycle length of *F. schultzei* is about 15 to 21 days at 30°C. Pupation takes place in the soil.

Adult colour varies from pale yellow to dark brown. Like all species of the genus *Frankliniella*, the pronotum bears four pairs of well-developed setae of practically equal length: one at the anterior angle, one at the anterior edge and two at the posterior angle. There is also a pair of small anteromarginal setae between the two posteromarginal pronotum setae.

The antennae have eight segments: I and II are brown, although I may sometimes be paler than II; the lower parts of III and IV are paler; V to VIII are brown. The interocellar setae III are within the ocellar triangle between the two posterior ocelli on a line between their anterior margins. They are approximately one and a half times as long as the longest of the postocular setae. There is a continuous row of setae on the first vein of the anterior wings.

This very active insect is found on young cotton plants (Delattre, 1957) where it causes deformation similar to that of *T. tabaci*, and also on flowerbuds, flowers and leaves. In Argentina (Barral and de Stacul, 1969) and Paraguay (Prudent, 1990), *F. schultzei* is the main pest at the beginning of the cotton season and destroys a large number of emerging plants. In Iran (Bournier and Couilloud, 1969), Sudan and Egypt, it only infests flowers and leaves in the season. It may also cause slight wounding in young bolls, but this damage does not seem to affect the maturation and quality of the seed cotton.

Frankiniella occidentalis (Pergande, 1895) (Western Flower Thrips)

This insect, known as the western flower thrips, originated in the western part of the USA (California) and Mexico and has progressively spread to the whole of the USA over the past decade. It damages cotton crops mainly in North America but was introduced into Europe about 1985 (Bournier and Bournier, 1987), and has gradually spread to all countries. In Spain, in particular, it has been reported on cotton. Like *T. palmi*, it will probably continue to spread in coming years. It has already been reported in South Africa, the Canary Islands, Kenya, Reunion and Martinique.

The colour of the females varies, ranging from pale yellowish-white with brown spots on the abdominal tergites in the summer forms, to dark brown in the winter forms; all intermediate colours are observed, with almost complete disappearance of the brown spots in the paler specimens. The antennae have eight segments: I, II and VI to VIII are dark brown with I being paler than II; III to V are dark with only the base being paler. The wings are slightly smoky; wing and body setae are brown. Ocellar setae III are halfway between the anterior ocellus and the posterior ocelli in a band bounded by the edge of the latter. The main postocular seta is almost as long (48–50 μm) as interocellar seta III (60 μm) (Fig. 17.2).

Males are distinctly smaller than the female and are usually pale. The *F. occidentalis* cycle lasts for 18 days at 26°C and about 35–45 days at 15°C. Pupation generally takes place in the soil and occasionally on the plant. Reproduction may be sexual, when the progeny is mainly female (seven to 15 females to one male) or parthogenetic when only males are produced. A single female can lay up to 100 eggs during its life, which can be 30 to 70 days according to the temperature.

Like *T. tabaci* and *F. schultzei*, *F. occidentalis* can infect vegetable crops with tomato spotted wilt virus. The damage in cotton is on young flowerbuds, flowers and leaves. Race (1965) reported that in New Mexico, *F. occidentalis* could lower seed cotton yields when there were more than five nymphs per plant during the first four weeks after sowing. In the USA, *F. fusca* and *F. tritici* – in addition to *F. occidentalis* – may cause damage to cotton (Rummel and Quisenberry, 1979).

Caliothrips impurus (Priesner, 1927)

Previously called *Heliothrips impurus, Hercothrips fummipennis* or *Hercothrips impurus*, this is a common species in Africa on various crops (onion, peanut, vegetables). It is particularly harmful to cotton in the Sudan (Bagnall and Cameron, 1932), where it is often combined as in Egypt with *C. sudanensis.*

Although the genera *Thrips* and *Frankliniella* belong to the Thripinae subfamily, the genus *Caliothrips* belongs to the Panchaetothripinae subfamily, which is characterized by a pattern of reticulated sculptures on the body and eight-segment antennae (VII and VIII are long and slender).

Caliothrips impurus has an overall dark brown colour. Antenna segments I, II and VI to VIII are dark brown throughout, whereas III to V are only dark brown in the centre. The legs are also dark except for the femur–tibia joint zone, the distal part of the tibia and the tarses, which are all pale. The wings are brown throughout except for two narrow, slightly paler bands at one quarter and three quarters of their length; the cilia at the anterior edge are also paler in these places. The lateral third of the abdominal tergites is reticulated with more or less elongated polygons containing dark ribs which are punctiform to a varying extent.

The life cycle lasts for 14 to 40 days, depending on the temperature. Reproduction is generally sexual but arrhenotokic parthenogenesis exists. There are usually five to six times as many females as males. Pupation generally takes place in the soil but may occur on the plant itself.

Most of the damage caused to cotton crops by *C. impurus* (Cameron, 1932) is sustained by the leaves. Young leaves are a preferential target; in cases of severe infestation they display serious necrosis, wither and fall .

Control

Various methods can be used to control Thysanoptera populations in cotton according to the type of farming (rainfed or irrigated) and the date of infestation in relation to the growth stage of the cotton. Heavy rainfall limits both leaf and soil infestation by thrips. Flooding of the fields after rainfall or by irrigation destroys a large proportion of pupal populations in the soil.

In some countries, such as the Sudan, early sowing can reduce attacks at emergence when damage is greatest, but if cotton plants emerge when nearby weeds are withering, the thrips migrate to the cotton and can destroy up to 50% of the young plants. In contrast, if cotton crops emerge at the same time as the weeds or earlier when irrigation is used, the plants will have grown sufficiently to withstand damage when the thrips arrive. Thrips are strongly attracted by onions and garlic so plantations of these near the cotton fields provide indirect protection as trap crops.

Biological Control

Thrips populations are checked by certain species of predatory thrips; the most common – *Franklinothrips vespiformis* – is very widespread and also preys on spider mites, leaf-hoppers and whiteflies. Other predator species include *Scolothrips hartwigi* Priesner in Africa and *Scolothrips acariphagus* in Central Asia. *Frankliniella occidentalis* is a predator on *Tetranychus urticae* in the USA (Trichilo and Leigh, 1986). Other predators on Thysanoptera include Anthocoridae, Lygaeidae and Neuroptera (especially Chrysopidae), ladybirds and mites.

Parasitoids are fairly rare on thrips. The best known are Hymenoptera of the Eulophidae family (genus *Ceranisus*) and the Trichogrammatidae family (genus *Megaphragma*). *Ceranisus* sp. parasitizes larval and nymphal stages whereas *Megaphragma* sp. parasitizes eggs. Several experiments have been carried out in recent years, some with encouraging results.

Chemical Control

Most damage caused by Thysanoptera is at the very beginning of the season when plants are not sufficiently developed to attract parasitoids and predators, so chemical control has to be used in almost all cases. As thrips suck the contents of cells, active ingredients with translaminar effects should be used as much as possible in chemical control. Seeds are treated or insecticide is applied to sowing rows in the regions where attacks by thrips are very early.

Aerial spraying is used to control later attacks of the vegetative parts of cotton plants. Soil applications can sometimes also reduce nymphal populations and thus complement aerial treatment.

Numerous insecticides have been used for seed treatment including acephate, aldrin, dieldrin, lindane, isophenphos, carbosulfan, furathiocarb, thiodicarb and imidachloprid. The latter has given a very good result in Paraguay (Prudent, 1990) on *F. schultzei* and *C. braziliensis*. Soil treatment includes granules of aldicarb, oxamyl, carbofuran, carbosulfan, lindane, diazinon and disulfoton.

Sprays of methomyl, endosulfan, acephate, dimethoate, monocrotophos, diazinon, dichlorvos, methamidophos, profenophos and imidachloprid have been used, with the latter giving excellent control of *T. palmi* which is particularly difficult to control.

References

Bagnall, R.S. and Cameron, W.P.L. (1932) Description of two species of *Hercothrips* injurious to cotton in the British Sudan and of an allied species of grass. *Annals and Magazine of Natural History* (10)10, 412–419.

Baloch, A.A. and Soomro, B.A. (1980) Preliminary studies on plant profile and population dynamics of insect pests of cotton. *Turkiye Bitki Koruma Dergisi* 4, 203–217.

Barral, J.M. and Velasco de Stacul, M. (1969) Determinacion de las especies de trips en cultivos de la region Centro-Chaquena, con especial referencia al algodon. *Revista de Investigaciones Agropecuarias* 6, 83–94.

Bournier, A. (1983) *Les Thrips, Biologie, Importance Agronomique*. INRA, Paris, 128pp.

Bournier, A. and Bournier, J.P. (1987) L'introduction en France d'un nouveau ravageur: *Frankliniella occidentalis* (Pergande). *Phytoma*, No. 388, May 1987, 14–17.

Bournier, A.and Couilloud, R., (1969) Les Thrips du cotonnier en Iran. *Coton et Fibres Tropicales* 24, 211–218.

Bournier, J.P. (1983) Un insecte polyphage: Thrips palmi Karny, important ravageur du cotonnier aux Philippines. *Coton et Fibres Tropicales* 38, 286–289.

Cameron, W.P.L. (1932) Thrips affecting the cotton crop in Gezira in 1930. *Bulletin of the Wellcome Tropical Research Laboratory, Entomology Section* No. 34, 32–37 (Khartoum, January 1931, published 1932).

Dahiya, A.S. and Ram Singh (1982) Bio efficacy of some systemic insecticides against jassid, thrips and white fly attacking cotton. *Pesticides* 16(12), 13–14.

Delattre, R. (1957) Note sur quelques deformations et aberrations du cotonnier. *Coton et Fibres Tropicales* 12, 335–350.

Ghabn, A.A.A.S (1948) Contribution to the knowledge of the biology of *Thrips tabaci* in Egypt. *Bulletin de la Societé Fouad 1er d'Entomologie* 32, 123–174.

Klein, M., Franck, A. and Rimond, D. (1986) Proliferation and branching of cotton seedlings; the suspected cause: *Thrips tabaci*, the influence on yields, and test to reduce damages. *Phytoparasitica* 14, 25–37.

Pan, Q.M., Wu, T.S. and Zhang, L.S. (1987) Dynamics and control of *Thrips tabaci* in cotton fields mulched with plastic films. *Plant Protection* 13(4), 25.

Passlow, T. (1971) *Thrips tabaci* as a seedling problem in Queensland cotton. *Queensland Journal of Agricultural and Animal Sciences* 28(2/3), 137–151.

Prudent, P. (1990) Efficacite de divers insecticides appliques en traitement des semences, contre les thrips du cotonnier au Paraguay. Rapport annuel IRCT, 1990 (not published).

Race, S.R. (1965) Importance and control of western flower thrips, *Frankliniella occidentalis*, on cotton. *Bulletin New Mexico Agricultural Experiment Station* No. 497, 25pp.

Rummel, D.R. and Quisenberry, J.E. (1979) Influence of thrips injury on leaf development and yield of various cotton genotypes. *Journal of Economic Entomology* 72, 706–709.

Shukurilloev, S. (1979) Thrips (Thysanoptera) components of the cotton agrobiocoenosis in Southern Tajikstan. *Izvestiya Akademii Nauk Tadzhikskoi SSR*, Series 3, Vol. 76, 28–34.

South, L. (1991) New thrips threatens south Florida. *American Vegetable Grower* April 1991, 30.

Trichilo, P.J. and Leigh, T.F. (1986) Predation on spider mite eggs by the western flower thrips, *Frankliniella cocidentalis*, an opportunist in a cotton agroecosystem. *Environmental Entomology* 15, 821–825.

18 Orthoptera

G.A. Matthews

International Pesticide Application Research Centre, Imperial College at Silwood Park, Buckhurst Road, Sunninghill, Ascot, Berkshire SL5 7PY, UK

Grasshoppers

Many species of non-migratory grasshoppers are recorded as feeding on cotton in Africa. Apart from members of three genera of long-horned grasshoppers (*Tettigoniidae*) of very minor importance, all these belong to the Acrididae, the most important and widespread of them being two species of *Zonocerus*, of which *Z. elegans* (Thnb.) (Plate VIII.1), often known as the elegant grasshopper or stinking grasshopper, is common throughout southern and eastern Africa, and *Z. variegatus* (L.) is distributed north of the equator, from Kenya to Sierra Leone. They have a wide range of host plants, including many common weeds, but they do not attack the Gramineae to any extent.

Description

Eggs 6 × 1.5 mm, brown, surrounded by masses of froth which harden to form a sponge-like packet about 25 mm long. Nymphs black, resembling the adult in general form, the body conspicuously striped and the appendages ringed with white or yellow. Adults 30–37 mm long, of typical grasshopper form, with a striking colour pattern, the saddle being olive-green, the elytra (which in *Z. elegans* often remain rudimentary) yellowish-green with red suffusions, and the rest of the body boldly patterned in black, yellow and orange. The adults exude drops of a yellowish, evil-smelling fluid when handled.

Biology and Damage

Each species has only one generation a year, development being arrested in the egg stage. Over the region extending from southern Africa to the equatorial regions (Sierra Leone, northern Zaire and Uganda), the eggs of both species are laid in loose soil in about March–April and hatch in October–December, and the nymphal and adult life then extends over a further four–seven months. The egg stage lasts for over six months and appears to require both a fairly high temperature and moist soil conditions for its completion. Thus, in southern Africa, eggs laid at the start of the cool, dry season do not hatch until the temperature rises and the first rain falls; in the equatorial regions, having a high rainfall in the second half of the year, hatching is retarded until the soil dries out and the temperatures rise. In the northern savannah areas, where *Z. variegatus* is the only species, the southern African pattern of events is repeated, with a difference of some six months on account of the reversal of the seasons, the eggs being laid in September–October and hatching in March–May. Some overlapping of the two patterns occurs north of the equator, thus accounting for the records from West Africa of two broods a year.

Both nymphs and adults are gregarious and sluggish; this may result in localized patches of severe defoliation, but it also facilitates their destruction by hand-picking. If local oviposition sites are known, farmers sometimes dig the area to expose the eggs to desiccation in the sun. Mortality of *Z. variegatus* is also caused by a fungus, *Entomophthora grylii* Fres., especially in the late nymphal and early adult stage after heavy rain (Chapman and Page, 1979).

Cotton is most severely damaged in the rainfed areas of the monsoon zone, where it is sown at the start of the single rainy season; the hatching of the nymphs then coincides with the early seedling stages of the crop. Where, as in Uganda, the cotton plants are fully grown by the time that hatching occurs, damage by defoliation is not serious.

In the rainfed cotton areas, damage by other species of grasshoppers is never more than sporadic, since there is normally an abundance of other food available at the time that cotton is growing. In semiarid areas, however, irrigated cotton may be attacked because it is the principal green food available after the brief rainy season is over. Thus, in the Sudan, *Cyrtacanthacris tatarica* (L.), which is not associated with the Gramineae, invades cotton from surrounding *Acacia* scrub at the end of the rains, and nymphs of the tree locust, *Anacridium moestum* (Serv.), which feeds normally on small trees and spends the cool, dry season gregariously in *Acacia* scrub, invade cotton fields when the plants are still young; swarms of adult *Anacridium* occasionally damage cotton. Adults and nymphs of *Pyrgomorpha cognata* Krauss migrate from weeds on fallow land at the end of the rains and invade the edges of cotton fields and there are numerous other species, of which the habits are not well known, that occasionally attack cotton.

Locusts

The migratory species of Acrididae that constitute the true locusts of Africa have a wide range of plants upon which they will feed, amongst which cotton is not particularly preferred, but they occasionally cause damage to the crop and this may sometimes be serious. The general appearance and habits of the locusts are well-known and only brief notes on them in relation to the cotton crop are given here.

Locusta migratoria migratorioides (R. & F.). The permanent breeding grounds of the African migratory locust are in the flood-plain of the Middle Niger, whence it may spread down into the cotton-growing regions of West and equatorial Africa. It displays a marked preference for monocotyledons and there are few records of its attacking cotton. It was reported as not touching this crop in northern Zaire, although hoppers have been known to feed on it in dry weather in Uganda and there is a record of severe damage to cotton in the Central African Republic.

Nomadacris septemfasciata (Serv.). Swarms of the red locust spread from its main breeding grounds, which are in the Lake Rukwa areas, into all the cotton-growing regions of Central, East and southern Africa during the outbreak of this species during the 1930s. *Nomadacris*, like *Locusta*, shows a preference for the Gramineae, but it will attack cotton, and severe damage, particularly to young plants, may occur, though this is usually localized, and losses, in the aggregate, are not large.

Schistocerca gregaria (Forsk.). The regular breeding grounds of the desert locust lie, for the most part, outside of Africa, in the Arabian Peninsula, but swarms occasionally reach the equatorial rainfed cotton areas of Africa, where they may cause a little damage to cotton, and the species sometimes causes serious damage to cotton in the Sudan. In that territory, the Gezira and the riverain cultivations of the Northern Province are invaded by young flying swarms in September, but fortunately little breeding takes place in the clay soils of the former. Swarms of both adults and hoppers can be serious in the Gash Delta, but the worst outbreaks have been in the Tokar Delta. The Red Sea littoral is regularly invaded by flying swarms from late September onwards, and these tend to concentrate in the cultivated crops of the Tokar flood area, which are conspicuous amongst the surrounding desert. Breeding occurs there during the winter rains and cotton is liable to attack throughout the winter and spring, although, as with the other species of locusts, cotton is not a preferred food, and the main object of the attack is the sorghum crops. Outbreaks can be effectively controlled using insecticides so serious damage from them is now unusual.

Crickets

Several species of true crickets (Gryllidae) are recorded from cotton in Africa, chiefly attacking the seedlings, and of these, *Brachytrupes membranaceus* (Dru.) is widespread throughout the rainfed cotton areas, where it is associated with sandy soils, in which it constructs underground burrows. It feeds on all soft parts of the plant and causes a certain amount of nuisance by clipping off seedlings early in the season. The mole cricket, *Gryllatalpa africana* P. de B. (Gryllotalpidae) has been reported as damaging cotton in Kenya and the Central African Republic.

Reference

Chapman, R.F. and Page, W.W. (1979) Factors affecting the mortality of the grasshopper *Zonocerus variegatus* in southern Nigeria. *Journal of Animal Ecology* 48, 247–70.

19 Isoptera

J.W.M. Logan

Natural Resources Institute, Central Avenue, Chatham Maritime, Chatham, Kent ME4 4TB, UK

Introduction

Termites (Isoptera) are generally minor pests of cotton but can cause localized major losses, particularly in the drier parts of Africa and the Indian subcontinent. They are social insects living in colonies which vary in size from a few hundred to hundreds of thousands depending on the species. Each colony consists of reproductives (either a pair or a group of males and females), workers, soldiers, immature termites and, at certain times of the year, flying reproductives (alates) which leave the nest to form new colonies. There are more than 2500 species of termite but only about 200 are considered to be pests of buildings, crops or forestry. Of these, only a few are pests of cotton (Table 19.1). The most important are *Hodotermes mossambicus* (Hagen)(Hodotermitidae), the harvester termite of southern and eastern Africa, and various species of *Microtermes*, *Odontotermes* and *Ancistrotermes* (Termitidae: Macrotermitinae) in Africa and Asia. *Hodotermes mossambicus* colonies live underground and forage on the surface, cutting living or dead plant material, usually grasses, into small sections and carrying it back to the nest to consume. The Macrotermitinae construct fungus gardens from their faeces. Symbiotic basidiomycete fungi of the genus *Termitomyces* break down the cellulose and lignin into carbohydrates which, with the fungi, are eaten by the termites. Some species of *Odontotermes* build mounds, but the other cotton-damaging Macrotermitinae live in completely subterranean nests consisting of many widely dispersed chambers linked by tunnels. The termites forage under the soil surface over a considerable area and attack plant material either from underneath, where it is in contact with the soil, or under cover of soil sheeting constructed by the termites. For further information on the biology of termites refer to Harris (1971), and Wood and Johnson (1986).

Table 19.1. Termites reported to damage cotton.

Species	Country	Reference
HODOTERMITIDAE		
Hodotermes mossambicus (Hagen)	Tanzania	Pearson (1958)
	Malawi	Pearson (1958)
	Zimbabwe	Mitchell (1972)
RHINOTERMITIDAE		
Coptotermes amanii (Sjöstedt)	Somalia	Harris (1971)
C. sjöstedti Holmgren	Congo	Harris (1971)
TERMITIDAE		
TERMITINAE		
Amitermes sp.	Tanzania	Harris (1971)
Eremotermes paradoxalis	Pakistan	Akhtar and Shahid (1989)
Microcerotermes spp.	East Africa	Harris (1971)
M. diversus Silvestri	Arabia	Harris (1971)
	Yemen Arabic Republic	El Bashir *et al.* (1981)
M. parvus (Haviland)	Zimbabwe	Mitchell (1972)
MACROTERMITINAE		
Ancistrotermes sp.	Zimbabwe	Brettel (1975)
A. equatorius Harris	Uganda	Harris (1971)
A. latinotus (Holmgren)	Southern Africa	Harris (1971)
	Zimbabwe	Mitchell (1972)
Macrotermes sp.	Zimbabwe	Mitchell (1972)
Microtermes spp.	East Africa	Harris (1971)
	Zimbabwe	Brettel (1975), Mitchell (1972)
M. aluco (Sjöstedt)	West Africa	Harris (1971)
M. kasaiensis (Sjöstedt)	Zambia	Harris (1971)
M. mycophagus (Desneux)	Pakistan	Akhtar and Shahid (1989)
M. najdensis Harris	Sudan	Tiben *et al.* (1990)
	Yemen Arabic Republic	Wood *et al.* (1987)
M. obesi Holmgren	Pakistan	Akhtar and Shahid (1989)
M. sudanensis (=*subhyalinus* Silvestri)	Burkina Faso	Ouattara *et al.* (1977)
M. thoracalis (Sjöstedt)	Sudan	Harris (1971)
M. unicolor Snyder	Pakistan	Akhtar and Shahid (1989)
Odontotermes spp.	China	Harris (1971)
	Zimbabwe	Brettel (1975) Mitchell (1972)
O. latericius (Haviland)	Zimbabwe	Mitchell (1972)
O. obesus (Rambur)	India	Harris (1971)
NASUTITERMITINAE		
Trinervitermes biformis (Wasmann)	India	Harris (1971)
Termites (unspecified)	Sudan	Crowther and Barlow (1943), Tarr (1960)
	Texas	Crowther and Barlow (1943)

Identification

Samples collected from damaged cotton for identification should include soldiers as mandible shape is an important distinguishing feature (Fig. 19.1). Soldiers are readily distinguished from the workers by their prominent mandibles

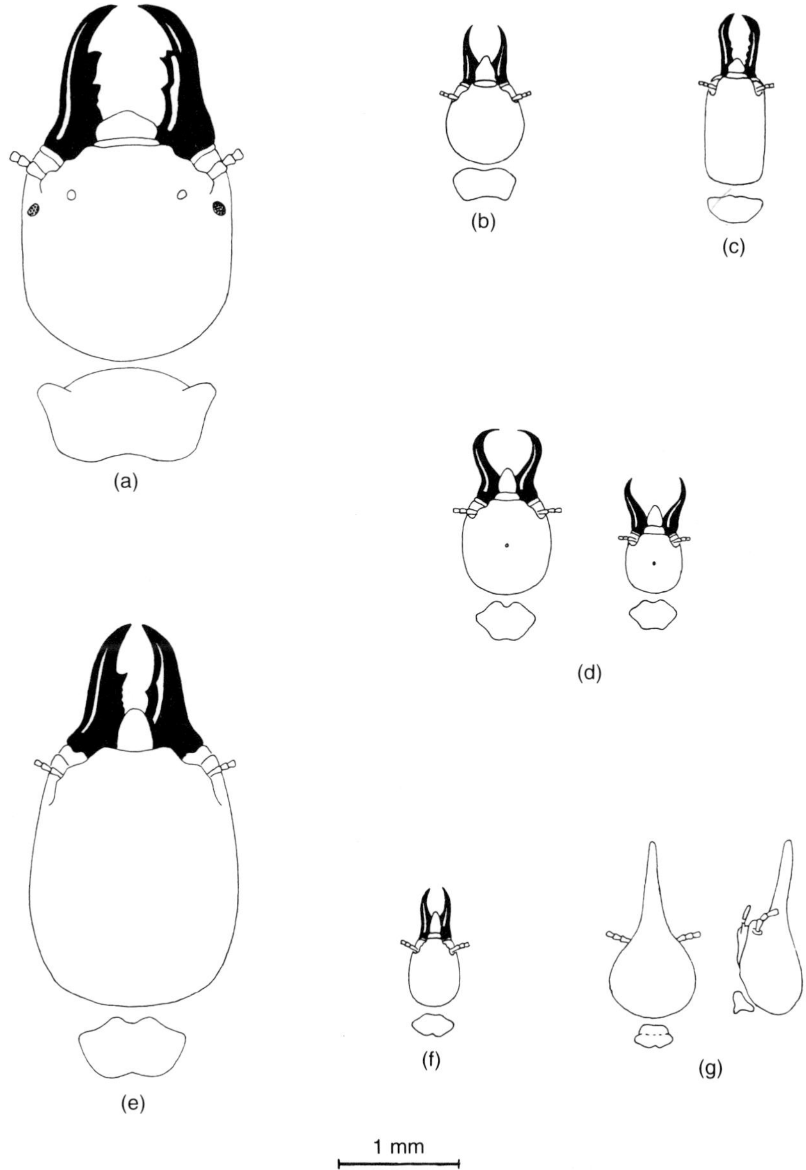

Fig. 19.1. Heads and mandibles of the soldier castes of cotton-damaging termites: (a) *Hodotermes mossambicus*; (b) *Coptotermes* sp.; (c) *Microcerotermes* sp.; (d) *Ancistrotermes* sp. major and minor soldiers; (e) *Odontotermes* sp.; (f) *Microtermes* sp.; (g) *Trinervitermes* sp.

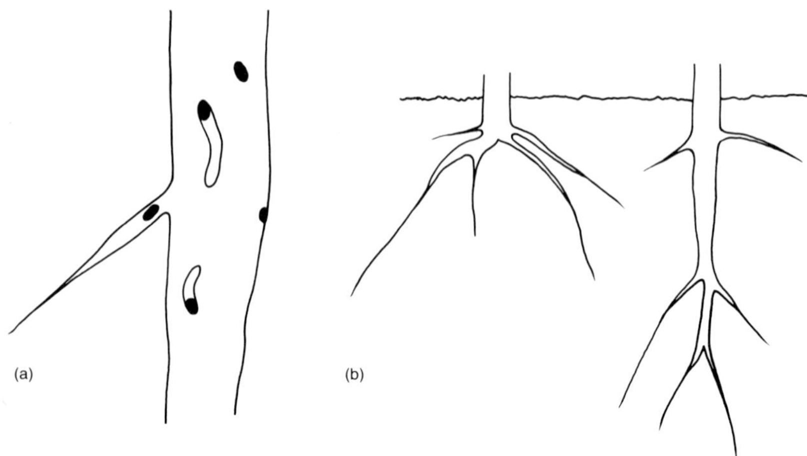

Fig. 19.2. Termite damage to cotton roots: (a) surface damage and tunnels caused by termites (*Microtermes* spp.); (b) forked root (left) due to termite attack on the developing taproot compared with normal root (right) (after Crowther and Barlow, 1943).

although in small termites, such as *Microtermes* and *Ancistrotermes*, the soldier mandibles are not so obvious. Soldiers of the Nasutitermitinae have distinctive flask-shaped heads and the mandibles are vestigial or absent. The keys in Pearce *et al.* (1993), Mitchell (1980) and Chotani (1980) can be used to identify most termite pests to genus and many of those of southern Africa (Mitchell, 1980) and India (Chotani, 1980) to species.

Damage

Hodotermes mossambicus attacks seedling cotton in Tanzania and Malawi, cutting it off at ground level and may defoliate older plants (Pearson, 1958; Harris, 1971). Damage can be considerable, in some cases causing complete loss of the seedlings (Pearson, 1958; G.K.C. Niyrendra, personal communication, 1990).

The Macrotermitinae, however, attack cotton at all stages of growth. It is reported that seeds in India are eaten by termites between sowing and germination (Butani, 1972), but generally, attack is on growing plants (Crowther and Barlow, 1943; Butani, 1972; Gupta and Sharma, 1977; Tiben *et al.*, 1990; Wood *et al.*, 1987). In Sudan and Yemen, mortality due to termites increased toward the end of the season (Tiben *et al.*, 1990), but in Zimbabwe, J.H. Brettell (personal communication, 1990) found mortality was confined to plants up to 50–70 days old. Termites penetrate the roots then tunnel up into the stem replacing the tissue with soil. In young plants the first sign of attack is wilting during the day followed shortly by death. If the plant is dug up, tunnels in the root or stem are evident, usually containing termites (Fig. 19.2). The well-developed root system of older plants may help them survive the attack. Lateral roots can take over the function of the severed taproot, giving a distinctive forked root (Fig. 19.2). Such plants lack support and may fall over, soiling the bolls and making harvest more difficult (Crowther and Barlow, 1943; Wood *et al.*, 1987; Akhtar and Shahid, 1989; Tiben *et al.*, 1990).

In the Tokar Delta of Sudan, *Microtermes najdensis* Harris also attack the developing bolls at the top of the plant by constructing earth covered runways up the stem (Tiben *et al.*, 1990). *Microtermes* sp. in the Lower Shire in Malawi and near Kadoma, Zimbabwe, also forage on the outside of the stem, occasionally tunnelling into the stem, particularly if it is already damaged (Logan, unpublished, 1990; J.H. Brettell, personal communication, 1990).

Losses Due to Termites

Damage is often assessed as the number of plants attacked by termites. For example, Brettell (1975) recorded 2900 plants per hectare killed by termites in Zimbabwe, and 20–40% of the plants were attacked in trials in the Gezira irrigation scheme, Sudan (Crowther and Barlow, 1943; Tarr, 1960). However, this method can overestimate the overall effect of the damage. Plants adjacent to those killed may benefit from the improved spacing to produce more or larger bolls (Wood *et al.*, 1987). Loss of up to 30% of plants does not affect the yield if the mortality occurs evenly (Crowther and Barlow, 1943). On the other hand, yield losses may occur in the surviving plants due to secondary attack by pathogenic fungi (Tarr, 1960), or to sublethal root damage which reduces the growth of the plants (Akhtar and Shahid, 1989) and can reduce yields by 2–17% compared with healthy plants (Crowther and Barlow, 1943).

Yield losses of between 7.1% and 50% were associated with *Microtermes najdensis* in the Yemen Arabic Republic and in the Tokar Delta of Sudan (El Bashir *et al.*, 1981; Wood *et al.*, 1987; Tiben *et al.*, 1990). Insecticide treatment increased yields of seed cotton by over fourfold and lint by nearly threefold in India (Bhatnagar, 1962). Generally, losses elsewhere are low and termites are considered only a minor pest although locally damage may be severe (Harris, 1971; Butani, 1972; Mitchell, 1972; Akhtar and Shahid, 1989).

Control

Chemical

Hodotermes mossambicus can be controlled by baits made of chopped hay or lucerne treated with carbaryl or sodium fluosilicate placed near the foraging holes. This method may also control *Trinervitermes* spp. which have similar foraging habits.

Control of termites attacking the roots of cotton has generally relied on the use of the cyclodiene group of organochlorine insecticides, such as aldrin, dieldrin, heptachlor and chlordane, to form a protective barrier around the roots (Harris, 1971; Sands, 1973) Other organochlorines, such as BHC and DDT, have been used in the same way. Applied to the furrow, planting hole or as a seed dressing, they persist in the soil and give protection through to harvest. They are cheap and therefore can be cost-effective (Brettell, 1975; Wood *et al.*, 1987; Tiben *et al.*, 1990). However, increasing concern over hazards to human health and the environment has led to their use being strictly controlled in agriculture and, in many countries, they are no longer available for use in cotton.

The organochlorine insecticides have been largely replaced as termiticides in the building industry by chlorpyrifos, isofenphos and permethrin which persist under buildings for between five and 15 years (Mauldin *et al.*, 1987). However in agricultural soils they are broken down much more rapidly. Persistence of these insecticides can vary between a few days and several months depending on many factors, including initial concentration, soil type, temperature, humidity and previous exposure to the pesticide (Whitney, 1967; Naitam and Sukhani, 1984; Agnihotri *et al.*, 1986; Felsot, 1989). Most currently available insecticides probably have insufficient persistence to protect cotton through to harvest but there is little published information.

When isofenphos was applied as granules at 5.0 kg ha^{-1}, 10% remained in the soil at harvest (Wood, 1982). Damage was reduced and yields increased (Wood *et al.*, 1987). Brettell (personal communication, 1990) found that 1.5 kg a.i. ha^{-1} endosulfan was as effective as aldrin in reducing plant mortality where damage was limited to young cotton. This insecticide may not be as effective if damage occurs at the end of the season. The use of carbaryl was also reported to control damage and increase yields in India (Bhatnagar, 1962). More research is required to establish the efficacy of these or other insecticides under a variety of field conditions before recommendations for control can be made.

Persistence of short-lived insecticides can be enhanced by incorporating them into an inert matrix from which they are released at a slow but steady rate throughout the life of the crop. Such controlled-release granules have been shown to control subterranean termite attack on groundnuts (Logan, 1988) and *Eucalyptus* (Mitchell, 1989) and are currently under test on cotton in Zimbabwe (J.H. Brettell, personal communication, 1990).

Non-chemical Control

Non-chemical methods, cultural, biological, resistance and the use of botanical insecticides, for the control of termites in crops, including cotton, are reviewed by Logan *et al.* (1990).

Losses due to subterranean and harvester termites can be reduced by planting excess seeds per station and thinning surviving plants later (Matthews, 1989). Subterranean termite attack has been shown to be influenced by irrigation, organic manures, mulches, cultural practices and crop rotation (Crowther and Barlow, 1943; Shinde, 1971) and it may be possible to use these factors to reduce losses (Logan *et al.*, 1990). Wounds, caused by the wind rubbing the plant against the soil or wind-blown sand, predispose the plant to termite attack (Harris, 1971; Tiben *et al.*, 1990). In the Tokar Delta in Sudan, nurse crops of sorghum and millet are planted to protect cotton against wind damage and therefore may reduce termite damage.

Termite resistance in cotton has not been investigated, although Sen-Sarma

(1986) noted that indigenous cotton in India was less susceptible than American and Egyptian varieties.

A wide range of mammals, amphibians, reptiles and arthropods prey on termites as specialist or opportunistic predators but, in view of the high production rate of termite colonies, it is unlikely that this predation can be significantly enhanced to achieve more effective control. Biological control of termites by bacteria, fungi and nematodes has been investigated but generally was unsuccessful in the field, possibly because the termites seal off parts of the colony containing sick or dying individuals (Pearce 1987; Logan and Abood, 1990).

Termites are usually minor pests of cotton and in most cases the cost of providing a prophylactic treatment each season cannot be justified. However, the increasing spread of agriculture into marginal areas and the growing of cotton on degraded soils is likely to result in stressed plants which are more susceptible to attack and the problem may get worse, justifying further research.

References

Agnihotri, N.P., Jain, H.K. and Gajbhiye, V.T. (1986) Persistence of some synthetic pyrethroid insecticides in soil, water and sediment. *Journal of Entomological Research (New Delhi)* 10, 147–151.

Akhtar, M.S. and Shahid, A.S. (1989) Termite populations and damage in cotton fields at Quadar, Pur, Multan Pakistan (Isoptera). *Sociobiology* 15, 349–359.

Bhatnagar, S.P. (1962) Insecticidal trials against termites infesting cotton plants under unirrigated (barani) conditions in Rajasthan. In: *Termites in the Humid Tropics.* UNESCO, Paris, 223pp.

Brettell, J.H. (1975) Soil pest trials. In: *Annual Report of the Cotton Research Institute, Gatooma, Rhodesia, 1973–1974.* Cotton Research Institute, Gatooma, Rhodesia, pp. 136–138.

Butani, D.K. (1972) Combating pests and diseases of cotton. *Indian Farmers' Digest* 4, 54–55.

Butani, D.K. (1973) Les insectes ravageurs du cottonier. XVI. Les principaux problemes parasitaires du cotonier en Inde. *Coton et Fibres Tropicales* 28, 259–268.

Chotani, O.B. (1980) *Termite Pests of Agriculture in the Indian Region and their Control.* Technical Monograph No. 4, Zoological Survey of India, pp. 1–84.

Crowther, F. and Barlow, H.W.B. (1943) Tap-root damage of cotton, ascribed to termites, in the Sudan Gezira. *Empire Journal of Experimental Agriculture* 11, 99–112.

El Bashir, S., Khairalla, A.I. and El Kateb, B. (1981) *Crop Damage Due to Termites in the Tihama Region of the Yemen Arab Republic.* Ministry of Agricultural Research Service, Yemen Arab Republic, 12pp.

Felsot, A.S. (1989) Enhanced biodegradation of insecticides in soil: implications for agroecosystems. *Annual Review of Entomology* 34, 453–476.

Gupta, V.S. and Sharma S. (eds) (1977) *Package of Practices. Kharif 1977.* Hayana Agricultural University, Hissar, India, 126(+40)pp.

Harris, W.V. (1971) *Termites: Their Recognition and Control,* 2nd edition. Longmans, London, 186pp.

Logan, J.W.M. (1988) *Overseas Development Natural Resources Institute and the International Crops Research Institute for the Semi-Arid Tropics collaborative project for the control of termites in groundnuts 1986–87. Final Report.* Overseas Assignment Report, Project A1402. Overseas Development Natural Resources Institute, Chatham, UK, 47pp.

Logan, J.W.M. and Abood, F. (1990) Laboratory trials on the toxicity of hydramethylnon

(Amdro; AC 217,300) to *Reticulitermes santonensis* Feytaud (Isoptera: Rhinotermitidae) and *Microtermes lepidus* Sjöstedt (Isoptera: Termitidae). *Bulletin of Entomological Research* 80, 19–26.

Logan, J.W.M., Cowie, R.H. and Wood, T.G. (1990) Non-chemical control of termites in crops and forestry. A review. *Bulletin of Entomological Research* 80, 309–330.

Matthews, G.A. (1989) *Cotton Insect Pests and Their Management.* Longman, Harlow, Essex, 199pp.

Mauldin, J., Jones, S. and Beal, R. (1987) Viewing termiticides. *Pest Control* 55, 46–59.

Mitchell, B.L. (1972) Termite survey of Rhodesia. *Rhodesian Agricultural Journal* 69, 39.

Mitchell, B.L (1980) Report on a survey of the termites of Zimbabwe. *Occasional Papers of the National Museum of Rhodesia. B Natural Science* 6, 187–323.

Mitchell, M.R. (1989) Comparison of non-persistent insecticides in controlled release granules with a persistent organochlorine insecticide for the control of termites in young *Eucalyptus* plantations in Zimbabwe. *Commonwealth Forestry Review* 64, 281–294.

Naitam, N.R. and Sukhani, T.R. (1984) Effect of ecological conditions on the persistence of carbofuran and isofenphos in the soil and sorghum seedlings. *Indian Journal of Entomology* 46, 452–459.

Ouattara, S., Jolivet, P. and Parys, E.V. (1977) *Seconde Liste des Insectes et des Plantes-hotes en Haute-Volta et dans les Regions Limitrophes*, 2nd edition. Bobo-Dioulasso, Haute-Volta, Projet de Renforcement de la Protection des Plantes en Haute-Volta, 107pp.

Pearce, M.J. (1987) Seals, tombs, mummies and tunnelling in the drywood termite *Cryptotermes* (Isoptera; Kalotermitidae). *Sociobiology* 13, 217–226.

Pearce, M.J., Bacchus, S. and Logan, J.W.M. (1993) What termite? A guide to identification of termite pest genera in Africa. Technical Leaflet No. 4, Natural Resources Institute, Chatham, 19pp.

Pearson, E.O. (1958) *The Insect Pests of Cotton in Tropical Africa.* Commonwealth Agricultural Bureau, London, 355 pp.

Sands, W.A. (1973) Termites as pests of tropical food crops. *Pest Articles and News Summaries* 19, 167–177.

Sen-Sarma, P.K. (1986) Economically important termites and their management in the oriental region. In: Vinson, S.B. (ed.) *Economic Impact and Control of Social Insects.* Praeger, New York, pp. 69–102.

Shinde, C.B. (1971) Termites, a dangerous insect of agriculture. *Farmer and Parliament* 6, 23.

Tarr, S.A.J. (1960) The effects of fungicide–insecticide seed treatments on emergence, growth and yield of irrigated cotton in the Sudan Gezeira. *Annals of Applied Biology* 48, 591–600.

Tiben, A., Pearce, M.J., Wood, T.G., Kambal, M.A. and Cowie, R.H. (1990) Damage to crops by *Microtermes najdensis* (Isoptera: Macrotermitinae) in irrigated semi-desert areas of the Red Sea coast. 2. Cotton in the Tokar Delta region of Sudan. *Tropical Pest Management* 36, 296–304.

Whitney, W.K. (1967) Laboratory tests with Dursban and other insecticides in soil. *Journal of Economic Entomology* 60, 68–74.

Wood, T.G. (1982) Trials and demonstrations of termite control in cotton and other crops in the Tihama, Yemen Arab Republic. Unpublished Report. FAO, Rome, 29pp.

Wood, T.G. and Johnson, R.A. (1986) The biology, physiology and ecology of termites. In: Vinson, S.B. (ed.) *Economic Impact and Control of Social Insects.* Praeger, New York, pp. 1–68.

Wood, T.G., Bednarzik, M. and Aden, H. (1987) Damage to crops by *Microtermes najdensis* (Isoptera: Macrotermitinae) in irrigated semi-desert areas of the Red Sea coast 1. The Tihama region of the Yemen Arab Republic. *Tropical Pest Management* 33, 142–150.

20 Acari – Leaf-Feeding Mites

J. Gutierrez

Institut Français de Recherche Scientifique pour le Développement en Coopération (ORSTOM), BP 5045, 34032 Montpellier Cedex 1, France

Mites living on cotton belong to three large superfamilies of phytophagous mites, i.e. Tetranychoidea, Tarsonemoidea and Eriophyoidea. Representatives of these groups each have a particular morphology and cause characteristic damage.

The Tetranychoidea, which are subdivided into Tetranychidae or spider mites and Tenuipalpidae or false spider mites, are coloured, the females are visible with the naked eye and measure 0.3–0.5 mm. They puncture cells of the epidermis and parenchyma and absorb the contents. The damage takes the form of spots of variable intensity on leaves that can eventually cause drying and defoliation.

Tarsonemoidea have an oval translucent bright body whose largest dimension varies from 0.1–0.3 mm. The phytophagous species mainly live on young leaves that have not yet unfolded, which are particularly sensitive to substances contained in the injected saliva. They cause leaf deformation and can even stunt the whole plant. The only species incriminated causes bronzing on leaf undersurfaces and limb margin deformations. As the leaves age, they may split or crack open.

Eriophyoidea have a vermiform body and only two pairs of legs. They are small, with a body length ranging from 0.1–0.25 mm. They have a waxy white colour and generally can only be seen with a stereomicroscope. Unlike tetranychid mites, eriophyid mites do not kill the leaves they attack. On cotton, they either cause velvet-like hair to appear on leaves and petiole or induce the appearance of galls and branch deformation. They have a very narrow range of host plants and in most cases they are only collected on a single plant species. They are capable of transmitting virus infections.

In these three superfamilies, reproduction is characterized by male-producing parthenogenesis (arrhenotoky), i.e. males are born from an unfertilized haploid egg, females from a diploid egg. Some species reproduce by female-producing parthenogenesis (thelytoky) and in this case the life cycle only comprises individuals with 2 N chromosomes.

Only tetranychid and tarsonemid mites have a severe impact on cotton farming in the world. Tetranychidae outbreaks tend to occur in arid regions, whereas Tarsonemidae cause more problems in humid regions that are very cloudy during the period of cotton cropping. The importance of the damage is considerably increased because of the destruction of the natural predators of these mites which occurs as a consequence of insecticide treatment designed to control a major cotton pest.

TETRANYCHOIDEA

Tetranychidae

The species and their distribution

Thirty-six tetranychid species (Table 20.1) are reported to be pests of cotton in different parts of the world but only nine of them have real economic significance: they belong to the genera *Oligonychus* and *Tetranychus*.

Mites of the genus *Oligonychus* colonize the upper leaf surface, while those of the genus *Tetranychus* feed on the under surface (Plate VIII.2). In the latter case the injury shows on the upper surface as whitish areas where feeding has occurred.

Oligonychus gossypii (Zacher), originally described on cotton in Togo, is widespread in all tropical Africa and is also known in Central America and Brazil. This relatively large species is dark red in colour. It is very common on many host plants in West Africa and appears to be sensitive to acaricides.

Tetranychus desertorum Banks, the desert spider mite, has been recorded in the southern USA, and Central and South America. The females are carmine in colour. Cases of resistance to organophosphates have been reported (Cranham and Helle, 1985).

Tetranychus gloveri Banks is known throughout tropical America. The name *Tetranychus tumidus* Banks has been wrongly used for this species by many authors. *T. tumidus* is the water hyacinth spider mite, while *T. gloveri* infests many plants (Boudreaux, 1979). The females are carmine and the freshly laid eggs are colourless.

Tetranychus lombardinii Baker and Pritchard, first described on cotton in Mozambique, has subsequently been collected on cotton plantations in Malawi, Zimbabwe and South Africa (Meyer and Rodrigues, 1966; Meyer, 1987). The females are dark red with a dark spot on each side of the body.

Tetranychus ludeni Zacher, although widespread throughout the tropics, is known as a cotton pest in Mozambique, South Africa and Australia. The females are carmine with reddish legs. Some strains of this species may be resistant to organophosphates (Cranham and Helle, 1985).

Tetranychus neocaledonicus Andre (=*Tetranychus cucurbitae* Rahman and Sapra = *T. equatorius* McGregor). The vegetable mite has a wide host range and is

Table 20.1. Tetranychidae reported to be pests of cotton, their alternative host plants, and their world distribution.

Species	Alternative host plants	Geographic distribution	References
Petrobia latens (Muller)	Monocotyledons	Europe, North Africa, South Africa, North America, Australia	Baker and Pritchard, 1953; Pritchard and Baker, 1955
Eutetranychus africanus (Tucker)	Citrus, frangipani, castor bean, breadfruit	Egypt, Mozambique, South Africa, Madagascar, Mauritius, Burma	Attiah, 1967; Meyer and Rodrigues, 1966; Meyer, 1987
Eutetranychus citri Attiah	Citrus	Egypt	Yousef *et al.*, 1976
Eutetranychus orientalis (Klein)	Citrus, frangipani, castor bean, papaya	Southern Asia, Middle East, Tropical Africa	Attiah, 1967; Jeppson *et al.*, 1975; Meyer, 1987
Eutetranychus palmatus Attiah	Date palm	Egypt	Yousef *et al.*, 1976
Allonychus littoralis (McGregor)	Avocado	Ecuador, Central America	McGregor, 1955
Eotetranychus falcatus Meyer and Rodrigues	Peanut, Hibiscus, Sida	Tropical Africa	Meyer and Rodrigues, 1966; Meyer, 1987
Eotetranychus smithi Pritchard and Baker	Rose, Rubus	USA, Japan, Madagascar	Caldwell, 1967
Mononychellus planki (McGregor)	Beans, peanut, soyabean	Brazil, Colombia, Puerto Rico	Flechtmann and Baker, 1970
Oligonychus andrei Gutierrez	Grewia	Madagascar	Gutierrez, 1967
Oligonychus coffeae (Nietner)	Tea, mango, frangipani, cassava, rose	Pantropical	Meyer and Rodrigues, 1966
Oligonychus intermedius Meyer	Dombeya, Grewia	South Africa, Malawi	Meyer, 1987
Oligonychus mangiferus (Rahman and Sapra)	Mango, rose	Pantropical	Mohamed, 1963

Table 20.1. Continued

Species	Alternative host plants	Geographic distribution	References
Oligonychus peruvianus (McGregor)	Grapes, willow, carob, avocado	California, Texas, Mexico, Guatemala, Venezuela, Trinidad, Peru	Estebanes and Baker, 1968
Oligonychus gossypii (Zacher)	Cassava, citrus, rose	Tropical Africa, Madagascar, Central America, Brazil	Baker and Pritchard, 1960; Jeppson *et al.*, 1975
Oligonychus mcgregori Baker and Pritchard	Ficus, Lonchocarpus	Central America, Mexico	Baker and Pritchard, 1953; Estebanes and Baker, 1968
Oligonychus stenoperitrematus (Ugarov and Nikolskii)	Maize, watermelon, thistle	Former USSR	Pritchard and Baker, 1955
Tetranychus amicus Meyer and Rodrigues	Banana, cassava, peanut	South Africa, Mozambique, Reunion, Mauritius	Meyer and Rodrigues, 1966; Gutierrez, 1974
Tetranychus canadensis (McGregor)	Apple, plum, rose, horsechestnut, poplar	Canada, USA	Pritchard and Baker, 1955
Tetranychus desertorum Banks	Leguminous forage crops, beans, Graminae, horseweed, eggplant, melons	Southern USA, Central and South America, China, Japan	Nickel, 1960; Wang, 1981
Tetranychus gigas Pritchard and Baker	No other known host	Texas and Arizona	Pritchard and Baker, 1955
Tetranychus gloveri Banks	Beans, eggplant, banana	Southern USA, Central and South America, Guam	Boudreaux, 1979
Tetranychus lombardinii Baker and Pritchard	Hibiscus, banana, Solanaceae	South and East Africa, Madagascar, Australia, Indonesia	Gutierrez and Schicha, 1983; Meyer, 1987

Table 20.1. Continued

Species	Alternative host plants	Geographic distribution	References
Tetranychus ludeni Zacher	Compositae, Cucurbitae, Leguminosae	Pantropical, in greenhouses in more temperate areas	Meyer and Rodrigues, 1966; Davis, 1968; Meyer, 1987
Tetranychus macfarlanei Baker and Pritchard	Beans, Hibiscus, pumpkin	India, Mauritius, Madagascar	Jose and Shah, 1986 and 1988
Tetranychus marianae McGregor	Castor bean, passionflower	Australia, Pacific Islands, Tropical America	Davis, 1968; Jeppson *et al.*, 1975; De Moraes *et al.*, 1987
Tetranychus neocaledonicus André	Polyphagous	Pantropical	Meyer and Rodrigues, 1966; Gutierrez, 1976
Tetranychus pacificus McGregor	Deciduous fruits, walnut, grapes, leguminous forage crops	Interior western USA, Mexico	Jeppson *et al.*, 1975
Tetranychus piercei McGregor	Banana, papaya, Palmae	Southeast Asia	Fauvel, personal communication
Tetranychus rooyenae Meyer	Hibiscus, Solanaceae	South Africa, Malawi	Meyer, 1987
Tetranychus schoenei McGregor	Deciduous fruits, beans	Eastern and southeastern USA	Jeppson *et al.*, 1975
Tetranychus tchadi Gutierrez and Bolland	Beans, soyabean	Chad, Senegal, Mali	Gutierrez and Etienne, 1981
Tetranychus turkestani Ugarov and Nikolskii	Polyphagous	Regions of the northern hemisphere with a temperate or mediterranean climate, South Africa	Uspenskii, 1978; Meyer, 1987
Tetranychus urticae Koch	Polyphagous	Cosmopolitan but more widespread in temperate climate	Jeppson *et al.*, 1975; Meyer, 1987
Tetranychus yusti McGregor	Beans, soyabean, Compositae, Leguminosae, Graminae	Southeastern USA, Mexico, Central America, Ecuador, Nigeria	McGregor, 1955; Matthysse, 1978
Tetranychus zambezianus Meyer and Rodrigues	Soyabean	Angola, Zimbabwe, Mozambique, Madagascar	Meyer and Rodrigues, 1966; Meyer, 1987

distributed throughout tropical regions. It is an important pest mite in India, Madagascar and Africa where it is common on cotton. The females are bright red with clearer legs. Numerous acaricides are effective against this mite and no resistant strain has been reported in the literature.

Tetranychus pacificus McGregor, the Pacific spider mite, has been recorded from British Colombia to California and Mexico. It is a pest of a wide variety of crops. Feeding females are greenish with dark spots on each side of the body. It is one of the most difficult tetranychid mites to control because its populations have developed resistance to several acaricides (Jeppson *et al.*, 1975).

Tetranychus turkestani Ugarov and Nikolskii (=*Tetranychus atlanticus* McGregor). The strawberry spider mite is usually collected in regions with a temperate or Mediterranean climate. It feeds primarily on low-growing hosts, less often on fruit trees. Females are straw coloured or greenish with dark spots on each side of the body. They can hibernate, the winter forms being bright orange. In California this species predominates in fields during late spring and early summer (Leigh, 1985). Cases of resistance to organophosphates have been reported (Cranham and Helle, 1985).

Tetranychus urticae Koch, the two-spotted spider mite is considered to be a complex that includes *Tetranychus cinnabarinus* (Boisduval) (Dupont, 1979). More than fifty synonyms have been used for this species, the best known being *T. telarius* L., *T. bimaculatus* Harvey, *T. arabicus* Attiah and *T. cucurbitacearum* (Sayed). *Tetranychus urticae* originates in the temperate zones but is frequently found in intensively farmed regions of the tropics. Its strains show resistance to most of the acaricide groups (Cranham and Helle, 1985). When obliged to survive on spontaneous vegetation in tropical regions, it is often less competitive than indigenous Tetranychidae, despite its large polyphagia. On the other hand its resistance to pesticides allows it to predominate in regularly treated areas. Summer females have two colour forms: carmine with a dark spot on each side of the body (*cinnabarinus* form) and green with similar spots (*urticae sensu stricto* form). Hibernating females are uniformly orange.

Other Tetranychidae mentioned by different authors cited in Table 20.1 have been collected on cotton and manage to develop on this plant without attaining outbreak proportions. These species are actually only found in limited areas near other host plants that are more favourable to their multiplication and constitute infestation reservoirs.

Tetranychus tchadi Gutierrez and Bolland (=*Tetranychus joanni* Meyer), which was first found on soybean and beans in Chad and Senegal, has been collected on cotton in Nigeria (Meyer, 1987) and was recently found on this plant in Mali. *Eotetranychus falcatus* Meyer and Rodrigues, first reported in South Africa, Mozambique, Zimbabwe, Malawi and Angola (Meyer, 1987) has been identified in Togo and Mali. *Tetranychus piercei* McGregor, which was described in Southeast Asia on different crops, was also recently identified on cotton in the Philippines (G. Fauvel, personal communication, 1991).

In terms of the large cotton farming regions in the world, *T. urticae* and *T. turkestani* are the most widespread in China, the former USSR and the Middle East (Uspenskii, 1978; Demillo, 1979; Cai, 1985). *Tetranychus urticae, T. ludeni* and *T. neocaledonicus* are most common in India (Gupta, 1985). In Africa, in addition to

T. urticae which is rampant in Egypt (Yousef *et al.*, 1976) and was introduced into South Africa (Meyer and Rodrigues, 1966) as well as into Benin, Ivory Coast and Senegal in intensively cultivated areas, cotton plantations are attacked by species from the surrounding spontaneous vegetation. These include *T. lombardinii, T. neocaledonicus* and *T. ludeni* in East Africa and South Africa (Meyer and Rodrigues, 1966; Duncombe, 1977) and *Oligonychus gossypii* and *T. neocaledonicus* in West Africa.

Tetranychus urticae is present in all regions of the USA, along with *T. turkestani* and *T. pacificus* in California (Leigh, 1985), *T. desertorum* in Texas (several authors cited by Nickel, 1960), and *T. gloveri* in Louisiana and all over the southeast (Jeppson *et al.*, 1975).

Life history (Fig. 20.1)

With the exception of *Petrobia latens* (Muller) which reproduces by thelytoky and is of secondary importance (Baker and Pritchard, 1953), all the species shown in Table 20.1 reproduce by arrhenotoky. In all cases, there are three active larval stages between the egg and the adult (larva, protonymph, deutonymph), alternating with three resting stages (protochrysalis, deutochrysalis and teliochrysalis). The total development time of *T. urticae* growing on cotton at a constant temperature of 25°C and a constant relative humidity of 50% is nine days for males and 9.2 days for females (Gutierrez, 1976).

Reared at 25°C, most females of the *Tetranychus* species mentioned (*T. desertorum, T. neocaledonicus, T. pacificus, T. turkestani* and *T. urticae*) live for three to four weeks and lay 50 to 100 eggs (several authors reviewed by Sabelis, 1985).

In tropical regions, a generation lives for 15 to 20 days depending on the

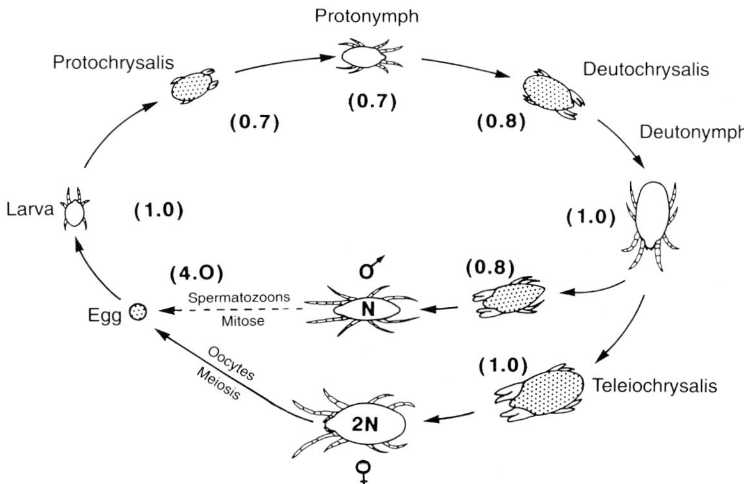

Fig. 20.1. Life history of a tetranychid mite. Figures in brackets indicate the length in days of the different development stages of *Tetranychus urticae* Koch bred at 25°C and RH 50%.

period of the year, meaning that there can be as many as ten successive generations of tetranychid mites during the period of cotton cropping (Gutierrez, 1976).

Dispersal and within-plant distribution

Tetranychid mites can crawl along plants and the ground. For long-distance dispersion they can be carried from one location to another on man, insects or birds but the main means of dispersion is probably wind: mites lower themselves from the host plant on silk threads, that serve as balloons or parachutes to carry them considerable distances.

Working on three species (*T. urticae, T. pacificus* and *T. turkestani*) in California, Carey (1982) demonstrated that on developing cotton plants, females found colonies about midway up the plant and that the majority of the mites occupied the fifth to the tenth mainstem leaves below the apical plant terminal. By late season they move into the plant terminal.

Damage and crop losses

Mechanical injury to epidermal and mesophyll cells leads to important water losses. Cell chloroplasts are damaged and the intensity of photosynthesis is reduced. Disturbance of metabolic processes results in decreased growth, flowering and cropping.

A comparison between physiological injuries caused by *T. turkestani, T. pacificus* and *T. urticae* shows that *T. turkestani* is the most injurious species and *T. pacificus* is somewhat more harmful than *T. urticae. Tetranychus turkestani* causes toxin-induced injury to cotton unlike the other two species (Brito *et al.*, 1986).

Authors who have carried out measurements of yield decreases due to tetranychid mites all agree that losses are severe, and that the economic significance of the damage increased with the earliness of the attack.

In the region of Sao Paulo in Brazil, outbreaks of *T. urticae* reduce the harvest by 17–25% and affect fibre quality (Oliveira and Calcagnolo, 1975). In Alabama, an early attack of the same species decreased the quality of the cotton and reduced the seed yield by 14–44% depending on the infestation level (Canerday and Arant, 1964a).

A late attack of *T. turkestani* in Alabama caused a seed loss of 13–22% (Canerday and Arant, 1964b). In Kazakhstan, an infestation by this species starting in the flowering stage, caused yield reductions of the order of 44% and losses even reached 62% when the mites attacked at the beginning of the season (Wehner, 1989).

In Zimbabwe, infestation by one of the three common species (*T. urticae, T. lombardinii, T. ludeni*) 14 weeks after germination causes damage of the order of 14%, but when it occurs six weeks after germination, losses reach 67% (Duncombe, 1977). This is partly due to the fact that the tolerance of cotton for Tetranychidae increases as the period of boll maturity approaches (Wilson *et al.*, 1987).

Natural enemies

Tetranychids are attacked by many predators, including insects and other mites.

Insects

These consist of Coleoptera, Thysanoptera, Diptera, Hemiptera, Neuroptera and Dermaptera. Only the first three orders comprise predators that are truly specific for mites. In the Coleoptera, these includes Staphylinidae of the genus *Oligota* (ten species mentioned by Chazeau, 1985) and Coccinellidae of the genus *Stethorus* (thirty-odd species mentioned by Chazeau, 1985). Out of the three families of Thysanoptera predators of mites, members of the family Thripidae are the most active in tropical countries. Diptera are mostly represented by Cecidomyiidae, whose larvae feed on tetranychid adults and all the developmental stages. Predation by Hemiptera is less frequent on cotton, since the Anthocoridae or Miridae attacking phytophagous mites are more rare in tropical countries than in the summer in temperate countries.

Insect predators are relatively indifferent to the host plant and their flying capacity allows them to focus on high concentrations of tetranychid mites. Their high sensitivity to pesticides causes their disappearance in cotton fields as soon as the first insecticides are applied, and for this reason they have not yet been considered for use in integrated biological control.

Acari

The mites which prey on tetranychid mites belong to the Bdellidae, Anystidae, Stigmaeidae and Cheyletidae families, but it is above all the Phytoseiidae which are the most widespread and effective. Sixty-three species of Phytoseiidae divided into 15 different genera have been recorded on cotton throughout the world (De Moraes *et al.*, 1986).

Numerous Phytoseiidae also show a resistance to insecticides, particularly to organophosphates. This feature is commonly made use of in integrated pest management programmes in temperate zone orchards and vineyards, and it could be employed with cotton plants in the warm regions. Although Phytoseiidae strains resistant to specific insecticide (e.g. organophosphates) have still not been introduced into cotton plantations, the impact of pesticide treatments on their survival is already taken into account when insecticides and acaricides are chosen.

Control

At present chemical application is the method most commonly employed to control outbreaks of Tetranychidae. Treatment ought only to be undertaken if infestation exceeds a threshold determined by sampling (Leigh, 1985).

Careful regular inspection of fields is necessary to detect the first appearance of an infestation, which may be confined to a few plants or localized patches within a field. Slight mottling at the base of leaves may be seen more easily than the mites. When detected early, it may be possible to 'spot' treat these areas and reduce the amount of acaricide needed. Unfortunately, prolonged use of an acaricide can lead to rapid selection of resistant populations, so acaricide rotation

Table 20.2. Tenuipalpidae reported to be pests of cotton, their alternative host plants, and world distribution.

Species	Alternative host plants	Geographic distribution	References
Brevipalpus californicus (Banks)	Polyphagous	Pantropical, particularly reported from Mozambique and Nigeria	Meyer, 1979
Brevipalpus obovatus Donnadieu	Polyphagous	Pantropical, particularly reported from Egypt, Mozambique and South Africa	Yousef *et al.*, 1976; Meyer, 1979
Brevipalpus phoenicis (Geijskes)	Polyphagous	Pantropical, particularly reported from Egypt, Mozambique and South Africa	Yousef *et al.*, 1976; Meyer, 1979
Raoiella indica Hirst	Coconut, Arecanut	India, Mauritius	Gupta, 1985

schemes have been devised (Duncombe, 1972, 1973) in which the use of a particular chemical group within a specified area is limited to a maximum of two seasons. Farmers then change to a different type of acaricide.

Field officers often have a tendency to favour the use of compounds which also have a considerable effect on other crop pests, whereas one ought to choose them according to the species present in the field. The establishment of an acaricide rotation scheme avoids selection of resistant populations (see Chapter 27).

When there are several species present, as in California (*T. urticae, T. pacificus* and *T. turkestani*), an individual treatment affects their relative abundance (Trichilo *et al.*, 1990).

In Brazil, Andrade *et al.* (1989) reported that dimethoate (400 g l^{-1} and 600 ml ha^{-1} and ethyl-chlopyrifos (480 g l^{-1} and 200 ml $^{-1}$) were very effective against *T. urticae* in the state of Parana. In São Paulo state, Gavioli *et al.* (1988) showed the effectiveness of abamectin (10 g ha^{-1} a.i.) against the same species.

In South Africa, Botha *et al.* (1988) successfully controlled *T. urticae* with pyrethroid insecticides such as fenvalerate (200 cm^3 a.i. ha^{-1}) and biphenthrin (300 cm^3 a.i. ha^{-1}), which were more effective than triazophos or hexythyazox.

Several studies of some cultivars show an important resistance to Tetranychidae (Leigh, 1985; Trichilo and Leigh, 1985; Sengonca *et al.*, 1986; Botha *et al.*, 1989), but these promising results have not yet been implemented.

Tenuipalpidae

False spider mites are smaller than Tetranychidae and move more slowly. Only four species (Table 20.2) have been reported on cotton, three belonging to the same genus, i.e. *Brevipalpus californicus* (Banks), *B. obovatus* Donnadieu and *B. phoenicis* (Geijskes), and a fourth, *Raoiella indica* Hirst.

Tenuipalpidae have a life cycle comparable to that of Tetranychidae and their reproduction is generally based on arrhenotoky. The species found on cotton comprise males and females, but males are extremely rare in the three *Brevipalpus* species, and reproduction of most strains living in the tropics occurs by thelytoky.

Brevipalpus californicus, B. obovatus and *B. phoenicis* are all polyphagous and cosmopolitan. They are very flat and a brick-red colour. The literature only reports their presence on cotton in Africa, but it is probable that they infest this plant in other parts of the world. They are sometimes collected in mixtures, and the species can only be distinguished after preparation and microscopic examination. They live on both surfaces of leaves near the midrib. The damage they cause is basically comparable to that of the Tetranychidae but much less widespread. Since they are sensitive to many insecticides and acaricides, their populations in cotton plantations are small and their economic impact is practically negligible.

Raoiella indica has only been reported by Gupta (1985) on cotton in India and its normal plant hosts are Palmaceae. Its presence on cotton is probably accidental in plantations located near palm trees.

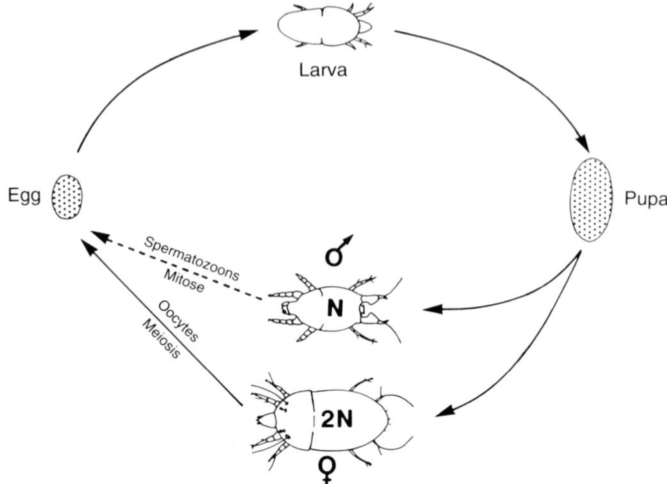

Fig. 20.2. Life history of a tarsonemid mite.

TARSONEMOIDEA

Tarsonemidae

Only one species has been recorded on cotton *Polyphagotarsonemus latus* (Banks), the yellow tea mite or the broad mite, previously named *Hemitarsonemus latus* (Banks). This mite is distributed throughout the tropics and is a greenhouse pest in temperate regions. It feeds on a wide variety of agricultural crops, ornamental and wild plants.

Life history (Fig. 20.2)

The developmental cycle of *P. latus* includes only egg, larva, pupa and adult stages. The oval and elongate eggs are attached to the leaf surface, and their upper surface is studded with longitudinal rows of small tubercles. The larvae are hexapodal and move and feed like adults. They pass without moulting into an inactive stage called a pupa before hatching into adults. At 26–30°C and high relative humidity, only four to five days are required to complete a generation (Schmitz, 1962), i.e. the species multiplies rapidly. The average egg deposition under such conditions is 3.5 eggs per day for six days.

Dispersion is improved by the male's habit of carrying away a female pupa, but longer-range dispersion is accomplished by wind and phoretic association with different insects (Flechtmann *et al.*, 1990).

Damage and crop losses

Lesions caused by *P. latus* appear as limb margin deformations and even perforations leading to limb lacerations. The disturbance of plant physiology takes the form of shortened internodes and reduced mean boll weight.

Polyphagotarsonemus latus is a pest of cotton in humid tropical regions, such as Central Africa (Schmitz, 1962), Uganda (Ingram, 1960), Ivory Coast (Vaissayre, 1982) and the state of São Paulo, Brazil (Oliveira and Calcagnolo, 1974).

Schmitz (1962) reported large population variations in successive years and indicated that damage was more severe when mites appeared early in the cotton-growing season. As soon as the blades of the first three fully-developed leaves at the top show marginal curling, and darker green colour, the plants lose a third of their productivity. In Brazil, Oliveira and Calcagnolo (1974) found that the yellow tea mite reduced cotton seed production by 11% over a whole plot. In Ivory Coast, Vaissayre (1982) reported that in years favourable to outbreaks of Tarsonemidae, cotton seed yields were reduced by 54%.

Control

The predators of Tarsonemidae are mainly Phytoseiidae mites and Hemiptera insects (Anthocoridae). The irregularity of tarsonemid attacks and their fast breeding rate precludes biological control, and consequently no particular research has been carried out in this field.

Organochlorine insecticides (endrin and endosulfan), which were used to control insect pests of cotton, also acted against *P. latus*, but repeated use of synthetic pyrethroids, commonly used now against bollworms has made it necessary to apply specific acaricide treatments against Tarsonemidae in regions where climatic conditions favour their multiplication.

After field trials of about 50 compounds, Vaissayre (1986) selected several insecticides that were quite effective against *P. latus*. These comprise several organophosphorus compounds (profenofos, chlorthiophos, triazophos, ethyl-chlorpyrifos, ethyl-azinphos), used at doses ranging from 250 to 450 g a.i. ha^{-1}, and abamectin at 10 g a.i. ha^{-1}. The capacity of the latter acaricide has also been reported in Brazil, in the state of São Paulo (Donatoni *et al.*, 1988; Gaviolo *et al.*, 1988).

The method of application has proved to be important, i.e complete coverage of the foliage is essential, especially for organophosphorus compounds. Treatments are generally applied between the second and third months after emergence, when the plants are particularly sensitive to infestation (Vaissayre, 1986).

No resistance to the recommended acaricides has yet been reported in this species. On the scale of the whole field, possible resistant strains may be diluted in large untreated populations living on plants in the immediate vicinity of the plantations.

Eriophyoidea

Four species from the family Eriophyidae are recorded on cotton, belonging to two subfamilies: the Eriophyinae and the Phyllocoptinae. These four mite species are restricted to the genus *Gossypium* and each has a peculiar association with cotton plants.

Eriophyinae

The body is wormlike and whitish. The shield has a narrow, basally flexible anterior projection over the rostrum.

Acalitus gossypii (Banks), the cotton blister mite, belongs to a genus that lacks the forefemoral seta and the foretibial seta. Females reach 225–250 μm in length. This species is known from tropical America. According to Mohanasundaram (1982), some Indian records in the literature may concern *Eriophyes puttarudriahi* (Channabasavanna). *Acalitus gossypii* produces hairy deformations on leaves and flowers. Leaves may be crumpled and shoots distorted. *Gossypium barbadense* L. (Sea Island cotton) is very sensitive to this mite, whereas *G. hirsutum* L. (Upland cotton) suffers little damage. Keifer (in Jeppson *et al.*, 1975) recommends combining field clear-up with cotton-free periods and also selecting mite-resistant cotton varieties for planting.

Eriophyes puttarudriahi (Channabasavanna), the Indian cotton blister mite, is distinguishable from the previous species by its foretibial seta typical of the genus *Eriophyes*, and also by the absence of longitudinal shield lines in the centre. Females reach 140–150 μm in length. *Eriophyes puttarudriahi* has only been recorded in India where it is collected on *Gossypium herbaceum* L. (Levant or Asiatic cotton). It causes hairy and feltlike outgrowths on tender shoots. The plants may be stunted and boll formation suppressed.

Phyllocoptinae

The body is rather fusiform, and the shield has a broad-based and rigid anterior lobe over the rostrum.

Heterotergum gossypii Keifer. Females of the cotton rust mite are light yellow and from 150–215 μm long. The abdomen has five narrow rings dorsally just behind the shield followed by 14 broad tergites bearing elongate microtubercles. The sternites(63 to 68) are completely microtuberculate. *Heterotergum gossypii* is only known from Brazil. It was described by Keifer (1955) from specimens collected on *Gossypium hirsutum* in the state of Rio Grande do Norte. This eriophyid mite bronzes the mature leaves and produces tip blighting of young leaves.

Abacarus gossypii Mohanasundaram. Females of the Indian cotton rust mite are whitish and 200–210 μm long. On the dorsum a longitudinal thanosomal trough is formed by a subdorsal ridge on each side with a central longitudinal ridge that ends before the trough ends. Known only from India this species was described by Mohanasundaram (1982) from specimens collected on *Gossypium arboreum* in Tamil Nadu. It produces white erineum patches on both sides of leaves.

Life history (Fig. 20.3)

In eriophyid mites, there are two active larval stages between the egg and the adult (first and second nymphal stages), alternating with two resting periods

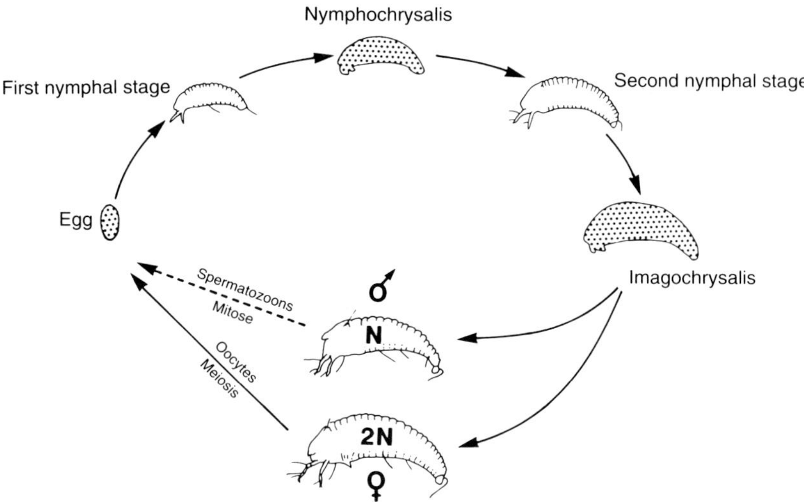

Fig. 20.3. Life history of an eriophyid mite.

(nymphochrysalis and imagochrysalis). The life histories of species living on cotton have never been studied, but the total development time is generally shorter than that of tetranychid mites reared under the same conditions. In tropical regions, reared female eriophyid mites were found to lay ten to 80 eggs during their two to four weeks of life (Jeppson *et al.*, 1975).

Natural enemies and control

Eriophyid mites are particularly attacked by other mites, i.e. Phytoseiidae, Tydeidae, Stigmaeidae and Tarsonemidae predators. Since species living on cotton cause damage that apparently has no economic impact, no particular measures have been taken against them. The differences in sensitivity observed between representatives of the genus *Gossypium* indicate that it may be possible to select resistant varieties.

References

Andrade, C.F.S., Habib, M.E.M. and Rossi, W.L. (1989) Acaricide efficiency of some chemicals against *Tetranychus urticae* Koch, 1836 (Acarina: Tetranychidae) under field conditions. *Ecossistema* 14, 163–171 (in Portugese).

Attiah, H.H. (1967) The genus Eutetranychus in U.A.R. with description of three new species (Acarina: Tetranychidae). *Bulletin de la Societé Entomologique d'Egypte* 51, 11–16.

Baker, E.W. and Pritchard, A.E. (1953) A guide to the spider mites of cotton. *Hilgardia* 22, 401–413.

Baker, E.W. and Pritchard, A.E. (1960) The tetranychoid mites of Africa. *Hilgardia* 29, 455–574.

Botha, J.H., Scholtz, A.J., Marais, A. and Berg, A.M. Van Den (1988) Preliminary tests on the contact toxicity of some pesticides to *Tetranychus urticae* females, with reference to the situation on field cotton. *Phytophylactica* 20, 277–279.

Botha, J.H., Greeff, A.I. and Scholtz, A.J. (1989) Preliminary screening of cotton plants for resistance to spider mite damage in South Africa. *Phytophylactica* 21, 379–383.

Boudreaux, H.B. (1979) Confusion of names for the spider mites *Tetranychus tumidus* and *T. gloveri*. In: Rodriguez, J.G. (ed.) *Recent Advances in Acarology*. Academic Press, New York, vol. 2, pp. 395–398.

Brito, R.M., Stern, V.M. and Sances, F.V. (1986) Physiological response of cotton plants to feeding of three *Tetranychus* spider mite species (Acari: Tetranychidae). *Journal of Economic Entomology* 79, 1217–1220.

Cai, S.H. (1985) Study on the occurence of *Tetranychus cinnabarinus* and its control target. *China Cotton* 5, 39–40 (in Chinese).

Caldwell, S.D. (1967) Cotton, a new host for the spider mite *Eotetranychus smithi*. *Journal of Economic Entomology* 60, 1169.

Canerday, T.D. and Arant, F.S. (1964a) The effect of spider mite populations on yield and quality of cotton. *Journal of Economic Entomology* 57, 553–556.

Canerday, T.D. and Arant, F.S. (1964b) The effect of late season infestations of the strawberry spider mite, *Tetranychus atlanticus*, on cotton production. *Journal of Economic Entomology* 57, 931–933.

Carey, J.R. (1982) Within-plant distribution of Tetranychid mites on cotton. *Environmental Entomology* 11, 796–800.

Chazeau, J. (1985) Predaceous insects. In: Helle, W. and Sabelis, M.W. (eds) *Spider Mites, Their Biology, Natural Enemies and Control*. Elsevier, Amsterdam, vol. 1B, pp. 211–246.

Cranham, J.C. and Helle, W. (1985) Pesticide resistance in Tetranychidae. In: Helle, W. and Sabelis, M.W. (eds) *Spider Mites, Their Biology, Natural Enemies and Control*. Elsevier, Amsterdam, vol. 1B, pp. 405–421.

Davis, J.J. (1968) Studies of Queensland Tetranychidae. 3. Records of the genus *Tetranychus*. *Queensland Journal of Agricultural and Animal Sciences* 25, 57–67.

Demillo, A.P. (1979) Ecological–biological characteristics of the harmfulness of spider mites on cotton in Tajikstan. In: Novozhilov, K.V. (ed.) *Proceedings of the All-Union Research Institute for Plant Protection*, Lenin Academy of Agricultural Sciences, Leningrad, 55–66 (in Russian).

De Moraes, G.J., McMurtry, J.A. and Denmark, H.A. (1986) *A Catalog of the Mite Family Phytoseiidae*. Embrapa – DDT, Brasilia, 353pp.

De Moraes, G.J., McMurtry, J.A. and Baker, E.W. (1987) Redescription and distribution of the spider mites *Tetranychus evansi* and *T. marianae*. *Acarologia* 28, 333–343.

Donatoni, J.L., Biondo, C.J., Geraldi, F.I., Raizer, A.J., Arashiro, F.Y., Clari, A.I. and Mariconi, F.A.M. (1988) Chemical acaricides for the control of the broad mite *Polyphagotarsonemus latus* (Banks, 1904) on cotton crop. *Anais da Sociedade Entomologica do Brasil* 17, 283–292 (in Portugese).

Duncombe, W.G. (1972) Red spider mite on cotton and its control. *Rhodesia Agricultural Journal* 69(1), 7–10.

Duncombe, W.G. (1973) The acaricide spray rotation for cotton. *Rhodesia Agricultural Journal* 70(5), 115–118.

Duncombe, W.G. (1977) Cotton losses caused by spider-mites (Acarina: Tetranychidae). *Rhodesia Agricultural Journal* 74, 141–146.

Dupont, L.M. (1979) On gene flow between *Tetranychus urticae* Koch, 1836 and *Tetranychus cinnabarinus* (Boisduval) Boudreaux, 1956 (Acari: Tetranychidae): synonymy between the two species. *Entomologia Experimentalis et Applicata* 25, 297–303.

Estebanes, G.M.L. and Baker, E.W. (1968) Aranas rojas de Mexico (Acarina: Tetranychidae). *Anales de la Escuela Nacional de Ciencias Biologicas Mexico* 15, 61–133.

Flechtmann, C.H.W. and Baker, E.W. (1970) A preliminary report on the Tetranychidae (Acarina) of Brazil. *Annals of the Entomological Society of America* 63, 156–163.

Flechtmann, C.H.W., Guerrero, J.M., Arroyave, J.A. and Constantino, L.M. (1990) A little known mode of dispersion of *Polyphagotarsonemus latus* (Banks). *VIII International Congress of Acarology*, Ceske Budejovice, Czechoslovakia. Aug. 6–11, 1990, Abstracts 43.

Gavioli, L.A., Gravena, S., Leao Netro, R.R. and Tozzati, G. (1988) Effect of abamectin, cyfluthrin and fenpropathrin on *Polyphagotarsonemus latus* (Banks, 1904) and *Tetranychus urticae* (Koch, 1836) and some natural enemies in cotton. *Ecossistema* 12, 66–77 (in Portugese).

Gupta, S.K. (1985) *Handbook Plant Mites of India.* Zoological Survey of India, Calcutta, 520pp.

Gutierrez, J. (1967) Contribution à l'étude morphologique et biologique de *Tetranychus neocaledonicus* André 1933 (Acarien – Tetranychidae) 'araignée rouge' du cotonnier à Madagascar. *Coton et Fibres Tropicales* 22, 183–195.

Gutierrez, J. (1974) Les espèces du genre *Tetranychus* Dufour (Acariens: Tetranychidae) ayant une incidence économique à Madagascar et dans les îles voisines. Compétition entre les complexes *Tetranychus neocaledonicus* André et *Tetranychus urticae* Koch. *Acarologia* 16(2), 258–270.

Gutierrez, J. (1976) Etude biologique et écologique de *Tetranychus neocaledonicus* André (Acariens: Tetranychidae). *Travaux et Documents ORSTOM Paris* 57, 1–173.

Gutierrez, J. and Etienne, J. (1981) Quelques données sur les Acariens Tetranychidae attaquant les plantes cultivées au Sénégal. *Agronomie Tropicale* 36, 391–394.

Gutierrez, J. and Schicha, E. (1983) The spider mite family Tetranychidae (Acari) in New South Wales. *International Journal of Acarology* 9, 99–116.

Ingram, W.R. (1960) The control of yellow tea mite, *Hemitarsonemus latus* (Banks), with DDT on cotton in Uganda. *Bulletin of Entomological Research* 51, 577–582.

Jeppson, L.R., Keifer, H.H. and Baker, E.W. (1975) *Mites Injurious to Economic Plants.* University of California Press, Berkeley, 614pp.

Jose, V.T. and Shah, A.H. (1986) Chemical control of the spider mite, *Tetranychus macfarlanei* B. & P., the pest of cotton. *Pesticides* 20(12), 19–23.

Jose, V.T. and Shah, A.H. (1988) Bionomics of the spider mite (*Tetranychus macfarlanei* Baker and Pritchard) injurious to cotton in Gujarat. *Gujarat Agricultural University Research Journal* 13(2), 12–18.

Keifer, H.H. (1955) Eriophyid studies XXIII. *The Bulletin Department of Agriculture State of California* 44, 126–130.

Leigh, T.F. (1985) Cotton. In: Helle, W. and Sabelis, M.W. (eds) *Spider Mites, Their Biology, Natural Enemies and Control.* Elsevier, Amsterdam, vol. 1B, pp. 349–358.

McGregor, E.A. (1955) Notes on spider mites (Tetranychidae) of Ecuador. *Revista Ecuatoriana de Entomologia y Parasitologia* 2(3–4), 365–377.

Matthysse, J.G. (1978) Preliminary report on mites collected from plants and animals in Nigeria. I. Mites from plants. *Nigerian Journal Entomology* 1(3) 57–70.

Meyer, M.K.P. (1979) The Tenuipalpidae (Acari) of Africa. With keys to the world fauna. *Department of Agricultural Technical Services. Republic of South Africa, Entomology Memoir* 50, 1–135.

Meyer, M.K.P. (1987) African Tetranychidae (Acari: Prostigmata) with reference to the world genera. *Department of Agriculture and Water Supply, Republic of South Africa, Entomology Memoir* 69, 1–175.

Meyer, M.K.P. and Rodrigues, M.C. (1966) Acari associated with cotton in Southern Africa (with reference to other plants). *Garcia de Orta* 13, 193–226.

Mohamed, I.I. (1963) Acarine mites occurring on cotton plants in Egypt. *Bulletin de la Societé Entomologique d'Egypte* 46, 511.

Mohanasundaram, M. (1982) Two new species of *Abacarus* (Acari; Eriophyidae) on economic hosts in Tamil Nadu. *Indian Journal of Acarology* 6(1–2), 9–13.

Nickel, J.L. (1960) Temperature and humidity relationships of *Tetranychus desertorum* Banks with special reference to distribution. *Hilgardia* 30, 41–100.

Oliveira, C.A.L. De and Calcagnolo, G. (1974) Acao do 'acaro branco' *Polyphagotarsonemus latus* (Banks, 1904) na depreciacao quantitativa e qualitativa da producao algodoeira. *Biologico* 40, 139–149.

Oliveira, C.A.L. De and Calcagnolo, G. (1975) Acao do acaro 'rajado' *Tetranychus urticae* (Koch, 1836) na depreciacao quantitativa da producao algodoeira. *Biologico* 41, 307–327.

Pritchard, A.E. and Baker, E.W. (1955) A revision of the spider mite family Tetranychidae. *Pacific Coast Entomological Society, San Francisco, Memoirs Series* 2, 1–472.

Sabelis, M.W. (1985) Reproduction strategies. In. Helle, W. and Sabelis, M.W. (eds) *Spider Mites, Their Biology, Natural Enemies and Control.* Elsevier, Amsterdam, vol. 1A, pp. 265–278.

Schmitz, G. (1962) L'acariose à *Hemitarsonemus* affection foliaire du cotonnier. *Publication INEAC Série scientifique* 99, 5–50.

Sengonca, C., Lababidi, M.S., Gerlach, S. (1986) The effect of different cotton varieties on the carmine spider mite, *Tetranychus cinnabarinus* Boisd. (Acari: Tetranychidae). *Plant Breeding* 97, 297–303.

Trichilo, P.J. and Leigh, T.F. (1985) The use of life tables to assess varietal resistance of cotton to spider mites. *Entomologia Experimentalis et Applicata* 39, 27–33.

Trichilo, P.J., Wilson, L.T. and Gonzalez, D. (1990) Relative abundance of three species of spider mites (Acari: Tetranychidae) on cotton, as influenced by pesticides and time of establishment. *Journal of Economic Entomology* 83, 1604–1611.

Uspenskii, F.M. (1978) The Turkestan cotton spider mite. *Zashchita Rastenii* 6, 47–48 (in Russian).

Vaissayre, M. (1982) Observations relatives à l'incidence économique de l'acariose à *Polyphagotarsonemus latus* (Banks) en culture cotonnière. *Coton et Fibres Tropicales* 37, 313–314.

Vaissayre, M. (1986) Lutte chimique contre l'acarien *Polyphagotarsonemus latus* (Banks) en culture cotonnière. *Coton et Fibres Tropicales* 41, 31–43.

Wang, H.F. (1981) Economic insect fauna of China (Acariformes: Tetranychoidea). *Science Press, Beijing* 23, 1–150 (in Chinese).

Wehner, F. (1989) Einfluss von Besiedlungstermin und Befallsdauer mit *Tetranychus turkestani* Ug. et Nic. auf Pflanzenwachstum, Ertrag und Qualität von Baumwolle in Mittelasien. *Beiträge zur Tropischen Landwirtschaft und Veterinärmedizin* 27, 203–209.

Wilson, L.T., Pickett, C.H., Leigh, T.F. and Carey, J.R. (1987) Spider mite (Acari: Tetranychidae) infestation foci: cotton yield reduction. *Environmental Entomology* 16, 614–617.

Yousef, A.T.A., El-Badry, E.A. and Heykal, I.H. (1976, publ. 1980) Mites inhabiting cotton and associated weeds in Egypt (Acarina). *Bulletin of the Entomological Society of Egypt* No. 60, 223–227.

III Pest Management

21 The Effects of Insect Attack on the Yield of Cotton

G.A. Matthews

International Pesticide Application Research Centre, Imperial College at Silwood Park, Buckhurst Road, Sunninghill, Ascot, Berkshire SL5 7PY, UK

Introduction

Pearson (1958) pointed out that it cannot be assumed that the mere abundance of an insect, or the physical evidence of its activities, are valid indications of its effect upon yield or quality of the end-products of the plant. An insect infestation has to be considered in relation to the state of the plant at the time and its subsequent history. The cotton plant, through its capacity for continued flowerbud production and vegetative growth is able to compensate for quite considerable damage, especially in varieties of an indeterminate growth habit, and so the damaging effect of an insect can depend very much on when the infestation occurs in relation to plant development and other factors, such as availability of moisture, nutrients and temperature.

Removal of large numbers of plants in the early stages of development may have very little effect on yield. In trials with one variety in one locality, a wide range of plant populations from 3 m^2 to 20 m^2 have shown few significant differences in yield, when considered over several seasons and over a wide range of different cotton growing areas. When seedling plant populations were reduced from the recommended seven plants per square metre to 0.7 plants per square metre to simulate a 90% loss due to termites, grasshoppers and other early season pests, there was still a high yield (>1600 kg seed cotton ha^{-1}) if plants were protected from subsequent bollworm infestations (Matthews *et al.*, 1972). In this trial seedlings were removed at random and this resulted in variations in the distribution of the remaining plants with small and large gaps along individual rows. It might be argued that not all losses are at random, but a more even spacing would be similar to the normal spacing trial in which few yield differences have been noted. The compensation with the indeterminate variety was primarily due to longer sympodia, on which late bolls could mature due to the favourable climatic conditions. More determinate varieties such as Tamcot are less likely to compensate in this way.

The later there is a total loss of plants, the extent to which compensation can occur will be reduced, and yield losses will be more closely related to the loss of plants. Losses of older plants are more likely to be due to diseases such as *Verticillium*, *Fusarium* or *Rhizoctonia* as well as bacterial blight *Xanthomonas malvacearum* than insects, although termites can damage mature plants. In a few areas there can be isolated plants lost due to stem girdling beetles such as *Alcidodes*, but many of the insects attacking stems remove only one of the monopodia.

Assessing yield losses due to artificial removal of leaves is much more difficult, as the plant can replace those destroyed by insects. Studies in the Sudan on cotton unprotected against insect attack showed that defoliation would have the most effect during flowering, since this would cut off the food supplies to the developing buds, flowers and bolls (Goodman, 1956, 1957). Since those studies very little additional experimental data has been published on effects of defoliation, although there have been attempts to establish economic or action thresholds for defoliation on which to base decisions on when to apply insecticides. In Egypt, the cotton leafworm *Spodoptera littoralis* can cause severe defoliation but 20% defoliation per week did not affect the yield (Russell *et al.*, 1993). Routine hand-collection of egg masses is used to control infestations. However, Hosny *et al.* (1986) suggested that spraying, ideally with a selective insecticide, could be delayed until there were more than 4000 egg-masses per feddan. In Uganda, the effect of tattering of leaves due to *Taylorilygus* was studied (Coaker, 1957; McKinlay and Geering 1957), but differences between treatments were generally not significant.

The effect on the plant of sap-feeding insects that attack the foliage might be regarded as comparable with defoliation by chewing insects, although any effect is more protracted compared with the immediate loss of leaf area. However, attacks by aphids, whiteflies and jassids are debilitating, causing a decrease in plant growth and in some cases severe stunting and death of the plant. Their severity depends very much on the effects of natural enemies or, in the case of jassids, the level of host plant resistance. The main problem with jassids is the transfer of a toxin into the leaves; and with aphids and whiteflies it is the production of honeydew, causing stickiness of open bolls allowing the growth of sooty moulds which interfere with the physiology of the leaves and cause discoloration of lint. The overall direct effect of whiteflies, restricting the supplies of carbohydrates reaching the maturing bolls, is to interfere with the lint-wall thickening and consequently with fibre strength and quality. Both of these effects have very significantly reduced the yield and quality of cotton in several areas such as the Sudan (Everleens, 1983).

The shedding of flowerbuds and part-grown bolls is a marked feature of the behaviour of the cotton plant. Flowerbuds can be shed at any stage from a very early 'match-head' to when the flower is opening. Apart from insect injury causing shedding, flowerbuds or young bolls are also shed for physiological reasons. Natural shedding occurs if the plant is stressed, for example if there is a drought during the season, and increases normally at the end of the season with moisture stress, lower temperatures and the approach of senescence. This shedding commences well before cessation of plant growth, as plants balance the leaf area with the number of bolls it can mature. Where there is continued adequate moisture, especially with irrigated cotton, natural shedding can be reduced in comparison with the typical rainfed cotton, which does not have a prolonged season. Conversely waterlogging can also increase natural shedding. The proportion of insect induced shedding can be significantly reduced by crop protection, but the total amount of shedding is a function of the interaction of the plant's physiological condition in relation to its physical environment. It is the time at which shedding occurs that influences the yield.

Studies on the effect of shedding, by artificial removal of buds over different periods and from particular positions on plants, may not provide a valid comparison with what occurs naturally (Dale, 1962). This is due to: (i) removal of a bud that in normal circumstances would not have shed; (ii) loss of nutrients that may have been recovered by the plant by retranslocation from the abscinding organ; (iii) artificial removal may affect the whole mechanism of growth; and (iv) most studies of disbudding occur over a definite period, rather than at a rate which varies in relation to soil moisture and other factors. Nevertheless, there is evidence that early disbudding increases vegetative and root growth (Dale, 1959), and that disbudding and shedding can increase flowerbud production. Thus a small loss of young flowerbuds during the early part of flowering has generally stimulated plant growth and can, under some circumstances increase the yield (Beeden, 1976). However, early shedding can delay the formation of the crop and this needs to be avoided as it is the earliest formed bolls which usually produce the best quality, most mature fibre. Generally studies have confirmed that protection of the plants between the eighth and 16th week after germination is crucial to obtaining the maximum yield, and that further protection later may be necessary if pest infestations persist. After the 20th week, serious crop loss is less likely as the bolls open, although late-season pests, such as the pink bollworm can still cause damage to seeds.

Shedding of the boll, if it occurs, almost invariably occurs within the first two weeks after the flower has opened; bolls that are damaged by pests or disease when they are older than this may abort, but no abscission layer is formed and they do not fall off the plant. Thus any insect induced damage which occurs after these have passed the stage at which they can be shed, represents an irreplaceable loss to the plant. The earlier that damage occurs, the greater the loss. Early infestations of cotton stainers transmitting *Ashbya gossypii* will cause increased damage to young bolls having a high moisture content, in contrast to late infestations feeding on bolls which are nearly mature. Any delay in the formation of the crop due to early insect damage causes the remaining bolls to mature much later when cotton stainer populations may have increased. However, late damage may not cause staining of the lint, and feeding by bollworms may result in only partial destruction of the boll. Any insect damage can introduce other pathogens that will cause boll rot. Care must be taken therefore in estimating losses, as the most serious fraction of the damage consists of bolls that have been rendered unpickable and thus never reach the market.

The main problem in identifying crop losses is to determine what is the actual financial loss, and what are the costs of appropriate control tactics so that an integrated pest management programme can be recommended that will be economically viable to the farmers and all those involved in the cotton industry, including the distributors of agrochemicals, pheromones and other inputs, the ginners and traders.

It must be recognized, however, that the economic relationship between levels of yield and pest management inputs will depend upon the type of farming system involved. Peasant farmers with limited resources and small cotton areas may find a relatively low input/output system of pest management economically preferable to one of high input/output required by large scale cotton growers with a high financial investment. Peasant farmers in particular may decide, as in Malawi, to apply fewer sprays than suggested in the recommendations. Farrington (1977) reported that these lower inputs of insecticide more closely approached the economic optimum for the farmer. It is thus important for research to ensure that the lower inputs are used to the best advantage of the farmer, and this will have implications for the timing and method of application and other aspects of the pest management programme.

References

Beeden, P. (1976) Ecology and Pest Status of *Heliothis armigera* (Hübner) (Lepidoptera, Insecta) in the Northern States of Nigeria. Unpublished PhD Thesis, University of London.

Coaker, T.H. (1957) Studies of crop loss following insect attack on cotton in East Africa. II. Further experiments in Uganda. *Bulletin of Entomological Research* 48, 851–866.

Dale, J.E. (1959) Some effects of the continuous removal of floral buds on the growth of the cotton plant. *Annals of Botany* 23, 636–649.

Dale, J.E. (1962) Fruit shedding and yield in cotton *Empire Cotton Growing Review* 39, 170–175.

Everleens, K.G. (1983) Cotton-insect control in the Sudan Gezira: analysis of a crisis. *Crop Protection* 1, 273–287.

Farrington, J.A. (1977) Research-based recommendations versus farmers' practices: some lessons from cotton spraying in Malawi. *Experimental Agriculture* 13, 9–15.

Goodman, A. (1956) The Tokar investigation. *Empire Cotton Growing Review* 33, 300–306.

Goodman, A. (1957) Insects and the yield of a cotton crop. *Empire Cotton Growing Review* 34, 13–20.

Hosny, M.M., Topper, C.P., Moawad, G.M. and El Soadany, G.B. (1986) Economic damage thresholds of *Spodoptera littoralis* Boisd. on cotton in Egypt. *Agricultural Research Review* 64, 1–11.

McKinlay, K.S and Geering, Q.A. (1957) Studies in crop loss following insect attack on cotton in East Africa. *Bulletin of Entomological Research* 48, 833–850.

Matthews, G.A., Rowell, J.G. and Beeden, P. (1972) Yield and plant development of reduced stands of cotton in Malawi. *Experimental Agriculture* 8, 33–48.

Pearson, E.O. (1958) *Insect Pests of Cotton in Tropical Africa.* Commonwealth Institute of Entomology, London.

Russell, D.A., Radwan, S.M., Irving, N.S., Jones, K.A. and Downham, M.C.A. (1993) Experimental assessment of the impact of defoliation by *Spodoptera littoralis* on the growth and yield of Giza '75 cotton. *Crop Protection* 12, 303–309.

22 Breeding for Insect Resistance

J.R. Gannaway

Texas Agricultural Experiment Station, PO Box 219, Lubbock, Texas 79401–9757, USA

Introduction

Plant breeders are constantly searching for new and improved genetic sources of resistance to insect pests. These new sources have been limited to variability present within the *Gossypium* spp., until the advent and rapidly expanding use of genetic materials through biotechnology. The tremendous expansion of research endeavours utilizing biotechnology has opened what seems to be infinite new avenues for improving crop plants, including the area of insect resistance.

Plant breeding has in the past and will continue in the future to be composed of a system of trade-offs and compromises. One breeder may choose to accept a slight yield reduction in order to incorporate one or several enhancements concerning cotton fibre quality. Another cotton breeder may, out of necessity, be forced to achieve a compromise between earliness of crop maturity and resistance to such diseases as *Verticillium* and/or *Fusarium* wilt. While yet another individual scientist may decide to compromise the degree of pubescence in an effort to achieve a smaller amount of very fine or pin trash in the lint following the ginning process. From these few examples, it becomes readily apparent that combining all of the desired traits into a single cultivar is highly unlikely. However, to make progress in the development of new cultivars, the number of desired traits remaining to be incorporated into our new and improved cultivars is being reduced.

In this account of the development of cotton resistant to various insect pests, there are often conflicting reports concerning various traits. No attempt will be made to explain these differences except to suggest that some of these varied responses may be due to different genetic backgrounds.

Pink Bollworm

One of the earliest reports concerning the damaging effects of pink bollworm, *Pectinophora gossypiella*, on cotton was in 1842 in India (Brown, 1927). This insect (see Chapter 4) occurs throughout the cotton-growing regions of the world and is of major importance in India, Pakistan, Egypt, China, the western United States, the Philippines, and numerous countries in South America and Africa.

One of the most effective methods to circumvent the excessive costs of control is the incorporation of genetically inherited traits that impart resistance to the pink bollworm. Traits that have been used to impart resistance are nectariless, glabrous, okra-leaf, earliness of crop maturity, reduced bract size, and bolls that exhibit a very rough, rugate surface. Resistance among several of the diploid species such as *Gossypium arboreum, G. herbaceum, G. thurberi, G. trilobum, G. armourianum* and *G. somalense* was reported by Brazell and Martin (1956) following their review of numerous reports concerning the pink bollworm. Several authors (Brazell and Martin, 1956; Reed and Adkisson, 1961; Adkisson *et al.*, 1962; Wilson and Wilson, 1975) have found that numerous races of *G. hirsutum* exhibit levels of resistance to the pink bollworm that could be significantly important and useful. Smith (1992) cites work by Wolcott (1927) and Harland (1929) noting that the pink bollworm exhibited a preference for the tetraploid *G. hirsutum* over the diploid *G. arboreum*, as well as several native cottons grown in Haiti. All of these studies utilizing various species of *Gossypium* and the racestocks of *G. hirsutum* suggest that a resistance factor or factors are inherent within these materials. However, it is not believed that any cultivars utilizing these sources of resistance have been developed for commercial production.

In numerous studies, researchers have observed that the adult moths of the pink bollworm and many other adult insects exhibit an egg-laying and feeding preference for cottons that have extrafloral nectaries (Lukefahr and Rhyne, 1960; Schuster *et al.*, 1976; Henneberry *et al.*, 1977; Adjei-Maafo and Wilson, 1983), but there are conflicting reports concerning the influence of the nectariless trait upon resistance to the pink bollworm. In 1960, Lukefahr and Rhyne were unable to determine the influence of the presence or absence of nectaries upon pink bollworm damage in a field trial. Reductions in damage to nectariless cottons have been reported by other investigators. El-Zik and Thaxton (1989) cite numerous works in which the nectariless trait resulted in less damage to the cotton crop. Niles (1980) also reviews the work of several scientists who reported a very positive impact of the nectariless trait. Flint *et al.* (1988) noted a 29% reduction in catches of male pink bollworm moths in a nectariless cultivar when compared with a nectaried cultivar.

The combination of the nectariless trait with other morphological characteristics has proven beneficial. Wilson and Wilson (1975) noted that when the nectariless and glabrous traits were combined, the number of damaging insects and associated damaged seed were reduced; both of these traits exhibit some influence independently, but when the two are combined in a single cultivar, the effect becomes additive.

The okra-leaf trait has been reported by Wilson and George (1982) to exhibit a detrimental effect upon pink bollworm populations. Senft (1986) reported that in two growing seasons, okra-leaf cottons had better yields than their normal-leaf counterparts by 7%. Attempts to determine the origin of the resistance factor associated with the okra-leaf trait were unsuccessful. Efforts to incorporate the resistance factor into commercial cultivars were only partially successful. Some of the experimental cottons that were produced suffered significantly less seed damage than did the parent normal-leaf cultivars; however, none of the experimental lines yielded as much as the normal-leaf parents. Wilson *et al.* (1986) postulated that the resistance factor associated with the okra-leaf trait might be related to increased temperatures observed with the plant structure. Boll and soil temperatures were slightly higher in okra-leaf cotton plots, but further investigation has determined that the temperature theory is an incomplete explanation. Wilson (1986) concluded that while okra-leaf stocks in certain genetic backgrounds exhibit some resistance to this pest and may yield equal to or better than normal-leaf lines, the inconsistent response of some okra-leaf genotypes makes comparison with normal-leaf genotypes over a range of environments necessary.

The combination of nectariless, okra-leaf and semi-smoothleaf traits was utilized to determine their impact upon resistance to the pink bollworm (Wilson, 1989). The semi-smoothleaf trait was included because it reduces the amount of leaf trash in mechanically-harvested cotton. Isolines combining these three traits were compared with control cultivars at three locations. All of the isolines (nectariless alone, nectariless in combination with okra-leaf, nectariless in combination with semi-smoothleaf and nectariless combined with okra-leaf combined with semi-smoothleaf) sustained less seed damage when compared with nectaried, normal-leaf cultivars and most were equal in lint yield and earliness.

Other factors that have been reported to impart resistance to the pink bollworm are earliness, short bracts and rough boll surface. El-Zik and Frisbie (1985) report that earliness of crop fruit-set with associated earliness of crop maturity have increased cotton's resistance to the pink bollworm. Zaman (1986) suggested that earliness in association with short bracts and small, rough-surfaced bolls as exhibited by the cultivar '231 Rosy' could be factors related to resistance to the pink bollworm.

The rapidly expanding world of biotechnology, as it relates to the use of foreign DNA and the development of transgenic cottons with resistance to the pink bollworm, is an area that will be discussed in another section of this chapter.

Boll Weevil

The boll weevil, *Anthonomus grandis*, has very few alternate hosts and feeds primarily on cotton pollen. Because of this near obligate feeding preference for cotton, this insect can be extremely devastating to commercial cotton production unless measures to control this pest are enacted. The author has observed fields that have been virtually destroyed by the boll weevil, and the few bolls that remained in the field were not worth harvesting due to lint discoloration and seed damage. If left unchecked, this insect has the destructive potential to eliminate cotton as a commercial crop in areas where cotton and the boll weevil are adapted.

Howard (1898) reported that the use of early maturing cultivars was an easily adapted and effective method of dealing with this pest. Fruit production in the early portion of the growing season is an escape mechanism still in use today (Fehr, 1987; El-Zik and Thaxton, 1989). Early maturing cultivars produce the majority of their fruit at the end of the first generation of weevils and prior to the onset of the second generation. Producers in areas where boll weevils are a pest strive to take advantage of this 'fruiting window' as an escape mechanism.

Walker *et al.* (1976) found that early maturing, rapid fruiting genotypes would produce bolls and escape damage from the first generation weevil if infestation levels did not exceed 50 overwintering females per hectare. Economic loss would be sustained if 148 or more female weevils per hectare were found. These scientists concluded that the 30 days of blooming prior to build-up of damaging insect levels would provide adequate bolls of sufficient age to escape serious damage.

A morphological character strongly associated with resistance to boll weevil is frego bract (Jones *et al.*, 1964; Jones, 1972). This mutated bract type was first observed in a commercial field of 'Stoneville 2B'. Niles (1980) describes this trait as having relatively narrow, elongated and twisted bracts which flare away from the fruiting form. The strapped-like appearance of this bract is easily identifiable from the normal type bract. Frego bract and the rolled-bract trait (Maredia, 1986) provide resistance to the boll weevil by reducing the desirability of cotton fruit as an oviposition site.

Reduction in weevil oviposition on frego bract cotton is a controversial issue. Reductions in oviposition ranging from 60% to 100% have been reported (Lincoln and Waddle, 1966; Jenkins *et al.*, 1969; Maxwell *et al.*, 1969; Jenkins and Parrot, 1971; Pieters and Bird, 1976; Jones *et al.*, 1987), but Schuster *et al.* (1981) noted no differences in oviposition punctures between frego bract and normal bract cottons. This controversy might be resolved if a feeding puncture could be accurately identified from an oviposition puncture. Smith (1992) discusses this problem.

The primary conclusion that can be drawn concerning the use of frego bract as a factor for boll weevil resistance concerns behaviour modification. Mitchell *et al.* (1973) reported that weevils increased their plant-to-plant movement eight-fold on the frego cotton, spent only 50% as much time at each puncture site on non-frego cotton, and significantly reduced the amount of time spent in flower-buds on cottons with frego bracts. This increased movement throughout the crop also increases the probability that the weevil will come into contact with some insecticide. Buds of frego bract cottons have a seven-fold increase in insecticide when compared with normal bract cottons (Jenkins, 1976).

While the frego bract trait may have some relative advantages concerning resistance to the boll weevil, there are some rather serious problems with increased susceptibility to plant bugs such as the tarnished plant bug, fleahopper and related species. The problems created by the increased susceptibility to these insects may override the positive effects of frego bract on boll weevils.

Incorporation of the okra-leaf trait into cotton has been shown to provide some level of resistance to the boll weevil. Andries *et al.* (1969) reported that cottons exhibiting the okra-leaf trait have approximately 40% less foliage than normal-leaf cottons. This reduction in foliage allows about 70% more light penetration through the crop canopy. Reddy (1974) found that this increased the temperature within the crop canopy while reducing the relative humidity. These two factors resulted in increased desiccation of flowerbuds and insects.

Numerous authors (Stephens, 1957; Wannamaker, 1957; Wessling, 1958; Stephens and Lee, 1961; Hunter, *et al.*, 1965; and El-Zik and Thaxton, 1989) have reported the positive influence that increased pubescence, primarily pilose, has on reducing boll weevil damage. The source of resistance is mechanical. The pilose line produces pubescence on the inside and outside of the bract. Almost all, if not all, highly pubescent cottons do not produce the pubescense on the inner side of the bract. The pilose trait allows for a 'velcro effect' as the bracts tend to stick together and protect the flowerbud. As the buds grow in size, this form of protection is diminished.

Several disadvantages are associated with the pilose and other highly pubescent traits. The bollworm complex is attracted to highly pubescent lines. Secondly, increased pubescence results in increased leaf and bract trash content in the lint which is highly undesirable to the textile industry.

Other traits that have been studied in search of resistance to the boll weevil include red plant colour, reduced androecium, increased gossypol level and oviposition suppression factors. With the exception of reduced androecium and oviposition suppression factors, these traits all tended to present mixed results (Smith, 1992)

Only recently a number of these primitive racestocks of *Gossypium hirsutum* L. have been tested for resistance. Maxwell *et al.* (1969) described an oviposition suppression factor which was found in Sea Island 'Seaberry' (*G. barbadense*). A similar suppression factor has been found in several of the racestocks of *G. hirsutum* with reductions in oviposition of up to 84% being observed (Jenkins, 1976). The United States Department of Agriculture currently has an ongoing programme to convert these photoperiodic racestocks into day-neutral types, with the goal of identifying numerous insect-resisting and other traits that will be of significant importance to the cotton industry. Programmes of this type, in concert with all of the other cotton breeding programmes throughout the world will provide cotton producers with constantly improved cultivars that will be less susceptible, and possibly immune, to insect pests.

Bollworms and Budworms

Numerous traits and factors have been noted as imparting varying levels of resistance to bollworms and budworms. The glabrous character has received extensive attention as a source of resistance to bollworms and budworms. Surfaces covered with pubescence are preferred oviposition sites for the bollworm (Smith,1992). Gillham (1963) showed that the bollworm preferred the pubescent cottons when compared with glabrous cottons, the glabrosity disrupting the egg-laying habits of bollworms (Fehr, 1987). El-Zik and Thaxton (1989) also reported that cottons exhibiting a reduced number of trichomes incurred much less egg-laying activity and subsequent damage from the bollworm–budworm complex. Similarly Ha (1987) found that the smoothleaf trait suppressed oviposition by *H. zea*. The inclusion of the glabrous factor in a breeding programme striving to achieve resistance to budworms and bollworms has been experienced by most cotton breeders involved in developing materials where these insects are a problem.

Breeders have also found that resistance can be enhanced by combining several other insect-resisting traits – nectariless, earliness of crop maturity, reduced branching, yellow and/or orange pollen, red leaf colour and numerous allelochemicals. Tang (1987) noted that bollworms did less damage to nectariless cotton, which exhibited a 5.6% higher boll retention with yield increases of 16.7–28.2% when compared to nectaried cotton. Jones *et al.* (1989) developed three strains of cotton which all possessed the frego bract and glabrous traits and exhibited good field resistance to the bollworm–budworm complex. One of these three strains was nectariless. The nectariless line had exceptionally high yield stability across environments with an associated lack of yield reduction in the presence of the bollworm–budworm complex. Alimukhamedov and Shvetsova (1988) noted that cottons resistant to the American bollworm (*H. armigera*) morphologically produced few hairs, underdeveloped nectaries and a plant type with limited branching. In both cage and field studies, Luo *et al.* (1988) utilized 100 cotton accessions characterized by three morphological traits (nectariless and glabrous, nectariless and pubescent, and nectaried with pubescence) in studies to determine their impact upon *H. armigera*. Lines possessing the nectariless and glabrous traits reduced the egg-mass of *H. armigera* by 52.5–78.7% in cage tests and by 45.9% in the field. The nectariless and pubescent materials decreased the egg-mass by 33.4% in field tests. In 1966, Lukefahr *et al.* reported that the nectariless–glabrous traits significantly reduced bollworms when compared with nectaried–pubescent materials. A tenfold reduction in trichomes resulted in a 60% reduction in egg population. Lukefahr (1977) postulated that reducing trichomes to less than 200 per square inch of leaf surface should reduce egg deposition and larval population by 50%. Numerous other authors have reported the positive effects of the nectariless trait with respect to the bollworms and budworms that attack cotton (Lukefahr *et al.*, 1965; Davis *et al.*, 1973; Laster and Meredith, 1974a; Schuster and Maxwell, 1974; Maxwell *et al.*, 1976). Reports that nectariless cotton has no effect on bollworms and budworms also exist (Lukefahr, 1977; Sosa *et al.*, 1981; Ha *et al.*, 1987). Speculation still exists among entomologists, agronomists and breeders concerning the effectiveness of the nectariless trait to impart resistance to bollworms and budworms. The ability of a single trait such as nectariless to impart resistance to an insect pest can be strongly influenced by the genetic background into which this trait is inserted, in addition to the plethora of other factors that impact the relationship between the host plant and the pest of concern.

El-Zik and Thaxton (1989) report that earliness of crop maturity and time of fruit set affect resistance to bollworms and budworms. In many areas of the cotton-producing regions of the world, the utilization of the escape mechanism offered by of 'short- season' cultivars is a cost-effective method for coping with late season pests (El-Zik and Frisbie, 1985). Jenkins *et al.* (1986) tested 13 cultivars and lines of Upland cotton for three years with insect pressure provided by *H. virescens*. Regression analyses indicated that approximately 65% of the resistance to *H. virescens* is associated with earliness and rapid fruiting. The authors also stated that additional unidentified factors were also involved.

The presence of a water-soluble substance, contained on the growth terminal of an Upland cotton line resistant to the cotton bollworm and tobacco budworm provided another avenue to explore in terms of breeding for resistance factors. In 1987, Bhardwaj *et al.* reported on a study in which the terminals of 'PD 695' and 'Coker 304' were excised and used in larval feeding studies. The larval diets were prepared using the washed and unwashed terminals of the 'PD 695' (resistant) and 'Coker 304' (susceptible) plants. Larval growth was found to be similar on diets prepared from 'Coker 304' terminals; but, significant reductions in larval growth were observed on diets prepared from unwashed terminals of 'PD 695' when compared with washed terminals. The authors suggest that the terminals of 'PD 695' produce a water-soluble substance which confers resistance to *H. zea* and *H. virescens*.

Smith (1992) presents an excellent review of the work that has been reported concerning the use of allelochemicals as sources of resistance to the bollworm–budworm complex. He also notes that due to the complex chemistry involved, the expense of chemical analyses and, in this author's personal opinion, the uncertainty of which allelochemicals will be most effective and the ease with which allelochemicals can be genetically manipulated, that no major cotton breeding programme in the United States is devoting a concentrated effort to breeding for specific allelochemicals found in the lysigenous glands of cotton. Stipanovich *et al.* (1988) reported the hemigossypolone and the heliocides H_1, H_2, H_3 and H_4 are all associated with resistance to *Heliothis* spp. and *Helicoverpa* spp. They report that these compounds are the major terpenoid aldehydes in leaves, with gossypol being the major terpenoid aldehyde in flowerbuds and essentially the only one in seed.

Kuznetsova (1986), in a laboratory feeding study utilizing fourth-instar larvae placed on bolls from three cultivars with high levels of gossypol and gossypol-type substances, three with low levels plus one that was unknown, observed very different feeding patterns. The cultivars with high levels of gossypol and gossypol- type substances were all characterized by the predominance of a behavioural pattern in which the larvae began feeding quickly, penetrating the boll and feeding without much further significant change in position, a form of feeding that is known to be linked with low nutrient values and the inhibitory effect of such compounds as gossypol and gossypol-related substances. On the other cultivars, the varied feeding patterns were relatively evenly represented. Kuznetsova concluded that the predominance of a particular feeding pattern represents an adaptive reaction by the larvae. Gossypol and related substances strongly influence this behavioural pattern by the larvae; however, one cannot conclusively state that other chemicals are not involved.

Prior to adopting the philosophy that allelochemicals, including gossypol and related compounds, will impart the resistance essential to combat such insects as the bollworm and budworm, one should consider the work of Montandon *et al.* (1986). The effects of glandular chemicals, gossypol being the primary one, on the survival and following development of the cotton leafworm,

Alabama argillacea, which is a specialist feeder, and *H. virescens*, a polyphagous species, were studied when fed the cotyledonary leaves of a glanded and glandless cotton. Individuals of *A. argillacea* survived equally well on either type; however, the weight of older larvae was significantly greater when fed on glanded leaves. The survival and pupal weight were reduced in *H. virescens* when fed on glanded leaves. The difference in response by these insects suggested that changes made in plant chemistry to enhance the resistance of plants against generalist feeders may not lessen, and could possibly increase, the impact of adapted specialist feeders on cotton. This research points out the importance of knowing and understanding the organisms with which we work. An incomplete understanding of the pest and the host could result in personnel involved with crop improvement creating a problem that could be as detrimental as the problem they are attempting to rectify.

The influence of pubescence and nectaries has been discussed as it relates to resistance to bollworms and budworms. However, this author feels that a brief discussion of how these two traits impact the effectiveness of beneficial predators is important. Treacy *et al.* (1987) conducted several studies in an attempt to determine the effect of trichome density on the predation of *H. zea* eggs by larvae of *Chrysopa rufilabris*. A no-choice greenhouse test showed that fewer eggs were destroyed on a heavily pubescent (pilose) cotton than on smoothleaf and hirsute cottons. Egg predation was greater on the smoothleaf cotton than on the hirsute cotton. Laboratory studies revealed that the leaf trichomes inhibited movement of *C. rufilabris* larvae over leaf surfaces. It is concluded that cotton leaf trichomes serve as mechanical barriers which reduce the mobility and the predatory effectiveness of *C. rufilabris*. In another study, Treacy (1986) found that extrafloral nectar was essential to the survival of *Trichogramma pretiosum* adults. This accounted for the greater amount of parasitism on *H. zea* eggs on nectaried than on nectariless plants. He also observed that *Chrysopa rufilabris* larvae destroyed equal numbers of eggs on nectaried and nectariless cottons.

As we continue to look at factors that impart resistance to various cotton insect pests, we will become acutely aware of the fact that the development of insect-resistant cotton is comprised of a system of trade-offs and compromise.

Thrips

Factors reported to impact resistance to thrips, *Frankliniella* spp. and *Thrips* spp. include plant pubescence, thick leaf epidermis and chemical content of leaves as it affects reproduction fitness. The pilose trait, controlled by the H_2 gene, is reported to confer resistance to thrips (El-Zik and Thaxton, 1989). Quisenberry and Rummel (1979) determined the degree of resistance to be expressed by differential reductions in leaf area. They report that genotypes with pubescent leaves may have some resistance to thrips pressure. Dense pubescence on very young leaves was reported as a trait imparting resistance to thrips by Ballard (1951). However, he also suggested that pubescence was not the only factor involved. No correlation was found between thrips resistance and pubescence by Egyptian scientists (Gawaad and Soliman, 1972). It was later reported by Gawaad *et al.* (1973) that a thick epidermis on the underside of the leaf was associated with resistance to thrips.

Trichilo and Leigh (1988) reported the reproductive fitness of thrips to be much lower when they fed on leaves of a cotton genotype resistant to spider mites (*Tetranychus urticae*), compared with leaf feeding on a susceptible genotype. The addition of pollen or mite eggs to the leaf diet significantly improved the reproductive fitness of the thrips. Mite eggs were not as nutritionally beneficial to the thrips as pollen. The addition of pollen to the leaf diet reduced the developmental time period, increased fecundity and resulted in a longer lifespan compared with leaf tissue alone. The effect of the pollen was much more pronounced on thrips feeding on leaves resistant to spider mites than on those that were susceptible. If pollen was made available, the thrips could completely overcome the negative effects associated with the resistant leaves.

Spider Mites

Spider mites are reported to be a problem requiring chemical control in 20 countries throughout the world (Ridgway, 1984). Simongulyan and Dzhunedzha (1988) conducted a study utilizing eight *G. hirsutum* cultivars and their F_1 hybrids mated in a diallele system. The traits measured were total leaf thickness and the thickness of the barrier layer. Only thickness of the barrier layer was found to impart resistance to spider mites (*Tetranychus urticae*). The two traits studied were found to be under polygenic control and could be used in a breeding programme to impart resistance. Alimukhamedov and Shvetsova (1988) evaluated 150 cultivars and breeding lines and found 14 that offered resistance to this pest. The resistant genotypes had a high content of protein, phenols and nucleic acids with a low content of carbohydrates, oil and thiols. Sugars predominated among the carbohydrates. Feeding on the resistant genotypes reduced the pests' fecundity and progeny weight.

A very comprehensive study seeking resistance to spider mites has been conducted by Schuster *et al.* (1972, 1973) and Schuster and Maxwell (1976). Numerous genotypes of *G. hirsutum* and *G. barbadense* were evaluated. No resistance was observed in any of the *G. hirsutum* lines; however, most of the *G. barbadense* materials evaluated imparted resistance.

The mechanism of resistance to this pest is still undefined. However, the studies cited here would suggest that they are both chemical and physical in nature.

Aphids

The aphid, *Aphis gossypii*, can cause severe damage to cotton (see Chapter 12), but often, producers will choose not to apply insecticides to remove these insects because they are considered to be a minor pest. Aphids have most often been ignored by plant breeders in host-plant resistance programmes. However, these attitudes can be changed, for example in the 1991 cropping season in the Texas

Southern High Plains production region, aphids became a very serious problem in the latter portion of the growing season, as they had become relatively resistant to several insecticides. There is now a renewed interest in host-plant resistance to this pest.

There is a paucity of information concerning research in this area. Traits reported to impact upon this pest are plant colour, pubescence, gossypol content, chemical make-up of the plant and stem-tip stiffness. Red plant colour is reported to confer resistance to aphids (El-Zik and Thaxton, 1989). Increased levels of leaf pubescence are reported to result in increased populations of aphids (Dunnam and Clark, 1939; Pollard and Saunders, 1956). However, Gawaad and Soliman (1972) showed no relationship between plant pubescence and aphid resistance. Gossypol is reported to be highly toxic to aphids (Bottger *et al.*, 1964). Alimukhamedov and Shvetsova (1988) and Sharipova (1987) report results indicating the importance of plant chemistry. Genotypes exhibiting resistance to aphids had a high content of protein, phenols and nucleic acids and a low content of carbohydrates, oil and thiols (Alimukhamedov and Shvetsova, 1988). Sharipova (1987), after screening several genotypes of *G. hirsutum* and *G. barbadense* for resistance, suggested that rate of aphid reproduction is the best indicator of resistance. This study implicated plant chemistry as impacting resistance.

The most interesting find concerning aphid resistance was by Kadapa *et al.* (1988). The stiffness of stem tips of ten cultivars ranging in aphid susceptibility was tested by determining the force required for a needle to penetrate the first two mainstem nodes. Susceptible cultivars required very little force for the needle to penetrate up to 1 mm. Stem tips of the tolerant to resistant strains required almost twice as much force to penetrate as susceptible lines. These more resistant stocks required forces from 50 to more than 100 g of force. At deeper levels of 2 mm or more, the tolerant genotypes were more than twice as hard to penetrate as the susceptible genotypes. These results indicate that the force required to insert the proboscis into the stems of tolerant genotypes is one of the main causes for aphid non-preference.

Plant Bugs

Plant bugs, *Lygus* spp., are a pest on crops other than cotton and have a relatively wide range of non-cultivated hosts. Where plant bugs are a problem in cotton production areas, producers can observe significant yield losses and delayed crop maturity.

The two species of plant bugs of most concern to cotton producers are *Lygus lineolaris*, tarnished plant bug, and *Lygus hesperus*. The nectariless trait has been reported to provide some resistance to these pests (El-Zik and Thaxton, 1989). Scott *et al.* (1988) assessed the effect of the nectariless trait on a field scale level in 1981 and 1982. The fields ranged in size from 16 to 60 hectares. The combined adult and nymphal populations were reduced 44% and 34% in the nectariless cotton, compared with its nectaried counterpart in 1981 and 1982, respectively. However, one of the major drawbacks noted in this study was the concomitant reduction in predatory insects. *Geocoris* spp., *Nabus* spp., *Orius* spp. and *Oxyopes* spp. were present in significantly reduced numbers in both years of this study.

The nectariless trait has been cited by numerous other researchers as reducing *Lygus* spp. populations (Schuster and Maxwell, 1974; Meredith and Laster, 1975; Schuster *et al.*, 1975; Benedict *et al.*, 1976, 1981; Schuster and Frazier, 1976; Lukefahr, 1977). Population reductions were reported as high as 60% for nymphs and adults, with egg lay reductions up to 90% for the nectariless cottons when compared with nectaried genotypes.

The glabrous trait has resulted in reduced numbers of the tarnished plant bug in several studies (Laster and Meredith, 1974b; Jenkins *et al.*, 1977; Meredith and Schuster, 1979). However, as was reported for the fleahopper, genotypes with reduced levels of pubescence actually sustain more damage than the pubescent types. Extremely high levels of pubescence as conferred by the H_2 gene have been reported to give increased levels of resistance to both species of *Lygus* (El-Zik and Thaxton, 1989). However, Tingey *et al.* (1975) noted that *L. hesperus* exhibited a preference for the pilose types compared to genotypes with normal levels of pubescence.

The problems encountered in the spinning mill when trying to use highly pubescent cottons spurred renewed interest in the effort to develop resistance to *Lygus* spp. in glabrous cottons. In the MAR programme at Texas A & M University, researchers (Bird *et al.*, 1983) report that they have broken the genetic linkage between increased sensitivity to *Lygus* spp. and the glabrous trait.

Frego bract and the rolled bract trait have been evaluated for resistance to *Lygus lineolaris* and *Lygus hesperus*. In certain genetic backgrounds, frego bract is associated with increased susceptibility to both *Lygus* spp. (El-Zik and Thaxton, 1989). Maredia (1986) reported that the rolled bract trait was much more sensitive to the tarnished plant bug than cotton genotypes with normal bracts.

Genotypes with increased levels of bud gossypol have been reported to adversely affect the development and behaviour of *Lygus* spp. (Tingey *et al.*, 1975; Schuster and Frazier, 1976). Tingey *et al.* (1975) postulated that since they observed significant reductions in growth rate of the insects on several genotypes, that a plant bug suppression factor was involved. Genotypes identified as exhibiting this suppression factor were four *G. barbadense* stocks and several *G. hirsutum* stocks, with cytoplasm from *G. longicalyx* and *G. harknessii*. As with almost any trait, conflicting reports exist. Meredith *et al.* (1979) reported no resistance to the tarnished plant bug when utilizing genotypes with cytoplasms from *G. barbadense, G. anomalum, G. arboreum, G. herbaceum* and *G. tomentosum*.

Milam *et al.* (1989), in a study to determine a method for selecting for resistant types and/or eliminating the susceptible lines, found that by growing segregating populations under high insect pressure provided the best method to identify the desired genotypes. As resistant materials are identified and further evaluated, Kitten *et al.* (1987) found that one effective method of determining resistance was to compare these genotypes in the presence and absence of insects. They expressed the yield of each genotype with insect pressure as a percentage of the yield of that genotype with the insects controlled. These percentages could also be compared with those of commercial cultivars for evaluation purposes.

Several traits have been discussed which, dependent upon genetic background and other circumstances, have offered varying levels of resistance. It becomes readily apparent that more research is needed to determine the impact of genetic background and the related effect of the various cotton cytoplasms.

Fleahoppers

The cotton fleahopper, *Pseudatomoscelis seriatus*, typically attacks cotton early in the growing season. The primary feeding site for this insect is very young flowerbuds. However, as this insect attempts to feed on these flowerbuds in the terminal growing area of the plant, they often damage or destroy the apical meristem resulting in the loss of apical dominance. When this type damage is incurred, numerous adventitious buds located on the main stem will initiate growth in an attempt to re-establish apical dominance. This results in an excessive amount of vegetative growth and is often referred to as 'crazy top' (Smith, 1992).

Niles (1980) cites research reports stating that fleahopper numbers were significantly reduced on cottons that were glabrous. This evidence appeared promising until it was observed that even though the insect numbers were reduced, damage was being incurred in varying degrees on glabrous cottons. It soon became apparent that glabrous cottons were more sensitive to fleahopper damage than more pubescent types (Walker and Niles, 1973; Niles *et al.*, 1974; Walker *et al.*, 1974; Schuster *et al.*, 1976). However, Bird *et al.* (1983) report that through the efforts in the multiadversity resistance (MAR) breeding programme, the linkage between increased sensitivity and glabrousness has been broken. The pilose trait, controlled by the H_2 gene, is reported to confer some level of resistance (El-Zik and Thaxton, 1989). This trait is also reported to impede the movement of fleahopper nymphs and possibly affect oviposition (Meredith and Schuster, 1979; Lidell *et al.*, 1986).

Nectariless cottons have been shown to exhibit varying levels of resistance to the fleahopper (El-Zik and Thaxton, 1989). Reports of significant reductions in fleahopper population have been cited by numerous authors (Laster and Meredith, 1974a; Meredith, 1976; Schuster *et al.*, 1976; Lidell *et al.*, 1986).

Ha (1987) studied the impact of five morphological traits (frego bract, nectariless, glandless, okra-leaf and smoothleaf) upon fleahopper populations. Comparisons of these traits in eight genetic backgrounds were made with normal controls. Fleahopper population was reduced 54% utilizing the smoothleaf trait in tests averaged over three years, two sites, genetic backgrounds and multiple sampling dates. Frego bract increased fleahopper numbers by 64%. El-Zik and Thaxton (1989) corroborate the increased susceptibility to fleahoppers in frego bract lines and also report that genetic background influences the degree of susceptibility. However, Ha (1987) reports that interactions of morphological traits with genetic backgrounds were not significant for fleahopper response.

In a study performed in Texas in 1983–1984, the nectariless trait was evaluated for providing resistance to the fleahopper (Lidell *et al.*, 1986). Nine experimental Upland cottons and three commercial cultivars were included in this study. Of the twelve genotypes studied, eight were nectariless, one was glabrous, one was pilose and two were hirsute with nectaries. These materials were evaluated under both high and low fleahopper population densities. At the low level of infestation, only the susceptible control suffered significant damage. Differences were much more varied at the high level of infestation. Resistance was observed in two genotypes, pilose and a hirsute cultivar with nectaries. Large and significant yield and earliness losses were observed in the nectariless and glabrous genotypes.

The single most important conclusion that can be drawn from this discussion of resistance to the fleahopper concerns genetic background. It is this author's opinion that plant breeders, all too often tend not to give as much credence to genetic background as we should. Most of the differences associated with differing genetic backgrounds are not phenotypically evident and therefore very easy to overlook or explain as an anomaly. These anomalies are often the materials being sought to identify and incorporate into cottons developed for commercial production.

Whiteflies

Two species of whiteflies *Bemisia tabaci* and *Trialeurodes abutilonea* are reported to attack cotton (see Chapter 13). The traits most commonly discussed concerning resistance to these pests are leaf shape and pubescence.

Leaf glabrosity, okra-leaf shape and open canopy are reported to be associated with resistance to the cotton whitefly, *Bemisia tabaci* (Ozgur *et al.*, 1988). Sippell *et al.* (1987) studied the effects of leaf shape and leaf hair density upon cotton whitefly population. These researchers used genotypes with leaf hair densities classed as low, medium and high, and combined those with three different leaf shapes (normal, okra and super okra). Resistance was observed in genotypes with low leaf hair density and two leaf shapes, okra and super okra. Cotton whitefly resistance was expressed as few numbers of adults per leaf and associated reductions in adults, scales and pupae per unit area. Honeydew, one of the causes of lint stickiness, was reduced to acceptable levels in resistant genotypes. Microclimatic conditions, primarily reductions in relative humidity and increased temperature, resulted in a more adverse situation for this pest. Similar results were reported by Ozgur and Sekeroglu (1986). The resistant genotypes they identified were glabrous, possessed smaller okra-leaf shape, exhibited an open canopy and were taller than the other genotypes. Butler *et al.* (1986) noted inconsistencies with respect to leaf shape resistance to cotton whiteflies. They also reported that a normal, hairy leaf cultivar had more adults and eggs than the semi-smooth and smoothleaf isolines. These researchers also removed the hair from one half of each of several leaves of this hairy leaf cultivar using an electric razor. One very interesting fact was that the unshaved half of these leaves contained 2.4 times as many adults and four times as many eggs as the half that had been shaved.

Studies concerning resistance of cotton to *Trialeurodes abuitilonea*, the bandedwing whitefly, suggest that leaf shape modification gives some increased resistance. Jenkins (1981) and Niles (1980) citing Lukefahr (1977) report that high levels of gossypol result in increases in population levels of thrips and the bandedwing whitefly.

Internal chemical characteristics of a cotton leaf are reported to affect feeding preference of *B. tabaci* (Hussein *et al.*, 1936). Berlinger *et al.* (1983) reported that this insect would discriminate between leaves with different pH values at the 0.25 level. The researchers noted that preference increased up to pH values of 7.25. The pH content of leaves of the vast amount of cotton germplasm has not been screened for leaf pH value. This could be a very valuable tool to be used in a host-plant resistance breeding programme that to date remains untapped.

Jassids

Jassids have been reported as a pest of numerous crops including cotton (see Chapter 14). Crops susceptible to jassids can be completely devastated by this insect. The development of cottons resistant to the jassid is a classic example, and one of the earliest successes, concerning host-plant resistance. Painter (1951) provides an excellent discussion concerning the development of jassid-resistant cottons.

Most researchers are in agreement that plant pubescence is a major source of resistance (Parnell *et al.*, 1949). However, with the detrimental effects associated with high levels of pubescence, researchers are continuing to look for other sources of resistance. Taneja *et al.* (1988) selected healthy and damaged leaves from seven cultivars and two hybrids that possessed varying degrees of resistance. These cottons were grown under identical conditions. The healthy and damaged leaves were analysed separately. Data revealed that susceptible cottons damaged by jassids exhibited decreased levels of protein, phosphorus, sulfate and potassium with concurrent increases in tannin, total phenols and silica. Undamaged leaves of resistant genotypes exhibited higher levels of total phenols, tannin and silica than those of susceptible cottons. These authors concluded that these three materials (total phenols, tannin and silica) are actively involved in the tolerance of cotton to jassids. Somewhat similar results were noted by Singh and Agarwal (1988). They reported that levels of nonreducing sugars, tannins, silica and free gossypol in the leaves were significantly negatively correlated with oviposition preference.

The preferred form of resistance to the jassid would be chemical in nature, especially if more than one compound is involved. If more than one chemical is involved, the longevity of the resistance will be increased tremendously. This would also enable the cotton industry to move away from the use of highly pubescent cultivars which tend to create problems for the cotton textile industry.

Biotechnology

The science of biotechnology is rapidly emerging and changing with unprecedented frequency. This science is providing means of incorporating genes from one species into another totally unrelated species. To date, the successes in transferring genes imparting resistance to pests have not been as rapidly forthcoming as initially proposed. However, progress is being made and the potential remains excellent.

One of the most discussed successes concerns the incorporation of *Bacillus thuringiensis* bacterium into cotton. There is no doubt that the inclusion of this bacterium provides excellent resistance to the Lepidopteran pests of cotton. However, the point of major concern to everyone involved concerns the development of resistance to this bacterium by the target species.

Will this resistance occur and, if so, when? If the source of pest control introduced into cotton utilizing the techniques afforded by biotechnology is controlled by a single gene, then it is this author's opinion that resistance will be forthcoming in a matter of only a few generations of the target pest.

One method of overcoming this rapid development of resistance would be to incorporate a minimum of two factors that are lethal to the insect pests. This technique would require the insect to mutate at more than one locus in order to be resistant to the introduced toxic factors. Since all of the toxic factors introduced into cotton would be lethal to the target species, an insect would be required to mutate at multiple sites simultaneously. The probability of this event occurring is much less than the probability of the insect mutating at a single site.

Another method frequently discussed concerning avoidance of resistance concerns the utilization of cultivars and hybrids composed of blends. These blends would be comprised of varying proportions of plants with and without the resistance trait(s). This technique would put much less pressure upon the target insect species to develop resistance in order to survive. The presence of plants in the population not possessing a toxin lethal to the targeted insect species would allow the wild type to exist. The maintenance of this wild type could possibly prolong the use of cultivars possessing newly introduced resistant factor(s).

These are only two possible methods that could be utilized to prolong the usefulness of traits imparting insect resistance. Human ingenuity will continually be postulating and testing new and innovative theories to avoid and/or overcome insect resistance. Cooperative research efforts between conventional plant breeders and biotechnologists will continue to provide cotton cultivars and hybrids that will allow the cotton production industry to flourish.

Summary

The majority of this chapter has discussed insect resisting traits in cotton as related to the primary pests of cotton (see Table 22.1). Berlinger (1986) points out that the mechanism of resistance to various pests in the same crop is oftentimes contradictory. It can be shown that glabrous cotton imparts resistance to pink bollworms, the *Heliothis* complex and whiteflies, while conversely, cotton containing increased levels of gossypol was resistant to the pink bollworm and the *Heliothis* complex. However, these cottons are shown to be very susceptible to attack from the whitefly. These situations concerning resistance and susceptibil-

Table 22.1. Host plant resistance traits in cotton.

	Pink bollworm	Bollworms and budworms	Boll weevil	Plant bugs	Jassids	Fleahoppers	Thrips	Spider mites	Aphids	Whiteflies
Nectariless	R(?)	R(?)		R		R				
Glabrous	R	R	N	S(?)	S	(?)	S	N	S	R
Hirsute		S	R	R	R	R	R		(?)	S
Pilose		S	R	R(?)	R	R	R	N	(?)	S
Okra-leaf	R(?)	N	R	N	N	N	N	N	(?)	R(?)
Fegro bract			R(?)	S		S				
Reduced bract-size	R		R(?)	S		S				
Red plant colour		R	(?)						R	
High gossypol	R	R	(?)	R	R	N	S	N	R	S
Tanninis		R			R		R			
Heliocides		R			R		R			
Oviposition suppression factor			(?)	R						
Rugate boll	R									
Early crop maturity	E	E	E							
Reduced branching		R								
Pollen colour (yellow & orange)		R								
Other allelo-chemicals		R			R		R	R	R	R
Thick leaf epidermis							R	R		
Stiffness of stem tips									R	

R = resistance; S = susceptible; E = escape; N = no effect; (?) = conflicting evidence or not verified.

ity in concert with the present and expected insect pressures, changing production practices, improved cropping systems, scientific advancements in the fields of genetics, both basic and applied, and biotechnology will provide plant breeders with unlimited opportunities and challenges to develop insect resisting cotton.

In striving to incorporate insect resisting traits into cotton for use in commercial production, several questions must be asked prior to this endeavour. Some of them are:

- What is (are) the insect(s) that is (are) causing economic losses?
- Do sources of resistance exist?
- What is the mode of inheritance of each insect resisting trait?
- What breeding methodology should be utilized to incorporate most effectively and efficiently the trait of concern?
- How will the incorporation of a given resistance factor impact other insect pests?
- Will the incorporation of a given insect resisting trait detract from the marketability and spinnability of cotton lint?
- Will incorporation of insect resistance into a cotton cultivar enhance the environmental soundness and production sustainability for the production area of concern?

As each plant breeder determines the answer to these and other pertinent questions, then a strategy can be developed to achieve the goal. An attempt to answer each of these questions for each scientist involved in cotton breeding throughout the world would be futile. However, an excellent review concerning the third and fourth questions is presented by Niles (1980). In this discussion on breeding techniques, the primary methods utilized will be those suited to a normally self-pollinating species with modifications based upon the mode of inheritance of the trait in question. Variations of standard breeding procedures will be utilized based upon the genetic material being used and the innovativeness of the plant breeder.

References

Adjei-Maafo, I.K. and Wilson, L.T. (1983) Factors affecting the relative abundance of arthropods on nectaried and nectariless cotton. *Environmental Entomology* 12, 349–352.

Adkisson, P.L., Bailey, C.F. and Niles, G.A. (1962) *Cotton Stocks Screened for Resistance to the Pink Bollworm, 1960–61.* Texas Agricultural Experiment Station Miscellaneous Publication 606.

Alimukhamedov, S. and Shvetsova, L. (1988) Immunity of cotton to pests. *Khlopok* 1988(4), 26.

Andries, J.A., Jones, J.E., Sloane, L.W. and Marshal, J.G. (1969) Effects of okra leaf shape on boll rot, yield and other important characters of Upland cotton, *Gossypium hirsutum* L. *Crop Science* 9, 705–710.

Ballard, W.W. (1951) Varietal differences in susceptibility to thrips injury in Upland cotton. *Agronomy Journal* 43, 34–44.

Benedict, J.H., Leigh, T.F. and Hyer, A. (1976) Host plant resistance in cotton to lygus bugs and other insect pests. *Proceedings of the Western Cotton Production Conference 1976*, 80–83.

Benedict, J.H., Leigh, T.F., Hyer A.H. and Wynholds, P.F. (1981) Nectariless cotton: Effect on growth, survival, and fecundity of lygus bugs. *Crop Science* 21, 28–30.

Berlinger, M.J. (1986) Host plant resistance to *Bemisia tabaci. Agriculture, Ecosystems and Environment* 17, 69–82.

Berlinger, M.J., Magaland, Z. and Benzioni, A. (1983) The importance of pH in food selection by the tobacco whitefly, *Bemisia tabaci. Phytoparasitica* 11, 151–160.

Bhardwaj, H.L., Weaver, J.B., Jr. and Severson, R.F. (1987) Presence of water-soluble materials on cotton terminals as related to bollworm (Lepidoptera: Noctuidae) resistance. *Journal of Agricultural Science* 109, 193–195.

Bird, L.S., El-Zik, K.M., Percy, R.G, Thaxton, P., Bendict, J.H., Reyes, L., Creelman, R.A., Clark, L.E., Heald, C.M. and Kappelman, A.J., Jr. (1983) Improved Multi-adversity Resistant (MAR) Cottons from MAR-4 Hybrid pool. *Texas Agricultural Experiment Station Progress Report* 4128, 11pp.

Bottger, G.T., Sheehan, E.T. and Lukefahr, M.J. (1964) Relation of gossypol content of cotton plants to insect resistance. *Journal of Economic Entomology* 57, 283–285.

Brazzel, J.R. and Martin, D.F. (1956) Resistance of cotton to pink bollworm damage. *Texas Agricultural Experiment Station Bulletin* 843.

Brown, H.B. (1927) *Cotton*. McGraw-Hill, New York.

Butler, G.D., Jr., Henneberry, T.J. and Wilson, F.D. (1986) *Bemisia tabaci* (Homoptera: Aleurodidae) on cotton: adult activity and cultivar oviposition preference. *Journal of Economic Entomology* 79, 350–354.

Davis, D.D., Ellington, J.V. and Brown, J.C. (1973) Mortality factors affecting cotton insects: 1. Resistance of smooth and nectariless characters in Acala cotton to *Heliothis zea*, *Pectinophora gossypiella*, and *Trichoplusia ni*. *Journal of Environmental Quality* 2, 530–535.

Dunnam, E.W. and Clark, J.C. (1939) The cotton aphid in relation to the pilosity of cotton leaves. *Journal of Economic Entomology* 31, 633–666.

El-Zik, K.M. and Frisbie, R.E. (1985) Integrated crop management systems for pest control and plant protection. In: Mandava, N.B. (ed.) *CRC Handbook of Natural Pesticides: Methods. Vol. I. Theory, Practice, and Detection*. CRC Press, Boca Raton, Florida, pp. 21–122.

El-Zik, K.M. and Thaxton, P.M. (1989). Genetic improvement for resistance to pests and stresses in cotton. In: Frisbie, R.E., El-Zik, Kamal M. and Wilson, T.L. (eds) *Integrated Pest Management Systems and Cotton Production*. John Wiley & Sons, New York, pp. 191–224.

Fehr, W.R. (1987) *Principles of Cultivar Development*. Macmillan, New York.

Flint, H.M., Curtice, N.J. and Wilson, F.D. (1988) Development of pink bollworm populations (Lepidoptera: Gelechiidae) on nectaried and nectariless Deltapine cotton in field cages. *Journal of Environmental Entomology* 17, 306–308.

Gawaad, A.A.A. and Soliman, A.D. (1972) Studies on *Thrips tabaci* Lindman. IX. Resistance of nineteen varieties of cotton to *Thrips tabaci* L. and *Aphis gossypii* G. *Zeitschrift für Angewändte Entomologik* 70, 93–98.

Gawaad, A.A.A., El-Gayer, A.S., Soliman, A.S. and Zaghlool, D.A. (1973) Studies of *Thrips tabaci* Lindman. X. Mechanism of resistance to *Thrips tabaci* L. in cotton varieties. *Zeitschrift für Angewändte Entomologik* 73, 251–255.

Gillham, F.E.M. (1963) A study in the response of bollworm, *Heliothis zea* (Boddie), to different genotypes of upland cotton. In: *Proceedings of the Beltwide Cotton Production Research Conference*, Memphis, Tennessee, pp. 80–88.

Ha, S.B. (1987) Effects of selected morphological traits in cotton on natural insect infestations, lint yield, lint percent, and fiber quality. *Dissertation Abstracts International, B (Sciences and Engineering)* 47, 3171B.

Ha, S.B., Young, J.H., Wilson, L.J. and Verhalen, L.M. (1987) Effects of morphological trait in cotton on natural infestations of the cotton fleahopper and bollworm. In: *Proceedings of the Beltwide Cotton Production Research Conference*, Memphis, Tennessee, p. 104.

Harland, S.C. (1929) A suggested method for the control of certain bollworms in cotton. *Empire Cotton Growing Review* 6, 333–334.

Henneberry, T.J., Bariola, L.A. and Kittock, D.L. (1977) Nectariless cotton: Effect on cotton leaf perforator and other insects in Arizona. *Journal of Economic Entomology* 70, 797–799.

Howard, L.O. (1898) Some miscellaneous results of the work of the Division of Entomology. *US Department of Agriculture, Division of Entomology Bulletin* 18.

Hunter, R.C., Leigh, T.F., Lincoln, C., Waddle, B.A. and Bariola, L.A. (1965) Evaluation of a selected cross-section of cottons for resistance to the boll weevil. *Arkansas Agricultural Experiment Station Bulletin* 700.

Hussein, M.A., Puri, A.N. and Trehan, K.N. (1936) Cell sap acidity and the incidence of whitefly *Bemisia gossypiperda* on cottons. *Current Science* 4, 486–487.

Jenkins, J.N. (1976) Boll weevil resistant cottons. In: *Boll Weevil Suppression, Management and Elimination Technology*. US Department of Agriculture, ARS-S-71, pp. 45–49.

Jenkins, J.N. (1981) Breeding for insect resistance. In: Frey, K.J. (ed.) *Plant Breeding II*. The Iowa State University Press, Ames, pp. 291–308.

Jenkins, J.N. and Parrott, W.L. (1971) Effectiveness of frego bract as a boll weevil resistance character in cotton. *Crop Science* 11, 739–743.

Jenkins, J.N., Maxwell, F.G., Parrott, W.L. and Buford, W.T. (1969) Resistance to the boll weevil (*Anthonomus grandis* Boh.) oviposition in cotton. *Crop Science* 9, 369–372.

Jenkins, J.N., McCarty, J.C., Jr. and Parrott, W.L. (1977) Inheritance of resistance to tarnished plant bugs in a cross of Stoneville 213 by Timok 811. In: *Proceedings of the Beltwide Cotton Production Research Conferences*, Memphis, Tennessee, p. 97.

Jenkins, J.N., Parrott, W.L., McCarty, J.C.. and Dearing, L. (1986) Performance of cottons when infested with tobacco budworm. *Crop Science* 26, 93–95.

Jones, J.E. (1972) Effects of morphological characters on cotton insects and pathogens. In: *Proceedings of the Beltwide Cotton Production Research Conference*, Memphis, Tennessee, pp. 88–92.

Jones, J.E., Newsom, L.D. and Tipton, K.W. (1964) Differences in boll weevil infestations among different biotypes and upland cotton. In: *Proceedings of the Beltwide Cotton Production Research Conference*, Memphis, Tennessee, pp. 48–55.

Jones, J.E., Dickson, J.I. and Beasley, J.P. (1987) Preference and nonpreference of boll weevil to selected cotton. In: *Proceedings of the Beltwide Cotton Production Research Conference*, Memphis, Tennessee, pp. 98–101.

Jones, J.E., Dickson, J.I., Graves, J.B., Pavloff, A.M., Leonard, B.R., Burris, E., Coldwell, W.D., Micinski, S. and Moore, S.H. (1989) Agronomically enhanced insect-resistant cottons. In: *Proceedings of the Beltwide Cotton Production Research Conference* Memphis, Tennessee, pp. 135–137.

Kadapa, S.M., Vizia, N.C. and Patil, N.B. (1988) A note on stem-tip stiffness in aphid tolerant cottons (*Gossypium hirsutum* L.) *Current Science, India* 57, 265–266.

Kitten, W.F., Bridge, R.R. and Laster, M.L. (1987) Response of advanced breeding lines of cotton to early season tarnished plant bug injury. *Mississippi Agricultural and Forestry Experiment Station Research Report* 12(2),1–5.

Kulkarni, S.N. and Raodeo, A.K. (1986) The incidence of sucking pest complex and its effect on plant growth and yield in cotton. *Indian Journal of Plant Protection* 14, 75–81.

Kuznetsova, T.L. (1986) Nature of feeding by larvae of the cotton moth on bolls of different cotton varieties. *Trudy Vsesoyuznogo Entomologicheskogo Obshchestva* 68, 129–134.

Laster, M.L. and Meredith, W.R. (1974a) Influence of nectariless cotton on insect pest populations. *Mississippi Agricultural and Forestry Experiment Station Research Highlights.* Information Sheet No. 1241, 2pp.

Laster, M.L. and Meredith, W.R., Jr. (1974b) Evaluating the response of cotton cultivars to tarnished plant bug injury. *Journal of Economic Entomology* 67, 686–688.

Lidell, M.C., Niles, G.A. and Walker, J.K. (1986) Response of nectariless cotton genotypes to cotton fleahopper (Heteroptera: Miridae) infestation. *Journal of Economic Entomology* 79, 1372–1376.

Lincoln, C. and Waddle, B.A. (1966) Insect resistance in frego-type cotton. *Arkansas Farm Research* 15, 5.

Lukefahr, M.J. (1977) Varietal resistance to cotton insects. In: *Proceedings of the Beltwide Cotton Production Research Conference*, Memphis, Tennessee, pp. 236–237.

Lukefahr, M.J. and Rhyne, C. (1960) Effects of nectariless cottons on populations of three lepidopterous insects. *Journal of Economic Entomology* 53, 242–244.

Lukefahr, M.J., Martin, D.F and Meyer, J.R. (1965) Plant resistance to five Lepidoptera attacking cotton. *Journal of Economic Entomology* 58, 516–518.

Lukefahr, M.J., Cowan, C.B., Pfrimmer, T.R. and Noble, L.W. (1966) Resistance of experimental cotton strain 1514 to the bollworm and cotton fleahopper. *Journal of Economic Entomology* 59, 393–395.

Luo, Y., Bu, L.Y., Liu, Z.Z., Li, Y.J. and Wang, J.Y. (1988) Effects of the resistance of nectariless and smooth leaf cotton on bollworms and its utilization. *China Cottons* 1988(3), 46–47.

Maredia, K.M. (1986) Evaluation of specific cotton cultivars for tarnished plant bug, *Lygus lineolaris* (PAB), resistance. *Dissertation Abstracts International, B (Sciences and Engineering)* 46(10), 3273B.

Maxwell, F.G., Jenkins, J.N., Parrott, W.L. and Buford, W.T. (1969) Factors contributing to resistance and susceptibility of cotton and other hosts to the boll weevil (*Anthonomus grandis Boh.*). *Entomologia Experimentalis et Applicata* 12, 801–810.

Maxwell, F.G., Schuster, M.F., Meredith, W.R., and Laster, M.L. (1976) Influence of the nectariless character in cotton on harmful and beneficial insects. *Symposia Biologica Hungarica* 16, 157–161.

Meredith, W.R. (1976) Nectariless cottons. In: *Proceedings of the Beltwide Cotton Production Research Conference*, Memphis, Tennessee, pp. 34–37.

Meredith, W.R. and Laster, M.L. (1975) Agronomic and genetic analysis of tarnished plant bug tolerance in cotton. *Crop Science* 15, 535–538.

Meredith, W.R., Jr. and Schuster, M.F. (1979) Tolerance of glabrous and pubescent cottons to tarnished plant bugs. *Crop Science* 19, 484–488.

Meredith, W.R., Jr., Meyer, V., Hanney, B. W. and Bailey, J.C. (1979) Influence of five *Gossypium* species cytoplasms on yield, yield components, fiber properties, and insect resistance in upland cotton. *Crop Science* 19, 647–650.

Milam, M.R., Jenkins, J.N., McCarty, J.C.., Jr. and Parrott, W.L. (1989) Breeding upland cotton for resistance to the tarnished plant bug. *Field Crops Research* 21(3–4), 227–238.

Mitchell, H.C., Cross, W.H., McGovern, W.L.M. and Dawson, E.M. (1973) Behavior of the boll weevil on frego bract cotton. *Journal of Economic Entomology* 66, 677–680.

Montandon, R., Williams, H.J., Sterling, W.L., Stipanovic, R.D. and Vinson, S.B. (1986)

Comparison of the development of *Alabama argillacea* (Hubner) and *Heliothis virescens* (F.) (Lepidoptera: Noctuidae) fed glanded and glandless cotton leaves. *Environmental Entomology* 15, 128–131.

Niles, G.A. (1980) Breeding cotton for resistance to insect pests. In: Maxwell, F.G. and Jennings, P.R. (eds) *Breeding Plants Resistant to Insects.* John Wiley & Sons, New York, pp. 337–369.

Niles, G.A., Walker, J.K. and Gannaway, J.R. (1974) Breeding for insect resistance. In: *Proceedings of the Beltwide Cotton Production Research Conference*, Memphis, Tennessee, pp. 84–86.

Ozgur, A.F. and Sekeroglu, E. (1986) Population development of *Bemisia tabaci* (Homoptera: Aleurodidae) on various cotton cultivars in Cukurova, Turkey. *Agriculture, Ecosystems and Environment* 17, 83–88.

Ozgur, A.F., Sekeroglu, E., Gencer, O., Gocmen, H., Yelin D. and Isler, N. (1988) Study of population development of important cotton pests in relation to various cotton varieties and plant phenology. *Doga, Turk Tarim ve Ormancilik Dergisi* 12(1), 48–74.

Painter, R.H. (1951) *Insect Resistance in Crop Plants.* Macmillan, New York, 520pp.

Parnell, F.R., King, H.E. and Ruston, D.F. (1949) Jassid resistance and hairiness of the cotton plant. *Bulletin of Entomological Research* 39, 539–575.

Pieters, E.P. and Bird L.S. (1976) Field studies of boll weevil resistant cotton lines possessing the okra leaf-frego bract characters. *Crop Science* 17, 431–433.

Pollard, D.G. and Saunders, J.H. (1956) Jassid resistant Sakeland hairiness in relation to other cotton pests. *Empire Cotton Growing Review* 33, 197–202.

Quisenberry, J.E. and Rummel, D.R. (1979) Natural resistance to thrips injury in cotton as measured by differential leaf area reduction. *Crop Science* 19, 879–881.

Reddy, P.S.C. (1974) Effects of three leaf shape genotype of *Gossypium hirsutum* L. and row types on plant microclimate, boll weevil survival, boll rot and important agronomic characters. PhD dissertation. Louisiana State University, Baton Rouge, 84pp.

Reed, D.K. and Adkisson, P.L. (1961) Short-day cotton stocks as possible sources of host plant resistance to the pink bollworm. *Journal of Economic Entomology* 54, 484–486.

Ridgway, R.L. (1984) Cotton protection practices in the USA and world. In: Kohel, R.J. and Lewis, C.F. (eds) *Cotton*, Agronomy Monograph No. 24. American Society of Agronomy, Crop Science Society of America, Soil Science Society of America, Madison, Wisconsin, pp. 265–287.

Schuster, M.F. and Frazier, J.S. (1976) Mechanisms of resistance to *Lygus* spp. in *Gossypium hirsutum* L. *Eucarpia/ OILB Host Plant Resistance to Insects and Mites.* Wageningen, Holland, pp. 129–135.

Schuster, M.F. and Maxwell, F.G. (1974) The impact of nectariless cotton on plant bugs, bollworms and beneficial insect. In: *Proceedings of the Beltwide Cotton Production Research Conference*, Memphis, Tennessee, pp. 86–87.

Schuster, M.F. and Maxwell, F.G. (1976). Resistance to two-spotted spider mite in cotton. *Mississippi Agricultural and Forestry Experiment Station Bulletin* 821, 13pp.

Schuster, M.F., Maxwell, F.G. and Jenkins, J.N. 1972. Antibiosis to two-spotted spider mite in Upland and American pima cotton. *Journal of Economic Entomology* 65, 1110–1111.

Schuster, M.F., Maxwell, F.G., Jenkins, J.N., Cherry, E.T., Parrott, W.L. and Holder, D.G. (1973) Resistance to two-spotted spider mite in cotton. *Mississippi Agricultural and Forestry Experiment Station Bulletin* 802.

Schuster, M.F., Lukefahr, M.J. and Maxwell, F.G. (1975). Impact of nectariless cotton on plant bugs and natural enemies. *Journal of Economic Entomology* 69, 400–402.

Schuster, M.F., Holder, D.G., Cherry, E.T. and Maxwell, F.G. (1976). Plant bugs and natural enemy insect populations on frego bract and smoothleaf cottons. *Mississippi Agricultural and Forestry Experiment Station Technical Bulletin* 75, 1–11.

Schuster, M.F., Anderson, R.F. and Cannon, C.E. (1981). Boll weevil oviposition on frego bract cotton. *Journal of Economic Entomology* 74, 346–349.

Scott, W.P., Snodgrass, G.L. and Smith, J.W. (1988) Tarnished plant bug (Hemiptera: Miridae) and predaceous arthropod populations in commercially produced selected nectariless cultivars of cotton. *Journal of Entomological Science* 23, 280–286.

Senft, D. (1986) Bollworms find okra-leaf cotton less tasty. *Journal of Agricultural Research* 34.

Sharipova, G.F. (1987) Resistance of cotton to cotton aphid. *Zashchita Rastenii* 6, 26–27.

Simongulyan, N. and Dzhunedzha, N. (1988) Inheritance of anatomical characters of the leaf. *Khlopok* 5, 45–47.

Singh, R. and Agarwal, R.A. (1988) Role of chemical components of resistant and susceptible genotypes of cotton and okra in ovipositional preference of cotton leafhopper. *Proceedings of the Indian Academy of Sciences* 97, 545–550.

Sippell, D.W., Biindra, O.S. and Khalifa, H. (1987) Resistance to whitefly (*Bemisia tabaci*) in cotton (*Gossypium hirsutum* L.) in the Sudan. *Crop Protection* 6, 171–178.

Smith, C.W. (1992) History and status of host plant resistance in cotton to insects in the United States. In: Sparks, D.L. (ed.) *Advances in Agronomy*, Vol. 48. Academic Press, San Diego, pp. 251–296.

Sosa, E.S., Silguero, J.F. and Wolfenbarger, D.A. (1981) Resistance of LA 17801 to *Heliothis* spp. in southern Tamaulipas. In: *Proceedings of the Beltwide Cotton Production Research Conference*, Memphis, Tennessee, pp. 80–81.

Stephens, S.G. (1957) Sources of resistance of cotton strains to the boll weevil and their possible utilization. *Journal of Economic Entomology* 50, 415–418.

Stephens, S.G. and Lee, H.S. (1961) Further studies on the feeding and oviposition of the boll weevil (*Anthonomus grandis*). *Journal of Economic Entomology* 54, 1085–1090.

Stipanovic, R.D., Altman, D.W., Begin, D.L., Greenblatt, G.A. and Benedict, J.H. (1988) Terpenoid aldehydes in Upland cottons: analysis by aniline and HPLC methods. *Journal of Agricultural and Food Chemistry* 36, 509–515.

Taneja, A.D., Sharma, J.C., Singh, D.P., Sharma, A.P and Kairon, M.S. (1988) Biochemical changes in *Gossypium hirsutum* cotton in relation to attack by Jassid (*Empoasca* spp.). *Journal of the Indian Society for Cotton Improvement* 13, 33–36.

Tang, C.M. (1987) Genetics and breeding of nectariless cotton. *China Cottons* 5, 11–12.

Tingey, W.M., Leigh, T.F. and Hyer, G.H. (1975) *Lygus hesperus*: Growth, survival, and egg laying resistance of cotton genotypes. *Journal of Economic Entomology* 68, 28–30.

Treacy, M.F. (1986) Role of cotton trichome density and extrafloral nectar in bollworm (Lepidoptera: Noctuidae) egg parasitism and predation. *Dissertation Abstracts International, B. (Sciences and Engineering)* 47(1), 60B.

Treacy, M.F., Benedict, J.H., Lopez, J.D. and Morrison, R.K. (1987) Functional response of a predator (Neuroptera: Chrysopidae) to bollworm (Lepidoptera: Noctuidae) eggs on smoothleaf, hirsute and pilose cottons. *Journal of Economic Entomology* 80, 376–379.

Trichilo, P.J. and Leigh, T.F. (1988) Influence of resource quality on the reproductive fitness of flower thrips (Thysanoptera: Thripidae). *Annals of the Entomological Society of America* 81, 64–70.

Walker, J.K. and Niles, G.A. (1973) Studies of the effect of bollworms and cotton fleahoppers on yield reduction in different cotton genotypes. In: *Proceedings of the Beltwide Cotton Production Research Conference*, Memphis, Tennessee, pp. 100–102.

Walker, J.K., Niles, G.A., Gannaway, J.R., Robinson, J.V., Cowan, C.B. and Lukefahr,

M.J. (1974) Cotton fleahopper damage to cotton genotypes. *Journal of Economic Entomology* 67, 537–542.

Walker, J.K., Gannaway, J.R. and Niles, G.A. (1976) Age distribution of cotton bolls and damage from the boll weevil. *Journal of Economic Entomology* 70, 5–8.

Wannamaker, W.K. (1957) The effect of plant hairiness of cotton strains on boll weevil attack. *Journal of Economic Entomology* 50, 418–423.

Wessling, W.H. (1958) Resistance to boll weevil in mixed population of resistant and susceptible cotton plants. *Journal of Economic Entomology* 51, 502–505.

Wilson, F.D. (1986) Pink bollworm resistance, lint yield, and lint yield components of okra-leaf cotton in different genetic backgrounds. *Crop Science* 26, 1164–1167.

Wilson, F.D. (1989) Yield, earliness, and fiber properties of cotton carrying combined traits for pink bollworm resistance. *Crop Science* 29, 7–12.

Wilson, F.D. and George, B.W. (1982) Effects of okra-leaf, frego-bract, and smooth-leaf mutants on pink bollworm damage and agronomic properties of cotton. *Crop Science* 22, 798–801.

Wilson, F.D., George, B.W., Fry, K.E., Szaro, J.L., Henneberry T.J. and Clayton, T.E. (1986) Pink bollworm (Lepidoptera: Gelechiidae) egg hatch, larval success, and pupal and adult survival on okra- and normal-leaf cotton. *Journal of Economic Entomology* 79, 1671–1675.

Wilson, R.L. and Wilson, D.F. (1975) *A Laboratory Evaluation of Primitive Cotton* (Gossypium hirsutum *L.*) *Races for Pink Bollworm Resistance.* United States Department of Agriculture, ARS-W-30.

Wilson, R.L. and Wilson, F.D. (1976) Nectariless and glabrous cotton: Effect on pink bollworm in Arizona. *Journal of Economic Entomology* 69, 623–624.

Wolcott, G.R. (1927) Haitian cotton and the pink bollworm. *Bulletin of Entomological Research* 18, 79–82.

Zaman, M. (1986) Relative abundance of the pink bollworm on different cultivars of cotton. *Pakistan Cottons* 30, 17–20.

23 Cultural Control

G.A. Matthews

International Pesticide Application Research Centre, Imperial College at Silwood Park, Buckhurst Road, Sunninghill, Ascot, Berkshire SL5 7PY, UK

Introduction

Prior to the introduction of modern synthetic pesticides, a number of cultural practices believed to minimize the impact of insect pests and pathogens had been devised. Many of these practices lapsed with the effective and easy-to-use pesticides, but there is now increasing concern about the adverse effects of pesticides on the environment, the widespread occurrence of pesticide resistance and the rising costs of pesticides. This is resulting in renewed attention being given to the role of cultural control in integrated pest management programmes.

Cultural control practices generally relate to the timing of agronomic operations, cropping system, crop growth and productivity and crop sanitation. Their effectiveness in achieving a significant and lasting control will depend largely on the area and range of crops and other host plants within an agroecosystem, dispersal potential and activity of natural enemies of the pest, and the ability of the pest to withstand adverse environmental conditions in the short term.

Cropping Practices

Traditional farming systems frequently depend on growing relatively small areas of different crops as an insurance against food crop failure. These crops, grown in close proximity and often as a mosaic, need to be weeded manually, so the total area of individual farms is determined by the availability of labour at the crucial period of crop establishment. There have been suggestions that mixed cultures, particularly relay- and intercropping which provide temporal and spatial ecological diversity, favour pest control, in contrast to the genetic and phenological uniformity of monocultures (Southwood and Way, 1970). Without any insecticides some intercrops yield more than the same crop grown as a monoculture. Cowpeas are a good example of this (Jackai *et al.*, 1985). Such intercrops provide a less attractive visual or olfactory stimulus to certain insect pests, or a less favourable environment for their development, hinder dispersal and enhance the diversity of natural enemies (Perrin, 1980). However mixed cropping systems, such as maize–cotton which are found in many countries, have produced low cotton yields, especially when tall food intercrops such as sorghum or maize overshadow the cotton, so plants remain stunted with delayed fruiting. If a relay system is adopted and farmers sow cotton later, climatic conditions may be less suitable, in which case low yields are inevitable. Small plants may attract fewer pests or completely escape some infestations, but relatively little damage can significantly decrease the yields. In practice, polyphagous pests such as *Helicoverpa* spp., *Aphis* spp. and *Bemisia* can readily migrate between crops and so increase the chances of crop damage unless sufficient populations of natural enemies can also be maintained.

Instead of the mosaic of crops within a field, the introduction of mechanical cultivations led to larger individual areas of crops. A considerable amount of research was done in the 1930s to determine the effects of different alternative crops grown alongside the cotton, to establish whether a cropping pattern could be established in which certain crops would act as trap crops. Parsons and Ullyett (1934) found that 20 rows of maize 130 m long attracted *Helicoverpa armigera* so that few eggs could be found on eight hectares of cotton nearby. As maize is only attractive to the *Helicoverpa* moths for 15–20 days during the tasselling period, and cotton remains attractive over a much longer period, a succession of maize sowings is generally required to reduce oviposition on cotton. In Queensland three successive sowings were successful unless the moth population was very high (Wells, 1942). Unfortunately in most areas and seasons, the timing and duration of the rain restricts which crops can be grown and when they can be sown. In the Nile delta, Egypt, irrigation does allow a succession of maize and other crops including tomatoes, so with *G. barbadense* there are normally very few *Helicoverpa* larvae on the cotton. Any interference with the balance between host plants could increase the risk of bollworm infestation, thus instead of being a trap crop, maize is more often a major source of *Helicoverpa* infestation on cotton. In Tanzania and Nigeria, damage to cotton has generally increased when farmers have increased sowings of maize rather then sorghum; natural enemies often being very effective on sorghum.

Availability of irrigation should allow greater flexibility in control of sowing dates, where temperature or other factors are not limiting. In northwest China, bollworm infestations are less serious in areas where cotton is sown earlier under a polythene strip before the maize crop. The presence of poplar trees as wind breaks (Fig. 23.1) provides a hibernation site for natural enemies, especially coccinellids which invade the cotton fields from wheat and other crops. Unfortunately, the choice of sowing dates on irrigation schemes is usually dictated by other management criteria such as the availability of tractors, seed-drills, water and labour, rather than crop protection. In consequence intensification of cropping often exacerbates the number of pests by providing a sequence of host plants, especially where temperatures allow pest populations to survive. In the Sudan Gezira, severe infestations of *Helicoverpa armigera* were of short duration and *G. barbadense* with a long flowering period compensated for the loss of buds. However, the change in the rotation providing a sequence of crops, wheat, sorghum and groundnuts, all of which are host plants, has allowed the pest to survive between cotton crops, and infestations to become more prolonged.

In the Sudan and in Australia, aerial application over large cotton fields has led to selection of pests resistant to insecticides. Apart from other factors concerned with the choice and timing of sprays, the treatment of extensive areas of one crop limit the impact of natural enemies. Furthermore, when other crops are also sprayed selection for resistance is intensified, as in Australia, where there are relatively few natural habitats within the farming areas suitable for some of the pest population to survive on wild host plants and so remain susceptible to the insecticides (see Chapter 27).

Although trap cropping alone has not been always sufficiently effective in controlling key pests, there is some justification for re-examining the use of trap crops, such as maize, in integrated pest management programmes, especially where some other control tactic can be superimposed. To be acceptable the trap crop itself should provide revenue to the farmer.

It may be easier to conserve or release natural enemies on the trap crop, where conditions may be more favourable for sustained biological control. Thus carefully timing cuts of alfalfa (*Medicago sativa* L.) in California can be used to keep *Lygus* away from cotton. However, if there is an interuption to irrigation and the alfalfa dries out, the *Lygus* readily disperse to the cotton areas. In Egypt, *Spodoptera littoralis* populations on cotton come from berseem clover (*Trifolium alexandrinum*), so irrigation of this crop is stopped in May to reduce the source of infestation and encourage migration of the natural enemies to the cotton fields. Interestingly, where cotton is grown in areas with sugar cane scattered in small areas on which *Trichogramma* has been released against borers, or the parasitoid

Fig. 23.1. Poplar trees alongside cotton in China.

occurs naturally, there has generally been less *Helicoverpa* infestation. Detailed investigation of this is needed, but it may be that the sugar cane provides a reservoir of parasitoids to reinfect the annual cotton crop.

Alternatively a pesticide, preferably with a selective action, may be applied to the trap crop to prevent populations of the pest migrating to the cotton. This has been attempted with small areas of early sown cotton as a trap crop baited with pheromone for boll weevil, *Anthonomus grandis*, adults emerging from diapause in Nicaragua (Swezey and Daxl, 1988). Similarly, a single insecticide application has been applied to safflower (*Carthamus tinctorius* L.) to prevent *Lygus* migrating to cotton. Selective spraying using an electrostatic spray to minimize contamination of the soil under crops should allow greater survival of some ground dwelling predators such as ants (Endacott, 1983). Heaps of cotton seed treated with insecticide have also been used as a bait to attract cotton stainers and reduce numbers surviving the dry season.

Close Season

One cultural control widely adopted by cotton farmers is the need to uproot and destroy crop residues at the end of the season and prohibit sowing the next crop before a particular date. The aim is to have, where possible, a two month period free of cotton. The close season was introduced in most countries to reduce the survival of pink bollworm, although in its absence, a close season was introduced into Zimbabwe in an attempt to control the red bollworm (McKinstry, 1938). This was not successful due to the survival of diapause pupae in the soil. However in the Lower Shire valley in Malawi, early sowing combined with a close season did reduce *Diparopsis* populations (Pearson, 1958). Later in Zimbabwe and also in Malawi, a failure to observe the close season resulted in an immediate infestation of pink bollworm presumably derived from neighbouring Mozambique (Matthews *et al.*, 1965).

A close season can only be effective if the pest has a limited host plant range and is unable to survive on the alternative host plants in the absence of cotton. Pink bollworm will feed on *Abelmoschus esculentus*, okra, as well as certain other *Hibiscus* spp. such as *H. dongolensis*, but its survival on these during the close season is low, provided okra is grown at the same time as cotton and the other hosts are removed from irrigation ditches and other habitats close to cotton fields. A cleaning of seed cotton at the ginneries is also essential.

The close season has been most difficult to observe in areas where the cotton stalks have provided a valuable fuel, so that even if they are uprooted and removed from the fields, the diapause larvae can survive. Thus in Egypt it has been suggested that the main infestation of pink bollworm occurs initially close to the villages, where the stalks are stored, although vegetables such as okra are also more likely in the same locality. Grazing of the plants after harvesting by goats and other animals will reduce the number of bolls and seed adhering to the stalks, and has been encouraged where insecticides have not been applied for several weeks before harvest. Ideally the stalks are shredded, ploughed in and buried to provide organic matter and improve soil structure. The greater the number of cultivations or irrigations during the close season, the fewer pink bollworm that can survive in the trash buried in the soil. In northwest China, where pink bollworm is not present, the stalks are harvested and used to manufacture hardboard (Fig. 23.2).

The close season has also been relevent to boll weevil control where it has been possible to grow short-season cottons. Early harvesting, followed by removal of crop residues before climatic conditions are suitable to induce diapause in the adult weevils, will reduce the overwintering population. This strategy has been particularly successful in parts of the USA, although such programmes met with considerable logistic difficulties (Summy *et al.*, 1988; Summy and King, 1992).

A number of other pests are very effectively controlled by the close season, and only appear as pests when there is a failure to observe the close season. In South Africa, *Apion soleatum* has increased in severity in recent years (Broodryk and Mansfield, 1985) necessitating strengthening of the legislation (Niles and Greeff, 1989).

Crop Rotation

Soil borne pests and pathogens, such as nematodes, *Fusarium* wilt and bacterial blight are known to increase where cotton is grown without a suitable crop rotation. Cultivation, irrigation and other practices for other crops can affect the survival of cotton pests, particularly those that overwinter in the soil, for example red bollworm, although most mortality is due to the activities of natural enemies, especially ants. Dispersal of moths emerging from overwintering pupae presents

Fig. 23.2. Harvesting cotton stalks in northwest China to make hardboard.

few problems of survival where cotton is grown in fields close to the previous season's cotton. Isolated patches of cotton have become infested from cotton over 10 km from the nearest cotton.

Fertilizer Practice

High yields of cotton require some fertilizer application, but excessive use of nitrogen is liable to produce rank growth and increase the severity of infestations due to whiteflies, jassids and aphids. Studies in the Sudan by Joyce (1961), Procter and Tigani (1963) and Jackson *et al.* (1973) have all indicated increases in jassids, which were considered to be due to the high N level in leaves. Similar results have been obtained with whitefly populations. Studies on other crops suggest that the imbalance between N and P is one of the factors that leads to a higher rate of reproduction. The increase in whiteflies was exacerbated where persistent broad spectrum insecticides were used and underleaf coverage was poor on the rank crops.

Shedding of buds, flowers and bolls and stunted growth with small reddish leaves may not be caused by an insect pest. In several areas in Africa this damage to the crop was due to boron deficiency, corrected by foliar applications of the trace element (Rothwell *et al.*, 1967).

Discussion

To improve crop protection by manipulating cropping practices in an established farming system will require sustained research to understand the changes in pest populations on the different crops over entire seasons. Present evidence indicates that the close season is of fundamental importance in restricting the populations of key pests such as *Pectinophora gossypiella* as well as many minor pests. There is a also clear evidence that individual fields of cotton should not be too large to avoid decimation of natural enemies when insecticides have to be applied. Fields adjacent to cotton should ideally be sown with crops that are not hosts of cotton pests, but in most areas there will be hosts of polyphagous pests within their flight range to cotton, so other components of integrated pest management need to be used as much as possible. However any changes in farming practices are more likely to be due to political, marketing or other factors, rather than deliberate attempts to modify pest populations.

References

Broodryk, S.W. and Mansfield, J. (1985) A new cotton pest in the Transvaal. *Die OTKaner* 29, 15.

Endacott, C.J. (1983) Non-target organism mortality – a comparison of spraying techniques. *Proceedings, 10th International Congress on Plant Protection* 2, 502.

Jackai, L.E.N., Singh, S.R., Raheja, A.K. and Wiedijk, F. (1985) Recent trends in the control of cowpea pests in Africa. In: Singh, S.R. and Rachie, K.O. (eds) *Cowpea Research, Production and Utilization.* John Wiley & Sons, New York.

Jackson, J.E. Burhan, H.O. and Hassan, H.M. (1973) Effects of season, sowing date, nitrogenous fertilizer and insecticide spraying on the incidence of insect pests on cotton in the Sudan Gezira.*Journal of Agricultural Science* 81, 491–505

Joyce, R.J.V. (1961) Some factors affecting numbers of *Empoasca lybica* (de Berg) (Homoptera: Cicadellidae) infesting cotton in the Sudan Gezira. *Bulletin of Entomological Research* 52, 287–305.

McKinstry, A.H. (1938) Major pests of cotton in Southern Africa. In: *Third Conference on Cotton Growing Problems.* Empire Cotton Growing Corporation, London, p.102.

Matthews, G.A. Tunstall, J.P. and McKinley, D.J. (1965) Outbreaks of the pink bollworm (*Pectinophora gossypiella* Saund.) in Rhodesia and Malawi. *Empire Cotton Growing Review* 42, 197–208.

Niles, G.A. and Greeff, M.S. (1989) *Ratoon Cotton in Perspective* Technical Communiction No. 217 Department of Agriculture and Water Supply, Republic of South Africa, 12pp.

Parsons, F.S. and Ullyett, G.C. (1934) Investigations on the control of the American and red bollworms of cotton in South Africa. *Bulletin of Entomological Research* 25, 349–381.

Pearson, E.O. (1958) *The Insect Pests of Cotton in Tropical Africa.* Commonwealth Agricultural Bureau, London, 355pp.

Perrin, R.M. (1980) The role of environmental diversity in crop protection. *Protection Ecology* 2, 77–114.

Proctor, J.H. and Tigani, el Amin (1963) Annual Report of the Gezira Agricultural Research Station, Sudan 1962–63.

Rothwell, A., Bryden, J.W., Knight, H. and Coxe, B.J. (1967) Boron deficiency of cotton in Zambia. *Cotton Growing Review* 44, 23–28.

Southwood, T.R.E. and Way, M.J. (1970) Ecological background to pest management. In: Rabb, R.L. and Guthrie, F.E. (eds) *Concepts of Pest Management,* North Carolina State University, Raleigh, pp. 6–28.

Summy, K.R. and King, E.G. (1992) Cultural control of cotton insect pests in the United States. *Crop Protection* 11, 307–319.

Summy, K.R., Cate, J.R. and Hart, W.G. (1988) Overwintering strategies of boll weevil in southern Texas: Reproduction on cultivated cotton. *Southwestern Entomologist* 13, 159–164.

Swezey, S.L. and Daxl, R.G. (1988) Area-wide suppression of boll weevil (Coleoptera: Curculionidae) populations in Nicaragua. *Crop Protection* 7, 168–176.

Wells, W.G. (1942) Progress Report of the Experiment Stations of the Empire Cotton Growing Corporation, 1940–41, p. 15.

24 Biological Control

D.J. Greathead

International Institute of Biological Control, Silwood Park, Buckhurst Road, Ascot, Berkshire SL5 7TA, UK

Introduction

Before the advent of powerful synthetic pesticides, biological control was used alongside cultural and chemical methods in a commonsense approach to pest control, using the best means available to control each component of the pest complex of a crop. After new synthetic chemical pesticides became widely available at the end of World War II, biological control suffered an eclipse until the limitations of reliance on pesticides began to become apparent. This resulted in increased attention being given to the integrated pest management (IPM) concept in combating the effects of overuse of chemicals and, concurrently, interest in biological control revived.

IPM, with emphasis on biological control, has been developed in some countries for a number of perennial cropping systems but some success has also been achieved in annual crops, notably in glasshouses but also in the field for rice in tropical Asia. In spite of much recent research, particularly in the USA, deliberate use of biological control as part of IPM has been less widely adopted in cotton. Although there are a number of successes, notably from China, the former USSR and Peru, they are not well documented.

IPM is sensitive to changes in the economics of pesticide use and the availability of new products which may appear to offer cheaper and simpler solutions in the short term. However, this is short-sighted as a return to more complex ecologically sound IPM eventually becomes necessary. IPM is also sensitive to changes in farming systems, crop varieties, advent of new pests and other environmental changes which alter the relative importance of individual pest species.

Although more than 1300 insect species are recorded from cotton (Hargreaves, 1948) the majority of these are of no economic importance, but wherever cotton is grown it is subject to damage by a range of insect pests which severely restrict yield. Most of these pests are native to the regions in which they are found, but some important pests are introduced over a large part of their present range. The most important pest to have spread is the pink bollworm, *Pectinophora gossypiella* (Saund.), which is found almost everywhere where cotton is grown. It is now believed to have originated in Malaysia and Indonesia (Malesia) and/or northern Australia. Other pests which have become widespread are *Nezara viridula* (F.) (probable origin tropical Africa) and *Bemisia tabaci* (Genn.) (probable origin Turkey–Iran–Pakistan). Few of the major pests are restricted to cotton, several have a number of hosts within the Malvales and many are polyphagous, entering the cotton crop from other crop hosts, e.g. *Helicoverpa* spp., *Lygus* spp. and *Jacoboasca* spp.

In any particular cotton growing region a range of species attacking leaves, flowers and bolls are limiting factors, and belong to several insect orders: Heteroptera, Homoptera, Lepidoptera and Coleoptera. Thus, control measures are likely to be needed at different stages in development of the crop, against several pests at any one time.

In order to combat pests, notably *P. gossypiella* which continues to breed in the seeds after harvest, and diseases, cotton is normally only grown as an annual. Strict sanitation and a close season are enforced as part of the control strategy. Thus, the cotton crop is a temporary habitat in which natural enemies do not have time to develop stable populations able to maintain pests at low densities. However, many pests enter the cotton crop from other crops grown in rotation, e.g. *Helicoverpa armigera* (Hb.) from sorghum and groundnuts in the Sudan, or from wild plants and weeds, especially those pests with other hosts among the Malvales. Pests with alternative hosts in the immediate neighbourhood of the crop may be expected to enter the crop with their natural enemies but there is evidence, in some instances, that the natural enemies are affected by the host plant on which the pest is feeding. For example *H. armigera* in East Africa suffers higher levels of mortality by parasitoids, and greater intensity of disease due to NPV infection, on maize than on cotton (Nyambo, 1990), and there is evidence from Egypt (reviewed by Greathead and Bennett, 1981) that levels of parasitism of *B. tabaci* are markedly affected by the host plant.

Cotton is often grown in large fields so that there is a tendency to monoculture with consequent lack of diversity. This and the other factors combine to make the environment of the cotton plant a difficult one in which to practise biological control: i.e. there are few introduced pests; the crop is grown as an annual; it has a diverse pest complex; there is an enforced close season; and in many areas the agroecosystem lacks diversity. Nevertheless, there is potential for incorporating biological control into IPM systems for cotton.

Biological Control

Biological control – the use of living organisms to control pests – can be applied in several ways, all of which have been attempted at some time for the control of pests of cotton with varying degrees of success.

Classical Biological Control

Hitherto classical biological control has been the most widely applied method of exploiting natural enemies and the one which has the most successes to its credit overall. It involves the introduction and permanent establishment of exotic natural enemies to achieve long-term depression of the population density of a pest. It has the advantages that once implemented no further effort or cost is involved in maintaining control over a large area without individual farmers being involved in extra work.

In most instances the targets have been introduced pests, which appear to be ideal because they usually arrive without the specific natural enemies which suppress them in their areas of origin. Although there are few such targets on the cotton crop, biological control introductions have been made against most of them.

Classical biological control has also been carried out successfully against native pests by seeking more efficient natural enemies from related species, or introducing natural enemies to fill gaps in the natural enemy spectrum affecting the different stages of the pest (Carl, 1982). It has been argued by Hokkanen and Pimental (1984) that natural enemies of related species are more effective biological control agents than natural enemies of the pest itself, because the latter become less virulent as they have developed homeostasis with their hosts through genetic feedback over the time that they have been associated. This does not seem to be true of insect parasitoids and predators (Legner, 1986; Waage and Greathead, 1988), but may be true of some pathogens (Briese, 1986), and should not be a constraint to introductions.

The earliest classical biological control introduction against a cotton pest was of an ant *Ectatomma tuberculatum* Ol., imported into Texas in 1904 from Guatemala for control of the boll weevil *Anthonomus grandis* Boh. It is perhaps fortunate that this polyphagous predator failed to survive the winter. Introductions of two braconid parasitoids from Peru in 1941–1945 also failed as did releases of *Bracon kirkpatricki* (Wilk.), a parasitoid of *Pectinophora gossypiella* in East Africa, which has bred well on the boll weevil in the laboratory (Clausen, 1978). The situation was reviewed and a systematic evaluation of parasitoids of other *Anthonomus* spp. from Central and South America was started. The first species to be released, *Catolaccus grandis* (Burks) from Mexico, a parasitoid from the larvae, did not become established but other species were to be tried (Cate, 1985) although no reports of further releases have been traced. However, *C. grandis* has been evaluated as an agent for augmentative release with encouraging results and is considered to merit further research because it is the only control agent to kill the larval stage (Summy *et al.*, 1992); but the major effort for boll weevil control in the USA is now directed at eradication (Chapter 9).

Unsuccessful attempts have also been made to control a cotton stainer, *Dysdercus andreas* L., in Puerto Rico (Clausen, 1978), *Alabama argillacea* (Hb.) in the Caribbean (Cock, 1985) and leafworms, *Spodoptera* spp., in several countries (Waterhouse and Norris, 1987). However, natural enemies imported into Peru during 1909–1912 for control of the newly introduced scale insect, *Pinnaspis strachani* (Cooley) are believed to have been effective but this programme is poorly documented (Clausen, 1978).

The greatest effort has been directed at the control of bollworms, chiefly *Pectinophora gossypiella*, *Heliothis virescens* (F.) and *Helicoverpa* spp. Numerous introductions of Old World parasitoids were made, chiefly in the USA, Mexico and the Caribbean, up to 1975 but only one, *B. kirkpatricki*, became established but it does not contribute to control (Clausen, 1978; Luck, 1981; Cock, 1985). Efforts were then made in the USA, Mexico and Greece to introduce parasitoids of *P. gossypiella* from its possible area of origin in NW Australia but these also failed (Legner and Medved, 1979). Likewise introductions into India from the USA appear to have been unsuccessful (Nagarkatti, 1982). Although there are reports of establishment of the species concerned, they were being released in augmentative biological control experiments at the time of the recoveries (Yadav and Patel, 1987).

More encouraging is the recent successful introduction of a European parasitoid, *Cotesia kazak* (Telenga) into New Zealand for control of *Helicoverpa armigera*, an immigrant species with few natural enemies, that infests principally tomatoes. This introduction has had a substantial impact on the survival of larvae to the later instars, when it becomes most damaging, but has not precluded the need for chemical control (Cameron, 1989). This result stimulated a new effort to obtain biological control of this pest in Western Australia but results are not yet available (Waterhouse and Norris, 1987). *C. kazak* has also been released recently in South Africa (van Hamburg *et al.*, 1992). In the Sudan, *Trichogramma pretiosum* Riley appears to have been successfully established on the Rahad Irrigation Scheme and is reported to be achieving substantial parasitism which gives hope that early season pesticide applications can be withheld (Munir *et al.*, 1992).

Biological Control by Augmentation

When classical biological control is not possible other techniques may be used to achieve short-term biological control. These all involve the culture and periodic release of biological control agents and are collectively referred to as biological control by augmentation of natural enemies.

In annual crops there is frequently insufficient time for a favourable balance to develop between natural enemies and pest species. Thus outbreaks may occur which are not suppressed until the crop has suffered severe damage, or natural enemies are unable to exert sufficient control in these temporary habitats to obviate the need for intervention. In these circumstances small inoculative releases of biological control agents early in the season may be used to establish a favourable balance between the pest and its natural enemies which will persist until harvest. Alternatively, larger augmentative releases may be required at intervals to increase the numbers of existing natural enemies so as to enable them to maintain the pest population below the economic threshold. Biological control agents may also be applied using inundative releases, as a biological pesticide, for a quick kill with no expectation of a continuing impact. The distinction between these different techniques is often blurred in practice and as they share a common technology they are best discussed together.

To be successful, augmentative biological control requires careful planning and attention to detail in the production and delivery of the control agents at the correct time and in the right quantity. Many attempts have failed on an operational scale, even if successful on an experimental scale. Nevertheless, augmentative biological control is frequently attempted because of its attraction to scientists who can set up production facilities which provide employment and impressive statistics for publication in reports. The constraints reviewed by Greathead (1988), Voegele *et al.* (1988) and King and Coleman (1989) are summarized in relation to the applicability of augmentative biological control in cotton fields.

- *Timing of releases.* Efficient monitoring of pest numbers in relation to crop and weather conditions is required to ensure that release of control agents is timed when the pest is at the susceptible stage, and the biological control agents will begin to search for hosts and not be diverted by unsuitable weather conditions. Thus monitoring programmes, which may be part of an IPM programme, need to include assessing the effect of the releases. Trained staff are needed to supervise releases and inform farmers so that they understand the biological processes involved.
- *Choice of agent.* It is critically important to choose species or strains of control agents which are effective on the target pest in the field, and not to rely on readily available or easily cultured material which may have a poor field performance.
- *Supplies of agents.* Living insect natural enemies are highly perishable so they cannot be stockpiled. Production must therefore be efficient, to ensure that the agents are produced in sufficient quantity as and when they are needed. This requires considerable investment in rearing facilities and trained dedicated staff. Since the agents are normally only required during a short season, production must be planned well in advance. An efficient distribution system and appropriate packaging is also essential, to ensure that agents arrive at the place where they are to be used in good condition and at the right time.
- *Quality control.* Insects cultured for long periods are prone to adapt to insectary conditions and lose behavioural characters which ensure efficient performance in the field. Therefore, care has to be taken to ensure a good genetic base for breeding stock, usually by periodic introduction of fresh material from the field and tests to monitor the quality of the product. Much can be done to obviate deterioration by designing rearing facilities in such a way that natural enemies are forced to search for hosts.
- *Competitive cost of the agents.* Most parasitoids and predators must be reared on their natural hosts and many larval parasitoids have a low fecundity, which makes production costs high even when the hosts can be reared in large numbers on artificial media. For these reasons most effort has been made to use egg parasitoids (e.g. see Plate VIII.3) which can be reared on alternative hosts, such as *Trichogramma* spp. on the eggs of flour moths, and Chrysopid predators which can be fed artificial media. Even so it has proved difficult, in

developed countries, to make unit costs of control comparable with chemical pesticide treatments.

- *Acceptance by farmers.* Because augmentative biological control requires monitoring, precision and the use of perishable agents, there must be a clear perceived advantage if farmers and agricultural authorities are to adopt the method. The advantages may include avoidance of pesticide resistance, lower cost or concern for the side effects on non-target organisms but, above all, there will need to be the assurance that the biological control method will not fail.

Overcoming these constraints is a considerable challenge which rarely has been met satisfactorily.

Augmentative biological control of cotton pests, particularly bollworms, has been attempted in a number of countries with very variable success. The most accessible and detailed reports are those from the USA. Mass rearing technology for egg-parasitoids, *Trichogramma* spp., was developed and field trials were carried out for *Heliothis virescens* and *Helicoverpa armigera* control. The feasibility of raising parasitism was demonstrated on field stations, and research has been carried out to improve the results using behaviour modifying chemicals to prevent parasitoid dispersal. Studies have also been made on larval and pupal parasitoids as potential control agents. However, biological control has not been adopted in current integrated pest management schemes in Texas and Arkansas (Cate, 1985). This work has been reviewed in a number of publications, notably by King and Coleman (1989) who imply that the reason for the failure to exploit the research is principally the high cost in relation to that of conventional chemical treatments. However, in the USA a major constraint to the use of biological control agents in cotton fields has been the need for chemical control of boll weevil. Even when there has not been a direct conflict, pesticide contamination of the crop environment reduces the efficiency of biological control agents.

In India trials are being carried out on experiment stations (Yadav and Patel, 1987; Dhandapani *et al.*, 1992) using applications of *Trichogramma* spp. and larval parasitoids, both native and imported, against the bollworm complex. In some of these, yield has been comparable to that obtained with chemical control and the costs were similar. On the other hand in Queensland trials with *Trichogramma* spp. did not provide adequate control (Twine and Lloyd, 1984).

In contrast some other countries apply *Trichogramma* spp. over large areas and report satisfactory control, but few details are available and it is often not clear which are the principal crops on which it is applied. For example, King *et al.* (1985) gathered estimates that over eight million hectares are treated in the former USSR, two million in China and lesser amounts in Mexico, Colombia and other countries. Matthews (1993, personal communication) visited Uzbekistan in 1993 and reports that in 1992 some 4000 kg of *Trichogramma pintoi* Voegele and smaller quantities of *Bracon hebetor* Say were applied to cotton fields for control of *H. armigera* and that biological control has largely replaced insecticides; applications have declined from 60,000 tons in 1975 to only 2000 tons now. Evidence of similarly impressive results have also recently become available for Colombia where insecticide applications for control of *Helicoverpa armigera*, *Heliothis virescens* and *Alabama argillacea* (Hb.) have been reduced from 19–22 in 1975–1977 to 0–2 in succeeding years when *Trichogramma* spp. have been applied (Garcia-Roa, 1990). King *et al.* (1985) suggest that the reason that *Trichogramma* spp. are used in these countries is that they are cost effective compared with alternative means of control. Thus, when there is a capability for cheap local production within the country, use of *Trichogramma* spp. is attractive when the alternative is relatively expensive imported pesticides which require foreign exchange for their purchase. Further, a level of control that is acceptable in one country may not be tolerable in another.

The constraints which apply to insects as agents for augmentative biological control, particularly for inundative use, are less severe for pathogens. Also they can be applied using conventional spraying equipment and so show more promise for short-term biological control in cotton crops.

Microbial Control

Chewing insects, Lepidoptera and Coleoptera, are affected by the full range of microorganisms pathogenic to insects (protozoa, fungi, bacteria, viruses). On the other hand, sucking insects, Heteroptera and Homoptera, are hosts to fewer pathogens, essentially fungi and nematodes, since infection by the other agents is principally through ingestion of spores deposited on the host plant. Thus, microbial agents have been investigated chiefly for the control of chewing pests on cotton.

Entomopathogenic nematodes are usually included as microbial agents since the constraints on their use and many of the problems of applying them are similar. They affect both chewing and sucking insects but are more dependent on high humidity for transmission, and so are most appropriate for the control of pests in soil and crops growing in wet conditions. For these reasons they are not promising agents for the control of pests affecting the aerial parts of the cotton plant and do not appear to have been considered seriously for use in this crop.

Early experiments with microbial agents, mainly fungi, for classical biological control were mostly unsuccessful, chiefly owing to inadequate understanding of the ecology of the pathogen – especially in relation to the mechanisms of infection and transmission. Experimental and theoretical studies have been reviewed by a number of authors (Fuxa and Tanada, 1987), and the implications for pest control have been examined (Franz, 1986). The conclusions are that microbial agents are unsuitable for classical biological control, unless there is an efficient dispersal mechanism and a degree of stability in the ecosystem. Further, the degree of virulence is critical in sustaining infection rates and avoiding violent fluctuations.

As a result, research is now concentrated on developing pathogens as biopesticides for which they have many advantages compared with arthropods, most importantly their small size and high multiplication rate. Some fungi and bacteria have the additional advantage that can be grown *in vitro* but protozoa and viruses must be grown in living tissue and are therefore expensive to produce in quantity. The pathogen must then be formulated to provide a stable product suitable for application as a spray or dust, and protected from inactivation in the field. Under tropical conditions ultraviolet radiation and high temperatures limit persistence, particularly of viruses. In this respect cotton fields are a particularly harsh environment. Notwithstanding, preparations of *Bacillus thuringiensis* and nuclear polyhedrosis viruses have been developed for use in cotton crops with some success on an experimental scale, against *Heliothis* spp. (Bell, 1982; McKinley, 1982), and *Spodoptera littoralis* (Boisd.) (McKinley *et al.*, 1989; Dhandapani *et al.*, 1992). See Chapter 25.

Experiments have been carried out to test the efficacy of a formulation of the pathogenic fungus *Beauveria bassiana* (Bals.) Vuill. as a biopesticide for boll weevil control in Texas. Field trials against emerging overwintered adults in 1989 and 1990 increased yields significantly over those on untreated controls but were only around two thirds of the yield of plots treated with chemical pesticides (Wright and Chandler, 1992).

This work has exposed problems in developing products with an adequate shelf life and effective protection from radiation so that they persist on the crop and retain their virulence. A further problem is that most pathogens, except *B. thuringiensis*, are slow to kill the pest and damage to the crop continues for some time after application. As a commercial proposition, biopesticides also have the disadvantage of relative specificity to one pest species, or a group of similar pests and so will have a restricted market. Products must also undergo expensive procedures to obtain registration as pesticides. These constraints make biopesticides unattractive to large multinational companies, while small companies have difficulty in bearing the risks and high development costs (Jutsum, 1988). However if effective low-cost production methods can be perfected for use on a cottage-industry scale in developing countries these constraints are less important. Thus, biological control of cotton pests using microbial agents may have promise as an alternative to chemical pesticides, especially in developing countries. However, there are many problems to be overcome before they can provide reliable control.

Conservation of Natural Enemies

Insect pests suffer mortality from parasitoids and predators occurring naturally in the crop environment. This mortality may be insufficient to prevent economic damage but can be useful in reducing the frequency of applying control measures. Many other insects are supressed by their natural enemies, and their potential as crop pests is only realized when careless use of pesticides destroys their natural enemies. Thus, *Bemisia tabaci* has only become a major threat to cotton production since the use of pesticides was intensified (Cock, 1986). Red spider mites have also increased in areas where an inappropriate insecticide has been applied too frequently. Therefore, efforts to reduce the frequency of pesticide applications, target them more accurately or to use more selective active ingredients will help conserve natural enemies and so improve natural control.

Plant breeding to incorporate characters into the cotton plant which favour natural enemies is another approach to biological control. Physical characters which have been shown to have an impact on the balance between pests and their natural enemies (Schuster and Calderon, 1986) include:

- hairiness, which inhibits movement of small arthropods, increases the frequency of attack by predators, but may also impede their movement and reduce searching efficiency;
- extra floral nectaries attract parasitoids and predators into the crop and so increase rates of parasitism and predation;
- bract shape affects the ability of parasitoids to locate bollworm larvae, narrow bracts enhance parasitism.

More subtle effects on pest population dynamics caused by changes in palatability of the host plant merit further investigation. Thus, chemical and physical characters which reduce survival or prolong development should increase parasitism and predation rates, and may enable natural enemies to prevent or delay outbreaks.

Changes in cultural practices may also have a significant beneficial effect on pest numbers. Weeds add diversity which provides nectar sources for parasitoids, alternative hosts for predators and shelter. For example, weeds growing around cotton fields in the Sudan Gezira provide reservoirs of whitefly parasitoids.

Crop rotation may be beneficial when it allows natural enemies to carry-over from one crop to another and so maintain high population densities, but not if the preceding crop is also a host to a serious pest. For example, the practice of growing sorghum and groundnuts in rotation with cotton in the Sudan aggravated the *Helicoverpa armigera* problem. On the other hand several investigations have shown that *H. armigera* is more heavily attacked by pathogens and parasitoids on some of its host plants than on others. As mentioned earlier, virus and bacterial disease epidemics are more frequent on maize than on cotton, but also on a weed, *Cleome* sp., in Tanzania (Nyambo, 1990). This weed and sorghum also appear to favour parasitism by insect parasitoids. These differences may be exploited in multiple-cropping farming systems practised by small farmers in many tropical countries.

The possibility of interplanting cotton with lucerne as a reservoir for the predatory mite, *Metaseiulus occidentalis* (Nesbitt), which preys on red spider mite, *Tetranychus urticae* Koch, in California has been tested over two years but provided successful control adjacent to the lucerne strips in only one of them (Corbett *et al.*, 1991).

Alterations in cultural practices to enhance biological control cannot be considered in isolation, and must be evaluated as part of farming system strategy to determine whether the changes will be of overall benefit to the farmer.

In large-scale intensive agriculture the opportunities are limited but for small farmers, in developing countries operating low input systems, there are many possibilities which have been extensively discussed in a number of reviews (Altieri, 1983; Way and Norton 1984). Wu (1986) provides an example from field investigations in China. Interplanting cotton with foodcrops and brassica oil crops enabled natural enemies to control *Aphis gossypii* Glover and the first generation of *H. armigera*. Subsequently control was maintained using *Trichogramma* sp. and NPV sprays.

Conclusions

The need to grow cotton as an annual crop and the diverse complex of pests which attack it in any particular region make it a difficult crop on which to practise biological control. Certainly, there is no benefit from biological control of single pests when pesticides, which must be applied to control others, also destroy the biological control agents. Thus, biological controls must be introduced as part of integrated pest management, acknowledging that pesticides will need to be used, at least in intensive farming systems.

Because few pests are introduced species the opportunities for classical biological control are limited. So far introductions against exotic pests have been of little benefit, but the possibilities are not exhausted. Exploration for the boll weevil's natural enemies in Mexico has provided more promising parasitoids (Cate, 1985). The exploration for the pink bollworm's natural enemies in the Indonesian region has been only superficial, and species may exist there which merit trial. Introductions against native pests have so far not been successful in achieving control. However, most of these were made because parasitoids were readily available, and not because careful analysis had shown that there were empty niches to be filled, or because the parasitoids had been shown to be superior to those already present. For example, Greathead and Girling (1982) drew attention to gaps in the natural enemy spectra affecting *Heliothis* spp. in several regions, which might be filled by introductions from other continents.

In spite of considerable research and favourable results from field trials, augmentation of insect natural enemies has not been adopted by farmers in the USA and India, although it is apparently effective in Uzbekistan and China. It is suggested that cost, and the complex logistics of producing a perishable product and delivering it on time, are against the adoption of this method. On the other hand, pathogens show more promise for use in this way, being more readily and cheaply produced and stored. However, so far, problems with formulation and stability in storage have made them unreliable in field use and unattractive to commercial production. These constraints are likely to be overcome and more research is required on strain selection for virulence against the target pests.

Probably the greatest potential in the short term lies in maximizing the impact of existing natural enemies through modifications in cultural practice and selectivity in necessary pesticide applications.

References

Altieri, M.A. (1983) *Agroecology. The Scientific Basis of Alternative Agriculture.* University of California Press, Berkeley.

Bell, M.R. (1982) The potential use of microbials in *Heliothis* management. In: *Proceedings of the International Workshop on* Heliothis *Management*, 15–20 November 1981, ICRISAT Center, Patancheru, India. International Crops Research Institute for the Semi-Arid Tropics, Patancheru, pp. 137–145.

Briese, D.T. (1986) Host resistance to microbial agents. In: Franz, J.M. (ed.) *Biological Plant and Health Protection.* Gustav Fischer Verlag, Stuttgart, pp. 233–256 (*Progress in Zoology* 32).

Cameron, P.J. (1989) *Heliothis armigera* (Hübner), tomato fruitworm (Lepidoptera: Noctuidae). In: Cameron, P.J., Hill, R.L., Bain, J. and Thomas, W.P. (eds) *Review of Biological Control of Invertebrate Pests and Weeds in New Zealand.* Technical Communication CIBC No. 10. CAB International, Wallingford, pp. 87–91.

Carl, K.P. (1982) Biological control of native pests by introduced natural enemies. *Biocontrol News and Information* 3, 191–200.

Cate, J.R. (1985) Cotton: status and current limitations to biological control in Texas and Arkansas. In: Hoy, M.A. and Herzog, D.C. (eds) *Biological Control in Agricultural IPM Systems.* Academic Press, New York, pp. 538–556.

Clausen, C.P. (ed.) (1978) *Introduced Parasites and Predators of Arthropod Pests and Weeds.* US Department of Agriculture, Handbook No. 480.

Cock, M.J.W. (ed.) (1985) *A Review of Biological Control of Pests in the Commonwealth Caribbean and Bermuda up to 1982.* Technical Communication CIBC No. 9. Commonwealth Agricultural Bureaux, Farnham Royal.

Cock, M.J.W. (ed.) (1986) Bemisia tabaci a *Literature Survey on the Cotton Whitefly with an Annotated Bibliography.* CAB, International Institute of Biological Control, Ascot, UK.

Corbett, A., Leigh, T.F. and Wilson, L.T. (1991) Interplanting alfalfa as a source of *Metaseiulus occidentalis* (Acari: Phytoseiidae) for managing spider mites in cotton. *Biological Control* 1, 188–196.

Dhandapani, N., Kalyanasunderam, M., Swamiappan, M., Sundra Babu, P.C. and Jayraj, S. (1992) Experiments on management of major pests of cotton with biocontrol agents in India. *Journal of Applied Entomology* 114, 52–56.

Franz, J.M. (ed.) (1986) *Biological Plant and Health Protection.* Gustav Fischer Verlag, Stuttgart (*Progress in Zoology* 32).

Fuxa, J.R. and Tanada, Y. (eds) (1987) *Epizootiology of Insect Diseases.* John Wiley & Sons, New York.

Garci-Roa, F. (1990) Effectiveness of *Trichogramma* spp. in biological control programs in the Cauca Valley, Colombia. *Colloques de l'INRA* No. 56, 197–199.

Glaser, I., Klein, M. and Nakache, Y. (1992) Comparison of entomopathogenic nematodes combined with antidesiccants applied by canopy sprays against three cotton pests (Lepidoptera: Noctuidae). *Journal of Economic Entomology* 85, 1636–1641.

Greathead, D.J. (1988) Crop protection without chemicals: pest control in the third world. *Aspects of Biology* 17, 19–28.

Greathead, D.J. and Bennett, F.D. (1981) Possibilities for the use of biotic agents in the control of the white fly, *Bemisia tabaci. Biocontrol News and Information* 2, 1–7.

Greathead, D.J. and Girling, D.J. (1982) Possibilities for natural enemies in *Heliothis* management and the contribution of the Commonwealth Institute of Biological Control. In: *Proceedings of the International Workshop on* Heliothis *Management, 15–20 November 1981*, ICRISAT Center, Patancheru, India, International Crops Research Institute for the Semi-Arid Tropics, Patancheru, pp. 147–158.

Hargreaves, H. (1948) *List of Recorded Cotton Insects of the World*, Commonwealth Institute of Entomology, London.

Hokkanen, H. and Pimental, D. (1984) New approach for selecting biological control agents. *Canadian Entomologist* 116, 1109–1121.

Jutsum, A.R. (1988) Commercial application of biological control: status and prospects. *Philosophical Transactions of the Royal Society, London* 318, 357–373.

King, E.G. and Coleman, R.J. (1989) Potential for biological control of *Heliothis* species. *Annual Review of Entomology* 34, 53–75.

King, E.G. and Powell, J.E. (1992) Propagation and release of natural enemies for control of cotton insect and mite pests in the United States. *Crop Protection* 11, 497–506.

King, E.G., Bull, D.L., Bouse, L.F. and Phillips, J.R. (1985). Biological control of bollworm and tobacco budworm in cotton by augmentative releases of *Trichogramma*. *Southwestern Entomologist*, *Supplement* No.8, 198pp.

Legner, E.F. (1986) Importation of exotic natural enemies. In: Franz, J.M. (ed.) *Biological Plant and Health Protection*, Gustav Fischer Verlag, Stuttgart, pp. 19–30.

Legner, E.F. and Medved, R.A. (1979) Influence of parasitic Hymenoptera on the regulation of pink bollworm, *Pectinophora gossypiella*, on cotton in the Lower Colorado Desert, *Environmental Entomology* 8, 922–930.

Luck, R.F. (1981) Parasitic insects introduced as biological control agents for arthropod pests. In: Pimental, D. (ed.) *CRC Handbook of Pest Management in Agriculture*, Vol. 2. CRC Press, Boca Raton, Florida, pp. 125–284.

McKinley, D.J. (1982) The prospects for the use of nuclear polyhedrosis virus in *Heliothis* management. In: *Proceedings of the International Workshop on* Heliothis *Management*, 15–20 November 1981, ICRISAT Center, Patancheru, India. International Crops Research Institute for the Semi-Arid Tropics, Patancheru, pp. 123–135.

McKinley, D.J., Moawad, G., Jones, K.A., Grzywacz, D. and Turner, C. (1989). The development of nuclear polyhedrosis virus for control of *Spodoptera littoralis* (Boisd.) in cotton. In: Green, M.B. and Lyon, D.J. de B. (eds) *Pest Management in Cotton*. Ellis Horwood, Chichester, pp. 93–100.

Matthews, G.A. (1993) Biological control takes precedence in Uzbekistan cotton. *Pesticide Outlook* (in press).

Munir, B., Abdelrahman, A.A., Mohammed, A.H. and Stam, P.A. (1992) Introduction of *Trichogramma pretiosum* Riley against *Heliothis armigera* (Hb.) in the Sudan. Proceedings, 3rd International Conference on Plant Protection in the Tropics, Genting Highlands, 20–23 March 1990, Vol. 5. Malaysian Plant Protection Society, pp. 70–73.

Nagarkatti, S. (1982) The utilization of biological control in *Heliothis* management in India. In: *Proceedings of the International Workshop on* Heliothis *Management*, 15–20 November 1981, ICRISAT Center, Patancheru, India, International Crops Research Institute for the Semi-Arid Tropics, Patancheru, pp. 159–167.

Nyambo, B. (1990) Effect of natural enemies on the cotton bollworm, *Heliothis armigera* Hübner (Lepidoptera : Noctuidae) in western Tanzania. *Tropical Pest Management* 36, 50–58.

Schuster, M.F. and Calderon, M. (1986) Interactions of host plant resistant genotypes and beneficial insects in cotton ecosystems. In: Boethal, D.J. and Eikenbary, R. D. (eds) *Interactions of Plant Resistance and Parasitoids and Predators of Insects*. Ellis Horwood, Chichester, pp. 84–97.

Summy, K.R., Morales-Ramos, J.A. and King, E.G. (1992) Ecology and potential impact of *Catolaccus grandis* on boll weevil infestations in the Lower Rio Grand Valley. *Southwestern Entomologist* 17, 279–288.

Twine, P.H. and Lloyd, R.J. (1984) Observations on the effect of regular releases of *Trichogramma* spp. in controlling *Heliothis* spp. and other insects in cotton. *Queensland Journal of Agricultural and Animal Science* 39, 159–167.

van Hamburg, H., Basson, N.C.J. and Broodryk, S.W. (1992) Integrated pest management of cotton pests: a compromise between biological and chemical control. *Plant Protection News* No. 29, 6–7.

Voegele, J., Waage, J. and van Lenteren, J. (eds) (1988) *Trichogramma and other Egg Parasites*. 2nd International Symposium Guangzhou, 10–15 November 1986. Institut National de la Recherche Agronomique, Paris (*Les Colloques de l'INRA* No. 43.).

Waage, J.K. and Greathead, D.J. (1988) Biological control: challenges and opportunities. *Philosophical Transactions of the Royal Society, London* 318, 111–128.

Waterhouse, D.F. and Norris, K.R. (1987) *Biological Control: Pacific Prospects.* Inkata Press, Melbourne.

Way, M.J. and Norton, G.A. (1984) Integrated crop protection. In: Hawksworth, D.L. (ed.) *Advancing Agricultural Production in Africa.* Commonwealth Agricultural Bureaux, Farnham Royal, pp. 212–218.

Wright, J.E. and Chandler, L.D. (1992) Development of a biorational mycoinsecticide: *Beauveria bassiana* conidial formulation and its application against boll weevil population (Coleoptera: Curculionidae). *Journal of Economic Entomology* 85, 1130–1135.

Wu, Q. (1986) Investigation on the fluctuations of dominant natural enemy populations in different cotton habitats and integrated application with biological agents to control cotton pests (in Chinese: English summary). *Natural Enemies of Insects* 8, 29–34.

Yadav, D.N. and Patel, R.C. (1987) Biological control of cotton pests with special reference to cotton bollworms. In: *Proceedings of the Seminar-cum-Sixth Workshop of Biological Control of Crop Pests and Weeds*, 29 June to 2 July 1987, Anand. Indian Institute of Horticultural Research, Bangalore, India, pp. 51–62.

25 Use of Baculoviruses for Cotton Pest Control

K.A. Jones

Natural Resources Institute, Central Avenue, Chatham Maritime, Chatham, Kent ME4 4TB, UK

Introduction

Virus diseases infecting insects have been recognized for over 80 years (Wahl, 1909). The first recorded attempt to control insect pests with entomopathogenic viruses was carried out by Lounsbury in 1913, who sprayed an aqueous mixture of virus infected *Colias electo* and *Heliothis 'obtectus'* larvae on to lucerne fields (Heimpel, 1967). It is in the last 30 years, however, that the study of the nature of insect pathogenic viruses and their use in pest control has blossomed.

Several types of viruses have been described from insects and these have been classified into several families and a number of other groups (Table 25.1). Pathogenic viruses have been described infecting several major and minor cotton pests (Table 25.2). The majority of these viruses are from one family, the Baculoviridae, which include the nuclear polyhedrosis viruses (NPV) and the granulosis viruses (GV).

The baculoviruses are unique to Arthropoda and within this phylum have been found mainly in insects. This family of viruses are therefore considered to be the most promising for pest control and the World Health Organization went as far as to recommend that baculoviruses were the only group of viruses considered to be safe for use in the field (WHO, 1973).

The use of baculoviruses for cotton pest control has several advantages over chemical insecticides. These include specificity, the viruses being harmless to non-target organisms including beneficial insects. To date no problems have been encountered with the development of field resistance to baculoviruses. They leave no toxic residues. They are generally cheaper than chemicals to develop and register as pesticides. Finally, baculoviruses can be produced at a local level using relatively simple technology.

Baculoviruses also have the advantage of being self-perpetuating in the environment; dead, infected insects releasing virus into the environment. The spread of disease through a pest population does occur naturally and has been

Table 25.1. Classification of viruses pathogenic to insects.

Virus group		Type of nucleic acid	Virion shape	Inclusion body	Main arthropod orders attacked	Other phyla attacked
Family: Baculoviridae						
Genus: Nucleopolyhedroviruses Nuclear polyhedrosis viruses	(NPV)	dsDNA	Rod/	+	Lepidoptera, Hymenoptera Diptera, Crustacea	None
Genus: Granuloviruses (Granulosis viruses)	(GV)	dsDNA	bacilliform	+	Lepidoptera	None
Unclassified dsDNA viruses						
Non-occluded rod-shaped viruses (*Oryctes*-type viruses)	–	dsDNA	Rod/ bacilliform	–	Lepidoptera, Coleoptera	None
Family: Reoviridae						
Genus: Cypovirus (Cytoplasmic polyhedrosis virus)	(CPV)	dsRNA	Icosahedral	+	Lepidoptera, Hymenoptera, Diptera, Coleoptera, Crustacea	Vertebrates, plants
Family: Ascoviridae	–	dsDNA	Bacilliform/ allantoid	+/–	Lepidoptera	–
Family: Polydnaviridae						
Genus: Ichnovirus	–	dsDNA	Rod/	–	Hymenoptera	None
Genus: Bracovirus	–	dsDNA	fusiform	–	Hymenoptera	None
Family: Poxviridae						
Subfamily: Entomopoxviridae Genera A, B & C	(EPV)	dsDNA	Ovoid	+	Lepidoptera, Orthoptera Coleoptera, Diptera	Vertebrates
Family: Iridoviridae						
Genus Iridovirus	(IV)	dsDNA	Icosahedral	–	Lepidoptera, Hymenoptera Coleoptera, Diptera Ephemeroptera, Hemiptera	Vertebrates
Family: Parvoviridae						
Genus: Densovirus	(DNV)	ssDNA	Icosahedral	–	Lepidoptera, Orthoptera Diptera	Vertebrates

Table 25.1. *Continued.*

Virus group		Type of nucleic acid	Virion shape	Inclusion body	Main arthropod orders attacked	Other phyla attacked
Family: Rhabdoviridae						
Sigma virus	(σ)	ssRNA	Bacilliform	–	Diptera	Vertebrates, plants
Family: Birnaviridae						
Drosophila X virus	–	dsRNA	Icosahedral	–	Diptera	Vertebrates
Family: Nodaviridae						
Nodavirus	–	ssRNA	Isometric	–	Diptera, Coleoptera, Lepidoptera	–
Family: Tetraviridae						
e.g. *Helicoverpa armigera* stunt virus	–	ssRNA	Icosahedral	–	Lepidoptera	–
Family: Picornaviridae						
e.g. Cricket paralysis virus	–	ssRNA	Icosahedral	–		
Drosophila C virus	–	ssRNA	Icosahedral	–	Lepidoptera, Hymenoptera	Vertebrates, plants
Gonometa virus	–	ssRNA	Icosahedral	–	Coleoptera, Orthoptera Diptera, Hemiptera	
Unclassified small RNA viruses						
e.g. Bee acute paralysis virus	–	ssRNA	Icosahedral	–		
Aphid lethal paralysis virus	–	ssRNA	Icosahedral	–		
Queensland fruitfly virus	–	ssRNA	Icosahedral	–		

ds = Double stranded; ss = single stranded.
Adapted from Evans and Harrap (1982), Hunter *et al.* (1984), Entwistle and Evans (1986) and Francki *et al.* (1991) with recent revisions.

exploited in pest control in some crops, especially forest. However, this epizootic effect cannot be relied upon for effective control of cotton pests and an 'insecticidal' approach has been universally adopted with baculoviruses for cotton pest control.

On the negative side, disadvantages compared with chemical insecticides include the slow speed of action of baculoviruses and generally short environmental persistence. The narrow host range of baculoviruses is also regarded as a disadvantage where a complex of insect pests are found; with a few notable

Table 25.2. Cotton insect and mite pests from which pathogenic viruses have been isolated.

Major pests	Virus group
Anomis flava	NPV
Anthonomus grandis	NPV, IV
Crytophlebia leucotreta	CPV, GV
Diparopsis watersi	NPV, CPV
Earias insulana	VD
Helicoverpa armigera	NPV, CPV, GV, IV
Helicoverpa zea	NPV, CPV, GV, IV, NOVD
Pectinophora gossypiella	NPV, CPV
Spodoptera littoralis	NPV, CPV, GV
Spodoptera litura	NPV, CPV, GV, IV, EPV
Sylepte derogata	IV
Tetranychus cinnabarinus	non-occluded virus
Minor/occasional pests	
Agrotis ipsilon	NPV
Agrotis segatum	NPV, CPV, GV
Alabama argillacea	NPV
Amsacta spp.	NPV
Aphis spp.	VD
Bacculatrix thurberiella	NPV
Diacrisia spp.	NPV, CPV, GV
Epilachna spp.	VD
Estigmene acrea	NPV, CPV, GV
Helicoverpa assulta	NPV
Helicoverpa puntigera	NPV
Heliothis virescens	NPV
Mythimna seperata	NPV, EPV
Planococcus citri	non-occluded virus
Pseudoplusia includens	NPV, small RNA virus
Spodoptera exigua	NPV, CPV, GV
Spodoptera ornithogalli	NPV
Trichoplusia ni	NPV

NPV = nuclear polyhedrosis virus; GV = granulosis virus; CPV = cytoplasmic polyhedrosis virus; IV = iridovirus; NOVD, non-occluded virus; EPV = entomopox virus; VD = unidentified virus disease; non-occluded virus = unidentified non-occluded virus.

Main reference sources Martignoni and Iwai (1981); Hill (1982).

exceptions (e.g. *Mamestra brassicae* and *Autographa california* NPVs) most baculoviruses infect only one or two host species. Commercial companies will also regard the difficulty in patenting naturally occuring microorganisms as a disadvantage. However, the increasing importance of environmental issues, problems of insecticide resistance and general lack of new groups of chemical insecticides has resulted in an increasing interest in the use of baculoviruses in cotton pest control.

Properties of Baculoviruses

The family Baculoviridae consist of double stranded DNA viruses that have bacilliform or rod-shaped virions which are contained within a protein inclusion body. The family is divided into two genera, the Nucleopolyhedrosis viruses (NPV) or nuclear polyhedrosis viruses and the Granuloviruses (GV) or granulosis viruses (a third genus the non-occluded rod-shaped viruses has recently been removed from the Baculoviridae). The occlusion bodies of NPVs may be polyhedral in shape and are known as polyhedral inclusion bodies (PIB). They contain one or many virions. The occlusion bodies of GVs are ovicylindrical and are known as capsules. They contain one, or rarely two, virions. The structure and biology of the baculoviruses have been described in detail in a volume edited by Granados and Federici (1986).

Baculoviruses have been isolated mainly from Lepidoptera, Hymenoptera, Diptera, and Crustacea. There are also reported isolations from Coleoptera, Neuroptera and Trichoptera. Infection of the host is normally by ingestion. The inclusion body dissolves in the insect mid-gut and the released virions enter the mid-gut epithelial cells. From here other susceptible cells are infected; in Lepidoptera these can include the fat body, hyperdermis, tracheal matrix and its muscles, sheaths lining muscles and nerve fibres, ganglia and pericardial cells (Benz, 1963). Infected cells disintegrate, liberating viral products and eventually disruption of the organs and death of the insect ensues. The resulting liquefied body contents and virus are contained within the fragile 'bag' of the larval skin which easily ruptures when disturbed, releasing infective virus into the environment.

Activity and Susceptibility

The LD_{50} of baculoviruses to insect pests of cotton varies with virus type and strain, as well as the species, strain, weight, instar and age of insect. McKinley (1985) reported that the LD_{50} of *Spodoptera littoralis* NPV to its host insect varied from 1.9 polyhedral inclusion bodies (PIB) per larva for first instar to 3546.2 PIB per larva for the sixth instar this equating to 6.0 and 8.1 PIB per mg body weight respectively. The LD_{50}s also varied within an instar, greater susceptibility occurring immediately after moulting and when feeding recommences following a moult. El-Nagar *et al.* (1983) working with a different strain of *S. littoralis* reported LD_{50}s ranging from 1.95×10^2 PIB per insect for the second instar to 1.4×10^5

PIB per insect for the fifth instar. Klein and Podoler (1978) noted a 12-fold difference in susceptibility to NPV between laboratory and field populations of *S. littoralis*. Similar variation has been found with *Heliothis* and *Helicoverpa* spp. Susceptibility decreased with increasing larval age and body weight. Thus the LD_{50} increased from 9.5 to 2294 PIB per insect when three- to eight-day-old *Helicoverpa zea* larvae were infected with *Heliothis* NPV (Allen and Ignoffo, 1969). Williams and Payne (1984) tested three NPVs against *Helicoverpa armigera*; a multiple enveloped and single enveloped NPV of *H. armigera* and single enveloped virus of *H. zea*. The LD_{50} values for six-day-old larvae were 1400, 670 and 640 PIB per larva, respectively, and for neonate larvae eight, 15 and five PIB per larva. Whitlock (1978), using a different strain of *H. armigera* NPV gives an LD_{50} value of 5×10^4 PIB for second and third instar larvae.

Ignoffo and Couch (1981) reported a 56-fold range of activity for 34 isolates of the NPV of *Heliothis* spp.; *Heliothis virescens* was approximately 30–50% more susceptible than *H. zea* to the *Heliothis* NPV, *H. virescens* was also reported to be five times more susceptible than *H. zea* to *Autographa californica* NPV.

The time taken from infection to death also varies with the virus type and dose, as well as insect species and age; however, mortality invariably takes several days to occur. Ignoffo and Couch (1981) reported that when one–six-day-old *H. zea* and *H. virescens* larvae were infected with NPV, two days were required for initial mortality regardless of the dose. The medium survival time (ST_{50}) of neonate *H. zea* larvae infected with NPV was 75.4 hours, the first mortality being observed after 66 hours (Hughes and Wood, 1986), the dose of NPV used in this study was not reported. The LT_{50} of NPV infection of *H. armigera* was reported to be 3.6 days for one-day-old larvae and 9.9 days for eight–nine-day-old larvae (Daoust, 1974). In a later study Whitlock (1978) used an NPV dose twice as high and recorded LT_{50}s of two–three days for one-day-old larvae and 14 days for eight-day-old larvae. McKinley (1985) found that the LT_{50} of neonate larvae to *H. armigera* NPV was reduced from 11.7 to 4.7 days as the virus dose was increased from 1.5×10^4 to 1.5×10^8 PIB. With *S. littoralis* the LT_{50} fell from 11.5 to 3.0 days as the virus dose was increased from 4.3×10^3 to 2.8×10^8 PIB.

The situation in the field is further complicated by environmental factors, e.g. temperature, which can influence the development of the disease and in some cases may initiate an inapparent or latent infection.

The development of resistance to baculoviruses has been studied in the laboratory with NPV against *H. zea* (Ignoffo and Allen, 1972) and a NPV and a GV against *H. armigera* (Whitlock, 1977). Neither study revealed any evidence for the development of resistance. Apparent development of resistance has been reported after selection experiments in the laboratory with insects other than cotton pests (e.g. Briese and Mende, 1983); Briese (1986), however, points out that no problems of resistance have been encountered with the use of baculoviruses in large-scale control programmes.

Other reported effects of baculovirus infection to cotton pests include a reduction in fecundity of adult moths that were infected in the larval stage (Abul-Nasr *et al*., 1979; Varges-Osuna and Santiago-Alverez, 1986) and a reduction in feeding of infected larvae (Eid *et al*., 1985). Feeding does continue, however, until shortly before death of the insect.

Table 25.3. Reported field trials and use of baculoviruses for pest control on cotton.

Insect pest	Virus	Country	Reference
Agrotis segatum	*A. segatum* GV	Europe	Yearian and Young (1982)
Alabama agillacea	*A. agillacea* NPV	Brazil	Andrade and Habib (1983)
	A. californica NPV	USA	Luttrell *et al.* (1981)
Amsacta spp.	*Amsacta* NPV	(Chad)	Atger (1970)
Anomis flava	*A. flava* NPV	(Australia), China	Bishop *et al.* (1978); Liang *et al.* (1981)
		(Mali)	Atger and Chevalet (1975)
Bacculatrix thurberiella	*A. californica* NPV	USA	Bell and Romine (1982)
Crytophlebia leucotreta	*C. leucotreta* GV	(Ivory Coast)	Labonne (1971)
Diparopsis watersi	*A. californica* NPV	Chad	Atger (1969)
Helicoverpa armigera	*H. armigera* NPV	Botswana, Chad, China, Egypt, India, Ivory Coast, S. Africa, Thailand, Uganda, former USSR, Zimbabwe, Ivory Coast	Coaker (1958); McKinley (1971); Angelini and Couilloud (1972); Cadout and Soubrier (1974); Roome (1975); Whitlock (1978); Maiorov *et al.* (1984); Abou Bakr *et al.* (1986); Wu (1986); Ketunuti and Prathomrut (1989); Anon. (1986)
	H. zea NPV	Australia, Chad	Cadout and Soubrier (1974); Room (1979)
		Zimbabwe	McKinley (1971)
	Mamestra brassicae NPV	N. Cameroon, Chad	Biache and Chaufaux (1986); Renou (1987)
Helicoverpa assulta	*M. brassicae* NPV	China	Liu *et al.* (1987)
Helicoverpa punctigera	*H. zea* (?) NPV	Australia, USA	Pinnock (1980); Ignoffo and Couch (1981)
Helicoverpa virescens	*H. zea* NPV	USA	Bell and Kanavel (1978)
	A. californica NPV	USA	Bell and Romine (1980)
Heliothis zea	*H. zea* NPV	Nicaragua, USA	Ignoffo and Couch (1981); Martinez and Swezey (1988)
Mythimna seperata	*M. seperata* NPV	China	McKinley (1982)
Pectinophora gossypiella	*A. californica* NPV	USA	Bell and Kanavel (1977)
Pseudoplusia includens	*P. includes* NPV	(Guatemala)	Livingston and Yearian (1972)
Spodoptera exigua	*S. exigua* NPV	(Chad), Thailand, USA	Atger (1970); Ketunuti (personal communication); Falcon (1974)
Spodoptera littoralis	*S. littoralis* NPV	Egypt, Israel	Klein and Podoler (1978); Topper *et al.* (1984)
Spodoptera litura	*S. litura* NPV	India, China	Hwang and Ding (1975); Jayaraj *et al.* (1981)
Spodoptera ornithogalli	*S. ornithogalli* NPV	USA	Falcon (1974)
Trichoplusia ni	*T. ni* NPV	(Brazil), Columbia, USA	Falcon (1974); Bellotti and Reyes (1980); Silva and Santos (1980)
	A. californica NPV	USA, Guatemala	Vail *et al.* (1976); Fuxa (1990)

Countries in brackets = potential of control reported, field trials not undertaken.

Field Use of Baculoviruses for Control of Insect Pests of Cotton

Control of insects on cotton with baculoviruses has been attempted or recommended in more than 20 countries and has been directed against some of the major insect pests; these have mainly been lepidoptera (Table 25.3). Results have been variable but in many cases control has been compared favourably with chemical insecticides.

Generally the current use of baculoviruses on cotton, can be split between the control of leaf-eating Lepidoptera and the control of bollworms. Control of leaf feeders has tended to be more consistent than control of bollworms; this can mainly be attributed to the small amount of surface feeding of bollworms resulting in a reduced probability of virus ingestion. Successful control requires accurate application to the site of the target insect as well as accurate timing of application. Unsuccessful control has often resulted when virus applications were made against unsuitably high populations of insects or against populations of older larvae. From the discussions on susceptibility above it should be realized that older larvae require more virus (mitigated to a degree by a greater intake of food and hence virus on contaminated plants) and a longer time to kill. In the period from infection to death late instar lepidopterous larvae can cause considerable damage. Thus with noctuid larvae the food comsumed during the sixth instar amounts to 50% or more of the total amount consumed during the larval stage. More effective control (Afifi and Mespah, 1990), as with insecticides, can therefore be achieved when baculoviruses are aimed at first and second instar larvae. However, successful control of later instar larvae has been obtained with leaf-feeders, in certain circumstances, as cotton can withstand a considerable amount of damage to the leaves, up to a 50% loss of leaf surface area, without reduction in yield (Anon., 1969; Hosny *et al.*, 1986; Russell *et al.*, 1993).

Control of Leaf Feeding Lepidoptera

A representative example of this group is control of *Spodoptera* spp. Successful control of *S. littoralis, S. litura* and *S. exigua* has been reported in several countries (Table 25.3). Topper *et al.* (1984) showed that application of NPV to control *S. littoralis* in Egypt was as effective in reducing damage as either the hand-collection of egg-masses (the current method of control) or the application of chemical insecticides. Virus was applied as an aqueous suspension at the rate of 5×10^{12} PIB ha^{-1}, in a volume of 140 l ha^{-1}, when egg-masses numbers had exceeded a threshold level of 4000 ha^{-1}. Using this threshold it was estimated that on average each field would need to be treated a total of 1.6 times. Recent field trials have indicated that the virus rate can be reduced to 1×10^{12} PIB ha^{-1} applied in 90 l water ha^{-1} (Jones *et al.*, 1994). A commercial preparation of *S. littoralis* NPV, called Spodopterin, has been produced in France by Calliope and has been field tested in Egypt. Dhandapani *et al.* (1992) report successful control of *S. litura* on cotton in India with no applications of *S. litura* NPV at a rate of 250 larval equivalents ha^{-1} (approx. 5×10^{11} PIB ha^{-1}).

In Columbia, application of *Trichoplusia ni* NPV to cotton was so successful in controlling *T. ni* larvae that it replaced all other pest control methods; the insect is now no longer considered to present a pest problem (Bellotti and Reyes, 1980).

Control of Bollworms

By far the greatest use of baculoviruses in cotton has been directed against the control of *Heliothis/Helicoverpa* spp. and the first commercially registered baculovirus was *Heliothis/Helicoverpa zea* NPV (often referred to as *Baculovirus heliothis*) manufactured by Sandoz as Elcar. Ignoffo and Couch (1981) and Yearian and Young (1982) have summarized much of the research that led to the development of Elcar. A review of the field trials shows that effective control of *H. zea* was obtained at virus rate of 2.5×10^{11} PIB ha^{-1} for light to moderate infestations and 1.125×10^{12} for heavy infestations. An average of three applications of virus in a season was assumed. However, Entwistle and Evans (1985) who summarized the criteria for successful control of bollworms (*H. zea* and *H. virescens*) in the USA concluded that higher doses of virus were required: to obtain increases in cotton yield the total amount of PIB applied needed to be $>5 \times 10^{12}$ PIB per hectare per season. At doses above this, yield increased linearly with virus dose. Generally, yearly doses of 1.5×10^{13} PIB ha^{-1} have been applied in five to 13 applications, there being some indication that better control was obtained with the greater number of applications (constant total virus dose) possibly due to poor persistence of the virus.

McKinley (1971) and Brettell (personal communication) found that NPV did not control heavy populations of *H. armigera* in Zimbabwe. In contrast, Ketunuti and Prathomrut (1989) demonstrated effective control of *H. armigera* on cotton in Thailand at a rate of $1.3–1.95 \times 10^{12}$ PIB ha^{-1}. They estimated that the use of NPV could replace four–six of the 15 or more insecticide applications to cotton. Good control of *H. armigera* using NPV as part of an integrated pest management system has been reported in Australia (Room, 1979), and there is large scale use of *Heliothis* virus on cotton in China. In the latter case, a total of 23 tonnes of viral pesticides were produced in Hubei province from 1983 to 1991 to protect 40,000 ha, 80% of which was cotton (Winstanley and Rovesti, 1993). A commercial formulation of *H. armigera* NPV, called Virin-HS, is produced in the former USSR by NPO (Nauchno-proizvodstvennoe Objedinenie) Vector (Novosibirsk, Siberia) (Anon., 1991). Variable control of *H. armigera* in India has been reported (e.g. Jayaraj *et al.*, 1989). Control of *H. armigera* has also been successful using *Mamastra brassicae* NPV and a commercial formulation of this virus called Mamestrin has been produced by Calliope (Anon., 1986). The recommended rate of application for this latter product is 1×10^{13} PIB ha^{-1} in a volume of five litres. In Thailand a cottonseed company (Yong Suwat Seeds Co. Ltd) has built a factory for production of *H. armigera* and *S. exigua* NPVs for control of these pests on local cotton fields.

Attempts to control pink bollworm (*Pectinophora gossypiella*) using *Autographa californica* NPV were not successful (Bell and Kanavel, 1975, 1977), possibly due to the limited external feeding of this insect. This virus has also been field tested as an experimental formulation, San 404 I WP developed by Sandoz, against *Heliothis* spp. but proved to be inferior to Elcar (Yearian and Young, 1982). Yearian and Young (1982), however, reported that interest in the use of *A. californica* NPV still remained in the southwestern and western United States where insecticide applications against *Lygus hesperus* often result in an upsurge in the populations of *S. exigua* and *Trichoplusia ni* which are both susceptible to the virus.

Persistence of Baculoviruses on Cotton

Frequency of application and effectiveness of control with baculoviruses is partly dependent on the persistence of the virus on the cotton plant. Baculovirus persistence has been reviewed by Jaques (1977) and Richards and Payne (1982). Entwistle and Evans (1985) reported that the average half-life of *H. zea* NPV on the upper surface of *G. hirsutum* was 12.8 ± 5.1hours and the three-quarter-life was 41.25 ± 21.96 hours. The three-quarter-life of *S. littoralis* NPV on the upper surface of cotton leaves (presumably *G. barbadense*) was 105.6 hours. Persistence on protected parts of the plant, for example on the undersurface of leaves, can be considerably longer. Thus McKinley *et al.* (1989) estimated that whereas on the upper leaf surface toward the top of the plant (*G. barbadense* in Egypt) the amount of infective virus would fall from 3850 to 41 PIB cm^{-2} in 48 hours, after the same period on the lower leaf surface the number would only fall to 570 PIB cm^{-2}. Similarly, Yearian and Young (1974) reported that, in the USA, the persistence of *Heliothis* NPV on the lower surfaces of cotton leaves (*G. hirsutum*) was considerably greater than on the upper surface. In contrast, McKinley (1971) found little difference in the persistence of *H. zea* NPV on the upper and lower leaf surfaces of *G. hirsutum* in Zimbabwe.

The loss of activity of baculoviruses on cotton results from simple physical loss of inclusion bodies from the plant surface and inactivation of the virus by environmental factors.

Physical Persistence of Inclusion Bodies

Persistence studies have often taken no account of the physical loss of inclusion bodies. Most studies that have been done have been laboratory based. Ignoffo *et al.* (1965) and Bullock (1967) demonstrated that *Heliothis* NPV was not washed from cotton leaves by artificial rain. McKinley (1985) also showed that purified deposits of *S. littoralis* NPV were not easily washed from cotton leaves. It can be concluded that normally baculoviruses become firmly attached to the plant surface, attachment probably involving electrostatic forces, Van der Waal's forces and chemical bonds between the plant surface and the inclusion bodies (Small, 1985), and moderate rainfall therefore results in little loss of purified inclusion bodies from the plant. Jones and McKinley (1987) and Jones (1988), however, demonstrated there was an appreciable loss of inclusion bodies of *S. littoralis* NPV from cotton leaves in Egypt. This was attributed to physical abrasion caused by the action of wind and sand which resulted in more than a 90% loss of purified inclusion bodies in a week. The rate of loss was slower from the undersurface of the leaves and with unpurified virus, in the latter case the percentage of inclusion bodies lost within a week ranged between nil and 80%.

Although not strictly physical loss, it is also worth mentioning here growth dilution as this represents a loss of inclusion bodies per unit area. For example, Jones and McKinley (1987) showed that the leaf area of *G. barbadense* (var. Giza 75) could increase by 100% in a seven day period. This increase occurred mostly with younger leaves on which egg-masses and young larvae of *S. littoralis* are normally found. Thus within a seven day period there is a 50% loss of inclusion bodies per unit area. Similar rates of increase in leaf area have been noted with *G. hirsutum* by Bahtt (1974), Wilson *et al.* (1983) and Jones (unpublished data).

Inactivation of Baculoviruses on Cotton

Environmental factors that may inactivate baculoviruses in the field are high temperatures, solar irradiation, and interaction with chemicals on the plant surface.

Effect of field temperatures

Baculoviruses generally require temperatures above 60°C for more than 10 minutes to be inactivated (Entwistle and Evans, 1985), at lower temperatures little effect has been noted over several days. Field temperatures are therefore unlikely to inactivate baculoviruses and this has been demonstrated for *Heliothis* NPV by MacFarlane and Keely (1969) and McLeod *et al.* (1977), and for *S. littoralis* NPV by McKinley (1985) and Jones (1988).

Solar irradiation

Baculoviruses are rapidly inactivated by exposure to UV radiation. The sensitivity of *Heliothis* NPV to solar radiation has been demonstrated by Bullock (1967), Ignoffo and Batzer (1971) and McLeod *et al.* (1977), and with *S. littoralis* NPV by El-Nagar and Abul-Nasr (1980), McKinley (1985), Jones and McKinley (1987), Jones (1988) and Jones *et al.* (1993a). Laboratory and field experiments have demonstrated that it is the wavelengths in the UV range between 290 nm and 320 nm, which are present in sunlight and reach the Earth's surface, that are mainly responsible for the inactivation caused by solar radiation (Witt and Stairs, 1975; Griego *et al.*, 1985; Jones *et al.*, 1993a). When exposed in the open to direct solar radiation the half-lives of the NPVs of *Heliothis* spp. and *S. littoralis* were 'less than one day' – probably only a few hours (in the USA) and less than one hour (in Egypt), respectively (Ignoffo and Couch, 1981; Jones, 1988). However, in the field the target insects for control are often found in shaded areas of the cotton plant such as the undersurface of the leaves, where the intensity of solar UV radiation may be as little as 1% of that found in direct sunlight (Jones, 1988). Thus, Jones (1988) demonstrated that inactivation by solar UV radiation was much less on shaded areas of the cotton plant within the canopy and on the undersurface of leaves. Unpurified virus has also been shown to be considerably more resistant to inactivation by solar radiation than purified deposits of virus (El-Nagar and Abul-Nasr, 1980; Jones, 1988).

Interactions with chemicals on the leaf surface of cotton

Cotton leaves have a strongly alkaline surface with pH values as high as ten or 11 being recorded on both *G. barbadense* and *G. hirsutum* (Harr *et al.*, 1980; Jones and McKinley, 1987; Jones, 1988). Salt secretions from epidermal glands account for the high pH values. The effects of the salts, and the resulting alkaline dew encountered on cotton leaves, on the infectivity of baculoviruses have been studied in some detail. Griffiths (1982) demonstrated that baculovirus inclusion bodies are rapidly dissolved at pH values of around ten and field observations have shown that *Heliothis* NPV is inactivated on cotton leaves protected from sunlight (Andrews and Sikorowski, 1973; McLeod *et al.*, 1977; Young *et al.*, 1977) with more than a 50% loss of activity in one to six days. Richards (1980) and Elleman and Entwistle (1985a) also demonstrated in the laboratory that *S. littoralis* NPV was inactivated on the leaves of *G. hirsutum* placed in the dark.

The inactivation of *Heliothis* NPV by cotton leaf exudates has been attributed to the dissolution of inclusion bodies on the leaf surface, particularly under conditions of repeated wetting and drying (Andrews and Sikorowski, 1973). However, Elleman (1983) and Elleman and Entwistle (1985b) concluded that exposure of *S. littoralis* NPV to leaf exudates from *G. hirsutum* resulted in the occlusion bodies becoming less soluble in dilute alkali, an effect probably resulting from the presence of cations (Mg^{2+}, Ca^{2+}, Na^{+} and K^{+}) in the exudates. This lower solubility accounted for the reduced infectivity as the virions would not be released in the insect mid-gut. However, Jones (1988) showed that *S. littoralis* NPV was not inactivated after seven days when placed on leaves of *G. barbadense* in Egypt and protected from sunlight. These contradictory results might be related to differences in the cotton species or differences in environmental conditions, such as degree of water stress to the plants, amount of dew and other factors. It is generally agreed that the effect of the leaf-surface exudates plays only a minor role in virus inactivation in the field when compared to the effects of solar UV radiation.

Production and Formulation of Baculoviruses for Application to Cotton

Production

At present the production of baculoviruses can only be effectively and economically achieved *in vivo* and therefore requires the large-scale rearing of susceptible insects or the collection of infected insects from the field. In the majority of cases, baculoviruses used in cotton pest control have been produced in laboratories or factories under controlled conditions. Shapiro (1982, 1986) has described the basic techniques required for the *in vivo* mass production of virus. In summary, insects are reared until an optimum stage and then infected with virus. The infected insects are then incubated until death or harvested just prior to death, the latter case resulting in reduced problems of bacterial contamination (Shapiro, 1986; McKinley *et al.*, 1989). The process is optimized by establishing the best age and size of insect for infection, the appropriate virus dose and optimum rearing conditions. Some evidence has been reported that virus harvested from live larvae was less virulent than that harvested from dead larvae (Ignoffo and Shapiro, 1978). This problem might be overcome by a low temperature incubation of live larvae after harvest to allow the virus infection to proceed to death of the insect, whilst still limiting the growth of contaminant microorganisms (McKinley *et al.*, 1989). Yields of baculoviruses are related to the body weight of the infected insects. The maximum yield of Lepidoptera infected with baculoviruses is around 1×10^{10} PIB per insect (Entwistle and Evans, 1985); however under pilot or commercial scale production yields are considerably lower. Thus whilst the maximum number of PIB recovered from dead *S. littoralis* larvae was 5.4×10^9, the level of production averaged over a period of several months was nearer to 5×10^8 PIB per larva (Jones *et al.*, unpublished).

Entwistle and Evans (1985) summarized the cost (materials and labour) of producing several different baculoviruses. The cost of producing 1×10^9 PIB of *H. zea* NPV was put at 0.6 US cents in 1969 and the purchase price of the same amount of Elcar in 1980 was 1.3 US cents (Ignoffo and Couch, 1981). This suggests, as Entwistle and Evans (1985) point out, that production costs are becoming relatively cheaper as a result of greater efficiency. As the production process tends to be labour intensive, costs can be much lower in developing countries. An estimation of the cost of producting *S. littoralis* NPV in Egypt in 1990 made on the same basis as above gives a figure of about 1.0 US cent for 1×10^9 PIB (Bickersteth, 1990); at current application rates this equates to \$10.00 per hectare per application. The inclusion of the capital cost of building a production plant and the cost of applying the virus would increase the cost to \$18.06 per hectare per application. This compares with \$17.80–42.00 for the standard pest control measures used in Egypt (Jones *et al.*, 1993b).

Before formulation the harvested insects need to be processed. Processing techniques have included purification through sucrose density gradients, slow speed centrifugation, precipitation of the virus or simple filtration. The virus suspension may then be dried, often employing freeze-drying or spray-drying techniques. This processing may affect storage properties. Thus Ignoffo and Couch (1981) reported that acetone-extracted virus was not as stable as aqueous, lyophilized samples. Spray drying of formulations of *H. zea* NPV produced a more active product with greater stability than the previous processing method of acetone precipitation (Ignoffo *et al.*, 1976).

Formulation

Formulation of baculoviruses, like chemical insecticides, is essential for consistent and effective control. Although novel methods of application of baculoviruses have been suggested such as the release of contaminated adults or the spreading of solid baits containing virus (Martignoni and Milstead, 1962; McKinley, 1980), the practical use of baculoviruses for cotton pest control has relied almost entirely on the application of virus sprays to the crop. Where these have been compared to the application of dusts or solid baits, sprays have proved to be more effective (Montoya *et al.*, 1966; Stacey *et al.*, 1977a).

Formulation of baculoviruses has been reviewed by Young and Yearian (1986). Formulation additives such as wetting and thickening agents, and anti-evaporants can be added to improve physical performance by aiding dispersion and suspension of the virus in the spray liquid, to improve the deposition of the virus spray on the target and to aid the spread of spray droplets on the target . Additives may also be included in a formulation to maximize persistence (e.g. stickers, UV protectants) and encourage ingestion of the virus (e.g. gustatory stimulants and attractants) or to improve potency (e.g. various chemicals, including chemical insecticides). Finally, with liquid concentrates of virus, anti-bacterial and anti-fungal agents can be added to prevent growth of contaminant microorganisms. Generally, the development of formulations for baculoviruses has not been systematic and the choice of additive tested seems mainly to have been based on what was easily available to the researcher. Moreover, the goals required for a formulation, e.g. required length of persistence, have not been clearly defined. Consequently a large range of additives have been tested with baculoviruses for use on cotton, with varying success. In many cases, however, the effectiveness of individual additives on the performance of the formulation has not been reported.

Surfactants that have been tested and included in field formulations include Triton X-100 (Ignoffo and Montoya, 1966; McKinley, 1971), Teepol (Topper *et al.*, 1984) and Etocas 30 (McKinley *et al.*, 1989).

Stickers that have been tested include commercial products, e.g. Pitsulin with *H. armigera* NPV in Thailand (Ketunuti, personal communication) and generally available materials such as molasses (McKinley, 1971; Roome, 1975; Jones, 1988), larval extract, whole milk (Dhandapani *et al.*, 1987) or egg albumin (Chunderwar, personal communication). Jones (1988) tested a range of natural gums, demonstrating the most effective to be gum guar; none, however, was completely effective in preventing physical loss of polyhedra in the field and no formulation was more effective than simply using unpurified virus.

The largest group of additives that have been tested are UV protectants and several have been identified that provide significant protection. These range from commercial products: Tinopal RBS 200 (McKinley, 1985), Eusolex 232 (Jones, 1988) and Shade (Yearian and Young, 1974; Young and Yearian, 1976); and specialized dyes and chemicals: indigo carmine (Jones, 1988), lignin sulphate (Yearian and Young, 1974), activated charcol (Ignoffo *et al.*, 1972) and encapsulation in titanium dioxide, carbon black or starch (Bull *et al.*, 1976; Ignoffo *et al.*, 1991); to locally available products: molasses (Roome, 1975; Jones, 1988), 'Robin Blue' shirt whitener (Chundewar, personal communication) and various clays (Jones, 1988).

Numerous feeding stimulants have also been tested, these have included many sugars or sugar-based products, including sucrose (Stacey *et al.*, 1980), fructose and glucose (Stacey *et al.*, 1977c), invertase sugars (Andrews *et al.*, 1975) and jaggery (raw sugar, Dhandapani *et al.*, 1987). Molasses has also been added as a feeding stimulant (Stacey *et al.*, 1977b; Pfrimmer, 1979), although Jones (1988, 1990) found that molasses did not encourage feeding by larvae of *S. littoralis*. Increase in cotton yields resulting from the addition of sugars and sugar-based products to baculovirus sprays may have resulted from physiological effects on the plants (Stacey *et al.*, 1977b). Other feeding stimulants have been based on plant products and extracts, including dried garbanzo bean (Potter and Watson, 1983a), cotton seed and soybean flour (Bell and Kanavel, 1978; Potter and Watson, 1983b), corn, pulverized wheat and cottonseed meal (Stacey *et al.*, 1977b) and water extracted cottonseed, maize silks, kernels and seeds (Bell and Kanavel, 1978). Wheats and powdered milk have also been used (Stacey *et al.*, 1977b). The importance of feeding stimulants in aiding ingestion, and hence effectiveness, of viral pesticides was demonstrated by the recommendation by Sandoz that Elcar should be used in combination with a commercially available feeding stimulant. In this case the recommended product was Gustol, which is based on soyabean meal. A similar product has been sold under the trade name Coax; this is based on cottonseed flour (5 parts), mixed with cottonseed oil (1 part), sucrose (2 parts) and 0.4 per cent Tween 80 (Smith *et al.*, 1978). Several workers have demonstrated that the addition of Gustol or Coax improves control of *Heliothis* spp. on cotton (e.g. Johnson, 1982; Renou, 1987). Coax has also been successfully used in combination with *S. littoralis* NPV in Egypt; virus formulated with Coax gave superior control over unformulated virus. However, this may simply be a result of increased UV protection or Coax acting as a sticker because similar results were obtained by the addition of molasses (Topper *et al.*, 1984). Potter and Watson (1984) found that, although the addition of Coax extended the activity of Elcar on cotton, it did not increase the mortality of larvae feeding on the cotton. Under the environmental conditions in Egypt, Coax was extremely difficult to suspend and often blocked the spray nozzles during application (Jones, unpublished).

Other chemicals have been added to enhance virus potency. An example of this is sodium tetraborate which has been combined with *S. littoralis* NPVs (McKinley, 1985) in laboratory tests. Also included in this category are conventional chemical insecticides. Many instances of additive or synergistic combinations have be cited, but in general, results have been very variable. Thus, virus combined with *Bacillus thuringiensis*, chloridimeform, carbaryl, endrin, dicrotophos, pyrethrum (+ piperonyl butoxide) or deltamethrin have all been reported to give improved control in the field over virus alone (Hafez *et al.*, 1970; Atger, 1971; Chapman and Ignoffo 1972; Anon., 1976, Mohamed *et al.*, 1983, Biache and Chaufaux, 1986). In contrast, other workers have reported that addition of *B. thuringiensis*, permethrin, methomyl, chlordimeform and carbaryl resulted in a minimal increase, or even reduced control (Pieters *et al.*, 1978; Luttrel *et al.*, 1979, 1982; Johnson 1982). It can be concluded that the compatability of baculoviruses with chemical insecticides is dependent on dose, insect age and timing of application, and results are therefore unpredictable. Preferably such combinations should be avoided. However, a few cases exist where combinations are recommended. Calliope, for example, produce a formulation of *M. brassicae* NPV containing a low dose of pyrethroid insecticide; field trials in Chad demonstrated NPV–deltamethrin combinations to be synergistic (Biache and Chaufaux, 1986). In Thailand virus is often mixed with chemicals when spraying *H. armigera* on cotton, the insecticide killing most individuals rapidly and the virus killing insecticide resistant individuals over several days. Recently *S. exigua* NPV has been added to these sprays to kill mixed populations.

Compatability with spray tank mixes has also been investigated. No adverse effect on *Heliothis* NPV has been demonstrated when mixed with thiabendazole, or benomyl, but chlorothalonil or fentin hydroxide were antagonistic (Mohamed *et al.*, 1983).

Although baculoviruses have mostly been applied using water as the carrier, several oil-carriers have been tested with a view to using ULV application. Actipron and cottonseed oil have been used with *S. littoralis* NPV and were found to have no adverse effect (Topper, 1984); in contrast Shellsol T was found to inactivate the virus, as did Petrol Spirit (McKinley, 1985). *Spodoptera littoralis* NPV has been successfully applied in the ICI Electrodyn oil carrier, McKinley (1985) also demonstrating that suspension in this carrier, isophorone, did not inactivate the virus. *Heliothis zea* NPV has been applied in several mineral oils (Smith *et al.*, 1978) and cottonseed oil (Falcon *et al.*, 1974) without adverse effect to the virus. This virus has also been applied in Multifilm buffer X (Ignoffo and Montoya, 1966) and phosphate buffer to provide protection against the effects of the high pH on the leaf surface, however, no improvement in control was demonstrated (Young and Yearian, 1976).

Formulations that have proved to be successful on cotton have normally contained several additives. However, some of the formulations that have proved to be effective in field trials are impractical in terms of ease of use or cost. For example, effective control of *S. littoralis* on cotton in Egypt was demonstrated using an aqueous formulation of highly purified NPV containing 10% molasses, 0.001% Tinopal RBS200 and 0.5% Teepol (Topper *et al.*, 1984). However, this formulation was impractical because the cost of purification of the virus was too high, accounting for about 75% of the total cost of production and formulation, and some of the formulation processes (e.g. the addition of the molasses) had to be carried out in the field. A formulation was therefore developed to overcome these problems; this consisted of 100 g freeze-dried, unpurified virus, 60 g china clay and 40 g of a 1:1 mixture of synthetic silica and wetting agent (Etocas 30), the wetting agent having been adsorbed to the silica (McKinley *et al.*, 1989). In this formulation the insect debris, etc., in the unpurified virus provided protection against solar UV radiation and acted as a sticker, the china clay provided additional UV protection and was a bulking agent, the silica aided flowability of the powder and the Etocas aided suspension of the powder and spray deposition. Because the product is a dry powder there is no problem of growth of contaminant microorganisms. Estimates of the field persistence of this formulation in Egypt concluded that infective virus would remain on the plant for several days or weeks (Jones, 1988).

Unfortunately, additives used in commercial formulations developed for cotton pest control are not generally known. Elcar was produced as a wettable powder formulation and included clay amongst the additives as a diluent. Ignoffo *et al.* (1976) reported that Elcar contained an anti-evaporant, UV protectant and was gustatory-acceptable. Virin-HS is produced as a 'finely divided powder'. Mamestrin and Spodopterin are both liquid concentrate formulations; the technical leaflet for the former states that the formulation contains UV protectants and adhesivity polymerizing agents.

Spray Application

The importance of application in effective control using baculoviruses cannot be over-emphasized. In the absence of a very effective feeding attractant, the virus needs to be applied to regions of the cotton plant where it will be encountered and ingested by the target insect.

Application of baculoviruses to cotton has almost exclusively relied on ground spraying. Most research on application of baculoviruses to cotton has been carried out with *H. zea* NPV and this has been summarized by Young and Yearian (1986). Stacey *et al.* (1977c) found that higher volume sprays (47 l ha^{-1}) gave superior control than low volume spray (9.6 l ha^{-1}). Higher volume sprays were also found to be more effective by Chapman and Ignoffo (1972), and against *S. littoralis* in Egypt by Topper (1984). Drop size was not found to affect control of *H. zea* (Stacey *et al.*, 1980). Both drop size and spray volume, however, may influence the efficiency of gustatory stimulants (Young and Yearian, 1986) and UV protectants (Killick, 1986). Nozzle arrangement and orientation did not influence control of *Heliothis* spp. in the USA (Stacey *et al.*, 1977c). In contrast, upwardly directed sprays, obtained by using a cotton tailboom attached to a knapsack sprayer, were needed to obtain effective control of *S. littoralis* in Egypt, this insect being mainly found on the lower surface of leaves during the early instars. Good underleaf cover was also obtained on cotton in Egypt with an experimental electrostatic sprayer (Cooper, 1983) and McKinley (1985) showed that the passage of *S. littoralis* NPV through this sprayer did not affect the infectivity of the virus. ICI have also successfully applied *S. littoralis* NPV to cotton in Egypt using the Electrodyn spraying system.

Registration of Baculovirus Insecticides

Registration of baculoviruses for pest control has been reviewed by Entwistle and Evans (1985), Betz (1986) and in a volume edited by Evans (1991). The requirements of the regulatory bodies tend be less than that required for chemicals. Elcar was the first viral pesticide to be registered, breaking new ground. Until this time registration requirements were unclear, or totally non-existent. Elcar underwent rigorous safety and specificity tests, and was finally granted an exemption from 'requirement of a tolerance' on cotton by the USA Environmental Protection Agency in 1973 prior to registration in 1975. This helped to increase public confidence in the safety of baculovirus-based pesticides. Despite this and a large amount of evidence on the safety and specificity of baculoviruses the amount of safety testing required is still substantial. Different countries do have different requirements, although as both the EEC and FAO have recently drawn up guidelenes on the registration of microbials that are broadly similar to each other and to USA and Eastern European requirements, some standardization may occur. Broadly, these regulatory requirements include a description of the product and formulation, and the provision of data on infectivity, allergenicity, toxicity to defined non-target organisms (including beneficial insects), carcinogenicity and teratogenicity, and levels of residues (simplified in some instances to persistence data). This is required for the active ingredient and formulation. These are less stringent than for chemical insecticides and Lisansky (1984) estimated that toxicological testing for a naturally occurring microbial insecticide would cost some £40,000 compared with £3,000,000 for a chemical insecticide; both figures would be considerably higher at 1994 costs. The USA has now adopted a tiered approach to the registration of microbial pesticides. Tier 1 mammalian toxicology tests are acute oral, acute dermal, acute inhalation, intraperitoneal injection, infectivity studies, dermal and eye irritation, hypersensitivity tests, immune responses and tissue culture evaluation. Environmental data requirements are avian oral, avian injection, wild mammal tests, freshwater fish and invertebrate tests, estuarine and marine animal tests, plant tests and non-target insect tests. Further longer term and more detailed studies are only required if the product fails one or more of these tests. At present it would seem that every virus strain would need to undergo this process of registration, however, many strains of each specific virus occur naturally and it could be argued that a simplified or truncated procedure might be suitable for closely related (and non-genetically engineered) virus strains.

In developing countries registration requirements are varied. India has recently drawn up regulations that are similar in nature to the FAO guidelines. Other countries rely on plant quarantine regulations or simply require data on the product's efficacy, especially if the virus occurs naturally in that country; this is the case for Egypt. In Thailand efficacy and levels of contaminant microorganisms are required. In many countries no regulations requirements have been considered. However, as more products are developed the regulatory authorities will need to address the problem. Until then FAO, EPA or EEC guidelines can be referred to.

Conclusions and Future Use

Baculoviruses have been shown to control effectively some major cotton pests. Only a few commercial products have been developed; in the case of Elcar this was scientifically successful but due to competition from the synthetic pyrethroids was not an economic success (Huber, 1986) despite the lower commercialization costs (estimated by Ignoffo (1979) to be two–five times lower than for chemical insecticides) and production was halted. Recently, however, renewed interest has been shown in the use of environmentally friendly pesticides and new products have been developed by Calliope and are being produced in the former USSR. Recently, the USDA announced a new programme for testing Elcar, employing a strategy of spraying alternative weed hosts of *Heliothis* rather than cotton being employed: in 1990 some 100 square miles were subject to this treatment (Bell, 1990). In Egypt a pilot-scale production plant has been established at the Ministry of Agriculture's Plant Protection Research Institute, using funds from the UK's Overseas Development Administration, for production of *S. littoralis* NPV (Jones, 1990), and a similar approach has been adopted in Thailand for *H. armigera* and *S. exigua* NPVs and has led to commercial production of the viruses. Small-scale commercial production of *H. armigera* NPV is carried out in India and large-scale local production is carried out in China.

Successful control, however, requires a consistent product subject to some sort of quality control, effective and simple formulations that are easy to use and well defined methods of application, preferably using standard equipment. Education of users is also necessary so that baculoviruses are not used to try and control unsuitably high pest populations, rather they should mainly be used to replace some chemical insecticide sprays at low to moderate populations. This requires effective pest scouting. Farmers also need to be educated to accept the presence of some live pest insects after application of the virus and also some damage to the crop.

Effective control has been demonstrated in several developing countries, where the economics of production make viruses an attractive alternative to chemicals. Local production of viruses for cotton pest control in developing countries is likely to be the most appropriate strategy for the future.

Future developments include the possible use of genetically engineered baculoviruses (Beckmann and Davidson, 1991). Already *B. thuringiensis* and other toxin genes have been engineered into baculoviruses in an attempt to increase speed of kill or to stop the feeding of infected insects. Several other genes are also being considered. However, such products may prove to be difficult to register, a problem not encountered with naturally occurring viruses. For this reason, where possible, the use of genetically engineered viruses should be avoided.

References

Abou Bakr, H.E., El-Husseini, M.M. and Tawfik, M.F.S. (1986) Crossinfectivity and anti-genic characteristics of two nuclear polyhedrosis viruses in larvae of *Heliothis armigera* (Hbn.) and *Spodoptera littoralis* (Boisd.). *Bulletin of the Entomological Society of Egypt, Economic Series* 14, 13–21.

Abul-Nasr, S.E., Ammar, E.D. and Abul-Ela, S.M. (1979) Effects of nuclear polyhedrosis virus on various developmental stages of the cotton leafworm *Spodoptera littoralis. Zeitschrift für Angewändte Entomologie* 88, 181–187.

Afifi, F.M.L. and Mesbah, I.I. (1990) Economic threshold of infestation with the cotton leafworm. *Spodoptera littoralis* (Boisd.) (Lepidoptera: Noctuidae), in cotton fields in Egypt. 1. Food consumption of larvae. *Arab Journal of Plant Protection* 8, 96–100.

Allen, G.E. and Ignoffo, C.M. (1969) The nucleopolyhedrosis virus of *Heliothis*: quantitative in vivo estimates of virulence. *Journal of Invertebrate Pathology* 13, 378–381.

Andrade, C.F.S. and Habib, M.E.M. (1983) Pathology of the nuclear polyhedrosis of the cotton worm *Alabama argillacea* (Hubner, 1818) (Lep., Noctuidae). *Revista de Agricultura, Brazil* 58, 269–290.

Andrews, G.L. and Sikorowski, P.P. (1973) Effects of cotton leaf surfaces on the nuclear polyhedrosis virus of *Heliothis zea* and *Heliothis virescens* (Lepidoptera: Noctuidae). *Journal of Invertebrate Pathology* 22, 290–291.

Andrews, G.L., Harris, F.A., Sikorowski, P.P. and McLaughlin, R.E. (1975) Evaluation of *Heliothis* nuclear polyhedrosis virus in a cotton seed oil bait for control of *Heliothis virescens* and *Heliothis zea* on cotton. *Journal of Economic Entomology* 68, 87–90.

Angelini, A. and Couilloud, R. (1972) Biological control measures against certain pests of cotton and a view of integrated control on the Ivory Coast. *Coton et Fibres Tropicales* 27, 283–289.

Anon. (1969) *Insect Pest Management and Control.* Publication 1695, National Academy of Sciences, Washington, DC, 500pp.

Anon. (1976) Observations on the nuclear polyhedrosis virus of the cotton bollworm and its field tests in bollworm control. *Acta Entomologica Sinica* 19, 167–172.

Anon. (1986). Mamestrin[R] a new biological insecticide. Technical leaflet, Calliope S.A., Beziers, France.

Anon. (1991) *Agrow* 128, 9.

Atger, P. (1969) A virus disease with nuclear localisation in *Diparopsis watersi* Roth. *Coton et Fibres Tropicales* 24, 205–206.

Atger, P. (1970). Note on the entomopathogenic microorganisms of cotton pests utilised or discovered by the I.R.C.T. *Coton et Fibres Tropicales* 25, 521–524.

Atger, P. (1971) New prospects for the control by viruses of *Heliothis armigera* Hbn (Lepidoptera: Noctuidae) a pest of cotton in Africa. *Comptes Rendus des Sciences de l'Academie d'Agriculture de France* 57, 1544–1548.

Atger, P. and Chevalet, Y. (1975) Brief notice of a viral epizootic in *Cosmophila flava* F. (Noctuidae) in cotton plantations of Mali. *Coton et Fibres Tropicales* 30, 371.

Bahtt, J.G. (1974) Leaf growth and leaf area index in morphologically contrasting varieties of cotton. *Cotton Growing Review* 51, 187–191.

Beckmann, R. and Davidson, S. (1991) Virus to control bollworm pests? *Rural Research* 152, 12–14.

Bell, M.R. (1990) Use of baculovirus for the control of *Heliothis* spp. in area-wide pest management programs. In: *Proceedings and Abstracts, Vth International Colloquium on Invertebrate Pathology and Microbial Control*, 20–24 August 1990. Society for Invertebrate Pathology, Adelaide, Australia, pp. 486–490.

Bell, M.R. and Kanavel, R.F. (1975) Potential of bait formulations to increase effectiveness of nuclear polyhedrosis virus agains pink bollworm. *Journal of Economic Entomology* 68, 389–391.

Bell, M.R. and Kanavel, R.F. (1977) Field tests of a nuclear polyhedrosis virus in a bait formulation for control of pink bollworms and *Heliothis* spp. in cotton in Arizona. *Journal of Economic Entomology* 70, 625–629.

Bell, M.R. and Kanavel, R.F. (1978) Tobacco budworm: development of a spray adjuvant to increase effectiveness of a nuclear polyhedrosis virus. *Journal Economic Entomology* 71, 350–352.

Bell, M.R. and Romine, C.L. (1980) Tobacco budworm: field evaluation of microbial control in cotton using *Bacillus thuringiensis* and a nuclear polyhedrosis virus with a feeding stimulant. *Journal of Economic Entomology* 73, 427–430.

Bell, M.R. and Romine, C.L. (1982) Cotton leafperforator (Lepidoptera: Lyonetiidae): effect of two microbial insecticides on field populations. *Journal of Economic Entomology* 75, 1140–1142.

Bellotti, A.C. and Reyes, J.A. (1980) Progress in microbial control (1975–1980): World activities: North, South and Central America. In: *Proceedings of a Workshop on Insect Pest*

Management with Microbial Agents: Recent Achievements, Deficiencies and Innovations. Cornell University, Ithaca, N.Y., p. 20.

Benz, G. (1963) A nuclear polyhedrosis virus of *Malacosoma alpicola* (Staundinger). *Journal of Insect Pathology* 5, 215–241.

Betz, F.S. (1986) Registration of baculoviruses as pesticides. In: Granados, R.R. and Federici, B.A. (eds) *The Biology of Baculoviruses*, vol. II *Practical Application for Insect Control.* CRC Press, Boca Raton, Florida, pp. 203–222.

Biache, G. and Chaufaux, J. (1986) Studies on the conditions for using baculoviruses in the control of noctuids. *Colloques de l'INRA* 34, 149–165.

Bickersteth, S. (1990) A socio-economic assessment of the use of viruses for control of cotton leafworm in Egypt. *NRI Overseas Assignment Report* No. R1634(R), 37pp.

Bishop, A.L., Blood, P.R.B., Day, R.E. and Evenson, J.P. (1978) The distribution of cotton looper (*Anomis flava* Fabr.) larvae and larval damage on cotton and its relationship to the photosynthetic potential of cotton leaves at the attack site. *Australian Journal of Agricutural Research* 29, 319–325.

Briese, D.T. (1986) Insect resistance to baculoviruses. In: Granados, R.R. and Federici, B.A. (eds) *The Biology of Baculoviruses*, vol. II *Practical Application for Insect Control*, CRC Press, Boca Raton, Florida, pp. 237–263.

Briese, D.T. and Mende, H.A. (1983) Selection for increased resistance to a granulosis virus in the potato moth, *Phthorimaea operculella* (Zeller) (Lepidoptera: Gelechiidae). *Bulletin of Entomological Research* 73, 1–9.

Bull, D.L., Ridgeway, R.L., House, V.S. and Pryor, N.W. (1976) Improved formulations of the *Heliothis* nuclear polyhedrosis virus. *Journal of Economic Entomology* 69, 731–736.

Bullock, H.R. (1967) Persistence of *Heliothis* nuclear-polyhedrosis virus on cotton foliage. *Journal of Invertebrate Pathology* 9, 434–436.

Cadout, J. and Soubrier, G. (1974) Use of a nuclear polyhedrosis in the control of *Heliothis armigera* (Hb.) (Lep. Noct.) in cotton plantations in Chad. *Coton et Fibres Tropicales* 29, 357–365.

Chapman, A.J. and Ignoffo, C.M. (1972) Influence of rate and spray volume of a nucleopolyhedrosis virus on control of *Heliothis* in cotton. *Journal of Invertebrate Pathology* 20, 183–186.

Coaker, T.H. (1958) Experiments with a virus disease of cotton bollworm *Heliothis armigera* (HBN). *Annals of Applied Biology* 46, 536–544.

Cooper, J.F. (1983) A comparison between the distribution of charged and uncharged spray droplets on young cotton plants. Unpublished report, NRI, Chatham, Kent, UK.

Daoust, R.A. (1974) Weight related susceptibility of larvae of *Heliothis armigera* to a crude nuclear polyhedrosis virus preparation. *Journal of Invertebrate Pathology* 23, 400–401.

Dhandapani, N., Jayaraj, S. and Rabindra, R.J. (1987) Efficacy of ULV application of nuclear polyhedrosis virus with certain adjuvants for control of *Heliothis armigera* (Hbn) on cotton. *Journal of Biological Control* 1, 111–117.

Dhandapani, N., Kalyanasundaram, M., Swamiappan, M., Sundara Babu, P.C. and Jayaraj, S. (1992) Experiments on management of major pests on cotton with biocontrol agents in India. *Journal of Applied Entomology* 114, 52–56.

Eid, M.A.A., El-Nagar, S., Salem, M.S. and Badawy, E. (1985) Effect of nuclear polyhedrosis virus ingestion on consumption and utilization of food by *Spodoptera littoralis* larvae. *Bulletin of the Entomological Society of Egypt. Economic Series* 13, 67–74.

Elleman, C.J. (1983) The Inter-relationship between a Baculovirus of *Spodoptera littoralis* and the Leaf Surface of *Gossypium hirsutum*, with Comparative Observations on *Brassica oleracea.* DPhil thesis, University of Oxford.

Elleman, C.J. and Entwistle, P.F. (1985a) Inactivation of a nuclear polyhedrosis virus on cotton by substances produced by the cotton leaf surface glands. *Annals of Applied Biology* 106, 83–92.

Elleman, C.J. and Entwistle, P.F. (1985b) The effect of magnesium ions on the solubility of polyhedral inclusion bodies and its possible role in the inactivation of a nuclear polyhedrosis virus of *Spodoptera littoralis* by the cotton leaf exudate. *Annals of Applied Biology* 106, 93–100.

El-Nagar, S. and Abul-Nasr, S.E. (1980) Effect of direct sunlight on the virulence of NPV of the cotton leafworm, *Spodoptera littoralis. Zeitschrift für Angewändte Entomologie* 90, 75–80.

El-Nagar, S., Tawfik, M.S. and Adelrahman, T.A. (1983) The susceptibility to nuclear polyhedrosis virus among laboratory populations of *Spodoptera littoralis. Zeitschrift für Angewändte Entomologie* 96, 459–463.

Entwistle, P.F. and Evans, F. (1985) Viral control. In: Kerkut, G.A. and Gilbert, L.I. (eds) *Comprehensive Insect Physiology, Biochemistry and Pharmocology*, vol. 12. Pergamon Press, Oxford, pp. 347–412.

Evans, H.F. (ed.) (1991) IOBC/WRPS Working group *'Insect Pathogens and Insect Parasitic Nematodes'*. Second European meeting 'Microbial Control of Pests', 6–8 March 1989, Rome. IOBC/WRPS Bulletin 14/1, pp. 1–32.

Evans, H.F. and Harrap, K.A. (1982) Persistence of insect viruses. In: Mahy, B.W.S., Minson, A.C. and Darby, G. (eds) *Virus Persistence*. Cambridge University Press, pp. 57–98.

Falcon, L.A. (1974) Insect pathogens: integration into a pest management system. In: *Proceedings of the Summer Institute on Biological Control of Plant Insects and Diseases*, University Press of Mississippi, Jackson, Mississippii, pp. 612–627.

Falcon, L.A., Sorensen, A. and Akesson, N.B. (1974) Pioneering research on aerosol application of insect pathogens. *California Agriculture* 28(4), 11–13.

Francki, R.I.B., Fauquet, C.M., Knudson, D.L. and Brown, F. (1991) *Classification and Nomenclature of Viruses*. Fifth report of the International Committee on Taxonomy of Viruses. Archives of Virology supplement 2.

Fuxa, J.R. (1990) Insect control with baculoviruses. In: *New Directions in Biological Control. Alternatives for Suppressing Agricultural Pests and Diseases*, ULCA Symposia on Molecular and Cellular Biology New Series, vol. 112. Alan Liss, New York, pp. 97–113.

Granados, R.R. and Federici, B.A. (eds) (1986) *The Biology of Baculoviruses*, vol. II *Practical Application for Insect Control*, CRC Press, Boca Raton, Florida.

Griego, V.M., Martignoni, M.E. and Clyacomb, A.E. (1985) Inactivation of nuclear polyhedrosis virus (Baculovirus subgroup A) by monochromatic radiation. *Applied and Environmental Microbiology* 49, 709–710.

Griffiths, I.P. (1982) A new approach to the problem of identifying baculoviruses. In: Kurstak, E. (ed.) *Microbial and Viral Pesticides*. Marcel Dekker, New York, pp. 527–583.

Hafez, M., Kamel, A.A.M., Mostafa, T.H. and Omar, E.E. (1970) Field tests of combinations of polyhedrosis virus suspensions and certain chemical insecticides for control of the cotton leafworm, *Spodoptera littoralis* (Boisd.) (Lepidoptera: Noctuidae). *Bulletin of the Entomological Society of Egypt, Economic Series* 4, 65–69.

Harr, J., Guggenheim, R., Boller, Th. and Oertli, J.J. (1980) High pH-values on the leaf surface of commercial cotton varieties. *Coton et Fibres Tropicales* 35, 379–384.

Heimpel, A.M. (1967) Progress in developing insect viruses as microbial agents. In: *Proceedings of the Joint NS-Japan Seminar on Microbial Control of Insect Pests*, 21–23 April 1967, Fukuoka, Japan. pp. 51–61.

Hill, D.S. (1982) *Agricultural Insect Pests of the Tropics and their Control*. Cambridge University Press.

Hosny, M.M., Topper, C.P., Moawad, G.M. and El-Saadany, G.B. (1986) Economic damage thresholds of *Spodoptera littoralis* (Boisd.)(Lepidoptera: Noctuidae) on cotton in Egypt. *Crop Protection* 5, 100–104.

Huber, J. (1986) Use of baculoviruses in pest management programs. In: Granados, R.R. and Federici, B.A. (eds) *The Biology of Baculoviruses*, vol. II, *Practical Application for Insect Control.* CRC Press, Boca Raton, Florida, pp. 182–202.

Hughes, P.R. and Wood, H.A. (1986) *In vivo* and *in vitro* bioassay methods for baculoviruses. In: Granados, R.R. and Federici, B.A. (eds) *The Biology of Baculoviruses*, vol. II *Practical Application for Insect Control.* CRC Press, Boca Raton, Florida, pp. 1–30.

Hunter, F.R., Crook, N.E. and Entwistle, P.F. (1984) Viruses on pathogens for the control of insects. In: Grainger, J.M. and Lynch, J.M. (eds) *Microbial Methods for Environmental Biotechnology*. Society of Applied Bacteriology, London, pp. 323–347.

Hwang, G.H. and Ding, T. (1975) Studies on the nuclear polyhedrosis-virus disease of the cotton leafworm, *Prodenia litura* F. *Acta Entomologica Sinica* 18, 17–24.

Ignoffo, C.M. (1979) The first viral pesticide: past, present and future. *Developments in Industrial Microbiology* 20, 105–115.

Ignoffo, C.M. and Allen, G.E. (1972) Selection for resistance to an NPV in laboratory populations of the cotton bollworm *Heliothis zea*. *Journal of Invertebrate Pathology* 20, 187–192.

Ignoffo, C.M. and Batzer, O.F. (1971) Microencapsulation and ultraviolet protectants to increase sunlight stability of an insect virus. *Journal of Economic Entomology* 64, 850–853.

Ignoffo, C.M. and Couch, T.L. (1981) The nucleopolyhedrosis virus of *Heliothis* species as a microbial insecticide. In: Burges, H.D. (ed.) *Microbial Control of Pest and Plant Diseases 1970–1980*. Academic Press, London pp. 329–362

Ignoffo, C.M. and Montoya, E.L. (1966) The effects of chemical insecticides and insecticidal adjuvants on a *Heliothis* nuclear polyhedrosis virus. *Journal of Invertebrate Pathology* 8, 409–412.

Ignoffo, C.M. and Shapiro, M. (1978) Characteristics of baculovirus preparations processed from living and dead larvae. *Journal of Economic Entomology* 71, 186–188.

Ignoffo, C.M., Chapman, A.J. and Martin, D.F. (1965) The nuclear-polyhedrosis virus of *Heliothis zea* (Boddie) and *Heliothis virescens* (Fabricius) III. Effectiveness of the virus against field populations of *Heliothis* on cotton, corn and grain sorghum. *Journal of Invertebrate Pathology* 7, 227–235.

Ignoffo, C.M., Bradley, J.R., Gilliland, F.R., Harris, F.A., Falcon, L.A., McGarr, R.L., Sikorowski, P.P., Watson, T.F. and Yearian, W.C. (1972) Field studies on stability of the *Heliothis* nucleopolydedrosis virus at various sites throughout the cotton belt. *Environmental Entomology* 1, 388.

Ignoffo, C.M., Hostetter, D.L. and Smith, D.B. (1976) Gustatory stimulant, sunlight protectant, evaporation retardant: three characteristics of a microbial insecticide adjuvant. *Journal of Economic Entomology* 69, 207–210.

Ignoffo, C.M., Shasha, B.S. and Shapiro, M. (1991) Sunlight ultraviolet protection of *Heliothis* nuclear polyhedrosis virus through starch-encapsulation technology. *Journal of Invertebrate Pathology* 57, 134–136.

Jaques, R.P. (1977) Stability of entomopathogenic viruses. *Miscellaneous Publications of the Entomological Society of America* 10, 99–116.

Jayaraj, S., Santharam, G., Narayanan, K., Sundararajan, K. and Balagurunathan, R. (1981) Effectiveness of the nuclear polyhedrosis virus against field populations of the tobacco caterpiller *Spodoptera litura* on cotton. *Andhra Predesh Agricultural Journal* 27, 26–29.

Jayaraj, S., Rabindra, R.J. and Narayanan, K. (1989) Development and use of microbial agents for control of *Heliothis* spp. (Lepidoptera: Noctuidae) in India. In: King, E.G. and Jackson, R.D. (eds) *Proceedings of the Workshop on Biological Control of* Heliothis*: Increasing the Effectiveness of Natural Enemies. 11–15 November 1985, New Celhi, India.* For Eastern Regional Office, US Department of Agriculture, New Delhi, India, pp. 484–503.

Johnson, D.R. (1982) Suppression of *Heliothis* spp. on cotton by using *Bacillus thuringiensis*, Baculovirus and two feeding adjuvants. *Journal of Economic Entomology* 72, 207–210.

Jones, K.A. (1988) Studies on the Persistence of *Spodoptera littoralis* Nuclear Polyhedrosis Virus on Cotton in Egypt. PhD thesis, University of Reading, 390pp.

Jones, K.A. (1990) Use of a nuclear polyhedrosis virus to control *Spodoptera littoralis* in Crete and Egypt. In: Casida, J.E. (ed.) *Pesticides and Alternatives*. Elsevier, Amsterdam, pp. 131–142.

Jones, K.A. and McKinley, D.J. (1987) The persistence of *Spodoptera littoralis* nuclear polyhedrosis virus on cotton in Egypt. *Aspects of Applied Biology* 14, 323–334.

Jones, K.A., Moawad, G., McKinley, D.J. and Grzywacz, D. (1993a) The effect of natural sunlight on *Spodoptera littoralis* nuclear polyhedrosis virus. *Biocontrol Science and Technology* 3, 189–197.

Jones, K.A., Westby, A., Reilly, P.J.A. and Jeger, M.J. (1993b) The exploitation of microorganisms in the developing countries of the tropics. In: Jones, D. (ed.) *Exploitation of Microorganisms*. Chapman & Hall, London, pp. 343–370.

Jones, K.A., Irving, N.S., Grzywacz, D., Moawad, G.M., Hussein, A.H., and Fargahly, A. (1994) Application rate trials with a nuclear polyhedrosis virus to control *Spodoptera littoralis* (Boisd.) on cotton in Egypt. *Crop Protection* (in press).

Ketunuti, U. and Prathomrut, S. (1989) Cotton bollworm larvae control by *Heliothis armigera* nuclear polyhedrosis virus. In: *Abstracts of the First Asia-Pacific Conference of Entomology*, 8–13 November 1989, Chang Mai, Thailand, The Secretariat APCE, 1989, Bangkok, p. 16.

Killick, H.J. (1986) UV protectants in viral control of insects in relation to spray droplet size. In: Samson, R.A., Vlak, J.M. and Peters, D. (eds) *Fundamental and Applied Aspects of Invertebrate Pathology*. The Foundation for the Fourth International Colloquium of Invertebrate Pathology, Wageningen, The Netherlands, pp. 628–623.

Klein, M. and Podoler, H. (1978) Studies on the application of a nuclear polyhedrosis virus to control populations of the Egyptian cottonworm *Spodoptera littoralis*. *Journal of Invertebrate Pathology* 32, 244–248.

Labonne, V. (1971) *Cryptophlebia leucotreta* (Meyr.) present in Senegal. *Coton et Fibres Tropicales* 26, 468.

Liang, D.R., Zhao, C.X. and Yang, F.L. (1981) Effects of the nuclear polyhedrosis virus of *Anomis flava* (Fabricus). *Insect Knowledge* 18, 65–67.

Lisansky, S.G. (1984) Biological alternatives to chemical pesticides. *World Biotechnology Report* 1, 455–466. Online Publications, Pinner, UK.

Liu, S.N., Wang, H.M., Xing, Z.P., Jiang, L.R. and Gao, Y.H. (1987) Isolation and characterization of a nuclear polyhedrosis virus from cabbage moth *Mamestra brassicae*. *Chinese Journal of Biological Control* 3, 84–86.

Livingston, J.M. and Yearian, W.C. (1972) A nuclear polyhedrosis virus of *Pseudoplusia includens* (Lepidoptera: Noctuidae). *Journal of Invertebrate Pathology* 19, 107–112.

Luttrell, R.G., Yearin, W.C. and Young, S.Y. (1979) Laboratory and field studies on the efficacy of selected chemical insecticide – Elcar (*Baculovirus heliothis*) Combinations against *Heliothis* spp. *Journal of Economic Entomology* 72, 57–60.

Luttrell, R.G., Yearian, W.C. and Young, S.Y. (1981) Microbial and chemical insecticides against the cotton leafworm. *Arkansas Farm Research* 30, 10.

Luttrell, R.G., Young, S.Y., Yearian, W.C. and Horton, D.L. (1982) Evaluation of *Bacillus thuringiensis* – spray adjuvant – viral insecticide combinators against *Heliothis* spp. (Lepidoptera: Noctuidae). *Environmental Entomology* 11, 783–787.

MacFarlane, J.J., Jr. and Keely, L.L. (1969) Heat effects on the infectivity of a nuclear polyhedrosis virus of *Heliothis* spp. *Journal of Economic Entomology* 62, 925–929.

McKinley, D.J. (1971) Nuclear polyhedrosis virus of the cotton bollworm in Central Africa. *Cotton Growing Review* 48, 297–303.

McKinley, D.J. (1980) The use of viruses in the control of *Spodoptera* species and prior safety testing. In: Lundholm, B. and Stakerud, M. (eds) *Environmental Protection and Biological Forms of Control.* Ecological Bulletin (Stockholm) 31, 75–80.

McKinley, D.J. (1982) The prospects for use of nuclear polyhedrosis virus on *Heliothis* management. In: *Proceedings of the International Workshop on Heliothis Management,* 15–20 November 1981, ICRISAT, Patancheru, A P, India. International Crops Research Institute for the Semi-Arid Tropics, Patancheru, pp. 123–135.

McKinley, D.J. (1985) Nuclear Polyhedrosis Virus of *Spodoptera littoralis* Boisd. (Lepidoptera: Noctuidae) as an Infective Agent in its Host and Related Insects. PhD thesis, University of London.

McKinley, D.J., Moawad, G.M., Jones, K.A., Grzywacz, D. and Turner, C. (1989) The development of nuclear polyhedrosis virus for the control of *Spodoptera littoralis* (Boisd.) in cotton. In: Green, M.B. and Lyon, D.J. de B. (eds) *Pest Management in Cotton.* Ellis Horwood, Chichester, pp. 93–100.

McLeod, P.J., Yearian, W.C. and Young, S.Y. (1977) Inactivation of *Baculovirus heliothis* by ultraviolet irradiation, dew and temperature. *Journal of Invertebrate Pathology* 30, 234–241.

Maiorov, V.I., Lopatkin, A.V., Bogachev, S.S., Bekishev, G.A. and Kononenko, A.P. (1984) Evaluation of the effectiveness of the nuclear polyhedrosis virus of the cotton moth *Heliothis armigera* under various conditions of its practical use. *Izvestiya Akademii Nauk Tadzhikskoi SSR* 2, 43–46.

Martignoni, M.E. and Iwai, P.J. (1981) A catalogue of viral diseases of insects, mites and ticks. In: Burgess, K.D. (ed.) *Microbial Control of Pests and Plant Diseases 1970–1980.* Academic Press, London, pp. 897–911.

Martignoni, M.E. and Milstead, J.E. (1962) Trans-ovum transmission of the nuclear polyhedrosis virus of *Colias eurytheme* Boisduval through contamination of the female genitalia. *Journal of Insect Pathology* 4, 113–121.

Martinez, R. and Swezey, S.L. (1988) Control of *Heliothis zea* (Boddie) larvae with a nuclear polyhedrosis virus (*Baculovirus heliothis*) in cotton, Leon, Nicaragua 1983. *Revista Nicaraguense de Entomologia* 2, 13–18.

Mohamed, A.I., Young, S.Y. and Yearian, W.C. (1983) Susceptibility of *Heliothis virescens* (F.) (Lepidoptera: Noctuidae) larvae to microbial agent–chemical pesticide mixtures or cotton foliage. *Environmental Entomology* 12, 1403–1405.

Montoya, E.L., Ignoffo, C.M. and McGarr, R.L. (1966) A feeding stimulant to increase effectiveness of, and a field test with a nuclear polyhedrosis virus of *Heliothis. Journal of Invertebrate Pathology* 8, 320–324.

Pfrimmer, T.R. (1979) *Heliothis* spp. control on cotton with pyrethroids, carbamates, organophosphates and biological insecticides. *Journal of Economic Entomology* 72, 593–598.

Pieters, E.P., Young, S.Y., Yearian, W.C., Sterling, W.L., Clower, D.F., Melville, D.R. and Gilliland, F.R., Jr. (1978) Efficacy of microbial pesticide chlordimeform mixtures for control of *Heliothis* spp. on cotton. *Southwestern Entomologist* 3, 237–240.

Pinnock, D.E. (1980) E. Australia. In: *Proceedings of a Workshop on Insect Pest Management with Microbial Agents. Recent Achievements, Deficiencies and Innovations,* 12–15 May 1980, Boyce Thompson Institute, Ithaca, NY, pp. 13–17.

Potter, M.F. and Watson, T.F. (1983a) Garbanzo bean as a potential feeding stimulant for use with a nuclear polyhedrosis virus of the tobacco budworm (Lepidoptera: Noctuidae). *Journal of Economic Entomology* 76, 449–451.

Potter, M.F. and Watson, T.F. (1983b) Laboratory and greenhouse performance of *Baculovirus heliothis* combined with feeding stimulants for control of neonate tobacco budworm. *Protection Ecology* 5, 161–165.

Potter, M.F. and Watson, T.F. (1984) Field persistence of Elcar *(Baculovirus heliothis)* applied in a bait formulation for control of tobacco budworm in Arizona cotton. *Journal of Agricultural Entomology* 1, 78–81.

Rabindra, R.J. and Jayaraj, S. (1988) Evaluation of certain adjuvants for nuclear polyhedrosis virus (NPV) of *Heliothis armigera* (Hbn.) on chickpea. *Indian Journal of Experimental Biology* 26, 60–62.

Renou, A. (1987) Achievements in the biological control of *Heliothis armigera* (Hbn), a pest of the cotton crop in North Cameroon. *Mededelingen van de Faculteit Landbouwwetenschappen, Rijksuniversiteit Gent* 52, 311–318.

Richards, M.G. (1980) An Investigation into the Inactivation of a Nuclear Polyhedrosis Virus of *Spodoptera littoralis* by Cotton Leaf Exudates. Unpublished thesis for MSc degree, University of London.

Richards, M.G. and Payne, C.C. (1982) Persistence of baculoviruses on leaf surface. In: *Invertebrate Pathology and Microbial Control.* Proceedings of the Third International Colloquium of Invertebrate Pathology, 6–10 September 1982, University of Sussex, Brighton, UK. Society for Invertebrate Pathology, pp. 296–301.

Room, P.M. (1979) A prototype 'on-line' system for management of cotton pests in the Naomi Valley, New South Wales. *Protection Ecology* 1, 245–264.

Roome, R.E. (1975) Field trials with a nuclear polyhedrosis virus and *Bacillus thuringiensis* against larvae of *Heliothis armigera* (Hb.) (Lepidoptera: Noctuidae) on sorghum and cotton in Botswana. *Bulletin of Entomological Research* 65, 507–514.

Russell, D.A., Radwan, S.M., Irving, N.S., Jones, K.A. and Downham, M.C.A. (1993) Experimental assessment of the impact of defoliation by *Spodoptera littoralis* on the growth and yield of Giza '75 cotton. *Crop Protection* 12, 303–309.

Santharam, G. and Jayaraj, S. (1987) Effect of host plants and site of application on the infectivity of nuclear polyhedrosis virus to *Spodoptera litura* larvae. *Journal of Biological Control* 1, 39–43.

Shapiro, M. (1982) *In vivo* mass production of insect viruses for use as pesticides. In: Kurstak, E. (ed.) *Microbial and Viral Pesticides.* Marcel Dekker, New York, pp. 463–492.

Shapiro, M. (1986) In vivo production of baculoviruses. In: Granados, R.R. and Federici, B.A. (eds) *Biology of Baculoviruses,* vol. II *Practical Application for Insect Control.* CRC Press, Boca Raton, pp. 31–61.

Silva, S.M.T. and Santos, W.J. (1980) The occurrence of natural enemies of *Trichoplusia ni* (Hubner, 1802) on cotton in the municipalities of Urai and Londrina (Parana), in the year 1979. *Anais da Sociedade Entomologia do Brasil* 9, 179–187.

Small, D.A. (1985) Aspects of the Attachment of a Nuclear Polyhedrosis Virus from the Cabbage Looper (*Trichoplusia ni*) to the leaf surface of cabbage *(Brassica oleracea)*. DPhil thesis, University of Oxford.

Smith, D.B., Hostetter, D.L. and Ignoffo, C.M. (1978) Formulation effects on application of a viral insecticide (*Baculovirus heliothis*). *Journal of Economic Entomology* 71, 814–817.

Stacey, A.L., Yearian, W.C. and Young, S.Y. (1977a) Efficacy of *Baculovirus heliothis* and *Bacillus thuringiensis* applied in dry baits. *Arkansas Farm Research* 26, 3.

Stacey, A.L., Yearian, W.C. and Young, S.Y. (1977b) Evaluation of *Baculovirus heliothis* with feeding stimulants for control of *Heliothis* larva on cotton. *Journal of Economic Entomology* 70, 779–784.

Stacey, A.L., Young, S.Y. and Yearian, W.C. (1977c) *Baculovirus heliothis* effect of selective placement of *Heliothis* on mortality and efficacy in directed sprays on cotton. *Journal of the Georgia Entomological Society* 12, 167–173.

Stacey, A.L., Yearian, W.C., Young, S.Y., Luttrell, R.G., Matthews, G.J. (1980) Field evaluation of *Baculovirus heliothis* on cotton by using selected application methods. *Journal of the Georgia Entomological Society* 15, 365–372.

Topper, C. (1984) Report on the research and development of nuclear polyhedrosis virus of *Spodoptera littoralis* 1979–1983. Unpublished report, ODA and Egyptian Academy of Sciences (3 volumes). Overseas Development Administration, London.

Topper, C., Moawad, G., McKinley, D., Hosny, M., Jones, K., Cooper, J., El-Nagar, S. and El-Sheik, M. (1984) Field trials with a nuclear polyhedrosis virus against *Spodoptera littoralis* on cotton in Egypt. *Tropical Pest Management* 30, 372–378.

Vail, P.V., Soo Hoo, C.F., Seay, R. and Ost, R. (1976) Experiments on the integrated control of a cotton and lettuce pest. *Journal of Economic Entomology* 69, 787–791.

Vargas-Osuna, E. and Santiago-Alvarez, C. (1986) Differential response of male and female *Spodoptera littoralis* (Boisduval) individuals to baculovirus infections. In: Samson, R.A., Vlak, J.M. and Peters, D. (eds) *Fundamental and Applied Aspects of Invertebrate Pathology*. Foundation of the Fourth International Colloquium of Invertebrate Pathology, Wageningen, The Netherlands, p. 145.

Wahl, B. (1909) Uber die polyederkrakheit der Nonne *Chymantria monacha L. Cerr. f. gesam. Forsrwesen* 35, 164–172, 212–215.

Whitlock, V.H. (1977) Failure of a strain of *Heliothis armigera* (Hubn.) (Noctuidae: Lepidoptera) to develop resistance to a nuclear polyhedrosis virus and a granulosis virus. *Journal of the Entomological Society of Southern Africa* 40, 251–253.

Whitlock, V.H. (1978) Dosage mortality studies of a granulosis and nuclear polyhedrosis virus of a laboratory strain of *Heliothis armigera*. *Journal of Invertebrate Pathology* 32, 386–387.

WHO (1973) *The Use of Viruses for the Control of Insect Pests and Disease Vectors*. Report of a joint FAO/WHO meeting on insect viruses. World Health Organization Technical Report Services no. 531, Geneva.

Williams, C.F. and Payne, C.C. (1984) The susceptibility of *Heliothis armigera* larvae to three nuclear polyhedrosis viruses. *Annals of Applied Biology* 104, 405–412.

Wilson, A.G., Desmarchelier, J.M. and Malafant, K. (1983) Persistence on cotton foliage of insecticide residues toxic to *Heliothis* larvae. *Pesticide Science* 14, 623–633.

Winstanley, D. and Rovesti, L. (1993) Insect viruses as biocontrol agents. In: Jones, D. (ed.) *Exploitation of Microorganisms*. Chapman & Hall, London, pp. 105–136.

Witt, D.J. and Stairs, G.R. (1975) The effects of ultraviolet irradiation on a baculovirus infecting *Galleria mellonella*. *Journal of Invertebrate Pathology* 26, 321–327.

Wu, Q.L. (1986) Investigation on the fluctuations of dominant natural enemy populations in different cotton habitats and integrated application with biological agents to control cotton pests. *Natural Enemies of Insects* 8, 29–34.

Yearian, W.C. and Young, S.Y. (1974) Persistence of *Heliothis* nuclear polyhedrosis virus on cotton plant parts. *Environmental Entomology* 3, 1035–1036.

Yearian, W.C. and Young, S.Y. (1982) Control of insect pests of agricultural importance by viral insecticides. In: Kurstak, E. (ed.) *Microbial and Viral Pesticides*. Marcel Dekker, New York, pp. 387–423.

Young, S.Y. and Yearian, W.C. (1976) Influence of buffers on pH of cotton leaf surface and activity of a *Heliothis* nuclear polyhedrosis virus. *Journal of the Georgia Entomological Society* 11, 277–282.

Young, S.Y. and Yearian, W.C. (1986) Formulation and application of baculoviruses. In: Granados, R.R. and Federici, B.A. (eds) *The Biology of Baculoviruses*, vol. II *Practical Application for Insect Control.* CRC Press, Boca Raton, pp. 157–179.

Young, S.Y., Yearian, W.C. and Kim, K.S. (1977) Effect of dew from cotton and soybean foliage on activity of *Heliothis* NPV. *Journal of Invertebrate Pathology* 29, 105–111.

26 Pheromones for the Control of Cotton Pests

D.G. Campion

Natural Resources Institute, Central Avenue, Chatham Maritime, Chatham, Kent ME4 4TB, UK

Introduction

Pheromones are chemicals produced by one organism which influence the behaviour of other members of the same species. The two classes of pheromones most common among insects are the sex pheromones which are generally produced by the female to attract males for the purpose of mating; and the aggregation pheromones which are produced by one or both sexes and bring both sexes together for feeding and reproduction. The former are common among the Lepidoptera and the latter among the Coleoptera.

In the laboratory certain pheromones can induce a physiological response at concentrations involving only tens of molecules because of the finely tuned sensory apparatus on insect antennae. These chemicals therefore represent the most powerful of biologically active substances and efforts are being made to utilize them in the service of pest control.

Sex pheromones have been identified for over 500 insect species (Arn *et al.*, 1986) and the number is likely to increase even more over the next few years (Inscoe *et al.*, 1990). The use of these chemicals in pest control has developed along three main pathways: monitoring of insect populations with pheromone-baited traps; control by mass trapping using large numbers of traps to reduce population levels; and control by mating disruption in which synthetic pheromone is used to permeate the atmosphere so that an insect will be unsuccessful in finding a mate.

Most work has been on sex pheromones of Lepidoptera and aggregation pheromones of Coleoptera, and to a lesser extent on epideictic pheromones which influence oviposition behaviour in Coleoptera, Diptera, Homoptera, Hymenoptera, Lepidoptera and Orthoptera, and alarm pheromones which stimulate escape and other defensive behaviour in Dictyoptera, Hemiptera and eusocial Hymenoptera.

The chemical components range from blends of aliphatic alcohols,

Table 26.1. Pheromone components of major cotton pests.

Insect species	Common name	Distribution	Pheromone components	References
Alabama argillacea	Cotton leafworm	North & South America, Caribbean	(*S*)-9-methyl nonadecane (*Z,Z*)-6,9-heneicosadiene (*Z,Z*)-6,9-tricosadiene	Hall *et al.*, 1993
Anomis texana	Lesser cotton leafworm	South America	(*S*)-7-methylheptadecane	Hall *et al.*, 1993
Anthonomus grandis	Mexican boll weevil	North & South America	(IR-*cis*)-1-methyl-2-(1-methylethenyl) cyclobutane ethanol (*Z*)-2-(3,3-dimethylcyclohexylidene) ethanol (*Z*)-(3,3-dimethylcyclohexyl dene) acetaldehyde (*E*)-(3,3-dimethylcyclohexylidene) acetaldehyde	Tumlinson *et al.*, 1971
Bucculatrix thurberiella	Cotton leaf miner	North & South America & Caribbean	(*Z*)-9-tetradecenyl nitrate (*Z*)-8-tridecenyl nitrate	Hall *et al.*, 1992
Diparopsis castanea	Red bollworm	Southeastern Africa	9,11-dodecadienyl acetate (80% *Z*, 20% *E*) 11-dodecenyl acetate	Nesbitt *et al.*, 1973, 1975
Earias insulana	Spiny bollworm	Africa, India, Mediterranean countries, Pakistan & SE Asia	(*E,E*)-10-12-hexadecadienal (*Z*)-11-hexadecenal	Hall *et al.*, 1980; Cork *et al.*, 1988
Earias vittella	Spotted bollworm	India, Pakistan, SE Asia & northern Australia	(*E,E*)-10,12-hexadecadienal (*Z*)-11-hexadecenal (*Z*)-11-octadecenal	Cork *et al.*, 1988
Heliothis armigera	American bollworm	Widespread, Old World tropical and subtropical regions	(*Z*)-11-hexadecenal (*Z*)-9-hexadecenal hexadecanal hexadecanol	Nesbitt *et al.*, 1979, 1980; Kehat *et al.*, 1980; Konyukhov *et al.*, 1983

Table 26.1. Continued.

Heliothis virescens	Tobacco budworm	North & South America & Caribbean	(*Z*)-11-hexadecenal; (*Z*)-9-tetradecenal; 5–8 trace components	Tumlinson *et al.*, 1975; Klun *et al.*, 1980
Heliothis zea	Corn earworm	North & South America & Caribbean	(*Z*)-11-hexadecenal; (*Z*)-9-hexadecenal; (*Z*)-7-hexadecenal, hexadecanal	Klun *et al.*, 1980
Pectinophora gossypiella	Pink bollworm	Worldwide	(*ZZ*)-7,11-hexadecadienyl acetate; (*ZE*)-7,11-hexadecadienyl acetate	Bierl *et al.*, 1974; Hummel *et al.*, 1973
Spodoptera frugiperda	Fall armyworm	North & South America & Caribbean	(*Z*)-9-tetradecenyl acetate; (*Z*)-11-hexadecenyl acetate; (*Z*)-7-dodecenyl acetate; (*Z*)-9-dodecenyl acetate	Mitchell *et al.*, 1985
Spodoptera littoralis	Egyptian cotton leafworm	Africa, Mediterranean Countries	(*Z*)-9,(*E*)-11-tetradecadienyl acetate; (*Z*)-9,(*E*)-12-tetradecadienyl acetate; Other minor components	Nesbitt *et al.*, 1973; Dunkelblum *et al.*, 1982
Spodoptera litura	Tobacco cutworm	India, Pakistan, SE Asia, China, Philippines, Japan, Australia, Pacific Islands	(*Z*)-9,(*E*)-11-tetradecadienyl acetate; (*Z*)-9,(*E*)-12-tetradecadienyl acetate	Tamaki *et al.*, 1973
Spodoptera sunia		America & Caribbean	(*Z*)-9-tetradecenyl acetate; (*Z*)-9,(*E*)-12-tetradecadienyl acetate; (*Z*)-9-tetradecenal; (*Z*)-11-hexadecenyl acetate	Bestmann *et al.*, 1988

Fig. 26.1 Pheromone trap for *Pectinophora*.

aldehydes, esters and epoxides as found in lepidopterous insects to alkenoic acids and aldehydes, branched alkenones, esters, monoterpene alcohols and aldehydes as in Coleoptera. A similarity of chemical structure is evident within the Lepidoptera, particularly for related species and even within a species there is some evidence for differences in the pheromone components according to locality. Thus it was shown that for *Spodoptera littoralis* there were marked differences in the pheromone blends from calling females originating from Greece, Egypt and Israel (Campion *et al.*, 1980). The chemical structures of the major cotton insects have been identified (Table 26.1).

Recent general reviews include those edited by Jutsum and Gordon (1989) and Ridgway *et al.* (1990). This chapter considers the use of pheromones for the control of just the cotton pests, which will illustrate the potential advantages of this unique group of chemicals in pest-control situations and will also highlight problems and difficulties that have been encountered in their evaluation and utilization.

Monitoring

Pheromones are now being used throughout the world as monitoring tools and trapping systems are available on a commercial basis.

Traps for use with pheromones are generally simple, cheap devices specific for a given species using a sticky surface, water or dry funnel to capture and retain the insects (Figs 26.1, 26.2; Plate VIII.4) (Wall, 1989). The pheromone dispenser controls the release of the pheromone and typically contains a milligram or less of pheromone which lasts for a month or more (Campion *et al.*, 1978; Weatherston, 1989).

Pheromone traps generally catch target pest species even when population levels are very low and can therefore be used qualitatively to provide an early warning of pest incidence. This may be of particular value if the pest occurs sporadically from year to year or where long-range migration is suspected.

Quantitative monitoring, whereby the numbers of moths caught in the traps are related to some predicted number of eggs, larvae or economic threshold, has not been so successful. This is attributable in part to the varying environmental influences on trap catch and the nature and dispersal of the attractant plume. Indeed, in some instances, potent pheromones have drawn in insects from outside the designated crop area, thus providing misleading information about the population density within the crop. Apart from these difficulties there has also been a general lack of precise knowledge about the pest status of the insect, in other words, the level of damage caused before treatment is required.

Both qualitative and quantitative pheromone monitoring studies have been undertaken for key cotton pests.

Fig. 26.2. Pheromone trap for *Anthonomus*.

Fig. 26.3. Twist-tie containing Gossyplure.

Spodoptera

Long distance movements have been demonstrated for several species of *Spodoptera* including *Spodoptera exempta* (Walker) (Rose *et al.*, 1985) and *Spodoptera exigua* (Hb.) (French, 1969). In the case of *S. littoralis* on cotton, an analysis of daily records of catch in a pheromone trap network distributed throughout the Delta region of Egypt for a two year period suggested that build-up of local populations was the most important factor, after which some localized redistribution occurred (Nasr *et al.*, 1984), a view supported from results from a similar study earlier conducted in Cyprus (Campion *et al.*, 1977). The related species *Spodoptera litura* (F) is a sporadic pest of cotton in Asia and Australia while in Latin America several other *Spodoptera* species play a similar role including *Spodoptera frugiperda* (J.E. Smith) and *Spodoptera sunia* (Guenée).

It has been shown that wind speed, relative humidity and temperature can significantly affect *S. littoralis* and *S. litura* moth catches to pheromone baited traps (Campion *et al.*, 1974; Nakamura 1976; Murlis *et al.*, 1982) and given a relatively mobile insect, quantitative relationships between trap catch and subsequent larval infestations have not been significant.

Helicoverpa and *Heliothis*

The highly mobile polyphagous habit of *Helicoverpa* spp. again suggests that whereas pheromone baited traps can be used to study insect migration and provide information or early warning of initial insect attack, they are unlikely to provide precise quantitative data on the levels of larval infestation on a field to field basis.

A pheromone trap network to monitor the area-wide movement and distribution of *H. armigera* has been established in the Indian subcontinent (Anon., 1987). Predictive models, incorporating pheromone trap catches of *H. virescens* and *H. zea* have been developed in the USA to provide early warnings of probable oviposition patterns on cotton crops, or as aids to scheduling scouting and other pest management activities on an area-wide basis (Hartstack *et al.*, 1978, 1983; Hendricks and Hartstack, 1978; Hartstack and Witz 1981; Lopez *et al.*, 1990).

The relationship, between trap catches and subsequent densities of immature stages or damage, has been variable to poor (Hartstack *et al.*, 1978; Tingle and Mitchell, 1981; Kehat *et al.*, 1982; Rothschild *et al.*, 1982; Johnson 1983; Kononenko *et al.*, 1986; Leonard *et al.*, 1989; Nyambo, 1989). The closest relationships have been when moth densities are low and at the beginning of the seasonal cycle.

For *H. armigera* trap position, both absolute and in relation to the crop, wind speed and wind direction, relative humidity and temperature, can all significantly affect trap catch (Rothschild *et al.*, 1982; Dent and Pawar, 1988). Moonlight and barometric pressure have also been shown to affect captures of *H. virescens* and *H. zea* in the USA (Hartstack *et al.*, 1978; Hendricks and Hartstack 1978), although in India the effect of moonlight on pheromone trap catch of *H. armigera* was minimal (Dent and Pawar, 1988). However the high levels of inter-trap variation, as observed for *H. armigera* in India, would negate the usefulness of catch data from individual traps as indices of action threshold.

Alabama

The cotton leafworm, *Alabama argillacea* (Hb), may occur at almost any time of the year throughout tropical America with the first appearance of the moths coinciding with the young cotton plants of the rainfed crop.

Its pheromone identified by Cork *et al.* (1992) and Hall *et al.* (1993) and evaluated in Peru (Almestar, 1990) is likely to be useful in the development of an early warning device for the detection of incipient insect attacks.

The pheromone of a related species *Anomis texana* (Riley) of localized importance on cotton in Southern Peru has also been identified and field tested (Cork *et al.*, 1992; Hall *et al.*, 1993).

Pectinophora

The pink bollworm, *Pectinophora gossypiella*, is an established cotton pest throughout the world. Extensive qualitative surveys have been conducted in the United States to detect the possible spread of the insect from Arizona and New Mexico into adjacent States (Kennedy, 1981).

Pink bollworm occurred for the first time in Northern Peru in 1983 and since then a countrywide pheromone trap network has been maintained to determine areas of localized infestation and to monitor the spread of the insect (J. Gonzalez, private communication).

In Pakistan, Ahmad (1979) monitored year-round pink bollworm occurrence with pheromone traps with a view to improving control of the insect through alterations in the time of cotton growing and insecticide application. Similar studies were conducted in India by Singh and Lather (1989).

The quantitative use of pheromone traps by relating moth catch to an economic threshold reduced treatments by one third to one half in southern California (Toscano *et al.*, 1974). Early to midseason thresholds for insecticide application were 12 to 15 moths per trap per night, reducing to 3.5 to four moths per night later in the season. Later work by Henneberry and Clayton (1982) showed that catches of male moths were strongly correlated with oviposition, proportion of infested bolls and numbers of larvae per boll. However, after insecticidal treatment catches of moths remained above the threshold, due probably to immigration and fresh emergence, and so scheduling treatments on the basis of moth catches was only practical for the initial treatment. Beasley *et al.* (1985) showed a positive correlation between the numbers of male moths caught three to four days before first flowerbud infestation; the numbers of larvae in bolls were also positively correlated to later flower and boll infestation.

In Israel, where all cotton fields (60,000 ha) have been monitored by pheromone traps for *Pectinophora* since 1975, Melamed and Shoham (1975) recommended spraying when trap catches reached eight moths a night, but Teich *et al.* (1977) lowered the threshold to five per night. Insecticide treatments are limited to the beginning of the season in order to control the adult moths. Usually no more than two sprays are applied, whereas formerly farmers sprayed 10–15 times each season (Shani, 1982).

Pectinophora pheromone trap catches were shown to be closely correlated with boll damage in experiments carried out on Sea-Island cotton in Barbados. Catches of eight–nine moths in a night represented a 10% level of boll damage ten days later (Ingram, 1980). According to Taneja and Jayaswal (1981), cotton in India should be sprayed 24–48 hours after catches of eight moths/trap/night, while Dhawan and Sidhu (1984) in Punjab, found that spraying at four moths per night gave the best results. Page *et al.* (1984), using gossyplure-baited traps in Queensland, found that mean trap catches of 10–20 moths per night were related to economically damaging attacks of the pink spotted bollworm *Pectinophora scutigera* (Holdaway).

Thus with the level of present knowledge it does not seem possible to have a universal basis for correlating moth catches in pheromone traps with subsequent larval damage and the values need to be determined locally.

The pheromone of *Pectinophora* was combined with that of the red bollworm *Diparopsis castanea* (Hmps.) to develop a dual monitoring system for *Pectinophora* and *Diparopsis* in the Shire Valley cotton growing region of Malawi, when it was demonstrated that catches of both species were unaffected by the release of pheromone mixtures (Marks, 1976, 1977).

Earias

The pheromone of the spiny bollworm, *Earias insulana* (Boisd.), was identified and synthesized by Hall *et al.* (1980). Preliminary studies conducted in Syria suggested that a significant relationship did exist between numbers of *Earias* moths in the pheromone traps and the number of *Earias*-infested bolls sampled after an interval of 10 days (Campion *et al.*, 1981). No significant relationship between trap catch and subsequent larval infestations was found during similar studies conducted in Egypt (B.R. Critchley, private communication).

The spotted bollworm, *Earias vittella* pheromone was identified and field tested in Pakistan (Cork *et al.*, 1988) where *E. vittella* coexists with the closely related spiny bollworm, *E. insulana* (Boisd.) and the pink bollworm *P. gossypiella* and species specific monitoring tools are therefore particularly important.

Bucculatrix

The pheromone of the cotton leaf miner *Bucculatrix thurberiella* (Busck) was identified by Hall *et al.* (1992) and has been successfully field tested in Arizona.

Anthonomus

The boll weevil, *Anthonomus grandis*, is probably the most serious pest of cotton on the American continent. The male insect after first feeding on the cotton plant emits a pheromone which causes the aggregation of both sexes, while later in the season by the time of flowering only female insects are attracted. A boll weevil pheromone trap is used both for early season detection and dispersal studies (Davich *et al.*, 1970; Ridgway *et al.*, 1971), as it is much more efficient in detecting weevils than manual surveys (Mitchell and Hardee, 1974; Merkl *et al.*, 1978). Experiments conducted to compare traps placed around the border of fields compared with traps sited within the fields indicated that at low population densities results were similar, but that at higher densities, particularly during midseason when boll weevil reproduction is occurring, those traps sited in the fields were more efficient (Lloyd *et al.*, 1981; Leggett *et al.*, 1988).

The boll weevil first appeared in NE Brazil as recently as 1983 and it spread so rapidly that it reduced the cotton area by more than 80% in only two years (M. Hodge, unpublished report, 1987). In 1985 it appeared simultaneously in the vicinity of São Paolo and in the State of Parana which borders Paraguay, and by 1988 it had consolidated itself in and around São Paolo and Maringa and had spread west and south (Dos Santos, 1989).

It seems inevitable that it will spread into Paraguay where an estimated 450,000 hectares of cotton are cultivated. To counter this threat a network of pheromone traps in 20 different localities with 50 traps in each area has been established (Whitcomb and Marengo, 1985). The possibility of creating a cotton-free buffer zone between Paraguay and Brazil to delay the entry and spread of the boll weevil into Paraguay was explored, using remote sensing imagery to map out cotton fields in order to locate the buffer zone correctly and to ensure the best placement of a monitoring trap network (Denore *et al.*, 1988).

The boll weevil pheromone has thus been a valuable tool for detecting new boll weevil infestations where the boll weevil does not normally occur, and for detecting reintroductions into areas from which the boll weevil has been eliminated in the USA in the eradication programmes.

Quantitative monitoring to aid decision making in relation to defined economic thresholds has been less successful. Studies have been conducted to investigate the relationship between trap captures, insecticide applications for overwintered boll weevil control, and numbers of boll-weevil-damaged flowerbuds (Rummel *et al.*, 1980). From data obtained over several years on cumulative trap captures and subsequent flowerbud damage, a trap index system was derived, based on the number of weevils captured over a six-week period, for predicting the results of treatment for weevil control (Benedict *et al.*, 1985). Monitoring with pheromone traps as a guide to application of insecticides for diapause boll weevil control and even for midseason applications is practised in some areas of the USA, but more work is required to relate boll weevil population levels at different times of the year in relation to varying trap catch (Ridgway *et al.*, 1985).

Mass-trapping

The concept of mass-trapping seems simple enough since a powerful, highly specific insect attractant, if deployed in traps, should catch a sufficiently large number of the target insect species to reduce population increase to economically acceptable levels. The immediate question arises as to what proportion of the wild population needs to be trapped to achieve such a result. For Lepidoptera at least where only males are trapped, it is generally assumed that trapping efficiency should be as high as 80–95% (Knipling and McGuire, 1966).

Part of the problem is to quantify the number of traps necessary per unit area to achieve control. This has ranged in practice from 1 to 700 ha^{-1}, the trap density selected either on an empirical basis, or as a result of computer simulation modelling. The upper limit is, however, fixed by the economic factors of trap costs and the subsequent maintenance costs of the trapping network.

Once a highly efficient trapping system has been achieved and an adequate trapping regime established, then the problem remains of assessing accurately the effects of the treatment. The problem is that the treatment areas must be either isolated or sufficiently large to reduce the possibility of gravid female entry; the need for large or single area treatment has in turn resulted in many cases of trials consisting of unreplicated blocks, so that statistically acceptable comparisons with control areas have not been possible (Campion, 1984). It is assumed that for cotton pests, those of monophagous habit and of a limited migratory potential such as *Pectinophora* would offer the best prospects for control by this method.

Several attempts have been made to control *Pectinophora* in the United States by mass-trapping. Initial trials were conducted before the pheromone was fully characterized and the results were unsuccessful or equivocal (Graham *et al.*, 1966; Guerra *et al.*, 1969). A reduction in the level of larvae infesting bolls was claimed following mass-trapping with gossyplure in an area of 36 ha at a trap density of 5 ha^{-1}, although some late-season insecticide applications were still necessary (Flint *et al.*, 1976). To get round the problem of adequate controls the infestation levels during the mass-trapping period were compared with those in the same area for several years prior to treatment. A similar method of assessment was used by Huber *et al.* (1979) in a much larger area of 3000 ha using trap densities ranging from 5 to 10 ha^{-1}. A reduction in the level of boll infestation was claimed as the result of the mass-trapping, although the levels of infestation both prior to and during the trials were too low to warrant conventional control practices.

A mass-trapping trial was conducted in Syria in an attempt to control the spiny bollworm, *E. insulana*. Pheromone baited funnel traps at a density of 4 ha^{-1} were deployed in a cotton area of 150 ha and maintained throughout the cotton growing season. Although overall control was not achieved, it was claimed that the level of boll infestation due to *Earias* was 50% lower at the centre of the treatment area compared with the level at the periphery (Elmosa, 1986). This suggests that immigration of insects into the treatment area had been an important factor.

The relatively mobile polyphagous feeding habit of *Spodoptera littoralis* presents a major problem with the mass-trapping approach for the control of this insect, since traps have to be deployed over an extensive area and not confined to fields of a specific crop.

In Israel, using trap densities of 1–2 ha^{-1} in areas of mixed cultivation of up to 3000 ha, successful control of *Spodoptera* was claimed on the basis of a reduction by 30–40% in the number of insecticides required to control the cotton insect pest complex throughout the season when compared with conventional treatment areas (Teich *et al.*, 1979). By 1981 more than 20,000 ha of Israeli cotton were similarly treated (Shani, 1982).

Mass-trapping trials were also conducted in Egypt where *S. littoralis* is presently controlled by the hand picking of egg-masses from the young cotton plants. Teams of small children go through the cotton crop once every three days to collect the egg masses which are counted and then destroyed (Bishara, 1934; Hosny 1980). An economic appraisal of the mass-trapping method by Gubbins and Campion (1982) suggested that provided it was technically feasible, it would be considerably cheaper than the system of egg-mass collection.

Mass-trapping trials, at a density of three and five traps ha^{-1} in areas of up to 600 ha and maintained throughout the cotton growing season, failed to reduce significantly the number of egg-masses collected in the treatment areas compared with those collected in areas without traps (Hosny *et al.*, 1979; Campion, 1983). Other assessments, which included a comparison of mating of tethered virgin female *Spodoptera* moths in pheromone treated and control areas (McVeigh *et al.*, 1983), and the mating frequency of male moths as determined by the colour of the male reproductive tract as described by Haines (1981), also indicated no suppression of mating activity as a result of the traps.

Such differences in the outcome of trials in Egypt compared with those in Israel may also be attributed to the varying role of the beneficial insects. The level of *Spodoptera* attack in Egypt as measured by egg-mass numbers can vary from year to year from a daily average of 10 ha^{-1} in a light infestation year to more than 10,000 ha^{-1} in an outbreak year, with the frequency of outbreaks about one year in ten. Even where considerable egg-mass numbers do occur they may be heavily attacked by predators and possibly egg-parasites (W.R.Ingram, private communication). As long as the egg-mass collectors are active, no insecticide sprays are used and so the beneficial insects are preserved. It is therefore possible that successes claimed for mass-trapping are the result of the beneficial insects, since in the presence of the traps insecticides are less likely to be used (Campion, 1983).

The influence of beneficial insects may also have influenced the outcome of mass-trapping trials in Indian cotton fields for control of both *S. litura* and *H. armigera* using trap densities of 5 ha^{-1} where success was claimed, as in Israel, on a similar reduction of insecticide applications in treatment areas of 175 ha compared with areas receiving conventional treatment (Patel *et al.*, 1986).

In any event the results emphasize the need for reliable assay procedures to determine the effectiveness of control strategies using pheromones in the absence of a direct killing effect on the larvae.

The polyphagous, mobile characteristics of *S. littoralis* may not be ideal for control by mass trapping but may be better suited to alternative forms of control using pheromones. The development of a lure and kill technique, combining the potent attractancy of the pheromone with the lethal effect of insecticide in discrete sources (McVeigh and Bettany, 1987), may be a more effective alternative (see below).

Prospects for control of the boll weevil, *Anthonomus grandis*, by mass trapping might seem more promising since at least early in the season both sexes are attracted. However in the USA, despite attempts made over a number of years, no clearly defined success for mass-trapping as a control technique has yet been achieved. From results reported by Lloyd *et al.* (1981) using mark, release, recapture techniques and computer simulation models, it is suggested that at trap densities of 14 ha^{-1} a high proportion of the female population would be caught. It is doubtful whether such findings can be translated into a commercially viable control strategy given such a large trap density.

A more recent series of experiments directed toward low-density populations indicated that two traps ha^{-1} eliminated boll weevils from 80% of fields where the populations were estimated to be fewer than 2.5 ha^{-1} (Leggett *et al.*, 1988). Such low-density trapping can have a role in the boll weevil eradication programme (see Chapter 9).

Mating Disruption

Sex pheromones may be used to control insect pests by causing communication disruption between the sexes, thus reducing the incidence of mating and thereby reducing subsequent larval infestations. It is assumed that an insect will be unsuccessful in locating a mate if the atmosphere is permeated with pheromone from a number of point sources.

Slow release formulation is essential for pheromones used in control strategies in order to prolong the release and efficacy of compounds which are otherwise highly volatile, and to provide stabilization of remaining material under field conditions. Several formulations have so far been used commercially.

- Hollow fibres, developed by Albany International (subsequently Scentry) in the USA and now sold under licence by Sandoz in Switzerland. Polyether or polyester fibres, measuring 1.5 cm in length with an inner diameter of 0.2 mm, are sealed at one end and contain about 250–275 μg of pheromone (Swenson and Weatherston, 1989).

- Laminate flakes, developed by the Hercon Division of the Health-Chem Corporation in the USA and now sold under licence by BASF in Germany. The flakes, consisting of two layers of plastic laminate in between which is sandwiched a porous layer impregnated with pheromone, can be varied in size to give different rates of release or numbers of point sources per unit area to suit different types of ground or aerial application equipment (Quisumbing and Kydonieus, 1982).
- Microcapsules, developed in the United Kingdom jointly by the Natural Resources Institute (NRI) and ICI Agrochemicals. The microcapsules, made of polyamide and polyurea, consist of tiny spheres with a median diameter of 2–3 μm. They are miscible in water and contain light-stable additives to prevent degradation of the capsule wall and the enclosed pheromone particularly in the presence of sunlight (Hall *et al.*, 1982).
- Twist-tie dispensers, developed by the Shin-Etsu Company in Japan. This formulation consists of a polyethylene tube, 10–20 cm long, containing the pheromone and a soft-wire stiffener (Flint *et al.*, 1985).
- A polyvinyl chloride (PVC) resin formulation developed jointly by the Natural Resources Institute and Agrisense-BCS Ltd which has provided good protection for the more labile pheromones containing both aldehydic and conjugated diene functionality (Cork *et al.*, 1989).

Of the five formulations listed, the hollow fibres and laminate flakes have to be applied with a glue in a special applicator to ensure adherence to the plant foliage; the microcapsules, being a water-based suspension, can be sprayed with conventional applicators without the need of any special adhesives, while the twist-tie and PVC dispensers can as yet only be applied by hand, tied or attached around the stem of the plant (Fig. 26.3).

The relative merits of sprayable and hand applied pheromone formulations must depend on circumstances. Relatively stable pheromones such as that for the pink bollworm, a 1:1 mixture of (*ZZ*) and (*ZE*)-7,11-hexadecadienyl acetate applied in hollow fibre or microencapsulated formulations, have a field persistence of two to three weeks. Much greater persistence of up to 100 days can be achieved in the field using the twist-tie or PVC resin formulations.

In 1978, the United States Environmental Protection Agency (EPA) granted the first registration of a sex pheromone product to the pink bollworm pheromone (Gossyplure) formulated in hollow fibres (Brooks *et al.*, 1979). Following this, approximately 20,000 ha of cotton were treated with Gossyplure in the southwestern USA, particularly in Arizona and Southern California. In 1979, the EPA granted an experimental use permit (EUP) to the laminate flake pheromone formulation of Gossyplure and a total of 192 ha of cotton was treated (Kydonieus and Beroza, 1981). Both of these formulations were tested in larger-scale trials in 1980 and a combined total of 50,000 hectares was treated in 1981.

Following fears that whitefly, *Bemisia tabaci* (Gennadium), and *Helicoverpa* spp. outbreaks were associated with insecticide treatments for *Pectinophora* control, farmers in the Imperial Valley of California voted in 1982 to use pheromones in the entire valley, on about 14,700 ha of cotton, to try to reduce the amount of insecticides they were using. A Cotton Pest Abatement District (CPAD) was established in which a farmer who did not use pheromones to control pink bollworm on his cotton could be fined. The programme was considered a success, with fewer applications of conventional insecticides being made, higher yields and only 5% of the crop damaged, compared with over 30% in conventionally treated neighbouring fields (Doane *et al.*, 1983). The mandatory programme was discontinued in 1984 with over 80% of the farmers expected to continue using pheromones without the need for a CPAD. Unfortunately this did not prove to be the case and many farmers reverted to using regular schedules of conventional insecticides.

More recently the twist-tie formulation of pink bollworm pheromone was successfully used in Arizona, initially in small-scale trials totalling 10 ha (Flint *et al.*, 1985), and later on a larger scale in California in 1985 (Staten *et al.*, 1987). Twist-ties were applied at a rate of 1000 ha^{-1} (78 g a.i. ha^{-1}) with an expected half-life of 58 days. Mean larval counts in the pheromone treated fields were lower than in the conventional insecticide treated fields while the average number of insecticide treatments was reduced by 40% as a result of the pheromone treatments. A limiting factor may be the necessity for hand application in a country where labour costs are high, but such costs can be offset against savings in insecticide applications (Baker *et al.*, 1990).

A large scale trial was conducted in Haryana, India in 1980 using aerially applied hollow fibre pheromone formulations in an area of 290 ha (Pawar *et al.*, 1981). However, despite claims of success from this trial no further large scale works have since been undertaken. Small scale trials using the microencapsulated pheromone formulation 'Pectone' with ground spraying equipment were reported to be successful (Sundaramurthy *et al.*, 1987). Further small scale trials using microencapsulated pheromone also applied with traditional ground spraying equipment were successfully conducted in China (Shu *et al.*, 1987), in that the level of control achieved or measured by boll damage assessments was comparable with that achieved using conventional insecticides.

The possibilities for control of the pink bollworm have been examined in Egypt using laminate flake, microencapsulated and hollow fibre formulations of the pheromone. Each formulation was applied aerially or by hand three to five times during the cotton season at rates of between 20 and 50 g ha^{-1} per season. Levels of control comparable with, and sometimes better than, a conventional insecticide application programme were achieved. It was also demonstrated that populations of beneficial insects were much higher in pheromone treated areas than in those treated with broad-spectrum insecticides (Critchley *et al.*, 1985; El-Adl *et al.*, 1988).

Pink bollworm pheromones were first used commercially in Egypt in 1984 and by 1986, the area under treatment with pheromones had risen to 25,000 ha. The use of pheromones on such a scale was unique in the developing world and had only been undertaken on a comparable scale in the United States (Campion *et al.*, 1989).

More recently 'twist-tie' pheromone formulation which requires only one application for season-long control of pink bollworm, have been successfully evaluated in large-scale trials in Egypt (McVeigh *et al.*, 1988). Only hand application is possible with this formulation, but this may be of particular relevance in developing countries where there is an availability of labour.

In the Egyptian Province of Dakahlia, honey collectors benefited greatly from the pheromone applications in 1987. Total honey production rose from zero in 1986 when only conventional insecticides were used on the cotton, to a total for the district of over 10,700 kg in the pheromone treatment areas in 1987. Average honey production rose from zero to 4.5 kg per hive between 1986 and 1987. This encouraging 'side benefit' was achieved with a programme of three pheromone applications, starting at the first flower stage, followed by two applications of insecticide (Moawad *et al.*, 1991).

In Egypt, full responsibility for controlling the cotton pest complex rests with the Ministry of Agriculture. Although an environmentally acceptable control strategy is favoured, possible risks of failure with a relatively new technique, coupled with a strong lobby from the private sector promoting insecticides, have resulted in a policy of caution and the need for further demonstrations of efficacy and, therefore, areas under pheromone treatment are not likely to increase dramatically in the short term.

In the neighbouring country of Israel the effectiveness of the microencapsulated and hollow fibre pheromone formulations has also been confirmed (Kehat *et al.*, 1986) and commercial applications undertaken on an area of 2000 ha.

In Peru, the use of pink bollworm pheromome formulations has been to delay application of broad-spectrum insecticides until at least 100 days post-sowing so as to avoid the resurgence of what are otherwise unimportant insects (Gonzalez, 1982). Susceptibility to attack by pink bollworm commences with flowering, at 50 days post-sowing. Protection of the plants at this critical stage by one or two aerially applied pink bollworm pheromone treatments has provided an ideal selective treatment (A. Almestar and J.E. Gonzalez, private communication).

Hollow-fibre, microencapsulated and 'twist-tie' formulations of pink bollworm pheromone were also used in trials conducted in Pakistan from 1985 to 1988. The microcapsules were applied by motorized knapsack sprayers, and the fibre and twist-tie formulations applied by hand. The early season control of *Pectinophora* by mating disruption permitted an average reduction of two insecticide applications otherwise required to control the cotton pest complex, particularly at the time of flower and fruit setting when beneficial insects are most numerous. The pheromone formulations, together with a mixture of selective and broad-spectrum insecticides in plots of 5 or 10 ha of cotton were compared with plots of cotton of similar size in the same locality treated with a conventional insecticide spray programme. Comparisons of numbers of infested bolls and estimated yields showed that levels of control achieved using the pheromone/insecticide combinations were equal in effect to conventional programmes of insecticide sprays (Critchley *et al.*, 1991).

In Pakistan, in addition to *Pectinophora* there are also two other species of bollworm, *Earias vittella* and *E. insulana* considered to be of major importance, sometimes early in the season. Thus, insecticides may have to be applied at an early stage of crop development and beneficial insects are the major casualties of these treatments. Under these circumstances there is a strong case for pheromone control of *Earias* spp. in addition to that of *Pectinophora*, thus avoiding the use of early season insecticide applications and preserving beneficial insects.

Multicomponent pheromones for both species have been identified and synthesized (Hall *et al.*, 1980; Cork *et al.*, 1988). They have a common main component, (*E,E*)-10,12-hexadecadienal, and trials in Pakistan have indicated that this single component, formulated together with pink bollworm pheromone can be used as a common disruptant for season-long pheromonal control of the bollworm complex.

Twist-tie, PVC resin and black hollow fibre formulations of the main component were evaluated (Critchley *et al.*, 1987; Qureshi and Ahmed, 1989; Chamberlain *et al.*, 1992). The twist-tie and PVC resin formulations were applied once only and the hollow fibres were applied three or four times at ten-day intervals. At the end of the season, numbers of bolls per plant and the estimated yield of seed cotton in plots that were treated with both the pink bollworm pheromone and the *Earias* pheromone were greater than in plots treated with pink bollworm pheromone only or plots treated with insecticides. *Earias* damage cannot be related to boll infestations alone as both species attack the flowerbuds, causing premature shedding. The early-season protection may therefore be the reason for the higher numbers of bolls and higher estimated yields in the pheromone treated plots.

Night observations showed a complete absence of moth flight activity and mating pairs of *Earias* in the pheromone-treated areas, unlike the situation in insecticide-treated areas where moth activity was readily seen (Chamberlain *et al.*, 1992).

A study on the mating disruption technique for the control of the red bollworm, *Diparopsis castanea*, a key pest of cotton in southeastern Africa, was undertaken by Marks *et al.* (1978, 1981). However the results were equivocal in the absence at that time of a stabilized microencapsulated pheromone formulation. The red bollworm is closely related to the Sudan bollworm, *Diparopsis watersi* (Roths.), a widespread pest on cotton in West Africa, and the pheromones are likely to be very similar (Beevor *et al.*, 1973), so that a pheromone formulation appropriate for both species may be commercially viable.

A programme on the integrated use of pheromones and selective insecticides to ensure the preservation of the beneficial insect fauna is in progress in the Caribbean island of Barbados. The twist-tie pheromone formulation was used for the control of pink bollworm while the insect growth regulators flufenoxuron and buprofezin were employed respectively to control *Alabama* and the white fly, *Bemisia tabaci*. During the course of this work pheromone lures of *Alabama* in funnel traps were used as a monitoring device to provide the basis for an economic threshold (J. Jones and B.R. Critchley, unpublished data).

Computer simulation models were developed in the United States based on observations using the laminate flake pink bollworm pheromone formulation. They indicated that mating disruption was most effective when early season applications were made which subsequently delayed population increase, while late season applications were ineffective (Stone and Gutierrez, 1986). Results of trials in Egypt using a microencapsulated pheromone formulation confirmed these observations (Critchley *et al.*, 1984) and at present a modification of the basic model is being adapted for Egyptian cotton varieties and conditions as a guide not only to pheromone use but to provide an action framework for a comprehensive integrated control programme (Russell and Radwan, 1993).

Much research has been devoted to establishing the feasibility of the mating disruption approach for control of *Helicoverpa* spp. The number and importance of up to seven components of the pheromone blends is still not clearly confirmed (Klun *et al.*, 1979; Tumlinson *et al.*, 1982). The functional groups of the major attractant components are aldehydes, and hence more reactive and difficult to stabilize in the field than the acetate functional groups of the pink bollworm pheromone. All the *Heliothis* species are polyphagous and the adult moths highly migratory. Perhaps for these reasons attacks of cotton by both Old World and New World species are often unpredictable.

Trials conducted in the United States with hollow-fibre and laminate flake formulations of the main components or analogues of the *H. virescens* and *H. zea* pheromones were either unsuccessful or gave equivocal results (McLaughlin and Mitchell, 1982). Application rates did not, however, exceed 10 g a.i. ha^{-1} and therefore, apart from anticipated problems of moth migration, inadequate concentrations of pheromone may also have been a contributing factor. This view receives some support from trials conducted in Egypt using a microencapsulated formulation of the main pheromone components of *H. armigera*. As measured by trap-catch suppression in the treated areas, complete mating disruption was achieved for a period of 22 days when applied at a rate of 100 g ha^{-1}, whereas at rates of 10 and 50 g ha^{-1} such activity persisted for only four–five days (Critchley *et al.*, 1986).

Twist-tie and PVC resin pheromone formulations have resolved the problem of stability under field conditions (Dunkelblum and Kehat, 1990; D.R. Hall, private communication). However trials conducted in Australia using the twist-tie pheromone formulation failed to suppress mating which was attributed to the immigration of insects into the treatment areas (K. Ogawa, private communication).

Spodoptera littoralis has been shown to have a limited migratory potential (Campion *et al.*, 1977; Nasr *et al.*, 1984), therefore, the prospects for its control by use of pheromones are greater. Work in Egypt with microencapsulated formulations of its pheromone showed that mating disruption of *S. littoralis* is possible, but only at rates of active ingredient of 40 to 80 g ha^{-1} per application and such rates were too high to be economically viable (Campion *et al.*, 1981). However, in Egypt during 1990 the twist-tie and PVC resin formulations of *Spodoptera* pheromone were further evaluated in a total area of 50 ha. The effectiveness of the treatments was evaluated using pheromone traps, tethered virgin female moths, bait traps, light traps, low light-level video studies and numbers and viability of egg-masses were compared in both the treatment area and a nearby comparable untreated area. These methods of assessment showed that the one application of either pheromone formulation was effective in preventing mating of *S. littoralis* adults and that adult activity in the area was reduced for a period of six weeks. Although the pheromone appeared to be still effective in disrupting mating at that time, the trial had to be abandoned due to an application of insecticide to control another pest (McVeigh, 1990). To confirm these promising results further much larger trials are in progress.

In the laboratory, high concentrations of the boll weevil pheromone can prevent mating (Huddleston *et al.*, 1977) while reduced oviposition was also achieved in the field when dispensed at the relatively high rate of 25 g ha^{-1} (Villavaso, 1982), a level which is considered uneconomic for practical usage.

In summary it is concluded that factors leading to a successful implementation of cotton pest control by mating disruption requires:

- a basically stable pheromone which is inexpensive to manufacture on a large scale;
- the availability of stable formulations and appropriate methods of application;
- an insect which is a key pest with a narrow range of alternative host plants;
- an insect with a limited migratory capability;
- a basic biological knowledge of the insect is already available;
- adequate measures are available both for measuring disruption and for estimating larval populations and crop yield;
- the economic importance of the pest insect is great enough to warrant commercial interest;
- there are convincing reasons for changing from the presently available pest strategy.

Lure and Kill

Once commercial hollow-fibre formulations of Gossyplure became available, some growers in the USA added small amounts of pyrethroid insecticides to the sticker used with it to reduce the amount of pheromone applied and lessen the cost of each application. It was claimed that such formulations applied at a rate of 6000–12,000 fibres ha^{-1} were more effective since it was assumed that pink bollworm moths would be attracted to the insecticide impregnated fibres and would be killed or incapacitated. Subsequent investigations substantiated this to a degree (Butler and Las, 1983; Beasley and Henneberry, 1984), but most work on the effect of insecticides on pheromone-mediated behaviours was examined in laboratory flight tunnels (Haynes and Baker, 1985; Haynes *et al.*, 1986).

Field observations in Egypt using night vision equipment (B.W. Bettany, unpublished data) showed that male *S. littoralis* moths will land on, and remain in contact with microencapsulated point sources of pheromone at densities of 500–1000 ha^{-1} for an average of three seconds (McVeigh and Bettany, 1987). At higher numbers of pheromone sources, trail masking or confusion occurs such that the frequency of moth arrival at the sources is greatly reduced, while at densities of 10,000 ha^{-1} and above no arrivals occurred at all; hence distribution density is critical.

Laboratory studies involved selection of insecticides for lure and kill using contact toxicity assays with special reference to speed of action, residual persistence, effects on mating behaviour, fertilization and egg viability. The pyrethroids were found to be the most suitable class of insecticides and Lambda-cyhalothrin was selected as the most promising for field evaluation, since it was shown that up to 100% of moths exposed to insecticide-spiked pheromone sources for 10 to 20 seconds will die within 24 hours, and the mating capability of survivors is seriously impaired (De Souza *et al.*, 1992).

The levels of pheromone required for this technique are not likely to exceed 2.5 g ha^{-1} per application and, therefore, would be economically acceptable. The levels of insecticide required would also be very low. Field studies in 1987 and 1988 showed that an insecticide concentration of 2000 ppm active ingredient (a.i.) exhibited good contact toxicity to attractant sources and residual persistence, and had no apparent 'repellency' to *S. littoralis* or interactions with the pheromone. The residual toxicity of microencapsulated cyhalothrin (2000 ppm a.i.) on cotton in the field was found to be maintained for at least fourteen days. Studies on the field persistence of microencapsulated pheromone on filter papers suggested that the short-lived control of mating achieved in concurrent lure and kill trials (three–five days) was attributed to photochemical degradation of the pheromone.

A PVC resin pheromone formulation applied at 500 point sources ha^{-1} was found to be much less subject to photodegradation, and this was reflected by an improved trap-catch suppression of 90–95% for nearly one month post-application. Night observations using video techniques showed that although moths approached the PVC resin sources, the rate of landing was much lower when compared with filter papers sprayed with microencapsulated pheromone. This seemed to preclude lure and kill as being the chief mode of action (B.W. Bettany, unpublished data). To determine whether lure and kill or mating disruption was the major factor in effecting control by field applications, trials directly comparing the effects of the PVC resin sources with and without insecticide were conducted at a point-source distribution of 500 ha^{-1}. The extent of mating and trap-catch suppression were similar, indicating that little advantage was achieved by the inclusion of the insecticide (Downham *et al.*, 1991).

The results at least for *Spodoptera* indicate little advantage for the lure and kill technique using pheromone insecticide combinations over mating disruption. The complicated interaction between the density of attractant point sources, the need to maximize the contact time of attracted moths to insecticide treated surfaces which must also vary in relation to changing meteorological factors does not support an immediate practical outcome for this technique.

Host plant derived attractants and feeding stimulants for the boll weevil (McKibben *et al.*, 1985; Dickens 1989), together with the synthetic pheromone in combination with an insecticide are also being investigated. Such an all encompassing attractant source may be more effective than pheromone alone, but the problems referred to above may also have to be taken into consideration.

The use of an early planted trap crop of cotton which can be sprayed with insecticides has long been devised as a means of control of overwintering boll weevil (Scott *et al.*, 1974). The placement of boll weevil pheromone lures within the trap crop has increased the effectiveness of this technique both in the United States, Nicaragua and Colombia (Gilliland *et al.*, 1976; Laboucheix and Gonzalez, 1987; Mendoza, 1989).

Conclusions

It may be concluded that where precise population estimates are not required, pheromone-baited traps have been successful in delineating the presence or relative absence of target cotton pests. Apart from pink bollworm, the problem of relating trap catch to subsequent economic damage has not so far been convincingly successful. This approach requires much painstaking work over prolonged periods, and relatively few studies are in progress. It is important that such work is continued and it is certain that further successes will be achieved in the future.

Mass-trapping programmes for the control of cotton pests have so far shown little promise. Improvement in trap design and pheromone characterization may improve the situation in the future particularly in the utilization of beetle pheromones where both sexes are attracted to the traps, instead of only males as occurs in the case of most lepidopterous pheromones. However, the problem of trap deployment and maintenance poses economic constraints in countries where labour costs are high.

The lure and kill approach may be a more viable and practical option but there are still basic problems to be resolved.

The future prospects for control by mating disruption are good for those insects of key pest status with a limited range of host plants apart from cotton. However, the apparently low migration potential of the cotton pests successfully controlled may merely indicate that these attempts have been limited in size and scope by the relative expense of the pheromones, and this has precluded treatment over wider areas. Increased confidence in the already available techniques should lead to increasing areas of the target pest species being treated, as is occurring for pink bollworm control.

There is evidence from laboratory studies that at high application rates of pheromone, close-range as well as long-range orientation between the sexes is inhibited, which would prevent mating at high population densities, while for certain species adverse effects on female moth behaviour have also been noted. Further behavioural studies are necessary to verify these observations under field conditions. However, if it is confirmed that high application rates are generally necessary for successful control by mating disruption then, in many cases, ways will have to be found to reduce the cost of these chemicals to make the technique commercially competitive with conventional pesticides. More work also needs to be done to integrate insect control by mating disruption with other control strategies in well-constructed integrated pest management programmes.

References

Ahmad, Z. (1979) Monitoring the seasonal occurrence of the pink bollworm in Pakistan with sex traps. *Plant Protection Bulletin*, FAO 27, 19–20.

Almestar, A. (1990) Monitoro de *Alabama argillacea* (Hb.), en algodonero con trampas feromonas sexuales. *Proceedings of the 23rd Convencion Nacional de Entomologia*, August 1990, Cusco, Peru, 5pp.

Anon. (1987) ICRISAT Annual Report, pp. 153–154.

Arn, H., Toth, M. and Priesner, E. (1986) *List of Sex Pheromones of Lepidoptera and Related Attractants*, 10BC–WPRS, 123pp.

Baker, T.C., Staten, R.T. and Flint, H.M. (1990) Use of pink bollworm pheromone in the Southwestern United States. In: Ridgway, R.L., Silverstein, R.M. and Inscoe, M.N. (eds) *Behaviour-Modifying Chemicals for Insect Management: Applications of Pheromones and Other Attractants*, Marcel Dekker, New York, pp. 417–436.

Beasley, C.A. and Henneberry, T.J. (1984) Combining gossyplure and insecticides in pink bollworm control. *California Agriculture* 38, 22–24.

Beasley, C.A., Henneberry, T.J., Adams, C. and Yates, L. (1985) Gossyplure baited traps as pink bollworm survey, detection, research and management tools in Southwestern desert cotton growing areas. *California Agricultural Experiment Station Bulletin* 1915, 15pp.

Beevor, P.S., Campion, D.G., Moorhouse, J.E. and Nesbitt, B.F. (1973) Cross-attractancy and cross-mating between the red bollworm *Diparopsis castanea* (Hmps.) and the Sudan bollworm *Diparopsis watersi* (Roths.) (Lep., Noctuidae). *Bulletin of Entomological Research* 62, 439–442.

Benedict, J.N., Urban, T.C., George, D. M., Segers, J.C., Anderson, D.J., McWhorter, G.M. and Zummo, G.R. (1985) Pheromone trap thresholds for management of overwintered boll weevils (Coleoptera: Curculionidae). *Journal of Economic Entomology* 78, 169–171.

Bestmann, H.J., Attygalle, A.B., Schwartz, J., Vostrowsky, O. and Knauf, W. (1988) Identification of sex pheromone components of *Spodoptera sunia* Guenée (Lepidoptera: Noctuidae). *Journal of Chemical Ecology* 14, 683–690.

Bierl, B.A., Beroza, M., Staten, R.T., Sonnet, P.E. and Adler, V.E. (1974) The pink bollworm sex attractant. *Journal of Economic Entomology* 67, 211–216.

Bishara, I. (1934) The cotton worm *Prodenia litura* F. in Egypt. *Bulletin of the Royal Society of Entomology, Egypt* 18, 288–418.

Brooks, T.W., Doane, C.C. and Haworth, J.K. (1979) Suppression of *Pectinophora gossypiella* with sex pheromones. In: *Proceedings of the 1979 British Crop Protection Conference, Pests and Diseases*, 1979, Brighton. Lavenham Press, Lavenham, Suffolk, pp. 854–866.

Butler, G.D., Jr. and Las, A.S. (1983) Predaceous insects: effects of adding permethrin to the sticker used in gossyplure applications. *Journal of Economic Entomology* 76, 1448–1451.

Campion, D.G. (1983) Pheromones for the control of insect pests in Mediterranean countries. *Crop Protection* 2, 3–16.

Campion, D.G. (1984) Survey of pheromone uses in pest control. In: Hummel, H.E. and Miller, T.A. (eds) *Techniques in Pheromone Research*. Springer-Verlag, New York, pp. 405–449.

Campion, D.G. (1989) Semiochemicals for the control of insect pests. In: BCPC Monograph No. 43, *Progress and Prospects in Insect Control*. pp. 119–127.

Campion, D.G., Bettany, B.W. and Steedman, R.A. (1974) The arrival of male moths of the cotton leafworm *Spodoptera littoralis* (Boisd.) (Lepidoptera: Noctuidae) at a new continuously recording trap. *Bulletin of Entomological Research* 64, 379–386.

Campion, D.G., Bettany, B.W., McGinnigle, J.B. and Taylor, L.R. (1977) The distribution and migration of *Spodoptera littoralis* (Boisduval) (Lepidoptera: Noctuidae), in relation to meteorology on Cyprus, interpreted from maps of pheromone trap samples. *Bulletin of Entomological Research* 67, 501–522.

Campion, D.G., Lester, R. and Nesbitt, B.F. (1978) Controlled release of pheromones. *Pesticide Science* 9, 434–440.

Campion, D.G., Hunter-Jones, P., McVeigh, L.J., Hall, D.R., Lester, R. and Nesbitt, B.F.

(1980) Modification of the attractiveness of the primary pheromone component of the Egyptian cotton leafworm *Spodoptera littoralis* (Boisd.) (Lepidoptera: Noctuidae) by secondary pheromone components and related chemicals. *Bulletin of Entomological Research* 70, 417–434.

Campion, D.G., McVeigh, L.J. and Hunter-Jones, P. (1981) The use of sex pheromones for the control of spiny bollworm *Earias insulana* and the American bollworm *Heliothis armigera* in cotton growing areas of Syria. FAO Unpublished Report, 10pp.

Campion, D.G., Critchley, B.R. and McVeigh, L.J. (1989) Mating disruption. In: Justum, A.R. and Gordon, R.F.S. (eds) *Insect Pheromones in Plant Protection*. John Wiley & Sons, Chichester, pp. 89–119.

Chamberlain, D.J., Critchley, B.R. and Campion, D.G., Attique M.R., Rafique, M. and Arif, M.I. (1992) Use of multi-component pheromone formulations for the control of cotton bollworms (Lepidoptera: Gelchiidae and Noctuidae) in Pakistan. *Bulletin of Entomological Research* 82, 449–458.

Cork, A., Chamberlain, D.J., Beevor, P.S., Hall, D.R., Nesbitt, B.F., Campion, D.G. and Attique, M.R. (1988) Components of female sex pheromone of spotted bollworm, *Earias vittella* F. (Lepidoptera: Noctuidae): identification and field evaluation in Pakistan. *Journal of Chemical Ecology* 14, 929–945.

Cork, A., Hall, D.R., Smith, J. and Jones, O.T. (1989) A new resin formulation for the controlled release of insect pheromones. *Proceedings of 16th International Symposium of the Controlled Release Society*, 6–9 August 1989, Chicago, American Chemical Society, Washington DC, pp. 5–6, 9–10.

Cork, A., Hall, D.R., Campion, D.G., Chamberlain, D.J. and Almestar, A.A. (1992) Novel sex pheromone components of lepidopterous cotton pests from South America. *Proceedings XIX International Congress of Entomology*, Beijing, China, 28 June–4 July 1992. Abstracts, p. 209.

Critchley, B.R., Campion, D.G., McVeigh, L.J., Hunter-Jones, P., Hall, D.R., Cork, A., Nesbitt, B.F., Marrs, G.J., Jutsum. A.R., Hosny, M.M. and Nasr, El Sayed A. (1983) Control of the pink bollworm, *Pectinophora gossypiella* (Saunders) (Lepidoptera: Gelechiidae), in Egypt by mating disruption using an aerially applied microencapsulated pheromone formulation. *Bulletin of Entomological Research* 73, 289–299.

Critchley, B.R., Campion, D.G., McVeigh, E.M., McVeigh, L.J., Jutsum, A.R., Gordon, R.F.S., Marrs, G.J., Nasr, El Sayed A. and Hosny, M.M. (1984) Microencapsulated pheromones in cotton pest management. In: *Proceedings of British Crop Protection Conference, Pests and Diseases*, November 1984, Brighton, pp. 241–245.

Critchley, B.R., Campion, D.G., McVeigh, L.J., McVeigh, E.M., Cavanagh, G.G., Hosny, M.M., Nasr, El Sayed A., Khidr, A.A. and Naguib, A.A. (1985) Control of the pink bollworm, *Pectinophora gossypiella* (Saunders) (Lepidoptera: Gelechiidae), in Egypt by mating disruption using hollow-fibre, laminate-flake and microencapsulated formulations of synthetic pheromone. *Bulletin of Entomological Research* 54, 329–345.

Critchley, B.R., McVeigh, L.J., McVeigh E.M., Cavanagh, G.G. and Campion, D.G. (1986) Integrated control of cotton pests in Egypt using pheromones and viruses. Pheromone studies 1979–1983. *Tropical Development and Research Institute Overseas Assignment Report*. R1310(R), 59pp.

Critchley, B.R., Campion, D.G., Cavanagh, G.G., Chamberlain, D.J. and Attique, M.R. (1987) Control of three major pests of cotton in Pakistan by a single application of their combined sex pheromones. *Tropical Pest Management* 33, 374.

Critchley, B.R., Chamberlain, D.J., Campion, D.G., Cavanagh, G.G., Attique, M.R., Ali, M. and Ghaffar, A. (1991) Integrated use of pink bollworm pheromone formulations and selected conventional insecticides for control of the cotton pest complex in Pakistan. *Bulletin of Entomological Research* 81, 371–378.

Davich, T.B., Hardee, D.D. and Alcala, J.M. (1970) Long range dispersal of boll weevils determined with wing traps baited with males.*Journal of Economic Entomology* 63, 1706–1708.

Denore, B.J., Beaumont, T.E., Pender, J., Campion, D.G., Critchley, B.R. and Marengo, R.M. (1988) Cotton area mapping using multitemporal satellite data integrated within a geographical information system applied to a cotton boll weevil control programme in Paraguay. *Proceedings of the IGARSS 1988 Symposium*, 13–16 September 1988, Edinburgh, Scotland. ESA Publications Divison, 581–582.

Dent, D.R. and Pawar, C.S. (1988) The influence of moonlight and weather on catches of *Helicoverpa armigera* (Hübner) (Lepidoptera: Noctuidae) in light and pheromone traps. *Bulletin of Entomological Research* 78, 365–377.

De Souza, K.R., McVeigh, L.J. and Wright, D.J. (1992) Selection of insecticides for lure and kill studies against *Spodoptera littoralis* (Lepidoptera: Noctuidae). *Journal of Economic Entomology* 85, 2100–2106.

Dhawan, A. K and Sidhu, A.S. (1984) Assessment of capture thresholds of pink bollworm moths for timing insecticidal applications on *Gossypium hirsutum* Linn. *Indian Journal of Agricultural Sciences* 54, 426–433.

Dickens, J.C. (1989) Green leaf volatiles enhance aggregation pheromone of boll weevil *Anthonomus grandis*. *Entomologia Experimentalis et Applicata* 52, 191–203.

Doane, C.C., Haworth, J.K. and Dougherty, D.G. (1983) Nomate PBW, a synthetic pheromone formulation for wide area control of the pink bollworm. In: *Proceedings of 10th International Congress of Plant Protection*, 20–25 November 1983, Brighton, Lavenham Press, Lavenham, Suffolk.

Dos Santos, W.J. (1989) The boll weevil in Brazil. *Technical Seminar at 48th Plenary meeting of the International Cotton Advisory Committee, Scottsdale, Arizona*, October 1989, pp. 26–28.

Downham, M.C.A., McVeigh, L.J., Bettany, B.W., De Souza, K.R., Russell, D.A., Campion, D.G., Hall, D.R., Adams, P.H. and Moawad, G.M. (1991) Development of a lure and kill strategy for control of the Egyptian cotton leafworm in Egypt, 1986–1989. Natural Resources Institute, unpublished report.

Dunkelblum, E., Kehat, M., Gothilf, S., Greenberg, S. and Sklarsz, B. (1982) Optimised mixture of sex pheromonal components for trapping of male *Spodoptera littoralis* in Israel. *Phytoparasitica* 10, 21–26.

Dunkeleblum, E.M. and Kehat, M. (1990) The practical use of sex pheromones in cotton fields in Israel. In: *Pheromones in Mediterranean Pest Management*, 10BC-WPRS Working Group. Use of Pheromones and Other Semiochemicals in Integrated Control, 10–15 September 1990, Granada, p. 13.

Dunkelblum, E.M., Gothilf, S. and Kehat, M. (1980) Identification of the sex pheromone of the cotton bollworm *Heliothis armigera* in Israel. *Phytoparasitica* 8, 209–211.

El-Adl, M.A., Hosny, M.M. and Campion, D.G. (1988 Mating disruption for the control of pink bollworm *Pectinophora gossypiella* in the Delta cotton growing area of Egypt. *Tropical Pest Management* 34, 210–214.

Elmosa, H. (1986) Prospects for using sex pheromone for the control of spiny bollworm in cotton growing in Syria. *Dirasat* 13, 165–174.

Flint, H.M., Smith, R.L., Bariola, L.A., Horn. D., Forey, D. and Kuhn, S.J. (1976) Pink bollworm: trap tests with gossyplure. *Journal of Economic Entomology* 67, 738–740.

Flint, H., Merkle, J.R. and Yamamoto, A. (1985) Pink bollworm (Lepidoptera: Gelechiidae); field testing a new polyethylene tube dispenser for gossyplure. *Journal of Economic Entomology* 78, 1431–1436.

French R.A. (1969) Migration of *Laphygma exigua* (Lepidoptera: Noctuidae) to the British Isles in relation to large-scale weather movements. *Journal of Animal Ecology* 38, 199–210.

Gilliland, F.R., Jr., Lambert, W.R., Weeks, J.R. and Davis, R.L. (1976) Trap crops for boll weevil control. In: Davich, T.B. (ed.) *Boll Weevil Suppression, Management and Elimination Technology*. US Department of Agriculture, Agricultural Research Service, Southern Region, New Orleans, pp. 41–44.

Gonzalez, J.E. (1982), Manual de evaluación y control de insectos y acaros del algodonero. *Fundación para el Desarrollo Algodonero, Boletín Técnico*, No. 1, second edition. Lima, Peru, 79pp.

Graham, H.M., Martin, D.F., Ouye, M.T. and Hardman, R.M. (1966) Control of pink bollworm by male annihilation. *Journal of Economic Entomology* 59, 590–593.

Guerra, A.A., Garcia, R.D., and Leal, M.P. (1969) Suppression of populations of pink bollworm in field cages with sex attractant. *Journal of Economic Entomology* 62, 741–742.

Gubbins, K.E. and Campion, D.G. (1982) Economic aspects of pheromone trapping techniques for the control of Egyptian cotton leafworm. *Outlook on Agriculture* 11, 62–66.

Haines, L.C. (1981) Changes in colour of a secretion in the reproductive tract of adult males of *Spodoptera littoralis* (Boisduval) (Lepidoptera: Noctuidae) with age and mated status. *Bulletin of Entomological Research* 71, 591–598.

Hall, D.R., Beevor, P.S., Lester, R. and Nesbitt, B.F. (1980) (E,E)-10,12-Hexadecadienal: a component of the sex pheromone of the spiny bollworm, *Earias insulana* (Boisd.) (Lepidoptera: Noctuidae). *Experientia* 36, 52–153.

Hall, D.R. Nesbitt, B.F., Marrs, G.J., Green, A. St.J., Campion, D.G. and Critchley, B.R. (1982) Development of microencapsulated pheromone formulations. In: Leonhardt, B.A. and Beroza, M. (eds) *Insect Pheromone Technology: Chemistry and Applications*. American Chemical Society Symposium Series No. 190, Washington DC, pp. 131–143.

Hall, D.R., Beevor, P.S., Campion, D.G., Chamberlain, D.J., Cork, A., White, R., Almestar, A. and Henneberry, T.J. (1992) Nitrate esters: novel sex pheromone components of the cotton leafminer *Bucculatrix thurberiella* Busck. (Lepidoptera: Lyonetiidae). *Tetrahedron Letters* 33, 4811–4814.

Hall, D.R., Beevor, P.S., Campion, D.G., Chamberlain, D.J., Cork, A., White, R., Almestar, A., Henneberry, T.J., Nandagopal, V., Wightman, J.A. and Ranga Rao, G.V. (1993) Identification and synthesis of new pheromones. In: *Proceedings of an OILB Meeting on Pheromone Technology in Europe and the Developing Countries*, Chatham, Kent (in press).

Hartstack, A.W. and Witz, J.A. (1981) Estimating field populations of tobacco budworm moths from pheromone trap catches. *Environmental Entomology* 10, 908–914.

Hartstack, A.W., Hollingworth, J.P., Witz, J.A., Buck, D.R. Lopez, J.D. and Hendricks, D.E. (1978) Relation of tobacco budworm catches in pheromone baited traps to field populations. *Southwestern Entomologist* 3, 43–51.

Hartstack, A.W., King, E.G. and Phillips, J.R. (1983) Monitoring and predicting *Heliothis* populations in Southeast Arkansas. In: *Proceedings of the Beltwide Cotton Production Research Conference*, Memphis, Tennessee, pp. 187–190.

Haynes, K.F. and Baker, T.C. (1985) Sublethal effects of permethrin on the chemical communications system of the pink bollworm moth *Pectinophora gossypiella*. *Journal of Chemical Ecology* 14, 1547–1560.

Haynes, K.F., Li, W.G. and Baker, T.C. (1986) Control of pink bollworm moth (Lepidoptera: Gelechiidae) with insecticides and pheromones (Attracticides): lethal and sub-lethal effects. *Journal of Economic Entomology* 79, 1466–1471.

Hendricks, D.E. and Hartstack, A.W. (1978) Pheromone trapping as an index for initiating control of cotton insects, *Heliothis* spp: a compendium. In: *Proceedings of the Beltwide Cotton Production Research Conference*, Memphis, Tennessee, pp. 116–120.

Henneberry, T.J. and Clayton, T.E. (1982) Pink bollworm of cotton *(Pectinophora gossypiella* (Saunders)): male moth catches in gossyplure-baited traps and relationships to oviposition, boll infestation and moth emergence. *Crop Protection* 1, 497–504.

Hosny, M.M. (1980) The control of insect pests in Egypt. *Outlook on Agriculture* 10, 204–225.

Hosny, M.M., Iss-Hak, R.R., Nasr, El-Sayed A., El-Deeb, Y.A., Critchley, B.R., Topper, C.P. and Campion, D.G. (1979) Mass trapping for the control of Egyptian cotton leafworm *Spodoptera littoralis* (Boisd.) in Egypt. In: *Proceedings of the British Crop Protection Conference, Pests and Diseases*, 19–22 November 1979, Brighton, pp. 395–400.

Hosny, M.M., Topper, C.P., Moawad, G.M. and Saadany, G.B. (1986) Economic damage thresholds of *Spodoptera littoralis* (Boisd.) (Lepidoptera: Noctuidae) on cotton in Egypt. *Crop Protection* 5, 100–104.

Huber, R.T., Moore L. and Hoffmann, M.P. (1979) Feasibility study of area-wide pheromone trapping of pink bollworm moths in a cotton insect pest management programme. *Journal of Economic Entomology* 72, 222–227.

Huddleston, P.M., Mitchell, E.B. and Wilson, N.N. (1977) Disruption of boll weevil communication. *Journal of Economic Entomology* 70, 83–85.

Hummel. H.E., Gaston, L.K., Shorey, H.H., Kaae, R.S., Byrne, K.J and Silverstein, R.M. (1973) Clarification of the chemical status of the pink bollworm sex pheromone. *Science* 181, 873–875.

Ingram, W.R. (1980) Studies of the pink bollworm, *Pectinophora gossypiella*, on Sea Island cotton in Barbados. *Tropical Pest Management* 26, 118–137.

Inscoe, M.N., Leonhardt, B.A. and Ridgway, R.L. (1990) Commercial availability of insect pheromones and other attractants. In: Ridgeway, R.L., Silverstein, R.M. and Inscoe, M.A. (eds) *Behaviour-Modifying Chemicals for Insect Management.* Marcel Dekker, New York, pp. 631–715.

Johnson, D.J. (1983) Relationship between tobacco budworm (Lepidoptera: Noctuidae) catches when using pheromone traps and egg counts in cotton. *Journal of Economic Entomology* 76, 182–183.

Jutsum, A.R. and Gordon, R.F.S. (1989) *Insect Pheromones in Plant Protection.* John Wiley & Sons, Chichester, 369pp.

Kehat, M., Gothilf, S., Dunkelblum, E. and Greenberg, S. (1980) Field evaluation of female sex pheromone components of the cotton bollworm, *Heliothis armigera. Entomologia Experimentalis et Applicata* 27, 188–193.

Kehat, M., Gothilf, S., Dunkelblum, E. and Greenburg, S. (1982) Sex pheromone traps as a means of improving control programmes for the cotton bollworm. *Environmental Entomology* 11, 727–729.

Kehat, M., Chen, C., Klein, Z. and Fishler, R.G. (1986) Determining the efficiency of sex pheromones in controlling the pink bollworm *Pectinophora gossypiella* (Saunders), Lepidoptera, Gelechiidae in cotton fields in Israel. *Israel Journal of Entomology* 20, 25–36.

Kennedy, J.W. (1981) Practical application of pheromones in regulatory pest management programmes. In: Mitchell, E.R. (ed.) *Management of Insect Pests with Semiochemicals: Concepts and Practice.* Plenum Press, New York, pp 1–11.

Klun, J.A., Plimmer, J.A., Bierl-Leonhardt, B.A., Sparks, A.N. and Chapman, O.L. (1979) Trace chemicals: the essence of sexual communication systems in *Heliothis* species. *Science* 204, 1328–1330.

Klun, J.A., Plimmer, J.R., Bierl-Leonhardt, B.A., Sparks, A.N., Primiani, M., Chapman,

O.L., Lee, G.H. and Lepone, G. (1980) Sex pheromone chemistry of female corn earworm moth. *Heliothis zea. Journal of Chemical Ecology* 6, 165–175.

Knipling, E.F. and McGuire, J.U., Jr. (1966) Population models to test theoretical effects of sex attractants used for insect control. *US Department of Agriculture, Information Bulletin* 308, 20pp.

Kononenko, A.P., Grichanov, I.Ya., Kirov, E.I. and Maiorov, V.I. (1986) Correlation between trap catches of adults of the cotton moth and abundance of the pre-adult stages. *Izvestiya Academii Nauk Tadzhikskoi SSR, Biologicheskirkh Nauk* 3, 51–55.

Konyukhov, V.P., Kovalev, B.G. and Sammar-Zade, N.R. (1983) Isolation and identification of the components of the sex pheromone of the corn earworm *Heliothis armigera* (Hb.). *Soviet Journal of Bioorganic Chemistry* New York 9, 782–787.

Kydonieus, A.F. and Beroza, M. (1981) The Hercon dispenser formulation and recent test results. In: Mitchell, E.R. (ed.) *Management of Insect Pests with Semiochemicals: Concepts and Practice.* Plenum Press, New York, pp. 445–453.

Laboucheix, J. and Gonzalez, D.F. (1987) Evaluation de l'efficacité du méthyl parathion vis-à-vis d'*Anthonomus grandis* Boheman en culture cotonnière au Nicaragua. *Coton et Fibres Tropicales* 42, 41–53.

Leggett, J.E., Dickerson, W.A. and Lloyd, E.P. (1988) Suppressing low level boll weevil populations with traps: influence of trap placement, grandlure concentration and population level. *Southwestern Entomologist* 13, 205–216.

Leonard, B.R., Graves, J.B., Burris, E., Pavloff, A.M. and Church, G. (1989) *Heliothis* spp. (Lepidoptera: Noctuidae) captures in pheromone traps: species composition and relationship to oviposition in cotton. *Journal of Economic Entomology* 82, 574–579.

Lloyd, E.P., McKibben, G.H., Knipling, E.F., Witz, J.A., Hartstack, A.W., Leggett, J.E. and Lockwood, D.F. (1981) Mass trapping for detection, suppression and integration with other suppression measures against the boll weevil. In: Mitchell, R.R. (ed.) *Management of Insect Pests with Semiochemicals: Concepts and Practice.* Plenum Press, New York, pp. 191–203.

Lopez, J.D., Shaver, T.N. and Dickerson, W.A. (1990) Population monitoring of *Heliothis* spp. using pheromones. In: Ridgway, R.L., Silverstein, R.M. and Inscoe, M.A. (eds) *Behaviour-Modifying Chemicals for Insect Management.* Marcel Dekker, New York, pp. 473–496.

McKibben, G.H., Thompson, M.J., Parrott, W.L., Thompson A.C. and Lusby, W.R. (1985) Identification of feeding stimulants for boll weevils from cotton buds and anthers. *Journal of Chemical Entomology* 11, 1229–1238.

McLaughlin, J.R. and Mitchell, E.R. (1982) Practical development of pheromones in *Heliothis* management. In: Read, W. and Kumble, V. (eds) *Proceedings of the International Workshop on* Heliothis *Management,* 15–20 November 1981, International Crops Research Institute for the Semi-Arid Tropics, Patancheru, AP, India. ICRISAT, Patancheru, pp. 309–318.

McVeigh, E.M., McVeigh, L.J. and Cavanagh, G.G. (1983) A technique for tethering female moths of *Spodoptera littoralis* (Boisduval) (Lepidoptera: Noctuidae) to evaluate pheromone control methods. *Bulletin of Entomological Research* 73, 441–446.

McVeigh, L.J. (1990) The use of pheromones in pest control in cotton and other field crops in the Mediterranean region. In: *Pheromones in Mediterranean Pest Management,* IOBC-WPRS Working Group: Use of Pheromones and Other Semiochemicals in Integrated Control, 10–15 September 1990, Granada, p. 16.

McVeigh, L.J. and Bettany, B.W. (1987) The development of a lure and kill technique for control of the Egyptian cotton leafworm, *Spodoptera littoralis.* In: OILB Working Group: Use of Pheromones and Other Semiochemicals in Integrated Control, Conference Proceedings 8–12 December 1986, Neustadt, Germany, pp. 59–60.

McVeigh, L.J., Campion, D.G., Critchley, B.R., Adams, P., Khidr, A.A. and Moawad, G.A. (1988) Large scale commercial application of pink bollworm pheromone in Egypt using twist-tie and microencapsulated formulations. ODNRI Overseas Assignment Report, Chatham, Kent, 45pp.

Marks, R.J. (1976) Field evaluation of gossyplure, the synthetic sex pheromone of *Pectinophora gossypiella* (Saund.) (Lepidoptera: Gelechiidae) in Malawi. *Bulletin of Entomological Research* 66, 267–278.

Marks, R.J. (1977) Assessment of the use of sex pheromone traps to time chemical control of red bollworm *Diparopsis castanea* Hampson (Lepidoptera: Noctuidae) in Malawi. *Bulletin of Entomological Research* 67, 575–587.

Marks, R.J., Nesbitt, B.F., Hall, D.R, and Lester, R. (1978) Mating disruption of the red bollworm of cotton *Diparopsis castanea* Hampson (Lepidoptera: Noctuidae) by ultra-low volume spraying with a microencapsulated inhibitor of mating. *Bulletin of Entomological Research* 68, 11–29.

Marks, R.J., Hall, D.R., Lester, R., Nesbitt, B.F and Lambert, M.R.K. (1981) Further studies on mating disruption of the red bollworm, *Diparopsis castanea* Hampson (Lepidoptera: Noctuidae) with microencapsulated mating inhibitor. *Bulletin of Entomological Research* 71, 403–418.

Melamed, Y. and Shoham, Ch. (1975) *Insect Scouting in Cotton Fields.* Centre for International Agricultural Cooperation, Rehovot, Israel, 17pp.

Mendoza, A. (1989) The organization of pest control in Colombia. In: *Technical Seminar at 48th Plenary Meeting of the International Cotton Advisory Committee*, Scottsdale, Arizona, October 1989, pp. 23–26.

Merkl, M.E., Cross, W.H. and Johnson, W.L. (1978) Boll weevil: detection and monitoring of small populations with in-field traps. *Journal of Economic Entomology* 71, 29–30.

Mitchell, E.B. and Hardee, D.D. (1974) In-field traps: a new concept in survey and suppression of low population of boll weevils. *Journal of Economic Entomology* 67, 506–508.

Mitchell, E.R., Tumlinson, J.H. and McNeil, J.N. (1985) Field evaluation of commercial formulations and traps using a more effective sex pheromone blend for the fall armyworm (Lepidoptera: Noctuidae). *Journal of Economic Entomology* 78, 1364–1369.

Moawad, D.G., Khidr, A.A., Zaki, M.M., Critchley, B.R., McVeigh, L.J. and Campion, D.G. (1991) Large scale use of hollow fibre and microencapsulated pink bollworm pheromone formulations integrated with conventional insecticides for the control of the cotton pest complex in Egypt. *Tropical Pest Management* 37, 10–16.

Murlis, J., Bettany, B.W., Kelley, J. and Martin, L. (1982) The analysis of flight paths of male Egyptian cotton leafworm moths, *Spodoptera littoralis*, to a sex pheromone source in the field. *Physiological Entomology* 7, 435–441.

Nakamura, N. (1976) The effect of wind velocity on the diffusion of *Spodoptera litura* (F.) sex pheromone. *Applied Entomology and Zoology* 11, 312–319.

Nasr, El-Sayed, A., Tucker, M. R. and Campion, D.G. (1984) Distribution of moths of the Egyptian cotton leafworm *Spodoptera littoralis* (Boisduval) (Lepidoptera: Noctuidae), in the Nile Delta interpreted from catches in a pheromone trap network in relation to meteorological factors. *Bulletin of Entomological Research* 74, 487–494.

Nesbitt, B.F., Beevor, P.S., Cole, R.A., Lester, R. and Poppi, R.G. (1973) Sex pheromones of two noctuid moths. *Nature, New Biology* 244, 208–209.

Nesbitt, B.F., Beevor, P.S., Cole, R.A., Lester, R. and Poppi, R.G. (1975) The isolation and identification of the female sex pheromones of the red bollworm moth, *Diparopsis castanea. Journal of Insect Physiology* 21, 1091–1096.

Nesbitt, B.F., Beevor, P.S., Hall, D.R. and Lester, R. (1979) Female sex pheromone components of the cotton bollworm *Heliothis armigera*. *Journal of Insect Physiology* 25, 535–541.

Nesbitt, B.F., Beevor, P.S., Hall, D.R. and Lester, R. (1980) (Z)-9-hexadecenal: a minor component of the female sex pheromone of *Heliothis armigera* (Hübner: Noctuidae). *Entomologia Experimentalis et Applicata* 27, 306–308.

Nyambo, B.T. (1989) Assessment of pheromone traps for monitoring and early warning of *Heliothis armigera* (Hübner), (Lepidoptera: Noctuidae) in the western cotton growing areas of Tanzania. *Crop Protection* 8, 188–192.

Page, F.D., Modnii, M.P. and Stone, M.E. (1984) Use of pheromone trap catches to predict damage by pink spotted bollworm larvae in cotton. In: Proceedings of the 4th Australian Applied Entomological Research Conference, 24–28 September 1984, Adelaide. *Pest Control: Recent Advances and Future Prospects*, South Australia Government Printer, Adelaide, pp. 68–73.

Patel, R.C., Yadav, D.N., Joshi, A.D. and Patel, K.A. (1986) A preliminary note on the impact of mass-trapping of *Heliothis armigera* and *Spodoptera litura* in cotton with sex pheromones. *Cotton Development* 15, 24–26.

Pawar, A.D., Prasad, J., Sharma, R.K., Yadav, K.R., Pickett, C.H., Doane, C.C., Brooks, T.W., Bajikar, M.R. and Baskaran, E. (1981) An operational field trial project in India for suppression of the cotton pink bollworm, *Pectinophora gossypiella* (Saunders), (Gelechiidae: Lepidoptera) employing gossyplure hollow fibre controlled release sex pheromone formulation. *Technical Report No. 1, issued by the Plant Protection Adviser to the Government of India*, N.H.IV, Faridabad, Haryana, India, 44pp.

Quisumbing, A.R. and Kydonieus, A.F. (1982) Laminated structure dispensers. In: Kydonieus, A.F. and Beroza, M. (eds) *Insect Suppression with Controlled Release Pheromone Systems*, vol. 1. CRC Press, Boca Raton, Florida, pp. 213–236.

Qureshi, Z.A. and Ahmed, N. (1989) Efficacy of combined sex pheromones for the control of three major bollworms of cotton. *Journal of Applied Entomology* 108, 386–389.

Ramaswamy, S.B., Randle, S.A. and Ma, W.K. (1985) Field evaluations of the sex pheromone of *Heliothis virescens* (Lepidoptera: Noctuidae) in cone traps. *Environmental Entomology* 14, 293–296.

Ridgway, R. L., Bariola, L.A. and Hardee, D.D. (1971) Seasonal movement of boll weevils near the High Plains of Texas. *Journal of Economic Entomology* 64, 14–19.

Ridgway, R.L., Dickerson, W.A., Brazzel, J.R., Leggett, J.F., Lloyd, E.P. and Planer, F.R. (1985) Boll weevil pheromone trap captures for treatment thresholds and population assessments. In: *Proceedings of the Beltwide Cotton Production Research Conference*, Memphis, Tennessee, pp. 138–141.

Ridgway, R.L., Silverstein, R.M. and Inscoe, M.N. (1990) *Behaviour-Modifying Chemicals for Insect Management. Applications of Pheromones and Other Attractants*. Marcel Dekker, New York, 761pp.

Rose, D.J.W., Page, W.W., Dewhurst, C.F., Riley, J.R., Reynolds, D.R., Pedgley, D.E. and Tucker, M.R. (1985) Downwind migration of the African armyworm moth *Spodoptera exempta*, studied by mark-and-capture and by radar. *Ecological Entomology* 10, 299–313.

Rothschild, G.H.L., Wilson, A.G.L. and Malafant, K.W. (1982) Preliminary studies on the female sex pheromones of *Heliothis* species and their possible use in control programmes in Australia. In: Reed, W. and Kumble, V. (eds), *Proceedings of the International Workshop on* Heliothis *Management*, 15–20 November 1981, ICRISAT Centre, Patancheru, AP, India, ICRISAT, Patancheru, pp. 319–327.

Rummel, D.R., White, J.R., Carrol, S.C. and Pruitt, G.R. (1980) Pheromone trap index system for predicting need for overwintering boll weevil control. *Journal of Economic Entomology* 73, 806–810.

Russell, D.A. and Radwan, S.M. (1993) Modelling pink bollworm mating disruption in Egyptian cotton. In: *Proceedings of an OILB Meeting on Pheromone Technology in Europe and the Developing Countries.* NRI, Chatham, UK (in press).

Scott, W.P., Lloyd, E.P., Bryson, J.O. and Davich, T.B. (1974) Trap plots for suppression of low density overwintered populations of boll weevils. *Journal of Economic Entomology* 67, 281–283.

Shani, A. (1982) Field studies and pheromone application in Israel. Third Israeli Meeting on Pheromone Research, 4 May 1982, Ben Gurion University of the Negev. *Phytoparasitica* 10, 135 (abstracts).

Shu, C., Cao, C. and Zhang, Y. (1987) Control of pink bollworm, *Pectinophora gossypiella*, by mating disruption with microencapsulated pheromone. *Chinese Journal of Biological Control* 3, 106–108.

Singh, J.P. and Lather, B.P.S. (1989) Monitoring of pink bollworm moths and larvae. *Indian Journal of Plant Protection* 17, 199–204.

Staten, R.T., Flint, H.M., Weddle, R.C., Quintero, E., Zarate, R.E., Finnell, C.M., Hernandes, M. and Yamamoto, A. (1987) Pink bollworm (Lepidoptera: Gelechiidae): large-scale field trials with a high rate gossyplure formulation. *Journal of Economic Entomology* 80, 1267–1271.

Stone, N.D. and Gutierrez, A.P. (1986) Pink bollworm control in Southwestern desert cotton. II A strategic management model. *Hilgardia* 54, 25–41.

Sundaramurthy, V.T., Natarajan, K., Basu, A.K., Gordon, R. and Campion, D.G. (1987) Management of the pink bollworm through its pheromone gossyplure. *National Symposium on Integrated Pest Control – Progress and Perspectives*, 15–17 October 1987, Association for Advancement of Entomology, Trivandrum, Kerala, India, pp. 70–71.

Swenson, D.W. and Weatherston, I. (1989) Hollow-fibre controlled-release systems. In: Justum, A.R. and Gordon, R.F.S. (eds) *Insect Pheromones in Plant Protection.* John Wiley & Sons, Chichester, pp. 173–197.

Tamaki, Y., Noguchi, H. and Yushima, T. (1973) Sex pheromone of *Spodoptera litura* (F.), (Lepidoptera: Noctuidae), isolation, identification and synthesis. *Applied Entomology and Zoology* 8, 200–203.

Taneja, S.L. and Jayaswal, A.P. (1981) Capture thresholds of pink bollworm moths on hirsutum cotton. *Tropical Pest Management* 27, 318–324.

Teich, I., Neumark, S. and Jacobson, M. (1977) The capture threshold of male pink bollworm moths with gossyplure and its effect on boll infestation and frequency of insecticidal treatment. *Journal of Environmental Science and Health* (A) 12, 423–430.

Teich, I., Neumark, S., Jacobson, M., Klug, J., Shani, A. and Waters, A.M. (1979) Mass trapping of males of Egyptian cotton leafworm *Spodoptera littoralis* and large scale synthesis of Prodlure. In: Ritter, F.J. (ed.) *Chemical Ecology, Odour Communication in Animals.* Elsevier/North-Holland Biomedical Press, Amsterdam, pp. 343–350.

Tingle, F.C. and Mitchell, E.R. (1981) Relationship between pheromone catches of male tobacco budworm, larval infestations and damage levels in tobacco. *Journal of Economic Entomology* 74, 437–440.

Toscano, N.C., Mueller, A.J., Sevacherian, V., Sharma, R.K., Nilus, T. and Reynolds, H.T. (1974) Insecticide applications based on hexalure trap catches versus automatic schedule treatments for pink bollworm moth control. *Journal of Economic Entomology* 72, 144–147.

Tumlinson, J.H., Gueldner, R.C., Hardee, D.D., Thompson, A.C., Hedin, P.A. and

Minyard, J.P. (1971) Identification and synthesis of the four compounds comprising the boll weevil sex attractant. *Journal of Organic Chemistry* 36, 2616–2621.

Tumlinson, J.H., Hendricks, D.E., Mitchell, E.R., Doolittle, R.E. and Brennan, M.M. (1975) Isolation, identification and synthesis of the sex pheromone of the tobacco budworm. *Journal of Chemical Ecology* 1, 203–214.

Tumlinson, J.H., Heath, R.R. and Teal, P.E.A. (1982) Analysis of chemical communication systems of Lepidoptera. In: Leonhardt, B.A. and Beroza, M. (eds) *ACS Symposium Series* 190. American Chemical Society, Washington DC, pp. 1–25.

Villavaso, E.J. (1982) Boll weevil: isolated field pest studies of the disruption of pheromonal communication. *Journal of the Georgia Entomological Society* 17, 347–350.

Wall, C. (1989) Evaluation and use of behaviour modifying chemicals. Monitoring and spray timing. In: Jutsum, A.R. and Gordon, R.F.S. (eds) *Insect Pheromones in Plant Protection*. John Wiley & Sons, Chichester, pp. 39–66.

Weatherston, I. (1989) Alternative dispensers for trapping and disruption. In: Jutsum, A.R. and Gordon, R.F.S. (eds) *Insect Pheromones in Plant Protection*. John Wiley & Sons, Chichester, pp. 249–278.

Whitcomb, W.H. and Marengo, R.M. (1985) Use of pheromones in the boll weevil detection and control programme in Paraguay. *Florida Entomologist* 69, 153–156.

27 Chemical Control

G.A. Matthews

International Pesticide Application Research Centre, Imperial College at Silwood Park, Buckhurst Road, Sunninghill, Ascot, Berkshire SL5 7PY, UK

Farmers have relied very considerably on chemicals to control the many different insects and mites that can infest cotton fields and adversely affect the yield and quality of seed cotton. Pearson and Maxwell Darling (1958) gave a warning when the new organic insecticides were being introduced in the 1950s that cotton farmers could have considerable problems, due to the destruction of parasitoids and predators and the risk that pests could become resistant to an insecticide, if too much reliance was put on chemical control. The need for close collaboration between entomologists and agronomists to assess the effects of insecticides on yields was stressed, and that particular care was needed in introducing 'dangerous' chemicals to small-scale farmers who were unfamiliar with the exacting operation of timely application of insecticides. Inevitably many problems have arisen, especially where excessive use of insecticides has occurred. Applications of DDT rapidly led to outbreaks of red spider mites in many areas and, in the USA, control of the boll weevil was followed by severe infestations of *Helicoverpa zea* and *Heliothis virescens*. Prolonged use of some insecticides has been one of the factors responsible for severe infestations of whiteflies and aphids causing downgrading of lint quality due to honeydew and sooty moulds.

In some areas repeated application of insecticides has led to pest populations becoming so resistant to chemical control that farmers were no longer able to grow the crop profitably and in consequence cotton production ceased in parts of Mexico (Adkisson, 1971) and in the Ord Valley in northern Australia (Basinski and Wood, 1987). In contrast, early warnings of the dangers of over reliance on chemical control were heeded in Peru, where farmers quickly appreciated the need to adopt an integrated pest management strategy and restrict the use of the organochlorine insecticides then available (Barducci, 1973).

Many have campaigned against the use of toxic chemicals especially the small-scale farmer who fails to wear protective clothing (Bull, 1982).

Over the decades 1950–1990 differences in how insecticides were used on cotton can be grouped as follows:

- countries with a wide range of insecticides available, usually applied on a regular schedule;
- countries with a limited range of insecticides recommended on a fixed schedule;
- countries with an integrated pest management programme in which timings of insecticide applications are related to crop monitoring.

Wide Range of Available Insecticides

In the USA in particular, the agrochemical companies, universities and USDA research staff have evaluated insecticides routinely to determine the dosage required to control each of the main cotton pests. At an annual cotton conference, the results of these trials and the relevant data submitted to the Environmental Protection Agency (EPA) for registration of products are discussed, and lists of those recommended and the dosages to be used for particular pests are published. Individual states vary the basic recommendations to suit local conditions. In general, growers have had a choice of several products and made their own decision about which insecticide should be used, although there has been preference for certain products. As individual insecticides often have a limited range of effectiveness, farmers have used mixtures or mixed their own 'cocktails' of emulsifiable concentrates as these were easy to dispense and mix directly with water in the spray tank. One widely available mixture contained DDT, methyl parathion and toxaphene. Methyl parathion was used to control the boll weevil, DDT was required for the bollworms and toxaphene was added to improve the overall efficiency of the mixture. In areas of severe bollweevil infestation sprays of this mixture were applied every five days. Similar programmes were recommended elsewhere particularly in Central America, and in parts of Nicaragua, as many as 35 sprays a season were applied (Falcon and Smith, 1973).

Growers initially believed that extra dollars spent on insecticide would provide a larger return on their investment. Since the effects of using the persistent organochlorines on birds and their accumulation in foodchains became apparent, use of DDT has declined. In the USA it was initially replaced by chlordimeform as a bollworm ovicide, but subsequent tests on this insecticide revealed that it was carcinogenic so the EPA withdrew registration. More recently the pyrethroids have been widely used, but many of the mistakes learnt by using the organochlorines have been repeated. In particular some countries, such as Thailand and Australia have already reported selection of pest populations resistant to the pyrethroids, which has led to the agrochemical industry setting up a group (Insecticide Resistance Action Committee – IRAC) to investigate ways of minimizing the risk of resistance (Jackson, 1989). Subsequently, an international group involving governments and university scientists was set up to explore ways of managing insecticide resistance.

In eastern Texas, late insecticide treatments with malathion or methyl parathion were applied in conjunction with early-maturing varieties to reduce the population of boll weevils entering diapause at the end of the season and reduce the numbers of weevils entering the following season's crop (Lacewell and Taylor, 1978). In some areas, such as Nicaragua this was followed by spraying a small area of cotton sown in advance of the main crop and baited with a Grandlure pheromone trap to attract the weevils that had survived the winter (Swezey and Daxl, 1988).

Limited Range of Insecticides on Fixed Schedule

In many of the developing countries, especially in Africa, cotton continued to be grown as a cash crop by small-scale farmers. Introduction of any new technology was dependent on government support, especially with the difficulties of instructing large numbers of individual growers how to use chemicals. To simplify their introduction, very simple calendar based schedules were devised; thus in Uganda four sprays of DDT (1.12 kg a.i. ha^{-1}) at 14 day intervals commencing six weeks from planting were recommended, principally to control bollworms (Davies, 1970). The DDT supplied as a 25% emulsifiable concentrate called 'Dudumaki' was subsidized to encourage growers to use it, but much of the chemical was applied to their food crops, and with other agronomic factors cotton yields remained low. Similar fixed schedules were introduced elsewhere in East (Reed, 1971) and West Africa (Lyon, 1971), with 10–14 day schedules based on DDT often mixed with HCH for control of sucking pests. Only a small proportion of the growers adopted these recommendations; thus in Nigeria, farmers were more concerned with the establishment of their food crops so most cotton was sown so late in July that there was much less benefit in applying insecticides (Norman *et al.*, 1974). In many areas the lack of water and drudgery was a strong deterrent to the use of knapsack sprayers. In West Africa, a rear-mounted horizontal boom (Cadou, 1959) made spraying easier, but still only some of the farmers adopted its use. Later ultra-low-volume spraying was introduced into these areas, and since 1975 its adoption in francophone countries has been effectively encouraged by the Compagnie Francais pour le Developpement des Fibres Textiles (CDFT), so that virtually all the cotton crops are protected with insecticides (Cauquil, 1987). By 1989, the sprays were principally a pyrethroid mixed with an organophosphate insecticide. Similarly in Tanzania, ultra-low-volume spraying had completely replaced conventional knapsack spraying by 1975 (Nyambo, 1989). The problem with the simplified fixed-schedule using the same dosage at each application is that the incidence of insect pests varies significantly between localities within a season and also from one season to the next. Thus while some farmers may benefit if a spray does coincide with a pest infestation, many may apply a spray when there is a low infestation or after serious damage has already occurred. Furthermore the farmer may lose a crop on which the investment in insecticides has already been made.

Restricted Range of Pesticides – Crop Monitoring

In Zimbabwe early insecticide trials, notably with endrin, have shown that significant yield increases could be obtained where the bollworms, including *H.*

armigera, have been known to cause complete crop loss. Laboratory evaluation of a range of insecticides, with particular attention given to insecticides of low toxicity to minimize possible hazards to the farmer, showed that carbaryl would be very effective against *D. castanea*, while DDT was clearly better against *H. armigera* (Matthews, 1966). Concurrent field trials confirmed that neither insecticide alone would give the optimum yield, as the relative importance of these two bollworms varied within and between seasons (Matthews, 1966). A mixture containing 40% of each active ingredient was made commercially, but for the small-scale farmer there would be little point in applying carbaryl during an infestation of *H. armigera*. To see if costs of insecticide could be minimized, it was decided to see if farmers could respond to changes in pest status and use either carbaryl or DDT according to data collected by routine crop monitoring. Dimethoate was recommended initially if aphid or red spider mite infestations warranted control.

Scouting crops had been advocated previously in the USA by Isely and Lincoln, but at that time their advice was followed only by a few farmers in Arkansas. Field trials were carried out in Malawi and Zimbabwe on farmers' fields and on average yields were at least doubled, the best yields being related to improved agronomic practices (Tunstall and Matthews, 1966). Recommendations were published (Tunstall *et al.*, 1961) and with the support of the extension services, production has continued to expand in Zimbabwe (Matthews, 1989). In Malawi, individual farmers have benefited from these recommendations (Gower and Matthews, 1972), but as Farrington (1977) pointed out some farmers applied fewer sprays in years of low rainfall when there was lower potential yield. However, only a small proportion of farmers have sprayed their crops due to lack of sufficient extension, marketing and other resources.

A number of changes have been made to the original recommendations, especially in Zimbabwe. Red spider mite was reported to be resistant to dimethoate and other organophosphate insecticides, especially on farms with irrigation and where vegetable crops were grown throughout the year, so the use of dimethoate was withdrawn and menazon, later replaced by pirimicarb, was substituted for aphid control and binapacryl for red spider mite. Subsequently, an acaricide rotation scheme was introduced in which the cotton area was divided into three zones. In each zone one type of acaricide can be used for two years before being moved to another zone (Duncombe, 1973). Binapacryl was not included in the rotation scheme, and recent reports indicate that dimethoate is still effective in most parts of the country.

Endosulfan was recommended as a replacement for DDT as it was less likely to induce outbreaks of *Tetranychus* spp. Later when the pyrethroids were introduced, their use was restricted to a nine-week period coinciding with the peak *Helicoverpa* infestation. This restriction on pyrethroid sprays is discussed in relation to resistance management strategies later. In Zimbabwe recommendations for the photostable pyrethroids cypermethrin and deltamethrin were withdrawn due to a rapid upsurge in mite populations (Brettell, 1986). At present the following five pyrethroids are registered for cotton in Zimbabwe – fenvalerate, lambacyhalothrin, biphenthrin, fluvalinate and flucythrinate. Critics of this approach to chemical control consider the system is too complex and is dependent on well supported extension services. Education of the farmer is essential and in Zimbabwe a training school was financed from a crop levy so that farmers or their assistants could learn to recognize the pest species and natural enemies, as well as how to decide when to spray and which insecticide to apply (Burgess, 1983). In Malawi, cotton scouts were employed, initially by the government, to monitor selected crops and relay their recommendations to all the growers within a certain area. Advice was provided to the scouts by a newsletter, with the details explained when the newsletter was distributed at their regular monthly meeting. Further information was given on regular radio programmes, in manuals, advisory leaflets and in posters. Later a pegboard (Fig. 27.1) was designed (Beeden, 1972), so that farmers could inspect their own fields and decide on the spray schedule without needing to write down any data. Many of the small-scale farmers have not yet been trained, but the recommendations are followed to some extent due to the control on the availability of insecticides and interchange of information between farmers.

Fig. 27.1. Pegboard used for scouting cotton.

The success of cotton in Zimbabwe since 1950 can be attributed to a sustained research programme, cooperation between the growers, extension services, the agrochemical industry and the research scientists; and an acceptable price for seed cotton. The other important factor was that other components of an IPM programme had been established before chemical control was introduced. Jassid resistant varieties had begun in the 1920s, followed in 1960 by resistance to bacterial blight. A closed season had been established in the 1930s, principally in an attempt to reduce infestations of *Diparopsis*, but its value in controlling *Pectinophora* was clearly demonstrated in 1961 when infestations were recorded for the first time in an area where uprooting had not been done (Matthews *et al.*, 1965).

Methods of Scouting

The correct timing of an insecticide application will depend on knowing which pests are present, changes in the population of pests and the extent to which a given population will affect yields at different times during a season. Crop monitoring of pests needs to be a relatively simple task that can be done regularly, so that decisions on control methods can be taken quickly when needed. Some of the earliest sampling was in the USA using a point sampling system in which the length of row examined to find 50 buds was measured at four–eight different sites within a field to assess crop development in terms of buds per hectare, and the proportion of buds punctured by the bollweevil is determined (Lincoln *et al.*, 1963). Bollworm damage was sampled in the same way (Lincoln and Phillips, 1979). Elsewhere lygus were sampled by taking 50 sweeps with a net across the crop.

In Africa, the need to change insecticide according to which bollworm species was present and their relative status in the crop led to scouting for bollworm eggs (Matthews and Tunstall, 1968). Recording the number of eggs rather than larvae was recommended to ensure that the farmer timed his sprays against the first instar larvae before they penetrated a bud or boll. Also scouting was quicker, as it avoided handling individual bracts too much while searching for larvae. The pale blue *Diparopsis* eggs and cream *Helicoverpa* eggs are readily seen on the plants so individual farmers could be trained to decide when to start sprays and change to another insecticide on the basis of action thresholds. It has been suggested that sampling of larvae would be preferable to make an allowance for egg parasitoids (Kfir and Van Hamburg, 1983), but sprays were generally not recommended until before the sixth week after germination. In some areas there may be sufficient numbers of parasitoids early in the season but they can be extremely low on the rainfed crops in semiarid areas and were less effective when bollworm populations increase rapidly.

Sampling of whole plants particularly at the end of the season can be very time consuming, so methods of sampling only a small sector of the plant have been devised. *Helicoverpa* often oviposits on the younger leaves and terminal bud so the upper part of the plant can be sampled. However, the most useful method of reducing the time required for crop monitoring is by using a sequential sampling system (Fig. 27.2) (Ingram and Green, 1972; Sterling and Pieters, 1979). If the number of pests exceeds a threshold after only a few plants have been sampled, a decision can be made quite quickly. The threshold may change during a season, especially when the yield potential is affected by other factors such as rainfall. Sampling of natural enemies has also been advocated to delay spray application if sufficient natural enemies are present (Hasse, 1986).

Counting some pests, such as aphids and mites can be particularly difficult so it is usual to have a scoring system in which the numbers are coded. For example zero if none is present, one if one–ten are counted, two if 11–30 and four if there are over 30 on a leaf. Whiteflies can be sampled by a modified beating technique: a flat surface coated with a thin film of vegetable oil is held below the crop canopy and the plants are struck with a stick; insects knocked down are caught on the oily surface, counted and then the surface wiped clean with an oil-soaked cloth ready for the next sample (Butler *et al.*, 1986).

In other examples, it is the amount of damage that is assessed. The percentage of buds with flared bracts or bolls damaged by bollworm is used to indicate when to treat a crop; thus in Egypt 100 bolls collected at random are cut open to determine the level of infestation of *Pectinophora.* According to Salama (1983) spraying starts if 10% of the green bolls are infested, but a lower threshold of 3–5% has been used.

For the small-scale farmer, the pegboard referred to above provides a simple means of making a decision based on an individual inspection of a crop. Other farmers can now use a computer database to store such information, so that decisions can be based on trends in pest populations and assessments of crop development. Interactive with a central computer database, the SIRATAC program in Australia enabled those who subscribed to the on-line system to compare routine field assessments and meteorological data with a model of the

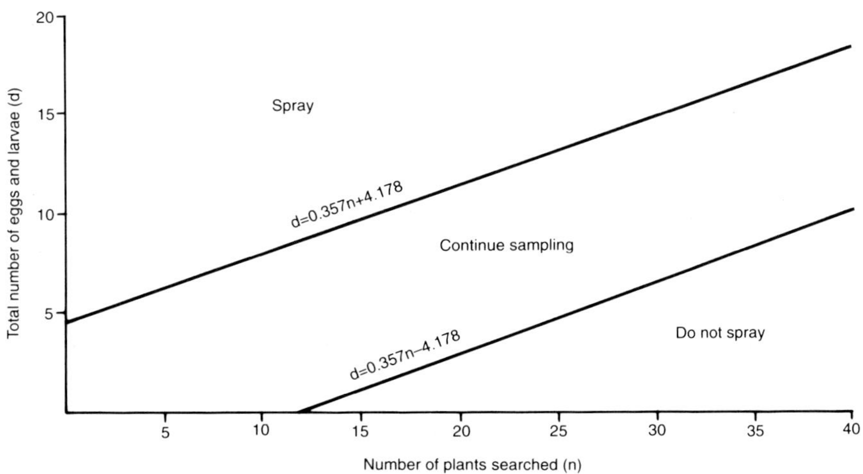

Fig. 27.2. Sequential sampling graph for use when the number of eggs and larvae is counted.

anticipated crop development, but the system was with irrigated cotton (Hearn and da Roza, 1985).

Efforts have been made to use pheromone traps to sample pest populations to assist in timing control operations. In most cases the pheromone trap provides an excellent indication of when a pest is entering the area in which the crop is grown, but there is not always a correlation between trap catches and the level of infestation in the crop. Marks (1977) tried to use dicastalure to sample *Diparopsis* in Malawi, but differences in plant growth on individual fields and other factors affected oviposition, so crop monitoring was still required. Similarly in the USA pheromone traps have only been suitable as a guide to when crop monitoring was required.

Resistance Management Strategies

Continued use of one or more insecticides has resulted in an increasing number of insect species becoming resistant to them, at least over part of their geographical range. Resistance to DDT in *Helicoverpa* led to the demise of the cotton crop in the Ord Valley in Australia. Subsequently detailed studies were carried out in Australia following the rapid onset of resistance to pyrethroid insecticides (Gunning *et al.*, 1984; Cox and Forrester, 1992). Similar problems have arisen in Thailand (Ahmad and McCaffery, 1988) and in the USA (Riley, 1988). Prolonged use of monocrotophos in the Sudan Gezira led to the selection of resistant whiteflies (Dittrich *et al.*, 1985). Much more study is needed to determine the optimum method of combating the resistance problem, but in the interim a number of strategies have been suggested to try and delay the onset of resistance or contain it, where it has already occurred.

Resistance management strategies

- Do not use same insecticide group or mixture throughout season.
- Only apply insecticide according to crop monitoring data using the highest effective action threshold, i.e. minimize number of applications.
- Use low dosages aimed only at early instars so deposits have less persistence.
- Favour insecticides with inherently less persistence.
- Avoid slow release formulations.
- Make local rather than area-wide treatments, thus individual small-scale farmers treating their crops on different days are less likely to select a resistant population compared to a similar large area of cotton treated quickly by aerial treatment.
- Leave refugia, i.e. untreated areas (see section on electrostatics).
- Consider suppression of detoxification mechanism by using synergist with insecticide.
- Use pesticides in a rotation scheme – e.g. acaricide rotation scheme in Zimbabwe.

As indicated previously, the lack of control and subsequent escalation of the number of applications led to the total failure of the cotton crop in the Ord Valley in Australia. There was, therefore, considerable concern when resistance to pyrethroids was reported in the Emerald district of Queensland. Anxious to avoid a repetition of the situation in the Ord, a pragmatic strategy was rapidly introduced with the support of growers and the agrochemical industry. Under this scheme pyrethroids were to be restricted to a 45-day period, later reduced to a 35-day period (10 January–13 February) for all crops, and more recently to a 28-day period. These insecticides were also used on other crops such as sorghum, so selection for resistance had been spread over several generations (Forrester and Cahill, 1987). The aim was to limit selection of a resistant population to one generation and hope that by the following year the frequency of resistance genes had decreased sufficiently for the insecticide to remain effective over a critical period of crop development and during the peak infestation of *Helicoverpa* (Fig. 27.3) (Sawicki *et al.*, 1989). Other changes have been made including the use of piperonyl butoxide as a synergist in the second spray, and allowing no applications of endosulfan after the pyrethroid sprays. *Bacillus thuringiensis* may be added to either endosulfan or methomyl for early season bollworm control (Forrester *et al.*, 1993). In Africa the situation is rather different, as there are still vast areas of uncultivated land where weeds and wild plants, and other crops such as maize, can support a population of *Helicoverpa* in the absence of insecticides.

In Egypt serious difficulties in controlling *Spodoptera littoralis* in the 1960s was due to prolonged use of the same insecticide. Insecticide use was reduced by the return to the use of children to collect egg-masses on leaves, but in the spray programme, aimed principally against *Pectinophora*, four insecticide groups are used; namely an insect growth regulator such as diflubenzuron mixed with an organophosphate has been used for the first spray that also controls any late infestation of *Spodoptera* (Dutton and Komblas, 1989), followed by a pyrethroid, a carbamate and finally an organophosphate spray. There is no indication that such

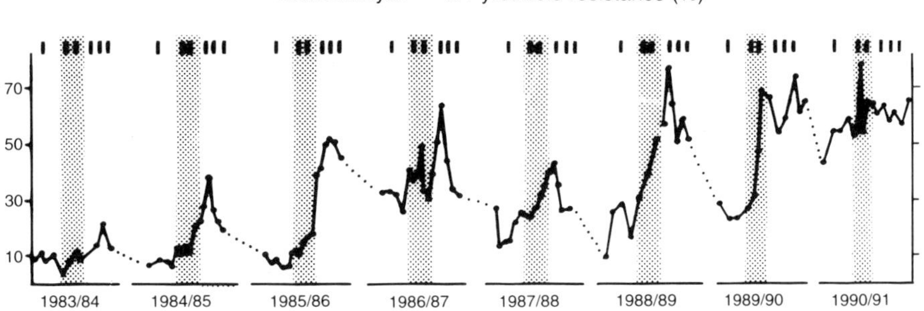

Fig. 27.3. Weekly pyrethroid resistance in *Heliothis armigera* from the Namoi and Gwydir river valleys of northern New South Wales, Australia, for the eight seasons since the introduction of a curative Resistance Management Strategy (for Stages I, II & III). Results expressed as the percentage of larvae (reared from field collected eggs) surviving the fenvalerate discriminating dose (0.2 µg per 30–40 mg larva). Stage II pyrethroid window 42 days duration to 1988–1989 season, thereafter 35 days. Piperonyl butoxide introduced into commercial use in 1990–1991.

Table 27.1. Examples of differences between countries in dosages of insecticides recommended.

Insecticide	Dosage (g a.i. ha^{-1}) recommended in	
	Zimbabwe*	Australia†
endosulfan	500(300‡)	720
fluvalinate	50	105
profenofos	200§	750
thiodicarb	412	525

* plants over 0.9 m high, rates are reduced on smaller cotton.
† most common used field rate (Cox and Forrester, 1992) for *Heliothis* control.
‡ lower rate applied if mixed with molasses.
§ only recommended as an acaricide in Zimbabwe.

a programme delays resistance, and according to Sawicki *et al.* (1989) some bioassays have indicated pyrethroid resistance. The use of broad spectrum insecticides also has an adverse effect on the natural enemies and this has resulted in serious outbreaks of aphids.

Dosage of Insecticide

Many insecticide trials on cotton have been aimed at determining the dosage of a particular product needed to control an established infestation to satisfy registration requirements. Trials have also examined season-long programmes of an individual company's products. In these trials, a relatively high volume of dilute spray has been applied on a fixed schedule irrespective of pest populations or the stage of the crop. Results of these trials on small plots can only give an indication of whether a particular insecticide is effective, due to large interplot effects. With trials in a number of different countries, manufacturers tend to build in an insurance or 'overkill' to achieve effective control under a wide range of conditions. The considerable increase in leaf area during the period of crop growth requiring protection suggests that the dosage should be related to leaf area index, but this is seldom considered. Where the dosage has been related to the size of plants and timed to control early instars, the dosage can be significantly reduced, as shown in Zimbabwe (Table 27.1). Apart from reducing the direct costs to the farmers, the lower dosage has less impact on natural enemies and there is less risk of residue in the crop. Of particular importance is the reduced risk of selecting a resistant population by having a heavy deposit persisting in the crop environment over a long period.

There have been relatively few trials to investigate how to devise a sustainable programme based on changes in pest populations. Crop monitoring of pest populations should be a standard feature of any integrated pest management programme. The original concept was to time applications according to an economic threshold, but this is difficult to determine if the final yield is unpredictable, especially with rainfed crops, and the price that the farmer will receive for his cotton has not been established. However, the farmer has to make a decision so there is a need for trials to establish an action threshold, even though the threshold may not be very precise. An increase in pest numbers above a threshold, or even a prolonged infestation or just below an initial arbitrary threshold, together with knowledge of previous crops in a locality, should be sufficient information to determine when to use an insecticide. Further trials will be needed to refine the 'action threshold'. In Zimbabwe, a threshold of six *H. armigera* eggs per 24 plants (0.25 eggs per plant) was selected to allow time for growers to get their equipment ready to spray within two–three days of the threshold being exceeded. Provided the crop can be treated immediately it is possible to delay until there are 0.5 eggs per plant. Detecting large *H. armigera* larvae is too late, as excessive damage will have already occurred and a higher dosage of insecticide would be required. Accepting that rapid growth occurs during insect infestations between weeks 6 and 16 after germination, it is important to be able to protect the new young growth, so more frequent application of a dosage optimized for the first–third instars is more effective than applying the much larger dosages that would be needed to kill fourth or fifth instar larvae, or if the spraying interval was increased to 14 days. However, a longer interval between sprays is possible once plants have reached their peak growth; in Zimbabwe this was usually at 15–16 weeks after germination.

In the USA recent recommendations have also advised scouting for bollworm eggs, instead of the traditional method of scouting for 'flared squares', i.e. damaged buds or young bolls (Bacheler and Bradley, 1989). Scouting is generally at weekly intervals, except from six weeks after sowing when twice-weekly crop inspection is essential in areas where *Helicoverpa* is a key pest.

Future for Insecticides

Adoption of resistance management strategies is clearly crucial to maintain the effectiveness of existing chemical groups for as long as possible. The number of new insecticides with insecticidal activity has decreased, and the usefulness of a new molecule can be limited if cross-resistance already occurs. This happened with pyrethroid resistance following rapidly where DDT had been used extensively. Unfortunately insect growth regulators and microbial insecticides such as *Bacillus thuringiensis* have had limited success on cotton. While some leafeaters such as *Spodoptera* and *Alabama* ingest deposits on leaves, bollworms feed within the protection of bracts, so are not exposed to a toxic deposit. If a frego bract variety were grown, perhaps their effectiveness against bollworms might be enhanced. Similarly okra-leaf cultivars might improve spray penetration. Genetically engineered cultivars of some crop plants such as tobacco are now being developed to contain genes to express insect toxins such as the delta-endotoxin from *Bacillus thuringiensis* (Vaeck *et al.*, 1987). Promising results have also been obtained with the cowpea trypsin inhibitor, particularly combined with other insect resistance genes (Boulter *et al.*, 1990). The extent to which this research will produce new cotton cultivars that will require less spray application is still not known (see Chapter 22), but insect pests of cotton have evolved to overcome the presence of gossypol and have developed resistance to insecticides so it is likely that resistence to genetically engineered plants will occur (Gould *et al.*, 1992) and that spraying will continue to be needed, especially if a variety provides a fibre quality specifically required by the spinning industry. In contrast to making plants more resistant to pest infestation, there have been efforts to grow cultivars which are glandless to provide a protein-rich flour suitable for human consumption. In many cases these glandless cottons can suffer damage from insects not normally regarded as pests. However in West Africa large areas of glandless cotton have been grown and required insecticide protection (Hau and Richard, 1986).

Complex insecticide management strategies can be followed by growers provided there is close liaison and cooperation between them, the agrochemical companies and the extension and research services. Training the scouts and how they interpret the data is essential. In Australia adoption of a pragmatic IRM was prompted for fear of a repeat of the complete loss of an industry as in the Ord Valley, while in Zimbabwe the virtual demise of the cotton industry in the 1950s has also stimulated growers to adopt research-led recommendations.

Application of Insecticides

The size of farms growing cotton can range from less than a hectare to several thousands of hectares, so a range of equipment is needed to satisfy each farmer's requirements. In planning a control programme, certain features are common to all farmers. Certain pests can immigrate into cotton fields over a wide area within a very short period, for example oviposition by *Helicoverpa* can increase significantly throughout large areas of the Sudan over one–two nights (Haggis, 1982). These eggs can hatch in two–three days, so larval damage can be significant if control action is not immediate. Rain or irrigation cycles can make soil too wet for access with ground equipment or wash-off deposits from the previous spray. Time must also be allowed for maintenance of the application equipment. When all these factors are considered, the farmer is advised to have sufficient equipment to treat his fields within three days of any one week, especially as rapid growth may necessitate a weekly spray schedule to protect new foliage and young buds on a well-grown crop.

Effectiveness of Different Methods

Where trials have compared different application techniques using the same insecticides and time of treatment, significant differences have seldom been obtained. Thus in Central Africa, a series of research station trials compared knapsack, tractor and animal-drawn sprayers (Tunstall *et al.*, 1965), knapsack and ultra-low-volume applications (Matthews, 1973), ultra-low volume with oil-based formulations and very-low volume with water-based formulations (Mowlam *et al.*, 1975) or electrostatically charged sprays (Nyirenda, 1986; Matthews, 1990). In these trials with relatively small plots, the treatments can be well supervised and timing of application is optimal. For the farmer, the choice of method of application is more related to the time that is needed to treat his crops. The quickest method which allows the farmer to devote more time to other duties will be the most suitable, provided other criteria such as capital investment, labour required, ease of maintenance, and reliability of the equipment are acceptable.

Knapsack Spraying (Plate VIII.5)

The majority of small-scale farmers in many developing countries have continued to use knapsack sprayers on the basis of their relatively low capital cost and versatility, despite many problems associated with their use (Matthews, 1992). Unless the nozzle on a lance is directed very carefully spray distribution is poor. Underleaf coverage is negligible if the nozzle is directed down over the crop, as the top leaves act as 'umbrellas' that restrict coverage of the sites of insect feeding, thus control of *Bemisia* in particular can be poor (Matthews, 1986). An operator moving the lance up and down between the rows walks directly into the treated foliage and gets heavily contaminated (Sutherland *et al.*, 1990). A rear-mounted 'tailboom' was developed to improve coverage at all stages of crop growth (Plate VIII.5) (Tunstall *et al.*, 1961) and its use also reduced operator contamination, as the operator walked away from the spray (Tunstall and Matthews, 1965). The nozzles were directed upwards at an angle of 45° so that in addition to good upper-surface leaf coverage, some droplets were deposited on the under-surfaces and the petioles on which bollworms moved to find buds or bolls. In this way deposits were also more protected from the effects of rainfall. The nozzles were also directed back at 10° so that spray was projected away from

Fig. 27.4. Sachets used for packaging insecticide in Malawi.

the operator, and nozzles were less obstructed by foliage. By maintaining a constant concentration of insecticide in the spray, the dosage was increased in proportion to the spray volume which was determined by the number of nozzles used. As the number of positions of the nozzle was adjusted according to plant height, and indirectly to the size of the canopy, the dosage of insecticide was approximately related to the leaf area index. Cone nozzles were selected rather than fan nozzles so that spray droplets were projected at more angles relative to the leaves. Water supplies to the farmer from streams or irrigation ditches are often poor and may contain sand and other particles which are liable to cause nozzle blockages, so a nozzle orifice diameter of 0.678 mm was required (throughput of 195 ml per minute at 2.8 bar pressure), and protected by a 50 mesh filter. The abrasive sand particles suspended in the water will erode the small orifice, so calibration needs to done regularly and the nozzle tips replaced if output has increased significantly (Beeden and Matthews, 1975).

Correct spray concentration was achieved by providing farmers with pre-packed polyethylene sachets containing sufficient wettable powder insecticide for one knapsack tank load (Fig. 27.4). The idea of a water-soluble sachet kept dry inside an outer protective waterproof container until needed and then just dropped into the sprayer without opening was considered in the 1970s, but the technology of water soluble plastics was not sufficiently advanced at that time, and early trials led to blocked nozzles. Now this form of packaging is increasing in importance.

Efficient distribution of the spray using the tailboom resulted in much lower dosages being recommended than elsewhere. Instead of emphasizing spray distribution, the horizontal boom (Cadou, 1959) used in West Africa was designed to increase the work rate.

Unfortunately the greatest disadvantage of knapsack sprayers is the drudgery of manually pumping up to 200 litres of water per hectare, in addition to the problem of obtaining and transporting sufficient water for each spray application. Many farmers had considerable difficulty in finding water near to their fields or had to carry it from the nearest borehole. This often meant that an unacceptably high proportion of time was needed each week for knapsack spraying. Although regarded as a universal sprayer, the lever-operated knapsack is best kept solely for insecticide or fungicide spraying by cotton farmers as it is extremely difficult to remove all traces of herbicide even with careful washing. Thus there is a risk of contaminating cotton with 2,4-D or similar hormone type chemical if it has been applied with the equipment.

Spinning Disc Sprayers

Constraints on water supplies for farmers in the semiarid areas led to research on ultra-low-volume application using hand-carried, battery-operated spinning disc sprayers from which a cloud of droplets, formed at the periphery of the disc, is allowed to drift downwind (see Plate VIII.6). With the correct choice of droplet size (50–100 μm VMD), air turbulence mixes the spray cloud with air within the top of the canopy. Much of the spray is impacted on the upper leaves facing windward, but a proportion of spray filters through the canopy to soil level. Swath width depends on the height of the nozzle above the crop and wind speed. In practice wind speed is very variable, and trials in Malawi (Matthews, 1973) indicated that swaths should not be too wide, otherwise there will be areas of poor coverage during periods of low wind speed, when there is little overlapping of deposits from successive swaths across a field. Another factor against wide swaths is that wind direction is seldom at right angles to the row direction so rows should never be in line with the predominant wind direction. Operator contamination is minimized by holding the spinning disc downwind and moving progressing upwind for successive swaths. In Malawi, the recommendation was to decrease the swath width in relation to the height of plants, thus the dosage increased for the larger canopy in much the same way that had been used with the knapsack sprayer fitted with a tailboom. Using a flow rate of 0.5 ml s^{-1}, the volume rate increased from 0.8 to 2.5 l ha^{-1} during the season (Matthews, 1971). Elsewhere most farmers were advised to use a fixed swath width usually five rows applying three litres per hectare. ULV spraying has been adopted most successfully in the French-speaking countries of Africa where, supported by cotton growing companies, almost all farmers protect their crops, and overall yields now exceed 1000 kg seed cotton ha^{-1}. However, with the high cost of oil-based formulations, the trend has been either to use less volume per hectare or use the spinning disc to apply conventional formulations diluted in water at very low volumes (i.e. 5–20 l ha^{-1}). In Zimbabwe, molasses at 17–25% is added to the spray to reduce the effect of evaporation on droplet size. This has enabled their farmers to apply about five litres of spray per hectare. In other countries, the volume rate has generally been 10–15 l ha^{-1} without any adjuvant. In the Cameroons, farmers are also assessing the use of conventional formulations following unreliable results when the volumes of oil-based sprays were reduced to 1 l ha^{-1} and the smaller droplets were more prone to dispersal by convective air movement away from the cotton fields. Recently, trials to assess water-based sprays have been started but there have also been changes in frequency and dosage of the insecticide applications.

Electrodynamic Spraying

Concern about dependence on wind velocity to distribute ULV spray droplets led to consideration of providing an extra force, an electrostatic force, to assist deposition of small droplets. While studies have been reported on charging a number of types of nozzles, including spinning discs, the development of an electrodynamic nozzle has provided a low energy method of producing a charged spray (Coffee, 1979). The system has been developed so that the small-scale farmer has a battery-operated high voltage generator inside a 'stick' at the end of which is mounted a pre-packaged container with integral nozzle, referred to as a 'Bozzle'. Used in a similar manner to the ULV spinning disc sprayers, the 'Electrodyn' sprayer nozzle is held approximately 0.4 m above the downwind inter-row, while the operator moving progressing upwind, walks along alternate rows (i.e. two row swaths). On young cotton the nozzle can be held closer above each row of plants to reduce contamination of the soil with insecticide. Extremely low flow rates are possible, so volume application rates can be as low as 0.25 l ha^{-1}, but usually 0.5–1.0 l ha^{-1} are applied. Application of these ultra-low volumes has restricted the range of pesticides to those that are sufficiently active at low dosages, can be formulated in the oil based/solvent mixture and provide the correct resistivity, i.e. electrical properties for droplet formation.

As with the ULV spinning disc sprayers, the 'Electrodyn' sprayer eliminates the need for water and, using oil-based formulations, the deposits are more rainfast. Additional advantages are the lower power requirement, the pre-packaging of the pesticide, and improved droplet deposition, especially on the under-surfaces of leaves. Like other spray techniques where the nozzle is above the crop, penetration through the crop canopy is poor, especially when compared with ground spraying using tailboom equipment. However, provided the canopy is not too tall and lush, adequate control of certain major pests, including *Pectinophora gossypiella* has been maintained, presumably due to persistent deposits affecting the moth stage. As distribution of spray is restricted, there is the possibility of more untreated refugia occurring within the crop canopy where natural enemies, especially general predators such as ants and spiders, can survive the insecticide treatment. Thus some survival of natural enemies should provide better integration of chemical and biological control, especially as it is also possible to have narrow strips of treated cotton intercropped with legumes, such as cowpea and other crops to provide crop diversification.

Electrodynamic spraying has been introduced into several countries in Africa and South America. In some areas only a limited number of insecticides was available but the range has increased, as farmers need to have a range of products to respond to crop monitoring decisions and to follow resistant management strategies. With this sprayer response to an economic or action threshold can be very rapid as the farmer does not need to collect water or prepare his spray.

Mechanized Ground Spraying

In Egypt, a motorized pump at the side of the field has been used with a portable hose carried through the cotton by a team of assistants. A single operator with a variable cone nozzle walked through the crop directing spray to either side. The high-volume technique is inefficient in the use of insecticide and labour which gets heavily contaminated by the spray. In other areas such as Thailand motorized mistblowers have been used to treat cotton, but these have not been widely used elsewhere due to difficulties in keeping the 2-stroke engines maintained.

In Spain and some other countries conventional tractor-mounted sprayers, fitted with a horizontal boom have been used. The main problem with conventional tractor equipment is the mechanical damage incurred by plants straddled by the tractor. However, plants can compensate for much of the damage if the tractor is used along the same pathways in the same direction for each application. In Zimbabwe it was more economic to grow cotton along the pathways than to have a crop such as groundnuts with a lower canopy.

In the USA, many of the ground sprayers are custom-built high clearance self-propelled sprayers, often with nozzles only along a horizontal boom. In some cases, drop-legs are used to direct spray laterally to the canopy. Only in Zimbabwe have tailbooms, similar to those used on knapsack sprayers, been adapted to fit tractor sprayers (Tunstall *et al.*, 1965). An air sleeve to direct a curtain of air downwards to assist penetration of spray has been used in cotton to improve control of *Bemisia tabaci* (Hadar, 1991). Similar studies in Zimbabwe used large slowly rotating fans mounted horizontally to create a turbulent downwards current of air into the canopy (Gledhill, 1971). Successful use of these air-assisted sprayers also requires attention to droplet size and formulation, as large droplets are effectively filtered by the upper leaves.

Aerial Application

Aerial insecticide treatments have been extensively used in the USA, former USSR, Central America, Colombia, Sudan, Egypt and Australia. In most countries conventional hydraulic nozzles, for example cone or whirl-jet types, are mounted on a boom, but rotary atomizers such as the 'Micronair' equipment have been used increasingly in Australia, Sudan and Southern Africa (Fig. 27.5). Adjustment of the blade settings of Micronair units gives some control of the droplet spectra (Parkin and Siddiqui, 1990). In general, penetration of crop canopies with aerial treatments is poor (Uk and Courshee, 1982). Serious

Fig. 27.5. Aerial spraying.

problems of *Bemisia* have undoubtedly been partly due to poor distribution of insecticide, with very low deposition on the undersurface of leaves, and the lower part of the canopy. Endosulfan has been added to other insecticides, such as amitraz to enhance the activity of the whiteflies (Peregrine, 1989). Attempts to improve distribution in cotton fields have included the use of helicopters, on the basis of increased downwash from the rotor, but at the speeds needed for economic operation, spray distribution from a helicopter is similar to a fixed-wing aircraft. Choice of the droplet spectrum and nozzle positions across the boom are important factors to achieve good distribution across successive swaths.

The main difficulty with aerial treatment is the need to avoid hot convective air movement which exacerbates evaporation of water-based sprays and reduces the proportion of insecticide deposited on the crop. Changes to less volatile ULV formulations improves the logistics of aerial operations, but application of small droplets is feasible only in extensive cotton areas, where drift is predominantly into neighbouring cotton fields. Studies by Uk (1987) in the Sudan showed a narrow swath at 06.00 hours with a high risk of drift; at 09.00 hours the swath was wider and with more wind deposition in the crop canopy was improved with less airborne drift 100 m downwind; but by midday and later in the day the thermals affected spray recovery so deposits were much more variable. Johnstone and Johnstone (1977) recommended that spraying should cease if the temperature difference between wet- and dry-bulb thermometer readings exceeds 4.5°C. Generally with water-based sprays the volume median diameter (VMD) of the spray droplets should be 200 μm, while less volatile sprays can be applied with smaller 100 μm VMD sprays. Attempts to improve the efficiency of aerial application have led to the use of molasses as an evaporation retardant in Zimbabwe, to spraying at night with cooler temperatures in Australia, and to using higher volumes of spray in the USA.

While aerial application enables large areas to be treated rapidly (Joyce, 1975), it is pertinent to note that many serious problems associated with resistance to insecticides or poor control of whiteflies and aphids have occurred in areas with extensive aerial application. An exception has been in Thailand where motorized mistblowers are commonly used. Factors other than method of application are more important in selection of resistant populations, but the extent of treatments clearly reduces the number of 'refugia' for natural enemies. Downwind 'drift' to adjacent crops may also exacerbate selection of resistant populations of polyphagous pest species such as *Helicoverpa*.

Granule Application

Control of *Bemisia* has also been attempted using tractor-mounted applicators to distribute aldicarb granules and cover them with soil to reduce the risk of goats or other animals ingesting the granules (Carlson and Mohamed, 1986). Unfortunately, large irrigated fields were not always accessible to tractor equipment, so areas which had to be left untreated were a source of reinfestation when the systemic activity declined. Granule treatment had to be before the plants were too tall otherwise the applicator would cause serious damage later in the season.

Seed Treatment

Young cotton seedlings can be exposed to insect and pathogen infestations. In an integrated pest management programme spraying such small plants with insecticides is unwarranted, but protection over a few weeks can be achieved using seed coatings of systemic chemicals, such as disulfoton, monocrotophos, phorate, acephate and more recently imidacloprid (Elbert *et al.*, 1990), but with some of these insecticides particular care is needed with subsequent handling of seed because of their toxicity. Uptake of the insecticide is better if the crop is irrigated. Halloin (1986) gives a comprehensive review of treatment of cotton seeds, using equipment described by Jeffs and Tuppen (1986). In many countries acid delinted seed is used as it provides better quality seed and removes the risk of pink bollworm infestation from diapause larvae.

Conclusions

Chemical control of cotton insect pests has provided considerable benefits through higher yields of better quality seed cotton. Nevertheless as major users of insecticides, cotton farmers have been confronted with many difficulties, of which insecticide resistance has become the dominant problem. Chemical control is not a simple system of applying any insecticide at any time, but must evolve into a regulated system of conserving resources by judicious use of selected products in as efficient a manner as possible. This will require greater attention to improved timing of treatment to minimize dosages, and more attention to accurate application.

References

Adkisson, P.L. (1971) Objective uses of insecticides in agriculture. In: Swift, J.E. (ed.) *Agricultural Chemicals – Harmony or Discord for Food, People, Environment.* Proceedings of the Symposium of the University of California, Division of Agricultural Science, Sacramento, California, pp. 43–51.

Ahmad, M. and McCaffery, A.R. (1988) Resistance to insecticides in a Thailand strain of *Heliothis armigera* (Hubner)(Lepidoptera: Noctuidae). *Journal of Economic Entomology* 81, 45–48.

Arnold, A.J. and Pye, B.J. (1981) Electrostatic spraying of crops with the APE80. *Proceedings, 1981 British Crop Protection Conference, Pests & Diseases.* BCPC Publication, Thornton Heath, Surrey, pp. 661–666.

Bacheler, J.S. and Bradley, J.R. (1989) Evaluation of bollworm action thresholds in absence of the boll weevil in North Carolina. The egg concept. In: *Proceedings of the Beltwide Cotton Production Research Conference*, Memphis, Tennessee, pp. 308–311.

Barducci, T.B. (1973) Ecological consequences of pesticides used for the control of cotton insects in Canete Valley, Peru. In: Farvar, M.T. and Milton, J.P. (eds) *The Careless Technology*. Stacey, London, pp. 423–438.

Basinski, J.J. and Wood, I.M. (1987) Kimberley Research Station. In: Basinski, J.J., Wood, I.M. and Hacker, J.B. (eds) *The Northern Challenge – a History of CSIRO Crop Research in Northern Australia*. Research Report 3, CSIRO, Division of Tropical Crops and Pastures. CSIRO, Melbourne.

Beeden, P. (1972) The pegboard – an aid to cotton pest sampling. *PANS* 18, 43–45.

Beeden, P. and Matthews, G.A. (1975) Erosion of cone nozzles used for cotton spraying. *Cotton Growers Review* 52, 62–65.

Boulter, D., Edwards, G.A., Gatehouse, A.M.R., Gatehouse, J.A. and Hilder, V.A. (1990) Additive protective effects of different plant-derived insect resistance genes in transgenic tobacco plants. *Crop Protection* 9, 351–354.

Brettell, J.H. (1986) Some aspects of cotton pest management in Zimbabwe. *Zimbabwe Agricultural Journal* 83, 41–46.

Bull, D. (1982) *A Growing Problem; Pesticides and the Third World*. OXFAM, Oxford.

Burgess, M.W. (1983) Development of cotton pest management in Zimbabwe. *Crop Protection* 2, 247–250.

Butler, G.D., Henneberry, T.J. and Hutchinson, W.D. (1986) Biology, sampling and population dynamics of *Bemisia tabaci*. In: Russell, G.R. (ed.) *Agricultural Zoology Review* 1, 167–195.

Cadou, I. (1959) Une rampe portative individuelle pour le pulverisation a faible volume. *Coton et Fibres Tropicales* 14, 47–50.

Carlson, G.A. and Mohamed, A. (1986) Economic analysis of cotton-insect control in the Sudan Gezira. *Crop Protection* 5, 348–354.

Cauquil, J. (1987) Cotton-pest control: a review of the introduction of ultra-low volume (ULV) spraying in sub-Saharan French-speaking Africa. *Crop Protection* 6, 38–42.

Coffee, R.A. (1979) Electrodynamic energy – a new approach to pesticide application. *Proceedings of the British Crop Protection Conference, Pests & Diseases*, 19–22 November 1979, Brighton. BCPC Publication, Thornton Heath, Surrey, pp. 777–789.

Cox, P.G. and Forrester, N.W. (1992) Economics of insecticide resistance management in *Heliothis armigera* (Lepidoptera: Noctuidae) in Australia. *Journal of Economic Entomology* 85, 1539–1550.

Davies, J.C. (1970) Effects of spraying and cultural practices on cotton in Uganda. *Experimental Agriculture* 6, 65–78.

Dittrich, V., Hassan, S.O. and Ernst, G.H. (1985) Sudanese cotton and the whitefly: a case study of the emergence of a new primary pest. *Crop Protection* 4, 161–176.

Duncombe, W.C. (1973) The acaricide spray rotation for cotton. *Rhodesian Agricultural Journal* 70, 115–118.

Dutton, R. and Komblas, K.V. (1989) Twenty years of use of chlorpyrifos in cotton in Egypt. In: Green, M.B. and Lyon, D.J. de B. (eds) *Pest Management in Cotton*. Ellis Horwood, Chichester, pp. 178–190.

Elbert, A. Overbeck, H., Iwaya, K. and Tsuboi, S.(1990) Imidacloprid, a novel systemic nitromethylene analogue insecticide for crop protection. *Proceedings of the British Crop Protection Conference, Pest and Diseases*, Brighton, 1990. BCPC Publication, Thornton Heath, Surrey, pp. 21–28.

Falcon, L.A. and Smith, R.F. (1973) *Guidelines for Integrated Control of Cotton Insect Pests*. FAO, Rome.

Farrington, J. (1977) Research-based recommendations versus farmers' practices: some lessons from cotton spraying in Malawi. *Experimental Agriculture* 13, 9–15.

Forrester, N.W. and Cahill, M. (1987) Management of insecticide resistance in *Heliothis armigera* (Hubner) in Australia. In: Ford, M.G., Hollomon, D.W., Khambay, B.P.S.

and Sawicki, R.M. (eds) *Biological and Chemical Approaches to Combating Resistance to Xenobiotics.* Ellis Horwood, Chichester.

Forrester, N.W., Cahill, M., Bird, L.J. and Layland, J.K. (1993) Management of pyrethroid and endosulfan resistance in *Helicoverpa armigera* (Lepidoptera: Noctuidae) in Australia. *Bulletin of Entomological Research Supplment No. 1.* International Institute of Entomology, London, 132pp.

Gledhill, J.A. (1971) A comparison of some tractor-mounted sprayers used in cotton pest control. In: Matthews, G.A. (ed.) *Cotton Insect Control*, Proceedings of the Cotton Insect Control Conference, Blantyre, Malawi, pp. 150–164.

Gould, F., Martinez-Ramirez, A., Andersen, A., Ferre, J., SIlva, F.J. and Moar, W.J. (1992) Broad-spectrum resistance to *Bacillus thuringiensis* toxins in *Heliothis virescens. Proceedings of the National Academy of Sciences of the USA* 89, 7956–7990.

Gower, J. and Matthews, G.A. (1972) Cotton development in the southern region of Malawi. *Cotton Growers Review* 48, 2–18.

Gunning, R.V., Easton, C.S., Greenup, L.R. and Edge, V.E. (1984) Pyrethroid resistance in *Heliothis armiger* (Hubner) (Lepidoptera: Noctuidae) in Australia. *Journal of Economic Entomology* 77, 1283–1287.

Hadar, E. (1991) Development criteria for an air-assisted ground crop sprayer. *BCPC Monograph* 46, pp. 23–26.

Haggis, M.J. (1982) Distribution of *Heliothis armigera* eggs on cotton in the Sudan Gezira: spatial and temporal changes and their possible relation to weather. In: Reed, W. (ed.) *International Workshop on* Heliothis *Management.* ICRISAT, Hyderabad, India.

Halloin, J.M. (1986) Treatment of cottonseeds. In: Jeffs, K.A. (ed.) *Seed Treatment.* BCPC Publication, Thornton Heath, Surrey, pp. 201–215.

Hasse, V. (1986) Introducing integrated plant protection to cotton farmers in the Philippines. *Second International Conference on Plant Protection in the Tropics*, Malaysian Plant Protection Society, Kuala Lumpur, p. 262.

Hau, B. and Richard, G.(1986) Results of the first large-scale growing of glandless varieties in the Ivory Coast. *Coton et Fibres Tropicales* 41, 100–101.

Hearn, A.B. and Roza, da G.D. (1985) A simple model for crop management applications for cotton (*Gossypium hirsutum* L.). *Field Crops Research* 12, 49–69.

Ingram W.R. and Green, S.M. (1972) Sequential sampling for bollworms on raingrown cotton in Botswana. *Cotton Growers Review* 49, 265–275.

Jackson, G.J. (1989) Insecticide resistance – the challenge of the decade. In: Green, M.B. and Lyon, D.J. de B. (eds) *Pest Management in Cotton.* Ellis Horwood, Chichester, pp. 27–30.

Jeffs, K.A. and Tuppen, R.J. (1986) Application of pesticides to seeds. In: Jeffs, K.A. (ed.) *Seed Treatment.* BCPC Publication, Thornton Heath, Surrey, pp. 17–50

Johnstone, D.R. and Johnstone, K.A. (1977) Aerial spraying of cotton in Swaziland. *PANS* 23, 13–26.

Joyce, R.J.V. (1975) Sequential aerial spraying of cotton at ULV rates in the Sudan Gezira as a contribution to synchronised chemical application over the area occupied by a pest population. *Proceedings of the 5th International Congress on Agricultural Aviation,* Cranfield, International Agricultural Aviation Centre, pp. 47–54.

Kfir, R. and Van Hamburg, H. (1983) Further tests of threshold levels for the control of cotton bollworms (mainly *Heliothis armiger*). *Journal of the Entomological Society of South Africa* 46, 49–58.

Lacewell, R.D. and Taylor, C.R. (1978) Economic implications of alternative boll weevil control strategies. In: *The Boll Weevil: Management Strategies.* Southern Cooperative Series Bulletin No. 228.

Lincoln, C.G. and Phillips, J.R. (1979) Point sample scouting. In: *Economic Thresholds and*

Sampling of Heliothis *Species on Cotton, Corn, Soybeans and Other Host Plants*. Southern Cooperative Series Bulletin 231, pp. 102–104.

Lincoln, C.G. Dowell, G.C. Boyer, W.P. and Hunter, R.C. (1963) The point sample method of scouting for boll weevil. *University of Arkansas Bulletin*, No. 666.

Lyon, D.J.de B. (1971) Timing of insecticide applications on cotton in the northern states of Nigeria. *Cotton Growers Review* 48, 281–296.

Marks, R.J. (1977) Assessment of the use of sex pheromone traps to time chemical control of the red bollworm *Diparopsis castanea* Hampson (Lepidoptera: Noctuidae) in Malawi. *Bulletin of Entomological Research* 67, 575–587.

Matthews, G.A. (1966) Investigations of the chemical control of insect pests of cotton in Central Africa. *Bulletin of Entomological Research* 57, 69–91.

Matthews, G.A. (1971) Ultra-low volume spraying of cotton – a new application technique. *Cotton Handbook of Malawi* Amendment 2/71, Agricultural Research Council of Malawi.

Matthews, G.A. (1973) Ultra-low volume spraying of cotton in Malawi. *Cotton Growers Review* 50, 242–267.

Matthews, G.A. (1986) Overview of chemical control. In: *Bemisia tabaci – A Literature Survey on the Cotton Whitefly with an Annotated Bibliography*. CAB International Institute of Biological Control, Ascot, Berkshire.

Matthews, G.A. (1989) *Cotton Insect Pests of Cotton and their Management*. Longman, Harlow, Essex.

Matthews, G.A. (1990) Changes in application techniques used by the small-scale cotton farmer in Africa. *Tropical Pest Management* 36, 166–172.

Matthews, G.A. (1992) *Pesticide Application Methods*, 2nd edn. Longman, London.

Matthews, G.A. and Tunstall, J.P. (1965) Aerial and ground spraying for cotton insect pest control in Rhodesia. *Cotton Growers Review* 42, 180–192.

Matthews, G.A. and Tunstall, J.P. (1968) Scouting for pets and the timing of spray application. *Cotton Growers Review* 45, 115–127.

Matthews, G.A., Tunstall, J.P.and McKinley, D.J. (1965) Outbreaks of pink bollworm *(Pectinophora gossypiella* Saund.) in Rhodesia and Malawi. *Cotton Growers Review* 42, 197–208.

Mowlam, M.D., Nyirenda, G.K.C. and Tunstall, J.P. (1975) Ultra-low volume application of water-based formulations of insecticides on cotton. *Cotton Growers Review* 52, 360–370.

Norman, D.W. Hayward, J.A. and Hallam, H.R. (1974) An assessment of cotton growing recommendations as applied by Nigerian farmers. *Cotton Growers Review* 51, 266–280.

Nyambo, B.T. (1989) Assessment of electrodynamic spraying as an alternative to ultra-low volume spraying in western Tanzania. *Crop Protection* 8, 417–421.

Nyirenda, G.K.C. (1986) Studies of the Effects of Insecticide Application on Cotton in Malawi. PhD Thesis, University of London.

Parkin, C.S. and Siddiqui, H.A. (1990) Measurement of drop spectra from rotary cage aerial atomisers. *Crop Protection* 9, 33–38.

Pearson, E.O. and Maxwell Darling, R.C. (1958) *The Insect Pests of Cotton in Tropical Africa*. Commonwealth Institute of Entomology, London.

Peregrine, D.J. (1989) The role of amitraz in cotton pest control. In: Green, M.B. and Lyon, D.J. de B. (eds) *Pest Management in Cotton*. Ellis Horwood, Chichester, pp. 191–199.

Reed, W. (1971) Comparison of insecticide sprayed and unsprayed cotton at Ukiriguru, western Tanzania. *Cotton Growers Review* 48, 200–209.

Riley, S.L. (1988) Pyrethroid resistance in *Heliothis virescens* and resistance management:

US perspective. *Proceedings of the Beltwide Cotton Production Research Conference*, Memphis, Tennessee.

Salama, H.S. (1983) Cotton-pest management in Egypt. *Crop Protection* 2, 183–191.

Sawicki, R.M., Denholm, I., Forrester, N.W. and Kershaw, C.D. (1989) Present insecticide-resistance management strategies in cotton. In: Green, M.B. and Lyon, D.J. de B. (eds) *Pest Management in Cotton.* Ellis Horwood, Chichester, pp. 31–43.

Smith, R.K. (1989) The 'Electrodyn' sprayer as a tool for rational pesticide management in smallholder cotton. In: Green, M.B. and Lyon, D.J. de B. (eds) *Pest Management in Cotton.* Ellis Horwood, Chichester, pp. 227–246.

Sterling, W.L. and Pieters, E.P. (1979) Sequential sampling. In: *Economic Thresholds and Sampling of* Heliothis *Species on Cotton, Corn, Soybeans and Other Host Plants.* Southern Cooperative Series Bulletin 231, pp.. 85–101.

Sutherland, J.A., King, W.J., Dobson, H.M., Ingram, W.R., Attique, M.R. and Sanjrani, W. (1990) Effect of application volume and method on spray operator contamination by insecticide during cotton spraying. *Crop Protection* 9, 343–350.

Swezey, S.L. and Daxl, R.G. (1988) Area-wide suppression of boll weevil (Coleoptera: Curculionidae) populations in Nicaragua. *Crop Protection* 7, 168–176.

Tunstall, J.P. and Matthews, G.A. (1961) Cotton insect control recommendations for 1961–62 in the Federation of Rhodesia and Nyasaland. *Rhodesian Agricultural Journal* 58, 233–258.

Tunstall, J.P. and Matthews, G.A. (1965) Contamination hazards in using knapsack sprayers. *Cotton Growers Review* 42, 193–196.

Tunstall, J.P. and Matthews, G.A. (1966) Large-scale spraying trials for the control of cotton insect pests in Central Africa. *Cotton Growers Review* 43, 121–139.

Tunstall, J.P., Matthews, G.A. and Rhodes, A.A.K. (1961) A modified knapsack sprayer for the application of insecticides to cotton. *Cotton Growers Review* 38, 22–26.

Tunstall, J.P., Matthews, G.A. and Rhodes, A.A.K. (1965) Development of cotton spraying equipment in Central Africa. *Cotton Growers Review* 42, 131–145.

Uk, S. (1987) Distribution patterns of aerially applied ULV sprays by aircraft over and within the cotton canopy in the Sudan Gezira. *Crop Protection* 6, 43–48.

Uk, S. and Courshee, R.J. (1982) Distribution and likely effectiveness of spray deposits within a cotton canopy from fine ultra-low volume spray applied by aircraft. *Pesticide Science* 13, 529–536.

Vaeck, M., Reynaerts, A., Hofte, H., Jansens, S., de Beuckleer, M., Dean, C., Zabeau, M. van Montagu, M. and Leemans, J. (1987) Transgenic plants protected from insect attack. *Nature* 328, 33–37.

28 Pest Management Systems

J.D. Mumford[1] and G.A. Norton[2]

[1]Department of Biology, Imperial College at Silwood Park, Buckhurst Road, Sunninghill, Ascot, Berkshire SL5 7PY, UK; [2]Cooperative Research Centre for Tropical Pest Management, University of Queensland, Brisbane, Queensland 4072, Australia

The Cotton Pest System

The development of pest problems on cotton is primarily determined by technical and ecological changes in the environment. However, pest problems may be aggravated by a complex of political, social and economic factors, which may also limit solutions. Similarly, the opportunities and constraints for cotton farmers to adopt pest management strategies are influenced by these same factors (Table 28.1). Knowledge about the key components in Table 28.1 is needed to understand how cotton pest problems develop, why farmers adopt the practices they do and how cotton pest management can best be improved.

Table 28.1 also shows that the components affecting pest problems in cotton lie in different systems. Agricultural and environmental policy issues, for instance, can affect the price of lint, the availability of credit and the regulations of insecticide use. These are part of a national system and involve decision makers at a national level. Clearly, decisions made at this level can affect the price that cotton farmers receive and their ability and incentive to purchase particular insecticides. At another level in the system, components in the village system such as the number and accessibility of extension agents and the availability and

Table 28.1. Components of the cotton pest system which contribute to pest development.

Agriculture, social and environment policies
Economic forces – input costs and output prices
Research and extension capability
Farming systems – land tenure, water use, mechanization, etc.
Pest status – climate, natural enemies, etc.
Agronomic practices – varieties, cropping patterns, etc.
Pesticide and application equipment availability
Social and traditional practices

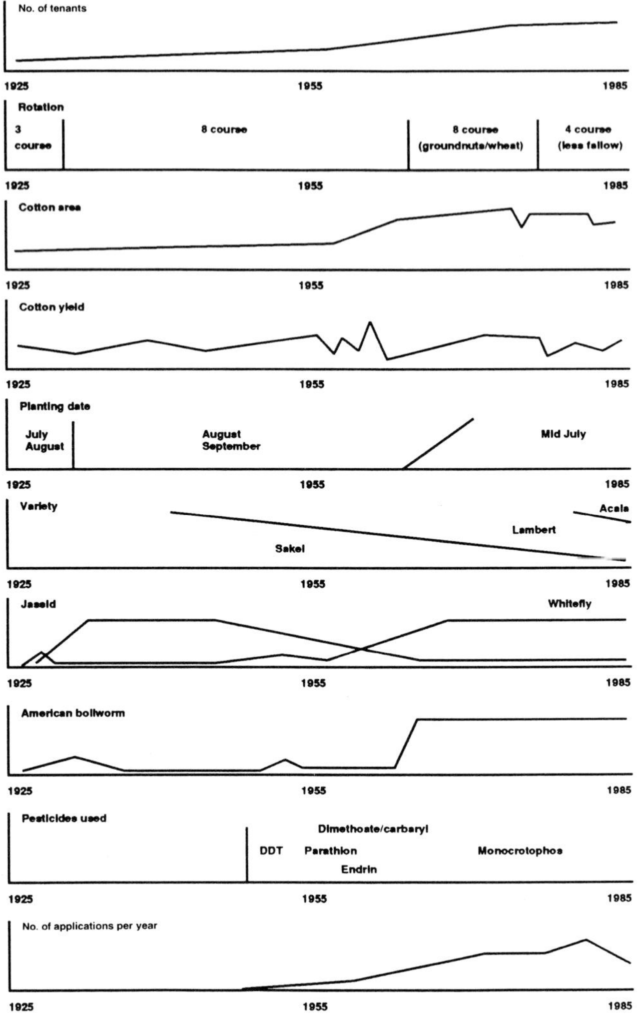

Fig. 28.1. Historical profile of pest problems in the Sudan Gezira (based on Griffiths, 1984).

cost of local labour can have a direct influence on the type of pest management practised. The farming system has a further impact on pest development and control, as do factors at the field level, such as agronomic and microclimatic features of the crop.

Thus, many interactions which can affect a farmer's pest problems occur both within and between these different levels of the system. For instance, decisions made on crop breeding policy at a national level can, in time, result in new varieties becoming available at the village level, which might then be adopted by farmers and so influence pest favourability and control. Thus, as well as interactions between these different levels, there is also a temporal dimension to consider, producing dynamic changes at the farm level following earlier changes at other levels in the overall system. While this complexity can be daunting, there are a range of systems analysis techniques that can help identify the key factors involved and provide a better appreciation of the problem.

An illustration of this is provided by an historical analysis carried out by Griffiths (1984) on developments in the Sudan Gezira from 1925 to 1984. Some examples of this analysis are presented in the historical profile shown in Fig. 28.1.

This figure collates a number of factors likely to have had a significant influence on the development of cotton pest problems in the Gezira over this period. This includes policy changes which affected cost sharing agreements and hence the profitability of cotton growing for the tenants, and agronomic changes, such as the shift in cotton varieties grown and changes in the 'official' rotation practice. All of these conditions had an influence on changes in pest status and in pesticide use over the period.

The recent problems associated with whitefly are likely to have resulted from several other developments. In the early 1960s, official changes in rotational practices to include additional crops, and the increased area cropped per season, caused more severe *Heliothis* problems. These led to an increased use of pesticide, particularly broad-spectrum insecticides such as DDT and monocrotophos (Fig. 28.1), which in turn contributed to the increased whitefly problem by reducing natural enemies and stimulating whitefly fecundity.

The changes in crop rotations in the early 1970s may also have had other, less direct effects on the whitefly problem. For instance, since the Gezira Board would not allow tenants to grow fodder crops, such as lubia, tenants irrigated cotton after picking in order to provide forage for their animals. This meant that cotton was available as a host plant for whitefly for a longer period, undoubtedly contributing to its increased pest status. Other changes, such as increased reliance on aerial spraying in the 1970s, may have added a further factor by not giving good penetration of the crop canopy to kill whitefly. Insecticide resistance also developed following the greater use of a wider range of insecticides.

Thus, as illustrated by the Gezira example, this type of analysis serves several purposes:

- it provides a rigorous framework for identifying the various causes (or hypotheses) of current pest management problems in cotton;
- it gives indications of changes in the system that will influence the ability and incentive for improved pest management;
- it provides a basis for predicting future developments at various levels that can affect pest status and management in the future, and so contributes to decision making on policy, research and development issues.

Descriptive analysis techniques, therefore, can be particularly useful in helping to identify strategic research priorities. Since this type of analysis is often aimed at designing research programmes where they have been nonexistent or inadequate in the past, much of the analysis must be subjective. This can be obtained by establishing a consensus of experienced opinion on the subject, through literature reviews and interviews. Structured workshops, meetings involving research and extension scientists, policy makers and farmers, provide an excellent means of achieving this.

An example of the output provided by such a workshop, carried out in Nantong City in China, is shown in Table 28.2. It provides an illustration of the factors that need to be included in a detailed specification of research and information requirements, in this case for a cotton pest management advisory system. Because of the emphasis on problem definition, inter-disciplinary workshops can also reveal other constraints to improved pest management, with consequent implications for appropriate action. Other constraints to improved pest management are likely to exist in agricultural policy, extension services and on the farm itself. Thus, the way in which decisions are made and the factors that affect this decision making process need to be studied as an integral part of the effort to improve pest management.

Systems analysis and decision analysis techniques can also address the more immediate pest problems facing cotton farmers, and other decision makers.

Decision Making

As we have seen above, decision making in various sectors, including policy, research and extension, can influence the long term strategy of cotton pest management. The demise of cotton growing in parts of Mexico, Peru and Australia, partly as a result of escalating pest control costs, brings home the importance of taking a long term strategic view of the problem. However, in the shorter term, the form of pest management actually practised in cotton crops will be determined by other decision makers and by the various factors that influence their decision making. Typically, there are the cotton growers themselves, ranging from those with large areas of irrigated cotton to those with only a few hectares or less of rainfed cotton. Alternatively, the person who makes pest management decisions may be a government official, as in Egypt, a state farm manager, as in parts of China, responsible for large tracts of cotton, or a pest management consultant employed by the farmer in the USA.

Whoever the decision maker is it is crucial to understand the decision maker's situation, and the key factors that influence his pest management decisions if these decisions are to be improved. Understanding the decision making process may provide ideas on how to achieve improvements, by targeting research on critical information gaps, for example. It can also give indications about how new technological options arising from research and development are likely to be received by farmers, and particularly, whether they are likely to be widely adopted. Such predictions should, of course, be followed up by extensive on-farm trials and liaison with research and extension personnel in the field so that the decision making process is monitored as it changes with new situations.

Surveys of cotton growers are an obvious way to obtain information on their current pest management practices and why they are carried out. For example, a survey of cotton growers in Zambia carried out by Javaid *et al.* (1987a,b) found that while virtually all growers were aware of the major insect pests, some of them applied more insecticide than was recommended, either because they over-estimated the potential losses caused by pests, or because they considered pesticides as a cheap insurance. This same survey also found that while many farmers inspect their crops to decide on the timing of pesticide application, they used their own rather than the recommended scouting method. The differences between farmers' practices and research based recommendations raises several possible responses. A misperception of the problem could be corrected through an extension campaign, research could be targeted at solutions that meet the farmers' objectives better, and recommendations could be reviewed to see whether farmers, who have their feet on the ground and their livelihoods at stake, may have developed a better way of solving their pest problem.

While this understanding of pest management decisions at the farm level is critically important, it is also important to examine the decisions at other levels.

Table 28.2. A specification for a computerized advisory system for late-season pest management produced in a workshop in Jiangsu Province, China (Zhang *et al.*, 1989).

General objectives

- A computer based advisory system should be developed at the county level, the lowest level of decision making at which computers are likely to be available
- The system should concentrate on bollworms and plant bugs, but other pests should also be considered in less detail
- The system should predict the time and size of the next peak of pink bollworms and hoppers, and based on this prediction (along with estimates of other minor pests) control recommendations and other information would be provided to advisers

Specific problems to be addressed

- Estimation of pest damage to cotton in the county
- Selection of appropriate control recommendations
- Allocation of staff for monitoring/spraying
- Allocation of pesticide supply

System outputs

- Extent of area needing treatment
- Timing of treatment
- Number of treatments
- Type of treatment (choice of pesticide)
- Identification of high-risk villages/fields

General data requirements for the decision support system

- Population development data (pink bollworm and bugs) for past seasons
  - timing (date and relative to crop development)
  - abundance
- Weather records for past seasons (daily through season)
  - temperature
  - rainfall
- Trials data on pesticide efficacy
- Crop growth through past seasons
  - timing of critical stages (buds, flowers, bolls)
  - number and proportion of buds, flowers, bolls
- Thresholds for each pest
- Current treatment suggested for each pest

Table 28.3. Decision options and criteria for cotton pest management in the Nantong City region of Jiangsu Province (Zhang *et al.*, 1989).

Level	Question	Factors
County and township	Target pest	weather soil other crops pest, natural enemies density period of occurrence
	Area to be treated	severity of attack chemicals available level of farmers' skill
	Timing of treatment	chemical type(s) pest, natural enemies numbers co-occurrence of other pests
	Number of applications	severity of attack residue period safety period chemical used occurrence of other pests
	Chemical type to recommend and supply	chemical availability target pest(s) pest stage price of chemical weather conditions toxicity (efficacy/safety)
	Spray method to recommend	type of chemical machinery availability
Village	Monitoring programme	local experience advice from county/township
	Target fields	local monitoring village discussions advice from township/county
	Timing of sprays	local monitoring effectiveness of previous spray level of local industry (affects labour)
Farm	Immediate control decision	field scouting village discussions advice from township/county
	Chemical choice	chemical availability effectiveness of control
	Application	cost of application cash availability technical ability machinery availability labour available

These decisions vary considerably from country to country. An example of different levels of pest management decisions, and the factors that are considered when making them in China is shown in Table 28.3. More strategic decisions are made at higher levels, and these appear as factors affecting local tactical decisions.

The decisions facing a particular decision maker can be expressed in terms of a decision profile, as shown in Figure 28.2 for the decision problem of a cotton farmer in South Africa. Along with the season profile of the crop there are points at which pest management decisions must be made. These relate to a weekly scouting schedule, and as the season progresses, decisions must be made each week about different scouting methods and control actions. Both action thresholds and control options change with time, for instance the red spider mite threshold increases as the crop grows and there is a limited period when pyrethroid insecticides can be used.

Decision making in integrated pest management (IPM) should be based on the economic threshold principle that control should only be initiated when the benefits exceed the costs (Mumford and Norton, 1984). While this is well recognized and agreed in theory, in practice many problems limit the use of this

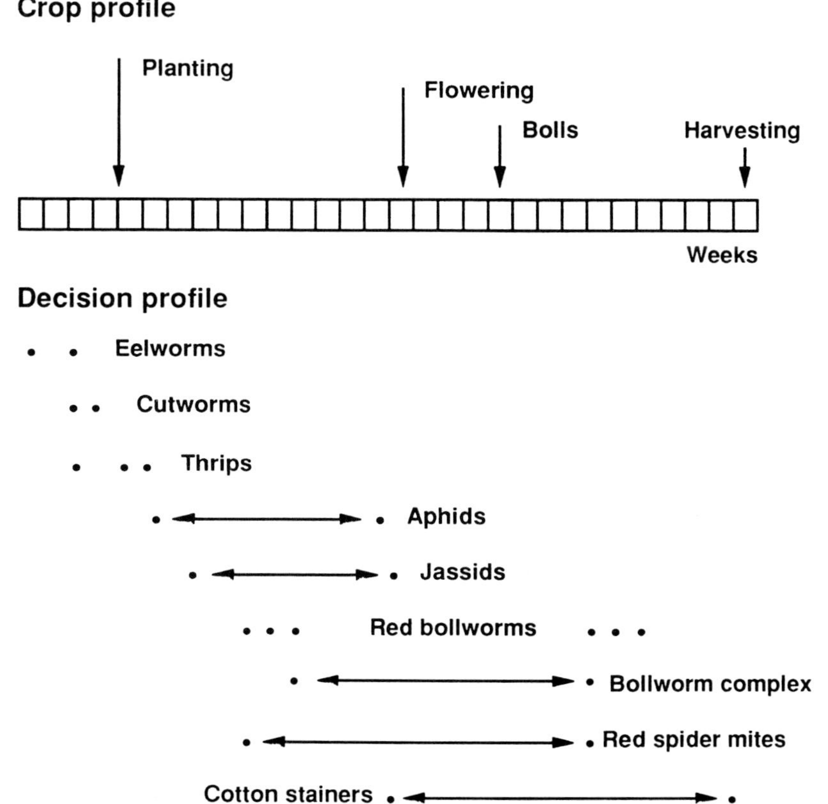

Fig. 28.2 Decision profile for cotton and pest management decisions (based on Mumford and van Hamburg, 1985).

principle by farmers. For the economic threshold principle to be implemented, two types of information must be available: sampling and prediction. Sampling of both crop (density and growth for yield/phenology estimates) and pests (scouting for loss/timing) is needed. Four types of prediction are required: yield potential of crop; loss (damage function for pests); performance of control options (control function); price (crop and control).

In making recommendations to farmers, the adviser must match control options to farmers' objectives, taking into account their particular expectations of yield, loss, control cost, and control performance. These choices can be based on sampling information and predictions, or on a routine schedule.

This information is not always easy to obtain, and it is particularly difficult to quantify these decision inputs when there are complexes of pest species. Further problems arise if broader issues are considered. For instance, there may be external factors, such as costs associated with pesticide resistance in other cotton pests or vectors of human disease living near cotton fields, operator safety or environmental contamination which affect large numbers of people as a group, rather than just individual decision makers. Farmers are not in a position to take these factors into consideration, therefore government intervention is necessary to ensure that their decisions do not go against the general welfare.

In some countries, such as Egypt and the Sudan, cotton pest control is conducted by a central authority over very large areas and the scale of decisions (such as widespread aerial spraying) makes it difficult to base them on sampling and prediction in individual fields. One spray sortie by an aircraft may cover fields owned by up to 100 small farmers.

Scouting provides information on the size of pest populations (and hence damage) and on the timing of susceptible or damaging stages of the pest populations and the crop. Scouting can be done either within the season on the crop (for bollworms) or outside the cotton season (for overwintering boll weevils). There are numerous examples of sampling cotton pests reported, including sequential sampling for *Heliothis* (Nachapong *et al.*, 1989); egg sampling for *Heliothis* (Sheng, 1985); boll weevil (Herzog and Lambert, 1984; Alvarez and Martinez, 1985); whitefly (Musuna, 1986); thrips (Mabbett *et al.*, 1984); and general pest complexes (CCGA, 1988).

Many studies have shown that control based on scouting is worthwhile (Brittain, 1983; Hatcher *et al.*, 1984; Lacewell and Masud, 1989), although most of these studies have not considered the extra risks that may be associated with waiting to spray on a threshold. One of few cases in the literature in which scouting has not proved to be beneficial is in Tanzania (Nyambo, 1989a,b), where schedule spraying was better. Thresholds were either too low or too high, sampling was often not reliable and pests were frequently present in high numbers, so either high costs or excessive damage often resulted. Threshold based control is only advisable in cases where accurate sampling methods are available, where pests do not regularly occur in high numbers and where control actions can be taken quickly and effectively in response to the scouting. Schedule spraying may be a more practical and economical alternative in cases where sampling is not feasible or where pest attack is regularly very serious. It may also be more attractive to the small-scale farmer where there is little extension input.

The value of an action threshold approach over routine calendar spraying must be related to the objectives of cotton growers and the potential for the information to change their actions. The relative performance of economic threshold and calendar spraying programmes can be measured in comparative field trials, but unless these trials are conducted over many years they do not reveal the riskiness of a strategy. While it may be impractical to conduct long-term comparative trials, it is possible to compare strategies based on the subjective opinion of people with considerable experience. Table 28.4 shows the consensus view of five commercial and research entomologists concerned with cotton pests in South Africa on the relative performance of several control strategies. The table is in the form of a pay-off matrix, which shows the cost of each strategy under three states of nature. The average returns can be compared, but the pay-off matrix also allows risk to be considered, so actions could be judged by their likely outcomes under adverse conditions for risk averse objectives, or under favourable conditions for risk takers. For instance, the higher threshold of eight larvae per 24 plants is generally better than a threshold of five, but in high attacks (24% of the time) it is worse.

Cotton Pest Models

Many computer models have been developed for cotton and cotton pests, which are relevant to sampling (sample size and distribution), prediction (population size, damage and timing) and simply better understanding of ecological processes involved in the cotton agroecosystem. Some of these models could also be incorporated into computerized advisory systems to help decision makers choose suitable control options.

A list of recent examples of cotton pest models is given in Table 28.5, indicating a variety of approaches and methods. The purpose of these models is also different. Some attempt to obtain 'explanations' of observed events, for example, Baumgartner *et al.* (1986) and Wallach *et al.* (1988). Others are concerned with crop loss and crop compensation to pest damage, for example Mangel *et al.* (1985). Some try to provide specific help in decision making, such as assessing strategic control by the use of mating disruption by pheromones (Stone and Gutierrez, 1986a,b) and using day-degree models of *Heliothis* development to determine the best time for augmentative release of *Trichogramma* spp. (Witz *et al.*, 1985). An extensive discussion of cotton pest models has been edited by Frisbie *et al.* (1989b).

Cotton Pest Decision Support Systems

Decision support systems are generally computer programs that provide help to decision makers by simulating the decision processes of human advisers to produce recommendations. Such systems have a number of advantages over human advisers, especially where human expertise is scarce. Once it has been developed a decision support system can be used repeatedly by large numbers of

Table 28.4. Pay-off matrix based on a consensus of experienced opinion about various threshold based strategies on irrigated cotton in the Transvaal, South Africa (Mumford and van Hamburg, 1985).

Strategies		Bollworm level and probability LOW *p*=0.36	MEDIUM *p*=0.40	HIGH *p*=0.24	Expected cost R ha^{-1}
Weekly	Sprays:	14–0–2	14–0–2	14–0–2	
sprays	% Loss:	–2%	–4%	–6%	
	R Loss:	408	448	488	443
ET eggs	Sprays:	3–1–1	4–2–1	5–3–1	
12/24	% Loss:	–0%	–1%	–2%	
plants	R Loss:	120	191	285	188
ET larvae	Sprays:	1–1–1	2–2–2	2–3–2	
5/24	% Loss:	–0%	–0.5%	–1%	
plants	R Loss:	74	158	196	137
ET larvae	Sprays:	0–1–0	1–1–1	1–2–1	
8/24	% Loss:	–0%	–2%	–9%	
plants	R Loss:	28	114	282	123
No spray	Sprays:	0–0–0	0–0–0	0–0–0	
	% Loss:	–5%	–33%	–70%	
	R Loss:	100	660	1400	636

Sprays: endosulfan–pyrethroid–triazophos, respectively, per season
R Loss: loss in Rand ha^{-1} based on a R2000 ha^{-1} expected crop value
Approximate exchange rate US$1.00 = R2.50

people who might otherwise not have access to one of the few extension advisers. The programs give consistent advice, and in many cases can explain the advice they give, helping to improve the expertise of their users. The rules used for computerized decision support systems can also be developed for other media, such as pegboards, manuals, etc., which would make them relevant to even the least developed cotton growing areas. In developing decision support systems, structuring the problem also provided valuable benefits in identifying key information gaps leading to specifications for research priorities (see also Table 28.3).

Decision support systems can be used to advise on any management decision, for example early season cotton aphid control in China (Table 28.6). The table illustrates examples of the factors that advisers consider when they are making recommendations. The decision support program asks the user to give information about these factors, in the classes shown, and is programmed with the help of an experienced cotton adviser to give appropriate recommendations for each possible combination of factors. Some or all of these factors may be considered in each case. As an example, if cotton is past the seedling stage and aphid density is less than 100 and the crop had previously been sprayed, then no spray would be advised. For very unusual combinations the system could be

Table 28.5. Some recent examples of cotton pest models.

Pest	Type of model	Place	Reference
Cutworm	Emergence timing	Former USSR	Gulamov *et al.* (1987)
Aphids	Population simulations	China	Xia and Sterling (1987); Zhang *et al.* (1987); Luo and Shen (1985)
Boll weevil	Day-degree, spring emergence and survival	USA	Stone *et al.* (1990)
	Cotton growth and weevil populations interact, based on supply and demand for nutrient and energy inputs	Brazil	Gutierrez *et al.* (1991a, b)
Heliothis	CIM-HEL, initial population emerging	USA	Brown *et al.* (1990)
	Heat units, soil temperature determines early season development	USA	Rummel and Hatfield (1988)
	Behavioural model related to adult emergence and oviposition choice	Sudan	Topper (1987)
	Population simulation using historical data	Israel	Wallach *et al.* (1988)
	MOTHZV, early season population used to predict timing and size of later population	USA	Witz *et al.* (1985); Hartstack and Witz (1983)
	MOTHZV-2, similar to above but with predators included	USA	Ables *et al.* (1983)
	HELDMG/GOSSYM, simulates damage and crop growth	USA	Thomas (1989)
Pectinophora	Diapause timing by temperature and day length	Brazil, USA	Gutierrez *et al.* (1986)
	Insect, crop and management models to test insecticide control strategies	USA	Stone *et al.* (1986b); Stone and Gutierrez (1986a,b)
	Climate based timing and damage models	China	Liu (1984); Liu *et al.* (1986)

Table 28.5. *Continued.*

Pest	Type of model	Place	Reference
Lygus	Population simulations	USA	Fleischer and Gaylor (1988)
	Crop damage	USA	Mangel *et al.* (1985)
Whitefly	Life-table models	Sudan	Baumgartner *et al.* (1986); Arx *et al.* (1983 a,b)
Pest complexes	TEXCIM, fleahopper/bollworm decision model for management strategies	USA	Masud and Benedict (1990); Breene *et al.* (1989); Legaspi *et al.* (1989); Thomas *et al.* (1988)
	CIM/COTCROP/CIM-BW/CIM-HEL, combined boll weevil, bollworm and crop model to test management strategies	USA	Brown *et al.* (1983)
	Cotton growth (day-degree) and simulated square damage	China	Bai *et al.* (1989)
	Cotton yield	China	Sheng and Hopper (1988)
	GOSSYM/COMAX, expert system and cotton growth model	USA	Hargett (1988)
	CIM/COTGAME, interactive training model for cotton management	USA	Akbay *et al.* (1988)
	SIRATAC, yield, management model	Australia	Hearn *et al.* (1981); Brook and Hearn (1983)
	COTSIM, combined crop/pest model	USA	Frisbie *et al.* (1989a)
	Socio-economic model of IPM extension benefits	USA	Napit *et al.* (1988)
	Simulation model to compare control strategies	USA	Simpson and Parvin (1983)

programmed to advise that an extension agent should visit the field to double check the recommendation.

Some decision support systems are specific to cotton pest problems only (Transvaal cotton pest system (Mumford and van Hamburg, 1985); SYSTEX for pest identification in the USA (Stone *et al.*, 1986a)). The Transvaal cotton pest system deals with the nine pests that have recommended controls available (nematodes, cutworms, thrips, aphids, jassids, red bollworms, American (and other) bollworms, red spider mites and cotton stainers). It was a prototype system intended for use in cotton cooperatives and pesticide dealerships, where farmers would be given access to a computer on a counter in the shop. The program provides diagnosis, control advice, calculations of pesticide volumes and costs for individual areas of cotton, advice on scouting techniques, pest identification and application volumes. The program is easily updated for new costs and pesticide rates. It ensures that all options are considered when making a pesticide decision and specifically prevents some recommendations being given, such as the use of pyrethroids outside the permitted crop stages. While this program is based on quite limited research inputs, it reflects the recommendations currently given to farmers by commercial and extension personnel.

Like many such systems, this one has only reached a prototype stage, partly through concern over who would or should take responsibility for the advice given by the computer in the absence of a human adviser to check the result. While this fear has not prevented the publication and use of numerous booklets of pest management instructions, the use of a computer is novel and still arouses some distrust. However, as more research data are available to add reliability to such systems, more prototypes are designed and thoroughly tested, and farmers and extension personnel become more familiar with computers such pest advisory programs are likely to become more widely used.

Table 28.6. Examples of inputs and recommendations used for a prototype cotton aphid decision support system in Jiangsu Province, China (Zhang *et al.*, 1989).

Factors considered	Classes		
Cotton stage	Seedling	Later	
Aphid density per 100 plants	>300	101–300	$<=100$
Humidity	$>80\%$	$<=80\%$	
Precipitation in the past 5 days	>20 mm	$<=20$ mm	
Planting method	transplanted	direct seeded	
Natural enemies (ratio to aphids)	more than about 1/60 to 1/100	less than about 1/60 to 1/100	
Previous spray on the crop	yes	no	

Information about these factors leads the expert system to one of four recommendations:

Do not spray
Reassess later after further monitoring
Spray once as soon as possible
Spray twice at a given interval

Other decision support systems are intended to give advice on more general aspects of cotton management, including pests (COTAFLEX (Frisbie *et al.*, 1989a); CALEX (Plant *et al.*, 1987); COMAX/GOSSYM (McKinion and Lemmon, 1985)), and these may form part of Integrated Cotton Management Systems.

Integrated Cotton Management Systems

Most decision makers do not make decisions about cotton pests in isolation. There is, therefore, a need to integrate advisory systems covering all aspects of crop management (including pest management as one component). There has been considerable emphasis on developing computerized management systems in the USA and Australia (e.g. COTAFLEX (Frisbie *et al.*, 1989a,c) and SIRATAC (Brook and Hearn, 1983)). The complexity of the total package and the difficulty of fitting the component models and computerized advisory systems together in practical systems has been a problem in the development of these systems. In some cases the poor economic returns from cotton have not justified such refined management. An alternative to an overall computerized crop management system would be to develop a series of advisory modules in different media (i.e. pegboards for field scouting, picture manuals for pest symptoms, printed time sheets for season plans, forms for budgeting and growth charts for yield prediction). In either form, there remains a great need for research and development in integrated cotton management systems throughout the world.

Constraints to Introduction of Cotton Pest Management Systems

The constraints to the introduction of improved cotton pest management systems vary greatly because of the range of management systems for cotton around the world. However, a number of general problems arise in all cases:

- few practical control options (due to cost or complexity);
- poor biological knowledge on which to base an economical, effective system;
- poor targeting and delivery to decision makers.

In many cases, computerized management systems can help to solve some of these problems. Computers are very useful for processing complex sets of information and decision support systems, for instance, can simplify the decision process for the farmer or adviser. While computers will not reduce the cost of individual control options, they can help to reduce the overall cost of pest control by making more efficient use of available options. Models can help to organize biological information and provide understanding of ecological processes, and both biological and decision models can help to provide direction for research and development activities. Decision models, in particular, can help to target information (since they are based on careful analysis of decision processes) and to deliver it to users. Well thought-out management programmes that are not based directly on computers may also help in many of these areas, if they are based on good descriptive and qualitative analyses of both the biological and decision problems.

References

Ables, J.R., Goodenough, J.L., Hartstack, A.W. and Ridgway, R.L. (1983) Entomophagous arthropods. *US Department of Agriculture, Agriculture Handbook*, No. 589, pp. 103–127.

Akbay, K.S., McClendon, R.W. and Brown, L.G. (1988) COTGAME: Cotton insect pest management simulation game. *Applied Engineering in Agriculture* 4, 201–206.

Alvarez, R.J.A. and Martinez, W.O. (1985) Evaluation of four sampling methods for *Anthonomus grandis* in cotton. *Revista Instituto Colombiano Agropecuario* 20, 131–137.

Arx, R. von., Baumgartner, J. and Delucchi, V. (1983a) Developmental biology of *Bemisia tabaci* (Gen..) (Sternorrhyncha, Aleyrodidae) on cotton at constant temperatures. *Mitteilungen der Schweizerischen Entomologischen Gesellschaft* 56, 389–399.

Arx, R. von., Baumgartner, J. and Delucchi, V. (1983b) A model to simulate the population dynamics of *Bemisia tabaci* (Genn.) (Ster. Aleyrodidae) on cotton in the Sudan Gezira. *Zeitschrift für Angewändte Entomologie* 96, 341–363.

Bai, L.X., Cao, C.Y., Zhang, Y.X. and Shu, C. (1989) A preliminary study on the dynamic model of cotton growing in relation to the pest management in the coastal region of Jiangsu. *Scientia Agricultura Sinica* 22, 7–14.

Baumgartner, J., Delucchi, V., Arx, R. von and Rubli, D. (1986) Whitefly (*Bemisia tabaci* Genn., Stern.: Aleyrodidae) infestation patterns as influenced by cotton, weather and *Heliothis*: hypotheses testing by using simulation models. *Agriculture, Ecosystems and Environment* 17, 49–59.

Breene, R.G., Hartstack, A.W., Sterling, W.L. and Nyffeler, M. (1989) Natural control of the cotton fleahopper, *Pseudatomoscelis seriatus* (Reuter) (Hemiptera, Miridae), in Texas. *Journal of Applied Entomology* 108, 298–305.

Brittain, J. (1983) Extension IPM programs in South Carolina. *Proceedings of the Congress of Plant Protection*, 1983, Brighton 3, 20–25.

Brook, K.D. and Hearn, A.B. (1983) Development and implementation of SIRATAC: a computer based cotton management system. *Proceedings of the First National Conference on Computers in Agriculture*, University of Western Australia, Perth, pp. 222–240.

Brown, L.G., McClendon, R.W. and Jones, J.W. (1983) Cotton and insect management simulation model. *US Department of Agriculture, Agriculture Handbook*, No. 589, pp. 437–479.

Brown, L.G., McClendon, R.W. and Akbay, K.S. (1990) Goal programming for estimating initial influx of insect pests. *Agricultural Systems* 34, 337–348.

CCGA. (1988) *Cotton Handbook*. Commercial Cotton Growers' Association, Harare, Zimbabwe.

Fleischer, S.J. and Gaylor, M.J. (1988) *Lygus lineolaris* (Heteroptera: Miridae) population dynamics: nymphal development, life tables, and Leslie matrices on selected weeds and cotton. *Environmental Entomology* 17, 246–253.

Frisbie, R.E., Crawford, J.L., Bonner, C.M. and Zalom, F.G. (1989a) Implementing IPM in cotton. In: Frisbie, R.E., El-Zik, K.M. and Wilson, L.T. (eds) *Integrated Pest Management Systems and Cotton Production*. Wiley, New York 437pp.

Frisbie, R.E., El-Zik, K.M. and Wilson, L.T. (eds) (1989b) *Integrated Pest Management Systems and Cotton Production*. Wiley, New York, 437pp.

Frisbie, R.E., El-Zik, K.M. and Wilson, L.T. (1989c) The future of cotton IPM. In: Frisbie, R.E., El-Zik, K.M. and Wilson, L.T. (eds) *Integrated Pest Management Systems and Cotton Production.* Wiley, New York, 437pp.

Griffiths, W.T. (1984) A Review of the Development of Cotton Pest Problems in the Sudan Gezira. Unpublished MSc Thesis, University of London.

Gulamov, M.I., Mukhitdinov, S.M. and Pasekov, V.P. (1987) Some problems of forecasting the abundance dynamics of insects using simulation modelling. *Zhurnal Obshchei Biologii* 48, 839–844.

Gutierrez, A.P., Pizzamiglio, M.A., Dos Santos, W.J., Villacorta, A. and Gallagher, K.D. (1986) Analysis of diapause induction and termination in *Pectinophora gossypiella* in Brazil. *Environmental Entomology* 15, 494–500.

Gutierrez, A.P., Dos Santos, W.J., Villacorta, A., Pizzamiglio, M.A., Ellis, C.K., Carvalho, L.H. and Stone, N.D. (1991a) Modelling the interaction of cotton and the cotton boll weevil. I. A comparison of growth and development of cotton varieties. *Journal of Applied Ecology* 28, 371–397.

Gutierrez, A.P., Dos Santos, W.J., Pizzamiglio, M.A., Villacorta, A., Ellis, C.K., Fernandes, C.A.P. and Tutida, I. (1991b) Modelling the interaction of cotton and the cotton boll weevil. II. Boll weevil (*Anthonomus grandis*) in Brazil. *Journal of Applied Ecology* 28, 398–418.

Hargett, J. (1988) Insect control for earliness: a producer's viewpoint. *Proceedings of the Beltwide Cotton Production Conference*, Memphis, Tennessee, pp. 30–31.

Hartstack, A.W. and Witz, J.A. (1983) Models for cotton insect pest management. *US Department of Agriculture, Agriculture Handbook* No. 589, pp. 359–381.

Hatcher, J.E., Wetzstein, M.E. and Douce, G.K. (1984) An economic evaluation of integrated pest management for cotton, peanuts and soybeans in Georgia. *Research Bulletin, College of Agriculture Experiment Stations, University of Georgia* No. 318, 28pp.

Hearn, A.B., Room, P.M., Thomson and Wilson, L.T. (1981) Computer-based cotton pest management in Australia. *Field Crops Research* 4, 321–332.

Herzog, G.A. and Lambert, W.R. (1984) A new scouting technique for sampling boll weevil reproduction following the use of insect growth regulators. *Southwestern Entomologist* 6, 27–32.

Javaid, I., Zulu, J.N., Matthews, G.A. and Norton, G.A. (1987a) Cotton insect pest management on small scale farms in Zambia – I. Farmers' perceptions. *Insect Science and its Application* 8, 1001–1106.

Javaid, I., Zulu, J.N., Matthews, G.A. and Norton, G.A. (1987b) Cotton insect pest management on small scale farms in Zambia – II. Training and sources of advice. *Insect Science and its Application* 8, 1007–1015.

Lacewell, R.D. and Masud, S.M. (1989) Economic analysis of cotton IPM programs. In: Frisbie, R.E., El-Zik, K.M. and Wilson, L.T. (eds) *Integrated Pest Management Systems and Cotton Production.* Wiley, New York, 437pp.

Legaspi, B.A.C., Jr., Sterling, W.L., Hartstack, A.W., Jr and Dean, D.A. (1989) Testing the interactions of pest-predator-plant components of the TEXCIM model. *Environmental Entomology* 18, 157–163.

Liu, B.Z. (1984) The dynamic models of fruits of cotton plant and the preliminary study on interaction between cotton fruits and pink bollworm. *Contributions from Shanghai Institute of Entomology* 4, 85–96.

Liu, S.Y., Zhang, S.F. and He, B.J. (1986) Studies on the life tables of the natural populations of cotton pink bollworm. *Pectinophora gossypiella* (Saunders). *Scientia Agricultura Sinica* 2, 65–71.

Luo, Z.Y. and Shen, X.Y. (1985) Preliminary investigation on population dynamics computer simulation and forecast of cotton aphid (*Aphis gossypii*) in the Sheshan field, Shanghai suburb. *Contributions from Shanghai Institute of Entomology* 5, 55–66.

Mabbett, T.H., Nachapong, M., Monglakul, K. and Mekdaeng, J. (1984) Distribution on cotton of *Amrasca devastans* and *Ayyaria chaetophora* in relation to pest scouting techniques for Thailand. *Tropical Pest Management* 30, 133–141.

McKinion, J.M. and Lemmon, J.E. (1985) Expert systems for agriculture. *Computers and Electronics in Agriculture* 1, 31–40.

Mangel, M., Stefanou, S.E. and Wilen, J.E. (1985) Modelling *Lygus hesperus* injury to cotton yields. *Journal of Economic Entomology* 78, 1009–1014.

Masud, S.M. and Benedict, J.H. (1990) Economic decision criteria for fleahopper and bollworm management in cotton: Texas Coastal Bend. *Texas Agricultural Experiment Station Bulletin* No. 1644, 16pp.

Mumford, J.D. and Norton, G.A. (1984) Economics of decision making in pest management. *Annual Review of Entomology* 29, 157–174.

Mumford, J.D. and van Hamburg, H. (1985) *A Descriptive Analysis of Cotton Pest Management in South Africa.* Plant Protection Research Institute, Pretoria. 84pp.

Musuna, A.C.Z. (1986) A method for monitoring whitefly, *Bemisia tabaci* (Genn.), in cotton in Zimbabwe. *Agriculture, Ecosystems and Environment* 17, 29–35.

Nachapong, M., Leggs, D.E., Kittiboonya, S. and Wangboonkong, S. (1989) Validation of computer simulated presence-absence sequential sampling plans for the cotton bollworm (*Heliothis armigera* (Hubner)) in cotton. *Thai Journal of Agricultural Science* 22, 293–302.

Napit, K.B., Norton, G.W., Kazmierczak, R.F., Jr. and Rajotte, E.G. (1988) Economic impacts of extension integrated pest management programs in several states. *Journal of Economic Entomology* 81, 251–256.

Nyambo, B.T. (1989a) Assessment of pheromone traps for monitoring and early warning of *Heliothis armigera* Hubner (Lepidoptera: Noctuidae) in the western cotton-growing areas of Tanzania. *Crop Protection* 8, 188–192.

Nyambo, B.T. (1989b) Use of scouting in the control of *Heliothis armigera* in cotton in WCGA in Tanzania. *Crop Protection* 8, 310–317.

Plant, R.E., Wilson, L.T., Zelinski, L., Goodell, P.D. and Kerby, T. (1987) CA-LEX/COTTON: an expert system based management aid for California cotton growers. *Proceedings of the Beltwide Cotton Production Research Conference*, Memphis, Tennessee, pp. 203–205.

Rummel, D.R. and Hatfield, J.L. (1988) Thermal-based emergence model for the bollworm (Lepidoptera: Noctuidae) in the Texas high plains. *Journal of Economic Entomology* 81, 1620–1623.

Sheng, C.F. (1985) Economic thresholds of the second generation of cotton bollworm in North China. *Acta Entomologica Sinica* 28, 382–389.

Sheng, C.F. and Hopper, K.R. (1988) Harvesting models and pest management in cotton. *Environmental Entomology* 17, 755–763.

Simpson, E.H., III and Parvin, D.W., Jr. (1983) Impact of alternative cotton insect management strategies on producer income in Mississippi. *US Department of Agriculture, Agriculture Handbook* No. 589, pp. 481–496.

Stone, N.D. and Gutierrez, A.P. (1986a) Pink bollworm control in southwestern desert cotton. I. A field-oriented simulation model. *Hilgardia* 54, 1–24.

Stone, N.D. and Gutierrez, A.P. (1986b) Pink bollworm control in southwestern desert cotton. II. A strategic management model. *Hilgardia* 54, 25–41.

Stone, N.D., Coulson, R.N., Frisbie, R.E. and Loh, D.K. (1986a) Expert systems in entomology: three approaches to problem solving. *Bulletin of the Entomological Society of America* 32, 161–166.

Stone, N.D., Gutierrez, A.P., Getz, W.M. and Norgaard, R. (1986b) Pink bollworm control in southwestern desert cotton. III. Strategies for control; an economic simulation study. *Hilgardia* 54, 42–56.

Stone, N.D., Rummel, D.R., Carroll, S., Makela, M.E. and Frisbie, R.E. (1990) Simulation of boll weevil (Coleoptera: Curculionidae) spring emergence and overwintering survival in the Texas Rolling Plains. *Environmental Entomology* 19, 91–98.

Thomas, W.M. (1989) Modelling within-plant distribution of *Heliothis* spp. (Lepidoptera: Noctuidae) damage in cotton. *Agricultural Systems* 30, 71–80.

Thomas, W.M., Roof, M.E. and Jones, R.G. (1988) Predicting *Heliothis* spp. oviposition peaks using TEXCIM. *Journal of Agricultural Entomology* 5, 253–256.

Topper, C.P. (1987) Nocturnal behaviour of adults of *Heliothis armigera* (Hubner) (Lepidoptera: Noctuidae) in the Sudan Gezira and pest control implications. *Bulletin of Entomological Research* 3, 541–554.

Wallach, D., Kletter, E., Sachs, Y. and Ungar, E.D. (1988) Towards optimal pest control in cotton. In: Rymon, D.J. (ed.) *Optimal Yield Management.* Avebury, Aldershot, UK, pp. 195–206.

Witz, J.A., Hartstack, A.W., King, E.G., Dickerson, W.A. and Phillips, J.R. (1985) Monitoring and prediction of *Heliothis* spp. *Southwestern Entomologist* 8, 56–70.

Xia, X.Y. and Sterling, W.L. (1987)Computer simulation of cotton aphid population dynamics. *Acta Phytophylactica Sinica*, 14, 151–156.

Zhang, L.X., Niu, X.T. and Wu, M. (1987) A dynamic simulation model of the biological control system of cotton aphids. *Proceedings of the 1st International Conference on Agricultural Systems Engineering*, 11–14 August 1987, Changchun, China, pp. 456–464.

Zhang, X.X., Norton, G.A. and Mumford, J.D. (1989) Report on cotton pest problems in Nantong City. Silwood Centre for Pest Management, Imperial College, Ascot, United Kingdom, 29pp.

Index

Page numbers of main sections of text are shown in **bold**. There are separate listings for acari (mites), countries, insecticides, insects, parasitoids, predators and countries.